D0151286

MASONRY DESIGN AND DETAILING

For Architects, Engineers, and Builders

MASONRY DESIGN AND DETAILING

For Architects, Engineers, and Builders

CHRISTINE BEALL, AIA

Prentice-Hall, Inc., Englewood Cliffs, New Jersey 07632

Library of Congress Cataloging in Publication Data

Beall, Christine.
Masonry design and detailing.

Bibliography: p.
Includes index.
1. Masonry. I. Title.
TA670.B43 1984 693'.1 83-21105
ISBN 0-13-559153-8

Editorial/production supervision
and interior design: *Mary Carnis*
Manufacturing buyer: *Anthony Caruso*

Printed in the United States of America

10 9 8 7 6 5 4 3 2 1

ISBN 0-13-559153-8

Prentice-Hall International, Inc., *London*
Prentice-Hall of Australia Pty. Limited, *Sydney*
Editoria Prentice-Hall do Brasil, Ltda., *Rio de Janeiro*
Prentice-Hall Canada, Inc., *Toronto*
Prentice-Hall of India Private Limited, *New Delhi*
Prentice-Hall of Japan, Inc., *Tokyo*
Prentice-Hall of Southeast Asia Pte. Ltd., *Singapore*
Whitehall Books Limited, *Wellington, New Zealand*

To Annie

CONTENTS

PREFACE xvii

PART I: INTRODUCTION

1 HISTORY AND DEVELOPMENT OF MASONRY TECHNOLOGY 1

1.1 Development 1
1.2 Decline 3
1.3 Revival 6
1.4 Modern Masonry 6

2 RAW MATERIALS AND MANUFACTURING PROCESSES 9

2.1 Clay Masonry 9

2.1.1 Composition 10
2.1.2 Clay Types 10
2.1.3 Material Preparation 11

2.1.4 Manufacturing 11
2.1.5 Forming 13
2.1.6 Drying 16
2.1.7 Glazing 16
2.1.8 Burning 16
2.1.9 Drawing and Storage 18

2.2 Concrete Masonry 18

2.2.1 Aggregates 18
2.2.2 Cements 20
2.2.3 Admixtures 20
2.2.4 Manufacturing 20
2.2.5 Material Preparation 21
2.2.6 Forming 21
2.2.7 Curing 22
2.2.8 Surface Treatment 24
2.2.9 Cubing and Storage 26

2.3 Mortar and Grout Materials 26

2.3.1 Cements 26
2.3.2 Lime 28
2.3.3 Masonry Cements 28
2.3.4 Sand 29
2.3.5 Water 29
2.3.6 Admixtures 29

PART II: MASONRY PRODUCTS AND ACCESSORIES

3 CLAY PRODUCTS 30

3.1 Brick 30

3.1.1 Sizes and Shapes 35
3.1.2 Special-Purpose Bricks 39
3.1.3 Glass Block 42
3.1.4 Hollow Brick 43

3.2 Structural Clay Tile 45

3.2.1 Loadbearing Wall Tile 45
3.2.2 Non-Loadbearing Tile 47
3.2.3 Facing Tile 47
3.2.4 Ceramic Glazed Facing Tile 49
3.2.5 Screen Tile 49

3.3 Ceramic Veneer (Terra Cotta) 49

3.4 Properties and Characteristics 52

3.4.1 Compressive Strength 52
3.4.2 Transverse Strength 52
3.4.3 Absorption 54
3.4.4 Durability 55
3.4.5 Expansion Coefficients 56
3.4.6 Fire and Thermal Resistance 56
3.4.7 Acoustical Characteristics 56
3.4.8 Colors and Textures 57

4 Cementitious Masonry Units 58

4.1 Concrete Brick 58
4.2 Sand-Lime Brick 60
4.3 Gypsum Block 60
4.4 Cast Stone 60
4.5 Cellular Concrete Block 60
4.6 Concrete Block 60

4.6.1 Coring 62
4.6.2 Grading 63
4.6.3 Moisture Content 64

4.7 Unit Types 66

4.7.1 Decorative Units 66
4.7.2 Glazed Units 71
4.7.3 Paving Units 72

4.8 Properties and Characteristics 73

4.8.1 Unit Strength 73
4.8.2 Absorption 74
4.8.3 Volume Changes 74
4.8.4 Fire, Sound, and Heat Resistance 75
4.8.5 Colors 76

5 NATURAL STONE 77

5.1 Geological Characteristics 77
5.2 Properties 78
5.3 Production 79
5.4 Building Stone 84

5.4.1 Granite 84
5.4.2 Limestone 84
5.4.3 Marble 86
5.4.4 Slate 87
5.4.5 Sandstone 87

5.5 Selecting Stone 88

6 MORTAR AND GROUT 91

6.1 Mortar Properties 91

6.1.1 Workability 92
6.1.2 Retentivity and Flow 92
6.1.3 Bond Strength 94
6.1.4 Compressive Strength 95
6.1.5 Extensibility and Volume Change 95
6.1.6 Durability 96
6.1.7 Efflorescence 96

6.2 Mortar Classification 96

6.2.1 Portland Cement-Lime Mortars 97
6.2.2 Masonry Cement Mortars 97

6.3 Mortar Types 97

6.3.1 Type M Mortar 99
6.3.2 Type S Mortar 100
6.3.3 Type N Mortar 100
6.3.4 Type O Mortar 100
6.3.5 Type K Mortar 100
6.3.6 Refractory Mortars 100
6.3.7 Chemical-Resistant Mortars 101
6.3.8 Extra-High-Strength Mortars 102

6.4 Grout 103

7 MASONRY ACCESSORIES 105

7.1 Horizontal Reinforcement 106

7.2 Anchors, Ties, and Fasteners 107

7.2.1 Ties 107
7.2.2 Anchors 110
7.2.3 Fasteners 113

7.3 Movement Joints 115

7.4 Flashing Materials 115

PART III: BUILDING SYSTEMS AND APPLICATIONS

10 ENVIRONMENTAL CHARACTERISTICS 117

8.1 Single-Wythe Walls 118

8.1.1 Hollow Tile 118
8.1.2 Brick and Block 118
8.1.3 Gypsum Block 118
8.1.4 Glass Block 118

8.2 Multi-Wythe Walls 119

8.2.1 Solid Masonry Walls 120
8.2.2 Solid Walls of Hollow Units 121
8.2.3 Cavity Walls 122
8.2.4 Metal Wall Ties 123

8.3 Fire Resistance Characteristics 124

8.3.1 Fire Tests 124
8.3.2 Fire Resistance Ratings 125
8.3.3 Steel Fireproofing 129

8.3.4 *Construction Classifications* 131
8.3.5 *Compartmentation* 131
8.3.6 *Fire Insurance Rates* 134

8.4 Thermal Properties 134

8.4.1 *Thermal Inertia* 138
8.4.2 *Heat-Gain Calculations* 138
8.4.3 *Heat-Loss Calculations* 142
8.4.4 *The* M-*Factor* 144

8.5 Added Insulation 147

8.5.1 *Granular Fills* 148
8.5.2 *Rigid Board Insulation* 148
8.5.3 *Foams* 149
8.5.4 *Vapor Barriers* 150
8.5.5 *Insulation Location* 150

8.6 Passive Solar Design 151

8.6.1 *Shading Devices* 151
8.6.2 *Direct-Gain Solar Heating* 153
8.6.3 *Thermal Storage Walls* 153
8.6.4 *The Trombe Wall* 155

8.7 Acoustical Properties 161

8.7.1 *Sound Ratings* 161
8.7.2 *Sound Absorption* 162
8.7.3 *Sound Transmission* 162
8.7.4 *STC Ratings* 164
8.7.5 *Code Requirements* 166

8.8 Differential Movement 166

8.8.1 *Temperature Movement* 167
8.8.2 *Moisture Movement* 167
8.8.3 *Elastic Deformation* 167
8.8.4 *Plastic Flow* 169
8.8.5 *Effects of Differential Movement* 169
8.8.6 *Movement Joints* 170

9 NON-LOADBEARING CONSTRUCTION 171

9.1 Partition Walls 172

9.2 Exterior Walls 175

9.2.1 *Veneer Design and Attachment* 175
9.2.2 *Brick Veneer* 176
9.2.3 *Concrete Brick and Concrete Block Veneer* 180
9.2.4 *Stone Veneer* 182
9.2.5 *Movement Joints* 186
9.2.6 *Panel Walls* 188
9.2.7 *Curtainwalls* 191

9.3 Screen Walls and Garden Walls 194

9.4 Fireplaces 203

10 PLAIN AND PARTIALLY REINFORCED LOADBEARING MASONRY 207

10.1 Bearing Wall Systems 208

10.1.1 Plain Masonry 208
10.1.2 Partially Reinforced Masonry 209
10.1.3 Axial Load Distribution 210
10.1.4 Foundations 211
10.1.5 Lateral Load Distribution 211
10.1.6 Diaphragms 212
10.1.7 Shear Walls 213
10.1.8 Floor-Wall Connections 213
10.1.9 Differential Movement 215

10.2 Structural Design 217

10.2.1 Empirical Design 218
10.2.2 Rational Analysis 225
10.2.3 Codes 225
10.2.4 Materials 226
10.2.5 General Design Requirements 226
10.2.6 Masonry Strength 227

10.3 Brick Masonry 229

10.3.1 Eccentricity 230
10.3.2 Slenderness 232
10.3.3 Cross-Sectional Area 232
10.3.4 Allowable Vertical Loads 232
10.3.5 Shear Walls 233

10.4 Concrete Unit Masonry 235

10.4.1 Slenderness 236
10.4.2 Effective Section 236
10.4.3 Eccentricity 237
10.4.4 Axial Loads 238
10.4.5 Shear Walls and Diaphragms 240
10.4.6 Flexural Strength 241

11 REINFORCED MASONRY 243

11.1 Basic Design Theory 243

11.1.1 Minimum Amount of Steel 244
11.1.2 Design Codes 245
11.1.3 Materials 245
11.1.4 Allowable Stresses in Reinforced Masonry 248

11.2 *ANSI A41.2* 248

11.2.1 Design of Columns 248
11.2.2 Design of Reinforced Walls 255

11.3 *NCMA-1970* 256

11.3.1 Design of Columns 257
11.3.2 Design of Reinforced Walls 258

11.4 *ACI-539* 259

11.5 *UBC-1982* 260

11.5.1 Design of Columns 261
11.5.2 Design of Reinforced Walls 261

11.6 *BIA-1969* 262

11.6.1 Design of Columns 262
11.6.2 Design of Reinforced Walls 263

11.7 A Design Example 264

11.8 High-Risk Design 268

11.8.1 Building Layout 269
11.8.2 Shear Walls and Diaphragms 269
11.8.3 Diagonal Tension 272
11.8.4 Building Code Requirements 272

12 BEAMS, LINTELS, AND ARCHES 277

12.1 *ANSI A41.2* 277

12.1.1 Shear and Diagonal Tension 278
12.1.2 Web Reinforcement 279
12.1.3 Bond Stress 279
12.1.4 Anchorage Requirements 280

12.2 *NCMA-1970* 281

12.3 *BIA-1969* 282

12.4 *UBC-1982* 282

12.5 Deep Wall Beams 283

12.6 Lintels 284

12.6.1 Load Determination 285
12.6.2 Steel Lintels 286
12.6.3 Concrete Masonry Lintels 288
12.6.4 Reinforced Brick Lintels 294
12.6.5 Precasting 297

12.7 Arches 297

12.7.1 Minor Arch Design 298
12.7.2 Graphic Analysis 302
12.7.3 Rotation 303
12.7.4 Sliding 304
12.7.5 Crushing 304
12.7.6 Thrust Resistance 305
12.7.7 Design Tables for Semicircular Arches 306
12.7.8 A Design Example 311
12.7.9 Major Arch Design 314
12.7.10 Arch Construction 314

13 RETAINING WALLS, BELOW-GRADE WALLS, AND POOLS 316

13.1 Retaining Walls 316

13.1.1 Lateral Earth Pressure 317
13.1.2 Surcharge 318

13.1.3 Overturning and Sliding 318
13.1.4 Drainage and Waterproofing 318
13.1.5 Expansion Joints 318
13.1.6 Footings 319
13.1.7 Materials 320
13.1.8 Concrete Masonry Cantilever Walls 320
13.1.9 Brick Cantilever Walls 320

13.2 Below-Grade Walls 324

13.2.1 Unreinforced Walls 328
13.2.2 Partially Reinforced Walls 328
13.2.3 Reinforced Walls 332
13.2.4 Footings 332
13.2.5 Material Requirements 332
13.2.6 Waterproofing 332

13.3 Swimming Pools 334

13.3.1 Materials 334
13.3.2 Design 334
13.3.3 Construction 336

PART IV: CONSTRUCTION PRACTICE

14 DETAILS AND WORKMANSHIP 338

14.1 Preparation of Materials 338

14.1.1 Mortar 338
14.1.2 Concrete Masonry Units 340
14.1.3 Brick 340

14.2 Construction and Workmanship 341

14.2.1 Modular Coordination 341
14.2.2 Control Joints 346
14.2.3 Expansion Joints 349
14.2.4 Movement Joint Locations 350
14.2.5 Differential Movement 351
14.2.6 Flexible Anchorage 353
14.2.7 Pattern Bonds 358
14.2.8 Mortar Joints 360
14.2.9 Laying Units 362
14.2.10 Weep Holes 363
14.2.11 Story Poles 363
14.2.12 SCR Masonry Process 364
14.2.13 Reinforcement and Accessories 364
14.2.14 Insulation 366
14.2.15 Grouting 366
14.2.16 Surface Bonding 368

14.3 Masonry Paving 371

14.3.1 Outdoor Paving 372
14.3.2 Bases 373
14.3.3 Setting Beds 374
14.3.4 Joints 376
14.3.5 Membrane Materials 378

14.3.6 Masonry Units 378
14.3.7 Paving Patterns 379
14.3.8 Brick Floors 381

14.4 Moisture Protection 384

14.4.1 Flashing 384
14.4.2 Material Selection 386
14.4.3 Waterproofing and Dampproofing 386
14.4.4 Condensation 388
14.4.5 A Design Example 390
14.4.6 Applied Finishes 394

14.5 Cold Weather Construction 400

14.6 Prefabricated Masonry 402

14.7 Masonry Restoration 403

14.7.1 Efflorescence 404
14.7.2 Cleaning Existing Masonry 407
14.7.3 Cleaning New Masonry 408
14.7.4 Tuckpointing 411
14.7.5 Replacing Masonry 411

15 ECONOMICS OF CONSTRUCTION 413

15.1 General Costs 413

15.2 Comparative Costs 414

15.3 Value Engineering 421

15.4 Design Decisions 425

16 SPECIFYING AND INSPECTING MASONRY CONSTRUCTION 428

16.1 Methods of Specifying Masonry 428

16.2 Guide Specifications 429

Section 04200 Unit Masonry 429
Section 04400 Stone Veneer 438
Section 04520 Masonry Restoration 441

16.3 Inspecting Masonry Construction 444

16.3.1 Materials 445
16.3.2 Construction 446
16.3.3 Workmanship 446
16.3.4 Protection and Cleaning 447

APPENDIX A: GLOSSARY 449

APPENDIX B: NOTATIONS AND SYMBOLS 471

APPENDIX C: ASTM REFERENCE STANDARDS 473

APPENDIX D: MASONRY ORGANIZATIONS 476

BIBLIOGRAPHY 479

INDEX 483

PREFACE

This handbook addresses a broad range of aesthetic, technical, and environmental considerations. In addition to the engineering aspects of design, technical information on energy and sound control, maintenance, life-cycle costing, and workmanship is included. My goal has been to assemble and correlate existing industry information into a single, concise, and complete reference aimed at architects and other design professionals.

Two major sources of information have been publications of the Brick Institute of America and the National Concrete Association, especially the BIA "Technical Notes" series and the NCMA "TEK Bulletins." A bibliography of detailed sources is given at the back of this book as well as a list of national and regional masonry organizations through which design assistance can be obtained.

I would like to thank Excy Johnston, AIA, for preparing the sketches in Figs. 1-1, 2-11, and 15-4; Gregg Borchelt, P.E., of the Masonry Institute of Houston/Galveston for his technical editing of Chapters 10, 11, and 12 and for the preparation of sample problems to accompany the text; Bernie Beall for typing the rough draft; and Kathy Cogburn for putting the text on computer disc.

Photographs and many technical illustrations, charts, graphs, and tables have been provided through the courtesy of the Brick Institute of America, the National Concrete Masonry Association, the American Concrete Institute, the Masonry Institute of America, and the Portland Cement Association.

Christine Beall, AIA
Austin, Texas

1

HISTORY AND DEVELOPMENT OF MASONRY TECHNOLOGY

The history of man is preserved in the things he built—in the temples, fortresses, sanctuaries, and cities he constructed. Man's earliest efforts at shelter were limited to the materials at hand. The trees of a primeval forest, the mud of a river valley, the rocks and cliffs of a mountain range afforded primitive opportunity for protection, security, and defense. The stone and brick of skeletal architectural remains date back to the temples at Ur built in 3000 B.C., the early walls of Jericho of 8000 B.C., and the vaulted tombs at Mycenae of the fourteenth century B.C. The permanence and durability of these masonry materials safeguarded the prehistoric record of achievements, and preserved through centuries of war and natural disaster the traces of human development from cave dweller to modern builder. Indeed, *the history of man is the history of his architecture, and the history of architecture is the history of masonry.*

1.1 DEVELOPMENT

Stone is the oldest, most abundant, and perhaps most important *raw building material* of prehistoric and civilized man. Stone formed his defense in walls, towers, and embattlements. He lived in buildings of stone, worshiped in stone temples, and built roads and bridges of stone. Man began to form and shape stone when tools had been invented that were hard enough to trim and smooth the irregular lumps and broken surfaces. Stone building was then freed from the limitations of monolithic slab structures like those at Stonehenge and progressed through the shaped and fitted

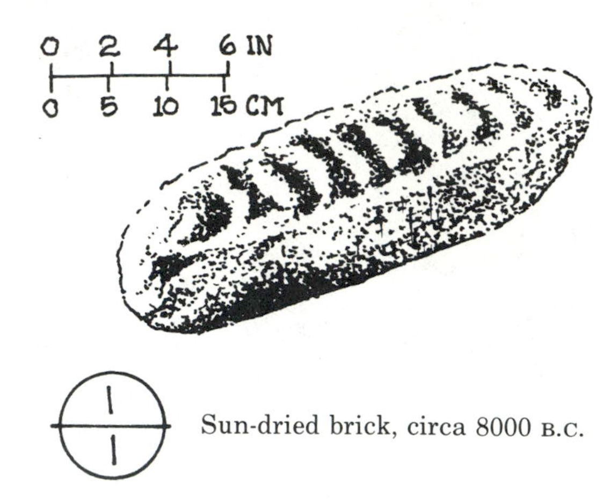

Sun-dried brick, circa 8000 B.C.

blocks of the Egyptians to the intricately carved columns and entablatures of the Greeks and Romans.

Brick is the oldest *man-made* building material, invented almost 10,000 years ago. Its simplicity, strength, and durability led to extensive use, and gave it a dominant place in history alongside stone.

Rubble stone and mud bricks as small, easily handled materials, could be stacked and shaped to form enclosures of simple or complex design. Hand-shaped, sun-dried bricks, reinforced with such diverse materials as straw and dung, were so effective that kiln-fired bricks did not appear until the third millenium B.C., long after the art of pottery had demonstrated the effects of high temperatures on clay. Some of the oldest bricks in the world, taken from archaeological digs at the site of ancient Jericho, resemble long loaves of bread, sometimes with a bold pattern of Neolithic thumbprint impressions on their rounded tops (*see Fig. 1-1*). The use of wooden molds did not supplant hand-forming techniques until the early Bronze Age, around 3000 B.C.

Perhaps the most important innovations in the evolution of architecture were the development of masonry arches and domes. Throughout history, the arch was the chief means of overcoming the span limitations of single blocks of stone or lengths of timber, making it possible to bridge spaces once thought too great. Early forms were only approximations of true arching action and were generally false, corbeled arches. True arches carry their loads in simple compression to each abutment, and as long as the joints are roughly aligned at right angles to the compressive stress, the precise curve of the arch is not critical.

Excavation of ruins in Babylonia exposed a masonry arch believed to have been built around 1400 B.C. Arch construction reached a high level of refinement under the Romans, and later developments were primarily in the adoption of different profiles. Islamic and Gothic pointed arches led to the design of groined vaults, and eventually to the zenith of cathedral architecture and masonry construction in the thirteenth century.

Simple dome forms may actually have preceded the true arch because, like the corbeled arch, they could be built with successive horizontal rings of masonry, and required no centering. These domes were seen as circular walls gradually closing in on themselves rather than as rings of vertical arches. Barrel vaults were built as early as the thirteenth century B.C., and could also be constructed without centering if one end of the vault was closed off.

Initial exploitation of the true dome form took place from the mid first century A.D. to the early second century, under the reigns of Nero and Hadrian. The Pantheon dome exerts tremendous outward thrusts counteracted by massive brick walls encircling its perimeter. Later refinements included the masonry squinch and pendentive, which were instrumental in the construction of the dome of the Florence Cathedral, and buttressing by means of half domes at the sides, as in the Church of Hagia Sophia in Constantinople.

1.2 DECLINE

Renaissance architecture produced few significant innovations in structural practice, since design was based primarily on the classical forms of earlier eras. The forward thrust of structural achievements in masonry essentially died during this period of "enlightenment," and masonry structures remained at an arrested level of development.

With the onslaught of the Industrial Revolution, emphasis shifted to iron, steel, and concrete construction. The invention of portland cement in 1824, refinements in iron production in the early nineteenth century, and the development of the Bessemer furnace in 1854 turned the creative focus of architecture away from masonry.

By the early twentieth century, the demand was for high-rise construction, and the technology of stone and masonry building had not kept pace with the developments of other structural systems. The Chicago School had pioneered the use of iron and steel skeleton frames, and masonry was relegated to secondary usage as facings, in-fill, and fireproofing. The Monadnock Building in Chicago (1891) is generally cited as the "last great building in the ancient tradition of masonry architecture" (*see Fig. 1-2*). Its massive structure—16 stories high, with stone and brick walls 6 feet thick at the base, supported on immense footings—seemed to prove that the medium was not suited to the demands of a modern industrialized society. Except for the revivalist periods following the 1893 World's Columbian Exposition and the "mercantile classicism" which prevailed for some time, a general shift in technological innovation took place, and skeleton frame construction began to replace loadbearing masonry.

During this period, only Antonio Gaudí's unique Spanish architecture manifested innovative masonry designs (*see Fig. 1-3*). His "structural rationalism" was based on economy and efficiency of form, using ancient Catalan vaulting techniques, parabolic arches, and inclined piers to bring the supporting masonry under compression. His work also included vaulting with hyperbolic paraboloids and warped "helicoidal" surfaces for greater structural strength. Gaudí, however, was the exception in a world bent on developing lightweight, high-rise building techniques for the twentieth century.

At this time, masonry was perceived as a compressive material with no tensile strength. Concrete construction was then at the same level of sophistication, but the introduction of ferroconcrete brought this material a step further by combining the advantages of iron or steel with a cementitious mix. What failed to occur was the widespread application of this new technique to masonry. While concrete technology developed rapidly into complex steel-reinforced systems, masonry research experienced a period of dormancy.

The first experimentally reinforced concrete building, the Eddystone Lighthouse (1774), was actually constructed of concrete *and* stone. However, the practice of using iron or steel as reinforcing was thereafter limited almost entirely to concrete. The few instances of reinforced brick masonry that can be noted historically took place in the early to mid nineteenth

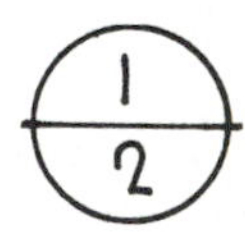

Monadnock Building, Chicago, 1891. Burnham and Root, architects. (*Photo courtesy of the School of Architecture Slide Library, the University of Texas at Austin.*)

century, but had disappeared by about 1880. Design with reinforced masonry was at that time intuitive or empirical rather than rationally determined, and rapid advances in concrete engineering quickly outpaced what was seen as an outmoded, inefficient, and uneconomical system. By the time the Monadnock Building was constructed, building codes recognized lateral resistance of masonry walls only in terms of mass, and this did indeed make the system uneconomical. Had the research in steel-reinforced concrete been applied equally and simultaneously to masonry, the history of modern architecture would be quite different.

(A)

(B)

(C)

(D)

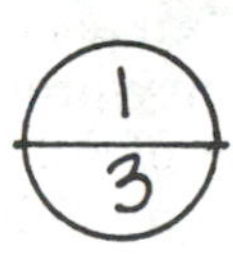

Gaudi's innovative masonry structures: (A) Inclined brick column, Colonia Guell Chapel; (B) Warped masonry roof, schools of the Sagrada Familia Church; (C) Thin masonry arch ribs, Casa Mila; and (D) Arch detail, Bell Esguard. (*Photos courtesy of the School of Architecture Slide Library, the University of Texas at Austin.*)

1.3 REVIVAL

In 1920, economic difficulties in India convinced officials that alternatives to concrete and steel structural systems must be found. Extensive research began into the performance of reinforced masonry walls, slabs, beams, and columns. This led not only to new systems of low-cost construction, but also to a basic understanding of the structural behavior of masonry. It was not until the late 1940s, however, that European engineers and architects began serious studies of masonry bearing wall designs—*almost 100 years after the same research had begun on concrete bearing walls*.

By this time, modern manufacturing techniques were capable of producing brick with compressive strengths in excess of 8000 psi, and modern portland cement mortars had exhibited strengths as high as 2500 psi. Extensive testing of some 1500 wall sections generated the scientific laboratory data necessary to develop a rational design method for masonry. These studies produced the first reliable, mathematical analysis of a very old material, freed engineers from the constraints of empirical data, and allowed formulation of rational theories of economical design. It was found that no new techniques of analysis were required, merely the application of accepted engineering principles normally used on other systems. The working stress formulas and calculations for reinforced concrete systems are often applicable to masonry as well.

The development of recommended practices in masonry design and construction took place in this country during the decade of the 1950s, and resulted in publication of the first engineered masonry building code in 1966. Continued research throughout the following two decades brought about refinements in testing methods and design procedures, and led to the adoption of engineered masonry structural systems by all of the major building codes in the United States. Laboratory and field tests have also identified and defined the physical properties of masonry and verified its excellent performance in fire control, sound attenuation, and thermal resistance.

Masonry construction today includes not only quarried stone and clay brick, but a host of other manufactured products as well. Concrete block, cast stone, structural clay tile, terra cotta veneer, glass block, mortar, grout, and metal accessories are all a part of the mason's trade. In the dozen or so definitions of masonry promulgated by recognized authorities, this group of materials is often expanded to include concrete, plaster, tile, or precast concrete. However, the most prevalent and the historically accepted definition of masonry is "an assemblage of small building units held together with mortar." Masonry products may be mechanically produced or hand made, and the variety of materials, colors, and textures available, combined with its structural capabilities, make it applicable to many types of architectural expression.

Masonry is currently enjoying a resurgence in popularity among the design professions, but is still widely misunderstood and often seen only as a decorative finish requiring other means of structural support. We need only refer to our architectural histories to find ample documentation to the contrary. The materials themselves can long outlive the people who produce them, and modernization of design and construction techniques have opened new horizons for the imaginative architect who wishes to leave a legacy of lasting beauty and performance.

1.4 MODERN MASONRY

Modern masonry may take one of several forms. Structurally, it may be divided into loadbearing, non-loadbearing, and veneer construction. Walls may be of single- or multi-wythe design. They may also be solid masonry, solid walls of hollow units, or cavity walls. Finally, masonry may be reinforced, partially reinforced, or plain, and either empirically or analytically designed.

Loadbearing masonry supports its own weight as well as the dead and live loads of the structure, and all lateral wind and seismic forces. Non-loadbearing masonry (including veneers) also resists lateral loads and may support its own weight for the full height of the structure, or be wholly supported by the structure at each floor. Solid masonry is built of solid units or fully grouted hollow units in multiple wythes with the collar joint between wythes filled with mortar or grout. Solid walls of hollow units have open cores in the units, but grouted collar joints. Cavity walls have two or more wythes of solid or hollow units separated by an open collar joint or cavity (*see Fig. 1-4*).

1-4 Types of masonry construction.

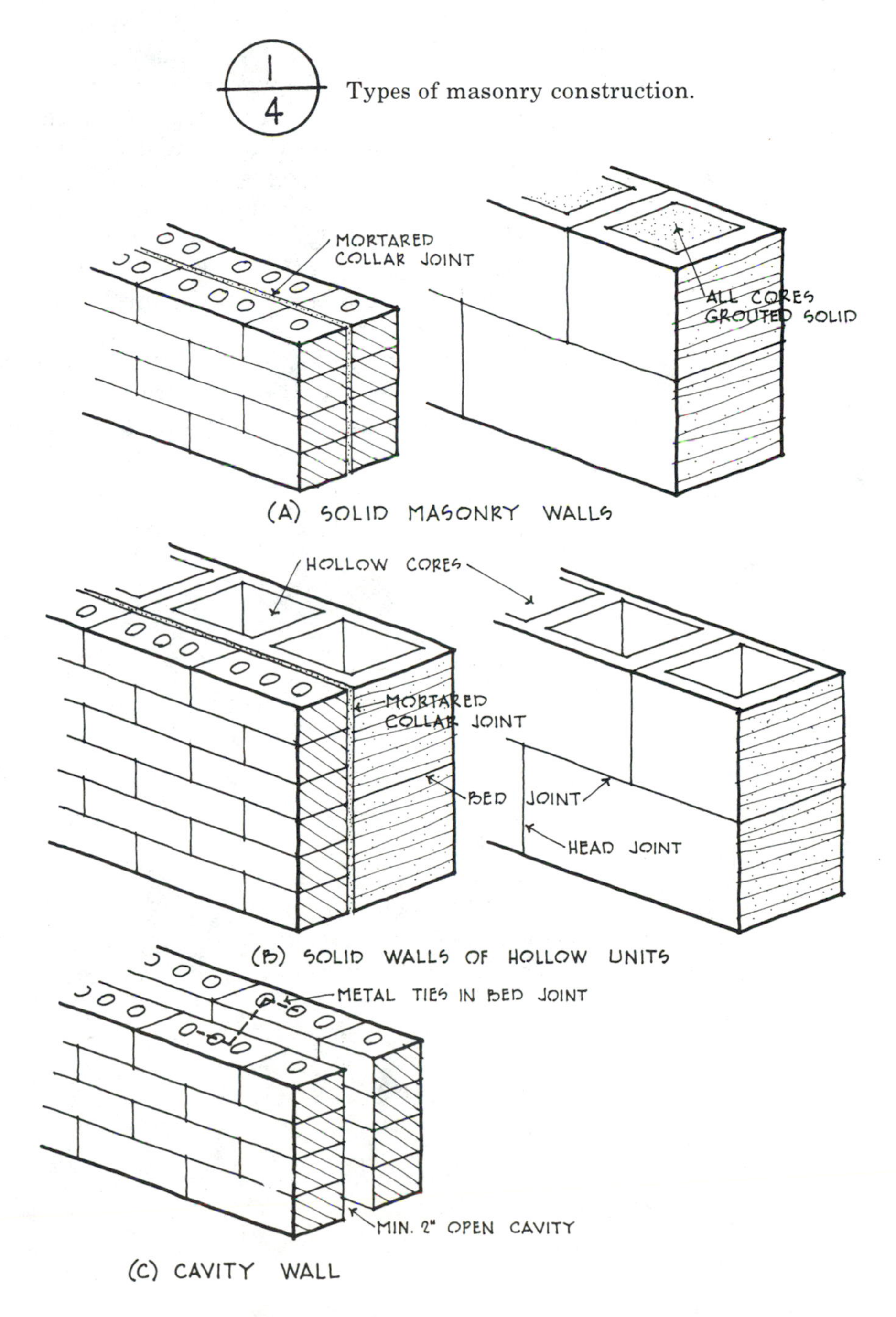

Empirical designs are based on arbitrary limits of height and wall thickness. Rational, engineered designs, however, are based on an analysis of the loads and the strength of the materials used in the structure. Standard calculations are used to determine the actual compressive, tensile, and shear stresses, and the masonry designed to resist these forces. Plain masonry can still be designed by empirical methods, but is applicable only to low-rise structures with modest loads. In plain masonry, shear and flexural stresses are resisted only by the bond between mortar and units. Although it is very strong in compression, it is weak in tension and shear. Small lateral loads and overturning moments are resisted by the weight of the wall.

Where lateral loads are higher, flexural strength can be increased by incorporating reinforcing steel in the member. Hollow units and cavity walls are both suited to this type of construction. Steel reinforcing bars are placed in the cores or cavity and grouted in place. The hardened grout binds the masonry and the steel together so that they act as a single element in resisting loads. Requirements for fully reinforced masonry are based on code-required minimums for area of steel. Partially reinforced masonry has no minimum area requirement, but instead selectively locates the reinforcement only in areas where design analysis indicates that tensile stress is developed. The minimum area requirements for reinforced masonry evolved primarily for seismic loading conditions. In most cases where earthquake or high wind forces are not a factor, partially reinforced masonry is most economical, and is applicable even for high-rise buildings of 20 stories or more.

Modern masonry is vastly different from the traditional construction of previous eras. Its structural capabilities are still being explored as continuing research develops greater understanding of its structural behavior.

2

RAW MATERIALS AND MANUFACTURING PROCESSES

The quality and characteristics of modern masonry products are directly and exclusively determined by the raw materials and methods of manufacture used in their production. Architects need not concern themselves with the esoterics of clay mineralogy or the details of plant automation in order to design and build successfully with brick and concrete block. However, a basic introduction to this aspect of masonry will aid in understanding the finished products and how they may best be used to perform specific design functions.

2.1 CLAY MASONRY

Clay, the raw material from which brick, structural clay tile, and architectural terra cotta are made, is the most plentiful natural substance used in the production of any building product. The American Society for Testing and Materials defines it as "an earthy or stony mineral aggregate consisting essentially of hydrous silicates of alumina—plastic when sufficiently pulverized and wetted, rigid when dry, and vitreous when fired to a sufficiently high temperature."

Clay is the end product of the chemical alteration over long periods of time of the less stable minerals in rock. This chemical weathering produces minute particles that are two dimensional or flake shaped. The unique plastic characteristics of clay soils are a result of the enormous amount of surface area inherent in this particle size and shape. A 1-in. cube of rock, for instance, if reduced to clay particle size, would yield a total surface area of approximately 1 acre. A thimble full of clay has about the same surface area as five trucks of gravel. The natural affinity of clay soils and moisture result in cohesiveness and plasticity from the surface tension of very thin layers of water between each of these minute particles. It is this plasticity which facilitates the molding and shaping of moist clay into usable shapes.

For the architect, the importance of understanding clay characteristics and methods of manufacture is their relationship to finished appearance and physical properties. Color depends first, on the composition of the raw material and the quantitative presence of metallic oxides. Second, it is an indication of the degree of burning to which the clay has been subjected. Lighter colors for a given clay are normally associated with underburning. They may also be indicative of high porosity and absorption along with decreased strength, durability, and resistance to abrasion. On the other hand, the very dark colors produced from the same clay are a result of overburning. This indicates that the units have been pressed and burned to a very high compressive strength and abrasion resistance, with greatly reduced absorption and increased resistance to freezing and thawing.

Most of the brick used in building construction falls between these two extremes. Since clay composition is the first determinant of brick color, lightness or darkness cannot be used as an absolute indicator of physical properties for brick made from different raw materials. It can, however, assist generally in the evaluation and selection of brick to meet specific program or environmental requirements.

2.1.1 Composition

Clays are complex materials composed basically of compounds of silica and alumina with varying amounts of metallic oxides and other minor ingredients and impurities. Metallic oxides act as fluxes to promote fusion at lower temperatures, influence the range of temperatures in which the material matures, and give burned clay the necessary strength for structural purposes. The varying amounts of iron, calcium, and magnesium oxides also influence the color of fired clay.

Based on chemical composition and the type of oxides present, clays may be divided into two different classes: (1) calcareous clays and (2) noncalcareous clays. While both are hydrous aluminum silicates, the calcareous clays contain around 15% calcium carbonate, which produces a yellowish color when fired. The noncalcareous clays are influenced by feldspar and iron oxide. The oxide may range from 2 to 25% of the composition, causing the clay to burn from buff to a salmon or red color as the amount increases.

If lime is present in a clay, it must be finely crushed or divided to eliminate large lumps. Lime becomes calcined in the burning process and later slakes or combines with water when exposed to the weather, so that any sizable fragments will expand and possibly chip or spall the brick.

2.1.2 Clay Types

There are three different types of clay which, while possessing similar chemical compositions, have different physical characteristics. Surface clays, shales, and fire clays are common throughout the world, and result from slight variations in the weathering process.

Surface clay occurs quite close to the earth's surface, and is characterized by a high oxide content, varying from 10 to 25%. Surface clays are the most accessible and easily mined, and therefore the least expensive.

Shale is clay that has been subjected to high pressure under natural conditions, causing a hardening and layering effect. It is very dense and harder to remove from the ground than other clays, and as a result, is more costly. Like surface clay, shale contains a relatively large percentage of oxide fluxes.

Fire clay is found at much greater depths than surface clay or shale. It tends to have fewer impurities and more uniform chemical and physical properties. The oxide content of fire clays is comparatively low, in the range of 2 to 10%. The reduced percentage of oxide fluxes raises the softening point of fire clay much higher than surface clay and shale, thus giving refractory qualities and the ability to withstand very high temperatures. This characteristic makes fire clays ideal for use in the production of brick and tile for furnaces, flue liners, ovens, and chimney stacks. The relatively small quantity of oxides also causes the clay to burn a very light brown or buff color, approaching white.

Clay is particularly suited to the manufacture of brick and structural tile. It has plasticity when mixed with water, so it can be molded or formed into the desired shapes; it possesses sufficient tensile strength to maintain those shapes after the dies or molds are removed; and its particles react to high temperatures by means of chemical and structural fusion.

2.1.3 Material Preparation

It is common practice for brick plants to mine more than one clay pit simultaneously. Since the raw clay is not always uniform in quality and composition, two or more kinds of material from different pits or from remote locations within the same pit are mixed together. This blending allows the manufacturer economically to minimize much of the natural variation in chemical composition and physical properties. Blending also produces a higher degree of product uniformity, helps control the color of the final unit, and permits some leverage in providing raw material suitable for specific types of brick or special product requirements.

Large lumps of clay are first washed to remove stones, soil, or excessive sand, and are then crushed into smaller pieces. The clay is then ground in hammer mills or in "dry pans," where huge wheels or rollers pulverize and then thoroughly mix the powdered clay. Particle size is controlled by an inclined, vibrating screen which retains the coarser particles and returns them to the grinder for further processing. The finer material is taken by conveyor to a storage bin or directly to the forming machine or pug mill for tempering and molding.

2.1.4 Manufacturing

After preparation of the raw clay, the manufacture of fired brick is completed in four additional stages: *forming, drying, burning*, and *drawing and storage* (*see Fig. 2-1*). Although raw materials may vary depending on local sources, this basic procedure is standard throughout the industry. Individual plants may introduce slight modifications to fit their particular material or method of operation, but the basic process is the same. Differences that may occur are in the brick molding techniques. In ancient

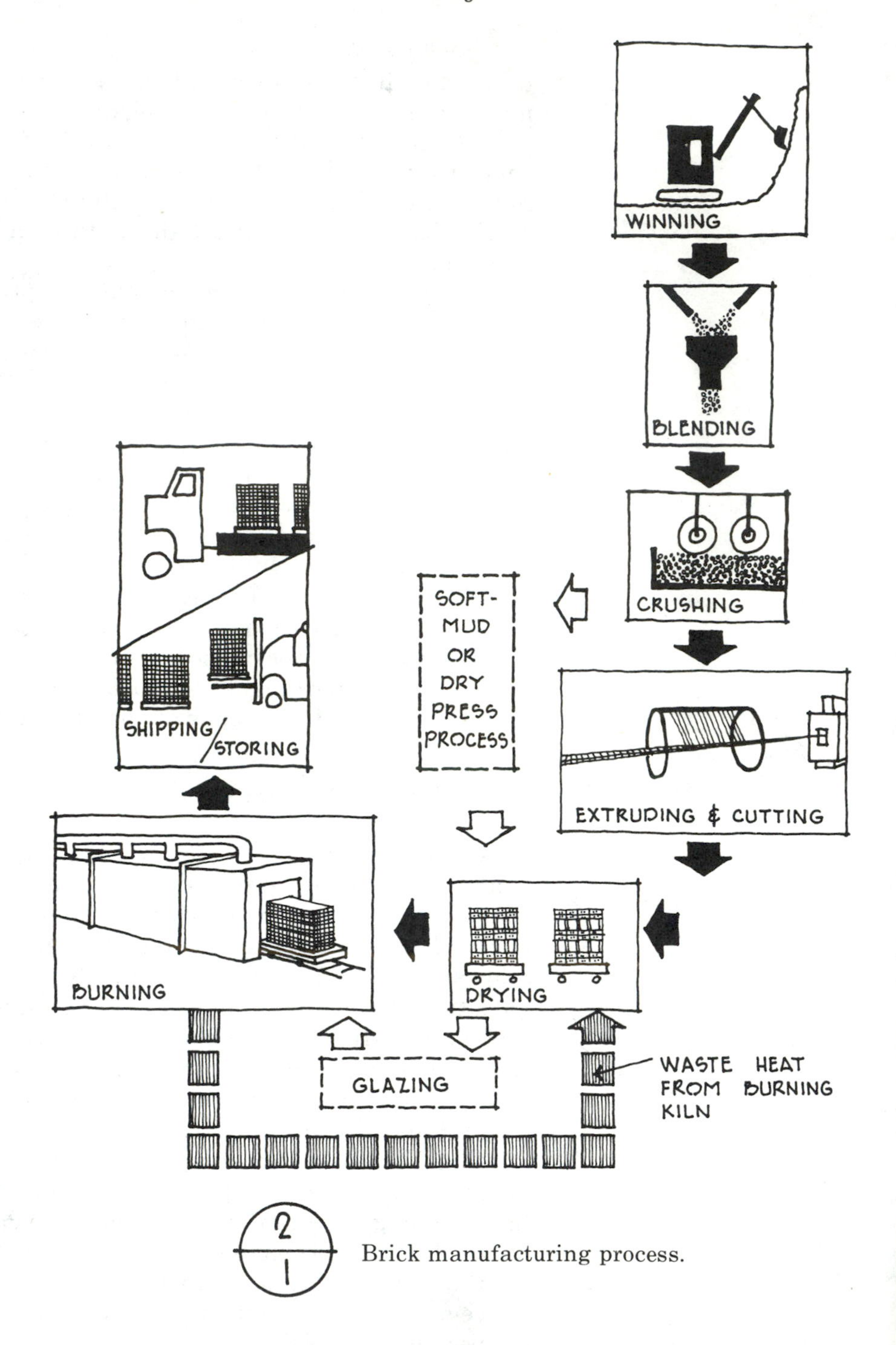

2/1 Brick manufacturing process.

as well as recent history, molding was performed exclusively by hand. Since brick-making machines were invented in the late nineteenth century, however, most of the structural clay products manufactured in the United States are machine made. At the present time, three principal methods are used in forming brick. The *stiff-mud*, *soft-mud*, or *dry-press* method may be used, depending on the suitability of the raw clay to the various methods of handling.

2.1.5 Forming

The first step in each of the forming methods is tempering, which produces a homogeneous, plastic mass. Tempering is done in a pug mill, where the clay is thoroughly mixed with a measured amount of water. The prescribed amount of water varies according to the forming method to be used.

The most common method is *stiff-mud extrusion*. A minimum amount of water, generally 12 to 15% moisture by weight, is mixed with the dry clay to make it plastic. After thorough mixing in the pug mill, the tempered clay goes through a deairing process in a partial vacuum. This removes air pockets and bubbles to increase workability and plasticity and produce greater strength. The clay is then forced by an auger through a steel die in a continuous extrusion or column of the desired size and shape. At the same time it is cored to reduce the weight of the unit and to facilitate drying and firing. Automatic cutting machines using thin wire attached to a circular steel frame cut the extruded clay into pieces either the height (side cut) or length (end cut) of the brick (*see Fig. 2-2*). Since clays shrink during both drying and burning, die sizes and cutter wire spacing must be carefully calculated to compensate. Texturing attachments may be affixed to roughen, score, scratch, or otherwise alter the smooth skin of the brick column as it emerges from the die (*see Fig. 2-3*). After cutting, clay slurry of contrasting color or texture may be rolled onto the brick surface to produce different visual effects.

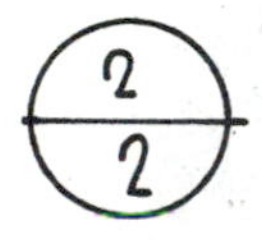

Wire-cutting extruded, stiff-mud brick. (*Photo courtesy BIA.*)

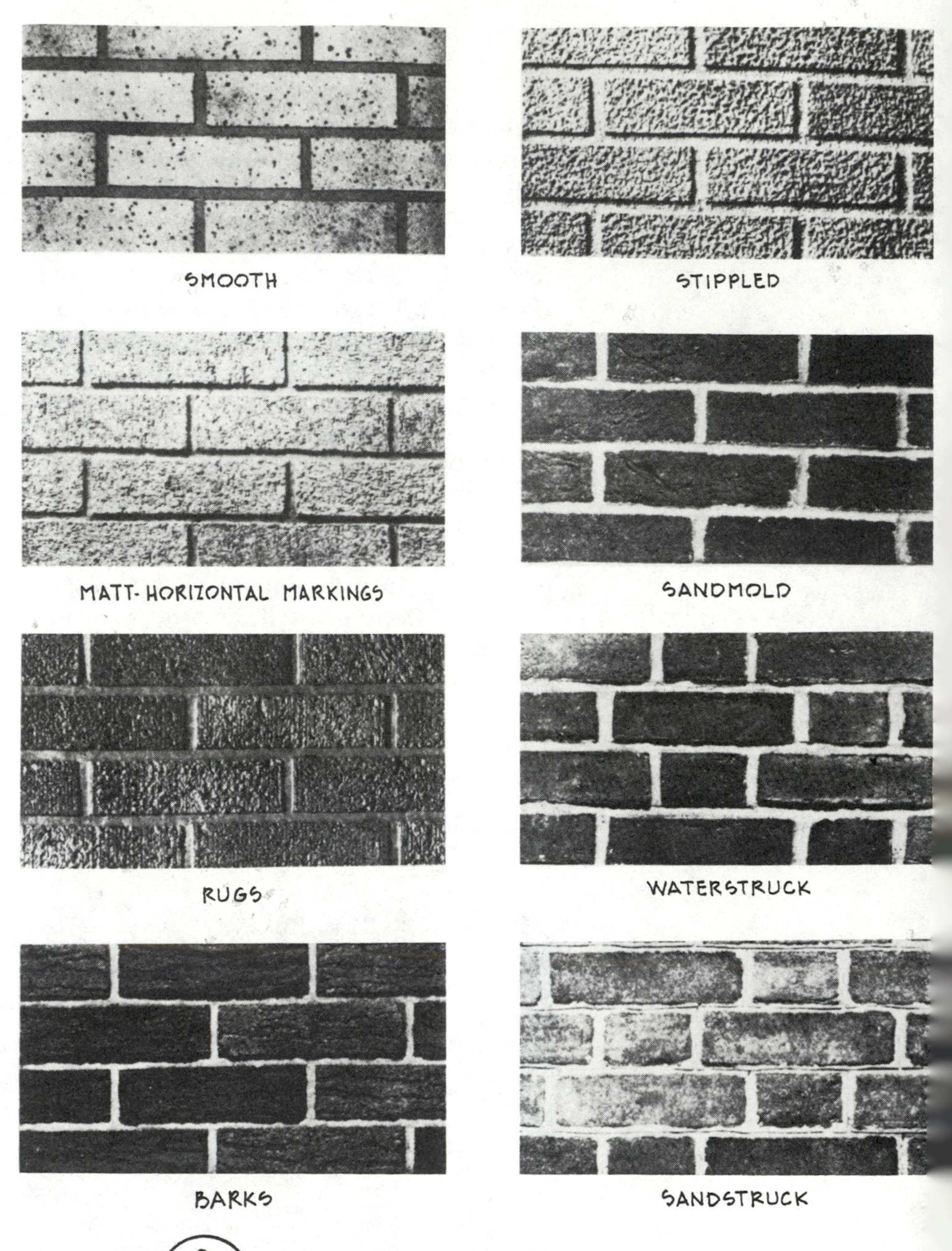

2-3 Typical clay brick textures. (*Photo courtesy BIA.*)

An off-bearing conveyor belt moves the "green" or wet bricks pas inspectors, who remove imperfect units and return them to the pug mill Satisfactory units are removed from the conveyor and stacked either b hand or machine on dryer cars in a prescribed pattern which will allow fre flow of drying air and kiln gases for burning. The stiff-mud process produce the hardest and most dense of the machine-made bricks, and delivers th highest volume of production.

The *soft-mud method* is the oldest method of production, and was use exclusively up until the nineteenth century (*see Fig. 2-4*). All handmad brick are formed by this process. Automated machinery can accomplish th molding more uniformly and efficiently than hand work, and is now widel used. The soft-mud process is particularly well suited for clays that contain too much natural water for the extrusion method. The clay is mixed with a 20 to 30% moisture content, approximately twice that of the stiff-mud clays. The soft, plastic mass is pressed into wooden molds by hand or machine. To prevent the clay from sticking, the molds are lubricated with sand or water. The resulting "sand-struck" or "water-struck" bricks have a unique appearance characterized by a rough, sandy surface, or a relatively smooth surface with minor variations from the texture of the individual molds (*see Fig. 2-3*). (Manufacturers often simulate the attractive rustic look of handmade brick by tumbling and roughening extruded brick.) In addition to having an attractive appearance, soft-mud units are more economical to lay because less precision is required, and masons can achieve a higher daily output.

The *dry-press method*, although it produces the most accurately formed brick, is no longer widely used. Dry-press bricks are costly to produce and install because of the labor involved in laying up such precision units. Clays of very low natural plasticity are required. A maximum of 10% water is added in tempering, and the relatively dry mix is pressed into steel molds by hydraulic plungers exerting a force of 500 to 1500 psi to form the unit.

Artist's engraving of a colonial brick-making operation. (*Courtesy BIA.*)

2.1.6 Drying

Green clay units coming from the molding or cutting machines may contain as much as 30% free moisture, depending on the forming process used. Before burning can begin, most of this excess moisture must be evaporated, either naturally by air drying or artificially in dryer kilns. The open sheds once used for air drying were affected by weather conditions, and the evaporation process could take anywhere from 7 days to 6 weeks. Modern brick plants now use separate dryer kilns supplied with waste heat from the exhaust of the firing kilns. Drying time has been reduced to a period of 24 to 48 hours, depending on the amount of moisture present. Temperatures normally range from 100 to 400°F, and must be carefully regulated, along with humidity, to prevent sudden changes which could cause excessive cracking or warping of the units.

2.1.7 Glazing

Although not common to all manufacturers, glazing is a highly specialized, carefully controlled procedure used in the production of decorative brick. *Ceramic glazes* are most widely used, and are normally applied at the end of the drying period. Chemical compounds are sprayed on the units before burning, and then subjected to normal firing temperatures to fuse the glaze to the body. They can produce a matte or high-gloss finish in a wide variety of colors. *Clay-coated glazes* produce a dull, nonreflecting, vitreously applied surface in softer tones than ceramic glazes. *Low-fired glazes* are used to obtain colors that cannot be produced at high firing temperatures. They are applied after the bricks have been burned to maturity, and then require refiring at lower temperatures. A *salt glaze* is produced by applying a vapor of sodium-iron silicate to the brick while it is at maximum firing temperature. The transparent finish shows the natural color of the fired brick under a lustrous gloss.

2.1.8 Burning

After excess moisture has been removed from the clay units and desired glazes, if any, have been applied, the bricks are ready for burning. This is one of the most specialized and critical steps in the manufacture of clay products. Burning is accomplished by controlled firing in a kiln to achieve ceramic fusion of the clay particles and hardening of the brick. Since so many of the properties of brick and clay tile depend on the method and control of firing, the development over the years of more sophisticated kilns has been instrumental in improving the quality and durability of clay masonry.

Originally, bricks were cured by sun drying. This permitted hardening by evaporation, but did not achieve the chemical fusion necessary for high strength. The first high-temperature firing of clay brick was done in scove kilns heated by hardwood fires and later by gas- and oil-fueled furnaces. Since the heat source was at the bottom of the kiln, and could be controlled only marginally, uneven firing resulted in hard-burned brick nearest the fire and soft, underburned brick at the top of the kiln. About 85% of the inexpensive, handmade brick imported from Mexico to the southwestern United States every year is characterized by this same uneveness o burning. Fires used there to heat the kilns are fueled by diverse combus tibles, and lack of control over the temperature results in a large percentage of underfired brick. Scove kilns were eventually replaced by beehive kilns. Heat was provided by more precisely controlled gas and oil fires in separate fireboxes, and could now be applied by a system of ducts from both the bottom and the top of the kiln. This distributed the heat more evenly and resulted in more uniform firing of the brick. However, the excessive time required for burning in a "periodic" kiln of this nature yields only a limited quantity of bricks.

Most plants now use continuous tunnel kilns, which are the latest and most sophisticated equipment for precisely controlled firing temperatures (*see Fig. 2-5*). The bricks travel through various temperature zones on flat rail cars. The clayware was first stacked on these cars for drying and then moved automatically into the first stage of the burning kiln.

Essentially, burning consists of subjecting the brick units to gradually increasing temperatures until fusion chemically alters the structure of the clay. The modern burning process consists of six phases, which are accomplished in the dryer kiln and in the preheating, firing, and cooling chambers of the burning kiln. The drying and evaporating of excess moisture is called the water-smoking stage. This is the initial preheat and may be done in separate dryers, or in the forward section of the burning kiln. This exposure to relatively low temperatures of up to 400°F begins the gradual, controlled heating process. Dehydration, or removal of the remaining trapped moisture, requires anywhere from 300 to 1800°F, oxidation from 1000 to 1800°F, and vitrification 1600 to 2400°F. It is only within this final temperature range that the silicates in the clay melt and fill the voids between the more refractory materials binding and cementing them together to form a strong, dense, hard-burned brick. The actual time and exact temperatures required throughout these phases varies according to the

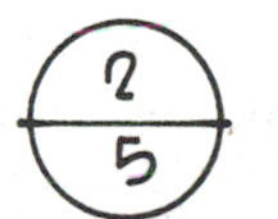

Modern tunnel kiln.

fusing characteristics and moisture content of the particular clay. Near the end of the vitrification phase, a reducing atmosphere may be created in which there is insufficient oxygen for complete combustion. This variation in the process is called flashing, and is intended to produce different hues and shadings from the natural clay colors. The final step in the firing of brick masonry is the cooling process. In a tunnel kiln, this normally requires up to 48 hours, as the temperatures must be very carefully and gradually reduced to avoid cracking and checking of the bricks.

2.1.9 Drawing and Storage

The removal of the bricks from the kiln is called drawing. The loaded flatcars leave the cooling chamber and are placed in a holding area until the bricks reach room temperature. They are then sorted as necessary, bound with metal bands into cubes equaling 500 standard-size bricks, and moved to storage yards or loaded directly onto trucks or railcars for shipment.

2.2 CONCRETE MASONRY

The development of modular concrete masonry was a logical outgrowth of the discovery of portland cement, and was in keeping with the manufacturing trends of the Industrial Revolution. Although the first rather unsuccessful attempts produced very heavy, unwieldy, and poorly adaptable units, the molding of ingredients such as quicklime and sand, or powdered lime and aggregates, into large blocks promised a bright new industry. With the invention and patenting of various block-making machines, unit concrete masonry began to have a noticeable effect on building and construction techniques of the late nineteenth and early twentieth centuries.

Concrete masonry is made from a relatively dry mix of cementitious materials, aggregates, water, and occasionally special admixtures. The material is molded and cured under controlled conditions to produce a strong, finished block that is suitable for use as a structural building element. Both the raw materials and the method of manufacture influence strength, appearance, and other critical properties of the block and are important in understanding the diversity and wide-ranging uses of concrete masonry products.

2.2.1 Aggregates

The aggregates in concrete block and concrete brick account for as much as 90% of their composition. The characteristics of these aggregates therefore play an important role in determining the properties and economy of the finished unit. Aggregates may be evaluated on the basis of (1) hardness, strength, and resistance to impact and abrasion; (2) durability in resisting freeze-thaw cycles; (3) uniformity in gradation of particle size; and (4) absence of foreign particles or impurities. A consistent range of fine and coarse particle sizes is necessary to produce a mixture that is easily workable and a finished surface that is dense and impervious. The presence of impurities such as silt and organic matter will impair strength and possibly cause surface blemishes.

There are two categories of aggregates used in the manufacture of concrete masonry: *normal-weight or heavyweight aggregates*, and *lightweight aggregates*. Early concrete masonry units were, for the most part made with the same heavyweight aggregates as those used today. Well-graded sand, gravel, crushed stone, and air-cooled slag are combined with other ingredients to produce a block that is heavy, strong, and fairly low in water absorption.

Efforts to make handling easier and more efficient led to the introduction of lightweight aggregates as a substitute for sand and gravel. Pumice, cinders, expanded slag, and other natural or processed aggregates are often used, and the units are sometimes marketed under special trade names, such as Haydite. Testing and performance have proven that lightweight aggregates generate other benefits that are of even greater importance than weight to masonry construction. Thermal, sound, and fire resistance are all affected, as well as color and texture. Lightweight aggregates increase the thermal and fire resistance of concrete masonry, but sound ratings are lower because of reduced density.

Colors resulting from the mix of aggregate and cement may range from white, to buff or brownish tones, to dull grays. Special colors may be produced by the use of selected crushed stones.

Surface textures depend on the size and gradation of aggregates. Classification of surface effects is only loosely defined as open or tight, with either fine, medium, or coarse texture (*see Fig. 2-6*). Although interpretation

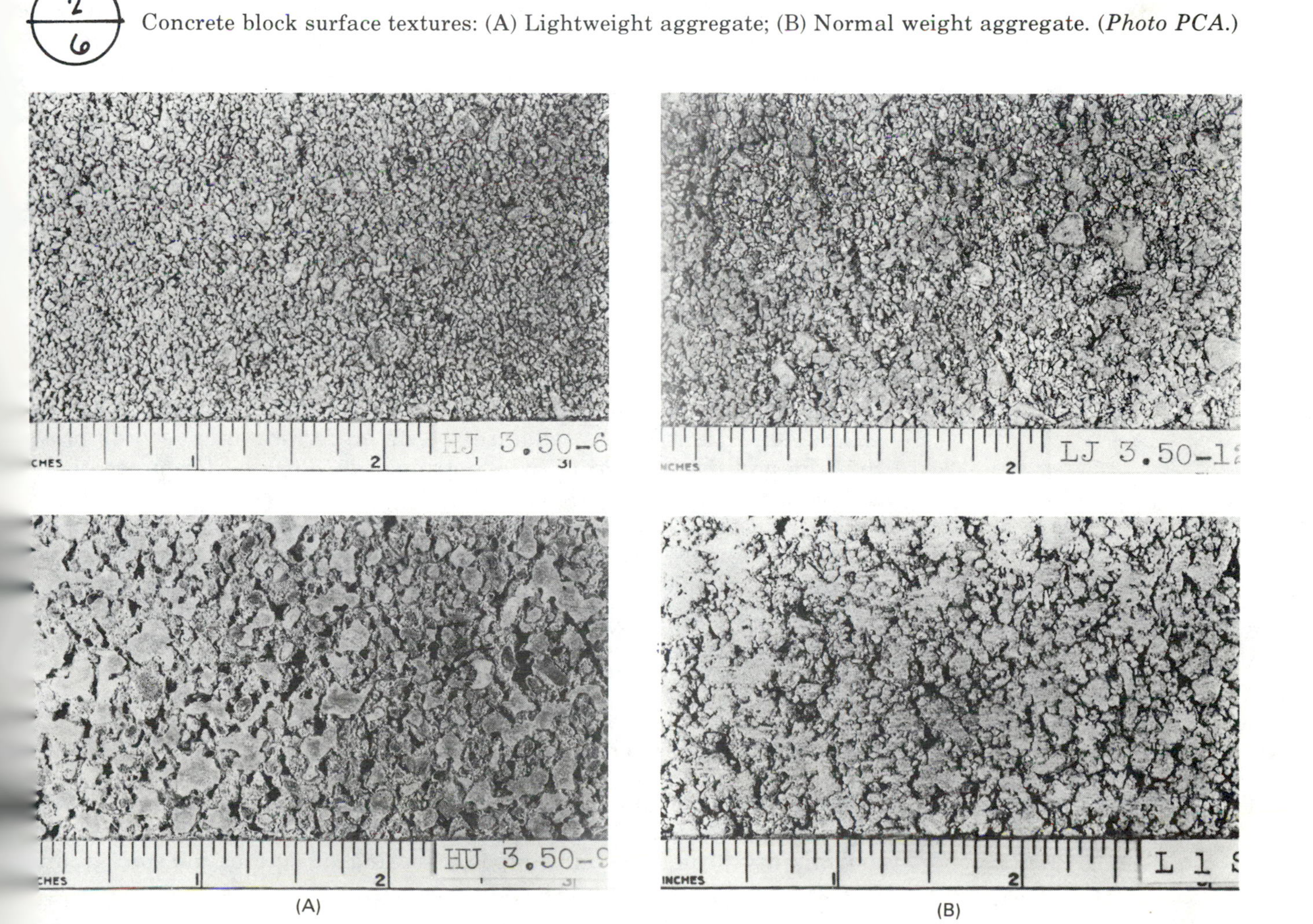

Concrete block surface textures: (A) Lightweight aggregate; (B) Normal weight aggregate. (*Photo PCA.*)

of these groups may vary in different geographical regions, an open surface is generally characterized by numerous large voids between the aggregate particles. A tight surface has few pores or voids of the size easily penetrated by water or sound. Fine textures are smooth, and consist of small, very closely spaced granular particles. Coarse textures are large grained and rough, and medium textures are, of course, intermediate. Both coarse and medium textures provide better sound absorption than the smoother faces, and are also recommended if the wall is to be plastered.

In general, local availability will determine the exact selection of aggregates, and the American Society for Testing and Materials (ASTM) has developed standards to regulate quality and composition. Within the limits of the required structural properties of the masonry, the architect may select different aggregates to serve nonstructural functions required by building type, occupant use, or aesthetics.

2.2.2 Cements

The cementitious material in concrete masonry is normally Type I, all-purpose portland cement. Type III, high-early-strength cement, is sometimes used to provide sufficient early strength to avoid distortion during the curing process. The air-entraining counterparts of these two cements (Types IA and IIIA) are sometimes used to improve the molding and off-bearing characteristics of the green units, and to increase resistance to weathering cycles. Air entrainment, however, will cause some strength reduction.

2.2.3 Admixtures

Admixtures marketed chiefly for use in concrete have shown few beneficial or desirable effects in the manufacture of concrete masonry. Air entrainment facilitates compaction and the close reproduction of the contours of the molds, but increased air content always results in lower compressive strengths. Calcium chloride accelerators speed the hardening or set of the units, but tend to increase shrinkage. Water-repelling agents reduce capillary action and the rate of water and moisture absorption of the units, but their exact effect is not known. ASTM specifications do not allow the use of any admixtures unless laboratory tests or performance records prove that the additives are in no way detrimental to the performance of the masonry.

Special colors can be produced by using pure mineral oxide pigments but this procedure is not widely recommended. Integral coloring of this nature is expensive, and although it can produce exotic hues, the danger of nonuniformity is much greater than when using colored aggregates. Natural aggregate colors are more durable, and more easily duplicated in the event of future additions to a building.

2.2.4 Manufacturing

The concrete masonry manufacturing process consists of six standard phases of production: (1) receiving and storing raw materials, (2) batching and mixing, (3) molding unit shapes, (4) curing, (5) cubing and storage, and (6) delivery of finished units (*see Fig. 2-7*).

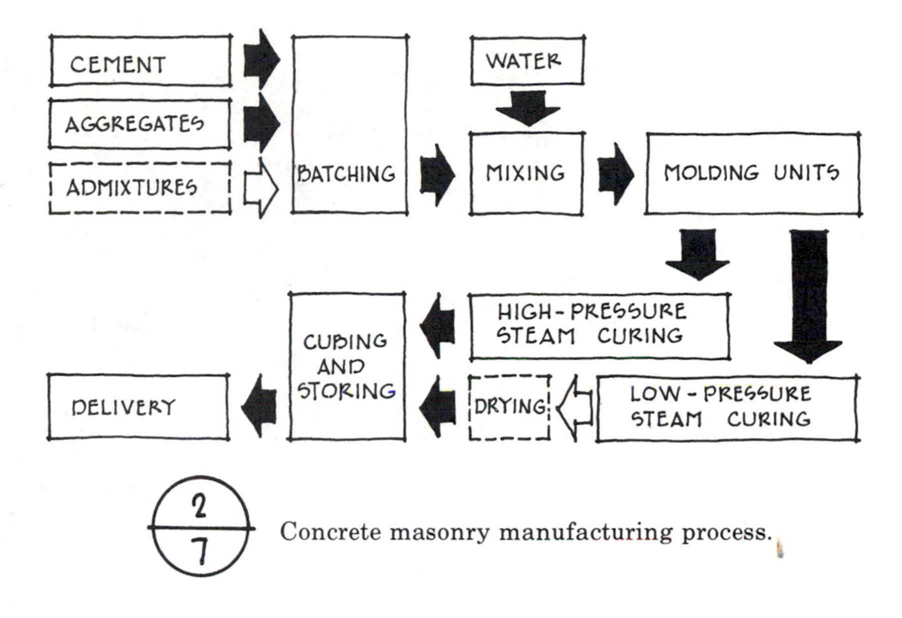

2-7 Concrete masonry manufacturing process.

2.2.5 Material Preparation

Materials are delivered in bulk quantities by truck or rail. Dissimilar aggregates are stored separately and later blended for different block types. Mixes will vary depending on aggregate weight, particle characteristics, and water absorption properties. Ingredients must be carefully regulated so that consistency in texture, color, dimensional tolerances, strength, and other physical properties is strictly maintained. Batching may be based on either weight or volume proportions, although weight batching is more commonly used.

The mixes are normally quite dry and would be classified as zero-slump concrete. Special high-strength units are made with more cement and water, but still have no slump. In the production of some special units, the batching is changed so that the mix will slump within controlled limits when the unit is removed from its mold. The unpredictable roll in texture produces the appearance of a handmade adobe brick.

2.2.6 Forming

Early block production consisted of hand-tamping the concrete mix into wooden molds with crude implements. A two-man team could turn out about 80 blocks a day. By the mid-1920s, automatic machines using mechanical action and gravity could produce as many as 3000 blocks a day. Today, units are molded with a combination of mechanical vibration and hydraulic pressure, and production is typically in the neighborhood of 1000 units per hour.

Custom concrete masonry units are often designed by architects. Special forms or textures can lend themselves to certain decorative motifs or to the expression of larger building shapes and masses (*see Fig. 2-8*). The economics of custom units generally requires that the volume of the order offset the expense of producing the molds. Most manufacturers now stock a series of typical customized units.

(A)

(B)

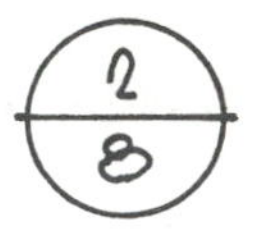

Paul Rudolph's Crawford Manor Apartments in New Haven, Connecticut (A). Customized masonry units used (B). (*Photos courtesy National Concrete Masonry Association.*)

2.2.7 Curing

Freshly molded blocks are lightly brushed to remove loose aggregate particles, then moved to a kiln or autoclave for accelerated curing.

The normal 28-day curing cycle for concrete is not conducive to mass production of unit masonry. As early as 1908, controlled tests were performed on experimental products cured by steam. In addition to hastening the hydration process, steam curing increases compressive strength, helps control shrinkage, and aids in uniformity of performance and appearance. Two types of steam curing are used in the industry: low-pressure and high-pressure.

About 80% of the block manufactured in the United States is produced by *low-pressure steam curing*. The first phase in the curing process is the holding or preset period of 1 to 3 hours. The units are allowed to attain initial hardening at normal temperatures of 70 to 100°F before they are exposed to steam. During the heating period, saturated steam is injected to raise the temperature to a maximum of 190°F. The exact time duration and temperature span recommended by the American Concrete Institute (ACI) depends on the composition of the cementitious materials and the type of aggregate used. Once maximum temperature is reached, the steam is shut off and a soaking period begins. Blocks are held in the residual heat and moisture for 12 to 18 hours or until the required strengths are developed. An artificial drying period may be used, with the temperature in the kiln raised to evaporate moisture.

The entire cycle is generally accomplished within a 24-hour period. Compressive strengths of 2- to 4-day-old units cured by low-pressure steam are approximately 90% of ultimate strength compared with only 40% for blocks of the same age cured by 28-day moist sprinkling. Steam-cured units are also characterized by a generally lighter color.

A variation of the low-pressure steam method adds a carbonation phase. Carbon dioxide is introduced into the drying atmosphere, causing irreversible shrinkage. Preshrinking decreases volume changes caused by atmospheric moisture conditions and reduces shrinkage cracking in the wall. Carbonation also increases tensile and compressive strength, hardness, and density.

High-pressure steam curing improves the quality and uniformity of concrete blocks, speeds production, and lowers manufacturing costs. Curing takes place in an autoclave kiln 6 to 10 feet wide and as much as 100 feet long (*see Fig. 2-9*).

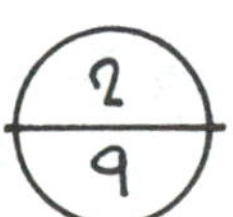

Loading racks of fresh units into an autoclave for high-pressure steam curing. (*Photo PCA.*)

A typical curing cycle consists of four operating periods: preset, temperature rise, constant temperature and pressure, and rapid pressure release. The low-heat preset period hardens the masonry sufficiently to withstand the high-pressure steam. The temperature rise period slowly brings both pressure and temperature within the autoclave to maximum levels where they remain constant for 5 to 10 hours. Temperature is actually the critical curing factor. Pressure is used as a means of controlling steam quality. Rapid pressure release or "blow-down" causes quick moisture loss from the units without shrinkage cracks. For normal-weight aggregates, the cycle produces relatively stable, air-dry blocks soon after removal from the autoclave. Lightweight blocks may require additional drying time to reach this same air-dry condition.

Blocks cured by autoclaving undergo different chemical reactions from those cured at low pressure. They are more stable and less subject to volume change caused by varying moisture conditions. The improved dimensional stability has greatly relieved the problem of control joints and shrinkage cracking in completed wall assemblies.

2.2.8 Surface Treatment

Concrete blocks are sometimes finished with ceramic, organic, or mineral *glazes*. These special finishes are applied after curing, and then subjected to heat treatment. The facings vary from epoxy or polyester resins to specially treated glass silica sand, colored ceramic granules, mineral glazes, and cementitious finishes. The treated surfaces are resistant to water penetra-

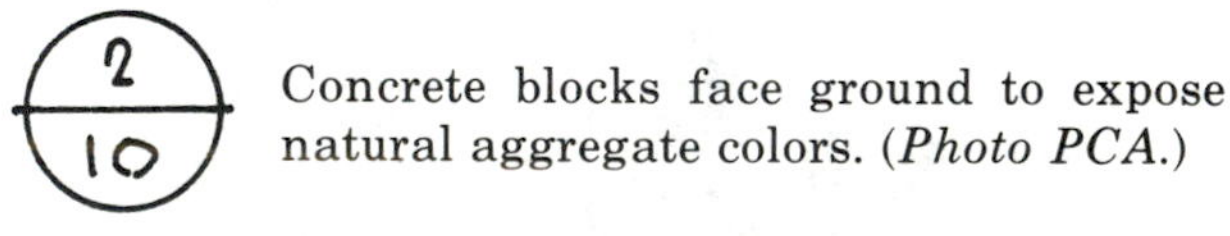

Concrete blocks face ground to expose natural aggregate colors. (*Photo PCA.*)

tion, abrasion, and cleaning compounds, and are very durable in high-traffic areas.

Surface *textures* are applied to hardened concrete blocks in a number of ways. Grinding the unit face produces a smooth, polished finish that highlights the aggregate colors (*see Fig. 2-10*). Ground faces can be supplementally treated with a wax or clear sealer. Sandblasting a block face exposes the underlying aggregate, adding color, texture, and dimension. Split-face units are produced by splitting ordinary blocks lengthwise. Solid units produce a rough stone appearance, while cored units are used to make split-ribbed block (*see Fig. 2-11*).

Detail of a veneer wall using split face and split rib units to create a pattern (Avner Nagger, architect).

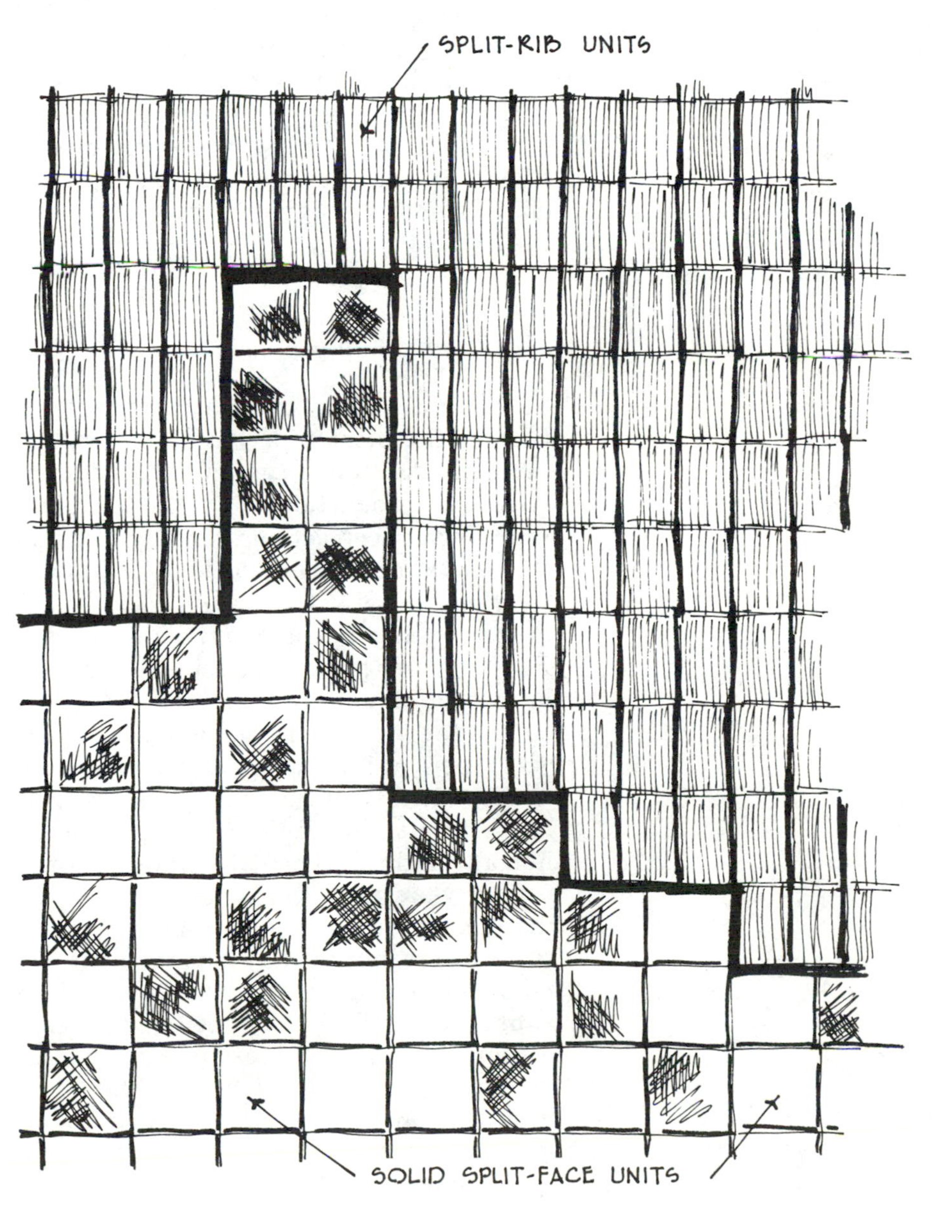

2.2.9 Cubing and Storage

Once the masonry units have been cured and dried, and any additional surface treatments have been completed, the blocks are removed from the steel curing racks and assembled in "cubes." The cubes are moved to a storage yard where, depending on the curing method used, they may remain in inventory anywhere from a few days to several weeks before they are shipped to a job site.

2.3 MORTAR AND GROUT MATERIALS

Mortar may account for as little as 7% of the volume of a masonry wall, but the role that it plays and the influence that it has on performance are far greater than the proportion indicates. The selection and use of various mortar ingredients directly affects the performance and bonding characteristics of masonry. It is important to be aware of the materials available and the effects they may have on the overall integrity of the masonry.

The principal components of masonry mortar and grout are cement, lime, sand, and water. Each of these constituents is essential in the performance of the mix. Cement gives the mortar strength and durability. Lime adds workability, water retentivity, and elasticity. Sand acts as a filler and contributes to economy and strength, and water imparts plasticity. To produce high-quality mortar, each of these ingredients must be of the highest quality.

2.3.1 Cements

Since its discovery in the early nineteenth century, portland cement has become the most widely used material of its kind. Portland cement is a carefully controlled combination of lime, silica, alumina, and iron oxide. Although production of portland cement is a lengthy and complicated procedure, it consists principally of grinding the raw materials, blending them to the desired proportions, and burning the mix in a rotary kiln until it reaches incipient fusion and forms clinkers. These hardened pellets are ground with gypsum, and the fine powder is then bagged for shipment (*see Fig. 2-12*). When mixed with water, portland cement undergoes hydration, an actual change in the chemical composition of the ingredients which causes the mass to harden and set.

There are five types of portland cement, each with different physical and chemical characteristics. Since the bonding properties required for mortar are significantly different from the qualities called for in concrete, not all of these types are suitable for masonry construction. For most ordinary mortars, Type I, all-purpose cement, is most widely used. In some instances, such as masonry catch basins or underground drainage structures where mortar may come in contact with sulfates in the soil, Type II portland cement can be used to resist chemical attack. A more common substitute for Type I is Type III, high-early-strength cement. This mixture attains ultimate compressive strength in a very short period of time, and generates greater heat during the hydration process. For use in cold-weather construction, these properties aid in keeping the wet mortar or grout from freezing and permit a reduction in the period of time required for protection against low temperatures.

Air-entraining portland cement, Types IA, IIA, IIIA, and so on, are

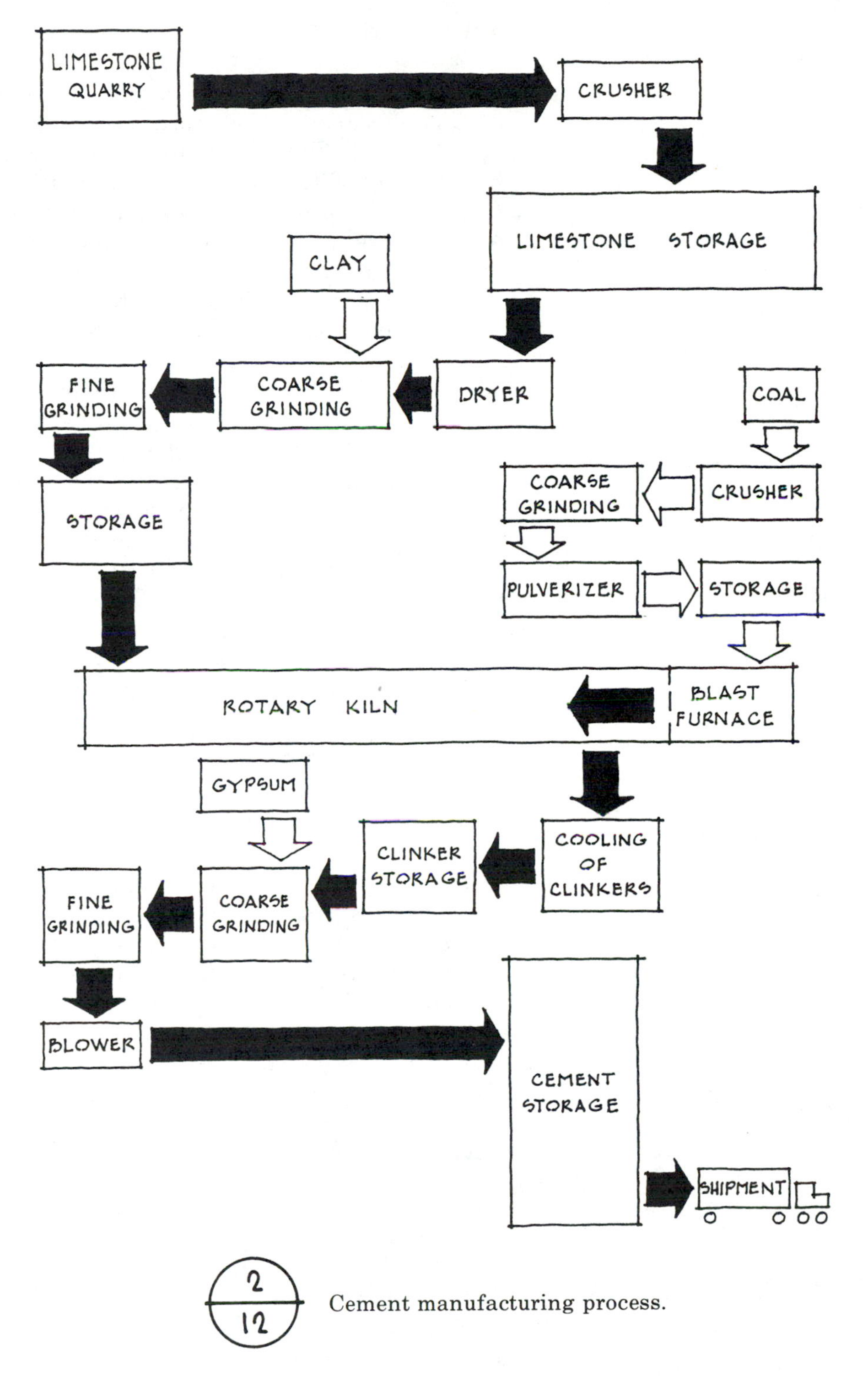

Cement manufacturing process.

made by adding selected chemicals to produce minute, well-distributed air bubbles in the hardened concrete or mortar. Increased air content improves workability, increases resistance to frost action and the scaling caused by chemical removal of snow and ice, and enhances moisture, sulfate, and abrasion resistance. Air-entrained mixes are not as strong as ordinary portland cements, and are detrimental in that they reduce the bond between mortar and masonry units. These mixes are not recommended by industry experts, particularly for engineered structural masonry.

2.3.2 Lime

Lime is perhaps the most valuable ingredient of masonry mortar and grout because of the many beneficial effects it has.

The term "lime" when used in reference to building materials means a burned form of lime derived from the calcination of sedimentary limestone. Powdered, hydrated lime is the most common and convenient form used today. Of the two grades of hydrated lime produced, only Type S is suitable for masonry work.

Mortar without lime is stiff and unworkable, high in compressive strength, but weak in bond and other required characteristics. Lime adds plasticity to the mortar, enabling the mason to spread it smoothly and fill joints completely, improving both productivity and workmanship. The plastic-flow quality of lime helps the mortar to permeate tiny surface indentations, pores, and irregularities in the masonry units and develop a strong physical bond. Lime also improves water retention. The mortar holds its moisture longer, resisting the suction of dry, porous masonry which can cause mortar to "pancake" before it has bonded to the units. Lime has low efflorescence potential because of its relatively high chemical purity. Its slow setting quality allows retempering. It undergoes less volume change or shrinkage than other mortar ingredients. It contributes to mortar integrity and bond by providing a measure of autogenous healing, the ability to combine with moisture and carbon dioxide to reconstitute or reknit itself if small cracks develop. In short, good-quality lime is an absolutely essential ingredient of good mortar.

2.3.3 Masonry Cements

In recent years, proprietary mixes of cement and workability agents or "masonry cements" have become popular with masons because of their convenience and good workability. However, the ASTM specification governing masonry cements places no limitations on chemical composition, and the ingredients as well as the properties and performance vary widely among the many brands available. Although the exact formula is seldom disclosed by the manufacturer, masonry cements generally contain equal parts of portland cement and ground, inert limestone. The use of inert limestone particles in lieu of hydrated lime reduces this ingredient to the capacity of a nonessential filler. Limestone is a carbonate form of calcium with completely different properties from lime. The absence of a suitable form of lime greatly impairs the quality and performance of mortar. Some masonry cement mixes contain as much as 30% entrained air, which significantly reduces strength and bond. Industry experts recommend only 6 to 15% air content.

Because of the wide variation of ingredients and proportioning, the properties of masonry cements cannot be accurately predicted solely on the basis of compliance with ASTM standards. They must be established through performance records and laboratory tests. Without substantiating evidence of adequate performance characteristics, some authorities discourage the use of masonry cements per se. Brick Institute of America (BIA) standards for recommended practice stipulate that without such laboratory test data, masonry cement mortars may not be used in structural or loadbearing masonry.

2.3.4 Sand

Sand aggregate accounts for a large proportion of the ingredients in mortar. Manufactured sands have sharp, angular grains, while natural sands obtained from banks, pits, and riverbeds have particles that are smoother and more round.

For use in mortar, sand must be clean, sound, and well graded according to requirements set by ASTM standards. Sand particles should always be washed and treated to remove foreign substances. Silt can cause mortar to stick to the trowel, and can impair proper bond of the cementitious material to the sand particles. Clay and organic substances reduce mortar strength and can cause brownish stains varying in intensity from batch to batch.

The sand in masonry mortar and grout acts as a filler. The cementitious paste must completely coat each particle to lubricate the mix. Sands that have a high percentage of large grains produce voids between the particles, and will make harsh mortars with poor workability and low resistance to moisture penetration. When the sand is well proportioned of both fine and coarse granules, the smaller grains fill these voids and produce mortars that are more workable and plastic. If, on the other hand, the percentage of fines is too high, more cement is required to coat the particles thoroughly, and the mortar will be weaker and more porous. Many commercially available sands do not meet ASTM gradation requirements for mortar. Such shortcomings may be easily and inexpensively corrected by the addition of the deficient fine or coarse sands.

2.3.5 Water

Water for masonry mortar must be clean and free of harmful amounts of acids, alkalis, and organic materials. Whether the water is drinkable is not in itself a consideration, as some drinking water contains appreciable amounts of soluble salts, such as sodium and potassium sulfate, which can contribute to efflorescence. If necessary, laboratory analysis of the water supply should be used to verify suitability.

2.3.6 Admixtures

Although admixtures are often used with some success in concrete construction, they can have adverse effects on the properties and performance of masonry mortar and grout. Recommended specifications do not incorporate, nor in fact even recognize, admixtures of any kind. Conventional portland cement–lime mortars do not require additives.

A variety of proprietary materials are available which claim to increase workability or water retentivity, lower the freezing point, and accelerate or retard the set. Although they may produce these effects, they can also reduce compressive strength, impair bond, contribute to efflorescence, increase shrinkage, or corrode metal accessories and reinforcing steel. As a rule, if admixtures are used to produce or enhance some special property in the mortar, the specifications should require that laboratory tests establish the effects on strength, bond, volume change, durability, and density.

3

CLAY PRODUCTS

Clay as a raw material is most valued for its ceramic characteristics. When subjected to high firing temperatures in a kiln, the silicates in clay melt, fusing the particles to a density that approaches vitrification. The resulting strength and weather resistance make brick, tile, and terra cotta among the most durable of building materials.

3.1 BRICK

There are many different shapes, sizes, and types of brick. The first distinction made by ASTM standards is between building brick and facing brick, based on appearance of the unit.

Building brick (sometimes called common brick) is used primarily as a structural material or backing for other finishes, where strength and durability are of more importance than appearance. Under ASTM C62, Standard Specification for Building Brick, grading is based on physical requirements and directly related to durability and resistance to weathering (*see Fig. 3-1*).

Grade SW (severe weathering) is used where a high degree of resistance to frost action is required and where conditions of exposure indicate the possibility of freezing when the unit is permeated with water. Grade SW is recommended for below-grade installations in moderate and severe weathering areas, and for horizontal or nonvertical surfaces in all weathering conditions. Grade MW (moderate weathering) may be used in geographical areas subject to freezing only when the brick will not be exposed to water permeation. Grade MW is more commonly used in moderate and negligible

TABLE A GRADE REQUIREMENTS FOR BRICK EXPOSURES			
	Weathering index		
Exposure	*Less than 50*	*50 to 500*	*500 and greater*
In vertical surfaces			
In contact with earth	MW	SW	SW
Not in contact with earth	MW	MW	MW
In other than vertical surfaces			
In contact with earth	SW	SW	SW
Not in contact with earth	MW	SW	SW

TABLE B PHYSICAL REQUIREMENTS						
	Minimum compressive strength (brick flatwise), gross area (psi)		*Maximum water absorption by 5-hr boiling (%)*		*C/B Maximum saturation coefficient* *	
Designation	*Average of 5 brick*	*Individual*	*Average of 5 brick*	*Individual*	*Average of 5 brick*	*Individual*
Grade SW	3000	2500	17.0	20.0	0.78	0.80
Grade MW	2500	2200	22.0	25.0	0.88	0.90
Grade NW	1500	1250	No limit	No limit	No limit	No limit

*The saturation coefficient is the ratio of absorption by 24-hr submersion in cold water (C) to that after 5-hr submersion in boiling water (B).

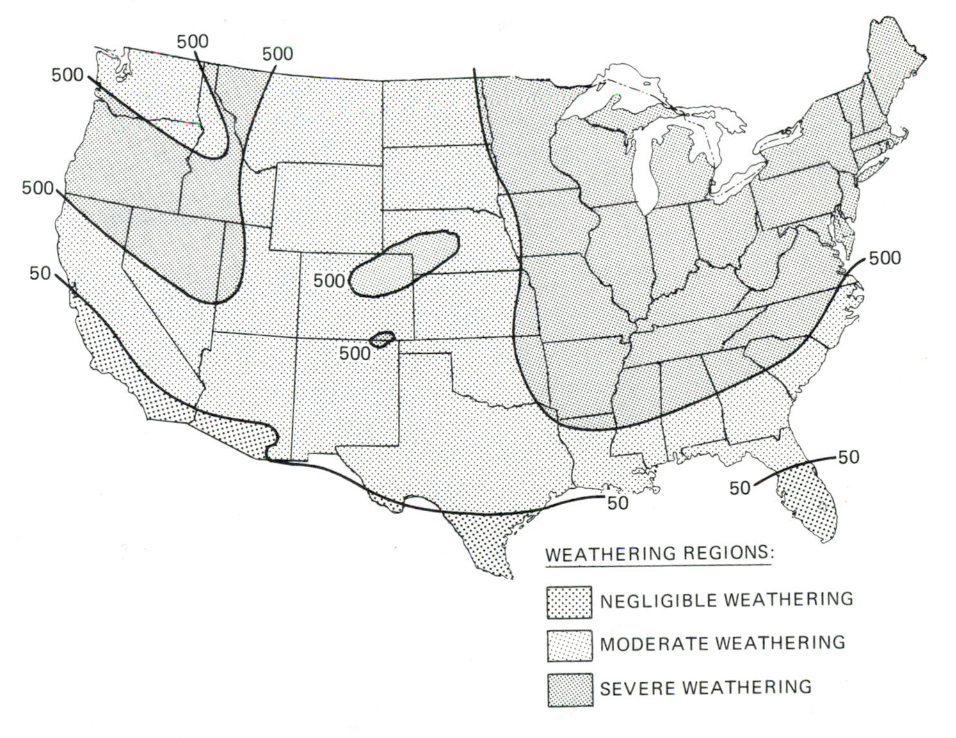

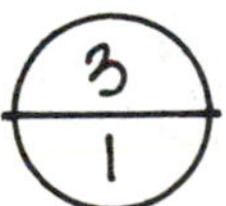

Requirements for building brick, from ASTM C62. (*Copyright, American Society for Testing and Materials, 1916 Race Street, Philadelphia, Pa. 19103. Reprinted, with permission.*)

weathering regions for above-grade vertical installations. Grade NW (no weathering) is permitted only for vertical applications in interior work where there will be no weather exposure.

Moisture enters the face of a brick by capillary action. When present in sufficient quantity and for an extended time, water will penetrate through the brick and approximate the laboratory condition (resulting from 24-hour submersion in cold water) defined as "permeated." Permeation may easily occur in units exposed in parapet walls, horizontal surfaces, and retaining walls, but is unlikely for ordinary exterior wall exposures if the brick is suitably protected by flashings and/or overhanging eaves. Under most circumstances, permeation of the brick in building walls would result only from defective workmanship or faulty drainage.

Face brick is used for exposed areas where appearance is an important design criteria. These units are normally selected for color, dimensional tolerances, uniformity, surface texture, and limits on the amount of cracks and defects. ASTM C216, Standard Specification for Facing Brick, covers Grades SW and MW, which correspond to the same physical and environmental requirements as those for building brick. Within each of these grades, face brick may be produced in three specific types based on aesthetic effect. Type FBX is for general use in exposed exterior and interior applications where a high degree of mechanical perfection, narrow color range, and minimum size variation are required. Type FBS permits wider color ranges and greater variations in size. Type FBA is manufactured with characteristic architectural effects, such as distinctive irregularity in size, color, and texture of the individual units (*see Fig. 3-2*). Extruded, stiff-mud brick may be produced in any of the three types by increasing the amount of texturing and roughening the units receive after leaving the die. Dry-press brick normally falls well within the strict tolerances required for Type FBX, but is not widely used because of the higher labor costs associated with laying up such precision units. Soft-mud brick, including hand-molded units, are rustic in appearance and meet the specifications only for Type FBA. Both labor economy and distinctive appearance make FBA brick very popular with architects and builders.

Requirements for face brick, from ASTM C216. (*Copyright, American Society for Testing and Materials, 1916 Race Street, Philadelphia, Pa. 19103. Reprinted with permission.*)

TABLE A MAXIMUM PERMISSIBLE EXTENT OF CHIPPAGE FROM THE EDGES AND CORNERS OF FINISHED FACE OR FACES INTO THE SURFACE

	Chippage (in.) in from:	
	Edge	*Corner*
Type FBX	$\frac{1}{8}$	$\frac{1}{4}$
Type FBS (smooth)*	$\frac{1}{4}$	$\frac{3}{8}$
Type FBS (rough)†	$\frac{5}{16}$	$\frac{1}{2}$
Type FBA	As specified by the purchaser	

*Smooth texture is the unbroken natural die finish.
†Rough texture is the finish produced when the face is sanded, combed, scratched, or scarified or the die skin on the face is entirely broken by mechanical means such as wire-cutting or wire-brushing.

TABLE B PERCENTAGES OF SHIPMENT THAT MAY BE ALLOWED CHIPPAGE OVER MAXIMUM PERMISSIBLE IN TABLE A			
		Chippage (in.) in from:	
	Percentage Allowable	*Edge*	*Corner*
Type FBX	5	$\frac{1}{4}$	$\frac{3}{8}$
Type FBS (smooth)	10	$\frac{5}{16}$	$\frac{1}{2}$
Type FBS (rough)	15	$\frac{7}{16}$	$\frac{3}{4}$
Type FBA	As specified by the purchaser		

TABLE C TOLERANCES ON DIMENSIONS		
	Maximum permissible variation from specified dimension, plus or minus (in.)	
Specified dimension (in.)	*Type FBX*	*Type FBS*
3 and under	$\frac{1}{16}$	$\frac{3}{32}$
Over 3 to 4, incl.	$\frac{3}{32}$	$\frac{2}{16}$
Over 4 to 6, incl.	$\frac{2}{16}$	$\frac{3}{16}$
Over 6 to 8, incl.	$\frac{5}{32}$	$\frac{4}{16}$
Over 8 to 12, incl.	$\frac{7}{32}$	$\frac{5}{16}$
Over 12 to 16, incl.	$\frac{9}{32}$	$\frac{3}{8}$

TABLE D TOLERANCES ON DISTORTION		
	Maximum permissible distortion (in.)	
Maximum face dimension (in.)	*Type FBX*	*Type FBS*
8 and under	$\frac{1}{16}$	$\frac{3}{32}$
Over 8 to 12, incl.	$\frac{3}{32}$	$\frac{1}{8}$
Over 12 to 16, incl.	$\frac{1}{8}$	$\frac{5}{32}$

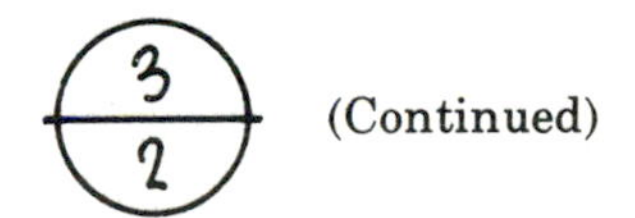

(Continued)

Used brick are sometimes specified by architects because of their weathered appearance and broad color range. In many instances, these specimens are not totally in compliance with accepted standards of durability for exposed usage. Sources for salvaged masonry are generally buildings at least 30 to 40 years old constructed of solid masonry walls with hard-burned brick on the exterior and inferior "salmon" brick as backup. Since the color differences used in originally sorting and selecting the brick become obscured with exposure and contact with mortar, salmon brick may inadvertently be used for an exterior exposure, where they can undergo rapid and excessive deterioration. Building code requirements may vary regarding the use of salvaged brick, and should be consulted prior to their selection and specification.

Imported Mexican brick give a distinctive, handcrafted quality to masonry. They also lack uniformity in compliance with standards of durability. Officials at the Brick Institute of Texas estimate that as much as 85% of the imported brick sold in that state each year are found to be substandard in water absorption, weathering, and compression tests. The abbreviated burning period and low firing temperatures typical of some Mexican brick plants produce units that are extremely soft and porous, causing severe maintenance problems even in relatively dry climates (*see Fig. 3-3*).

3/3 Substandard brick. (*Photos courtesy Brick Institute of Texas.*)

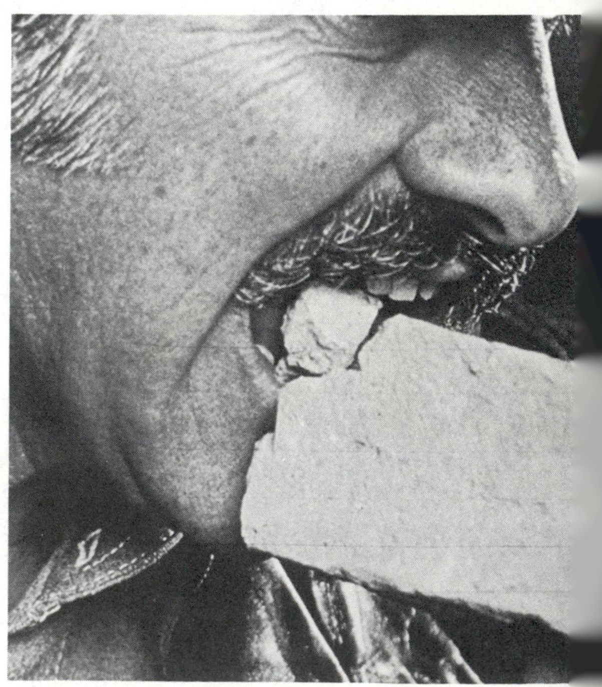

Underburned, salmon brick are not acceptable under any building code for use in areas exposed to weather. Some standards do make allowance under certain circumstances for unburned clay products such as adobe brick. Unless they are protected with a waterproofing agent, however, sun-dried brick are susceptible to moisture penetration, which can result in damage or disintegration. Commercially available units treated with emulsified asphalt have been tested and approved for use by some local authorities. Traditional blends of clay with straw or fiber reinforcing that have not been treated or certified must either be used in a completely sheltered location or receive a protective plaster or stucco coating.

3.1.1 Sizes and Shapes

Masonry unit sizes and shapes have proliferated over the last 5000 years to meet various regional standards and requirements throughout the world. Even within the United States, unit dimensions may vary from one area to the next and be further confused by different names for the same size of unit. At one time, there were only three commonly used brick sizes: "standard," Norman, and Roman. Industry demand has increased that number substantially. Brick is now available in thicknesses or bed depths ranging from 3 to 12 in., heights from 2 to 8 in., and lengths of up to 16 in. Production includes both nonmodular and modular sizes conforming to the 4-in. grid system of structural and material coordination. Some typical units are illustrated in *Fig. 3-4. Figure 3-5* lists several of the modular sizes, their recommended joint thicknesses, and coursing heights.

For clarity in specifying brick, best practice requires identification first by dimensions, then by name. It is also recommended that actual dimensions be used, listed thickness × height × length. Nominal dimensions may vary from actual sizes by the thickness of mortar joint with which the unit was designed to be used. Firebrick, however, are laid without true mortar beds, and sizes given should always be actual dimensions. Mortar joint thicknesses are determined by the type and quality of the unit. In general, glazed brick are laid with a $\frac{1}{4}$-in. joint, face brick with a $\frac{3}{8}$- or $\frac{1}{2}$-in. joint, and building brick with a $\frac{1}{2}$-in. joint.

The bricks in *Fig. 3-4* show a variety of core designs. Although they are typical of commercially available products, the corings vary with the manufacturer, and are not necessarily typical for or limited to the particular size with which they are shown. These design modifications have been developed over the years to facilitate, among other things, ease of forming, ease of handling, and improved grip and mortar bond. The oldest pattern is an identation or "frog" producible only by dry-press or soft-mud processes. Originally conceived as a scheme for reducing the weight of a solid unit, this depression provided a space for identification by early craftsmen, who would write the name of the reigning monarch during the time of construction. This practice has since aided archeologists in dating ancient buildings. Still in use today, the "frog" is now often stamped with the name of the brick manufacturer or date of production. Extruded brick are made with a series of holes cored through the unit which, for "solid brick" as defined by ASTM, may not exceed 25% of the area in the bearing plane. In addition to the cores, a $\frac{3}{4} \times \frac{3}{4}$-in. notch may be cut in one end of 6-in.-thick units to serve as a window jamb slot. Roman brick are made in double form and broken into two units on the job site, leaving a rough, exposed edge.

The trend in development of different brick sizes has been toward modular coordination and toward slightly larger dimensions. Most modern

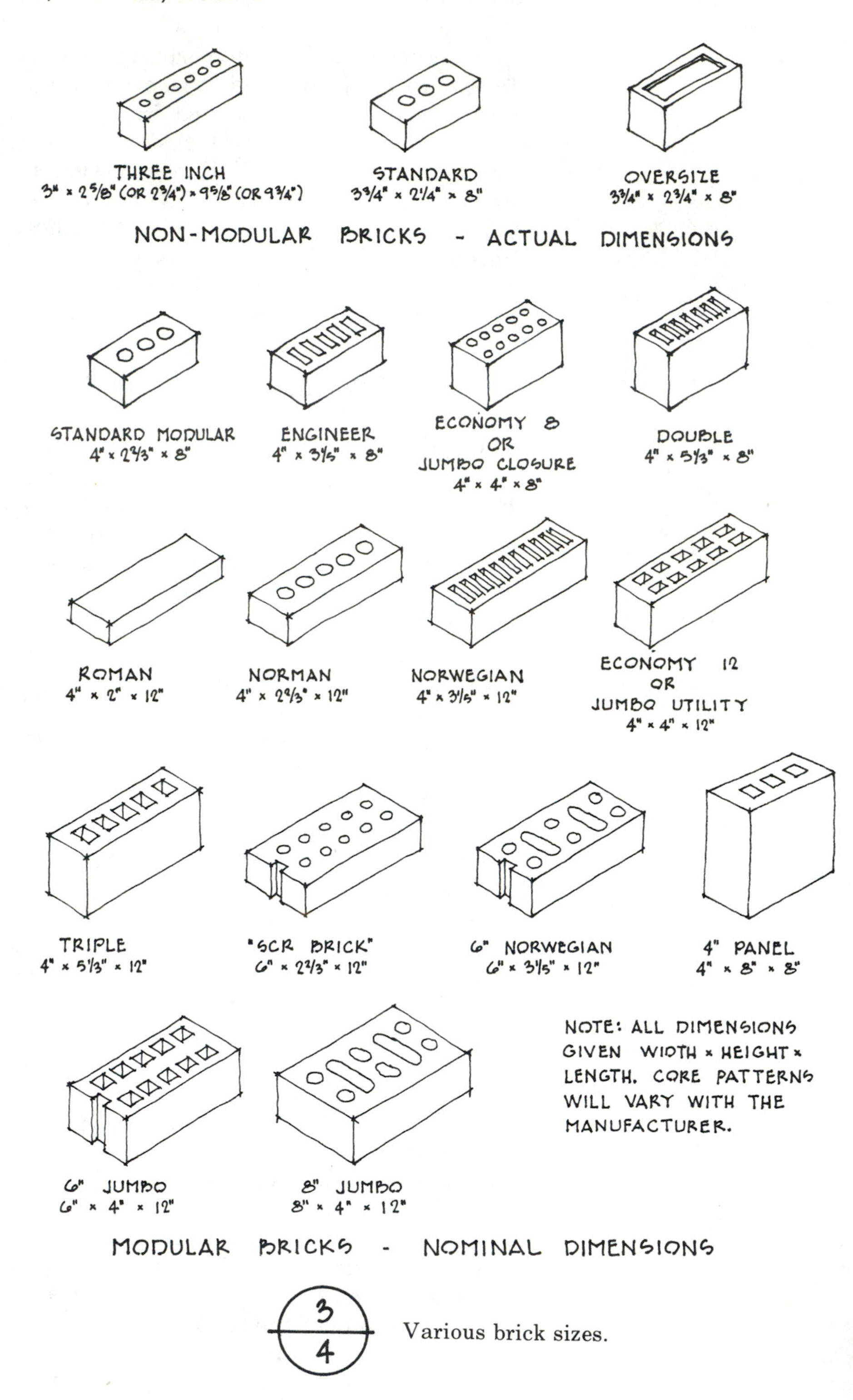

3-4 Various brick sizes.

masonry products, including clay tile and concrete block, are designed for connection at 8- or 16-in. course heights. For example, two courses of concrete block with mortar joints will equal 16 in. vertically, while three five, or six courses of various size brick, and two, three, or four courses of clay tile equal the same height. This permits horizontal mechanical connection between the facing and backup elements of a multi-wythe wall. One of the first oversize brick units was introduced by the Brick Institute of

America (BIA). The SCR brick was developed for use in single-wythe, 6-in. masonry walls, which are now permitted by many building codes for one-story construction. Larger brick sizes have also increased labor production. Although a mason can lay fewer of the large units in a day, the square

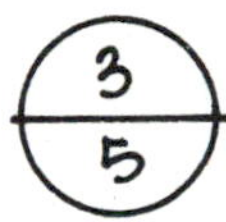

Modular brick sizes and coursing heights. (*From Brick Institute of America*, Technical Note 10B, *BIA, McLean, Va.*)

	Sizes							
	Nominal dimensions (in.)				Manufactured dimensions (in.)			
Unit designation	T	H	L	Joint thickness (in.)	T	H	L	Modular coursing (in.)
Standard Modular	4	$2\frac{2}{3}$	8	$\frac{3}{8}$	$3\frac{5}{8}$	$2\frac{1}{4}$	$7\frac{5}{8}$	3C = 8
				$\frac{1}{2}$	$3\frac{1}{2}$	$2\frac{1}{4}$	$7\frac{1}{2}$	
Engineer	4	$3\frac{1}{5}$	8	$\frac{3}{8}$	$3\frac{5}{8}$	$2\frac{13}{16}$	$7\frac{5}{8}$	5C = 16
				$\frac{1}{2}$	$3\frac{1}{2}$	$2\frac{11}{16}$	$7\frac{1}{2}$	
Economy 8 or Jumbo Closure	4	4	8	$\frac{3}{8}$	$3\frac{5}{8}$	$3\frac{5}{8}$	$7\frac{5}{8}$	1C = 4
				$\frac{1}{2}$	$3\frac{1}{2}$	$3\frac{1}{2}$	$7\frac{1}{2}$	
Double	4	$5\frac{1}{3}$	8	$\frac{3}{8}$	$3\frac{5}{8}$	$4\frac{15}{16}$	$7\frac{5}{8}$	3C = 16
				$\frac{1}{2}$	$3\frac{1}{2}$	$4\frac{13}{16}$	$7\frac{1}{2}$	
Roman	4	2	12	$\frac{3}{8}$	$3\frac{5}{8}$	$1\frac{5}{8}$	$11\frac{5}{8}$	2C = 4
				$\frac{1}{2}$	$3\frac{1}{2}$	$2\frac{1}{4}$	$11\frac{1}{2}$	
Norman	4	$2\frac{2}{3}$	12	$\frac{3}{8}$	$3\frac{5}{8}$	$2\frac{1}{4}$	$11\frac{5}{8}$	3C = 8
				$\frac{1}{2}$	$3\frac{1}{2}$	$2\frac{1}{4}$	$11\frac{1}{2}$	
Norwegian	4	$3\frac{1}{5}$	12	$\frac{3}{8}$	$3\frac{5}{8}$	$2\frac{13}{16}$	$11\frac{5}{8}$	5C = 16
				$\frac{1}{2}$	$3\frac{1}{2}$	$2\frac{11}{16}$	$11\frac{1}{2}$	
Economy 12 or Jumbo Utility	4	4	12	$\frac{3}{8}$	$3\frac{5}{8}$	$3\frac{5}{8}$	$11\frac{5}{8}$	1C = 4
				$\frac{1}{2}$	$3\frac{1}{2}$	$3\frac{1}{2}$	$11\frac{1}{2}$	
Triple	4	$5\frac{1}{3}$	12	$\frac{3}{8}$	$3\frac{5}{8}$	$4\frac{15}{16}$	$11\frac{5}{8}$	3C = 16
				$\frac{1}{2}$	$3\frac{1}{2}$	$4\frac{13}{16}$	$11\frac{1}{2}$	
SCR brick	6	$2\frac{2}{3}$	12	$\frac{3}{8}$	$5\frac{5}{8}$	$2\frac{1}{4}$	$11\frac{5}{8}$	3C = 8
				$\frac{1}{2}$	$5\frac{1}{2}$	$2\frac{1}{4}$	$11\frac{1}{2}$	
6-in. Norwegian	6	$3\frac{1}{5}$	12	$\frac{3}{8}$	$5\frac{5}{8}$	$2\frac{13}{16}$	$11\frac{5}{8}$	5C = 16
				$\frac{1}{2}$	$5\frac{1}{2}$	$2\frac{11}{16}$	$11\frac{1}{2}$	
6-in. Jumbo	6	4	12	$\frac{3}{8}$	$5\frac{5}{8}$	$3\frac{5}{8}$	$11\frac{5}{8}$	1C = 4
				$\frac{1}{2}$	$5\frac{1}{2}$	$3\frac{1}{2}$	$11\frac{1}{2}$	
8-in. Jumbo	8	4	12	$\frac{3}{8}$	$7\frac{5}{8}$	$3\frac{5}{8}$	$11\frac{5}{8}$	1C = 4
				$\frac{1}{2}$	$7\frac{1}{2}$	$3\frac{1}{2}$	$11\frac{1}{2}$	

footage of wall area completed is greater, less mortar is required, and jobs progress faster.

It should be noted that "standard" brick produced before 1946-1947, when modular coordination was adopted, had actual heights of $2\frac{1}{4}$ in. (designed to lay up three courses to 8 in.). This size is still widely available so that in renovation or restoration work, coursing heights can be effectively matched.

In addition to the common rectangular cut, brick may be formed in many special shapes for specific job requirements. Some of the more commonly used items include square and hexagonal pavers, bullnose and stair tread units, caps, sills, special corner brick, and wedges for arch construction (*see Fig. 3-6*). Unique custom shapes may be available on

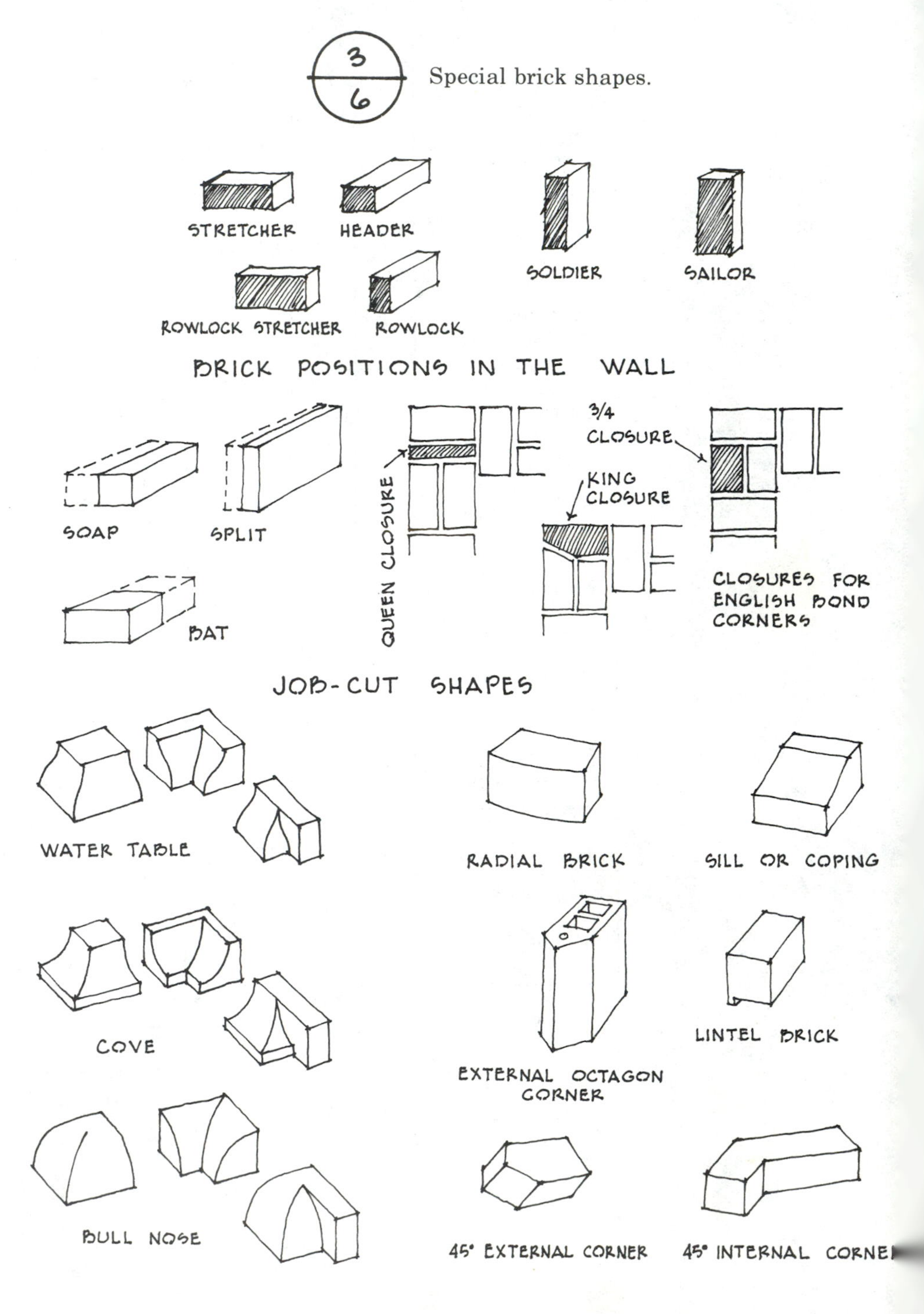

3-6 Special brick shapes.

request from some manufacturers, but can be expensive to produce depending on the size of the order. Job-cut shapes must often be made for corners or other locations where a full brick length may not fit. These special units may be called half or bat, three-quarter closure, quarter closure, queen closure, king closure, and split.

The most unusual examples of customized masonry are sculptured pieces handcrafted from the green clayware before firing. The unburned units are firm enough to allow the artist to work freely without damage to the brick body, but sufficiently soft for carving, scraping, and cutting. After execution of the design, the units are returned to the plant for firing and the relief is permanently set in the brick face (*see Fig. 3-7*).

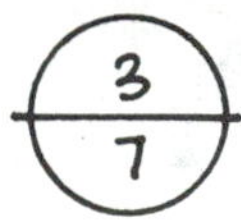

Sculptured brick panels at Loew's Anatole Hotel, Dallas. Beran and Shelmire, architects; Mara Smith, sculptor. (*Photo courtesy BIA.*)

3.1.2 Special-Purpose Bricks

Special-purpose bricks serve many functions in architecture and industry. Refractory brick or *firebrick*, for instance, are used in furnaces, chimney stacks, fireboxes, and ovens. The fire clay from which they are made has a much higher fusing point than that of ordinary clay or shale. Once the initial kiln firing has been accomplished, firebrick are extremely resistant to high temperatures without cracking, decomposition, or distortion. Firebrick are normally heavier and softer than other units and are produced in a slightly larger size ($4\frac{1}{2} \times 2\frac{1}{2} \times 9$ in.), to be laid with a thin coating of fire clay mortar in lieu of standard mortar joints. Fire clays typically burn to a white or buff color.

Refractory brick of different chemical composition are covered in a series of ASTM standards, and are graded according to fusion temperature, porosity, spalling strength, resistance to rapid temperature changes, thermal conductivity, and heat capacity. Some commonly used types of refractory brick are alumina brick, chrome brick, magnesite brick, and silica brick. The highly specialized nature of refractory design requires consultation with manufacturers to assure correlation between design needs and product specifications.

Glazed brick are fired with ceramic coatings which fuse to the clay body in the kiln and produce an impervious surface in clear or color, matte or gloss finish. Standards for glazed brick are outlined in ASTM C126, with requirements covering compressive strength, imperviousness, chemical resistance, crazing, and limitations on distortion and dimensional variation. They may be used for exposed interior and exterior walls and partitions, but are not recommended for copings or other horizontal surfaces. Units are manufactured in Grade S (select) and Grade SS (select sized, or ground edge), where a high degree of mechanical perfection, narrow color range, and minimum variation in size are required. Units may be either Type I, single-faced, or Type II, double-faced (opposite faces glazed). Type II are generally special-order items and are not widely used. Glazed brick are commonly available in "standard," oversize, Norman, and modular sizes, and in stretchers, jambs, corners, sills, and other supplementary shapes (*see Fig. 3-8*).

The naturally high abrasion resistance of ceramic clay products makes them very durable as paving materials. *Paving brick* are unique in color, pattern, and texture and are often specified as a wearing surface for roadways, walks, patios, drives, and interior floors. ASTM C902, Standard Specification for Pedestrian and Light Traffic Paving Brick, lists specific physical requirements. The BIA recommends a minimum average com-

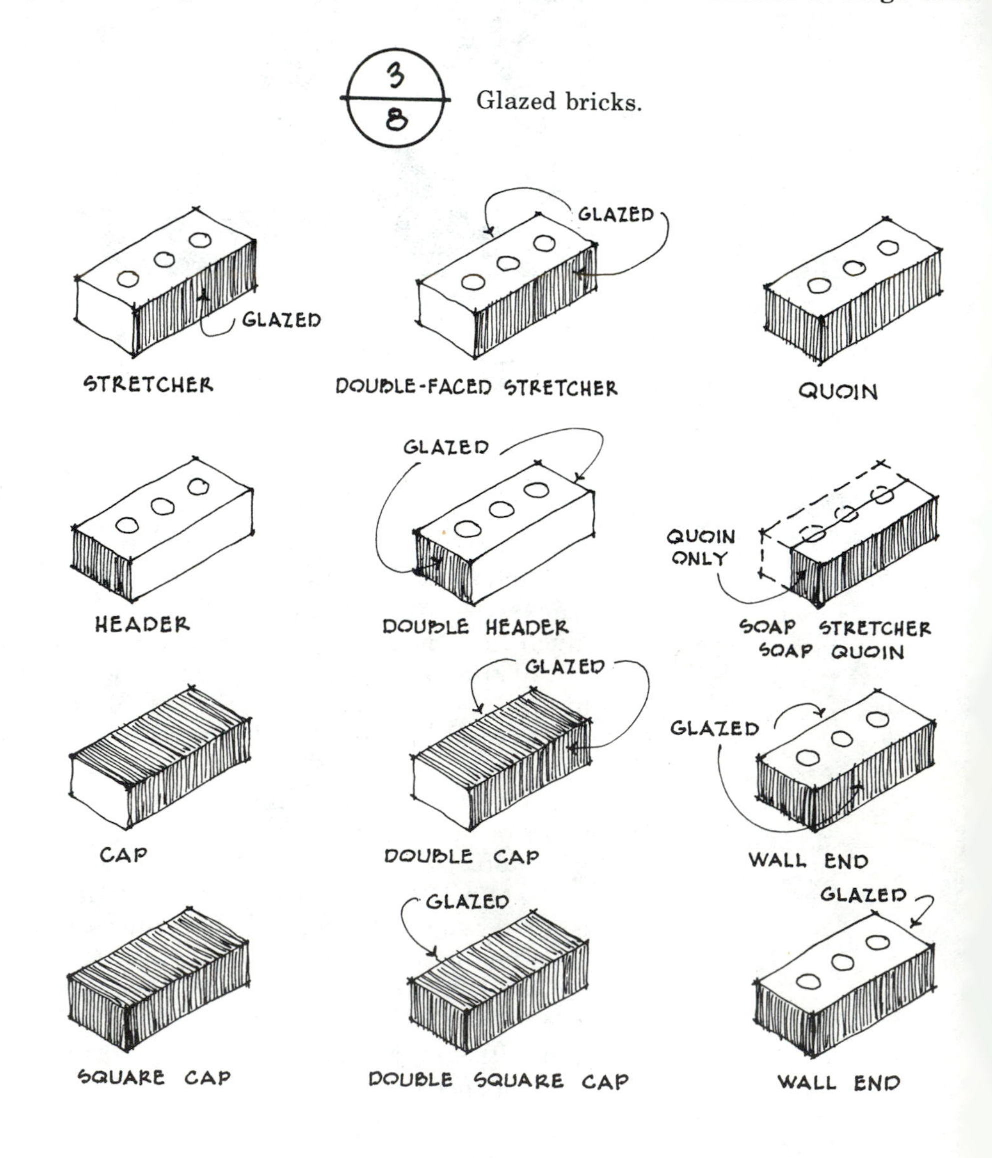

Glazed bricks.

pressive strength of 8000 psi for extruded brick and 4500 psi for molded brick; a maximum average cold-water absorption of 8%; and a maximum saturation coefficient of 0.78. For exterior paving, resistance to freezing and thawing in the presence of moisture is even more important than resistance to abrasive wear. Dense hard-burned units, graded SW in accordance with ASTM C62 or C216, withstand both types of abuse and should be specified for all residential and nonindustrial flooring and paving. Used brick are not recommended because nonuniformity in physical characteristics can cause spalling, flaking, pitting, and cracking.

The paving appearance depends largely on color, size, texture, and bond pattern. Paving brick are usually uncored and designed to be laid flat, but cored brick may be used if placed on edge. Colors may range from reds to buffs, grays, and browns. Surface textures include smooth, velour, and rough, slip-resistant finishes. Standard or round-edge pavers are available in rectangular as well as square and hexagonal shapes (*see Fig. 3-9*).

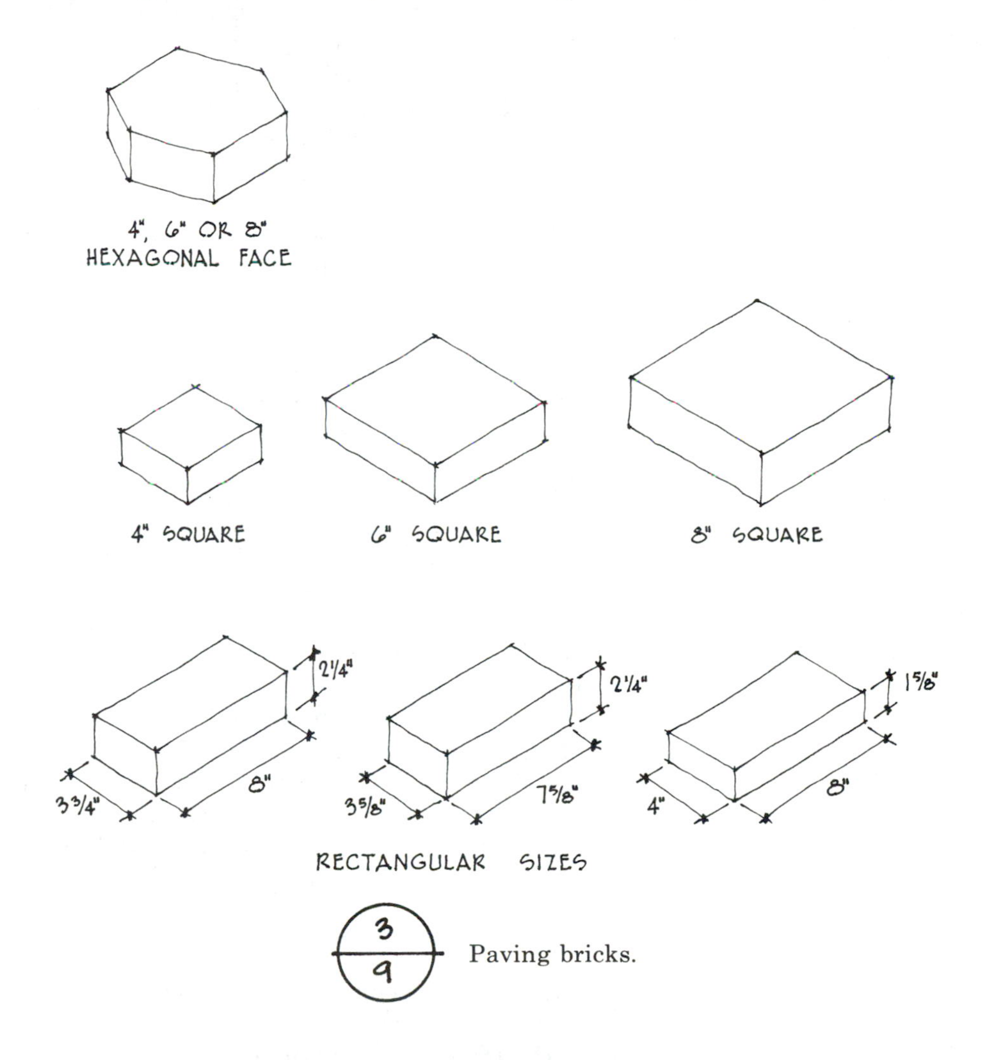

Paving bricks.

Many industrial operations, such as foundries, steel mills, refineries, and breweries, require flooring materials that are resistant to vibration, impact, heavy vehicular traffic, thermal shock, and chemical attack. *Industrial floor brick* have been used successfully in these applications because of their dense structure, chemical stability, hardness, and "non-dusting" characteristics. Four basic types of units are described in ASTM C410, and are classified on the basis of absorption, solubility in sulfuric acid, and modulus of rupture. Type T provides high resistance to thermal shock and mechanical impact, but also has relatively high absorption. Type H has a lower percentage of absorption, but offers only moderate resistance to chemicals and thermal shock. Type M should be used where low absorption is required. Both Type H and Type M offer only limited mechanical shock resistance, but are highly resistant to abrasion. Type L provides minimum absorption and maximum chemical and abrasion resistance, but limited resistance to thermal and mechanical shock. Jointing material for industrial floor brick should be portland cement mortar or grout or, when required, chemical resistant mortar.

Most well-burned clay masonry, including conventional face brick, building brick, structural clay tile, and ceramic glazed units, have excellent resistance to chemicals and chemical agents. In some installations, however, such as waste treatment facilities, dairies, chemical plants, refineries, and food processing plants, extraordinary resistance may be required. *Chemical-resistant brick* or acid-proof brick are machine-made, uncored, kiln-fired units made specifically for this purpose. They are strong, free of laminations, burned to vitrification to close all pores, and sufficiently rough in texture to ensure complete and intimate bond with the mortar. Conditions of temperature and acidity and the absorption rate of the unit are the primary factors governing material selection for use in corrosive environments. Determination of the nature and severity of exposure will dictate which of the two types of units covered in ASTM C279 should be used. Type H is used where thermal shock (or rapid temperature change) is the primary service factor and absorption is not critical. Type L provides the opposite characteristic of minimum absorption where thermal shock is not critical. The basis of the classification is the unit's modulus of rupture, absorption, and solubility in sulfuric acid. Chemical-resistant brick performs satisfactorily in the presence of mild alkalis and all acids except hydrofluoric. In instances where strong alkalis or hydrofluoric acid and its salts are present, a special "carbon brick" is required. Chemical-resistant mortars must be used with these units to assure an effective installation (see Chapter 6).

Standards for sewer and manhole brick are set out in ASTM C32. Two grades for each usage are identified. For *sewer brick*, Grades SS and SM distinguish between the amounts and velocities of abrasive materials carried. Grade SS is lower in absorption and offers greater erosion resistance. *Manhole brick* are graded on their ability to withstand freezing action rather than abrasion. Grade MS provides a high and uniform degree of resistance, while Grade MM offers only moderate and nonuniform resistance. These brick may be used in drainage structures for the conveyance of sewage, industrial wastes, and storm water, and for related structures such as manholes and catch basins.

3.1.3 Glass Block

Glass block is considered a masonry material since it is laid up in cement mortar and uses the same type of joint reinforcement as other units. Although they are not made from clay, glass blocks do share some common characteristics with burned clay products. Both contain silicates as a primary raw ingredient and the glass units, like brick, undergo vitrification when subjected to a heat process. They are available in a variety of sizes and in both solid and hollow form. Decorative blocks are produced in clear reflective, or color glass with smooth, fluted, etched, or rippled texture Functionally, glass block is used to diffuse or direct light for different illuminating requirements, and provide a high level of energy efficiency for glazed areas. Compressive strengths range from 400 to 600 psi. The current popularity of glass block masonry has saved this material from relative obscurity, and greatly increased domestic production in a variety of new shapes and styles (*see Fig. 3-10*).

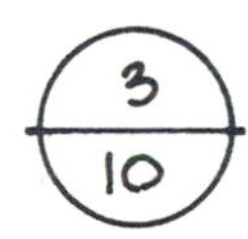

Alfred C. Glassel School of Art, Houston. Morris/Aubry, architects. (*Photo courtesy Houston Museum of Fine Arts.*)

3.1.4 Hollow Brick

One of the traditional distinctions made between different clay masonry products is based on the definition of brick as "solid" (core area of less than 25%) and clay tile as "hollow" (more than 25% cored area). However, during the 1970s, new standards were developed for "hollow brick" with a greater core area than that previously permitted for brick, but less than that allowed for tile.

The trend toward larger unit sizes led to production of jumbo brick in $8 \times 4 \times 12$-in. dimensions as early as the 1920s. In the southeastern part of the United States, this in turn prompted experimentation with greater coring as an effective means of reducing the weight and production costs of such large units. Originally made and marketed under a number of different proprietary names and specifications, these hollow brick are now classified in ASTM C652 for standardization of physical properties (*see Fig. 3-11*). Sometimes referred to as through-the-wall units, hollow brick may be laid with opposite faces exposed. They offer considerable economy in speed and construction of masonry walls while maintaining the aesthetic appeal of conventional multi-wythe systems.

ASTM C652 covers hollow brick with core areas between 25% and 40% of the gross cross-sectional area in the bearing plane (i.e., a minimum of 60% solid). The two grades listed correspond to the same measure of durability as that used for building brick and face brick: Grade SW and Grade MW. Types HBX, HBS, and HBA are identical to face brick types FBX, FBS, and FBA. Another type, HBB, is for general use in walls and partitions where color and texture are not a consideration and greater variation in size is permissible (as for building brick or common brick). Hollow brick are used for both interior and exterior construction in much the same way as solid brick. Sizes range from $4 \times 2\frac{1}{4} \times 12$ in. to $8 \times 4 \times 16$ in.

TABLE A PHYSICAL REQUIREMENTS						
	Compressive strength (hollow brick in bearing position) gross area, min, (psi)		*Water absorption by 5-hr boiling, max. (%)*		*C/B Saturation coefficient, max.*	
Designation	*Average of 5 brick*	*Individual*	*Average of 5 brick*	*Individual*	*Average of 5 brick*	*Individual*
Grade SW	3000	2500	17.0	20.0	0.78	0.80
Grade MW	2500	2200	22.0	25.0	0.88	0.90

TABLE B REQUIRED COMPRESSIVE STRENGTH FOR SATURATION COEFFICIENT WAIVER*	
Percent solid: $\frac{\text{net area}}{\text{gross area}} \times 100$	*Compressive strength (hollow brick in bearing position) gross area, min. (psi)*
75	8000
70	7750
65	7500
60	7250

*Interpolate between given values.

Property requirements from ASTM C652

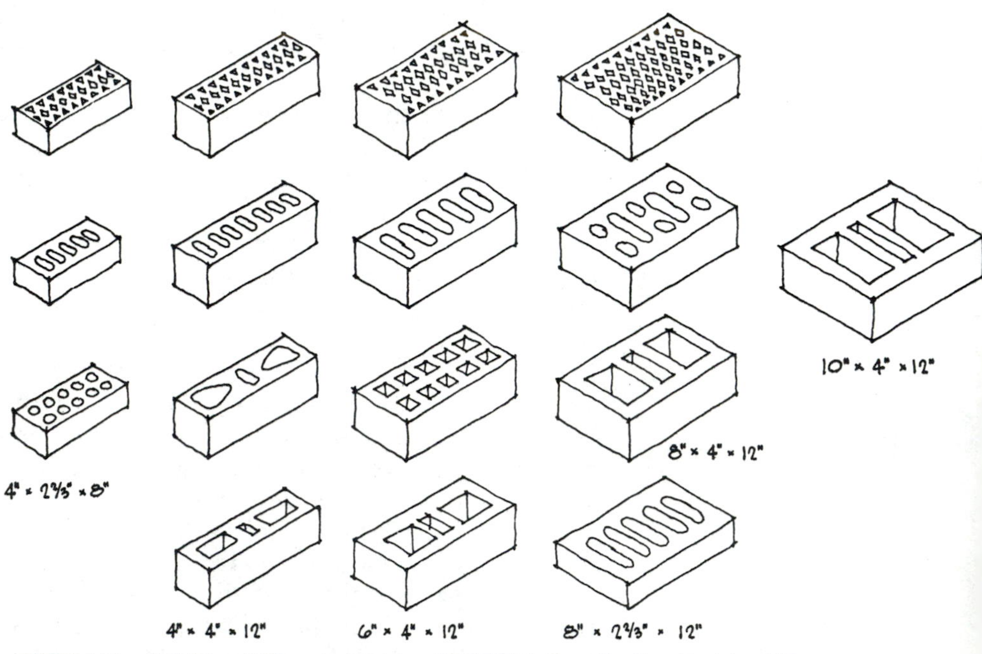

TYPICAL SIZES AND CORING PATTERNS FOR HOLLOW BRICK (LESS THAN 75% SOLID)

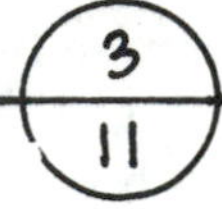

Hollow brick. *Copyright, American Society for Testing and Materials, 191 Race Street, Philadelphia, Pa. 19103. Reprinted, with permission.)*

3.2 STRUCTURAL CLAY TILE

Structural clay tile is the most recently developed of clay masonry products, first produced in this country in 1875. Up until that time, most buildings were constructed with solid masonry walls of brick or stone in multi-wythe form as loadbearing systems. With the invention and mastery of structural steel framing, a need arose for lightweight backing materials for the facing masonry used to clad these skeleton frames. Clay tile satisfied the demand and added elements of economy and fire resistance.

Although use has fallen off considerably since the mid-twentieth century, clay tile is still available through a limited number of manufacturers, and is often required for restoration/retrofit work. It may serve both as structural, facing, and backup elements in construction. Tile, like brick, is made of clay that is molded and then burned in a kiln to ceramic fusion. Clay tile may be used with cells either horizontal (side construction) or vertical (end construction), for both loadbearing and non-loadbearing applications. "Structural" tile is distinguished from clay wall tile and flat clay floor tile by its ability to carry load and support its own weight. The numerous types of tile used today are classified by function. *Structural clay loadbearing wall tile* and *structural clay non-loadbearing tile* may be used in the construction of walls and partitions where a finish coat of plaster or other material will be applied or where appearance is not a primary concern. These units are the equivalent of building brick, and are considered principally utilitarian in nature. *Structural clay facing tile* and *ceramic glazed facing tile* may be loadbearing or non-loadbearing, but are distinguished from the above on the basis of finish, much the same as facing brick is distinguished from building brick.

3.2.1 Loadbearing Wall Tile

The physical requirements for structural clay loadbearing wall tile are governed by ASTM C34, and the units are divided into two grades based on compressive strength and resistance to frost action in the presence of water (*see Fig. 3-12*). The higher grade, LBX, is suitable for areas exposed to weathering. Grade LB is limited to unexposed areas unless protected by at least 3 in. of stone, brick, or other masonry. Requirements for strength and absorption, as indicators of durability, are thus more stringent for grade LBX. In either case, the tile carries the structural load, the live load, the weight of the facing material, and its own weight. Loadbearing tile may also

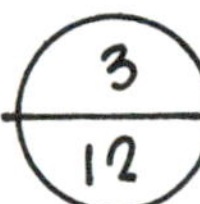

Physical requirements for structural clay loadbearing wall tile, from ASTM C34. (*Copyright, American Society for Testing and Materials, 1916 Race Street, Philadelphia, Pa. 19103. Reprinted, with permission.*)

Grade	Maximum water absorption by 1-hr boiling* (%)		Minimum compressive strength (based on gross area)† (psi)			
			End-construction tile		Side-construction tile	
	Average of 5 tests	Individual	Average of 5 tests	Individual	Average of 5 tests	Individual
LBX	16	19	1400	1000	700	500
LB	25	28	1000	700	700	500

*The range in percentage absorption for tile delivered to any one job shall be not more than 12.
†Gross area of a unit determined by multiplying the horizontal face dimension of the unit as placed in the wall by its thickness.

be used as a backing in composite wall construction with facing tile, brick, or other masonry units. In this instance, the two or more wythes of the wall are bonded together structurally so that the tile bears an equal share of the superimposed load, but does not support the facing itself.

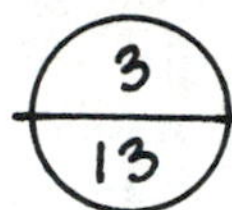

Clip and angle tile sizes. (*From Harry C. Plummer,* Brick and Tile Engineering, *Brick Institute of America, McLean, Va., 1962.*)

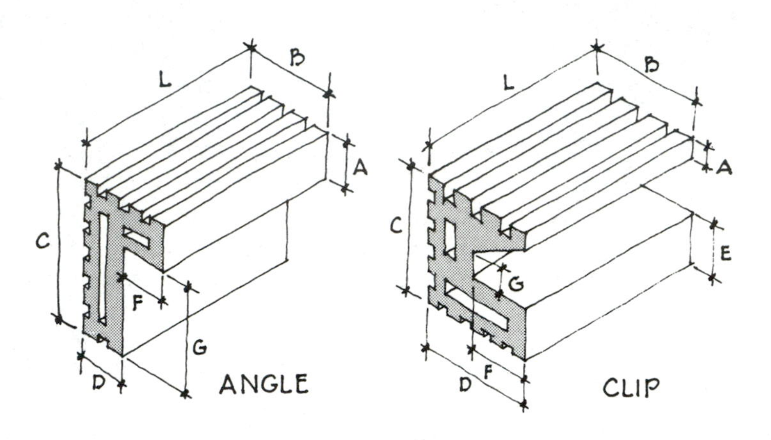

TYPICAL CLIP AND ANGLE TILE SIZES									
Unit number	*Outside measure (sq ft)*	*Weight per piece (lb)*	*A*	*B*	*C*	*D*	*E*	*F*	*G*
G20	0.65	13	$\frac{3}{4}$	$3\frac{3}{8}$	$3\frac{3}{8}$	$3\frac{3}{4}$	2	2	$\frac{3}{4}$
G25	0.70	14	$\frac{3}{4}$	$3\frac{7}{8}$	4	$4\frac{1}{2}$	2	$2\frac{1}{2}$	$\frac{13}{16}$
G30	0.75	15	$\frac{7}{8}$	$4\frac{1}{4}$	$4\frac{1}{4}$	$4\frac{3}{4}$	2	3	$\frac{7}{8}$
G35	0.80	16	$\frac{3}{4}$	$4\frac{3}{4}$	$4\frac{1}{4}$	$5\frac{1}{2}$	2	$3\frac{1}{2}$	1
G40	0.85	17	$\frac{3}{4}$	$4\frac{3}{4}$	$4\frac{1}{4}$	6	2	4	1
G45	0.90	18	$\frac{3}{4}$	$4\frac{3}{4}$	$4\frac{1}{4}$	$6\frac{1}{2}$	2	$4\frac{1}{2}$	1
G46	0.95	19	$\frac{3}{4}$	$4\frac{3}{4}$	$4\frac{3}{4}$	$6\frac{5}{8}$	2	$4\frac{5}{8}$	$1\frac{1}{2}$
G50	0.95	19	$\frac{3}{4}$	$4\frac{3}{4}$	$4\frac{1}{4}$	7	2	5	1
G55	1.00	20	$\frac{3}{4}$	$4\frac{3}{4}$	$4\frac{1}{2}$	$7\frac{1}{2}$	2	$5\frac{1}{2}$	$1\frac{1}{4}$
G60	1.05	21	1	$5\frac{1}{2}$	$4\frac{5}{8}$	8	2	6	$1\frac{3}{8}$
G61	1.05	21	1	$4\frac{1}{2}$	$4\frac{5}{8}$	8	2	6	$1\frac{3}{8}$
G70	1.16	23	1	6	$4\frac{7}{8}$	9	2	7	$1\frac{1}{2}$
G71	1.16	23	1	$4\frac{1}{2}$	$4\frac{7}{8}$	9	2	7	$1\frac{1}{2}$
G75	1.33	29	2	$7\frac{1}{2}$	$6\frac{3}{8}$	$9\frac{1}{2}$	2	$7\frac{1}{2}$	$2\frac{3}{8}$
G80	1.40	32	2	8	$6\frac{7}{8}$	10	2	8	$2\frac{7}{8}$
L23	0.60	12	2	4	5	2	—	2	3
L26	0.90	18	2	4	$8\frac{1}{2}$	2	—	2	$6\frac{1}{2}$
L43	0.65	13	2	6	5	2	—	4	3
L46	0.95	19	2	6	$8\frac{1}{2}$	2	—	4	$6\frac{1}{2}$

3.2.2 Non-Loadbearing Tile

There are three types of structural clay non-loadbearing tile covered by ASTM C56; partition, furring, and fireproofing tile. The standards include only one grade, NB, and one physical property specification, which limits the rate of water absorption. Maximum allowable absorption is 28%, and the maximum range of absorptions for tile delivered to any one job may not exceed 12%. *Partition tile* is restricted to non-loadbearing interior partitions. *Furring tile* is used to line the inside surface of exterior walls and provides an insulating air space, a moisture barrier, and a surface suitable for plastering. Partition and furring tile may be used to fireproof structural steel members, but for some applications around beams and girders, special shapes of *fireproofing tile* are required to conform to the profile of the steel. Clip and angle shapes have been devised for this purpose and, when used in conjunction with conventional rectangular tiles, provide a simple means of complete coverage (*see Fig. 3-13*).

Tile that will be plastered has a special surface texture that is ribbed or wire cut (sometimes called universal finish). Both textures provide good bond between the plaster and the clay body (*see Fig. 3-14*).

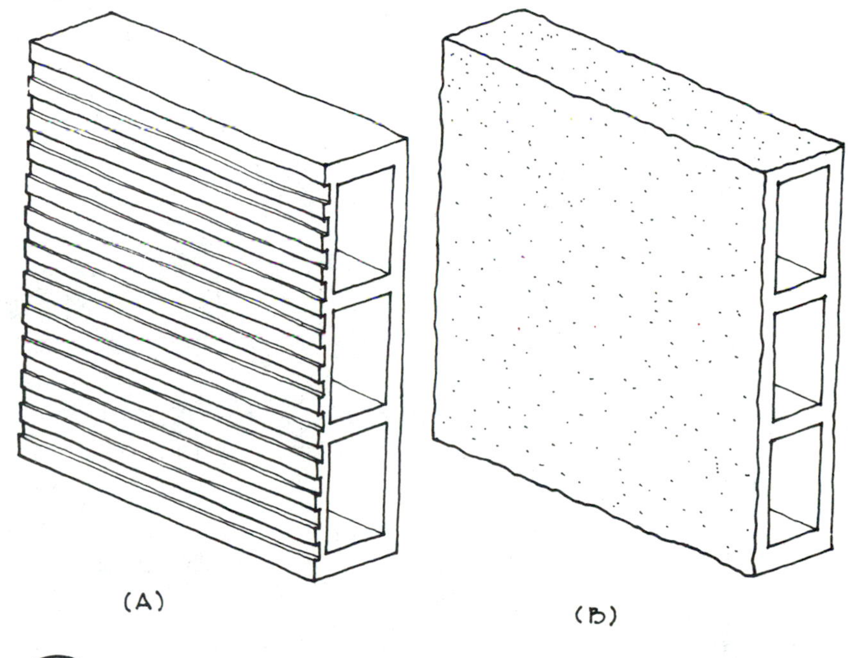

Scored (A) and wire cut or universal finish (B) on structural clay tile units.

3.2.3 Facing Tile

Facing tile combines the loadbearing capacity of ordinary clay tile with a permanent, finished surface suited for architectural applications. These natural-color unglazed tile are covered in ASTM C212. Two classes of tile are defined based on face shell and web thickness. "Standard" tile are general-purpose units for exterior or interior locations. "Special duty" tile have heavier webs and shells designed to increase resistance to impact and moisture penetration. Aesthetic factors are designated the same as for face brick. Type FTX tile have a smooth finish for general use in interior and exterior applications requiring minimum absorption, easy cleaning, and resistance to staining. They provide a high degree of mechanical perfection,

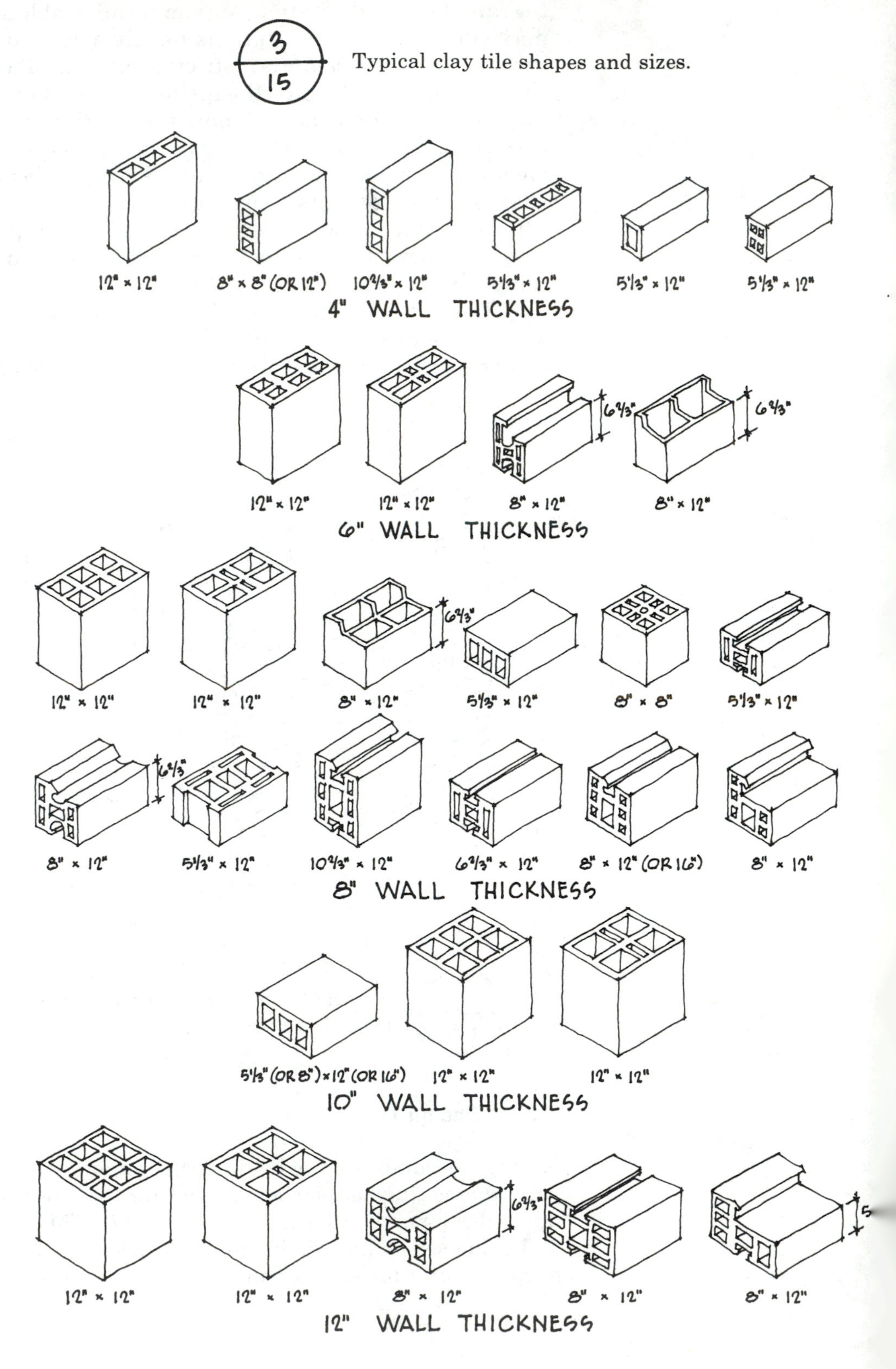

Typical clay tile shapes and sizes.

narrow color range, and minimum variation in face dimensions. Type FTS may have a smooth or rough texture, are suitable for interior and exterior construction where moderate absorption and moderate variation in face dimensions are permissible, and may be used where minor defects in surface finish are not objectionable. ASTM C212 lists compressive strength and absorption, and sets limits on chippage, dimensional variation, and face distortion (*see Fig. 3-15* for sizes and shapes available).

3.2.4 Ceramic Glazed Facing Tile

Glazed units are also of loadbearing quality, but have an impervious finish in either clear or color glaze. Physical requirements are outlined in ASTM C126, which also governs glazed brick. Grade and type classifications are identical for both products. Grade S (select) units are used with comparatively narrow mortar joints. Grade SS (selected sized, or ground edge) are used where variation of face dimension must be very small. Both grades may be produced in either Type I, single-faced units, where only one finished face will be exposed; or Type II, double-faced, where two opposite faces will be exposed. Standards cover compressive strength, absorption rate, number of cells, shell and web thickness, dimensional tolerances, and properties of the ceramic finish, including imperviousness, chemical resistance, and crazing.

The shapes of all structural tile units are controlled by the dies through which the plastic clay is extruded. The relative ease with which various designs can be produced led to the development of a large number of sizes and patterns. Through a process of standardization, this number has been reduced to only the most economical and useful units. Development and acceptance of the criteria for modular coordination encouraged refinement aimed at correlation with other manufactured masonry products. Structural clay tile are designed for use with $\frac{1}{4}$, $\frac{3}{8}$, or $\frac{1}{2}$-in. mortar joints, while facing tile uses only $\frac{1}{4}$-in. joints. Nominal dimensions, as for brick, include this thickness and are multiples of the 4-in. module or fractions of a multiple of that module (i.e., three courses of $5\frac{1}{3}$-in.-high tiles equals 16 in.).

Since glazed clay tile is not as easily cut and trimmed in the field as some other masonry materials, a larger variety of special shapes is required to facilitate door and window openings, headers, corner situations, and so on. In addition to full-size stretcher units, shapes include half-lengths, half-heights, corner and jamb units, as well as sills, caps, lintels, cove bases, and coved internal corners (*see Figs. 3-16 and 3-17*). Some manufacturers prepare shop drawings from the architectural plans to show actual tile shapes and locations. If the project is laid out with modular dimensions, very few (if any) extraordinary special shapes will be required and job-site cutting and waste will be minimum.

3.2.5 Screen Tile

Clay masonry solar screens have always found wide acceptance whether constructed of screen tile or of standard units ordinarily used for other purposes. Screen tile are available in a variety of shapes and patterns and in all colors of glazed and unglazed clay masonry (*see Fig. 3-18*). Lighter colors, because of greater reflectivity, provide brighter interiors. Darker colors absorb more of the sun's heat and light and give greater protection from its harsh rays. Screen tile are covered by ASTM C530.

AVAILABLE SIZES		
Series	*Nominal face dimensions, inches*	*Nominal thickness, inches*
PREFIX 6T	$5\frac{1}{3} \times 12$	2,4,6,8
4D	$5\frac{1}{3} \times 8$	2,4,6,8
4S	$2\frac{2}{3} \times 8$	2,4
8W	8×16	2,4

NOMENCLATURE	
Prefix: Denotes face size	6T, 8W, etc.
Number: Denotes horizontal or vertical axis and bed depth	bullnose, stretcher, quoin, etc.
Suffix: Denotes return, reveal, back face, and right- or left-hand shape	
Example: 6T24CR — $5\frac{1}{3} \times 12$; Horizontal bullnose; Vertical bullnose; 4" bed, 4" return; Right-hand unit	

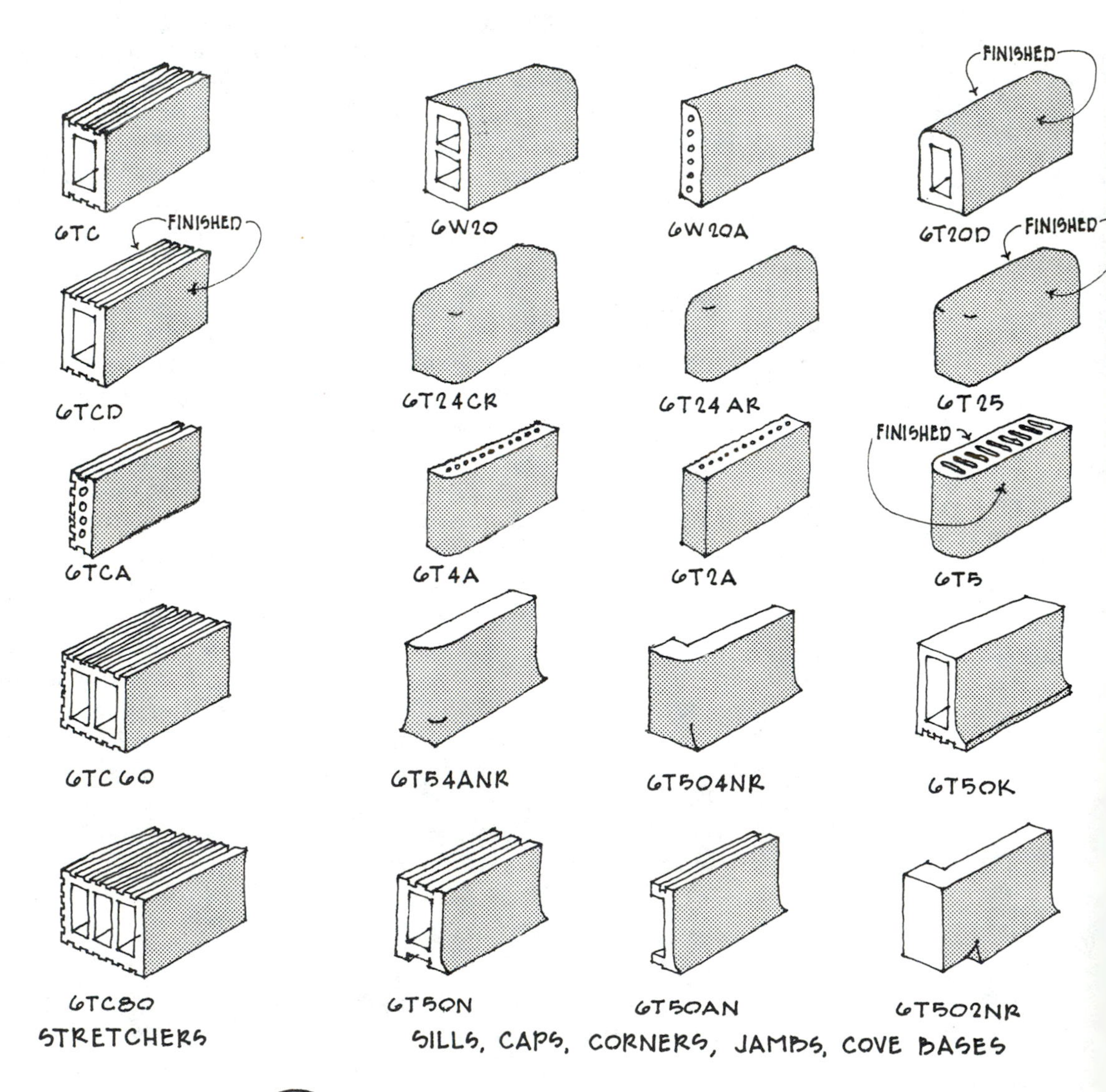

3/16 Glazed facing tile shapes.

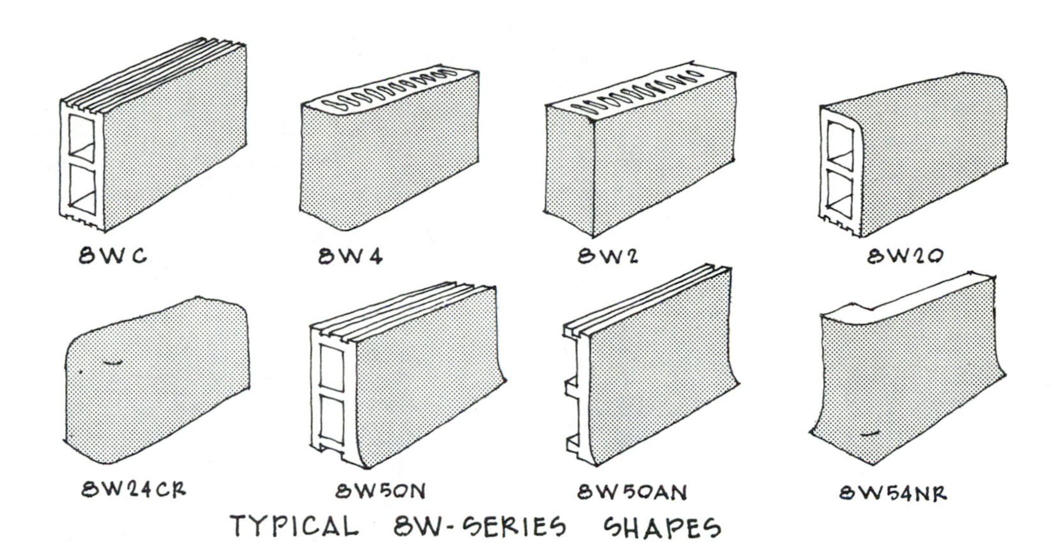

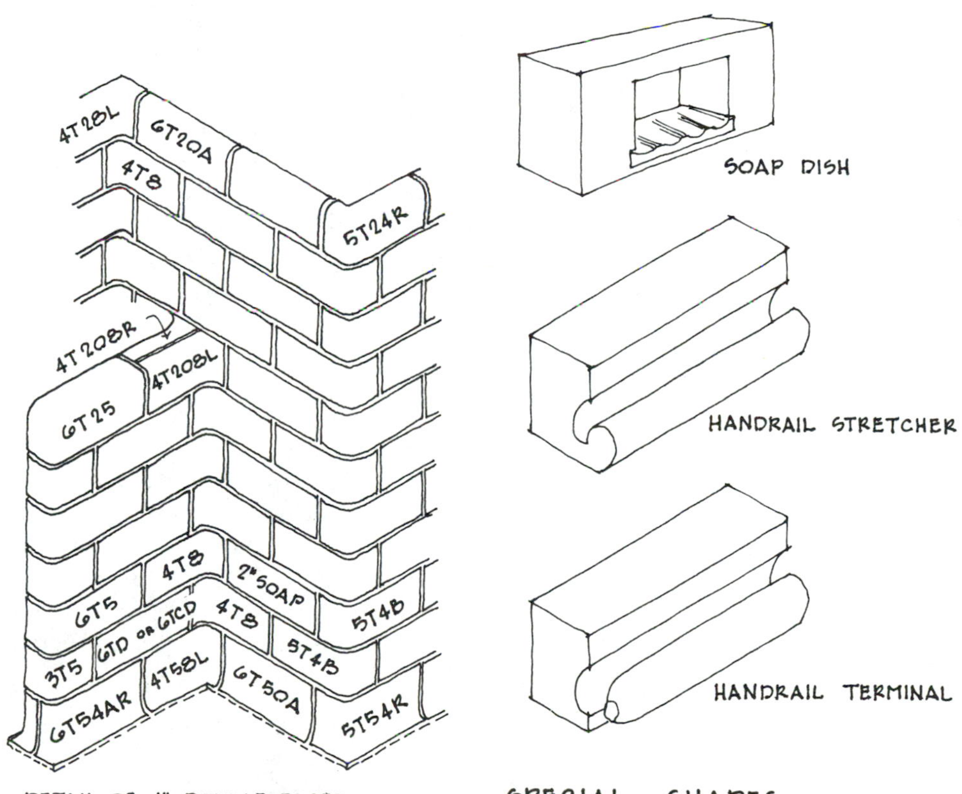

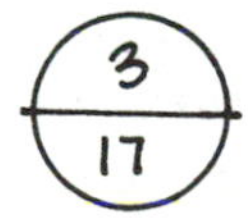

Glazed facing tile shapes.

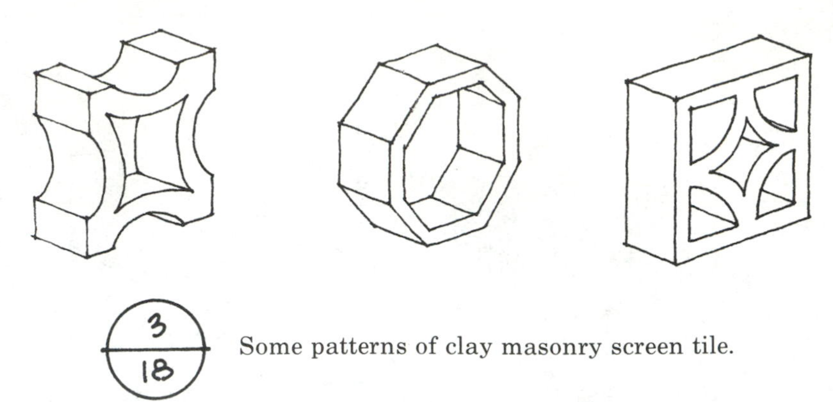

Some patterns of clay masonry screen tile.

3.3 CERAMIC VENEER (TERRA COTTA)

Architectural terra cotta has been used as a decorative veneer for centuries. The name itself, which means "fired earth," dates from Roman antiquity. Hand-molded slabs with either plain or sculptured surfaces are still produced in the traditional manner. Modern methods of production now offer machine-extruded units as well. These mechanically fabricated pieces, usually referred to as ceramic veneer, may have a smooth-ground surface, or may be beveled, scored, scratched, or fluted. Both the hand- and machine-made types may be glazed in clear, monochrome, or polychrome colors and in matte, satin, or gloss finishes. Both are custom products, and are more in demand today for restoration work than for new construction.

The backs of veneer slabs may be flat or ribbed, depending on the method of attachment to be used. Adhesion-type veneer is bonded to the backup material with mortar and requires no supplementary support. Its thickness is limited to $1\frac{1}{4}$ in. by most current building codes; its maximum face dimension to 36 in.; its area to 720 sq in.; and its weight to 15 lb/sq ft. Anchored-type veneer, usually $1\frac{5}{8}$ in. or greater in thickness, is attached by mortar bond *and* metal ties. Size and area limitations vary among building codes.

3.4 PROPERTIES AND CHARACTERISTICS

Physical properties and characteristics of masonry units are important to the architect only insofar as they affect performance and appearance of the finished wall or structure. The major building codes in the United States rely primarily on ASTM standards and requirements of the American Nationa Standards Institute (ANSI) for minimum property specifications. Thes deal mainly with compressive strength, absorption, and saturation coef ficients as indicators of acceptable performance (*see Fig. 3-19*).

3.4.1 Compressive Strength

The compressive strengths of brick and tile are usually based on gross area Extruded brick generally have higher compressive strength and lowe absorption than those produced by the soft-mud or dry-press processes. Fc a given clay and method of manufacture, higher compressive strength an lower absorption are also associated with higher burning temperatures. Th minimum compressive strength values listed in *Fig. 3-19* are substantiall exceeded by most manufacturers. Actual compressive strengths of cla masonry units are usually higher than concrete. For standard run bric strengths typically range from 1500 to 22,500 psi, with the majority of uni produced being in excess of 4500 psi (*see Fig. 3-20*).

Unit	Grade	Minimum compressive strength, gross area (psi)		Maximum water absorption by 5-hr boiling (%)		C/B Maximum saturation coefficient	
		Average of 5 tests	Individual unit	Average of 5 tests	Individual unit	Average of 5 tests	Individual unit
Building brick (ASTM C62)	SW	3000	2500	17.0	20.0	0.78	0.80
	MW	2500	2200	22.0	25.0	0.88	0.90
	NW	1500	1250	No limit		No limit	
Facing brick (ASTM C216)	SW	3000	2500	17.0	20.0	0.78	0.80
	MW	2500	2200	22.0	25.0	0.88	0.90
Hollow brick (ASTM C652)	SW	3000	2500	17.0	20.0	0.78	0.80
	MW	2500	2200	22.0	25.0	0.88	0.90

Unit	Grade	End construction		Side construction		By 1-hr. boiling, %	
		Av.	Ind.	Av.	Ind.		
Structural clay load bearing tile (ASTM C34)	LBX	1400	1000	700	500	16.0	19.0
	LB	1000	700	700	500	25.0	28.0
Structural clay facing tile (ASTM C212)	FTX					9.0	11.0
	FTS					16.0	19.0
	Standard	1400	1000	700	500		
	Special-duty	2500	2000	1200	1000		

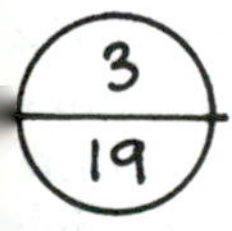

Minimum physical requirements of clay masonry products.

Compressive strength, flatwise		Modulus of rupture	
Range (psi)	Percentage of production within range	Range (psi)	Percentage of production within range
21,001–22,500	0.46	2101–3450	6.95
19,501–21,000	0.69	1951–2100	3.00
18,001–19,501	0.46	1801–1950	2.74
16,501–18,000	2.04	1651–1800	7.57
15,001–16,500	1.49	1501–1650	8.34
13,501–15,000	3.71	1351–1500	5.34
12,001–13,500	4.76	1201–1350	7.12
10,501–12,000	7.78	1051–1200	10.55
9,001–10,500	8.61	901–1050	10.44
7,501–9,000	11.92	751–900	13.60
6,001–7,500	15.47	601–750	11.74
4,501–6,000	16.81	451–600	7.52
3,001–4,500	17.97	301–450	4.35
1,501–3,000	7.46	151–300	0.37
0–1,500	0.36	0–150	0.37
Total percent	99.99		100.00

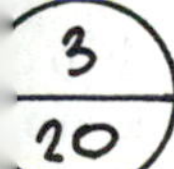

Actual compressive strengths of bricks produced in the United States. (*From Brick Institute of America,* Principles of Clay Masonry Construction, *BIA, McLean, Va., 1973.*)

3.4.2 Transverse Strength

The transverse strength of a brick acting as a beam supported at both ends is called the *modulus of rupture*. Tests at the National Bureau of Standards (NBS) indicate minimum values for well-burned brick to be in excess of 500 psi, with a maximum average of 2,890 psi. There is no general rule, however, for converting values of compressive strength to transverse strength, or vice versa. Tensile and shearing properties of burned brick have not been widely tested, but data from NBS indicate that *tensile strength* normally falls between 30 and 40% of the modulus of rupture and *shear values* from 30 to 45% of the net compressive strength. Tensile strength of structural clay tile is quite low and usually will not exceed 10% of the compressive strength. The *modulus of elasticity* for brick ranges from 1,400,000 to 5,000,000.

3.4.3 Absorption

The weight of burned clay products ranges from 100 to 500 lb/cu ft. Variations may generally be attributed to the process used in manufacturing and burning. Increased density and weight result from fine grinding of raw materials, uniform mixing, pressure exerted on the clay as it is extruded, deairing, and hard or complete burning. (Manufacturers are now placing organic materials such as sawdust in the clay body. These materials burn in the firing process, reducing fuel requirements. The resultant brick is lighter in weight, with little effect on strength or absorption.) The extrusion process produces very dense brick and tile characterized by high strength and a small percentage of voids. Since properties of absorption are also affected by the method of manufacture and degree of burning, these factors indicate fairly close relationships among total absorption, weight, density, and compressive strengths. With few exceptions, hard-burned units are highest in strength and density and lowest in absorption.

The absorption of a brick or clay tile is defined as the weight of water taken up by the unit under given test conditions, and is expressed as a percentage of the dry weight of the unit. Since highly absorptive brick exposed to weathering can cause a buildup of moisture in the wall, and since efflorescence is fostered by water in the wall, ASTM standards limit absorption to 17% for Grade SW and 22% for Grade MW units. Most brick produced in the United States have absorptions of only 4 to 10% (*see Fig. 3-21*).

An important property of brick that critically affects bond strength is the *initial rate of absorption, or suction*. High-suction brick absorb water from the mortar too quickly, impairing bond, proper hydration, and curing. Laboratory tests and field experience indicate that maximum strength and minimum water penetration occur with units having initial rates of absorption of less than 20 g/min at the time they are laid. Suction for commercially produced brick and tile may be as low as 1 or 2 g/min or as high as 60 g/min. Brick with high suction rates should be thoroughly wetted prior to installation, then allowed to "surface dry." Since suction can be controlled by this means, it is not covered in ASTM requirements. It should, however, be included in project specifications.

Saturation coefficient, or *C/B* ratio, is a measure of the relationship of two aspects of water absorption: the amount freely or easily absorbed and the amount absorbed under pressure. The former (C) is determined by the 24-hour cold water absorption test, and the latter (*B*) by the 5-hour boil absorption test. The ratio must be 0.78 or less to meet ASTM standards. The

Designation	Absorption by total immersion		Saturation coefficient, C/B (%)	Saturation by partial immersion, flat, 1 min (g)
	24-hr cold, C (%)	5-hr boil, B (%)		
A	1.9	3.5	0.53	10
B	9.4	13.4	0.70	33
C	14.6	16.9	0.86	112
D	11.3	15.1	0.74	25
E	10.2	14.7	0.69	38
F	13.8	18.7	0.74	42
G	5.4	7.8	0.69	12
H	3.3	6.0	0.54	6
J	9.3	13.5	0.68	31
K	6.8	13.4	0.54	27
L	3.5	7.6	0.44	9
M	1.6	1.9	0.73	3
N	7.4	8.5	0.87	16
O	3.6	6.6	0.55	20
P	1.5	2.8	0.51	6
R1	4.2	6.5	0.64	7
R2	2.1	4.7	0.45	3
R3	4.9	6.6	0.73	5
U	5.3	9.3	0.56	30
V	8.98	15.07	0.59	32
W	4.02	9.82	0.38	4
X	9.11	15.83	0.57	35

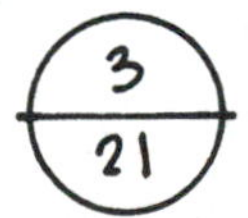

Absorption properties of various U.S. manufactured bricks. (*From Brick Institute of America,* Pocket Guide to Brick Construction, *BIA, McLean, Va.*)

C/B ratio determines the volume of open pore space remaining after free absorption has taken place. This is important under severe weathering conditions when a unit has taken in water that must have room to expand if frozen in order to avoid damage to the clay body. The theory does not apply to hollow masonry units or to certain types of deaired products. In those cases, strength and absorption alone are used as measures of resistance to frost action.

3.4.4 Durability

The durability of clay masonry usually refers to its ability to withstand freezing in the presence of moisture, since this is the most severe test to which it is subjected. Compressive strength, absorption, and saturation coefficient are evaluated together as an indication of freeze-thaw resistance since values cannot be assigned specifically for this characteristic.

Resistance to wear and abrasion are important aspects of durability for brick and tile paving, and for the lining of structures which will carry sewage, industrial waste, and so on. Abrasion resistance is also closely associated with the degree of burning, and ranges from underfired salmon brick at the low end to vitrified shale and fire clay at the opposite extreme.

The stronger the unit, and the lower the absorption, the greater this resistance will be. In salvaged brick or imported brick, underburned units are easily detected without sophisticated laboratory equipment or procedures. Extremely soft units are easily scratched or scored with a coin, cut with a knife, or even broken by hand (*see Fig. 3-22*). Brick and tile complying with ASTM standards, however, are high-quality products with proven records of performance in service.

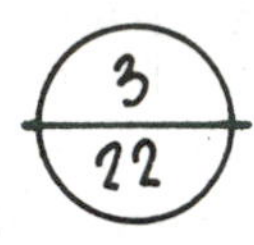

Underburned bricks are easily scratched or scored with a coin, cut with a knife, or broken by hand. (*Photos courtesy Brick Institute of Texas.*)

3.4.5 Expansion Coefficients

Clay products are among the most dimensionally stable of building materials. They do not shrink in any exposure, but do expand slightly under variable conditions of temperature and moisture. The *coefficient of thermal expansion* ranges from 2.5 millionths inch (0.0000025) per °F for fire clay units to 3.3 millionths inch (0.0000033) per °F for surface clay and shale units. This minute thermal expansion-contraction is usually absorbed by the flexibility of the mortared wall. Moisture expansion, however, is not reversible. Tentative test results have assigned a value of 0.0002 in./°F for the *coefficient of moisture expansion.* Pressure-relieving joints must be provided in the structure to permit this movement. Severe problems can develop when clay masonry expansion is restrained by concrete elements that have shrinkage potential. Jointing details must provide flexible anchorage to accommodate such differential movement.

3.4.6 Fire and Thermal Resistance

Masonry fire resistance and thermal qualities are both determined by mass The characteristics of the individual units are not considered, but ratings are established for finished wall assemblies. Detailed analysis of these properties is covered in Chapter 8.

3.4.7 Acoustical Characteristics

The high density of clay masonry predetermines its acoustical charac teristics. Although sound absorption is almost negligible, the heavy mas provides excellent resistance to the transmission of sound through walls. This suggests best use as partitions or sound barriers between areas of different occupancy. Where higher absorption is required in addition to sound isolation, special units are used. The SCR Acoustile was developed to offer 60 to 65% absorption. The unit is a structural facing tile with a perforated face shell. The adjacent cell(s) are factory-filled with a fibrous glass pad. The perforations may be round or slotted and arranged in random or uniform patterns. The tile itself is of loadbearing quality, may be glazed or unglazed, and otherwise exhibits the same properties and characteristics of structural clay facing tile manufactured in accordance with ASTM C212 or C126.

3.4.8 Colors and Textures

Brick and tile are available in an almost unlimited variety of colors and textures. They may be standard items or custom units produced for unique project requirements. Natural clay colors can be altered or augmented by the introduction of various minerals in the mix, and further enhanced by application of a clear, lustrous glaze. Ceramic glazed finishes range from the bright primary colors through the more subtle earth tones in solid, mottled, or blended shades. Glossy, matte, and satin finishes, as well as applied textures, add other aesthetic options (*see Fig. 2-3*).

4

CEMENTITIOUS MASONRY UNITS

Cementitious masonry units are hardened by chemical reactions rather than by ceramic fusion. This group includes concrete brick and block as well as sand-lime brick, cast stone, and cellular concrete block. The majority of these units are classified as solid and have less than 25% core area in relation to the gross cross section in the bearing plane. Concrete blocks, however, typically have 40 to 50% coring and are thus defined as hollow.

4.1 CONCRETE BRICK

Concrete brick are produced from a controlled mixture of portland cemen and aggregates in sizes, colors, and proportions similar to clay brick. The are governed by the requirements of ASTM C55 and can be loadbearing o non-loadbearing. Aggregates include gravel, crushed stone, cinders burned clay, and blast-furnace slag, producing both normal-weight an lightweight units. Coring or "frogging" may be used to reduce weight an improve mechanical bond.

Grading is based on strength and resistance to weathering. Grade N provides high strength and maximum resistance to moisture penetratio and frost action. Grade S has only moderate strength and minimum resistance to frost action and moisture penetration. Each grade may b produced as Type I, moisture-controlled, or Type II, non-moisture-con trolled, designated as N-I, N-II, S-I, or S-II. For moisture-controlled units the standards limit the moisture content at the time of constructio according to the relative humidity conditions typical of the site's geographic location. These requirements establish predictable shrinkage characteristics so that proper allowance can be made for control joints in the structure (*see Fig. 4-1*).

TABLE A MOISTURE CONTENT REQUIREMENTS FOR TYPE I CONCRETE BRICK

Linear shrinkage (%)	Moisture content, max. percent of total absorption (average of 3 concrete brick) — Humidity* conditions at job site or point of use: Humid	Intermediate	Arid
0.03 or less	45	40	35
From 0.03 to 0.045	40	35	30
0.045 to 0.065, max.	35	30	25

*Arid, average annual relative humidity less than 50%; intermediate, average annual relative humidity 50 to 75%; humid, average annual relative humidity above 75%.

TABLE B STRENGTH AND ABSORPTION REQUIREMENTS

Grade	Compressive Strength, min. (psi) (concrete brick tested flatwise) — Average gross area: Average of 3 concrete brick	Individual concrete brick	Water absorption, max. (average of 3 brick) with oven-dry weight of concrete (lb/ft³) — Weight classification: Lightweight (less than 105)	Medium weight (less than 125 to 105)	Normal weight (125 or more)
N-I N-II	3500	3000	15	13	10
S-I S-II	2500	2000	18	15	13

Requirements for concrete brick, from ASTM C55. (*Copyright, American Society for Testing and Materials, 1916 Race Street, Philadelphia, Pa. 19103. Reprinted, with permission.*)

Concrete mixes may be altered to produce a roll or slump when forms are removed, creating a unit similar in appearance to adobe brick. Color is achieved by adding natural or synthetic iron oxides, chromates, or other compounds to the mix. Manufacturers can produce solids, blends, or mingles. Synthetic colors are more easily controlled, but the earth tones obtained with natural iron oxides are more durable and easier to duplicate for future building additions. ASTM standards do not include color, texture, weight classification, or other special features. Such properties, if required, should be covered separately in the project specifications.

4.2 SAND-LIME BRICK

Calcium silicate brick, or sand-lime brick, are made with pure silica sanc and 5 to 10% hydrated lime, then steam cured in high-pressure autoclaves Sand-lime brick are used extensively in Europe, and most U.S. buildin codes permit their use in the same manner as clay brick. ASM C73, whic governs sand-lime brick, outlines grading standards identical to those fo building and face brick (Grades SW, MW, and NW). Compressive strength generally range from 2500 to 3000 psi, with a modulus of rupture of 450 psi Absorption rates from 7 to 10% are common. Sand-lime brick are a natura pearl gray, but other colors may be produced by the addition of minera oxides.

4.3 GYPSUM BLOCK

Gypsum block is used as a non-loadbearing partition material in 2, 3, 4, o 6-in. thicknesses. In addition to gypsum, ingredients include asbestos o vegetable fibers used as binders and reinforcing. Gypsum block is normall used as a base for plaster application. These units provide excellent fir resistance. As fireproofing around a steel column for instance, a 2-in thickness with plaster on one side has a rating of 4 hours. Properties ar specified under ASTM C52. Standards require minimum compressiv strengths of 75 psi when dry and 25 psi when wet to assure strength to carr the applied plaster and the dead load of the unit's own weight. Gypsum block should be laid only with mortar containing 1 part gypsum to 3 parts sand The availability of gypsum block units is limited.

4.4 CAST STONE

Cast stone is most widely used as an accessory for masonry construction i the form of lintels, sills, copings, and so on. Some manufacturers do produc simulated stone products designed for use as facing materials. The shape o the mold used for casting will determine the appearance of the unit. Surfac textures include shale, slate, or chipped sandstone. Color may be integrall mixed or surface applied to cure inseparably with the stone. Granite o marble chips are added to units that will be face ground to create a moc terrazzo. Cast stone veneer units are made in approximately 1-in. thick nesses with face dimensions 4×4 in., 4×8 in., 8×8 in., or 8×16 in.

4.5 CELLULAR CONCRETE BLOCK

Cellular concrete block has properties of thermal and sound insulation a well as nailability. The units weigh only one-fourth as much as norma concrete block, but are not made with lightweight aggregates. The mi contains sand, lime, and aluminum powder, and water is added to form slurry. Large steel vats are used as molds. A chemical reaction takes plac releasing hydrogen gas and generating heat, which causes the concrete t expand and set in cellular form. Smaller units are wire- or saw-cut from th large forms and curing is completed under steam pressure in autoclaves The extremely lightweight blocks have compressive strengths rangin from 450 to 1000 psi. They have high fire-resistance ratings and can b easily cut or sawed with ordinary woodworking tools. Cellular concret blocks are used for bearing walls in low-rise construction, as backing fo direct plaster application, as fireproofing, or as sound partitions. They ar laid up in ordinary cement-lime mortar.

4.6 CONCRETE BLOCK

Of the cementitious masonry products marketed in this country, concret block is the most familiar and most widely used. Aggregates determine th weight of the block and give different characteristics to the units. Ligh weight aggregates reduce the weight by as much as 20 to 45% with little sacrifice in strength. Specifications for aggregates are covered in ASTM C33 for normal-weight and C331 for lightweight materials. Weight classifications are based on density of the concrete and are subdivided as follows: normal-weight units are those whose concrete mix weighs more than 125 lb/cu ft; medium weight between 105 and 125 lb/cu ft; and two categories of lightweight, mixes weighing 85 to 105 lb/cu ft, and those weighing less than 85 lb/cu ft.

Some of the more commonly used aggregates are listed in *Fig. 4-2* together with the concrete unit weight and weight classifications. Exact individual unit weights depend on the coring design of the block and the percentage of solid volume to voids. An ordinary 8 × 8 × 16-in. unit weighs 40 to 50 lb when made from the more dense aggregates, and 25 to 35 lb when made from the lighter aggregates. Manufacturers can supply information regarding exact weight of their products, or the figures may be calculated if the percent of solid volume is known. Heavy or lightweight block can be used in any type of construction, but lightweight units have higher fire resistance, thermal insulation, and sound control properties. Choice of unit will depend largely on local availability and various design requirements of the building project. Three major kinds of concrete block are recognized: *hollow loadbearing concrete masonry units*, ASTM C90; *solid loadbearing concrete masonry units*, ASTM C145; and *non-loadbearing concrete masonry units*, ASTM C129.

Classification	*Aggregate*	*Unit weight of concrete (pcf)*	*Average weight of 8×8×16 unit (lb)*
Normal weight	Sand and gravel aggregate	135	44
	Crushed stone and sand aggregate	135	40
Medium weight	Air-cooled slag	120	35
Light weight	Coal cinders	95	28
	Expanded slag	95	28
	Scoria	95	28
	Expanded clay, shale, and slate	85	25
	Pumice	75	22

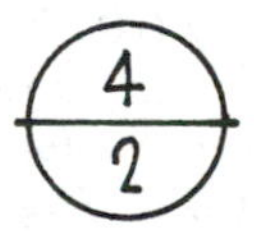

Weight variations of concrete with different aggregates.

Units defined as solid must have a minimum of 75% net solid area. Although the industry has standardized exterior dimensions of modular units, no such standardization exists for number and size of cores. Coring designs and percent of solid volume varies depending on the unit size, the equipment, and the methods of the individual manufacturers. For structural reasons, ASTM standards for loadbearing units specify minimum face shell and web thickness, but these stipulations do not apply for non-loadbearing units. Although minimum face shell and web thickness will not necessarily correspond to actual dimensions for all units, they can be used to estimate properties for preliminary design (*see Fig. 4-3*).

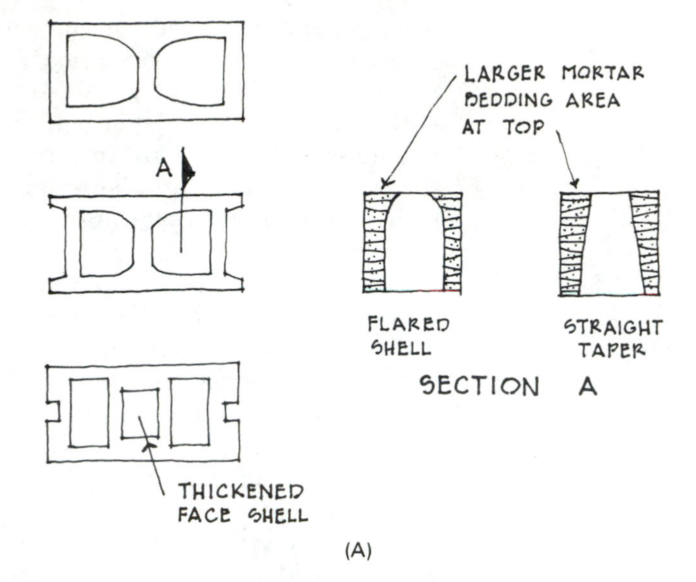

(A)

		Web thickness (WT)	
Nominal width, W, of units (in.)	*Face-shell thickness (FST), min. (in.)**	*Webs,* min. (in.)*	*Equivalent web thickness, min. (in./lin ft)†*
3 and 4	$\frac{3}{4}$	$\frac{3}{4}$	$1\frac{5}{8}$
6	1	1	$2\frac{1}{4}$
8	$1\frac{1}{4}$	1	$2\frac{1}{4}$
10	$1\frac{3}{8}$ $1\frac{1}{4}$‡	$1\frac{1}{8}$	$2\frac{1}{2}$
12	$1\frac{1}{2}$ $1\frac{1}{4}$‡	$1\frac{1}{8}$	$2\frac{1}{2}$

*Average of measurements on 3 units taken at the thinnest point.

†Sum of the measured thickness of all webs in the unit, multiplied by 12, and divided by the length of the unit.

‡This face-shell thickness (FST) is applicable where allowable design load is reduced in proportion to the reduction in thickness from basic face-shell thicknesses shown.

(B)

Concrete block coring: (A) Various concrete block core patterns; (B) Table of minimum thickness of face shells and webs (ASTM C90 units only). (*From National Concrete Masonry Association,* TEK Bulletin 36, *NCMA, Herndon, Va.*)

4.6.1 Coring

Block is produced in two- and three-core designs and with smooth or flanged ends (*see Fig. 4-3*). Two-core designs offer several advantages, including a weight reduction of approximately 10%, and larger cores for the placement of vertical reinforcing steel and conduit. In addition, the thickened area of the face shell at the center web increases tensile strength and helps to reduce cracking from drying shrinkage and temperature changes. End designs may be smooth or flanged, and some also have a $\frac{3}{4}$-in. mortar key or groove to assure weathertight joints. Smooth face ends must be used for corner construction, piers, pilasters, and so on. The cores of hollow units are usually tapered, with the face-shell thickness increasing from bottom to top of the unit. This facilitates form removal, and provides a larger bedding area for the mortar and a better grip for the mason. Minimum thickness required by ASTM standards refers to the narrowest cross section, not an average thickness of top and bottom. Since compressive strengths of hollow units are established on the basis of gross area, and fire-resistance ratings on equivalent solid thickness, these details of unit design become important in determining actual ratings for a particular unit (*see Fig. 4-4*).

Width (in.)	Gross volume [cu in. (cu ft)]	Minimum thicknesses		Three-core units		Two-core units	
		Shell (in.)	Web (in.)	Percent solid volume	Equivalent solid thickness (in.)	Percent solid volume	Equivalent solid thickness (in.)
$3\frac{5}{8}$	432 (0.25)	0.75	0.75	63	2.28	64	2.32
		1.00	1.00	73	2.66	73	2.66
$5\frac{5}{8}$	670 (0.388)	1.00	1.00	59	3.32	57	3.21
		1.12	1.00	63	3.54	61	3.43
		1.25	1.00	66	3.71	64	3.60
		1.37	1.12	70	3.94	68	3.82
$7\frac{5}{8}$	908 (0.526)	1.25	1.00	56	4.27	53	4.04
		1.37	1.12	60	4.57	57	4.35
		1.50	1.12	62	4.73	59	4.50
$9\frac{5}{8}$	1145 (0.664)	1.25	1.12	53	5.10	48	4.62
		1.37	1.12	55	5.29	51	4.91
		1.50	1.25	58	5.58	54	5.20
$11\frac{5}{8}$	1395 (0.803)	1.25	1.12	49	5.70	44	5.12
		1.37	1.12	51	5.93	46	5.35
		1.50	1.25	54	6.28	49	5.70
		1.75	1.25	57	6.63	52	6.05

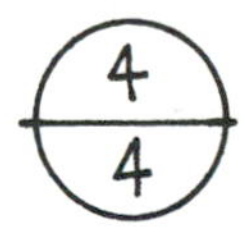

Volume characteristics of typical concrete masonry units (based on $7\frac{5}{8}$ in height × $15\frac{5}{8}$ in. length). (*From National Concrete Masonry Association, TEK Bulletin 2A, NCMA, Herndon, Va.*)

Hollow concrete masonry units (CMUs) are more widely used than solid units because of reduced weight, ease of handling, and lower cost. Most hollow blocks have core areas of 40 to 50%, leaving a net solid volume of 50 to 60%. Some concrete brick manufacturers have begun to capitalize on this economy by producing a cored "through-the-wall" unit. These units have increased thicknesses of 8 in., but maintain the face dimensions of brick. They may be classified as either solid or hollow depending on the percentage of voids created.

4.6.2 Grading

Loadbearing concrete masonry units are graded in exactly the same manner as concrete brick. Grade N units provide higher strength and resistance to frost action, and Grade S provide moderate strength and resistance to freezing. Durability under freeze-thaw conditions is taken as a measure of the absorption characteristics of the unit. Non-loadbearing units are not graded.

4.6.3 Moisture Content

For both solid and hollow, loadbearing and non-loadbearing block, a Type I designation represents moisture-controlled units, and Type II, non-moisture-controlled units (*see Fig. 4-5*). Since concrete shrinks with water loss, limits on unit moisture content were established to control the potentially damaging effects. When moist units are built into a wall and shrinkage is restrained, tensile and shearing stresses develop and may cause cracking. Allowable moisture content depends not only on the relative humidity of the

Requirements for concrete masonry, from ASTM C90, C129, and C145. (*Copyright, American Society for Testing and Materials, 1916 Race Street, Philadelphia, Pa. 19103. Reprinted, with permission.*)

TABLE A MOISTURE CONTENT REQUIREMENTS FOR HOLLOW LOADBEARING, SOLID LOADBEARING, AND HOLLOW NON-LOADBEARING UNITS, (ASTM C90, C145, AND C129).

	Moisture content, max. percent of total absorption (average of 3 units)		
	Humidity conditions at job site or point of use		
Linear Shrinkage (%)	*Humid**	*Intermediate†*	*Arid‡*
0.03 or less	45	40	35
From 0.03 to 0.045	40	35	30
0.045 to 0.065, max.	35	30	25

*Average annual relative humidity above 75%.
†Average annual relative humidity 50 to 75%.
‡Average annual relative humidity less than 50%.

TABLE B STRENGTH REQUIREMENTS FOR HOLLOW NON-LOADBEARING UNITS (ASTM C129).

	Compressive strength (average net area), min. (psi)
Average of 3 units	600
Individual unit	500

TABLE C STRENGTH AND ABSORPTION REQUIREMENTS FOR HOLLOW LOADBEARING UNITS (ASTM C90).

	Compressive strength, min. (psi)		*Water absorption, max. (average of 3 units) with oven-dry weight of concrete (lb/ft³)*			
	Average gross area		*Weight classification (lb/ft³)*			
			Light weight			
Grade	*Average of 3 units*	*Individual unit*	*Less than 85*	*Less than 105*	*Medium weight (less than 125 to 105)*	*Normal weight (125 or more)*
N-I N-II	1000	800	—	18	15	13
S-I* S-II*	700	600	20	—	—	—

*Limited to use above grade in exterior walls with weather-protective coatings and in walls not exposed to the weather.
Note: To prevent water penetration protective coating should be applied on the exterior face of basement walls and when required on the face of exterior walls above grade.

TABLE D STRENGTH AND ABSORPTION REQUIREMENTS FOR SOLID LOADBEARING UNITS (ASTM C145).

Grade	*Compressive strength, min. (psi)*		*Water absorption, max. (average of 3 units) with oven-dry weight of concrete (lb/ft^3)*			
	Average gross area		*Weight classification (lb/ft^3)*			
			Light weight		*Medium weight (less than 125 to 105)*	*Normal weight (125 or more)*
	Average of 3 units	*Individual unit*	*Less than 85*	*Less than 105*		
N-I N-II	1800	1500	—	18	15	13
S-I* S-II*	1200	1000	20	—	—	—

*Limited to use above grade in exterior walls with weather-protective coatings and in walls not exposed to the weather.

Note: To prevent water penetration, protective coatings should be applied on the exterior face of basement walls and when required, on the face of exterior walls above grade.

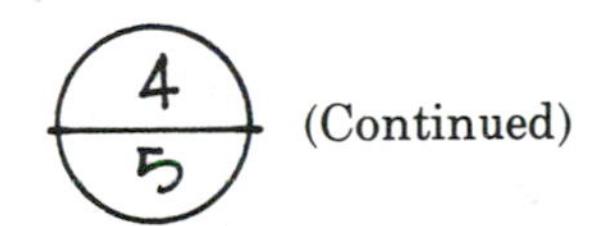

project location, but on the various shrinkage characteristics of the particular block. In order to equalize the expected drying shrinkage, units made from dense materials with lower shrinkage characteristics are allowed higher moisture contents than are lighter-weight units with higher shrinkage characteristics. The table in *Fig. 4-6* lists the average linear

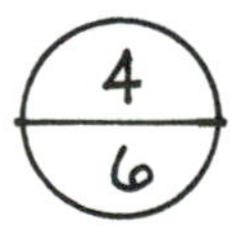

Actual shrinkage and permissible moisture content for various aggregate units.

Aggregate type	*Steam curing method*	*Average Linear shrinkage* (%)*	*Maximum moisture content, as percent of total absorption (average of 3 units) for average humidity conditions at job site†*		
			Humid	*Inter-mediate*	*Arid*
Normal weight	Autoclave low-pressure	0.019 0.027	45 45	40 40	35 35
Light weight	Autoclave low-pressure	0.023 0.042	45 40	40 35	35 30
Pumice	Autoclave low-pressure	0.039 0.063	40 35	35 30	30 25

*As determined by test, ASTM C246.

†From Table A, Fig. 4-5.

shrinkage for various aggregate types and the corresponding ASTM moisture values permitted.

4.7 UNIT TYPES

Concrete masonry units are governed by the same modular standards as clay masonry products. The basic concrete block size is derived from its relationship to modular brick. A nominal $8 \times 8 \times 16$-in. block is the equivalent of two bricks in width and length, and three brick courses in height. Horizontal ties may be placed at 8- or 16-in.-heights with either brick or structural clay tile. Nominal dimensions include a $\frac{3}{8}$-in. mortar joint, which is standard for concrete masonry. Concrete brick dimensions are the same as clay brick, but fewer sizes are generally available. Some variation in face size of standard concrete block stretcher units has been introduced to increase productivity on the job. Both the 12-in. high × 16-in.-long and the 8-in.-high × 24-in.-long units have 50% larger face area. To compensate for this additional size, lighter-weight aggregates are used to yield an 8-in.-thick unit weighing only 33 lb, (less than a normal-weight $8 \times 8 \times 16$-in. block). Each of the larger units can be laid as rapidly as a standard block, but covers 50% more wall area. These oversize units, however, are not typical. Size variation in most concrete block is limited to 2-in. incremental widths of 4 to 12 in., with a standard face size of 8×16 in. (*see Fig. 4-7*).

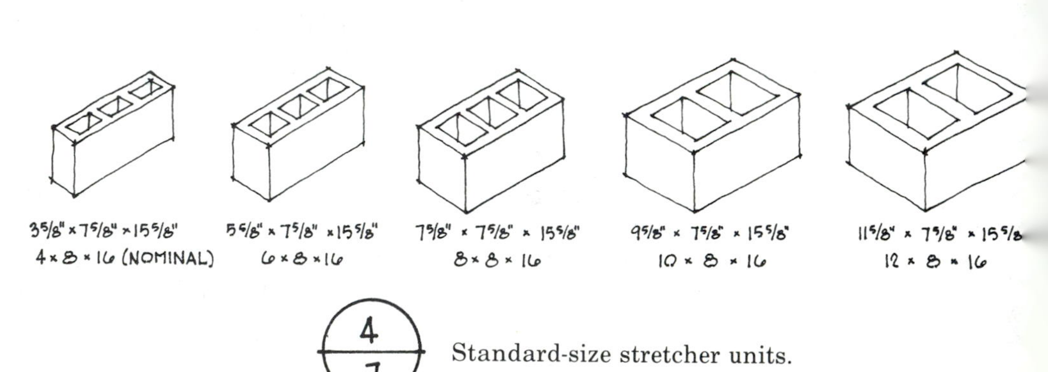

Standard-size stretcher units.

Half-lengths and half-heights are available for special conditions a openings, corners, and so on. A number of special shapes have beer developed for specific structural functions, such as lintel blocks, sasl blocks, pilaster units, and control joint blocks (*see Fig. 4-8*). Terminology i not fully standardized, and availability will vary, but most manufacturer produce and stock at least some of the more commonly used special items. Ir the absence of such shapes however, standard units can be field cut t accommodate many functions.

4.7.1 Decorative Units

Many decorative effects can be achieved through various CMU surfac treatments. Perforated screen block are available in several patterns an can be used as sun screens, ornamental partitions, and exterior soun baffles for damping low-frequency airborne noise (*see Fig. 4-9*). Ribbe grooved, and fluted faces laid in various bonding patterns can creat different effects scarcely resembling ordinary concrete block masonr

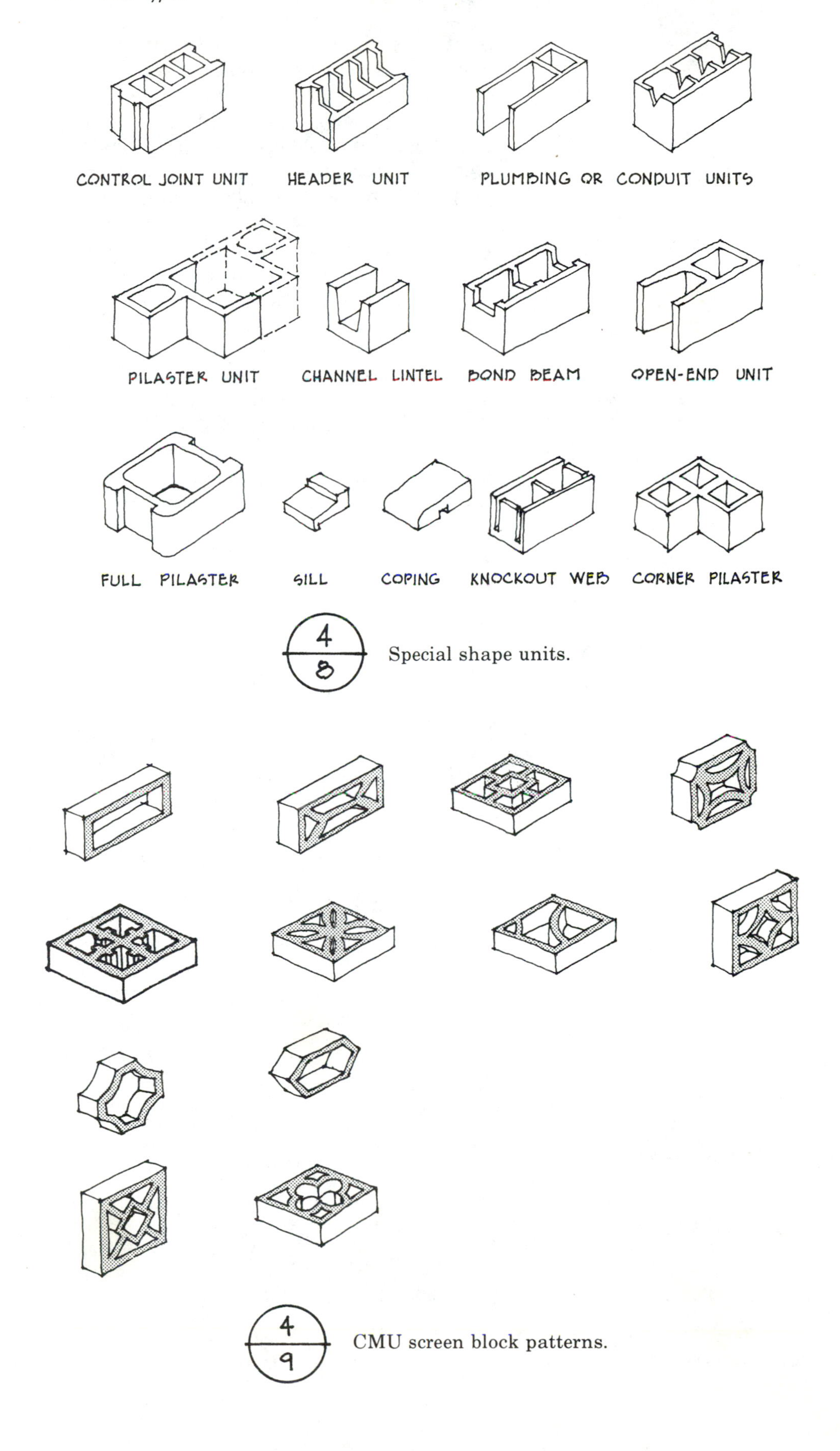

4/8 Special shape units.

4/9 CMU screen block patterns.

Split units with solid or ribbed surfaces simulate the effect of rough quarried stone. Surface grinding gives a smooth, almost polished appearance that enhances the color of the natural aggregate. Slump block units made with warm-tone cement and light-color aggregates give an irregular, "hand-made" impression. Customized designs have been created by many architects in shapes, patterns, and sculptured faces particularly suited to their projects, some with loadbearing capability and others strictly for veneer (*see Figs. 4-10 through 4-13*).

Since the selection of custom units over standard block is based primarily on appearance, the physical properties necessary for satisfactory performance may sometimes be taken for granted. Although a standard is currently being developed, at the present time there is no ASTM specification covering decorative units for exposure in exterior walls without protective coatings or facing materials. Most manufacturers have developed their own recommended specifications which are more stringent than those for conventional CMUs. Stricter requirements are essential to assure durability and imperviousness to water penetration when units are exposed to severe weather conditions with no surface protection. In the absence of specific minimum standards, empirical studies have shown that the strength and absorption requirements for normal-weight, Grade N concrete brick (ASTM C55) are applicable for decorative facing units where high strength and resistance to weathering are desired.

Custom concrete block. (*Courtesy National Concrete Masonry Association.*)

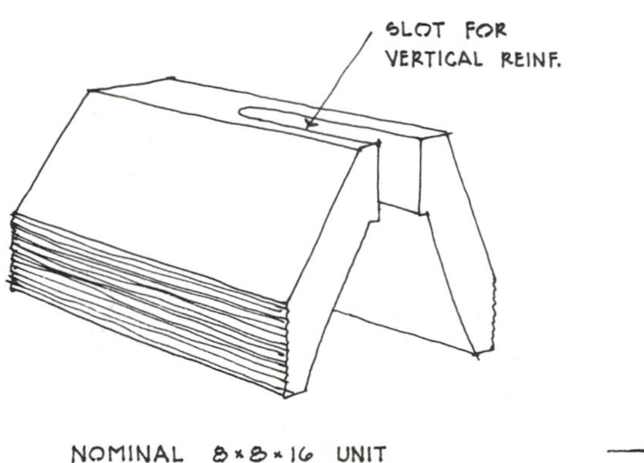

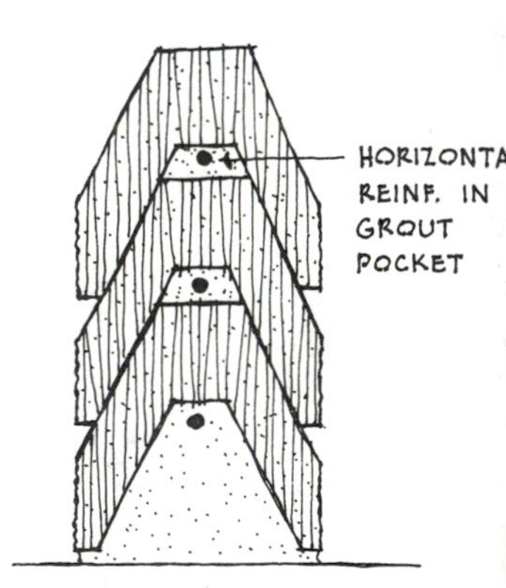

CUSTOM UNIT DESIGNED BY EDWARD HARDIN, ARCHITECT

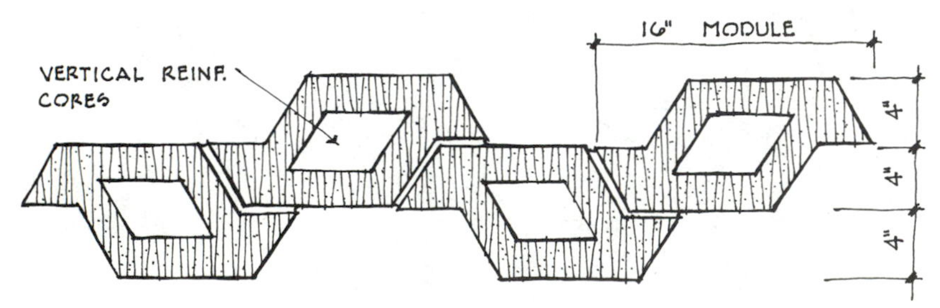

CUSTOM UNIT DESIGNED BY RICHARD E. CAMPBELL, ARCHITECT

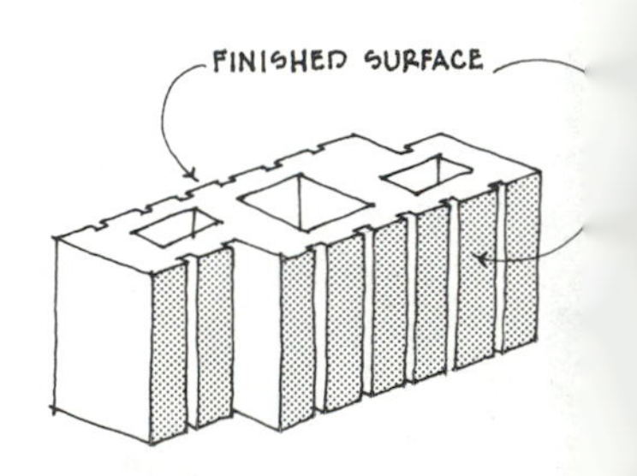

CUSTOM UNIT DESIGNED BY CHARLES W. DUEMMLING, ARCHITECT

4/11 Custom 12-rib split-face block, Coldspring New Town, Baltimore. Moshe Safdie, architect. (*Photo courtesy National Concrete Masonry Association.*)

(A)

(B)

4/12

Split rib and round fluted blocks: (A) Split-rib block: Guilderlan Reinsurance Bldg., Delmar N.Y. Howard Geyer Assoc., architects; (B Round-fluted block: Polytarp Office Bldg., Toronto, Canada, Boigon an Armstrong architects. (*Photos courtesy National Concrete Masonr Association.*)

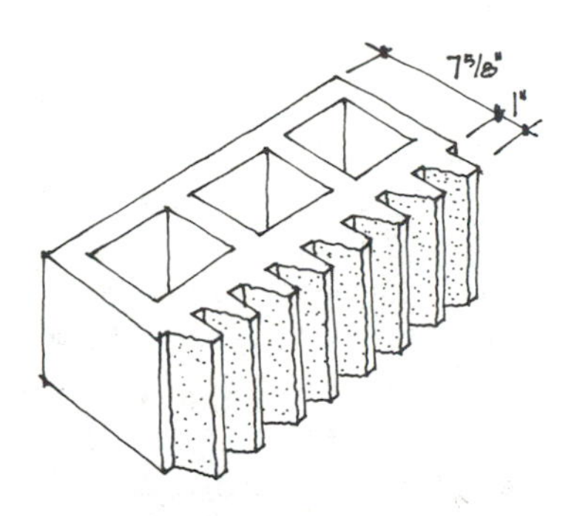

SPLIT-SAWTOOTH

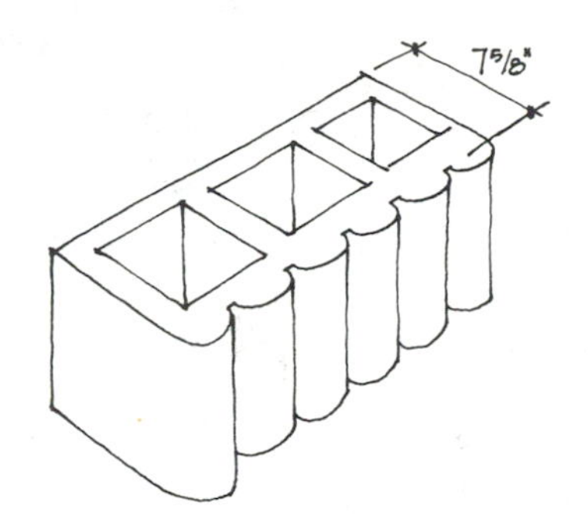

CURVED FLUTE

Two custom face designs by Paul Rudolph, architect. (*Photos courtesy National Concrete Masonry Association.*)

4.7.2 Glazed Units

Glazed surfaces may be applied to concrete brick or block as well as to sand-lime brick. Surfaces may consist of epoxy, polyester, ceramic, porcelainized, or mineral glazes, or cementitious finishes. All applied surfaces must meet the requirements of ASTM C744 in tests of imperviousness, abrasion, stain-, chemical-, and fire-resistance as well as crazing and adhesion of facing material to unit. Grade N, Type I, hollow loadbearing units are used and a thermosetting, resinous coating combined with specially treated silica sand, pigments, and/or ceramic colored granules is applied. The minimum requirements for both strength and abrasion are lower for glazed cementitious and concrete products than for glazed clay masonry units.

4.7.3 Paving Units

Two kinds of concrete masonry paving units are available for roadway and parking area surfacing. Solid interlocking units in a number of patterns provide a continuous topping over standard sand and gravel base materials. Open grid blocks permit grass to grow through the perforations while stabilizing the soil, protecting vegetation, and supporting vehicular traffic (*see Fig. 4-14*). Densely compacted units of 5000 to 10,000 psi compressive strength have a high resistance to moisture penetration and great durability in severe weathering conditions. CMU paving systems permit percolation of rainwater back into the soil despite the relative imperviousness of

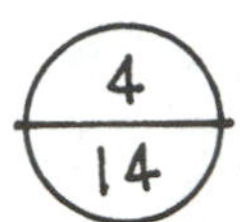

Concrete masonry paving units. (*From National Concrete Masonry Association*, TEK Bulletins 52, 75, 87, *NCMA, Herndon, Va.*)

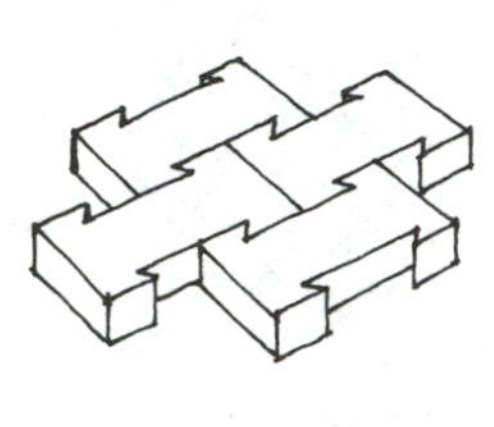

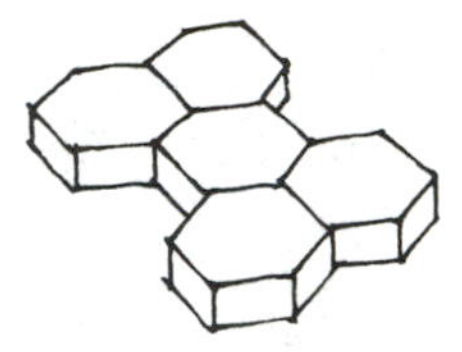

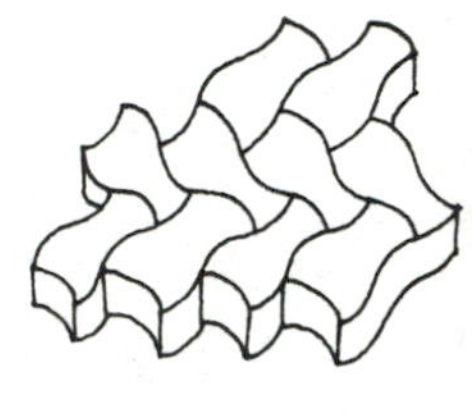

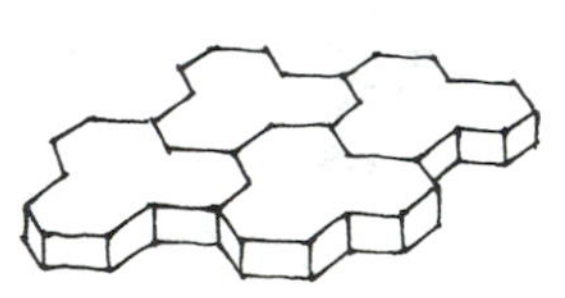

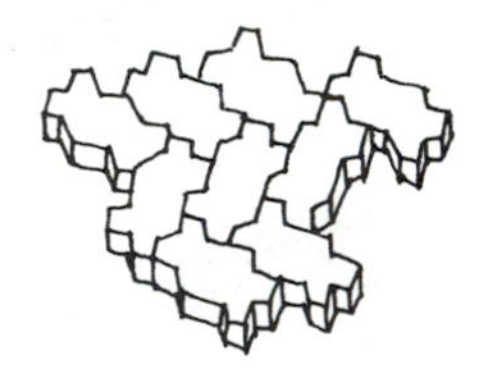

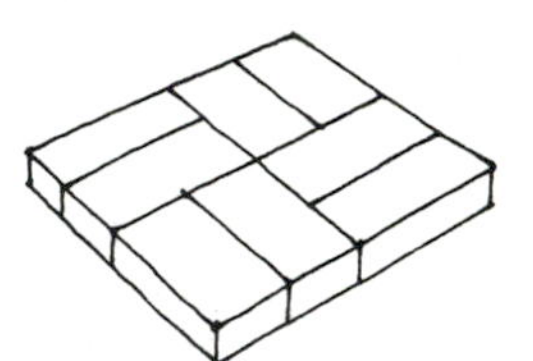

SOLID PAVERS

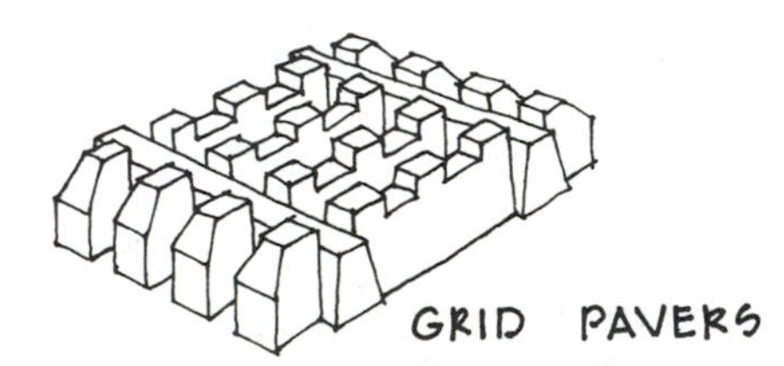

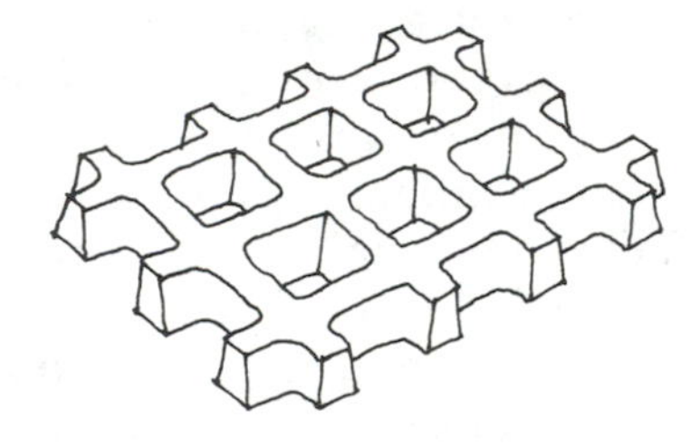

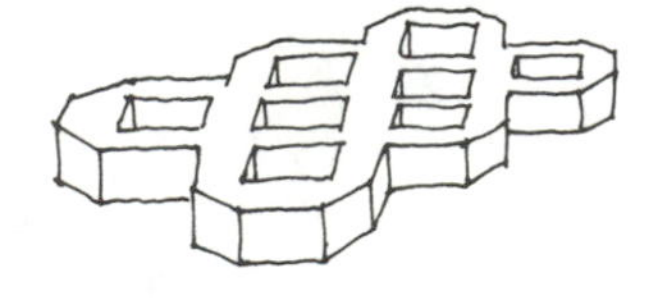

the unit itself. Prevention of excessive runoff is an important environmental consideration for standard installations as well as those where erosion and drainage of surrounding areas may be a problem. Grid units are generally 15 to 18 in. wide, about 24 in. long, and from $4\frac{1}{2}$ to 6 in. deep. Solid units are usually of proprietary design, and the sizes and shapes available will vary among manufacturers in different locations. Thicknesses range from $2\frac{1}{2}$ to 6 in., depending on the type of service and traffic load anticipated. Heavy-duty performance can be provided for industrial areas and roadways when speeds do not exceed about 40 mph.

The National Concrete Masonry Association (NCMA) has developed a specification, NCMA A-16-77, to govern properties and performance of solid concrete masonry pavers for vehicular traffic. Under these criteria, units with minimum compressive strengths of 8000 psi, and average absorption rates of less than 5%, have been found to give satisfactory performance in both laboratory and field tests.

4.8 PROPERTIES AND CHARACTERISTICS

CMU physical properties and characteristics fall into a number of structural, aesthetic, and functional categories. The two basic aspects, strength and absorption, have the greatest influence on overall performance. Compressive strength varies with the type and gradation of the aggregate, the water-cement ratio, and the degree of compaction in molding. In general, the lighter-weight aggregates produce slightly lower strength values and have increased rates of absorption (*see Fig. 4-15*).

Classification	*Aggregate*	*Compressive strength (gross area) (psi)*	*Water absorption (lb/cu ft of concrete)*	*Thermal expansion coefficient (per °F) $\times 10^{-6}$*
Normal weight	Sand and gravel	1200–1800	7–10	5.0
	Crushed stone	1100–1800	8–12	5.0
Medium weight	Air-cooled slag	1100–1500	9–13	4.6
Light weight	Coal cinders	700–1000	12–18	2.5
	Expanded slag	700–1200	12–16	4.0
	Scoria	700–1200	12–16	4.0
	Expanded clay, shale, and slate	1000–1500	12–15	4.5
	Pumice	700– 900	13–19	4.0

Effects of aggregate on concrete masonry properties.

4.8.1 Unit Strength

Aggregate particle size and gradation, as well as the amount of mixing water, affect compaction and consolidation and are important determinants of strength. Higher compressive values are associated with wetter mixes, but manufacturers must individually determine optimum water proportions to obtain a balance among moldability, handling, breakage, and strength. For special applications, higher-strength units may be obtained from the same aggregates by careful design of the concrete mix and slower curing. Gross strength ratings can thus be increased to as much as 2650 psi.

Because the core areas for hollow unit masonry are not standardized, strength values are always listed for gross area of the unit. For hollow masonry, net compressive strength is then computed on the percentage of hollow versus solid material in the block. If exact information regarding core area is not available or cannot be estimated from the table in *Fig. 4-4*, an average multiplier of 1.88 may be used for preliminary work. This figure represents units with 53% solid material and yields a net compressive strength of approximately 1800 psi from a gross area value of 1000 psi; 2250 psi net strength for 1200 psi gross, and so on.

Other CMU structural values can be estimated from compressive strength. *Tensile strength* generally ranges from 7 to 10% of compressive, *flexural strength* from 15 to 20%, and the *modulus of elasticity* from 300 to 1200 times the value in compression. For engineering calculations in reinforced masonry construction exact figures must be computed, but for general design purposes, these rules of thumb give a fairly accurate idea of the properties and capabilities of the block or brick being considered.

4.8.2 Absorption

Water absorption characteristics are an indication of durability in resistance to freeze-thaw cycles. Highly absorptive units, if frozen when permeated with water, can be fractured by the expanding ice crystals. A drier unit can accommodate some expansion into empty pore areas without damage. Minimum ASTM requirements differentiate between unit weights because of the effect of aggregate characteristics on this property. Absorption values are measured in pounds of water per cubic foot of concrete. They may range from as little as 4 or 5 lb/cu ft for heavy sand and grave materials, to 20 lb/cu ft for the most porous, lightweight aggregates.

Porosity influences other properties, such as thermal insulation and sound absorption. Increases in these characteristics are often accompanied by an undesirable increase in moisture absorption as well. Pore structure varies for different aggregates and material types and has varying influence on these values and their relationships to one another. Relatively large interconnected pores readily absorb air and sound as well as water and offer less resistance to damage from freezing. Unconnected or closed pores offer good insulating qualities, and reduced absorption of water and sound. A high initial rate of water absorption, or suction, adversely affect the bond between mortar and unit just as it does in clay masonry. Unlik brick, however, concrete products may not be prewetted at the job site t control suction because of the moisture shrinkage inherent to concrete. Suction can be controlled only through proper product specification b ASTM standards, and through the use of highly water retentive mortars t ensure the integrity of the bond.

4.8.3 Volume Changes

Volume changes in concrete masonry are caused by several thing Moisture shrinkage can be the most damaging because initial cycles of wetting and drying cause permanent shrinkage. Aged material has th ability to expand and contract in reversible movement. Controlling the initial moisture content at the time of construction is essential in preventing excessive shrinkage cracks in the finished wall. The different manufacturing techniques described in Chapter 2 bear significantly on this characteristic because of the variations in curing and drying methods. For a given aggregate, shrinkage tendencies due to moisture change can be reduced by as much as half by using high-pressure autoclave curing methods as opposed to low-pressure steam curing.

Small dimensional variations may occur due to changes in temperature. These changes, however, are fully reversible, and the units return to their original size after being heated and cooled through the same temperature range. Coefficients of thermal expansion (*see Fig. 4-15*) vary with different aggregates and are generally greater than values for clay masonry. As a result, provisions must be made for flexible anchorage and/or pressure-relieving control joints.

Volume changes are also caused by a natural chemical reaction called carbonation. Cured concrete absorbs carbon dioxide from the air, causing irreversible shrinkage. Preliminary tests indicate that under certain conditions, the magnitude of this change may nearly equal that of moisture shrinkage. Carbonation stages added to the normal manufacturing process can eliminate many field problems by effectively "preshrinking" the masonry and producing a more dimensionally stable unit.

4.8.4 Fire, Sound, and Heat Resistance

Fire resistance, thermal insulation, and acoustical characteristics are all related to the density of the product. *Fire-resistance* ratings are based on the rate of heat transmission through the unit and the rate of temperature rise on the opposite face rather than on structural failure because no such failure occurs. Ratings are calculated on the equivalent solid thickness exclusive of voids. For some aggregates and core designs, maximum 4-hour ratings can be obtained with ordinary 8-in. hollow units. *Thermal insulation* characteristics vary with aggregate type and density. Exact values may be easily determined from basic information. (Insulating qualities based on engineering calculations are discussed in Chapter 8.)

Acoustical characteristics may be subdivided into two categories: (1) sound absorption and reflectance, which depend primarily on surface texture; and (2) sound transmission, which is a function of density and mass. Normal-weight or heavyweight units have higher resistance to sound transmission. They will produce walls with higher STC ratings than those of lightweight units because of their resistance to diaphragm action. Absorption is higher for coarse, open-textured surfaces with large pore spaces. Reflectance is greater for tighter, closer grained, or painted surfaces with few, if any, open pores. CMUs can absorb from 18 to 68% of the sound striking the face of the wall, with lightweight units having the greater values. Specially designed block with slotted face shells provide high absorption by permitting sound waves to enter the cores, where their energy is dissipated through internal reverberation. Noise problems, particularly of middle- and high-frequency sounds, can often be controlled by these units, but they are proprietary products and may not be available in all locations.

4.8.5 Colors

CMU unit colors may be altered through the use of different aggregates, cements, or the integral mixing of natural pigments. Pearl grays, buffs, or even whites can easily be produced, offering great versatility within the generic product group. (Penetrating stains may also be applied to the finished wall to achieve a uniform color.)

5

NATURAL STONE

The earth's hard crust has undergone many changes throughout the millennia of geologic history. The stress and strains, the wearing away by atmospheric forces, by rain, wind, and heat, have produced a great variety of stones differing widely in appearance, but sharing similarities of composition. All stone is made up of one or more minerals of specific crystalline structure and definable chemical makeup. No two blocks of stone, however, even if quarried side by side, are identical in internal structure or physical and chemical composition.

5.1 GEOLOGICAL CHARACTERISTICS

As a natural, inorganic substance, stone can be categorized by form and geological origin. *Igneous rock* is formed by the solidifying and cooling of molten material lying deep within the earth or thrust to its surface by volcanic action. Granite is the only major building stone of this origin. *Sedimentary rock* such as sandstone, shale, and limestone is formed by waterborne deposits of minerals produced from the weathering and destruction of igneous rock. The jointed and stratified character of the formation makes it generally weaker than igneous rock. *Metamorphic rock* is either igneous or sedimentary material whose structure has been changed by the action of extreme heat or pressure. Marble, quartzite, and slate are all metamorphically formed.

Stone may also be classified by mineral composition. Building stone generally contains as the major constituent (1) silica, (2) silicates, or (3) calcareous materials. The primary silica mineral is quartz, the most abundant mineral on the earth's surface, and the principal component of granite. Silicate minerals include feldspar, hornblende, mica, and serpentine. Feldspar may combine with lime or potash to produce red, pink, or clear crystals. Hornblende, combining often with lime or iron, appears green, brown, or black. Mica, with iron or potash, produces clear crystals. Serpentine, in combination with lime, is generally green or yellow in color. The most common silicate building stone is also called serpentine after this mineral. Calcareous minerals include carbonates of lime and magnesia, such as calcite and dolomite, forming limestone, travertine, and marble.

5.2 PROPERTIES

Prior to the twentieth century, stone was the predominant material used in major building construction. It was not only the structural material, but also the exterior and interior finish, and often the flooring and roofing as well. The term "masonry" at one time referred exclusively to stonework, and the "architects" of medieval castles and cathedrals were actually stonemasons. Because of its massive weight and the resulting foundation requirements, stone is seldom used today as a structural element in contemporary architecture. It is, however, still widely used as a facing or veneer; in retaining walls, steps, walks, paths, and roads; as a floor finish; and is enjoying renewed popularity as a roofing material.

Despite their abundant variety, relatively few types of stone are suitable as building materials. In addition to accessibility and ease of quarrying, the stone must also satisfy the requirements of *strength*, *hardness*, *workability*, *porosity*, *durability*, and *appearance*.

The strength of a stone depends on its structure, the hardness of its particles, and the manner in which those particles are interlocked o cemented together. Generally, the denser and more durable stones are als stronger, but this is not always true. A minimum *compressive strength* o 5000 psi is considered adequate for building purposes, and the stones mos often used are many times stronger in compression than required by th loads imposed on them. Failures from bending or uneven settlement are no uncommon, however, since stone is much stronger in compression than i flexure or shear. Stones of the same type may vary widely in strength, thos from one quarry being stronger or weaker than those from another. Thus th average crushing strength of any type of stone may be misleading becaus of the wide variation in test results produced by stones within the sam classification. The table in *Fig. 5-1* illustrates the ranges typical for severa major types of stone. In modern building construction, *shearing strength* i stone is not nearly so important as compressive strength. The allowable uni stress of stone in shear should not be taken at more than one-fourth th allowable compressive unit stress. In *tension*, a safe working stress for ston masonry with portland cement mortar is 15 psi.

Hardness of stone is critically important only in horizontal planes suc as flooring and paving, but hardness does have a direct influence o workability. Characteristics may vary from soft sandstone which is easil scratched, to some stones which are harder than steel. Both strength an hardness are proportional to silica content. *Workability* in this instanc refers to the ease with which a stone may be sawed, shaped, dressed, carved, and will directly affect the cost of production. Workability decrease as the percentage of siliceous materials increases. Limestone, for instanc which contains little silica, is easily cut, drilled, and processed. Granit

Name	Type	Principal constituent	Weight (lb/cu ft)	Specific gravity	Compressive strength (psi)	Modulus of rupture (lb/sq ft)
Granite	Igneous	Silica	170	2.61–2.70	7,700–60,000	1430–5190
Marble	Metamorphic	Calcium carbonate	165	2.64–2.72	8,000–50,000	600–4900
Slate	Metamorphic	Calcium carbonate	170	2.74–2.82	10,000–15,000	6000–15,000
Limestone	Sedimentary	Calcium carbonate	165	2.10–2.75	2,600–28,000	500–2000
Sandstone	Sedimentary	Calcium carbonate	155	2.14–2.66	5,000–20,000	700–2300

Properties of common building stones (based on National Bureau of Standards Reports).

however, which consists largely of quartz, is the most difficult stone to cut and finish.

Porosity, the percentage of void content, affects the stone's absorption of moisture, thus influencing its ability to withstand frost action and repeated freeze-thaw cycles. Pore spaces are usually continuous and often form microscopic cracks of irregular shape. The method of stone formation, and the speed of cooling of the molten material, influence the degree and structure of these voids because of compaction and the possibility of trapped gases. Thus, sedimentary rock, formed in layers without high levels of pressure, is more porous than rock of igneous or metamorphic origin. Closely linked to this characteristic are grain and texture, which influence the ease with which stones may be split, and for ornamental purposes contribute to aesthetic effects as much as color.

Durability of stone, or its resistance to wear and weathering, is also considered roughly analogous to silica content. This is perhaps the most important characteristic of stone because it affects the life span of a structure. The stones traditionally selected for building construction have exhibited almost immeasurable durability compared to other building materials.

5.3 PRODUCTION

Stone is quarried from its natural bed by various techniques, depending on the nature of the rock. Drilling and splitting is the most basic, and the oldest method. With stratified material such as sandstone and limestone, the process is facilitated by natural cleavage planes, but also limited in the thickness of stone that can be produced. Holes are drilled close together along the face of the rock, and plugs and wedges then driven in with sufficient pressure to split the rock between holes. For stratified rock, holes are drilled only on the face perpendicular to the bed, but nonstratified material must be drilled both vertically and horizontally. Channeling machines are often employed on sandstone, limestone, and marble, but cannot be used with granite or other very hard stone. Wire saws are now used by most stone producers to cut a smoother surface, reduce the required mill finishing, and to subdivide large blocks of stone for easier transport, handling, and finishing.

The first stones cut from the quarry are large, with rough, irregular faces (*see Fig. 5-2*). These monolithic pieces are cut or split to the required rough size, then dressed at the mill with power saws and/or hand tools. Finished stone surface textures may vary from a rough rock face to a more refined hand-tooled or machine-tooled finish. For thin facings of marble or granite, gang saws cut several slabs from a block of stone at the same time. Although the sawing is a slow process, the surface it produces is so even that much work is saved in later dressing and polishing. Other saws, such as chat saws, shot saws, and diamond saws, are used to cut rough blocks of stone to required dimensions. Each type of saw produces a different surface texture.

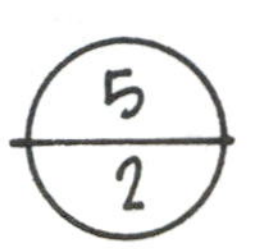

Quarrying stone. (*Photos courtesy Georgia Marble Co.*)

In addition to sawed finishes, stone may also be dressed with hand or machine tools. Planing machines prepare a surface for hammered finishes, for polished finishes, and for honed or rubbed finishes. A carborundum machine, used in place of a planer, will produce a very smooth finish. Honing is accomplished by rubbing the stone surface with an abrasive such as silicon carbide or sand after it has been planed. (A water spray is used to control dust.) Larger surfaces are done by machine, smaller surfaces and moldings by hand. Polished surfaces require repeated rubbing with increasingly finer abrasives until the final stage, which is done with felt and a fine polishing material. Only granite, marble, and some very dense limestones will take and hold a high polish. Power-driven lathes have been developed for turning columns, balusters, and other members that are round in section.

Hand-tooling is the oldest method of stone dressing. Working with pick, hammer, and chisels (*see Fig. 5-3*), the mason dressed each successive face of the stone, giving it the desired finish and texture. The drawings in *Fig. 5-4* illustrate the various steps in dressing the face, beds, and joints of a rough stone. Other hand-applied finishes include the bush-hammered, patent-hammered, pick-pointed, crandalled, and peen-hammered surface (*see Figs. 5-5 and 5-6*). Many of these finishes are now applied with pneumatic rather than hand tools, resulting in a more uniform surface.

Another finishing technique which produces a roughened surface is called flame cut, fire cut, or thermal finished. A natural gas or oxyacetelene flame is passed over a polished surface that has been wetted. The water that has been absorbed by the stone changes to steam and breaks off the surface, leaving an irregular finish. This finish can be selectively applied to portions of a stone surface to provide contrast.

Stone is used for masonry construction in many forms and is available

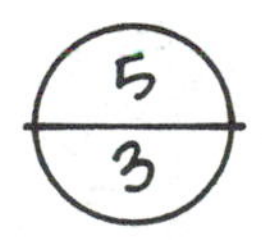

Traditional stone chisels. (*From Harley J. McKee,* Introduction to Early American Masonry, *National Trust for Historic Preservation and Columbia University, Preservation Press, Washington, D.C., 1973.*)

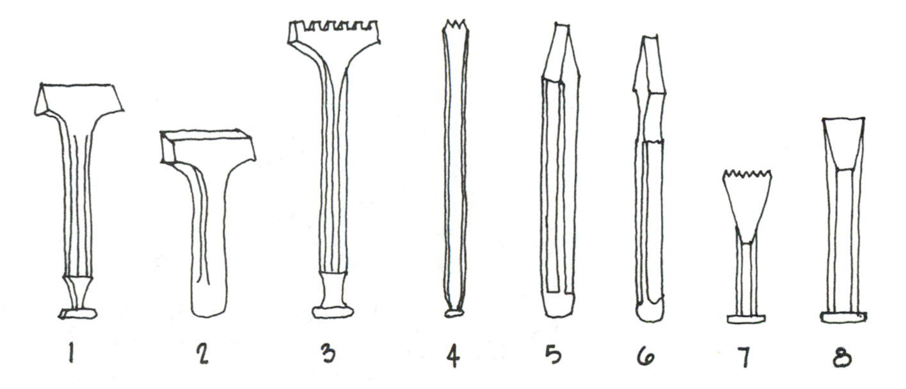

1. 2-3" WIDE DROVE CHISEL
2. 3½-4½" WIDE BOASTER OR BOLSTER TOOL
3. 19TH CENTURY TOOTH CHISEL
4. 16TH CENTURY ITALIAN TOOTH CHISEL
5. 19TH CENTURY NARROW CHISEL
6. SPLITTING CHISEL
7. 1¾", 7-TOOTH CHISEL
8. 1½" CHISEL

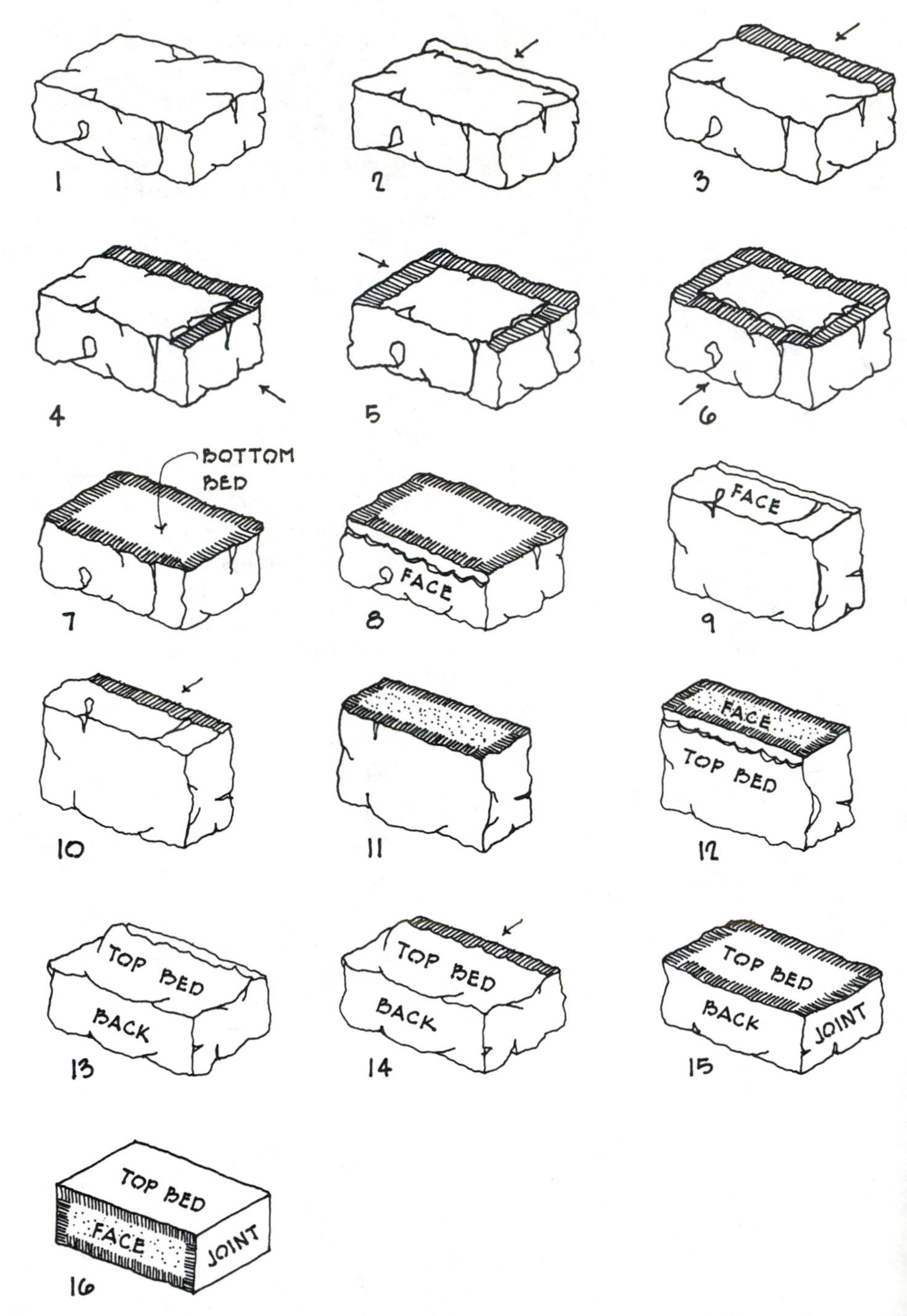

Various steps in hand-dressing the face, beds, and joints of a rough stone. (*From Harley J. McKee,* Introduction to Early American Masonry, *National Trust for Historic Preservation and Columbia University, Preservation Press, Washington, D.C., 1973.*)

commercially as (1) rubble stone, (2) dimension stone, and (3) flagston *Rubble* includes rough fieldstone and irregular stone fragments with least one good face. The stone may be either broken into suitable sizes, roughly cut to size with a hammer. *Dimension stone* (or cut stone) delivered from stone mills cut and dressed to a specific size, squared dimension each way, and to a specific thickness. Surface treatments inclu

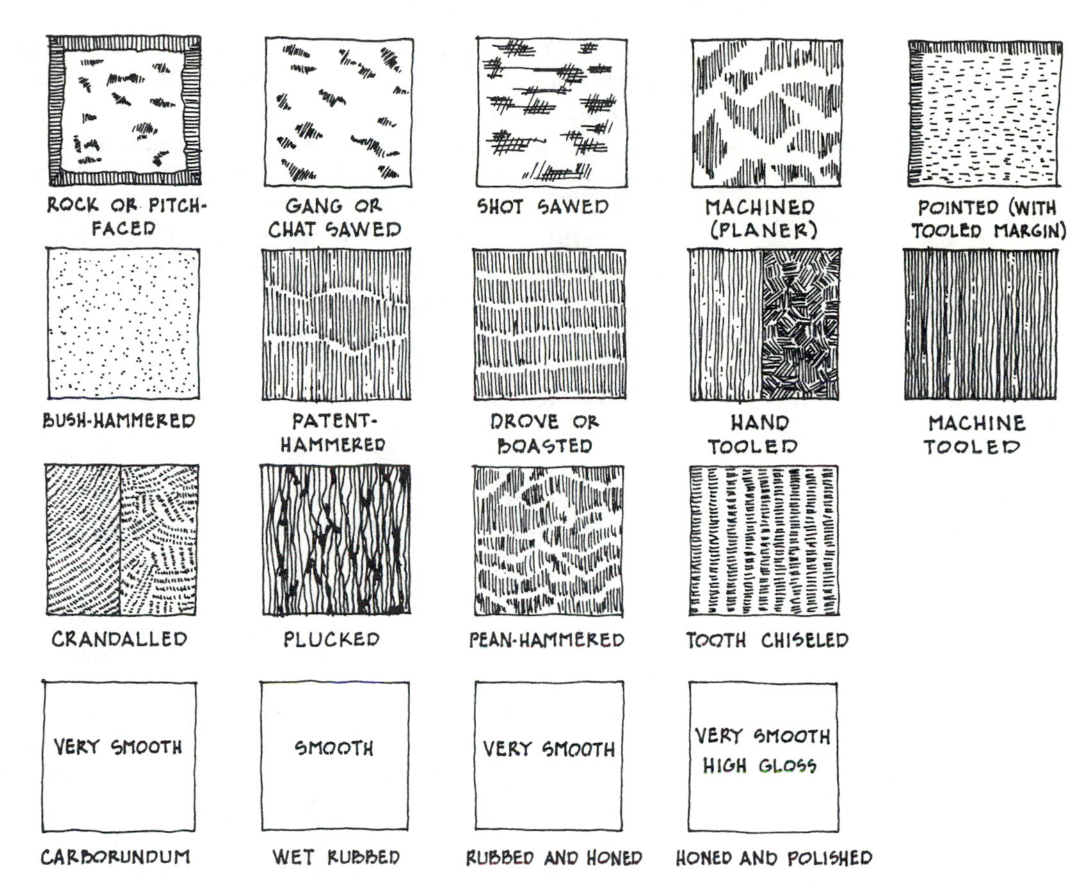

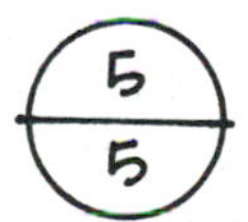

Stone surface finishes. (*From Charles G. Ramsey and Harold S. Sleeper,* Architectural Graphic Standards, *6th ed., ed. Joseph N. Boaz. Copyright © 1970 by John Wiley & Sons, Inc. Reprinted by permission of John Wiley & Sons, Inc.*)

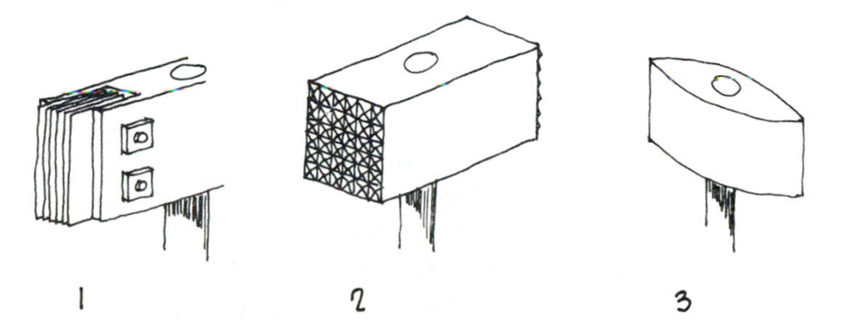

1. PATENT HAMMER (MAY HAVE 4 TO 8 BLADES)
2. BUSH HAMMER
3. PEEN HAMMER

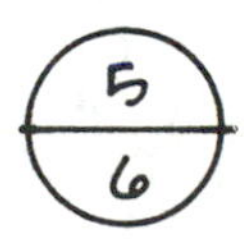

Stone hammers. (*From Harley J. McKee,* Introduction to Early American Masonry, *National Trust for Historic Preservation and Columbia University, Preservation Press, Washington, D.C., 1973.*)

a rough or natural split face, smooth, slightly textured, or polished finishes. Ashlar is a type of flat-faced dimension stone, generally in small squares or rectangles, with sawed or dressed beds and joints. Dimension stone is used for interior and exterior surface veneers, prefabricated panels, bearing walls, toilet partitions, arch stones, flooring, copings, stair treads, sills, and so on. *Flagstone* consists of thin slabs from $\frac{1}{2}$ to 2 in. thick in either squared or irregular shapes. Surfaces may be slightly rough, smooth, or polished. Flagstone is used on the exterior for walks, paths, and terraces, and on the interior as stair treads, flooring, coping, sills, and so on.

Some common types of stonework, as shown in *Fig. 5-7*, include rough fieldstone, random mosaic, coursed fieldstone, and cut-stone ashlar of various patterns.

5.4 BUILDING STONE

Some of the natural stones that satisfy the requirements of building construction are granite, limestone, sandstone, slate, and marble (*see Fig. 5-1*). Many others, such as quartzite and serpentine, are used locally or regionally, but to a much lesser extent.

5.4.1 Granite

Granite has been used as a building material almost since the inception of man-made structures. Because of its hardness, it was first used with exposed, hand-split faces. As tools and implements became more refined, the shapes of the stone also became more sophisticated. With the development of modern technology and improved methods of sawing, finishing, and polishing, granite was more readily available in the construction market and more competitive with the cost of other, softer stones.

Granite is an igneous rock composed primarily of quartz, feldspar mica, and hornblende. Colors vary depending on the amount and type o secondary minerals. Feldspar produces red, pink, brown, buff, gray, an cream colors, while hornblende and mica produce dark green or black Granite is classified as fine, medium, or coarse grained. It is very hard strong, and durable, and is noted for its hard-wearing qualities. Compressive strength may range from 7700 to 60,000 psi. While the hardness of th stone lends itself to a highly polished surface, it also makes sawing an cutting very difficult. A tolerance of $\pm\frac{3}{8}$ in. from nominal dimensions i generally required. Granite is often used for flooring, paneling, venee column facings, stair treads, flagstones, or in landscape application Veneer slabs may be cut as thin as $2\frac{1}{4}$ in., or the stone may be used in roughe form for ashlar work. Carving or lettering on granite, which was formerl done by hand or pneumatic tools, is now done by sandblasting, and ca achieve a high degree of precision.

5.4.2 Limestone

Limestone is a sedimentary rock which is durable, easily worked, ar widely distributed throughout the earth's crust. It consists chiefly calcium carbonate deposited by chemical precipitation or by the accumul tion of shells and other calcareous remnants of animals and plants. Ve few limestones consist wholly of calcium carbonate. Many contain ma nesium carbonates in varying proportions, sand or clay, carbonaceo

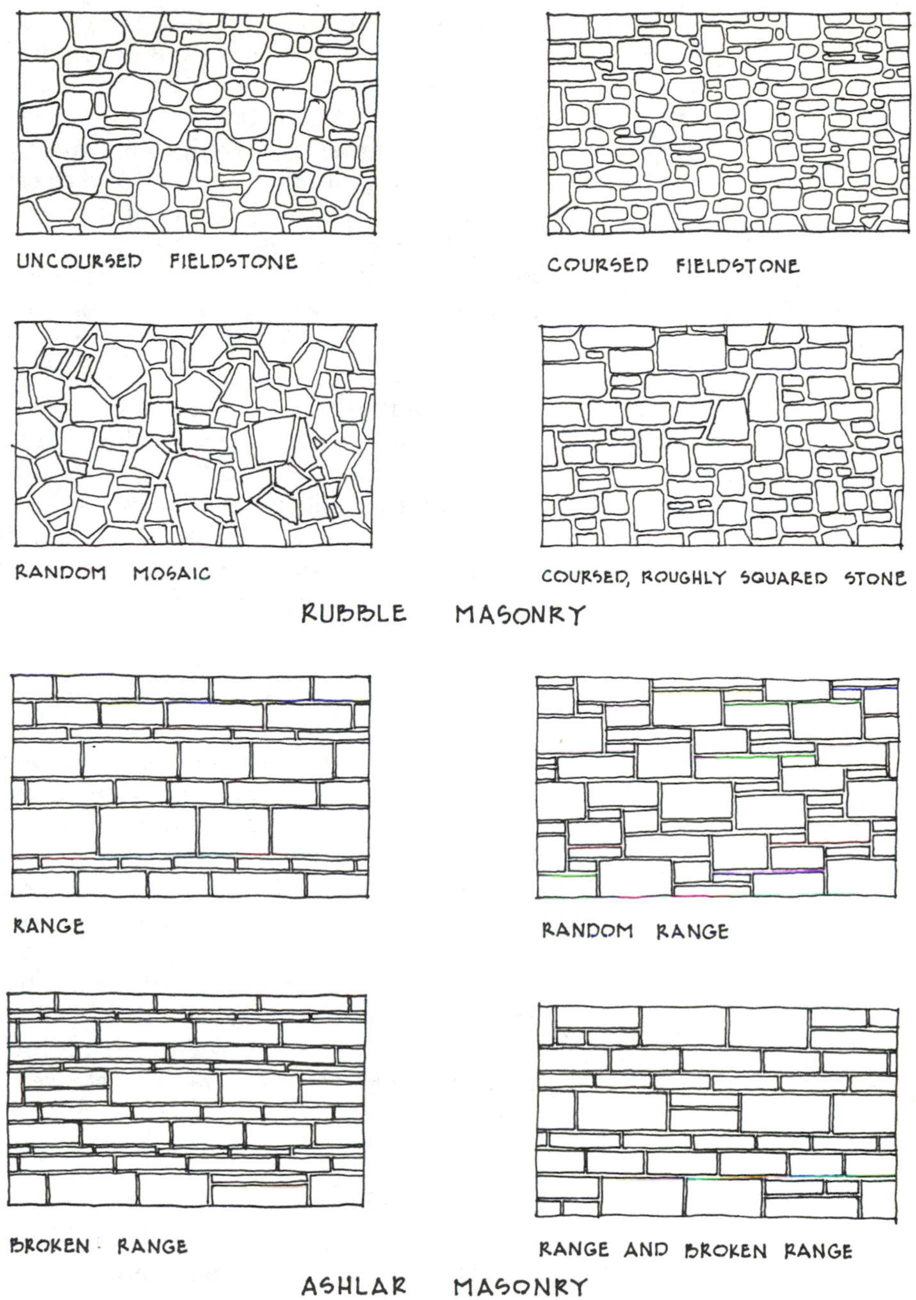

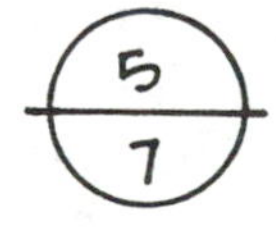

Some common types of stonework. (*From Charles G. Ramsey and Harold S. Sleeper*, Architectural Graphic Standards, *6th ed., ed. Joseph H. Boaz. Copyright © 1970 by John Wiley & Sons, Inc. Reprinted by permission of John Wiley & Sons, Inc.*)

matter, or iron oxides, which may color the stone. The most "pure" form is *crystalline* limestone, in which calcium carbonate crystals predominate, producing a fairly uniform white or light gray stone of smooth texture. It is highest in strength and lowest in absorption of the various types of limestone. *Dolomitic* limestone contains between 10 and 45% magnesium carbonate, is somewhat crystalline in form, and has a greater variety of texture. *Oolitic* limestone consists largely of small, spherical calcium carbonate grains cemented together with calcite from shells, shell fragments, and the skeletons of other marine organisms. It is distinctly noncrystalline in character, has no cleavage planes, and is very uniform in composition and structure.

The compressive strength of limestone varies from 2600 to 28,000 psi depending on the silica content, and the stone has approximately the same strength in all directions. Limestone is much softer, more porous, and has a higher absorption capacity than granite, but is a very attractive and widely used building stone. Although soft when first taken from the ground, limestone weathers hard upon exposure. Its durability is greatest in drier climates, as evidenced by the remains of Egyptian and Mayan monuments.

The impurities affect its color. Iron oxides produce reddish or yellowish tones as pink, buff, or cream, while organic materials such as peat give a gray tint. Limestone textures are graded as A, statuary; B, select; C, standard; D, rustic; E, variegated; and F, old Gothic. Grades A, B, C, and D come in buff or gray, and vary in grain from fine to coarse. Grade E is a mixture of buff and gray, and is of unselected grain size. Grade F is a mixture of D and E and includes stone with seams and markings.

Limestone is used as cut stone for veneer, caps, lintels, copings, sills, and mouldings, and as ashlar with either rough or finished faces. Naturally weathered or fractured fieldstone is often used as a rustic veneer on residential and low-rise commercial buildings. Veneer panels may be sliced in thicknesses ranging from 2 to 6 in. and face sizes from 3×5 ft to 5×14 ft. When the stone is set or laid with the grain running horizontally, it is said to be on its natural bed. When the grain is oriented vertically, it is said to be on edge.

Travertine is a porous limestone formed at the earth's surface through the evaporation of water from hot springs. It is characterized by small pockets or voids formed by trapped gases. This natural and unusual texturing presents an attractive decorative surface highly suited to facing materials and veneer slabs.

The denser varieties of limestone, including travertine, can be polished and for that reason are sometimes classed as marble in the trade. Indeed, the dividing line between limestone and marble is often difficult to determine.

5.4.3 Marble

Marble is a crystallized, metamorphosed form of noncrystalline limeston or dolomite. Its texture is naturally fine, permitting a highly polishe surface. The great color range found in marbles is due to the presence o oxides of iron, silica, mica, graphite, serpentine, and carbonaceous matter i grains, streaks, or blotches throughout the stone. The crystalline structure o marble adds depth and luster to the colors as light penetrates a sho distance and is reflected back to the surface by the deeper-lying crystal Pure marbles are white, without the pigmentation caused by mineral oxide Brecciated marbles are made up of angular and rounded fragmen embedded in a colored paste or cementing medium.

Marble often has compressive strengths as high as 50,000 psi, an when used in dry climates or in areas protected from precipitation, the stor is very durable. Some varieties, however, are decomposed by weathering exposure to industrial fumes, and are suitable only for interior work. Ov 200 imported and domestic marbles are available in the United States. Ea has properties and characteristics that make it suitable for different types of construction.

Marbles are classified as A, B, C, or D on the basis of working qualities, uniformity, flaws, and imperfections. For exterior applications, only group A, highest-quality materials should be used. The other groups are less durable, and will require maintenance and protection. Group B marbles have less favorable working properties than Group A, and will have occasional natural faults requiring limited repair. Group C marbles have uncertain variations in working qualities; contain flaws, voids, veins, and lines of separation; and will always require some repair (known as sticking, waxing, filling, and reinforcing). Group D marbles have an even higher proportion of natural structural variations requiring repair, and have great variation in working qualities.

Marble is available as rough or finished dimension stone, and is used extensively for wall and column facings, flooring, partitions, and other decorative surface work. Veneer slabs may be cut as thin as $\frac{7}{8}$ in. for surface areas up to 20 sq ft.

5.4.4 Slate

Slate is also a metamorphic rock, formed from argillaceous sedimentary deposits of clay and shale. Slates containing large quantities of mica are stronger and more elastic than clay slates. The texture of slate is fine and compact with very minute crystallization. It is characterized by distinct cleavage planes permitting easy splitting of the stone mass into slabs $\frac{1}{4}$ in. or more in thickness. Used in this form, slate provides an extremely durable material for flooring, roofing, sills, stair treads, and facings. Its average compressive strength is in the range of 10,000 to 15,000 psi, with tensile values as high as 8000 psi.

Small quantities of other mineral ingredients give color to the various slates. Carbonaceous materials or iron sulfides produce dark colors such as black, blue, and gray; iron oxide produces red and purple; and chlorite produces green tints. "Select" slate is uniform in color and more costly than "ribbon" slate, which contains stripes of darker colors.

5.4.5 Sandstone

Sandstone is a sedimentary rock formed of sand or quartz grains. Its hardness and durability depend primarily on the type of cementing agent present. If cemented with silica and hardened under pressure, the stone is light in color and very strong and durable. If the cementing medium is largely iron oxide, the stone is red or brown, and is softer and more easily cut. Lime and clay are less durable binders subject to disintegration by natural weathering. When first taken from the ground, sandstone contains large quantities of water, which make it easy to cut. When the moisture evaporates, the stone becomes considerably harder.

Sandstones vary in color from buff, pink, and crimson to greenish brown, cream, and blue-gray. It is traces of minor ingredients such as feldspar or mica which produce the range of colors. Both fine and coarse textures are found, some of which are highly porous and therefore low in durability. The structure of sandstone lends itself to textured finishes, and to cutting and tooling for ashlar and dimension stone in veneers, moldings, sills, and copings. Sandstone is also used in rubble masonry as fieldstone.

Flagstone or bluestone is a form of sandstone split into thin slabs for flagging.

5.5 SELECTING STONE

Stone for building construction is judged on the basis of (1) appearance, (2) economy, (3) durability, (4) strength, and (5) ease of maintenance. Design and aesthetics will determine the suitability of the color, texture, aging characteristics, and general qualities of the stone for the type of building under consideration. Colors may range from dull to brilliant hues, and from warm to cool tones. Textures may vary from coarse or rough to fine and dense. Some stones, such as marble and granite, are typically used on structures of grandeur and importance, whereas others, such as rubble sandstone, are more often seen on smaller, more picturesque buildings (*see Figs. 5-8 and 5-9*). Limestones are generally considered in the broader range of commercial and utilitarian applications. Some stones, such as granite, will soften very slowly in tone and outline, and will retain a sharp edge and hard contour indefinitely. Others mellow in tone and outline, becoming softer in shape without losing their sense of strength and durability.

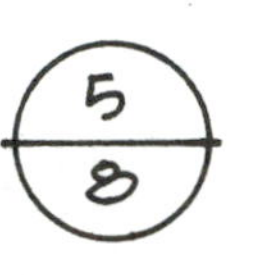

Field stone veneers.

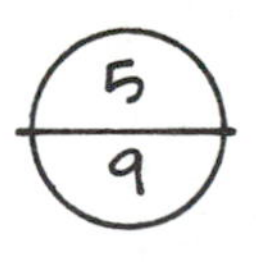

Veneer slabs and coursed dimension stone.

Elaborately carved ornaments and lettered panels require stones of fine grain to produce and preserve the detail of the artist's design.

The costs of various stones will depend on the proximity of the quarry to the building site, the abundance of the material, and its workability. In general, stone from a local source will be less expensive than those which must be imported; stone produced on a large scale less expensive than scarce varieties; and stone quarried and dressed with ease less expensive than those requiring excessive time and labor.

In terms of practicality and long-term cost, durability is the most important consideration in selecting building stones. Suitability will depend not only on the characteristics of the stone, but also on local environmental and climatic conditions. Frost is the most active agent in the destruction of stone. In warm, dry climates, almost any stone may be used with good results. Stones of the same general type such as limestone, sandstone, and marble differ greatly in durability based on softness and porosity. Soft, porous stones are more liable to absorb water and to flake or disintegrate in heavy frosts, and may not be suitable in the colder and more moist northern climates.

For flooring, abrasion resistance of the stone must also be considered. If two or more varieties of stone are used, the abrasion resistance should be approximately the same, or uneven wear will result. Only stones highly resistant to wear should be used on stair treads.

The compressive strength of stone was of great importance when large buildings were constructed of loadbearing stone walls and foundations. Today however, stone is more often used as a thin veneer over steel, concrete, or unit masonry structures, or as loadbearing elements only in low-rise structures. In these applications, the compressive loads are generally small, and nearly all of the commonly used building stones are of sufficient strength to maintain structural integrity.

6

MORTAR AND GROUT

Although mortar accounts for as little as 7% of the total volume of masonry, it influences performance far more than this proportion indicates. Functionally, it not only binds the individual elements together, but also seals against air and moisture penetration and bonds with steel reinforcing, metal ties, and anchor bolts to join the building components structurally. For engineered construction and loadbearing applications, mortar strength and performance are as critical as unit strength and workmanship.

Egyptian architects of the twenty-seventh century B.C. witnessed the first known use of masonry mortar, when a mixture of burned gypsum and sand was used in the construction of the Great Pyramid at Giza. Greek and Roman builders later added or substituted lime or crushed volcanic materials, but it was not until the nineteenth century development of portland cement that mortar became a true structural component equal in strength to the masonry units it bonded together.

6.1 MORTAR PROPERTIES

During this century, portland cement has become the principal ingredient of most mortar and grout mixes. Since it is also the principal ingredient of concrete, many designers assume that methods and materials for producing strong, durable concrete also apply to mortar technology. However, both laboratory tests and performance records disprove this assumption. The most important physical property of concrete is compressive strength, but the requirements for mortar and grout center primarily around bonding strength and durability. These qualities are influenced by two distinct sets of properties which interact to affect overall performance: (1) important *properties of the plastic mortar* include workability, water retentivity, initial flow, and flow after suction, and (2) critical *characteristics of the hardened mortar* are bond strength, durability, and extensibility, as well as compressive strength.

6.1.1 Workability

Workability significantly influences most other mortar characteristics. Workability is not precisely definable in quantitative terms because there are no definitive tests or standards for measurement. A "workable" mortar has a smooth, plastic consistency, is easily spread with a trowel, and readily adheres to vertical surfaces. Well-graded, smooth aggregates enhance workability as do lime, air entrainment, and proper amounts of mixing water. The lime imparts plasticity and increases the water-carrying capacity of the mix. Air entrainment introduces minute bubbles that act as lubricants in promoting flow of the mortar particles, but 12 to 15% maximum air content should be stipulated to minimize the reduction of bond strength. Unlike concrete, mortar requires a *maximum amount of water* for workability, and retempering should be permitted within the first $2\frac{1}{2}$ hours after mixing.

Variations in unit materials and in environmental conditions affect optimum mortar consistency. Mortar for heavier units must be more dense to prevent uneven settling after unit placement or excessive squeezing of mortar from the joints. Warmer summer temperatures require a softer, more moist mix to compensate for evaporation. Although workability is easily recognized by the mason, the difficulty in defining this property precludes a statement of minimum requirements in mortar specifications.

6.1.2 Retentivity and Flow

Other mortar characteristics that influence general performance, such as aggregate grading, water retentivity, and flow, can be accurately measured by laboratory tests and are included in ASTM Standards. *Water retentivity* allows mortar to resist the suction of dry masonry units and maintain moisture for proper curing. It is the mortar's ability to retain its plasticity in contact with absorptive masonry so that the mason can carefully align and level the units without breaking the bond. Less retentive mixes will "bleed" moisture, creating a thin layer of water between mortar and masonry unit and substantially decreasing bond strength (*see Fig. 6-1*). Highly absorptive clay units may be prewetted at the job site, but concrete products may not be moistened, thus requiring the mortar itself to resist water loss.

Under laboratory conditions, water retention is measured by flow tests, and is expressed as the ratio of *initial flow* to *flow after suction*. The flow test is similar to a concrete slump test, but is performed on a "flow table" that is rapidly vibrated up and down for several seconds.

Although they accurately reflect the properties of the mortar, laboratory values differ somewhat from field requirements. Construction mortars require initial flow values on the order of 130 to 150%. Laboratory mortars are required to have an initial flow of 100 to 115%. The amount of mixing

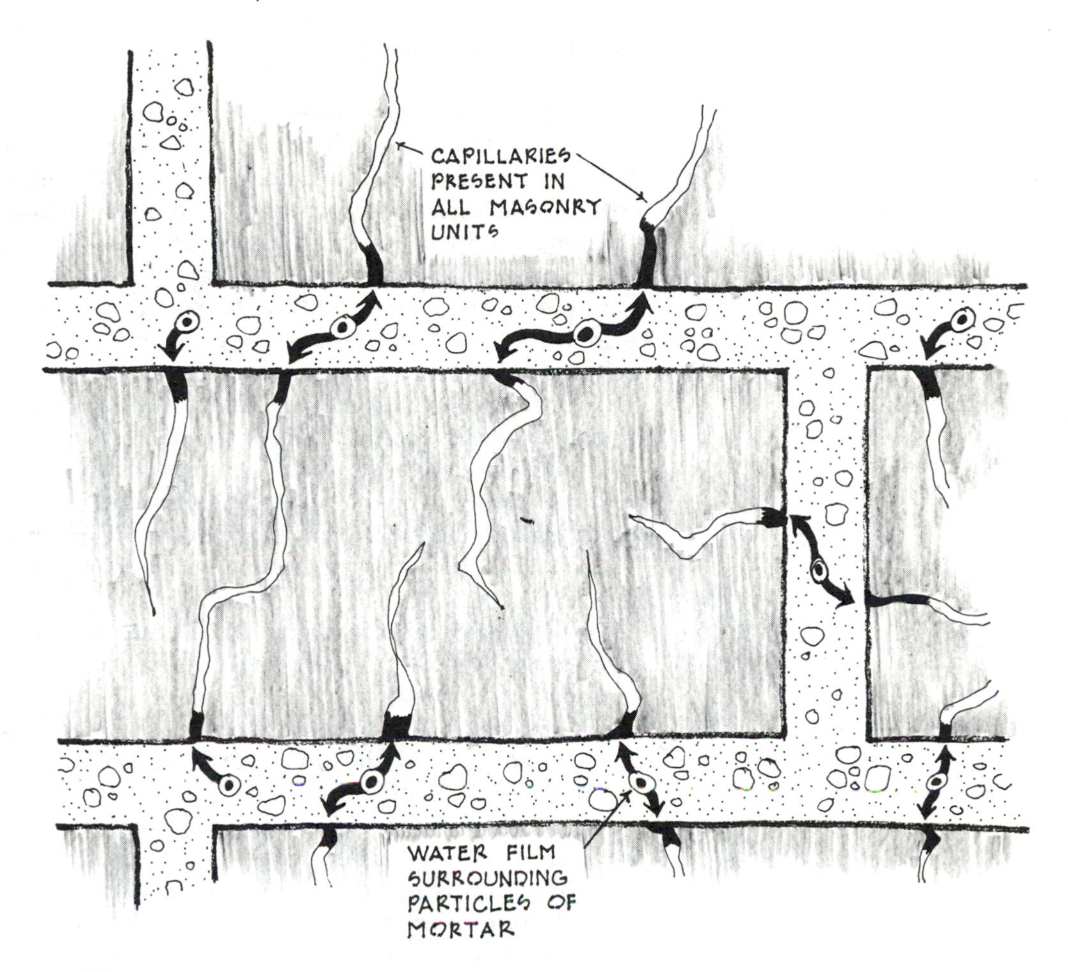

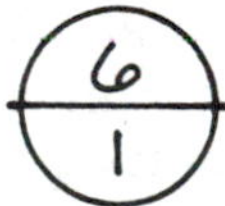

Exaggerated section showing capillary action of moisture being drawn out of mortar by dry units. (*Courtesy Acme Brick Co., Fort Worth, Tex.*)

water required to produce good workability, proper flow, and water retention are quickly and accurately adjusted by experienced masons. Results produced from assemblies prepared in the field reliably duplicate the standards set by laboratory researchers. Dry mixes lose water to the masonry units and will not cure properly. Excessively wet mixes cause units to float, and will decrease bond strength. The "proper" amount of mixing water is universally agreed upon as the maximum compatible with "workability," and workability is best judged by the mason.

Retempering (the addition of mixing water to compensate for evaporation) is acceptable practice in masonry construction. Since highest bond strengths are obtained with moist mixes having good flow values, a partially dehydrated mortar is less effective if the evaporated water is not replaced. Mortar normally begins to harden or set about $2\frac{1}{2}$ hours after initial mixing. After this point, retempering will decrease compressive strength by approximately 25%. ASTM standards require that all mortar be used within $2\frac{1}{2}$ hours and permit retempering as frequently as needed within this time period. Tests have shown that the decrease in compressive strength is minimal if retempering occurs only 1 to 2 hours after mixing. Mortar that is not used within $2\frac{1}{2}$ hours or that has begun to set should be discarded.

6.1.3 Bond Strength

The single most important property of mortar is bond strength, and it is critical that this bond be complete, strong, and durable. The mechanical bond between individual bricks, blocks, or stones unifies the assembly for integral structural performance, provides resistance to tensile stress, and seals against the penetration of moisture. The strength and extent of the bond are affected by many variables of material and workmanship. Complete and intimate contact between the mortar and the unit is essential, and workability influences the ease with which the mortar spreads and covers the surfaces. Rough units have a very porous surface that is highly receptive to the wet mortar and increases adhesion (*see Fig. 6-2*). The moisture content and suction of the units, the water retention of the mortar, and curing conditions such as temperature, relative humidity, and wind combine to influence the completeness and integrity of the mechanical and chemical bond. Voids at the mortar-to-unit interface offer little resistance to water infiltration and facilitate subsequent disintegration and failure if freezing occurs.

All other factors being equal, mortar bond strength increases as compressive strength increases, although the relationship has no direct proportions. Mortar with a laboratory compressive strength of 2500 psi develops tensile bond strength on the order of 50 to 100 psi. Increases in the cement content of the mix increases both these values. However, high cement-low lime mortars are stiff and do not readily penetrate porous unit surfaces, leaving voids and gaps which disrupt the bond. Increasing air content, or adding air-entraining ingredients, lowers both strengths.

Workmanship is also very important in bonding. Full mortar beds must be laid down by the mason to assure complete coverage of all contact surfaces. Once a unit has been placed and leveled, additional movement will break or seriously weaken the bond. The high water retention of cement-lime mortars allows more time for placing units on bed joints before evaporation or the suction of adjacent units alters the plasticity and flow of the mortar. In aligning the masonry, laboratory tests show that tapping the unit to level will increase bond strength 50 to 100% over hand pressure alone.

Because of the many variables involved, it is difficult to develop laboratory tests of bond strength that produce consistent results. Flexural

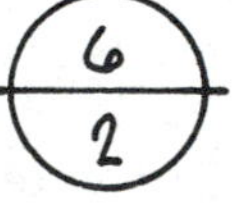

Enlargement showing the increased mechanical bond of mortar to a porous masonry unit surface.

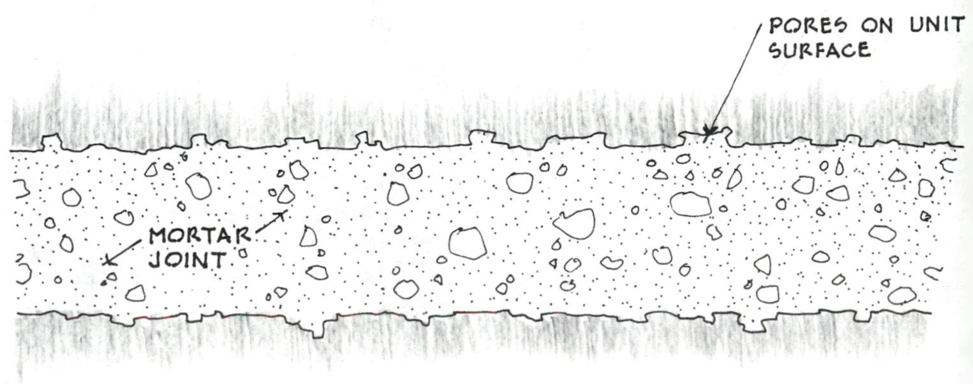

bond strength is presently measured by ASTM E518 or C952. (A new test method called the "bond wrench" is currently in the approval process with ASTM.) A simple field test to check adequacy of bond can be made by merely lifting a unit from its fresh mortar bed to determine if the mortar has fully adhered to all bedding surfaces. Good bond is indicated if the mortar sticks to the masonry unit and shows no air pockets or dry areas.

6.1.4 Compressive Strength

Masonry compressive strength depends on both the unit and the mortar. As with concrete, the strength of mortar is determined by the cement content and the water/cement ratio of the mix. Since plastic properties are often adjusted to achieve proper workability and flow, and since bond strength is ultimately of more importance, higher compressive values are sometimes sacrificed to increase or alter other characteristics. For loadbearing construction, building codes generally provide minimum allowable working stresses, and required compressive strengths may easily be calculated using accepted engineering methods. Strengths of standard mixes may be as high as 5000 psi, but need not exceed either the requirements of the construction or the strength of the units themselves. Although compressive strength is less important than bond, simple and reliable testing procedures make it a widely accepted basis for comparing mortars. Basically, compressive strength increases with the proportion of cement in the mix and decreases as the lime content is maximized. Increases in air entrainment, sand, or mixing water beyond normal requirements reduce strength values.

For two- and three-story construction, mortar strength is rarely a critical design factor since both the mortar and the masonry are usually much stronger than necessary. Compressive strength is important in engineered, loadbearing construction, but structural failure due to compressive loading is rare. More critical properties such as bond strength are given higher priority.

Although the compressive strength of masonry can be increased by using a stronger mortar, the improvement is not proportional. Tests indicate that wall strengths increase only about 10% when mortar strength increases 130%. In some instances, there are incentives other than economy which dictate using mortar with only the minimum required compressive strength. An unnecessarily hard, strong mortar will restrain concrete unit shrinkage and increase the amount of wall cracking as a result of the stress. A weaker mortar with higher lime content offers less opposition, permits greater movement, and more satisfactory performance as long as minimum requirements are met.

6.1.5 Extensibility and Volume Change

Two other important properties of hardened mortar are extensibility and volume change. *Volume changes* in mortar can result from the curing process, from cycles of wetting and drying, temperature change, or unsound ingredients that chemically expand. Available data indicate that expansion and contraction of masonry construction due to differential thermal volume change does not have a noticeable effect on performance. However, total volume change from different causes can sometimes be significant. Stronger mortars that are rich in cement and have less than 2 parts sand to

1 part cement can show substantial movement when exposed to alternate moist-dry conditions. Shrinkage during curing and hardening is greatest with higher water content. Volume changes caused by unsound ingredients such as reactive chemical compounds can cause disintegration of the masonry, but materials conforming to ASTM standards do not contain proprietary additives and should be free of this effect.

It is commonly believed that mortar shrinkage is significant, and that it is a primary cause of wall leaks. Research indicates, however, that maximum shrinkage across a mortar joint is minute, and is not in itself a cause of trouble. The most common leakage of masonry walls is through voids at the mortar-to-unit interface, where watertightness depends on a combination of good materials, workmanship, and design. The elastic properties of mortar, in fact, often counteract both temperature and moisture shrinkage. *Extensibility* is defined as the amount per unit length that a specimen will elongate, or the maximum unit tensile strain before rupture. This level is sufficiently high in mortar so that combined with the plasticity that lime gives, it allows slight movement without joints opening. For maximum resiliency (such as required in chimney construction), mortar should be mixed with the highest compatible lime proportion.

6.1.6 Durability

Durability is a measure of resistance to age and weathering, particularly to repeated freeze-thaw cycles. Mortars with high compressive strength can be very durable, but a number of factors other than strength affect this property. Ingredients, workmanship, volume change, flexibility, and the proper design and placement of expansion and control joints all influence durability and determine the maintenance characteristics of the construction. Although harsh environmental conditions and unsound ingredients can contribute to mortar deterioration, the most destructive factor is expansion of moisture in the wall by freezing. The bubbles introduced by air entrainment absorb the expansive forces of freezing water and provide good assurance against damage. They also decrease both the compressive and bond strength of the mortar. For this reason, air entrainment is not generally recommended. The best defense against freeze-thaw destruction is the elimination of moisture leaks at the joints with quality mortar ingredients and good bond.

6.1.7 Efflorescence

The same may be said regarding defense against efflorescence. Efflores- cence is the whitish deposit on exposed masonry surfaces caused by th leaching of soluble salts from within. If the mortar ingredients contain n soluble salts such as sodium or potassium sulfate, and if no moisture i present to effect leaching, efflorescence cannot occur. To minimize thi possibility, specify portland cements with low alkali content and lim which is a chemically pure ingredient. Good bond and tight mortar joint will then eliminate the presence of moisture, and both factors are unde control.

6.2 MORTAR CLASSIFICATION

The earliest known mortars consisted simply of lime, sand, and water. *Lim mortars* have low compressive strength and poor resistance to freeze-tha cycles, but do offer good workability and high water retention. Lim hardens only on contact with air, so curing occurs very slowly, over long periods of time, and at variable rates. In the past, slower methods of construction could accommodate this gradual hardening, but modern building techniques and faster-paced production have virtually eliminated the use of straight lime mortars. Since the latter part of the nineteenth century, portland cement has become the major ingredient of masonry mortars. Occasionally, it is used with sand and water only in what is called a *straight cement mortar.* Mixed in proportions of 1 part cement to 3 parts sand, these mortars harden quickly and consistently, exhibit high compressive strengths, and offer good resistance to freeze-thaw cycles, but are stiff and unworkable, and have low water retention and poor bond.

6.2.1 Portland Cement-Lime Mortars

A wider range of desirable properties and better construction results are achieved with combination portland cement-lime mortars. Lime is substituted for a certain amount of the portland cement, and is counted as part of the cementitious material in maintaining the 1:3 ratio with sand. In this type of mix, the cement contributes durability, high early strength, a consistent hardening rate, and high compressive strength, while lime adds workability, water retention, and elasticity. Cement-lime mortars produce highly satisfactory results, are predictable in performance, and meet ASTM requirements for all mortar types, including the high-strength mixes.

6.2.2 Masonry Cement Mortars

Proprietary masonry cements have become more widely used in recent years because of their convenience and the consistent, economical mixes they produce. Masonry cements have good plastic mortar properties, but these are often achieved through the use of workability agents and entrained air, often with adverse effects on compressive and bond strengths. Some products contain ground, inert limestone which is not a satisfactory substitute for calcined quicklime. Most masonry cements weigh about 70 lb per bag and contain approximately equal parts of portland cement and inert limestone by weight, plus additives. Because the exact proportions and ingredients of masonry cements are often not disclosed by the manufacturer, and are not limited by ASTM standards, predictable performance results between manufacturers are difficult to achieve. For this reason, ASTM standards permit masonry cements alone to be used only in lower-strength mortars for non-loadbearing construction. For reinforced masonry construction, portland cement must be added to the prepared masonry cement in order to produce adequate compressive strength.

6.3 MORTAR TYPES

ASTM C270, Mortar for Unit Masonry, outlines requirements for five different mortar types, designated as M, S, N, O, and K. Prior to 1954, mortar types were designated A-1, A-2, B, C, and D, but it was found that A-1 carried the connotation of "best" and that many designers consistently specified this type thinking it was somehow better than the others. To dispel this misunderstanding, the new, arbitrary letter designations were assigned so that no single mortar type could inadvertently be perceived as best for all purposes.

Compliance with ASTM C270 may be based on (1) strength and water retention or (2) on volume proportions (*see Fig. 6-3*). The ratio of cementitious material to aggregate may often be less than the minimum required by volume proportion since most mixes actually have higher strengths than the minimum required compressive strengths (*see Fig. 6-4*). This encourages preconstruction testing of sample mortar cubes to gain the economic advantage of meeting strength requirements with less cement. Some industry experts, however, do not recommend minimum property specifications since some mortars which meet the compressive strength and water retention standards may still be deficient in tensile bond strength. If ASTM C270 is referenced without indication as to the property or proportion standards, the proportion standard always governs. The volume proportions used are based on weights per cubic foot of materials as listed in *Fig. 6-5*.

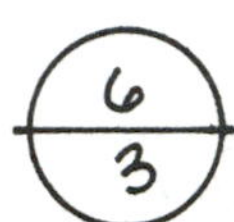

Mortar requirements by strength or by volume proportions, from ASTM C270. (*Copyright, American Society for Testing and Materials, 1916 Race Street, Philadelphia, Pa. 19103. Reprinted, with permission.*)

TABLE A COMPRESSIVE STRENGTH OF CUBES FOR LABORATORY PREPARED MORTAR

Mortar type	*Average compressive strength at 28 days (psi)*
M	2500
S	1800
N	750
O	350
K	75

TABLE B MORTAR PROPORTIONS BY VOLUME

Mortar type	*Parts by volume of portland cement, cement, or portland blast-furnace slag cement*	*Parts by volume of masonry cement*	*Parts by volume of hydrated lime or lime putty*	*Aggregate, measured in a damp, loose condition*
M	1	1	—	Not less than $2\frac{1}{4}$ and not more than 3 times the sum of the volumes of the cements and lime used
	1	—	$\frac{1}{4}$	
S	$\frac{1}{2}$	1	—	
	1	—	Over $\frac{1}{4}$ to $\frac{1}{2}$	
N	—	1	—	
	1	—	Over $\frac{1}{2}$ to $1\frac{1}{4}$	
O	—	1	—	
	1	—	Over $1\frac{1}{4}$ to $2\frac{1}{2}$	
K	1	—	Over $2\frac{1}{2}$ to 4	

Mortar type by proportions		Actual tested strengths in brick wallettes	
Type	Proportions	Compressive (psi)	Tensile bending (psi)
M	$1:\frac{1}{4}:3\frac{1}{2}$	3600	65
S	$1:\frac{1}{2}:4\frac{1}{2}$	3200	72
N	1:1:6	2800*	59
N	(Premixed masonry cement)	1500*	14
O	1:2:9	1600	20
K	1:3:12	Not tested	

*Note the difference between actual tested strength of Type N mortar mixed with portland cement and lime and that made with a premixed masonry cement and no added lime.

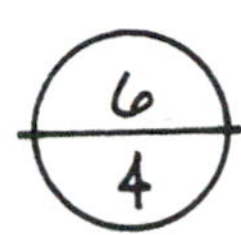

Actual strengths of mortars mixed by proportions. (*Courtesy Acme Brick Co., Fort Worth, Tex.*)

Material	Weight (lb/cu ft)
Portland cement	94
Portland blast-furnace slag cement	94 (approx.)
Masonry cement	Weight printed on bag
Hydrated lime	40
Lime putty	80
Sand, damp and loose	80 lb of dry sand

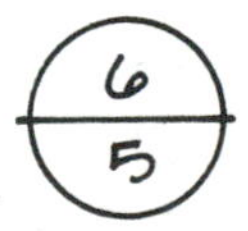

Weights of mortar materials, from ASTM C270. (*Copyright, American Society for Testing and Materials, 1916 Race Street, Philadelphia, Pa. 19103. Reprinted, with permission.*)

It should be emphasized that no single mortar type is universally suited to all applications. Variations in proportioning the mix will always enhance one or more properties at the expense of others. Each job must be evaluated individually to determine the highest-priority requirements and the mortar chosen accordingly. A mason or masonry contractor is not a good judge in mortar selection, because their bias is toward economy and workability. The architect or engineer is in a much better position to identify and assess all of the design criteria involved and to control the quality of the construction through the specifications. A good rule of thumb is to use the mortar with the least minimum compressive strength compatible with project requirements.

6.3.1 Type M Mortar

Each of the five basic mortar types has certain applications to which it is particularly suited and for which it may be recommended. Type M, for instance, is a high-strength mix that offers greater durability than other mortar types. It is recommended for both reinforced and unreinforced masonry which may be subject to high compressive loads, severe frost action, or high lateral loads from earth pressures, hurricane winds, or earthquakes. Because of its superior durability, Type M mortar may be used in structures below grade and in contact with the soil, such as foundations, retaining walls, walks, sewers, and manholes.

6.3.2 Type S Mortar

Type S mortar produces tensile bond values which approach the maximum obtainable with cement-lime mortar. It is recommended for structures subject to normal compressive loads but which require high flexural bond strength. Type S should also be used where mortar adhesion is the sole bonding agent between facing and backing, such as the application of adhesion-type ceramic veneer.

6.3.3 Type N Mortar

Type N is a good general-purpose mortar for use in above-grade masonry. It is well suited for masonry veneers and for interior walls and partitions. This "medium-strength" mortar represents the best compromise among strength, workability, and economy. When proportioned of portland cement, lime, and sand in a 1 : 1 : 6 ratio, actual laboratory compressive tests may show strengths as high as 2800 psi. The masonry cement-sand mix of the same designation will usually test no higher than 1500 psi. Conditions of workmanship, unit suction, and other variables do, of course, affect actual tested strength.

6.3.4 Type O Mortar

Type O is a "high-lime," low-strength mortar. It can be used in non-loadbearing walls and partitions, in exterior veneer that will not be subject to freezing in the presence of moisture, and in solid loadbearing walls where compressive stresses do not exceed 100 psi. Type O mortar is often used in one- and two-story residential work and is a favorite of masons because of its excellent workability and economical cost.

6.3.5 Type K Mortar

Type K mortar has a very low compressive strength and a correspondingly low tensile bond strength. It is suitable only for interior, non-loadbearing partitions carrying only their own dead weight. Type K mortar is very seldom used.

6.3.6 Refractory Mortars

In determining the requirements for mortar performance, two very specialized areas demand extensive and detailed project analysis. Refractory mortars may range from residential fireplace installations to extremely high-heat industrial boiler incinerators or steel pouring pits. For the lower heat residential jobs, ground fire-clay mortar is generally used in very thi joint applications. The fire bricks are actually dipped to get a thorough mortar coating, but no conventional mortar bed is laid. Increasing heat in the firebox ceramically fuses the mortar and seals the joints against heat penetration. ASTM C105, Ground Fire Clay as Mortar, covers material standards and properties for this use. For heavy-duty industrial service, ASTM C178 covers Air-Setting Refractory Mortar, which is designed to withstand intense and continuous heat. Using fire clay as a base, other materials, such as silica, chrome, silicon carbide, and alumina, may be added for special conditions. Manufacturers or suppliers should be consulted regarding design details and performance characteristics.

6.3.7 Chemical-Resistant Mortars

The field of chemical-resistant mortars is also highly specialized and complex in nature. Durability depends very heavily on proper mortar selection. Even with the use of chemical-resistant brick or tile, mortar may still be attacked by acids or alkalis, causing joint disintegration and loosening of the masonry units. There are few chemicals that do not attack regular portland cement mortars. Consequently, it is necessary to develop chemical resistance by means of admixtures or surface treatments. Special cements or coatings are available which will withstand almost all service conditions, but different types will react differently with various chemicals. The success of any particular treatment depends on local conditions, type and concentration of the chemical solution, temperatures, wear, vibration, type of subsurface, and workmanship. Joints should be as narrow as possible to minimize the exposed area and reduce the quantity of special material required. The selection of the optimum material for a particular installation must include the consideration of mechanical and physical properties as well as chemical-resistant characteristics (*see Fig. 6-6*).

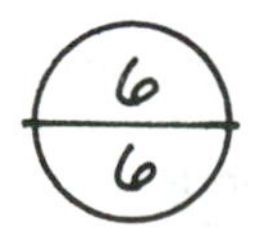

Physical and mechanical properties of chemical-resistant mortars. (*From Brick Institute of America*, Technical Note 32, *BIA, McLean, Va.*)

	Mortar type					
Property	*Sulfur*	*Silicate*	*Phenolic resin*	*Furan resin*	*Polyester resin*	*Epoxy resin*
Density (lb/cu ft)	138	125	112	95	125	95–125
Apparent porosity (%)	<1	>12	1	<1	<1	<1
Tensile strength (psi)	600	350	1200	1400	1400	1600
Compressive strength (psi)	6000	3500	10,000	12,000	12,000	16,000
Modulus of rupture (psi)	1400	750	1400	1600	1750	1800
Modulus of elasticity (10^6 psi)	1.2	1.0	0.6	1.0	1.5	1.3
Coefficient of linear expansion (10^{-6} in./in./°F)	18.2	6.0	8.0	11.5	18.0	14.0
Maximum service temperature (°F)	190	750–1600	350	375	250	250
Working time of mixed mortar at 70°F (min)	—	40	30–120	30	60	45
Final setting time (days)	—	3	3–4	2	2	2
Linear shrinkage on setting (%)	—	1.0	0.8	0.4	1.2	0

Several special types are available, including sulfur mortars, silicate mortars, phenolic resin mortars, and furan, polyester, and epoxy resin mortars. The capabilities and properties may often be altered by changing the formulations. For specific installations, full use should be made of available test procedures (*see Fig. 6-7*), and the engineering advice, services, and recommendations of manufacturing specialists in this field should be solicited.

Standard number	*Title*
ASTM C395	Chemical-Resistant Resin Mortars
ASTM C279	Chemical-Resistant Masonry Units
ASTM C466	Chemically Setting Silicate and Silica Chemical-Resistant Mortars
ASTM C287	Chemical-Resistant Sulfur Mortar
ASTM C413	Absorption and Apparent Porosity of Chemical-Resistant Mortars
ASTM C321	Bond Strength of Chemical-Resistant Mortars
ASTM C608	Brittle Ring Tensile Strength of Chemical Setting Silicate and Silica Chemical-Resistant Mortars
ASTM C267	Chemical Resistance of Mortars
ASTM C579	Compressive Strength of Chemical-Resistant Mortars
ASTM C396	Compressive Strength of Chemically Setting Silicate and Silica Chemical-Resistant Mortars
ASTM C306	Compressive Strength of Chemical-Resistant Resin Mortars
ASTM C580	Flexural Strength and Modulus of Elasticity of Chemical-Resistant Mortars
ASTM C531	Shrinkage and Coefficient of Thermal Expansion of Chemical-Resistant Mortars
ASTM C307	Tensile Strength of Chemical-Resistant Resin Mortars
ASTM C414	Working and Setting Times of Chemical-Resistant Silicate and Silica Mortars
ASTM C308	Working and Setting Times of Chemical-Resistant Resin Mortars
ASTM C386	Use of Chemical-Resistant Sulfur Mortars
ASTM C397	Use of Chemically Setting Chemical-Resistant Silicate and Silica Mortars
ASTM C398	Use of Hydraulic Cement Mortars in Chemical-Resistant Masonry
ASTM C399	Use of Chemical-Resistant Resin Mortars

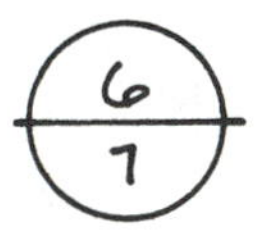

Specifications, tests, and practice standards for chemical-resistant mortars. (*From Brick Institute of America,* Technical Note 32, *BIA, McLean, Va.*)

6.3.8 Extra-High-Strength Mortars

Extra-high-strength mortars have been developed over the past few year for use with prefabricated masonry panels. Mortar additives greatl enhance the bonding, compressive, and tensile characteristics of portlan cement mortars. Recently, however, problems have occurred that indicat limited durability under some conditions of exposure. Because of the limite performance history of this mortar, caution should be exercised in i specification and use.

6.4 GROUT

Grout is a mixture of cementitious material and aggregate with enough water added to allow the mix to be poured into masonry cores and cavities without segregation. (It can be produced by adding excess water to mortars which contain a limited amount of lime.) ASTM C476 covers both fine and coarse mixtures determined on the basis of aggregate size and grading.

Selecting a fine or coarse grout is based on the size of the core or cavity as well as the height of the lift to be grouted. Some building codes and standards have different requirements for the relationship of maximum aggregate size to clear opening. For specific projects the governing standard should be checked. Generally, however, for fine grout in low-lift applications, the minimum space should be 2×3 in. cores or $\frac{3}{4}$-in. collar joints. In high-lift grouting where the smallest horizontal dimension is 3 in. a $\frac{1}{2}$-in. maximum coarse aggregate may be used, and in some instances with minimum 4-in. dimensions, $\frac{3}{4}$-in. aggregate is permitted.

Grout is an essential element of reinforced masonry construction. It must bond the masonry units and the steel together so that they perform integrally in resisting superimposed loads. In unreinforced, loadbearing construction, unit cores are sometimes grouted to give added area to resist loads. The fluid consistency of grout is important in determining compressive strength, in assuring that the mix will pour or pump easily and without segregation, and that it will flow around reinforcing bars and into corners and recesses without voids. ASTM C476 specifies grout proportions by volume, but does not indicate minimum strength or slump limits. (Some building codes do specify minimum 28-day compressive strength, so local standards should always be consulted.) Optimum consistency will depend on the absorption rate of the units as well as temperature and humidity conditions. Performance records indicate a desirable slump of 8 in. for units with low absorption, and as much as 10 in. for units with high absorption.

Depending on the amount of mixing water used, the mix proportions in *Fig. 6-8* will normally produce grouts with laboratory compressive strengths in nonabsorptive molds of 600 to 2500 psi at 28 days. However, actual field compressive strength usually exceeds 2500 psi because some of

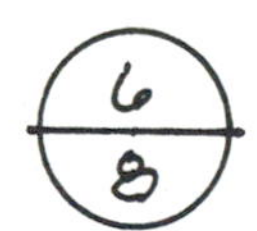

Grout requirements, from ASTM C476. *(Copyright, American Society for Testing and Materials, 1916 Race Street, Philadelphia, Pa. 19103. Reprinted, with permission.)*

Type	Grout proportions by volume			
	Parts by volume of portland cement or portland blast-furnace slag cement	Parts by volume of hydrated lime or lime putty	Aggregate, measured in a damp, loose condition	
			Fine	Coarse
Fine grout	1	0 to $\frac{1}{10}$	$2\frac{1}{4}$ to 3 times the sum of the volumes of the cementitious materials	—
Coarse grout	1	0 to $\frac{1}{10}$	$2\frac{1}{4}$ to 3 times the sum of the volumes of the cementitious materials	1 to 2 times the sum of the volumes of the cementitious materials

the mixing water is absorbed by the units, thus reducing the water/cement ratio and increasing the strength. When placed in the wall, the water/cement ratio of grout is rapidly reduced to a low value. The water absorbed by the units is retained for a period of time, thus providing a moist condition for optimum curing of the grout. Grout which may have a predicted 28-day strength of several hundred psi based on the water/cement ratio as mixed may prove to have an actual strength of 3000 to 4000 psi or higher when actual core samples are tested. This high strength is attributed to the immediate reduction of the water/cement ratio due to absorption, and to the optimum curing conditions provided within the wall.

ASTM standards do discourage and, in some instances prohibit, the use of admixtures in masonry grout. Particular caution must be exercised in the use of calcium chloride as an accelerating agent because of the damaging corrosive effect it has on steel reinforcement, metal ties, and anchors.

7

MASONRY ACCESSORIES

Accessory items are an important and integral part of masonry construction. Horizontal joint reinforcement, metal anchors, ties and fasteners, control joints, and flashing materials are common to the trade, and one or more of these items is usually necessary to complete a job. *Steel* is most often used for fabrication, but protective coatings must be applied to isolate the metal from the corrosive effects of wet mortar.

Corrosion of embedded metals is caused by direct oxidation or by galvanic currents. Galvanic corrosion requires the presence of moisture, which is, of course, available in wet mortar. This initial moisture can corrode the protective coating, which is then, in effect, sacrificed to protect the steel that it covers. Once the moisture has been consumed in the hydration process or has evaporated, strong, watertight joints will prevent infiltration of additional moisture, which could cause further corrosion.

Several nonferrous metals are also used for masonry accessories. *Copper* and copper alloys are essentially immune to the corrosive action of wet concrete or mortar. Because of this immunity, copper can be safely embedded in fresh mortar even under saturated conditions. Galvanic corrosion will occur if copper and steel items are either connected or in close proximity to one another. The presence of soluble chlorides will also cause copper to corrode. It is not advisable to use admixtures such as calcium chloride in mortar that will contain this or any other embedded metal.

Zinc is susceptible to corrosive attack. The corroded metal occupies a greater volume than the original material, and exerts expansive pressures around the embedded item. The film of zinc used to coat steel reinforcement, however, is so thin that the pressure is not sufficient to crack the masonry. The galvanized coating takes the initial corrosion to protect the steel that it covers. Calcium chloride accelerating agents are very damaging to zinc coatings and should not be used in reinforced or metal-tied masonry walls.

Aluminum is also attacked by fresh portland cement mortar and produces the same expansive pressures. Galvanic corrosion also occurs if aluminum and steel are embedded in the mortar in contact with one another. If aluminum is to be used in reinforced masonry, it should be electrically insulated by a permanent coating of bituminous paint, alkali-resistant lacquer, or zinc chromate paint. If the coating is not kept intact, chlorides can greatly accelerate corrosion. Because of the difficulty in assuring this protection, aluminum should not be embedded in, or come into contact with, mortar containing chlorides.

7.1 HORIZONTAL REINFORCEMENT

Horizontal joint reinforcement is used to control shrinkage cracking in CMU walls. It can also be used in lieu of masonry headers to bond multi wythe walls. It can bond intersecting walls, anchor masonry veneer, and assure maximum flexural wall strength.

Horizontal joint reinforcement is usually made of galvanized steel wire. Copper-coated wire is not used because electrochemical action destroys the welded joints. The four basic designs available are shown in *Fig 7-1*. Spacing of the welded lateral ties should not exceed 16 in. for deformed wire or 6 in. for smooth wire. If used as structural reinforcing for horizontal wall spans, the longitudinal chords must be of deformed wire.

Joint reinforcement is available in several wire diameters, and in standard lengths of 10 to 12 ft. Longitudinal wires should be 9 gauge or larger, welded to minimum 12-gauge cross wires. Fabricated widths are approximately $1\frac{5}{8}$ in. less than the actual wall thickness to assure adequate mortar coverage. The mortar between the tie and the wall face should be at least $\frac{5}{8}$ in. For cavity walls exceeding 50 ft between wall ends, control joints or expansion joints, ladder-type reinforcement will allow differential longitudinal movement between wythes. Ladder-type reinforcement should also be used in single-wythe concrete masonry walls. In cavity walls c

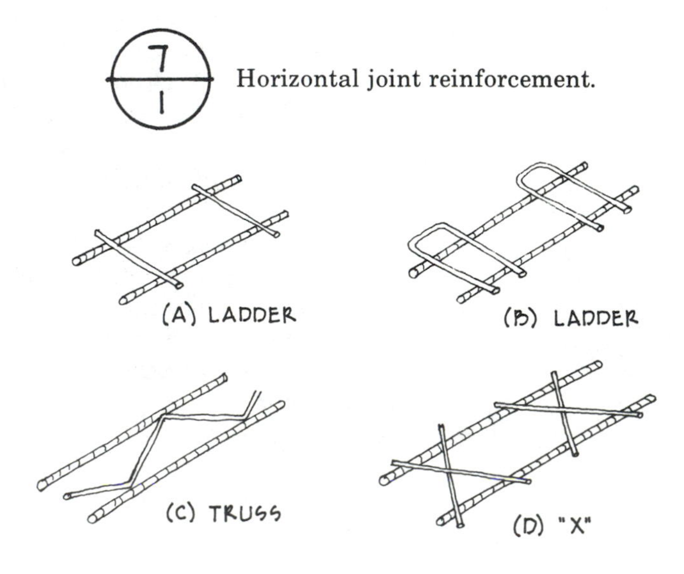

Horizontal joint reinforcement.

shorter length, and in solid or grouted walls of any length, any of the four types may be used. Wire that is not covered by mortar is high in tensile strength, but weak in shear. For maximum effectiveness, collar joints between wall wythes should be filled with grout.

7.2 ANCHORS, TIES, AND FASTENERS

Masonry *anchors* attach a veneer wall to its structural support: that is, another wall, a floor, a beam, or a column. *Ties* hold a multi-wythe masonry wall together, and *fasteners* attach other building elements to the masonry.

7.2.1 Ties

Truss-type joint reinforcement provides longitudinal strength in addition to lateral support between wythes. Individual *corrugated or wire ties* function only in the lateral direction, and provide intermittent rather than continuous support. There are several shapes and configurations, different wire gauges, and various sizes to suit the wall thickness (*see Fig. 7-2*).

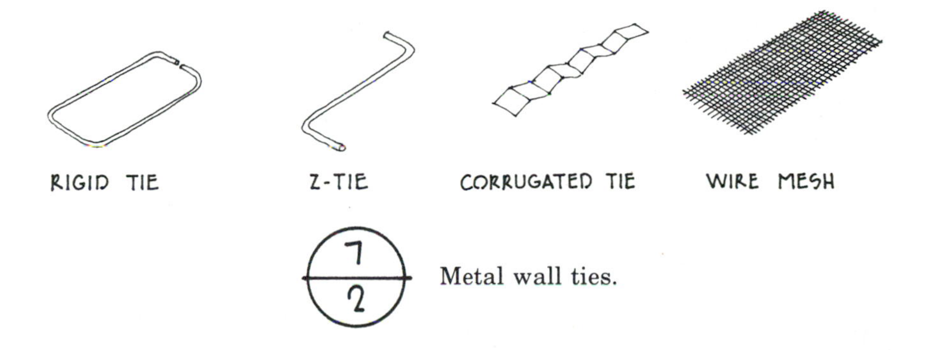

7/2 Metal wall ties.

Woven wire mesh is recommended only as a continuous lateral tie to hold masonry furring of maximum 2 in. thickness. Other types of ties can be used individually or with joint reinforcement (*see Fig. 7-3*). Only wire ties should be used in open cavity walls and grouted walls. These may be rigid for

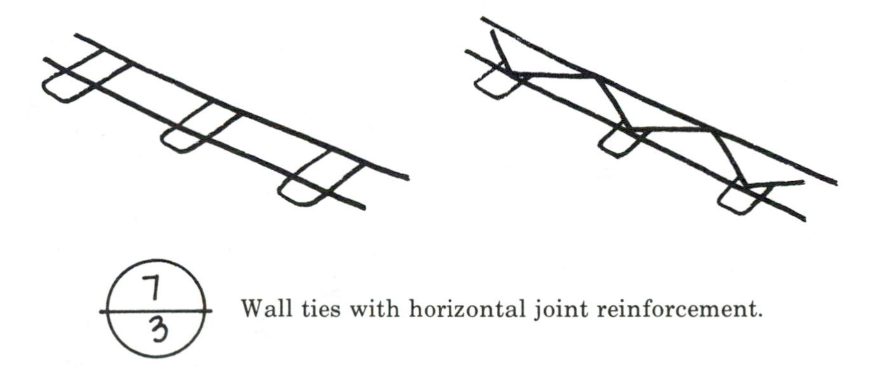

7/3 Wall ties with horizontal joint reinforcement.

laying in bed joints at the same height, or adjustable for laying in bed joints at different levels (*see Fig. 7-4*). The use of metal ties in lieu of masonry header courses increases the resistance of a wall to rain penetration, and increases ductility. Crimping the ties to form a drip is generally not recommended because of the reduction in strength caused by the deformation. Crimping is not permitted under some codes for certain types of wall

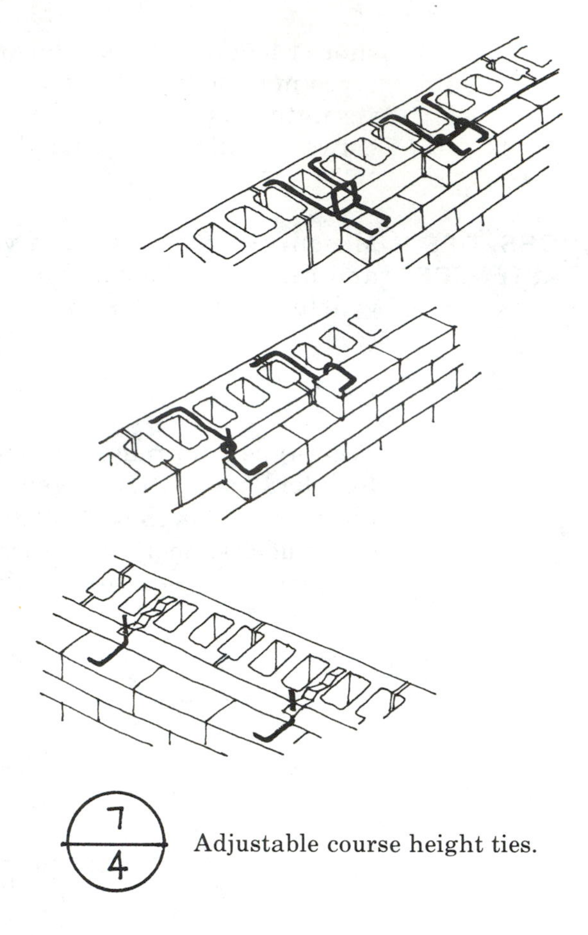

Adjustable course height ties.

construction. Ties should always be placed in the mortar bed, and n
directly on the masonry unit.

Recommended tie spacing is based on size and strength of the metal a
well as the configuration of the wall itself (*see Figs. 7-5 and 7-6*). Man
building codes prescribe maximum tie spacing as well. Ties should b
staggered vertically so that no two alternate courses form a continuous lin
Pull-out strengths for wire ties are based on the formula $p = 750\,(16.3d - 1$

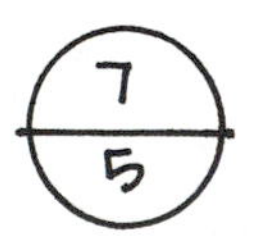

Properties of steel wire ties. (*From National Concrete Masonry Association*, TEK Bulletin 64, *NCMA, Herndon, Va.*)

Tie property	Wire gauge number			
	6	8	9	11
Diameter (in.)	0.1875	0.1620	0.1483	0.1205
Cross-sectional area (sq in.)	0.0276	0.0206	0.0173	0.0114
Radius of gyration (in.)	0.0469	0.0405	0.0371	0.0301
Allowable tensile load* (lb)	1,019	760	639†	273
Allowable mortar joint pull-out load‡ (lb)	515	410	350	240

*Based on yield strength 65,000 psi for 10 gauge and heavier, 40,000 for 11 gauge, and factor of safety of 1.67.

†Also applicable to individual corrugated sheet metal cross ties, 22 gauge, 1 in. wide.

‡Based on equation $p = 750\,(16.3\,d - 1)$, with factor of safety of 3.

Cavity width (in.)	Wall surface area, square feet per lateral tie for wire gauge number:			
	6	8	9	11
None (solid wall)	4.5	3.6	3.1	2.1
$3\frac{1}{2}$ or less	4.5	3.4	2.8	1.9
4	3.4	2.6	2.2	1.4
$4\frac{1}{2}$	2.7	2.0	1.7	1.1

Lateral ties in multi-wythe walls. (*From National Concrete Masonry Association*, TEK Bulletin 64, *NCMA, Herndon, Va.*)

where p is the pull-out strength of an individual wire tie from joints of Type M, S, or N mortar; and d is the tie diameter, in inches, where the diameter is between 0.17 and 0.19 in. Structural requirements of metal wall ties can be calculated by rational design methods. Particularly in the case of adjustable ties, and in loadbearing construction, it is recommended that engineering analysis be used to assure adequate strength and proper performance. Adjustable ties for cavity walls should be structurally designed for each condition of wind load, tie configuration, dimension, size, location, stiffness, embedment, modulus of elasticity of masonry, moment of inertia of each cavity wall wythe, and difference in level of connected joints.

Metal ties are typically made of several materials:

1. Cold-drawn steel wire, either plain (ASTM A82) or deformed (ASTM A496) for structural use as longitudinal wire in joint reinforcement.
2. Hard-drawn, copper-clad steel wire (ASTM B227, Grade 30 HS)—a copper sulfate bath gives the appearance of copper, but does not provide the 9-mil coating required by ASTM standards.
3. Hot-rolled steel wire (ASTM A510, Grade 1021) used for wire mesh.
4. Corrugated, hot-rolled steel sheets (ASTM A570, Grade E).
5. Electrogalvanized zinc-coated steel wire (ASTM A116).
6. Hot-dipped galvanized zinc coating (ASTM A153), applied after fabrication.

The most commonly used materials are uncoated steel, zinc-coated steel, and hot-dipped galvanized steel. Uncoated ties are not recommended.

Individual lateral wire ties may be rectangular or Z-shaped in lengths of 4, 6, or 8 in. (*see Fig. 7-2*). Z-ties should have at least a 2-in. 90° leg at each end. Rectangular ties should have a minimum width of 2 in. and welded ends if the width is less than 3 in. Either type may be used for solid masonry (core area less than 25%), but Z-ties are less expensive. Only rectangular ties should be used in ungrouted walls of hollow masonry. Corrugated steel ties should have a maximum wavelength of $\frac{3}{8}$ in., minimum amplitude of $\frac{1}{8}$ in., and minimum width of $\frac{7}{8}$ in. Thickness should be at least 22 gauge. Corrugated ties should be long enough to reach the outer-face-shell mortar bed of hollow units or the center of the mortar bed for solid units.

Wire mesh ties should be formed of unwelded, woven wire, 16 gauge or heavier. A minimum width of 4 in. is required and a $\frac{1}{2} \times \frac{1}{2}$ in. or finer mesh. Lengths may be field cut for convenience, and butt joints are acceptable.

7.2.2 Anchors

Masonry wall anchors provide connections that can resist compressive, tensile, and shear stresses. Rigid anchors resist all three types of loading, but flexible anchors, which may be needed to permit differential movement, do not resist shear.

Galvanized steel bolts and strips, either hooked or with cross bars, are used as *rigid anchors* (*see Fig. 7-7*). Hooked strips anchor intersecting walls at vertical intervals of 32 in. or less. They are typically $1\frac{1}{2} \times \frac{1}{4}$ in. in cross section and at least 24 in. long. A 1-in. 90° leg at each end is embedded in a grouted core or mortar filled joint. Bolt anchors are of several diameters and lengths. They are threaded at one end, and have a 1-in. 90° leg at the other for embedding in a grouted core or mortar joint.

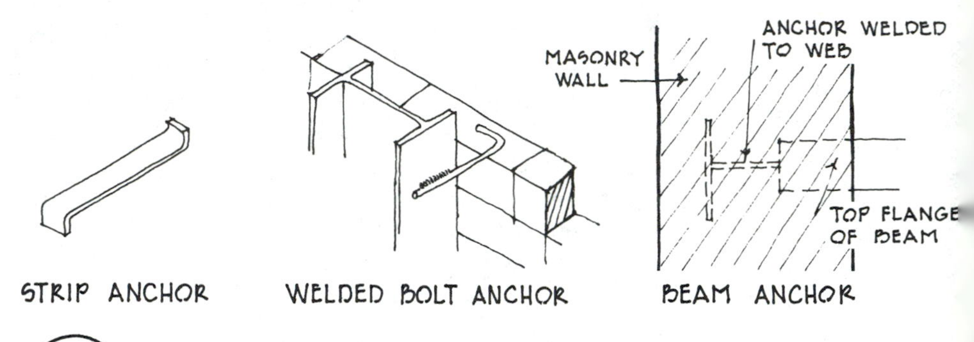

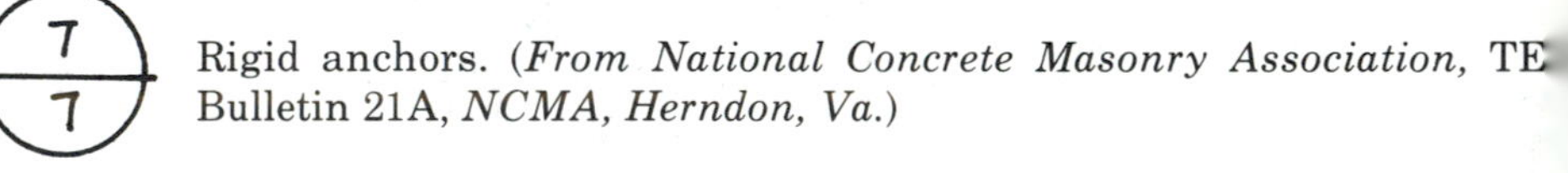

Rigid anchors. (*From National Concrete Masonry Association,* TE Bulletin 21A, *NCMA, Herndon, Va.*)

Flexible metal anchors may be of either wire or sheet metal. Th configuration of the wire to be embedded in mortar should be as shown i *Fig. 7-8*, or should have the equivalent pullout strength. This hooked le may be looped through a rod or notched plate welded to a steel structur member or bolted to a metal stud backing; through a sheet metal stri dovetailed to fit a slot in a concrete structural member; or through a met eye cast into the concrete. Flexible anchors, as in parts (A) and (D) of *Fi 7-8*, allow the wire end to remain perpendicular to the wall, reducing fatig loading, and providing better performance than corrugated anchor Pullout strength of any anchorage system should be verified before an ite is specified.

Corrugated sheet metal veneer anchors should meet the same physic requirements as corrugated ties ($\frac{3}{8}$-in. wavelength, $\frac{1}{8}$-in. amplitude, $\frac{7}{8}$-i width, and 22-gauge thickness). These anchors may be used only with sol or solidly grouted units where the distance between the veneer a supporting frame is 1 in. or less. One end of the anchor is nailed or screw directly to a wood or steel stud frame, and the other end is embedded in mortar joint (*see Fig. 7-9*). Performance is greatly reduced if the attachi nail or screw is not located exactly at the bend in the anchor. Corrugat dovetail anchors are fabricated to fit a dovetailed slot in a concre structural frame.

Only limited analytical research has been done on metal anchora

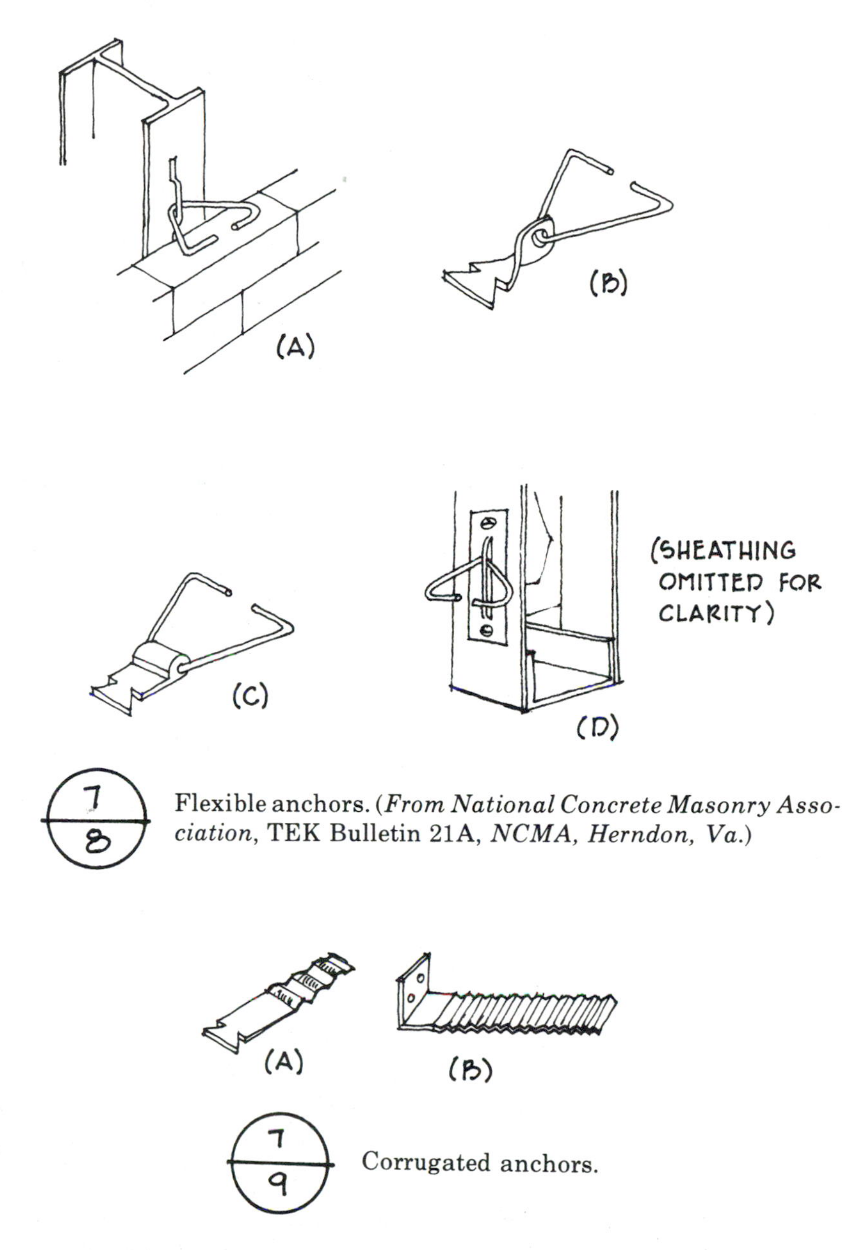

Flexible anchors. (*From National Concrete Masonry Association*, TEK Bulletin 21A, *NCMA, Herndon, Va.*)

Corrugated anchors.

systems. Most building codes give either required design load or maximum spacing for some types of anchors, but these are based only on empirical standards. Structural analysis indicates that the factors of safety for commonly used anchorage systems are generally lower than those required for the walls themselves. Conjectural allowable loads on flexible metal anchors are shown in *Fig. 7-10*. Better estimates may be made of the strength of rigid anchors (*see Fig. 7-11*).

Metal anchors for clay and concrete masonry are fabricated of the same materials as those used for horizontal joint reinforcement. Zinc- and copper-coated steel are most commonly used. In stone masonry, however, many metals can cause staining. Chromium-nickel stainless steel Types 302 and 304, and eraydo-alloy zinc are highest in corrosion and stain resistance. Hot-dipped galvanized zinc coatings are least resistant and although widely used, they are prohibited by some building codes. As stone anchors, copper, brass, and bronze may stain under some conditions. *Figure 7-12* shows some of the anchors used to attach stone veneer to various structural frames.

Form	Shape	Figure reference	Sheet metal width or wire gauge (in.)	(mm)	Conjectural allowable axial load lb
Sheet	Fully corrugated	7-9 (B)	1	25	80
Sheet	Dovetailed and corrugated	7-9 (A)	$\frac{7}{8}$	22	378
Sheet	Dovetailed and corrugated	7-9 (A)	1	25	460
Sheet	Dovetailed and corrugated	7-9 (A)	2	51	920
Wire	Dovetailed and twisted, looped, and hooked	7-8 (B)	No. 6	4.76	600
Wire	Dovetailed and twisted, looped, and hooked	7-8 (B)	No. 8	4.11	600
Wire	Dovetailed and twisted, looped, and hooked	7-8 (B)	No. 9	3.77	600
Wire	Dovetailed and twisted, looped, and hooked	7-8 (B)	No. 11	3.06	480
Wire	Looped and hooked	7-8 (A)	No. 6	4.76	500*
Wire	Looped and hooked	7-8 (A)	No. 8	4.11	500*
Wire	Looped and hooked	7-8 (A)	No. 9	3.77	500*
Wire	Looped and hooked	7-8 (A)	No. 11	3.06	273†
Wire	Dovetailed and twisted, looped, and hooked	7-8 (C)	No. 6	4.76	318‡

*Assumes No. 4 round tie-rod welded at 4-in. intervals.

†Limited by anchor tensile strength. No. 3 vertical rod welded at 4-in. intervals may be used.

‡Based on factor of safety of 3 in tension. No test data on compression available.

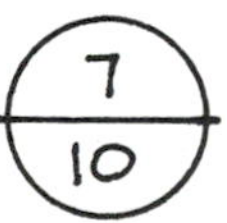

Conjectural allowable axial loads on flexible metal anchors. (*From National Concrete Masonry Association*, TEK Bulletin 21A, *NCMA, Herndon, Va.*)

TABLE A ESTIMATED ALLOWABLE SHEAR, TENSILE OR COMPRESSIVE LOAD ON ANCHOR BOLTS IN CONCRETE MASONRY OF SOLID UNITS OR SOLIDLY GROUTED HOLLOW UNITS		
Anchor diameter (in.)	*Minimum embedment (in.)*	*Allowable load (lb)*
$\frac{1}{4}$	4	270
$\frac{3}{8}$	4	410
$\frac{1}{2}$	4	550
$\frac{5}{8}$	4	750
$\frac{3}{4}$	5	1100
$\frac{7}{8}$	6	1500
1	7	1850
$1\frac{1}{8}$	8	2250

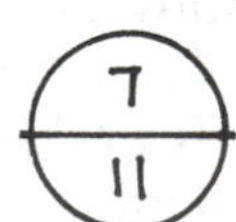

Allowable loads on rigid anchors. (*From National Concrete Masonry Association*, TEK Bulletin 21A, *NCMA, Herndon, Va.*)

TABLE B ESTIMATED ALLOWABLE TENSILE OR COMPRESSIVE LOAD ON ANCHORS GOVERNED BY CROSS BAR BEARING*		
Cross bar diameter (in.)	*Minimum cross bar length (in.)*	*Allowable tensile load (lb)*
$\frac{1}{4}$	8	540
$\frac{3}{8}$	8	820
$\frac{1}{2}$	8	1100
$\frac{5}{8}$	8	1500
$\frac{3}{4}$	10	2200
$\frac{7}{8}$	12	3000
1	14	3700
$1\frac{1}{8}$	16	4500

*These values are applicable when the tension or compressive strength of the anchor as controlled by shear is not exceeded.

7-11 (Continued)

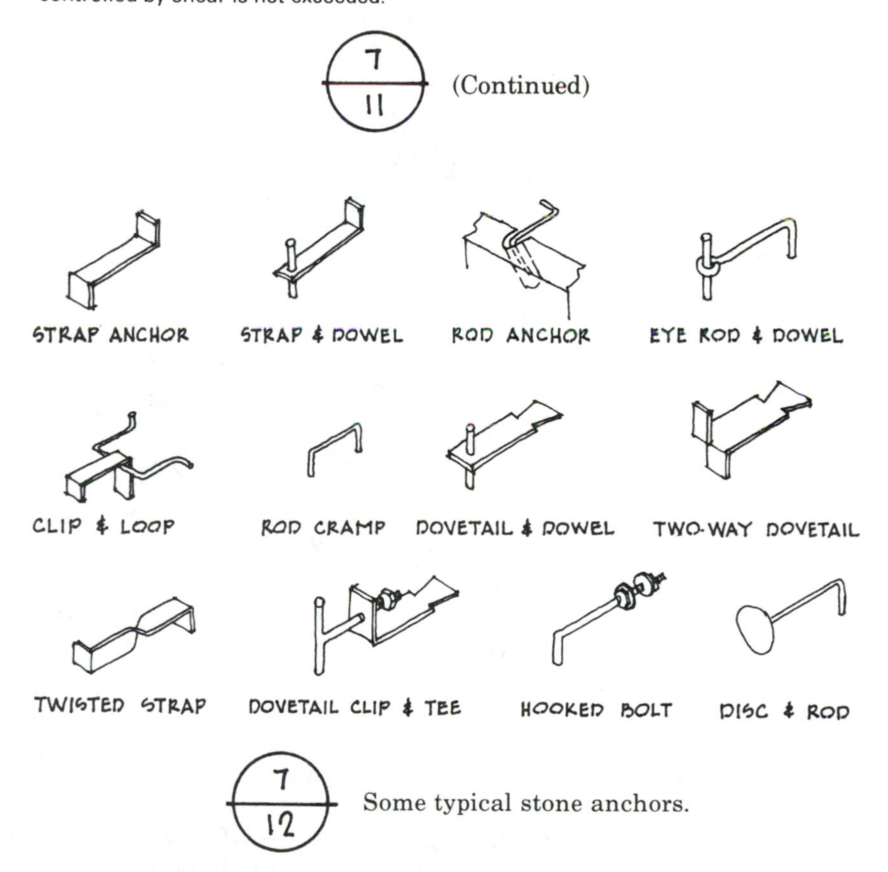

7-12 Some typical stone anchors.

7.2.3 Fasteners

Attaching fixtures or dissimilar materials to masonry requires some type of fastener. Most plugs, nailing blocks, furring strips, and so on, can be installed by the mason as the work proceeds. There are a variety of products and methods from which to choose, depending largely on the kind of fixture or material to be attached and the type of masonry involved (*see Fig. 7-13*).

The most common method of attaching wood trim items such as baseboards or chair rails is placing *wood nailing blocks* in the vertical

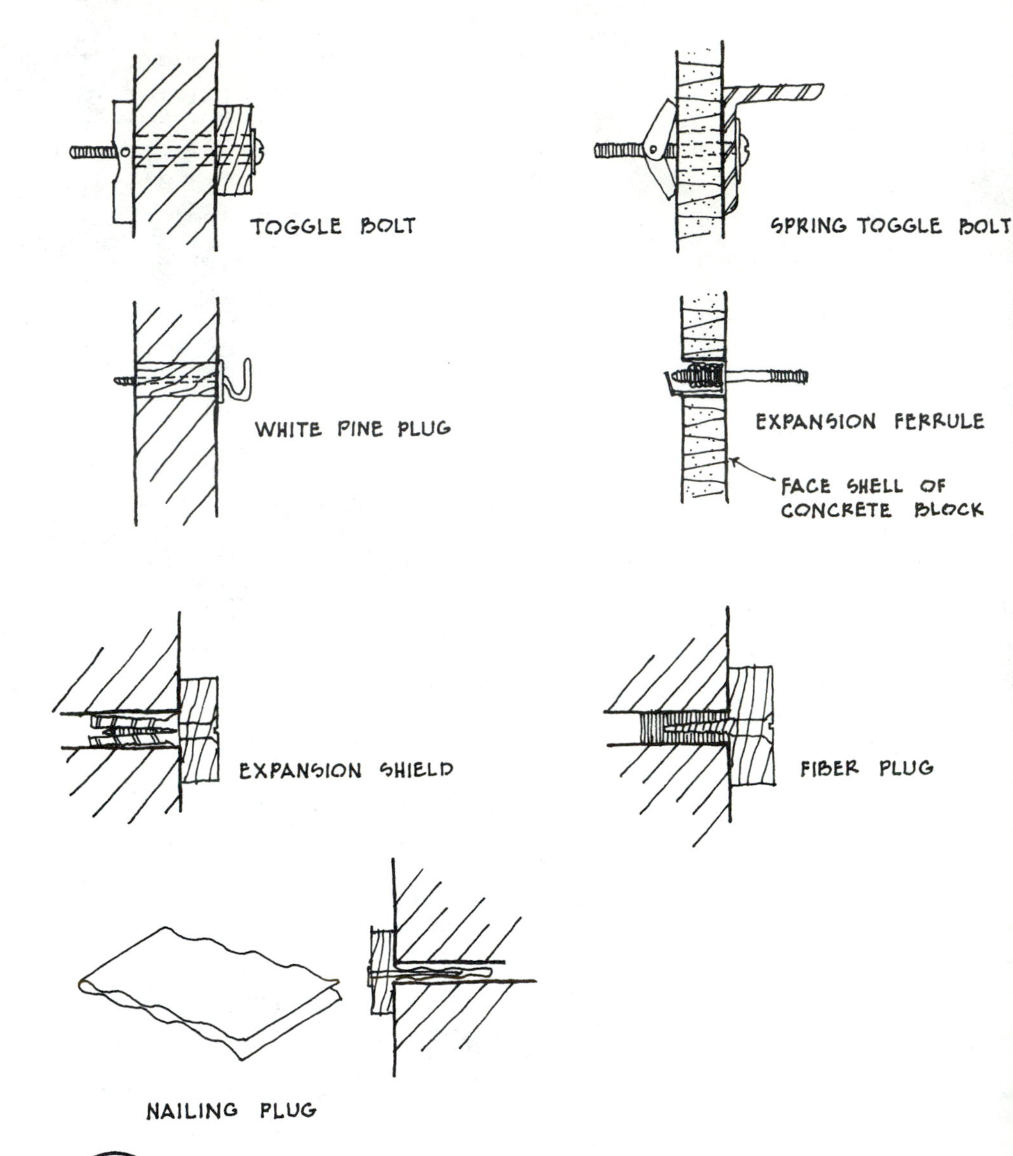

7/13 Masonry fasteners. (*From Brick Institute of America,* Technical Not *Vol. 2, No. 10, BIA, McLean, Va.*)

joints as the mason builds the wall. These blocks should be of seasone softwood creosoted to prevent shrinkage or rot. They should never be place in horizontal joints. *Galvanized metal nailing plugs*, with or withou fiberboard inserts, provide better construction and are easily set into th joints during construction. *Toggle bolts* and *double-threaded fasteners* ca be used only with hollow masonry units, and are installed after the wall i completed. *Wood plugs with threaded hooks* can be used with either solid o hollow masonry. The plug may be built into the wall or driven into a hol drilled after construction. *Plastic or fiber plugs* can also be used with soli or hollow units. They are placed in holes drilled into either the mortar joint or face shells of the masonry. *Expansion shields* and *wedge-type bolts* ma be used with solid or grouted masonry. Newer attachment methods includ pins or fasteners rammed or driven into solid masonry with a power tool c gun, and direct adhesive or mastic application.

Wood furring strips can be attached using nailing blocks, metal wall plugs, or direct nailing into mortar joints with case-hardened "cut nails" (wedge shaped) or spiral threaded masonry nails. Special anchor nails may be adhesively applied to the wall, or porous clay nailing blocks inserted into the bonding pattern (*see Fig. 7-14*). Metal furring strips are attached to the wall by tie wires built into the mortar joints or by special clips designed for this purpose.

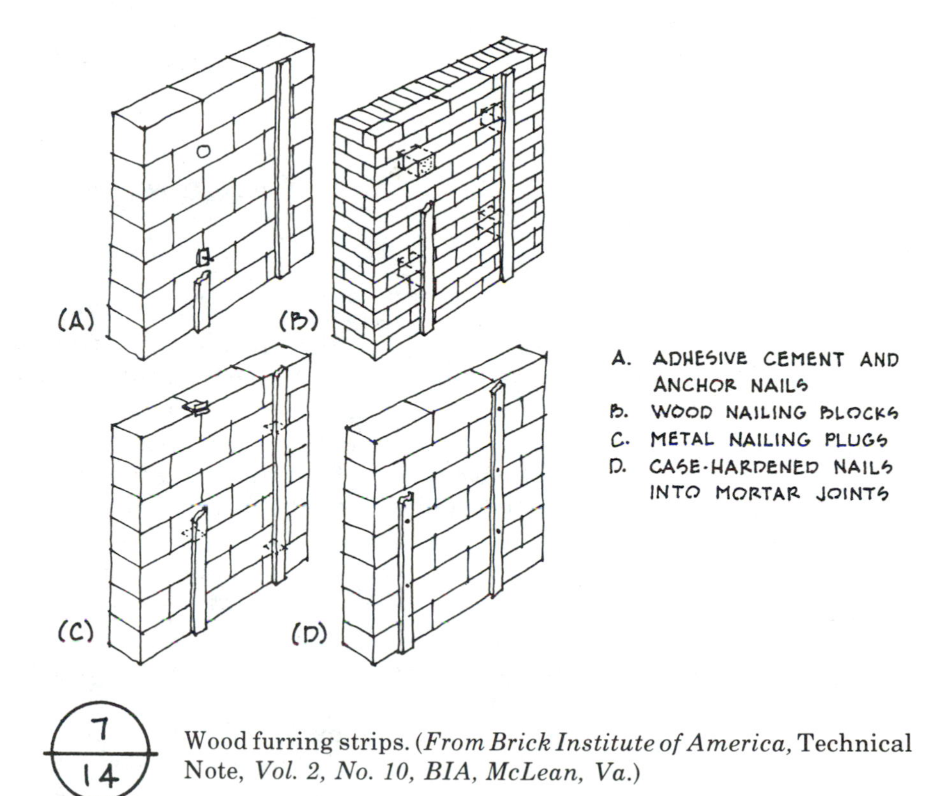

7-14 Wood furring strips. (*From Brick Institute of America,* Technical Note, *Vol. 2, No. 10, BIA, McLean, Va.*)

7.3 MOVEMENT JOINTS

Concrete masonry shrinkage and clay masonry expansion, along with thermal and moisture movement, are accommodated through special jointing techniques which allow movement without damage to the wall. Control joints for concrete masonry are designed as stress-relieving contraction points, and must extend completely through the masonry. Preformed, elastomeric fillers are available in widths to fit various wall thicknesses. The shear key and flanges provide a self-sealing joint, but exterior faces must be caulked to assure watertightness (*see Fig. 7-15*). Most of the rubbers used in fabricating control joint fillers have high durometer hardness and cannot absorb expansive movement. Softer materials such as neoprene rubber sponge are better suited for pressure-relieving joints in clay masonry walls, where expansion will compress the filler.

7.4 FLASHING MATERIALS

Masonry construction must include sheet flashing to divert penetrated moisture back to the exterior. Although they may be used in different situations, all flashing materials must be impervious to moisture and resistant to corrosion, abrasion, and puncture. In addition, they must be able to take and retain an applied shape to ensure proper performance after installation.

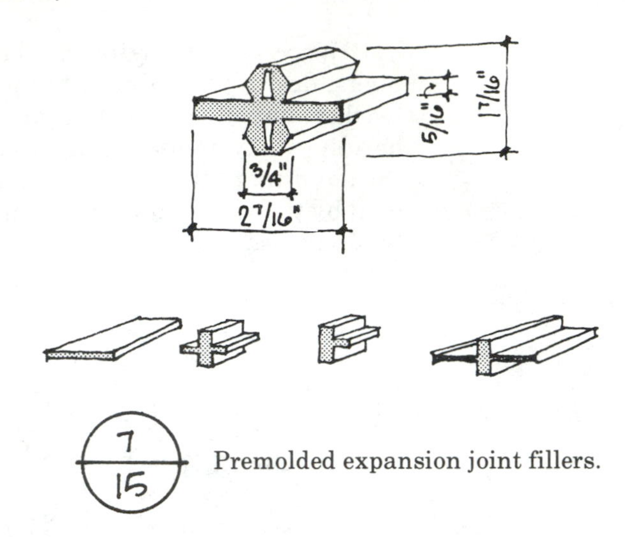

Premolded expansion joint fillers.

Galvanized steel is used in both residential and commercial construction, but is subject to corrosive attack from wet mortar unless covered with a bituminous coating. Exterior exposures require a 26-gauge thickness, and concealed installations require 28 gauge. *Stainless steel* flashings are highly resistant to corrosion, and provide good durability and workability without danger of staining adjacent masonry. For exposed locations, 28 gauge is recommended, and for concealed applications, 30 gauge. *Aluminum*, of course, is subject to corrosive damage from wet mortar and may not be used without a durable protective coating. Exposed aluminum should be 0.019 in. thick with 16,000 psi tensile strength, and concealed material 0.015 in. thick, 14,000 psi.

Copper sheet resists ordinary corrosive action, provides an excellent moisture barrier, and is easily shaped. Copper flashing can stain light-colored masonry unless it is coated with lead. A flashing weight of 16 oz/sq ft should be specified for exterior exposures, and 10 oz/sq ft for concealed locations. *Bituminous saturated fabric* and *plastic fabric* may not provide permanent protection since they are easily punctured during installation. Without careful workmanship to assure imperviousness, these materials are not recommended.

Combinations of materials are often used to provide adequate protection at lower cost by allowing thinner metal sections, or the use of otherwise corrodible metals. Sheet metals commonly used in *combination flashing* are 3-, 5-, or 7-oz copper, and aluminum of 0.004 or 0.005 in. thickness. Coatings may be bitumen, kraft paper, bituminous-saturated cotton fabrics, or glass fiber fabrics. These coated metals are suitable only for concealed installations.

8

ENVIRONMENTAL CHARACTERISTICS

Although masonry materials are often used for floor, ceiling, column, and beam construction, their most common use is in walls. Structural properties and design are discussed in Chapters 9 through 13, but some preliminary information is required regarding various wall types and their nonstructural characteristics.

The flexibility and diversity of masonry construction stem from the relatively small and heterogeneous components used to form mass building elements. Walls and partitions may be single- or multi-wythe in section and may combine different unit types for functional or aesthetic purposes. Standard brick, tile, and concrete masonry units can all be used in single-wythe loadbearing and non-loadbearing applications for low-, medium-, and high-rise buildings. In multi-wythe construction, the interior and exterior faces of the wall may be of the same material or of different units, such as clay brick/concrete block, brick/clay tile, concrete block/stone, and so on. One or both wythes may carry a superimposed load in bearing wall design, or they may serve respectively as curtain wall and panel wall sections in non-loadbearing construction. The properties of masonry walls in all instances are determined by the properties of the units and the mortar of which they are built. The nature of the material, and the quality and strength of the components will determine their suitability for various uses and applications.

The strength and structural capability of any of these wall types can be increased by adding steel reinforcement in bed joints, collar joints, or grouted cores (see Chapters 10 and 11).

8.1 SINGLE-WYTHE WALLS

Within the restrictions of height-to-thickness ratios prescribed by the model building codes (see Chapters 9 and 10), walls may be built with a single unit thickness of clay, concrete, glass, or gypsum masonry. Single-wythe walls of hollow units provide the options of grouting the core areas for greater mass and stability, or adding steel reinforcement for additional strength. Grouted, reinforced concrete block and hollow brick walls of a single 8 in. thickness can be used in loadbearing structures of 15 to 20 stories and more.

8.1.1 Hollow Tile

Hollow clay tile can be used in single-wythe construction of interior walls and partitions, and in some instances of exterior walls (*see Fig. 8-1*). Grading classifications—LB and LBX for loadbearing, and NB for non-loadbearing—will determine the type of unit selected. Facing tile and Type II glazed tile provide a finished surface on both faces of a single-wythe wall, with only one unit thickness for simplified construction. Standard structural tile designed to receive plaster applications can also be used in through-the-wall applications of one wythe. Type I glazed units are designed for finished exposure on only one side where the other wall face will be concealed or will receive a plaster finish.

8.1.2 Brick and Block

Hollow clay brick and solid units with a 6-in. bed depth are often used in single-wythe construction, and in some instances, codes permit 4-in. walls (*see Fig. 8-1*). Hollow brick sections are usually at least 8 in. thick. Hollow concrete blocks have decorative finishes on only one side. The opposite wall face must receive paint, plaster, gypsum board, or other materials if exposed to view. Single-wythe walls are widely used for veneer construction over wood frame, steel, concrete, or masonry structural backing. In some applications, they may also be loadbearing elements in themselves with only lateral bracing required for stability. Brick walls may of course be exposed on both sides without further finishing.

8.1.3 Gypsum Block

Gypsum blocks can be used only for interior, non-loadbearing partitions and are generally intended as a base for plaster application. Used primarily for lightweight fire and sound separation, thicknesses of 2, 3, 4, and 6 in. offer efficient and effective protection with relatively thin wall sections.

8.1.4 Glass Block

Glass block masonry is used for high-security glazing, and for glazed areas requiring light control and/or heat-gain reductions. The units are used only in single-wythe construction, and do not have loadbearing capabilities.

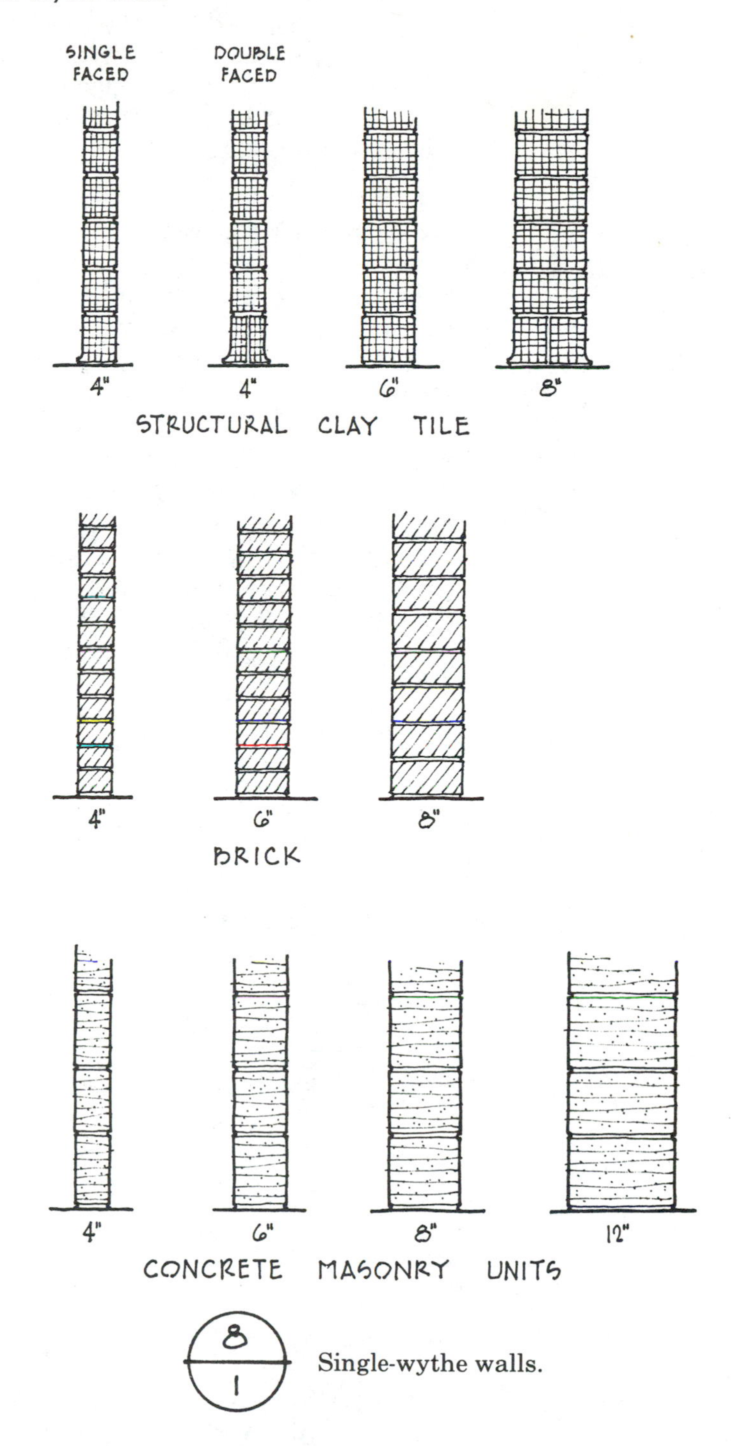

Single-wythe walls.

8.2 MULTI-WYTHE WALLS

For larger horizontal or vertical spans between stiffeners, or for greater resistance to fire, sound, and heat transmission, wall thicknesses are increased by adding additional wythes of masonry of the same type unit or of a different material. These wall types may be divided into (1) solid masonry walls, (2) solid walls of hollow units, and (3) cavity walls. A solid masonry wall is built of solid units, or of fully grouted hollow units laid contiguously with the collar joint between wythes filled with mortar or grout (*see Fig. 1-4*). A solid wall of hollow units is built of hollow clay or concrete masonry with mortared or grouted collar joints, but with core areas left void. A cavity wall consists of two or more wythes of hollow or solid units separated by an open collar joint or air space of at least 2 in. between two adjacent wythes.

8.2.1 Solid Masonry Walls

Solid masonry walls have been used in building construction throughout history. Strength, stability, and insulating value all depended on mass, and code requirements for empirically based, unreinforced bearing walls prescribed substantial thicknesses. The Monadnock Building in Chicago, completed in 1891, is 16 stories high with unreinforced loadbearing brick walls ranging in thickness from 12 in. at the top to 6 ft at the ground. At that time, wall wythes were bonded together with masonry headers as shown in *Fig. 8-2*. In 8-in. walls header courses extend the full width of the wall section, and moisture penetration from exterior to interior is facilitated, causing leakage and possible damage to the masonry. In most modern construction, masonry headers are replaced by metal wall ties grouted solidly in the bed joints to tie multiple wythes together. Today, solid

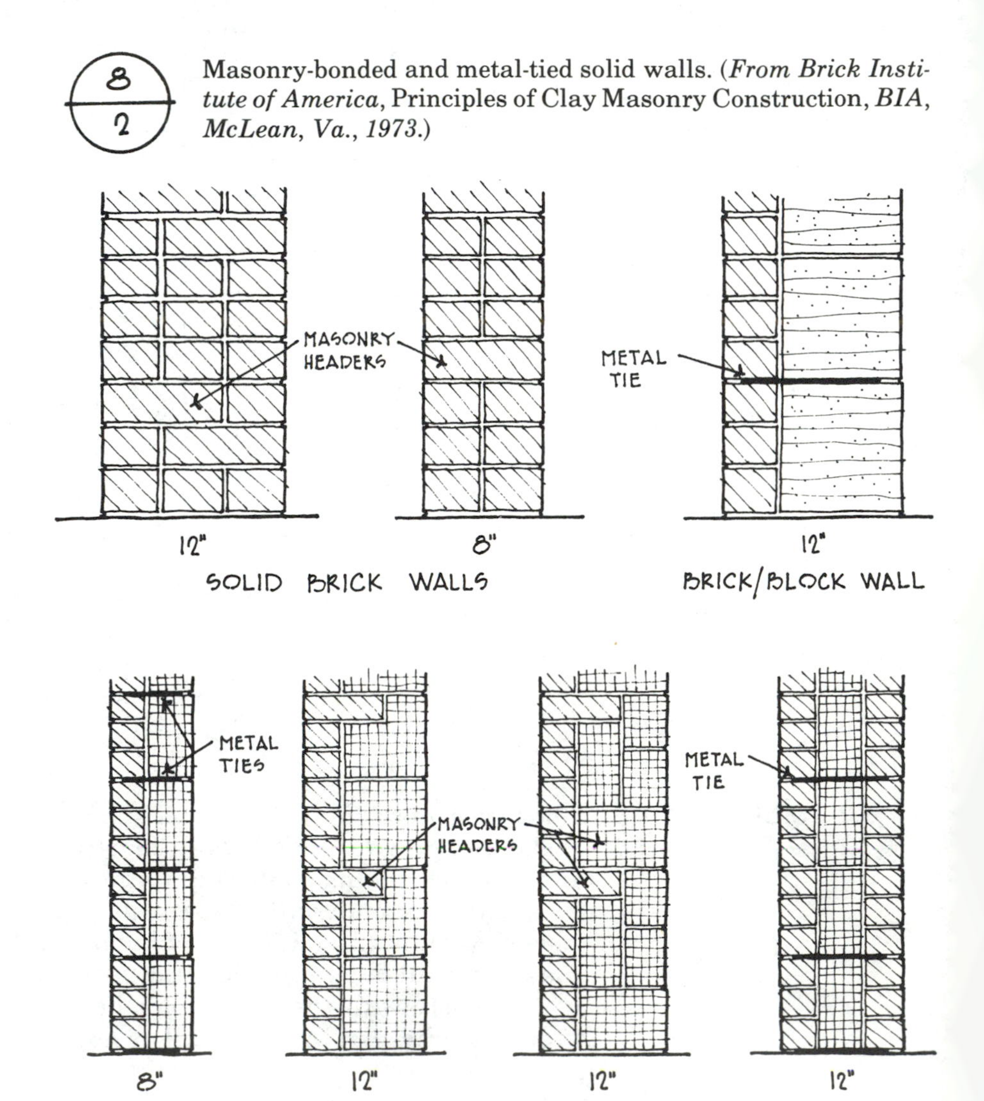

8-2 Masonry-bonded and metal-tied solid walls. (*From Brick Institute of America,* Principles of Clay Masonry Construction, *BIA, McLean, Va., 1973.*)

masonry walls do not generally exceed 12 in. in thickness except under special conditions or circumstances.

Multi-wythe walls of hollow tile, brick, or concrete units are often solidly grouted to increase strength or fire resistance. When treated in this manner, they are considered solid masonry walls. Composite walls of different materials such as brick/concrete block or concrete block/stone are also considered solid if both the collar joint between wythes and the cores of the hollow units are filled with mortar or grout. Typical material combinations include brick/tile, brick/block, brick/stone, block/stone, block/block, and occasionally block/tile.

8.2.2 Solid Walls of Hollow Units

Solid walls of hollow units consist of at least two wythes of material with no cavity or open collar joint. Construction may be of hollow clay tile, hollow CMUs, or hollow brick, or of stone or solid brick units in combination with any of these. If either wythe contains hollow masonry with open cores, the wall is considered to be of hollow units (*see Fig. 8-3*). Bonding can be accomplished with either masonry headers or metal wall ties.

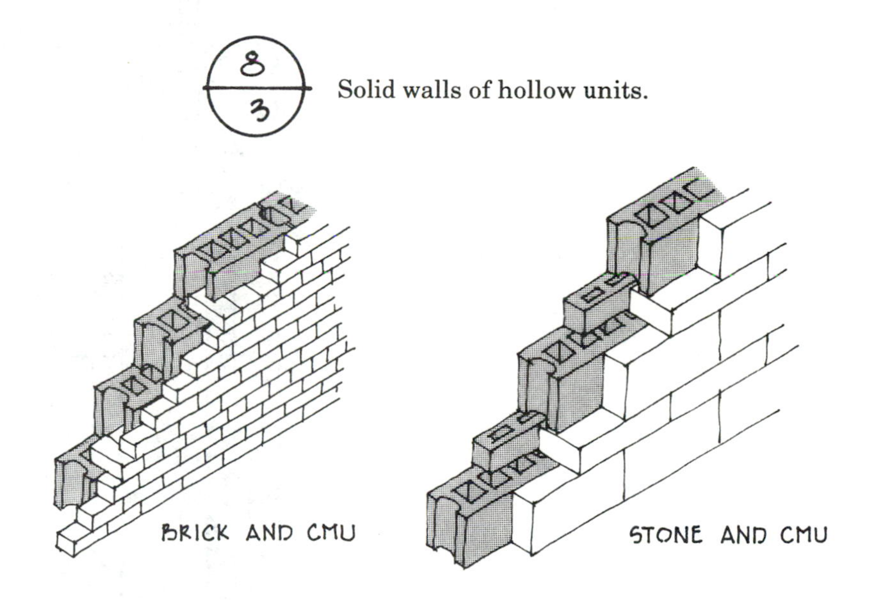

8-3 Solid walls of hollow units.

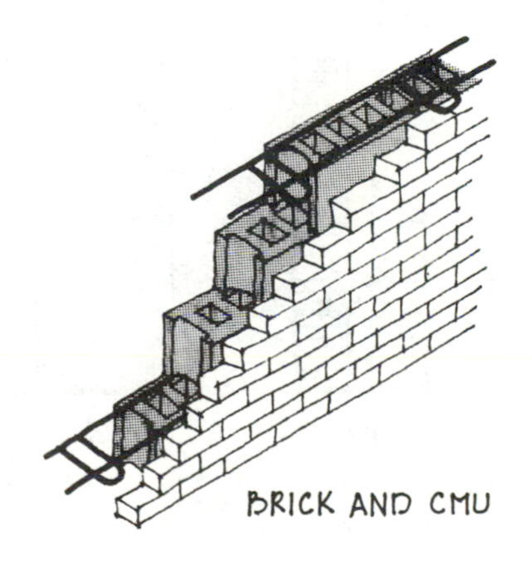

8.2.3 Cavity Walls

Cavity walls consist of two or more wythes of masonry units separated by an air space at least 2 in. wide. The wythes may be brick, clay tile, concrete block, or stone, anchored to one another with metal ties which span the open collar joint (*see Fig. 8-4*). The two wythes together can be designed as a unified loadbearing element, or the exterior wythe may be a nonstructural veneer. One of the major advantages of cavity wall construction is greater resistance to rain penetration resulting from the complete separation of the inner and outer wythes. This separation also increases thermal resistance by providing a dead air space, and allows room for additional insulating materials if desired.

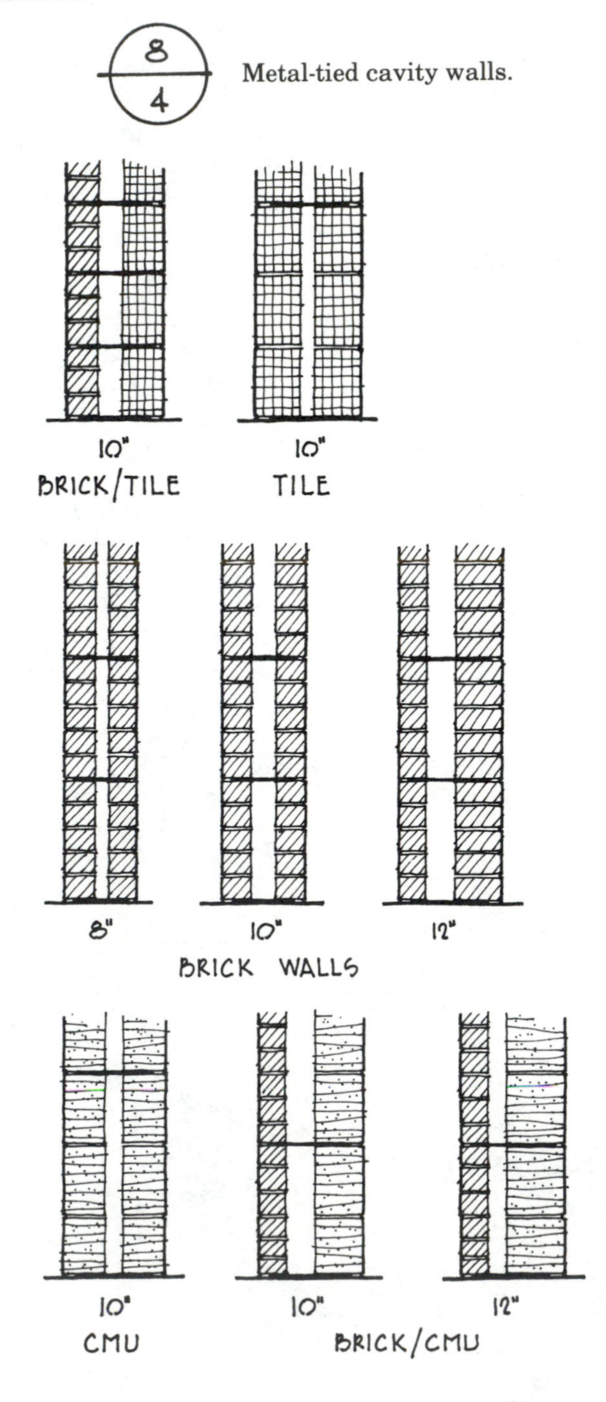

Metal-tied cavity walls.

Both sides of a cavity wall must act together in resisting wind loads and other lateral forces. The metal ties transfer these loads in compression, and must be solidly bedded in the mortar joints in order to perform properly. Crimped ties with a water drip in the center are not recommended because the weakened plane created can cause buckling of the tie and ineffective load transfer.

Cavity walls can prevent the formation of condensation on interior surfaces, so that plaster and other finish materials may be directly applied without furring. Insulation may be added in the wall cavity, including water-repellent vermiculite, silicone-treated perlite, or rigid boards. A vapor barrier or dampproof coating is usually required on the cavity face of the inner wythe unless waterproof insulation is used or the rigid insulation is held at least 1 in. away from the exterior wall.

In multi-wythe construction, a distinction is made between walls in which both the facing and the backup are loadbearing, and those in which only one wythe will carry the superimposed load. *Composite walls* are bonded with either masonry headers or metal ties so that both wythes act as a single element in resisting loads. In a *veneered wall*, the facing is attached but not structurally bonded, so that only the backup material is loadbearing. Clay brick and hollow tile, or brick and concrete block, are often used in this manner, with each material selected for optimum performance and economy.

8.2.4 Metal Wall Ties

Metal wall ties are generally recommended over masonry headers, for two reasons. Masonry headers increase the possibility of moisture penetration not only because of the continuity of the header through an 8-in. wall section, but also because of the difficulty of constructing some types of walls without breaking the mortar bond. If the mason is working from the exterior of a building, the backup units are naturally laid before the facing units. Brick headers are placed on top of the backup as the brick coursing reaches the proper height. With hollow units, this header is supported by two relatively thin strips of mortar. The weight of the next course of backup units can cause uneven settlement and joint separations through which moisture can seep (*see Fig. 8-5*). Metal wall ties eliminate this problem and simplify construction. They also provide the necessary flexibility for differential movement between the inner and outer wythes of masonry. Clay and concrete have different thermal and moisture expansion coefficients. Masonry bonding impairs movement in the wall and can create an eccentric load on the headers, rupturing the mortar bond at the exterior face.

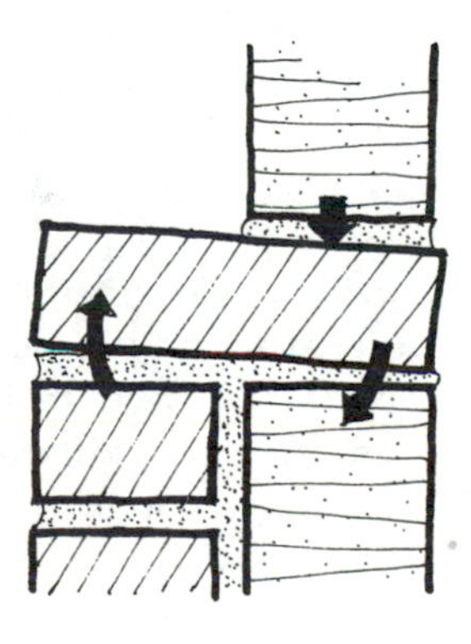

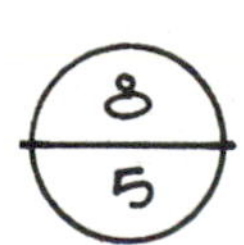

Rotation of brick from uneven settlement of heavy backup wythe. (*From Brick Institute of America*, Principles of Clay Masonry Construction, *BIA, McLean, Va., 1973.*)

8.3 FIRE RESISTANCE CHARACTERISTICS

Building fires are a serious hazard to life and property, and fire safety in construction is therefore a primary consideration of every building code authority. According to the National Fire Protection Association (NFPA), construction deficiencies are a major factor in large-loss fire experiences. NFPA records show that combustible construction is the predominant cause of conflagrations, particularly in areas of closely built wood-frame structures. Restricting the spread of fire is critical in reducing fire deaths and property loss. The overwhelming majority of U.S. fires are in residential buildings—apartments, hotels, and dwellings. Multifamily occupancies are particularly vulnerable because of the lack of physical separation between living units.

Fire regulations are concerned primarily with the safety of occupants, the safety of firemen, the integrity of the structure, and the reduction of damage. The overall risk is reduced when fire-resistant construction is used (1) to protect structural elements, and (2) to divide a building into compartments for the containment of fire.

The degree of fire protection offered by masonry construction was recognized long ago. In A.D. 1212 an ordinance was issued by royal proclamation requiring that all alehouses in London be built of masonry. After the great fire of 1666, which destroyed four-fifths of London, King Charles II decreed that the walls of all new buildings must be of masonry. Modern masonry construction has an excellent performance record in fire containment that is reflected in current building codes.

8.3.1 Fire Tests

Fire properties of building materials are divided into two basic categories: *combustibility* and *fire resistance*. Masonry is classified as noncombustible. Fire resistance is based on standard ASTM, NFPA, or National Bureau of Standards (NBS) fire endurance tests.

Specimens are subjected to controlled heat applied by standard time–temperature curve for a maximum of 8 hours and 2300°F. Wall assemblies must also undergo a hose stream test for impact, erosion, and thermal shock. Throughout the tests, columns and bearing walls are loaded to develop full design stresses. Within 24 hours after the testing is complete, bearing walls must also safely sustain twice their normal superimposed load. Fire resistance ratings, generally in 1- or $\frac{1}{2}$-hour increments, are assigned according to the elapsed time at which the test is terminated. The test is terminated when any one of three possible end-point criteria is reached: (1) an average temperature rise of 250°F or a maximum rise of 350°F is measured on the unexposed side of the wall; (2) heat, flame, or gases escape to the unexposed side igniting cotton waste samples; or (3) failure under the design load occurs (loadbearing construction only). The first two points concern containment of fire spread through the wall or section, while the third affects the structural integrity of the building. Despite this disparity, each of the criteria carries equal weight in determining the assigned fire rating.

Concrete and masonry fire ratings are almost invariably based on heat transmission. The temperature on the unexposed side has risen 250° F while the opposite face was heated to 2300°F. The structural integrity of th wall is maintained far beyond the time indicated by the fire rating. In a rea fire, maintaining structural integrity is critical to the safety of occupant and firefighters. It can also mean that property loss is restricted t superficial damage which can be readily repaired at far less expense than rebuilding a structure whose identical fire rating was based on structural failure.

8.3.2 Fire Resistance Ratings

Extensive fire testing has been done on masonry walls, and ratings are listed (1) by the National Bureau of Standards in its report BMS 92, (2) by the National Fire Protection Association in the *Fire Protection Handbook*, (3) by the Underwriters' Laboratories in its *Fire Resistance Index*, and (4) by the American Insurance Association in its publication *Fire Resistance Ratings*. The four model building codes most widely used in the United States list fire ratings that are taken from these reports or, in some instances, refer directly to the publications as reference standards. The *Standard Building Code* and the *Uniform Building Code* use tables listing the minimum thickness of a particular material or combination of materials required for ratings of 1, 2, 3, and 4 hours. The *National Building Code* references the Underwriters' Laboratories and American Insurance Association standards. The *Basic Building Code* refers directly to BMS 92 for rating requirements.

Ratings for brick and clay tile walls depend to some extent on the percent of cored area in the individual units. An 8-in. hollow tile obviously contains less mass than an 8-in. solid brick, and therefore offers less resistance to fire and heat. The rated wall assemblies listed by the National Bureau of Standards report and the American Insurance Association are illustrated in *Fig. 8-6*. These are typical of the ratings found in most model building codes.

The fire resistance of concrete masonry is a function of aggregate type and unit thickness. Walls and partitions of 1- to 4-hour ratings are governed by code requirements for actual or equivalent thickness computed on the percent of core area in the unit. Increasing the wall thickness or filling the cores with grout increases the rating. Units with less than 25% cored area are considered solid, and the actual thickness of the brick or block is used for calculations. Units with more than 25% coring are classified as hollow, and the equivalent thickness of solid material must first be computed in order to determine the fire rating. Since core size and design will vary, manufacturers' data are normally used to establish exact figures. A nominal 8-in. hollow unit reported to be 55% solid would be calculated as follows: *equivalent solid thickness* = *0.55 × 7.625 in. (actual thickness)* = *4.19 in.* (*see Fig. 8-7*).

Concrete masonry aggregates have a significant effect on the fire resistance characteristics of the units. Lightweight aggregates such as pumice, expanded slag, clay, or shale offer greater resistance to the transfer of heat in a fire because of their increased air content. Units made with these materials require less thickness to achieve the same fire rating as a heavyweight aggregate unit. The table in *Fig. 8-8* is taken from the *Standard Building Code*. It lists aggregate types and unit thicknesses that will satisfy certain fire rating requirements. A similar table is used in the *Uniform Building Code*.

The application of plaster to one or both sides of a clay or concrete masonry wall increases the fire rating of the member. For hollow units, the plaster thickness may be added to the equivalent solid thickness in calculating the rating classification.

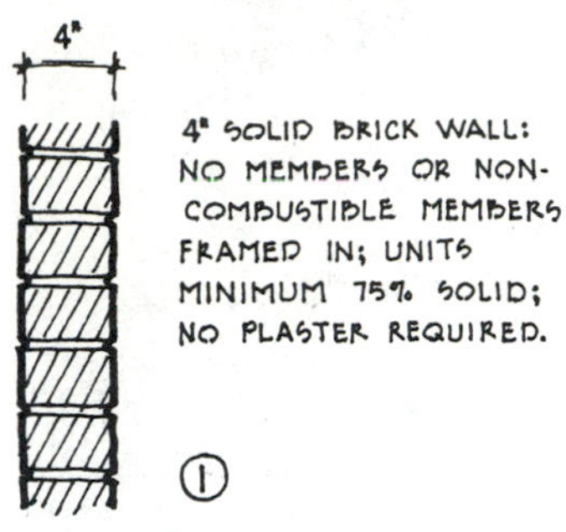

ONE-HOUR FIRE RATING

REFERENCES

① AMERICAN INSURANCE ASSOC., FIRE RESISTANCE RATINGS, DEC. 1964.

② NATIONAL BUREAU OF STANDARDS, FIRE RESISTANCE CLASSIFICATIONS OF BUILDING CONSTRUCTIONS, OCT. 1942.

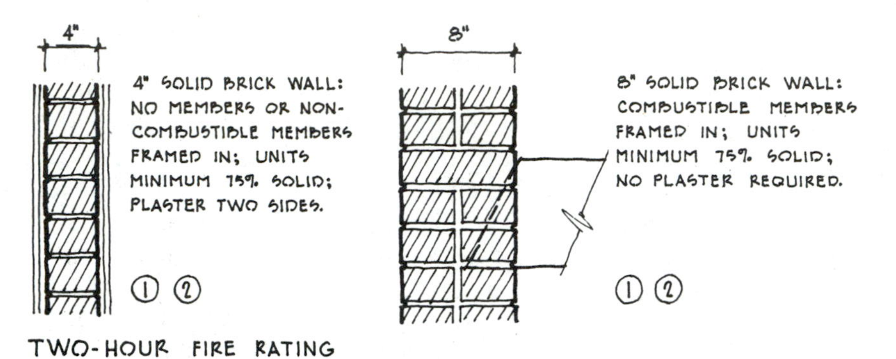

TWO-HOUR FIRE RATING

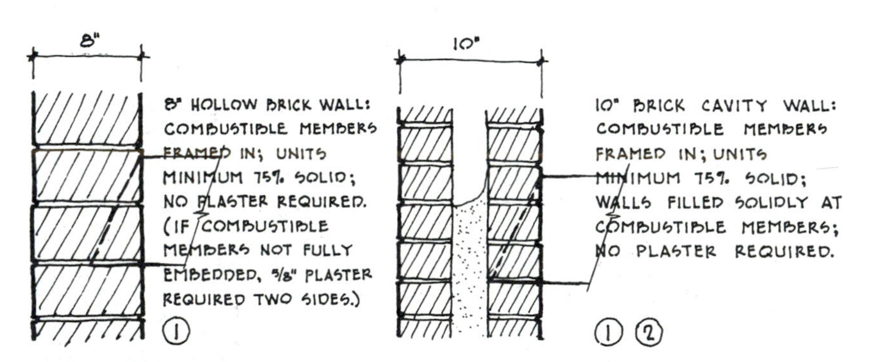

TWO-HOUR FIRE RATING

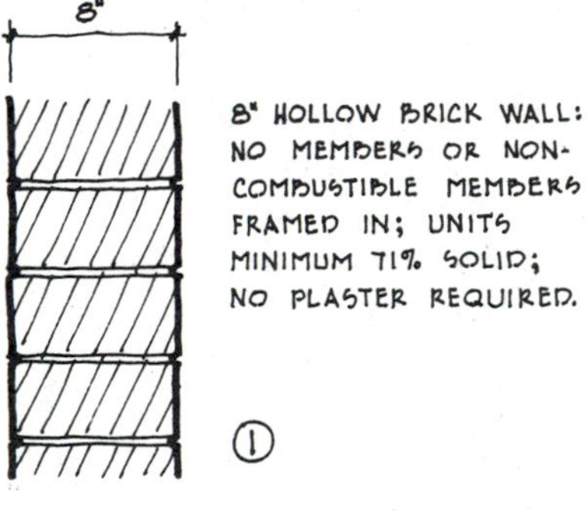

THREE-HOUR FIRE RATING

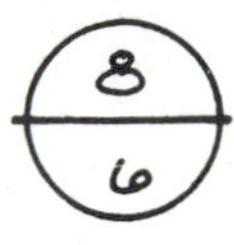

Fire resistance ratings. (*From Brick Institute of America*, Technical Note 16R, *BIA, McLean, Va.*)

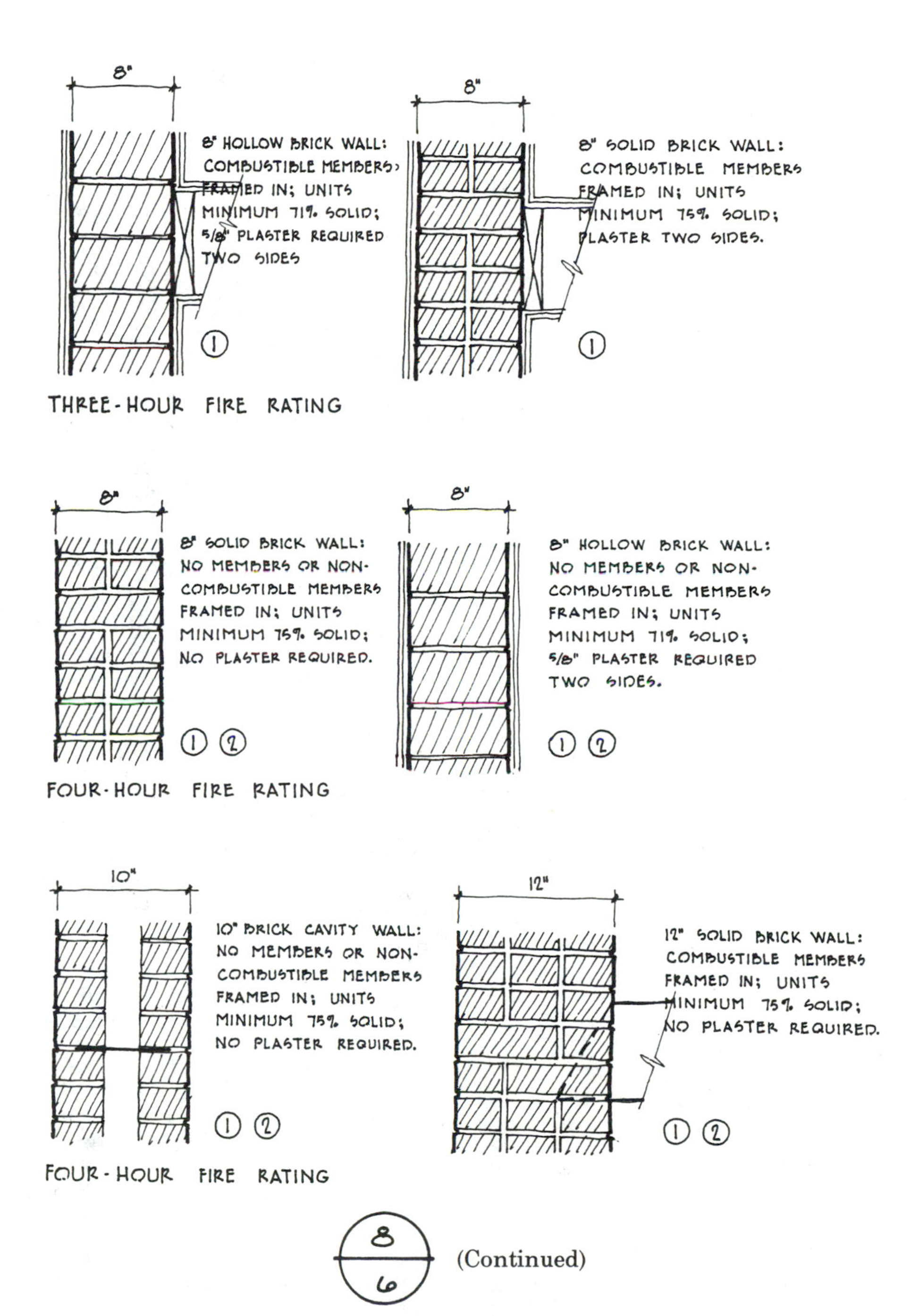
8"
8" HOLLOW BRICK WALL: COMBUSTIBLE MEMBERS FRAMED IN; UNITS MINIMUM 71% SOLID; 5/8" PLASTER REQUIRED TWO SIDES
1
8"
8" SOLID BRICK WALL: COMBUSTIBLE MEMBERS FRAMED IN; UNITS MINIMUM 75% SOLID; PLASTER TWO SIDES.
1
THREE-HOUR FIRE RATING
8"
8" SOLID BRICK WALL: NO MEMBERS OR NON-COMBUSTIBLE MEMBERS FRAMED IN; UNITS MINIMUM 75% SOLID; NO PLASTER REQUIRED.
1 2
8"
8" HOLLOW BRICK WALL: NO MEMBERS OR NON-COMBUSTIBLE MEMBERS FRAMED IN; UNITS MINIMUM 71% SOLID; 5/8" PLASTER REQUIRED TWO SIDES.
1 2
FOUR-HOUR FIRE RATING
10"
10" BRICK CAVITY WALL: NO MEMBERS OR NON-COMBUSTIBLE MEMBERS FRAMED IN; UNITS MINIMUM 75% SOLID; NO PLASTER REQUIRED.
1 2
12"
12" SOLID BRICK WALL: COMBUSTIBLE MEMBERS FRAMED IN; UNITS MINIMUM 75% SOLID; NO PLASTER REQUIRED.
1 2
FOUR-HOUR FIRE RATING
8
6

(Continued)

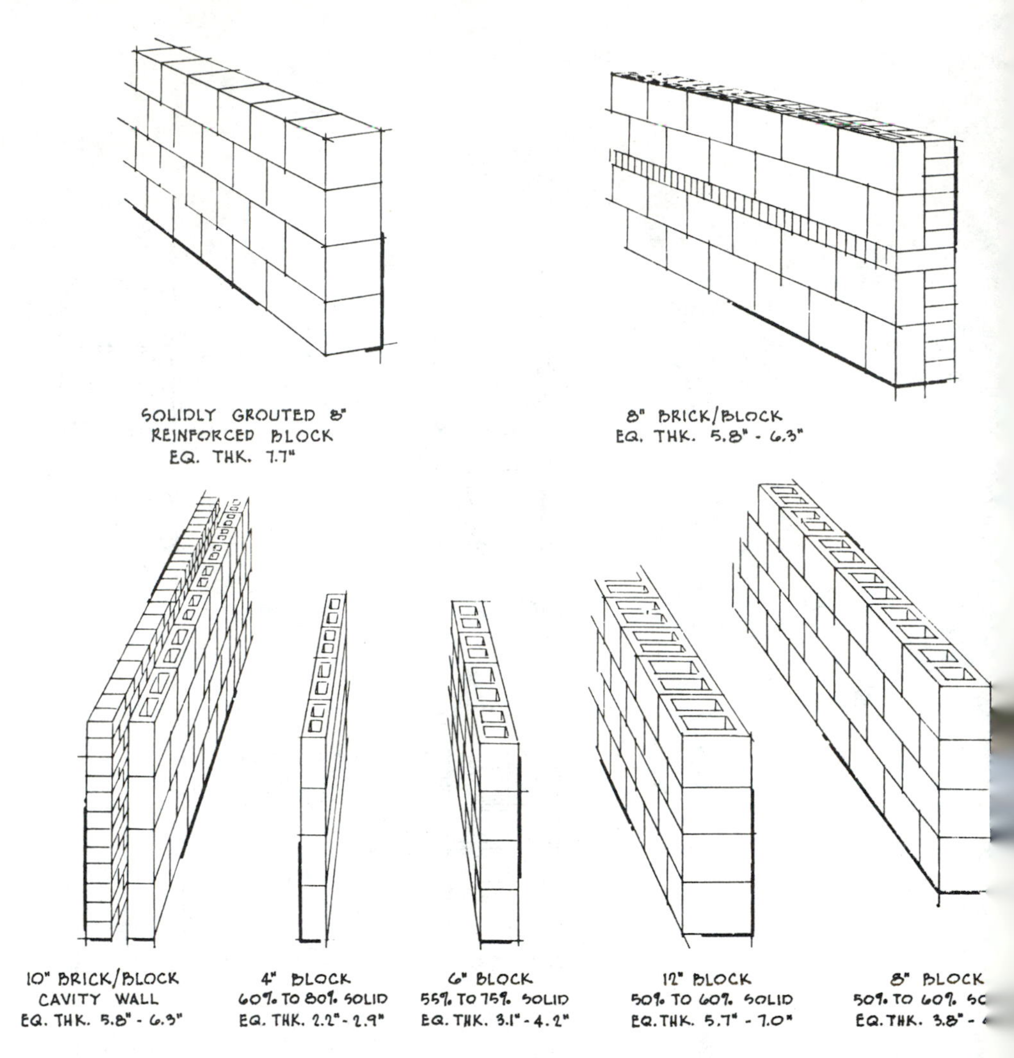

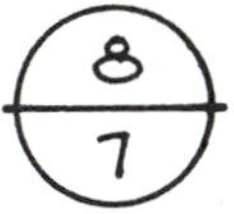

Equivalent thickness of solid and hollow masonry unit partitions. (*From National Concrete Masonry Association,* TEK Bulletin 6, *NCMA, Herndon, Va.*)

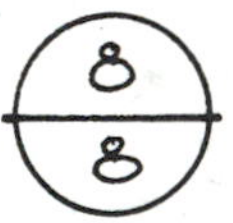

Fire ratings for various concrete block wall thicknesses. (*From Southern Building Code Congress,* Standard Building Code, *SBCC, Birmingham, Ala., 1982.*)

REQUIRED EQUIVALENT SOLID THICKNESS FOR CMU (IN.) FOR FIRE RESISTANCE RATINGS LISTED

	Members framed into wall—none or noncombustible			
Wall or partition assembly	*4 hr*	*3 hr*	*2 hr*	*1 hr*
Expanded slag or pumice aggregate	4.7	4.0	3.2	2.1
Expanded clay or shale aggregate	5.7	4.8	3.8	2.6
Expanded shale, clay, or slate aggregate (rotary kiln process)	5.35	4.5	3.95	2.6
Limestone, cinders, or unexpanded slag aggregate	5.9	5.0	4.0	2.7
Calcareous gravel aggregate	6.2	5.3	4.2	2.8
Siliceous gravel aggregate	6.7	5.7	4.5	3.0

8.3.3 Steel Fireproofing

Steel frame construction is vulnerable to fire damage and must be protected from heat and flame. Clay tile, brick, and concrete and gypsum block can all be used to fireproof steel columns and beams. Hollow tile units were originally manufactured for this purpose in the late nineteenth century. They offer effective and relatively lightweight protection. Fire test results from the National Bureau of Standards form the basis of modern code requirements for protection of steel structural elements. The table in *Fig. 8-9* is taken from the *Uniform Building Code* to show protective masonry coverings that are acceptable for various fire ratings. Comparisons may be made to the more detailed *Standard Building Code* requirements in *Fig. 8-10*. In some instances, a distinction is made for size of the structural steel member with smaller, lighter-weight sections requiring more protection because of their greater vulnerability to heat.

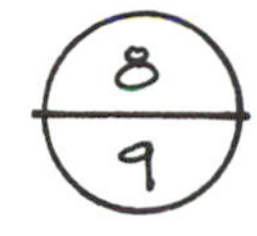

Uniform Building Code column protection requirements. (*Reproduced from the* Uniform Building Code, *1982 edition, copyright 1982, with permission of the publisher, the International Conference of Building Officials.*)

U.B.C. TABLE NO. 43-A—MINIMUM PROTECTION OF STRUCTURAL PARTS BASED ON TIME PERIODS FOR VARIOUS NONCOMBUSTIBLE INSULATING MATERIALS

Structural parts to be protected	*Item number*	*Insulating material used*	*Minimum thickness of insulating material for following fire-resistive periods (in.)*			
			4 hr	*3 hr*	*2 hr*	*1 hr*
Steel columns and all members of primary trusses	7	Clay or shale brick with brick and mortar fill	$3\frac{3}{4}$			$2\frac{1}{4}$
	8	4 in. hollow clay tile in two 2 in. layers; $\frac{1}{2}$ in. mortar between tile and column; $\frac{3}{8}$ in. metal mesh (wire diameter = 0.046 in.) in horizontal joints; tile fill	4			
	9	2 in. hollow clay tile; $\frac{3}{4}$ in. mortar between tile and column; $\frac{3}{8}$ in. metal mesh (0.046 in. wire diameter) in horizontal joints; Grade A concrete fill; plastered with $\frac{3}{4}$ in. gypsum plaster	3			
	10	2 in. hollow clay tile with outside wire ties (0.08 in. diameter) at each course of tile or $\frac{3}{8}$ in. metal mesh (0.046 in. diameter wire) in horizontal joints; Grade A concrete fill extending 1 in. outside column on all sides			3	
	11	2 in. hollow clay tile with outside wire ties (0.08 in. diameter) at each course of tile with or without Grade A concrete fill; $\frac{3}{4}$ in. mortar between tile and column				2
	12	Solid gypsum blocks with woven wire mesh in horizontal joints, laid with 1 in. mortar on flanges and plastered with $\frac{1}{2}$ in. gypsum plaster	$2\frac{1}{2}$	$2\frac{1}{2}$		
	13	Hollow gypsum blocks with $\frac{7}{8}$ in. wide 12-gauge metal clamps and woven wire mesh in horizontal joints; PL denotes $\frac{1}{2}$ in. gypsum plaster	$3\frac{1}{2}$ PL	$3\frac{1}{2}$ PL	3	3

FIRE RESISTANCE RATINGS FOR PROTECTED STEEL COLUMNS				
	Minimum nominal thickness outside of column (in.) for rating indicated			
Protective material	*4-hr*	*3-hr*	*2-hr*	*1-hr*
Hollow clay or shale tile[1] Unplastered	[2]	2[3]		2[4]
Plastered with $\frac{3}{4}$-in. sanded gypsum plaster (1:3 mix by volume)	2[5]			
Gypsum[1]—poured solid (re-entrant space filled) and reinforced with 4 × 4 in. wire mesh reinforcement wrapped around column, unplastered	2	$1\frac{1}{2}$	1	1
Gypsum (solid) block[1] Unplastered	4[6]		2[7]	
Plastered with $\frac{1}{2}$-in. sanded gypsum plaster	2[8]			
Gypsum (hollow) block[1] Unplastered			[9]	
Plastered with $\frac{1}{2}$-in. sanded gypsum plaster	3[10]			
Cinder concrete (hollow) block,[1] unplastered	3[11]			
Solid brick (clay or shale),[1] reentrant space filled with brick and mortar	$3\frac{3}{4}$	$3\frac{3}{4}$	$3\frac{3}{4}$	$2\frac{1}{4}$

Notes:

[1]The ratings indicated are applicable to steel columns 6 × 6 in. or larger.

[2]Structural steel columns protected with 2-in. hollow clay or shale tile, having wire mesh in horizontal joints, flanges covered with mortar or concrete, re-entrant space filled with concrete, shall have a fire-resistance rating of 4 hr if minimum area of solid material is not less than 225 sq in. For columns of less size two 2-in. layers of hollow clay or shale tile, $\frac{1}{2}$ in. mortar between tile and column, $\frac{3}{8}$ in. metal mesh in horizontal joints, hollow clay tile fill, shall be required for 4-hr rating.

[3]Hollow clay tile with outside wire ties (not less than No. 12 B&S gauge—0.08-in.-diameter steel wire tied around the outside of each course of tile at the middle) or with $\frac{3}{8}$ in. metal mesh in horizontal joints; limestone or trap rock concrete fill extending 1 in. outside column on all sides.

[4]Hollow clay tile with outside wire ties (not less than those prescribed in Note 3 herein), with or without concrete fill, $\frac{3}{4}$ in. mortar between column and tile.

[5]$\frac{3}{4}$-in. mortar between column and tile; $\frac{3}{8}$-in. metal mesh in horizontal joints, limestone concrete fill.

[6]$\frac{3}{8}$-in. metal mesh (or equivalent) ties in horizontal joints; 1-in. gypsum mortar on flange; poured gypsum fill, or re-entrant space filled with gypsum block and mortar.

[7]Same as prescribed in Note 6 herein—or $\frac{7}{8}$ in. × 12 gauge (or equivalent) metal clamps, at horizontal joints, set in holes drilled in blocks; 1-in. gypsum mortar on flange at horizontal joints only; re-entrant space not filled.

[8]Metal or wire lath, or mesh (or equivalent ties) in horizontal joints; $\frac{1}{2}$ in. mortar between column and block; poured gypsum fill, or re-entrant space filled with gypsum block and mortar or $\frac{7}{8}$ in. × 12 gauge (or equivalent) metal clamps, at horizontal joints set in holes drilled in blocks; re-entrant space not filled.

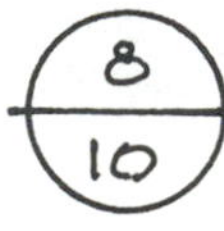

Standard Building Code column protection requirements. (*From Southern Building Code Congress*, Standard Building Code, *SBCC, Birmingham, Ala., 1982.*)

[9]$\frac{7}{8}$ in. × 12 gauge (or equivalent) metal clamps, at horizontal joints, set in holes drilled in blocks; re-entrant space not filled.

[10]Same as prescribed in Note 9 herein except that mortar is required between column flange and block.

[11]$1\frac{1}{4}$ in. mortar between column and block; re-entrant space filled with broken block and mortar.

(Continued)

8.3.4 Construction Classifications

Various types of construction are classified according to the degree of fire resistance they offer. Building codes distinguish at least four types: (1) Fire-Resistive construction, (2) Noncombustible construction, (3) Ordinary or Exterior Protected construction, and (4) Wood Frame construction. Within each of these classifications, specific fire ratings are required for various building components, such as interior and exterior bearing walls, columns and beams, floors, roofs, partitions, and elevator shafts. Although the nomenclature and number of classifications differs among the codes, the table in *Fig. 8-11* compares the fire ratings required for these construction types. (Detailed exceptions and specific occupancy requirements will reduce or increase ratings for some elements of the structure, so local codes should be consulted to verify exact requirements.)

Required fire resistance is based on fire hazard classification for different occupancies (i.e., residential, mercantile, industrial, and so on). High-hazard occupancies, and occupancies where life safety protection is critical, such as hospitals, high-rise residential or office buildings, and large assembly areas, require the greatest protection to assure control of spreading fires. Because of the fire-resistant characteristics of masonry structures, they easily satisfy the requirements for Fire-Resistive or Type I construction (*see Fig. 8-12*). Fire-Resistive construction is defined by NBS to include all buildings of noncombustible structure that will either withstand complete combustion of their contents without collapse or that have a general fire rating of 4 hours.

8.3.5 Compartmentation

A key element in fire control is compartmentation of a building to contain fire and smoke. Codes require that a building be subdivided by fire walls into areas related in size to the danger and severity of fire hazard involved. Fire walls must be constructed of noncombustible materials, have a minimum fire rating of 4 hours, and have sufficient structural stability under fire conditions to allow collapse of construction on either side without collapse of the wall. Masonry fire walls may be designed as continuously reinforced cantilevered sections. They are self-supporting without dependence on connections to adjacent structural framing. For additional lateral stability, free-standing cantilever walls may be stiffened by integral masonry pilasters with vertical reinforcing steel (*see Fig. 8-13*). Double fire walls can also be used, so that if the building frame on one side collapses,

REQUIRED FIRE RESISTANCE RATINGS FOR VARIOUS TYPES OF CONSTRUCTION (hr)																
	Fire-resistive				*Noncombustible*				*Ordinary*				*Wood frame*			
Building element	*NBC*	*UBC*	*SBC*	*BBC*	*NBC*	*UBC*	*SBC*	*BBC*	*NBC*	*UBC*	*SBC*	*BBC*	*NBC*	*UBC*	*SBC*	*BBC*
Exterior bearing walls	4[1]	4[7]	4[9]	4[18]	3[4]	1	2[11]	2	3[4]	4[7]	3[12]	2	3[6]	1	1[13]	$\frac{3}{4}$
Interior bearing walls	4	3	4	4[18]	2	1	1	2	2	1	1	$\frac{3}{4}$	N	1	N	$\frac{3}{4}$
Exterior non-bearing walls	3[2]	4[8]	2[10]	2[10]	3[2]	1[8]	2[10]	$1\frac{1}{2}$[10]	3[2]	4[8]	2[10]	2[10]	3[6]	1[16]	1[13]	$\frac{3}{4}$
Elevator shaft or stairways	2[3]	2	2	2	2[3]	1	2	2	2[3]	1	2	2	1	1	1	$\frac{3}{4}$
Partitions	N	1	1	N[14]	N	1	1	N[14]	N	1	1	N[14]	N	1	N[17]	N[14]
Columns and beams[15]	4	3	4	4	1	1	2	2	N[5]	1	1	$\frac{3}{4}$	N	1	1	$\frac{3}{4}$
Floors	3	2	$2\frac{1}{2}$	3	1	1	1	$1\frac{1}{2}$	N[5]	1	1	$\frac{3}{4}$	N	1	N	$\frac{3}{4}$
Roofs	2	2	$1\frac{1}{2}$	2	1	1	N	$\frac{3}{4}$	N[5]	1	1	$\frac{3}{4}$	N	1	N	$\frac{3}{4}$

Notes: NBC, 1967 *National Building Code* (American Insurance Association); UBC, 1967 *Uniform Building Code* (International Conference of Building Officials); SBC, 1965 *Southern Standard Building Code* (Southern Building Code Congress); BBC, 1965 *Basic Building Code* (Building Officials Conference of America). N, no fire rating required.

[1]Maximum percentage of window-to-wall area varies from 0% to 40% depending on horizontal separation. In business and residential occupancies, rating for bearing walls (exterior and interior) may be 3 hr.

[2]Pertains to a horizontal separation of 3 ft or less. Required rating and percentage of window-to-wall area varies with separation; if over 30 ft, no rating is required.

[3]In buildings 4 stories or more in height. In buildings less than 4 stories, the required rating is 1 hr.

[4]For horizontal separation over 3 ft, the required rating is 2 hr. Maximum percentage of window-to-wall area varies between 0 and 40% depending on separation.

[5]No rating; however, code specifies sizes of wood members.

[6]Pertains to horizontal separation of 3 ft or less and building area over 2500 sq ft. For 30-ft separation, no requirements.

[7]In office buildings, apartment buildings, etc., rating may be 2 hr where horizontal separation is 5 ft or greater.

[8]No rating required where horizontal separation is 50 ft or greater.

[9]Rating may be reduced to 2 hr where horizontal separation is 50 ft or greater.

[10]No rating required where horizontal separation is 30 ft or greater.

[11]Rating may be 1 hr outside first fire district.

[12]Rating may be 2 hr where horizontal separation is 30 ft or greater.

[13]Required only where horizontal separation is less than 3 ft.

[14]Corridor partitions and partitions separating tenant spaces required to have 3/4-hr rating.

[15]Applicable to columns and beams supporting loads from more than one floor or roof.

[16]No rating required where separation is 40 ft or greater in second fire district.

[17]Corridor partitions and partitions separating tenant spaces required to have 1-hr rating.

[18]In business and residential occupancies, rating for bearing walls may be 3 hr.

Comparative fire resistance requirements of the four model building codes. (*From Brick Institute of America*, Technical Note 16A, *BIA, McLean, Va.*)

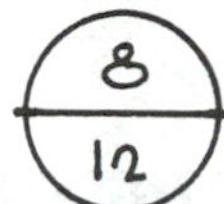

Compartmented concrete or masonry construction prevents the spread of fire. (*Photo PCA.*)

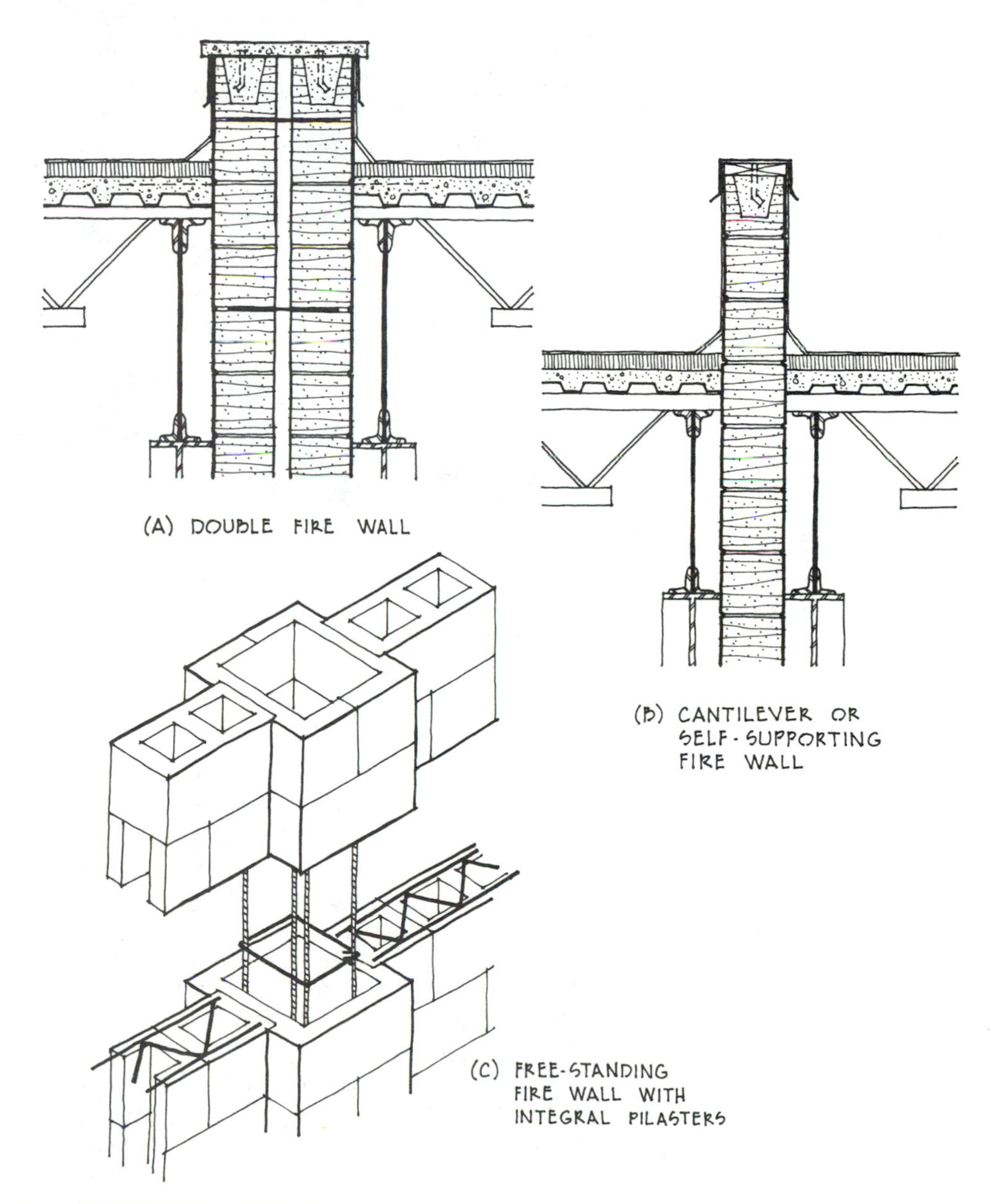

8/13 Masonry fire walls. (*From National Concrete Masonry Association,* TEK Bulletin 95, *NCMA, Herndon, Va.*)

half the wall can be pulled over while the other half still protects adjacent areas. Masonry walls also provide an airtight barrier against the spread of smoke and toxic fumes.

Fire walls are not extensively used in low-rise multifamily apartment units. Code requirements are much less stringent for two- and three-story residential occupancies than for high-rise buildings, and fires can quickly consume several adjacent units of combustible construction (*see Fig. 8-14*). Townhouse and zero-lot-line developments, however, are often required to have masonry fire walls separating units and as a result, statistics show that fire losses are greatly reduced for this building type.

Spread of fire through a building of combustible construction. (*Pho*
PCA.)

8.3.6 Fire Insurance Rates

Base rates for fire insurance are much lower for concrete and masonr buildings than for wood frame construction. Although exact premiums ar not determined until a structure is completed, the rates shown in *Figs. 8-1* and *8-16* are indicative of the relative cost savings.

8.4 THERMAL PROPERTIES

The thermal efficiency of a building material is normally judged by i resistance to heat flow. A material's *R-value* is a measure of this resistanc taken under laboratory conditions with a constant temperature differenti from one side to the other. This is called a *steady-state* or *static conditio*

Thermal resistance depends on the density of the material. By th measure, masonry is a poor insulator. Urethane insulation, on the oth hand, has a very high resistance because it incorporates closed cells pockets to inhibit heat transfer. (*R*-values of some typical building mat rials are listed in *Fig. 8-17*.)

The reciprocal of the *R*-value is the *U-value*, or the overall coefficient heat transmission. For estimating a building's heating and cooling requir ments, *U*-values are used in heat-loss and heat-gain calculations wi

Introductory Statement: When constructing apartments it can be very beneficial to subdivide the building into segments with fire walls. This reduces the potential loss and as a result has a direct influence on the insurance rates charged to the project. It should be noted that fire walls must be constructed of masonry or concrete construction to achieve insurance credits. There are also other construction requirements which must be met for fire walls. Your agent or the insurance services office should be contacted in regard to these construction details.

TABLE A FRAME*									
Occupancy— apartments:	*5–10 units*			*11–30 units*			*Over 30 units*		
Type of insurance	*Insurance amount coverage*	*Rate*	*Annual premium*	*Insurance amount coverage*	*Rate*	*Annual premium*	*Insurance amount coverage*	*Rate*	*Annual premium*
Building	200,000	0.90	1800	300,000	1.04	3120	500,000	1.09	5450
Contents	50,000	0.386	193	100,000	0.386	386	200,000	0.386	772
Extended coverage	250,000	0.312	780	400,000	0.312	1248	700,000	0.312	2184
Total annual premium			2773			4754			8406

*Buildings where the exterior walls are wood or other combustible materials, including construction where combustible materials are combined with other materials (such as brick veneer, stone veneer, wood, stucco on wood).

TABLE B JOISTED MASONRY*									
Occupancy— apartments:	*5–10 units*			*11–30 units*			*Over 30 units*		
Type of insurance	*Insurance amount coverage*	*Rate*	*Annual premium*	*Insurance amount coverage*	*Rate*	*Annual premium*	*Insurance amount coverage*	*Rate*	*Annual premium*
Building	200,000	0.74	1480	300,000	1.03	3090	500,000	1.08	5400
Contents	50,000	0.316	158	100,000	0.316	316	200,000	0.316	632
Extended coverage	250,000	0.312	780	400,000	0.312	1248	700,000	0.312	2184
Total annual premium			2418			4654			8216

*Buildings where the exterior walls are constructed of masonry materials such as adobe, brick, concrete, gypsum block, hollow concrete block, stone, tile, or similar materials, and where the floors and roof are combustible (disregarding floors resting directly on the ground).

TABLE C MASONRY NONCOMBUSTIBLE*									
Occupancy— appartments:	*5–10 units*			*11–30 units*			*Over 30 units*		
Type of insurance	*Insurance amount coverage*	*Rate*	*Annual premium*	*Insurance amount coverage*	*Rate*	*Annual premium*	*Insurance amount coverage*	*Rate*	*Annual premium*
Building	200,000	0.126	252	300,000	0.126	378	500,000	0.126	630
Contents	50,000	0.152	76	100,000	0.152	152	200,000	0.152	304
Extended coverage	250,000	0.061	153	400,000	0.061	244	700,000	0.061	427
Total annual premium			481			774			1361

*Buildings where the exterior walls are constructed of masonry materials, such as adobe, brick, concrete, gypsum block, hollow concrete block, stone, tile, or similar materials, with the floors and roof of metal or other noncombustible materials.

Fire insurance cost comparison for apartments, St. Louis County, Missouri. (*From Mason Contractors Association of Greater St. Louis,* Masonry Cost Guide and Related Technical Data, *MCA of Greater St. Louis.*)

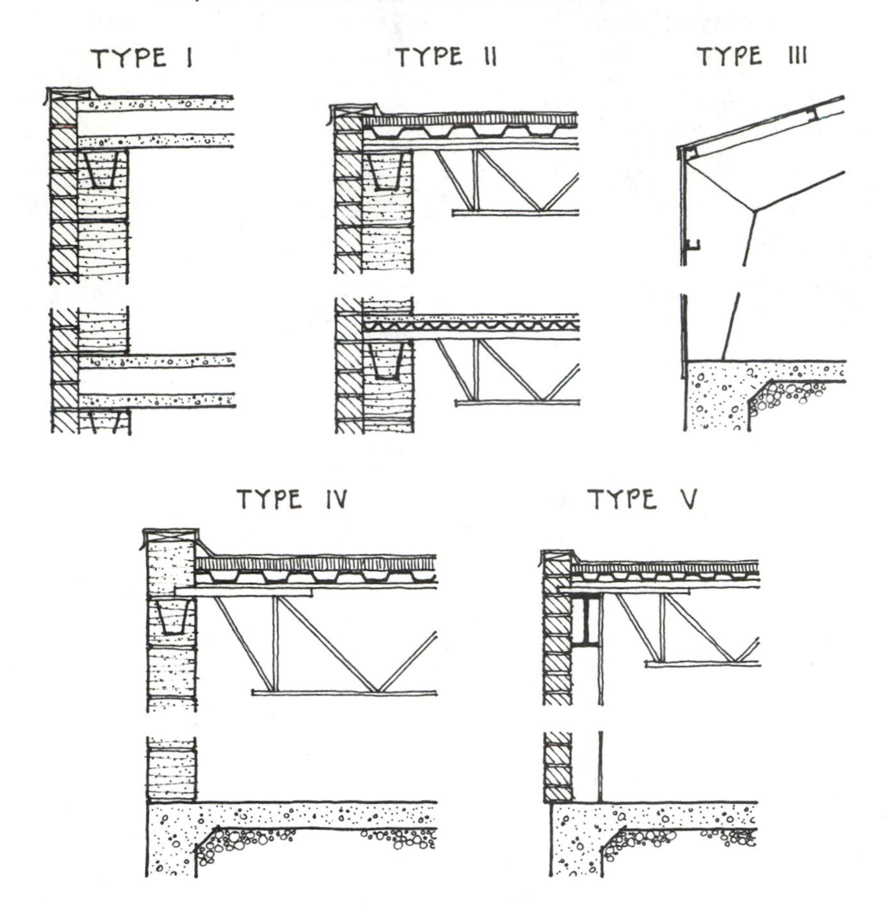

		12-in. masonry walls (brick faced) with roof of 2-hr fire-resistive rated		12-in. masonry walls (brick faced) with roof of minimum 0.22 gauge unprotected steel deck		Preengineered steel building walls and roof of unprotected steel		8-in. masonry walls (hollow concrete block) roof—minimum 0.22 gauge unprotected steel deck		4-in. brick veneer over steel stud, $\frac{1}{2}$-in. gypsum board each side—minimum 0.22 gauge steel deck roof	
Type of insurance	*Ins. amt. coverage*	*Rate*	*Ann. prem.*	*Rate*	*Ann. prem.*	*Rate*	*Ann. prem.*	*Rate*	*Ann. prem.*	*Rate*	*Ann. prem.*
Building	750,000	0.231	1,733	0.755	5,662	1.779	13,343	0.774	5,805	1.84	13,800
Contents	1,500,000	0.493	7,395	1.188	17,820	2.172	32,580	1.211	18,165	2.23	33,450
Extended coverage	2,250,000	0.114	2,565	0.114	2,565	0.350	7,875	0.114	2,565	0.232	5,220
Total annual premium			11,693		26,047		53,798		26,535		52,470

Note: Dimensions are 200 × 300 = 60,000 sq. ft. Divided with a standard 8-in. CMU fire wall. 1 story, no basement, 20 ft high, concret slab floor, exterior walls maximum of 20% glass windows. Location: St. Louis County, Missouri. Class 6 Fire Protection at risk; n exposure from adjacent buildings. All insulation listed by Underwriters' Laboratories as noncombustible, having a flame spread ratin under 25 and a smoke development rating under 200.

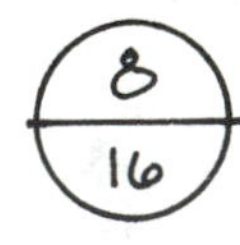

Fire insurance cost comparison for office-warehouse and storage buildings. (*From Mason Contractors Association of Greater St. Louis,* Masonry Cost Guide and Related Technical Data, *MCA of Greater St. Louis.*)

Description of material	Resistance (R) Per inch of thickness, 1/K	For thickness listed, 1/C
Air surfaces		
Inside—still air		0.68
Outside		
15-mph wind, winter		0.17
7.5-mph wind, summer		0.25
Air space		
$\frac{3}{4}$ to 4-in., winter		0.97
$\frac{3}{4}$ to 4-in., summer		0.86
Building board		
$\frac{3}{8}$-in. gypsum board		0.32
$\frac{1}{2}$-in. gypsum board		0.45
Plywood	1.25	
$\frac{1}{2}$-in. fiberboard sheathing		1.32
Insulating materials		
Batt or blanket		
2–2$\frac{3}{4}$-in.		7.00
3–3$\frac{1}{2}$-in.		11.00
5$\frac{1}{4}$–6$\frac{1}{2}$-in.		19.00
Expanded polystyrene	4.00	
Expanded polyurethane	5.88	
Vermiculite fill	2.27	
Perlite fill	2.70	
Masonry units		
Face brick, 130-lb density	0.11	
Sand and gravel aggregate concrete block		
4-in.		0.71
8-in.		1.11
12-in.		1.28
Lightweight aggregate concrete block		
4-in.		1.50
8-in.		2.00
12-in.		2.27
Siding		
$\frac{7}{16}$-in. hardboard		0.67
$\frac{1}{2}$-in. wood bevel		0.81
Aluminum or steel over sheathing		0.61

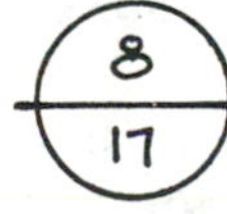

Thermal resistance of building materials.

specific outdoor design temperatures for winter and summer. These calculations (like the laboratory test conditions) assume a constant temperature differential between outdoor and indoor air, and do not take into account the diurnal cycles of solar radiation and air temperature. Since the sun does rise and set each day, the outdoor/indoor temperature differential continually fluctuates. *The static conditions on which R- and U-values are based do not actually exist outside the laboratory.* Building materials with heavy mass can react to temperature fluctuations, producing a dynamic thermal response that can differ substantially from heat flow calculations based solely on *U*-values. Research indicates that the actual measured rate of heat transfer for masonry walls is 20–70% less than steady-state calculation methods predict.

8.4.1 Thermal Inertia

Heat transfer through solid materials is not instantaneous. The time delay involving absorption of the heat is called *thermal lag*. Although most building materials absorb at least some heat, the higher density and greater mass of masonry cause it to have slower absorption and longer retention than other materials. The speed with which a wall will heat up or cool down is described as *thermal inertia*, and is dependent on wall thickness, density, specific heat, and conductivity.

For centuries people have known that buildings with massive masonry walls were more thermally stable than those of thin, lightweight construction. Until recently, however, there was no hard scientific evidence to explain this phenomenon. We have now come to understand that the transmission of heat through building walls is a dynamic process and that any method of calculating heat loss or heat gain that assumes it is static or steady state is not an accurate measure of performance.

Heat flows from hot to cold. As temperatures rise on one side of a wall heat begins to migrate toward the cooler side. Before heat transfer from on space to another can be achieved, the wall itself must undergo a temperatur increase. The amount of thermal energy necessary to produce this increas is directly proportional to the weight of the wall. Masonry is heavy enoug to store heat and substantially retard its migration. This characteristic i called *thermal storage capacity* or *capacity insulation*. One measure of thi storage capacity is the elapsed time required to achieve equilibrium betwee inside and outside wall surface temperatures. The midday solar radiatio load on the south face of a building will not completely penetrate a 12-in solid masonry wall for approximately 8 hours.

The effects of wall mass on heat transmission are dependent on th magnitude and duration of temperature differentials during the daily cycl Warmer climates with cool nights benefit most. Seasonal and climat conditions with fairly constant temperatures from day to night tend t diminish these benefits.

8.4.2 Heat-Gain Calculations

Thermal lag and capacity insulation are of considerable importance calculating heat gain when outside temperature variations are gre During a daily cycle, walls with equal *U*-values but unequal mass w produce significantly different peak loads. The greater the storage capacit the lower the total heat gain. Increased mass reduces actual peak loads in building, thus requiring smaller cooling equipment. Building envelopes with more thermal storage capacity will also delay the peak load until after the hottest part of the day when solar radiation through glass areas is diminished and, in commercial buildings, after lighting, equipment, and occupant loads are reduced. This lag time decreases the total demand on cooling equipment by staggering the loads.

Steady-state heat-gain calculations do not recognize the significant benefits of thermal inertia when they employ constant indoor and outdoor design temperatures. Computer studies completed by Francisco Arumi for the Energy Research and Development Administration and the National Concrete Masonry Association made close comparisons between static calculations and dynamic calculations. *Figure 8-18* shows the time–temperature curves derived from each method in calculating inside room temperature.

The attenuation of temperature amplitudes found with the dynamic response calculation graphically illustrates the actual effect that the thermal inertia of massive walls has on indoor comfort. Another study conducted by Mario Catani and Stanley E. Goodwin for the Portland Cement Association (PCA) and reported in the *Journal of the American*

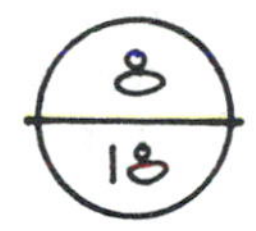

Francisco Arumi's comparison between a static and dynamic thermal calculation for a masonry wall. (*From Francisco N. Arumi,* Thermal Inertia in Architectural Walls, *National Concrete Masonry Association, Herndon, Va., 1977.*)

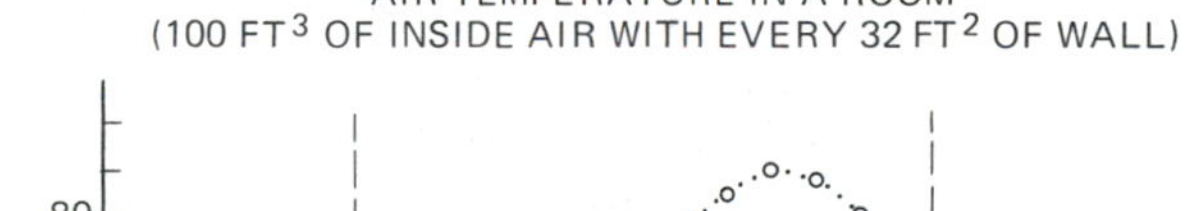

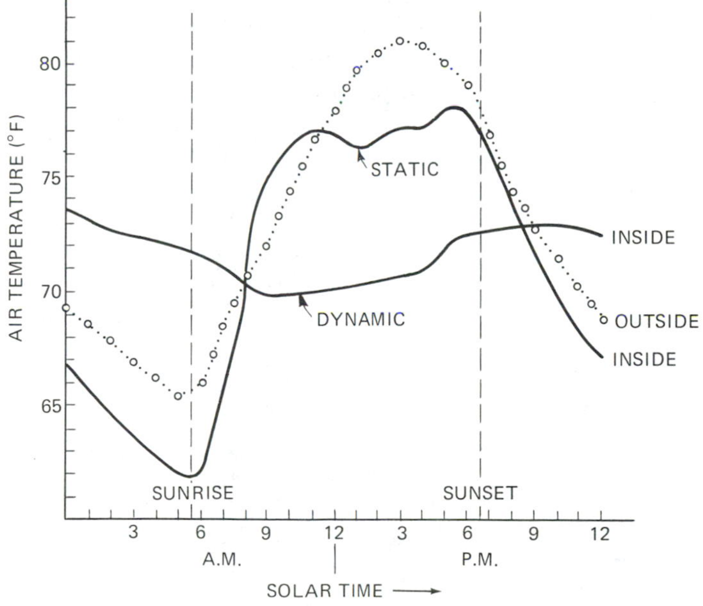

A DAY IN MAY IN AUSTIN, TEXAS

CONSTANTS OF THE CALCULATIONS

SN = 0.3 FT^{-1}
NO INTERNAL HEAT GENERATION
EAST ORIENTATION
SURFACE COLOR: BROWN

CLIMATE: 9-YEAR AVERAGE
AIR INFILTRATION = 4 VOL/HR

WALL: BRICK/CONCRETE BLOCK
C = 0.262 BTU/HR/FT^2/°F
γ = 2.98

Concrete Institute (February 1976) shows heat-gain comparisons for several wall types (*see Fig. 8-19*). Computer analysis using dynamic response methods showed that, with *U*-values equal, the peak heat gains of the lighter-weight walls were 38 to 65% higher than for the heavy walls. In comparisons of a model building with four alternative wall types, the same results were evident. Using dynamic analysis methods, two heavy concrete walls, a concrete tilt-up wall, and a metal building wall were studied to determine peak cooling loads (*see Fig. 8-20*). Results showed that the heavier walls were far superior in performance to the lightweight sections and that, despite a *U*-value that was 33% higher than the others, the peak loads for building A were 60 to 65% less than those for the lightweight construction.

The equation prescribed by the American Society for Heating, Refrigeration, and Air Conditioning Engineers (see *ASHRAE 90-75*) for calculating average thermal transmittance of the gross wall area does not make allowance for building mass. The formula is used merely to determine acceptable combinations of opaque and translucent materials in walls. Temperature conditions are assumed static and the performance evaluation of the wall is based solely on *U*-values. This type of heat-gain calculation can easily be compared to dynamic response methods and to actual field tests. In one series of tests conducted by the NCMA using National Bureau of Standards computer programs, two walls with the same *U*-value but o very different weights were measured for cooling load requirements. One building was of insulated wood frame construction weighing 8 lb/sq ft. Th second was insulated concrete masonry weighing 40 lb/sq ft. (The *U*-valu for both walls was 0.10.) When building size, roof construction, wall area percentage of glass area, and *U*-value are held constant, the ASHRAE formula would give identical results, indicating that both wall types would perform equally. However, actual measured results showed dramatic dif- ferences in performance. The masonry building required air-conditionin equipment capable of handling peak loads on the order of 34,000 Btu/hr The wood frame building showed peak requirements of approximatel

Catani and Goodwin's heat-gain curves for various wall types. (*From Mario Catani and Stanley Goodwin, "Heavy Building Envelopes and Dynamic Thermal Response,"* ACI Journal, *February 1976.*)

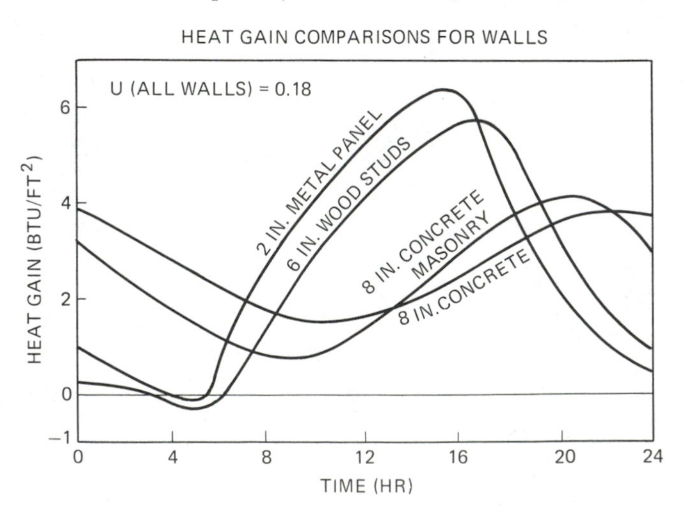

Average values of heat gain for four walls of a square building show that peak loads are increased from 38 to 65% as result of less mass.

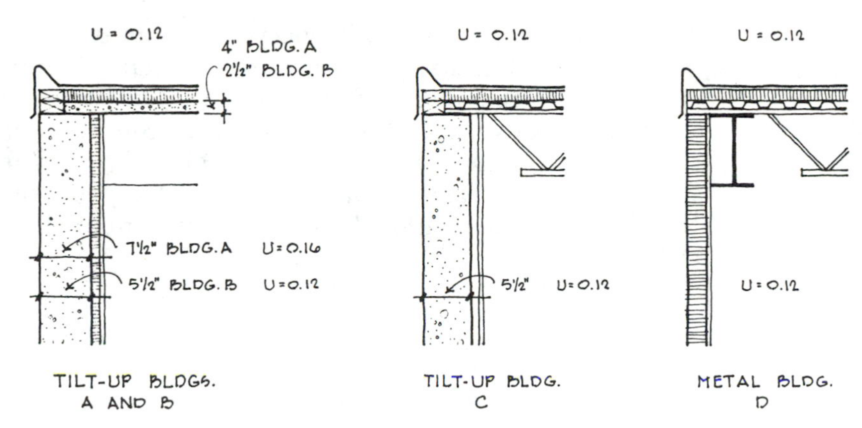

Four different building envelopes studied: two concrete, concrete and metal, and metal. Heavyweight Building A having walls with higher U-factors is used as a base for heat flow comparisons.

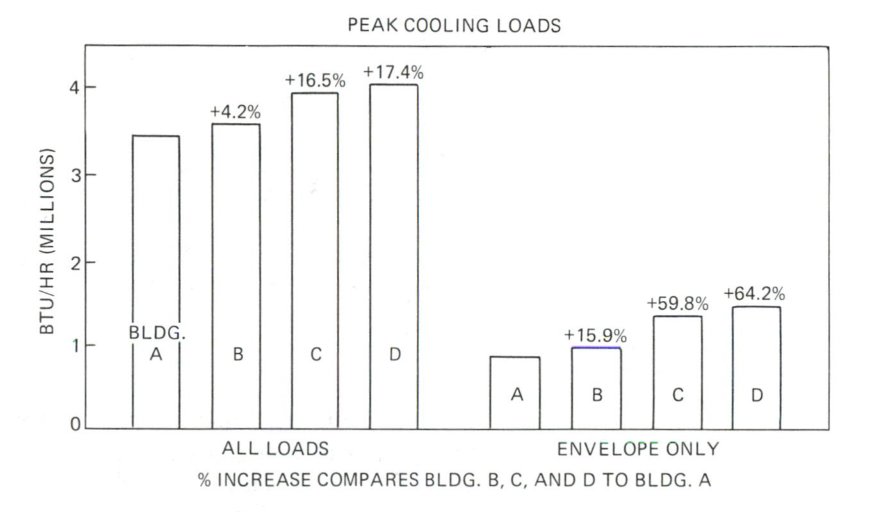

The percent increase compares Buildings B, C, and D to Building A. Because of heavier sections, peak cooling loads for Building A are reduced even though the wall U-factor is greater than for the other buildings.

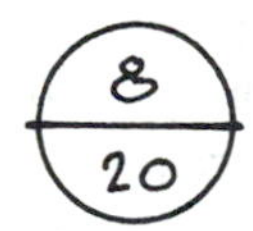

Catani and Goodwin's comparison of actual heat gain for various wall types. (*From Mario Catani and Stanley Goodwin, "Heavy Building Envelopes and Dynamic Thermal Response,"* ACI Journal, *February 1976.*)

40,500 Btu/hr—an increase of 6500 Btu/hr, or 20%. This drastic increase in heat gain is a direct result of reduced wall mass.

For high-rise and nonresidential building types, *ASHRAE 90-75* lists a different formula for heat-gain calculations that does consider wall mass and recognizes the benefits of thermal inertia and dynamic response. Lightweight walls are assigned an equivalent temperature difference (TD_{EQ}) of 44°F. This differential is reduced in increments to a low of 23°F for very heavy walls. These figures are used to modify effective *U*-values, and represent a theoretical improvement in thermal performance of as much as 48% for increased wall mass. These correction factors compare very well with research results and have been empirically tested to substantiate their validity.

NCMA reports other cooling load tests made using the NBS computer program. *U*-values of the walls, roof, and floor were held constant while the wall weight was varied from 10 lb/sq ft to 70 lb/sq ft. in 5-lb increments. The size of the required air-conditioning equipment varied inversely with the weight of the structure. The lightest-weight walls (10 lb/sq ft) required over 35,000 Btu/hour in air conditioning. The heaviest walls (70 lb/sq ft) required less than 25,000 Btu/hour. When the data are grouped in weight categories matching those of the equivalent temperature difference graph, the relationships are easily compared (*see Fig. 8-21*).

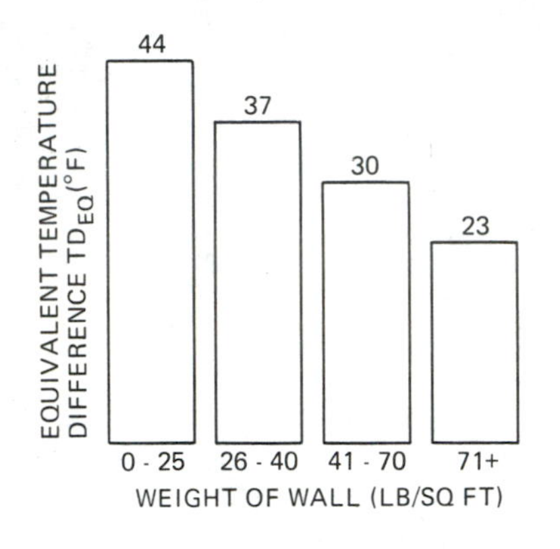

Effect of wall weight on heat transfer and air-conditioning load. (*From National Concrete Masonry Association*, TEK Bulletin 82, *NCMA, Herndon, Va.*)

In any climate where there are large fluctuations in the daily temper- ature cycle, the thermal inertia of masonry walls can contribute substan- tially to increased comfort and energy efficiency. The time lag created by delayed heat flow through the walls reduces peak cooling demands to a much greater extent than *U*-values indicate. The ASHRAE standard recognize, and extensive research demonstrates, that the steady-stat method for heat-gain calculations is not a valid assessment of the actua thermal performance of masonry walls. Dynamic response methods base on fluctuating outdoor temperatures will more accurately predict perfor- mance and permit more efficient sizing of cooling equipment tailored t actual peak load requirements.

8.4.3 Heat-Loss Calculations

In northern climates where heat loss is usually more critical than heat gai winter temperature cycles more nearly approximate static design cond tions because daily temperature fluctuations are smaller. There is stil however, significant advantage to be gained by using masonry walls wit thermal inertia. The methods developed by ASHRAE for measurin dynamic thermal response are more complicated for heat-loss calculatio than for heat gain, and require sophisticated computer programs.

The Catani and Goodwin study compared steady-state heat-lo calculations with dynamic analysis. They found that the predicted heat lo based on static conditions was 22% higher than the actual recorded loss f heavy walls, and 8% lower than the actual loss for lightweight walls. Usi three different wall types with the same U-value, they made a direct comparison of peak heating loads. It was found that, although the effects are not as dramatic for winter conditions, peak heating load requirements decreased as the weight of the building walls increased (*see Fig. 8-22*).

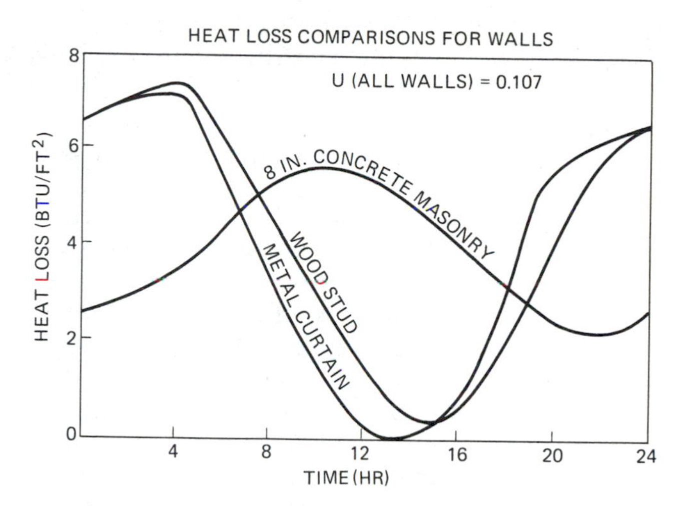

A computer program analysis shows that heat loss through walls with identical U-factors and configuration of insulation varies considerably due to mass. The effect of glass, occupants, and lights accentuates the difference.

Catani and Goodwin's heat-loss calculations for various wall types. (*From Mario Catani and Stanley Goodwin, "Heavy Building Envelopes and Dynamic Thermal Response,"* ACI Journal, *February 1976.*)

Test buildings have been used to validate computer programs for dynamic heat-loss calculations by comparing them to actual measured heating loads. The National Bureau of Standards conducted a series of tests on a full-scale building erected in its environmental chamber where both temperature and humidity can be controlled. The study also compared maximum heat flow rates predicted by the steady-state and dynamic methods with actual measured heat flow (*see Fig. 8-23*). Steady-state calculations were an average of 52% higher than measured results.

Test results of measured heat flow compared to steady-state and dynamic calculation methods. (*From National Concrete Masonry Association,* TEK Bulletin 58, *NCMA, Herndon, Va.*)

Test number	Insulation	Windows	*Maximum heat flow rate (Btu/hr)*		
			Steady-state method	*Response factor method*	*Measured heat flow*
1	None	Single glazed	15,135	11,558	11,372
2	Inside	Single glazed	4,470	2,814	2,748
3	Outside	Single glazed	4,748	3,047	2,811
4	Outside	Double glazed	4,499	2,525	2,700
5*	Outside	Double glazed	8,150	6,144	6,321

*Temperature limits changed to range of 10–70°F.

8.4.4 The M-Factor

The computer programs developed by ASHRAE and NBS for dynamic heat-loss calculations are so complex that they do not easily translate into a simple equation. Researchers recognized the need for a simpler method of calculation that would make the concept of thermal inertia more readily usable. In response to this need, the Masonry Industry Committee sponsored a study by the engineering firm of Hankins and Anderson that resulted in development of the *M*-factor, a simplified correction factor expressing the effects of mass on heat flow.

The *M*-factor is not a new calculation procedure, but is simply used to modify steady-state calculations to account for the effect of wall mass. The *M*-factor is a dimensionless correction factor. It is not a direct measure of the thermal storage capacity of walls. It is defined as the ratio of the cooling or heating load calculated by dynamic response methods to that computed with standard ASHRAE calculation methods.

The modifiers were plotted on a graph with variables of wall weight and number of degree-days (*see Fig. 8-24*). When the wall weight is very light, and in areas where the number of degree days is high (colder climates), the *M*-factors approach 1.0 (no correction). Ambient conditions in cold climates more closely approximate a steady-state condition and the traditional *U*-factor evaluation for heat loss is more accurate than for warmer regions.

The M-factors from the curves modify only heat-loss calculations and should not be used in cooling calculations. A study of heat-gain calculations

Thermal storage capacity correction graph for heat-loss calculations—*M*-factor curves. (*From Brick Institute of America*, Technical Note 4B, *BIA*, *McLean, Va.*)

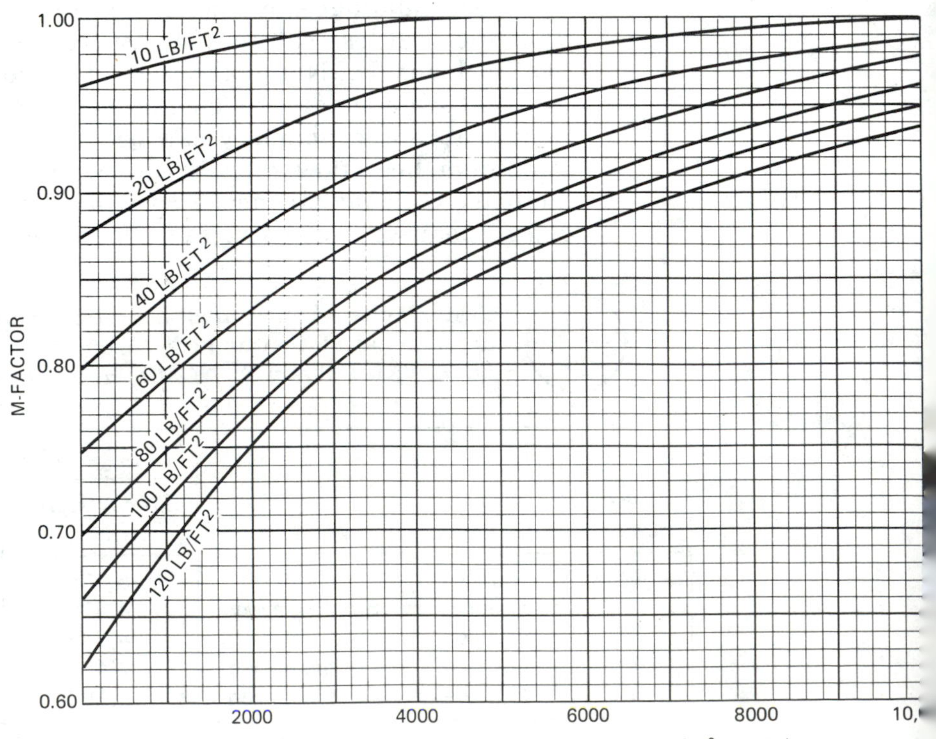

showed that dynamic analysis results were sufficiently close to those obtained using the TD_{EQ} modifier in the *ASHRAE 90-75* equation that no additional correction factor was needed for cooling load calculations.

Energy conservation codes may require a minimum heat-loss rate, a minimum R-value, or a maximum U-value. M-factors adjust the normal ratings of heavy walls, taking into account the effect of thermal inertia. They may be used (1) to modify heat losses determined by steady-state calculations, and (2) to determine effective U- and R-values for heavy walls.

As an example, consider a building project located in Memphis, Tennessee, with 3232 annual Fahrenheit degree-days (*see Fig. 8-25*), an outdoor design temperature of 21°F, and an indoor design temperature of

Annual degree-days for major U.S. and Canadian cities. (Data drawn from a publication of the U.S. Weather Bureau, covering the period 1931–1960.) (*From Brick Institute of America*, Technical Note 4B, *BIA, McLean, Va.*)

City and state	Yearly total	City and state	Yearly total
Birmingham, Alabama	2,551	Omaha, Nebraska	6,612
Anchorage, Alaska	10,864	Las Vegas, Nevada	2,709
Phoenix, Arizona	1,765	Reno, Nevada	6,332
Tucson, Arizona	1,800	Concord, New Hampshire	7,383
Little Rock, Arkansas	3,219	Trenton, New Jersey	4,980
Los Angeles, California	2,061	Albuquerque, New Mexico	4,348
Sacramento, California	2,502	Albany, New York	6,875
San Diego, California	1,458	Buffalo, New York	7,062
San Francisco, California	3,015	New York, New York	4,871
Denver, Colorado	6,283	Raleigh, North Carolina	3,393
Pueblo, Colorado	5,462	Bismarck, North Dakota	8,851
Hartford, Connecticut	6,235	Cincinnati, Ohio	4,410
Wilmington, Delaware	4,930	Cleveland, Ohio	6,351
Washington, D.C.	4,224	Oklahoma City, Oklahoma	3,725
Jacksonville, Florida	1,239	Portland, Oregon	4,635
Miami, Florida	214	Philadelphia, Pennsylvania	5,144
Orlando, Florida	766	Pittsburgh, Pennsylvania	5,987
Atlanta, Georgia	2,961	Providence, Rhode Island	5,954
Savannah, Georgia	1,819	Charleston, South Carolina	2,033
Honolulu, Hawaii	0	Columbia, South Carolina	2,484
Boise, Idaho	5,809	Sioux Falls, South Dakota	7,839
Chicago, Illinois	5,882	Knoxville, Tennessee	3,494
Springfield, Illinois	5,429	Memphis, Tennessee	3,232
Indianapolis, Indiana	5,699	Nashville, Tennessee	3,578
Des Moines, Iowa	6,588	Dallas, Texas	2,363
Topeka, Kansas	5,182	El Paso, Texas	2,700
Louisville, Kentucky	4,660	Houston, Texas	1,396
Baton Rouge, Louisiana	1,560	Salt Lake City, Utah	6,052
New Orleans, Louisiana	1,385	Burlington, Vermont	8,269
Portland, Maine	7,511	Richmond, Virginia	3,865
Baltimore, Maryland	4,654	Seattle-Tacoma, Washington	5,145
Boston, Massachusetts	5,634	Charleston, West Virginia	4,476
Detroit, Michigan	6,232	Milwaukee, Wisconsin	7,635
Minneapolis, Minnesota	8,382	Cheyenne, Wyoming	7,381
Jackson, Mississippi	2,239	Montreal, Canada	7,899
Kansas City, Missouri	4,711	Quebec, Canada	8,937
St. Louis, Missouri	5,000	Toronto, Canada	6,827
Great Falls, Montana	7,750	Vancouver, Canada	5,515

72°F. Two different wall types will be considered: an insulated wood frame wall weighing 8 lb/sq ft, and a masonry wall weighing 80 lb/sq ft. From *Fig. 8-24*, the M-factor for the frame wall is 1.0 and for the masonry wall 0.84. For the purpose of this example, both walls are assumed to have steady-state U-values of 0.12.

1. *Heat-loss calculation:* Using the standard heat-loss equation with the M-factor correction, the calculation for one square foot of the wood frame wall is

$$H_L = A \times U \times \Delta T \times M$$

$$= 1.0 \text{ sq ft} \times 0.12(72°\text{F} - 21°\text{F})\,1.0 = 6.12 \text{ Btu/hr/sq ft}$$

and for 1 sq ft of the masonry wall,

$$H_L = 1.0 \text{ sq ft} \times 0.12(72°\text{F} - 21°\text{F})\,0.84 = 5.14 \text{ Btu/hr/sq ft}$$

The thermal inertia of the heavy wall accounts for an approximate 16% decrease in heat loss. This reduced figure more accurately predicts performance of the masonry wall when compared to actual measured results in field tests.

2. *Effective U-values:* The steady-state U-values for both walls are assumed to be 0.12. To find the effective U-values, the formula $U_a = U(M)$ may be used for the frame wall:

$$U_a = 0.12(1.0) = 0.12$$

and the masonry wall:

$$U_a = 0.12(0.84) = 0.101$$

The better U-value for the masonry wall more accurately predicts its actua performance.

3. *Effective R-values:* The steady-state R-values for the two walls ar assumed to be 8.33. This can be adjusted for the effect of thermal inerti by the equation $R_a = R/M$. For the frame wall, the effective resistanc would be

$$R_a = \frac{8.33}{1.0} = 8.33$$

and for the masonry wall,

$$R_a = \frac{8.33}{0.84} = 9.91$$

This means simply that the thermal inertia of the masonry wall increas its effective R-value to a higher resistance of 9.91.

The M-factor is a simple means of quantifying the effect of therm inertia on heat-loss calculations without the aid of a computer. It permits more accurate prediction of dynamic thermal performance than stead state methods. The M-factors in *Fig. 8-24* are deliberately conservative. very cold climates, they give a credit of about 10% to a heavy wall, where t more detailed computer calculations indicate a much greater benefit. T results of some computer calculations for various wall weights are shown

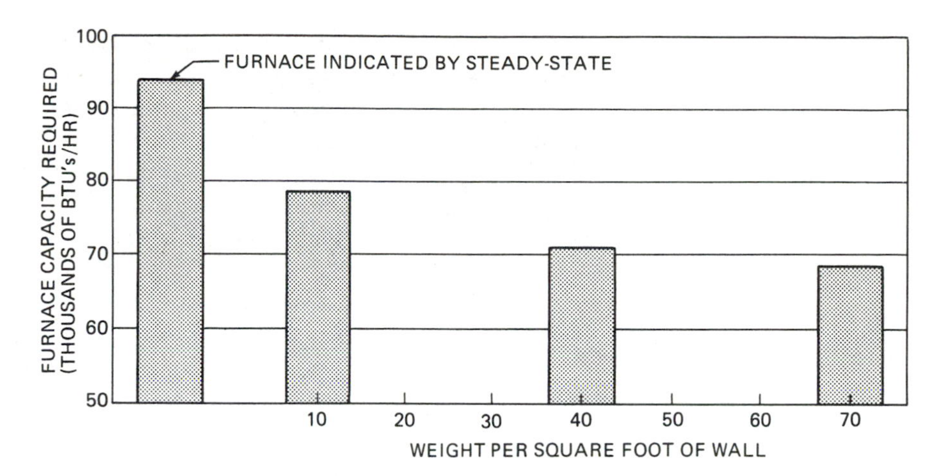

Size of furnace required for the heating load is reduced as the mass of the wall increases. (*From National Concrete Masonry Association*, TEK Bulletin 82, *NCMA, Herndon, Va.*)

Fig. 8-26. The difference between the static and dynamic methods was approximately 20% for the lightweight structure, and about 30% for the heaviest wall. The relationships of heating load to wall weight determined in this and other studies appear to validate the accuracy of the *M*-factor concept.

8.5 ADDED INSULATION

The thermal performance of masonry walls and their resistance to heat flow can be further improved by adding insulation (*see Fig. 8-27*). In severe winter climates where diurnal temperature cycles are of minimum ampli-

Insulation materials often used in masonry construction. (*From Brick Institute of America*, Technical Note 21A, *BIA, McLean, Va.*)

	*Properties**				
Material	*Density* *(lb/ft^3)*	*Thermal conductivity, k* *(Btu/hr/ft^2/°F)*	*Thermal resistance, R (per in.)* *(Btu/hr/ft^2/°F)*	*Permeability* *(perm-in.)*	*Vapor resistance, [ft^2/hr (in. Hg)]*
Granular fills					
Vermiculite (expanded)	5–9	0.44	2.27	62†	0.0163†
Perlite (expanded)	5–8	0.37	2.70	N.A.‡	N.A.‡
Rigid boards					
Expanded polystyrene, extruded	3.5	0.19	5.26	1.2	0.8333
Expanded polystyrene, molded beads	0.9–1.1	0.28	3.57	2.0–5.8	0.5–0.1724
Expanded polyurethane, extruded	1.5	0.16	6.25	0.4–1.6	2.5–0.625
Perlite aggregate	11	0.38	2.63	25	0.04
Rigid urethane	2	0.16§	6.25§	2	0.50
Cellular glass	9	0.35–0.44¶	2.86–2.44¶	0	Very high
Preformed fiberglass	4–9	0.21–0.26	4.76–3.86	Very high	Very low

Tabulated values are from varied sources. Designers should check with manufacturers and other sources for more precise values.

Material thickness is 2.5 in.

N.A., not available.

Based on *aged k*-factor.

From 0 to 90°F.

tude, the thermal inertia of brick and block walls can be complemented by the use of resistance insulation such as loose fill or rigid board materials. Hollow units can easily be insulated with loose fill or granular materials. Multi-wythe cavity walls and veneer walls over wood or metal frame construction have air spaces for granular fills or for rigid insulating boards. The proper selection of insulating materials for masonry walls depends on more than just thermal performance.

1. The insulation must not interfere with proper cavity wall drainage.
2. Thermal insulating efficiency must not be impaired by retained moisture from any source (e.g., wind-driven rain or vapor condensation within the cavity).
3. Granular fill materials must be able to support their own weight without settlement, to assure that no portion of the wall is without insulation.
4. Insulating materials must be inorganic, or be resistant to rot, fire, and vermin.
5. Granular insulating materials must be "pourable" in lifts of at least 4 ft for practical installation.

8.5.1 Granular Fills

Two types of granular fill insulation have been tested by researchers at the Brick Institute of America and found to comply with these criteria: *water repellent vermiculite* and *silicone-treated perlite* fill.

Vermiculite is an inert, lightweight, insulating material made from aluminum silicate expanded into cellular granules about 15 times thei original size. Perlite is a white, inert, lightweight granular insulating material made from volcanic siliceous rock expanded up to 20 times its original volume. Specifications for water-repellent vermiculite and silicone treated perlite are published by the Vermiculite Association and the Perlit Institute, Inc. Each of these specifications contains limits on density grading, thermal conductivity, and water repellency.

Cavity wall construction permits natural drainage of moisture o condensation. If insulating materials absorb excessive amounts of wate the cavity can no longer drain effectively. The insulation will act as vehicle by which moisture is transmitted across the cavity to the interic wythe. Untreated vermiculite and perlite will accumulate moisture.

Loose fill insulation is usually poured directly into the cavity from th bag or from a hopper placed on top of the wall. Pours can be made at an convenient interval, but the height of any pour should not exceed 20 f Rodding or tamping is not necessary and may in fact reduce the therma resistance of the material. The insulation in the wall should be protecte from weather during construction. Weep holes should be screened to preve the granules from leaking out or plugging the drainage path.

8.5.2 Rigid Board Insulation

Polystyrene, polyurethane, rigid urethane, cellular glass, preformed fib glass, and perlite board can all be used as insulating materials in cavi walls. They are classified physically as cellular or fibrous. Cellular insu tion includes polystyrene and polyurethane of open- and closed-cell cc struction. Fibrous insulation materials include fiberboards of wood, cane, vegetable fibers bonded with plastic binders. To make them moisture resistant, they are sometimes impregnated with asphalt. Fibrous glass insulation is made of nonabsorbent fibers formed into boards with phenolic binders, and surfaced with asphalt-saturated, fiberglass-reinforced material.

Water trapped in insulation can destroy its thermal insulating value. Water vapor can flow wherever air can flow—between fibers, through interconnected open cells, or where a closed-cell structure breaks down. With mean wall temperatures alternating above and below freezing, ice formation can break down the walls between cells. Repeated freeze-thaw cycles can progressively destroy closed cell insulation materials. Unimpaired cavity wall drainage is therefore important in the protection of the insulation board as well as the moisture protection of the interior spaces.

Some concrete block manufacturers produce units with built-in insulation installed at the plant prior to shipment. These inserts may be of polystyrene or polyurethane, and vary in shape and design for different proprietary products (*see Fig. 8-28*). Generally available in 6-, 8-, and 10-in. widths, these special units permit single-wythe construction of insulated walls with exposed masonry surfaces on each side, and better *U*-values than with loose fill insulation.

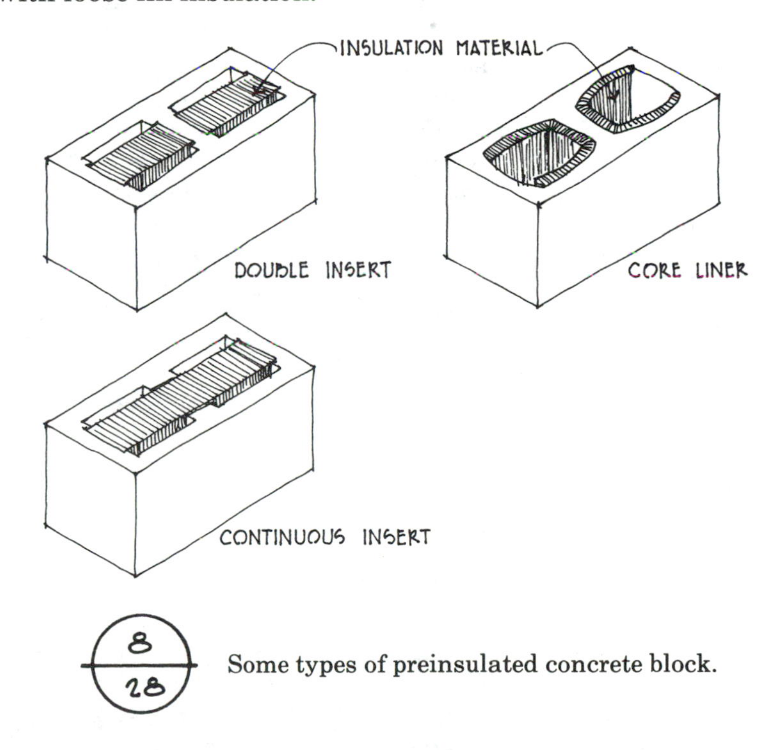

8-28 Some types of preinsulated concrete block.

There are probably as many different installation procedures as there are types of rigid board insulation. Generally however, rigid insulation is installed against the cavity face of the backup wythe. A minimum of 1 in. should be left between the cavity face of the exterior wythe and the insulation board to facilitate construction and allow for drainage of the cavity. Mechanical and/or adhesive attachment as recommended by the manufacturer is used to hold the insulation in place.

8.5.3 Foams

Urea-formaldehyde-based foams are not recommended for use in masonry walls.

8.5.4 Vapor Barriers

Under certain conditions of design, it may be necessary to add a vapor barrier to an insulated cavity wall to retard or stop the flow of water vapor. An acceptable barrier is one with a moisture vapor permeance of less than 1 perm. Vapor barriers may be in the form of bituminous materials, continuous polyethylene films, and so on. They may be attached to the insulation as part of the fabricated product, or they may be incorporated separately in or on the wall. For greatest effectiveness, barriers must be continuous and without openings or leaks through which vapor might pass. (See Chapter 14 for additional information regarding moisture control.)

8.5.5 Insulation Location

The most effective thermal use of massive construction materials is to store and reradiate heat. This means that insulation should be on the outside of the wall. *Figure 8-29* shows that the location of the insulation within the wall section has an effect on heat flux through the wall that is not accounted for by standard U-value calculations. In the thermal research conducted by the NBS and the NCMA, the effects of variable insulation location were studied. It was found that indoor temperatures were reduced by half when insulation was placed on the outside rather than the inside of the wall, and that the thermal storage capacity of the masonry was maximized. In cavity walls, performance is improved if the insulation is placed in the cavity rather than on the inside surface.

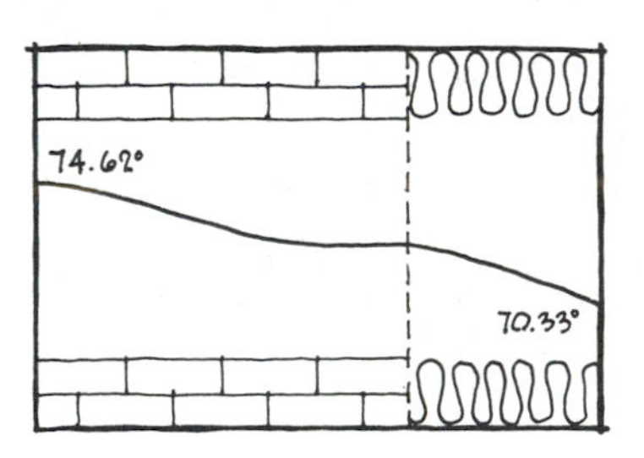

Francisco Arumi's temperature profiles for three different insulation locations. (*From* Energy Conservation through Building Design *edited by Donald Watson. Copyright 1979 by McGraw-Hill, Inc. Used with the permission of the publisher.*)

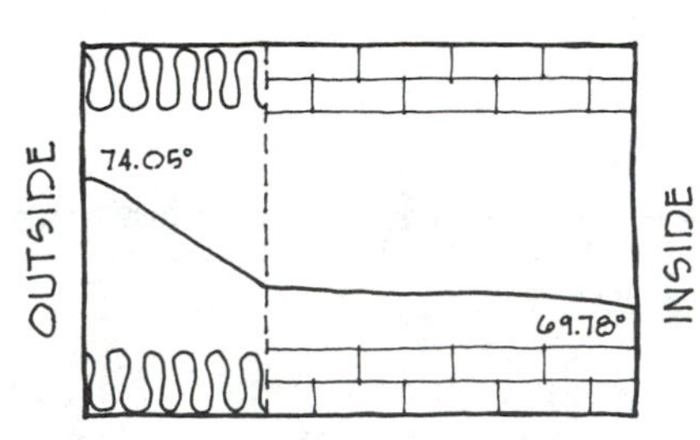

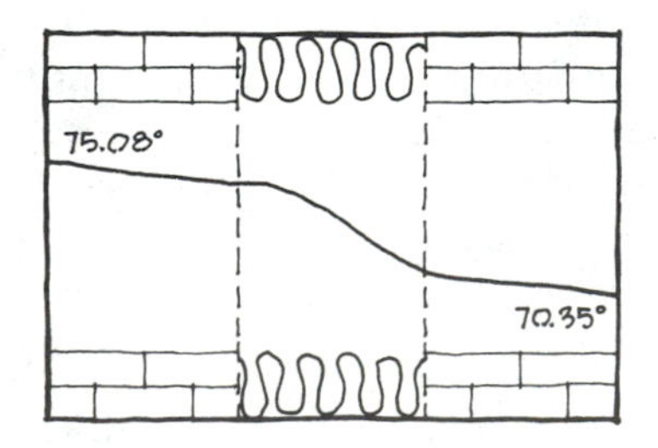

8.6 PASSIVE SOLAR DESIGN

Masonry construction can be used in several ways with passive solar design. It can provide a solar screen to shade glass areas on a facade, collect and distribute solar warmth in winter, and intercept excessive heat and radiation during the summer. In passive solar design, the buildings themselves collect, store, and distribute heat. A key element is the use of thermal mass—heavy materials that can absorb and reradiate large amounts of energy. Passive measures such as cross ventilation, evaporation, exhaustion of hot air by convection, and absorption of heat by thermal mass can provide up to 100% of a building's cooling needs in summer. Masonry is particularly cost effective in these applications because it simultaneously provides supporting structure, spatial definition, acoustical separation, fireproof construction, finished surfaces, and thermal storage.

Solar energy systems for buildings are divided into two categories: active and passive. Active systems generally use either flat-plate or concentrated collectors, heat storage tanks, pumps, heat exchangers, and extensive plumbing and electrical controls. Buildings need take no special form, and although building orientation is important, it need not be critical since solar collectors can often be oriented for optimum performance whatever the building's orientation. Passive buildings, on the other hand, *must* be oriented in relation to the seasonal and daily movements of the sun to maximize heat gain in the winter and to minimize solar loads in the summer. Solar heat gain through walls, windows, roofs, skylights, and other building elements can dramatically reduce winter energy requirements. If thermal energy flow is by natural means, such as radiation, conduction, and natural convection, and if solar energy contributes a significant portion of the total heating requirement, the building is considered a passive, solar-heated structure.

8.6.1 Shading Devices

Solar heat gain through windows can be as much as three times greater than heat loss because direct radiation is instantaneously transmitted to the building interior. The incident solar radiation received by a vertical surface often exceeds 200 Btu/hr/sq ft, and the annual operating cost of air-conditioning equipment attributed to each square foot of glass is considerable.

The desirability of direct solar heat is evaluated quite differently depending on location, climate, orientation, and time of day. Hot, arid regions generally require exclusion of solar radiation to prevent overheating, excessive air-conditioning loads, glare, or deterioration of materials. In other circumstances, it may be more desirable to ensure adequate sunlight, either for heat or purely for its psychological effect.

If sun control is necessary, the most efficient means is through the use of external shading devices. ASHRAE data indicate that exterior shading devices can reduce the instantaneous rate of heat gain by as much as 85%. Different orientations require different types of shading devices. Horizontal projections or overhangs work best on southerly orientations. Vertical fins are of little value on southern exposures, where the sun is high at midday. For easterly and westerly orientations, however, vertical fins work well. Horizontal elements are of little value here because low morning and afternoon sun altitudes negate their effect. Combination horizontal/vertical egg-crate devices work well on walls facing southeast, and are particularly effective for southwest orientations. Considered by some to give the best

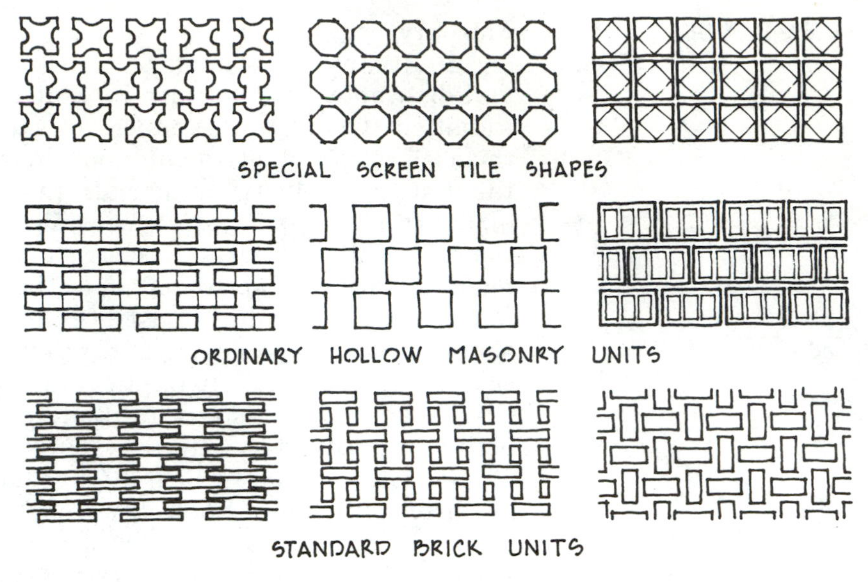

Variations of masonry solar screen designs. (*From Brick Institute of America*, Technical Note, *Vol. 11, No. 11, BIA, McLean, Va.*)

"all-around" shading, the egg-crate patterns are most advantageous in hot climates. Their high shading ratio and low winter heat admission can be undesirable in colder regions.

Clay or concrete masonry screens can be assembled in many decorative patterns, with either standard or custom units. Their shading characteristics are all of the egg-crate type (*see Fig. 8-30*). Masonry screens can be constructed in stack bond, running bond, or split bond (where the individual units are separated horizontally and the wall contains no vertical mortar joints). Standard concrete block or clay tile can be laid with cores perpendicular to the wall surface to create screen effects, or decorative units made expressly for this purpose can be used. Solid brick can be laid in split bond to give open screen patterns of various designs. The overall texture and appearance of the wall is affected by the size and shape of the units as well as the pattern in which they are assembled. Both glazed and unglazed units are available in a variety of colors. Lighter colors provide brighter interior spaces because of greater reflectivity. Darker colors reflect less light (*see Fig. 8-31*). Depending on orientation and latitude, small screen patterns can exclude much or all of the direct sun load.

The degree of shading provided by a masonry solar screen is a function of the shape, dimensions, and orientation of the openings. Standard sun path diagrams and shading masks can be used to compute time-shade cycles for openings of any shape, or to custom design a screen for a specific latitude and orientation. (See Ramsey and Sleeper's *Architectural Graphic Standards* and Olgyay and Olgyay's *Solar Control and Shading Devices* for more detailed information on solar screens.*)

Masonry screens can be used to reduce heat gain economically when building orientation cannot be easily adjusted. They can also be retrofitted to existing buildings to substantially reduce air-conditioning loads and lower overall energy consumption.

*Major references for which complete publishing information is not given in the text can be found in the Bibliography at the end of the book.

Material	Light reflectivity (%)
Unglazed clay masonry	
Cream manganese spot	52
Cream	50
Light buff	43
Light gray	40
Gray manganese spot	40
Golden buff	35
Red	30
Dark red	23
Ceramic glazed clay masonry	
White	83
Ivory	67
Sunlight yellow	65
White mottle	64
Coral	58
Cream glazed	51
Light gray	49
Green mottle	49
Cream mottle	49
Light green	46
Cream-tone salt glazed	44
Gray mottle	41
Ocular green	37
Tan	37
Blue	35
Buff-tone salt glazed	27
Black	5

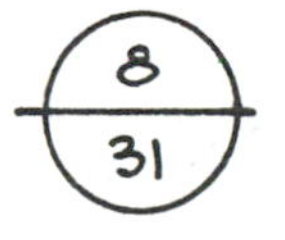

Reflectivity of colors. (*From Brick Institute of America,* Technical Note, *Vol. 11, No. 11, BIA, McLean, Va.*)

8.6.2 Direct-Gain Solar Heating

The simplest method of solar heating is *direct gain.* If a building is constructed of lightweight materials, solar radiation will heat its low thermal mass quickly and raise inside air temperatures above comfortable levels. At night, these buildings lose their heat just as rapidly, causing temperatures to drop again. Better designs allow sunlight to strike materials of high thermal mass that can store the heat and reradiate it at a later time (*see Fig. 8-32*). Contemporary materials include poured and precast concrete as well as masonry. When these materials with high heat storage capacity are used for walls, floors, and even ceilings, performance and efficiency is increased because the ratio of surface area to volume of mass is maximized.

8.6.3 Thermal Storage Walls

Another type of passive solar heating system uses the thermal storage wall concept. A glass-covered masonry or concrete wall is heated by direct radiation, stores the heat, and then reradiates it to the interior spaces. The

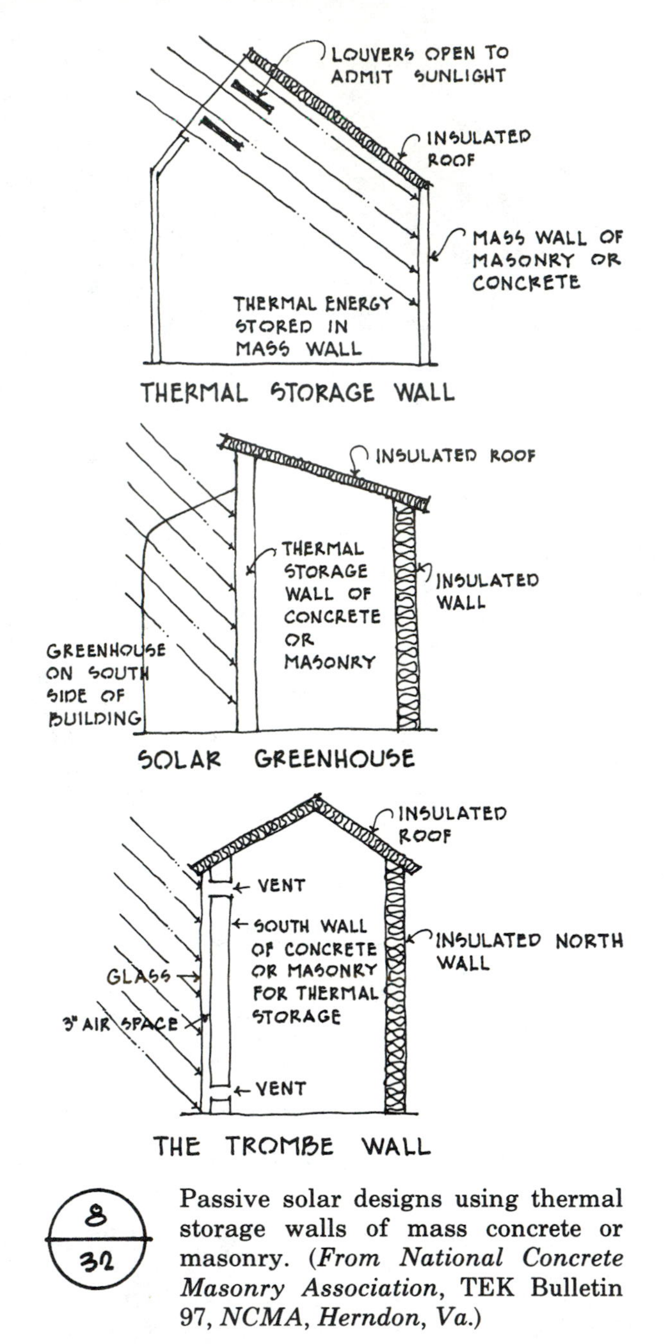

8/32 Passive solar designs using thermal storage walls of mass concrete or masonry. (*From National Concrete Masonry Association*, TEK Bulletin 97, *NCMA, Herndon, Va.*)

glass traps solar energy through a greenhouse effect. Sunlight strikes th mass wall, is converted to thermal energy, and stored. The storage mas becomes a radiant heat source, and creates natural convection currents tha help to distribute the heat. Buildings are most efficient when the glass are and thermal mass are properly sized and oriented for optimum exposur and are protected from heat loss by movable insulating panels or louver Thermal storage wall systems have much less temperature fluctuation tha do direct-gain systems, but do not usually achieve the same high initi interior temperatures.

Thermal mass walls have been built of concrete, water-filled contai ers, and masonry. The concept has been studied extensively in Europe ar the United States during the past 10 to 15 years. Proper design requir consideration of several factors, including heat capacity and thermal conductance. There must be sufficient heat storage capacity to absorb the solar energy entering the building through windows or skylights. The conductance of the wall must be such that it can store heat for a desired length of time and then release it into the room as required. When a mass storage wall is placed directly in front of the glass, optimum thermal conductance/heat storage capacity relationships must be achieved for efficient operation. If conductance values are too high, the wall loses too much heat during the "charging period," reducing the total amount of stored heat. If the conductance is too low, the wall reaches such high temperatures during the charging period that some heat is reradiated back through the glass and lost to the exterior. Several independent studies have indicated that the optimum thickness of a masonry thermal storage wall is 12 to 18 in. Heat storage capacity should be equivalent to 30 Btu/°F/sq ft of glass, or about 150 lb of masonry for every square foot of glazing.

Storage walls should be positioned to receive maximum solar radiation by direct exposure. South-facing walls are usually most effective. Special treatment, such as a dark exterior surface, roughness, grooves, or ribs, can improve convective and radiant heat exchange. At night, insulated louvers or movable insulated panels should be closed to prevent heat loss and to permit the wall to radiate heat into the interior.

8.6.4 The Trombe Wall

The most widely used type of massive thermal storage wall was developed by Felix Trombe in a series of experimental houses built in Odeillo, France. The Trombe wall consists of a glass-enclosed, heavy masonry wall painted black on its exterior south side to absorb the maximum amount of solar radiation. The glass and the wall are separated by a 3-in. air space. During the heating season, this air space is connected to the room by vents at the top and bottom of the wall. The heated air circulates into the room by thermal buoyancy currents. For summer operation, external vents are provided and the internal vents are closed (*see Fig. 8-33*). External venting may sometimes be combined with external insulation to provide additional cooling for the air space in front of the wall. Venting the wall to the interior will reduce temperature fluctuations and increase the maximum temperature reached in the living space. Vents with automatic or manual closures should be used so that the system does not reverse itself at night and create a heat loss. If controlled vents are not installed, movable insulation is essential to prevent heat losses at night.

In the earliest Trombe wall house, built in 1967, the wall thickness was approximately 2 ft. The primary heating is by radiation of heat passing through the wall, and by convection along its interior surface. About 30% of the energy is provided by a thermocirculation path which operates during the day by natural convection through the vents. The data shown in *Fig. 8-34* were taken over a 4-day period in December 1974 with the ambient temperature above freezing and variations in solar radiation ranging from sunny to cloudy conditions. The outside surface of the masonry wall (covered with one layer of glass) rose to 140-150°F during the day. The inside wall surface remained fairly constant at 85°F and provided radiant and convective heat to the room throughout the diurnal cycle as well as the alternately cloudy and sunny days. Additional data collected over a period of 1 year showed that about 70% of the total thermal energy required by the building was provided by passive solar collection in the masonry wall.

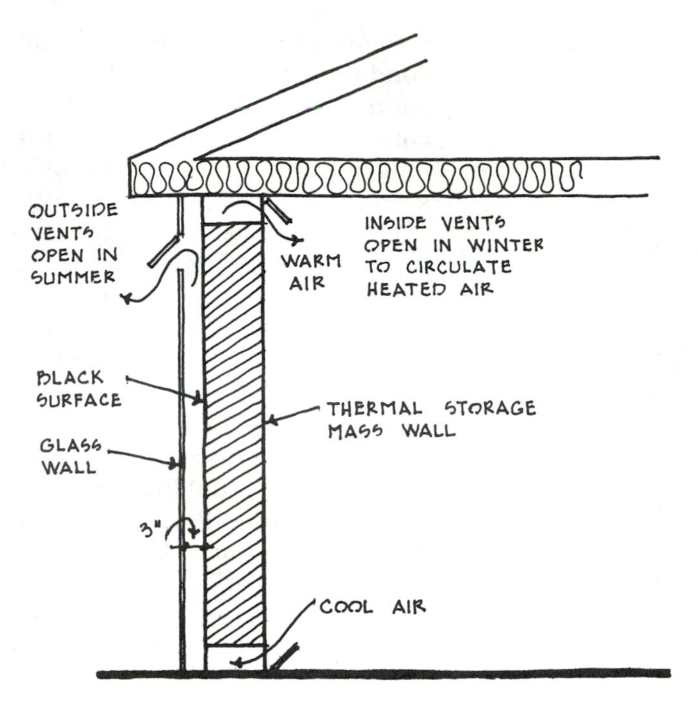

Detail of the Trombe wall. (*From S. V. Szokolay,* Solar Energy and Building, *The Architectural Press, Ltd., London, 1975.*)

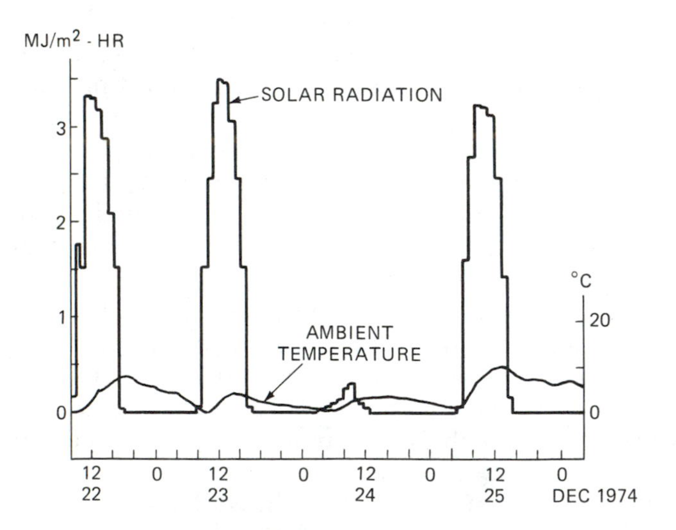

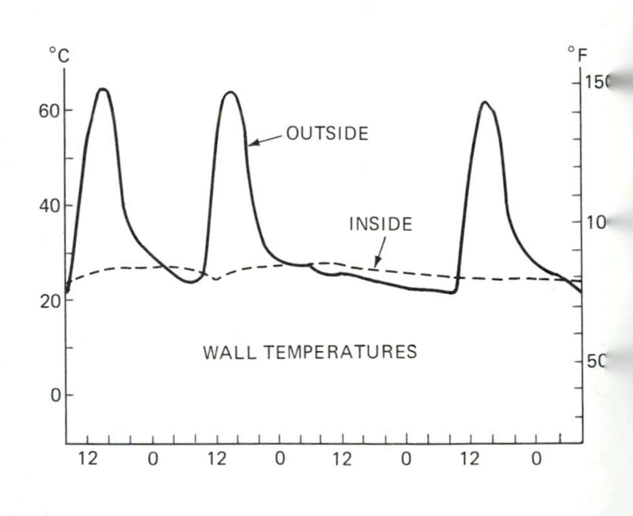

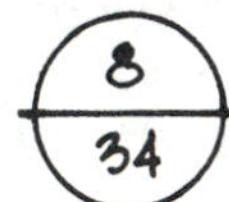

Data recorded on a test of the Trombe wall. (*From* Energy Conservation through Building Design *edited by Donald Watson. Copyright 1979 by McGraw-Hill, Inc. Used with the permission of the publisher.*)

Computer simulations were performed at the Los Alamos Scientifi Laboratory to test optimum wall thickness and performance characteristic under varying outdoor conditions. The graphs in *Fig. 8-35* show that th daily fluctuations felt on the inside wall surface are considerably differen for the 6-, 12-, and 24-in. thicknesses. However, the net annual therma contribution of the three walls was not markedly different. The 12-in. wal had the best overall performance, giving an annual solar heating contribu tion of 68%. Wall thickness was more important in reducing indoo

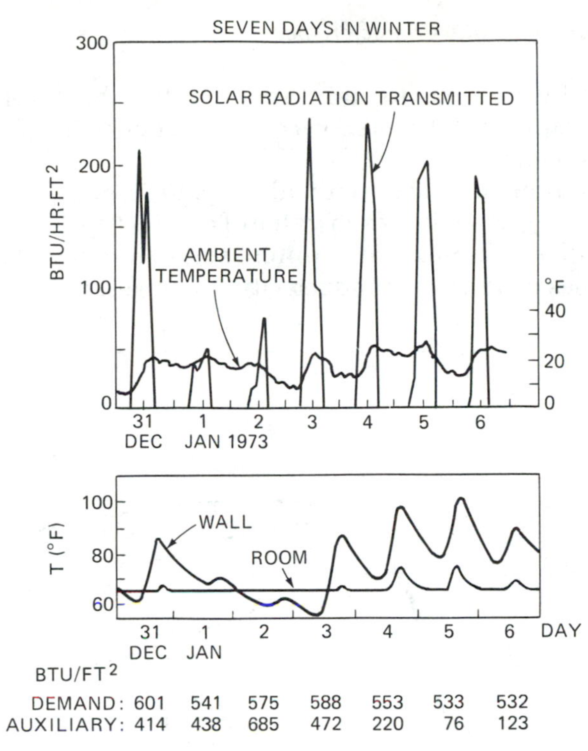

BTU/FT²

DEMAND:	601	541	575	588	553	533	532
AUXILIARY:	414	438	685	472	220	76	123

(A) TEMPERATURES TESTED DURING TYPICAL WINTER CONDITIONS

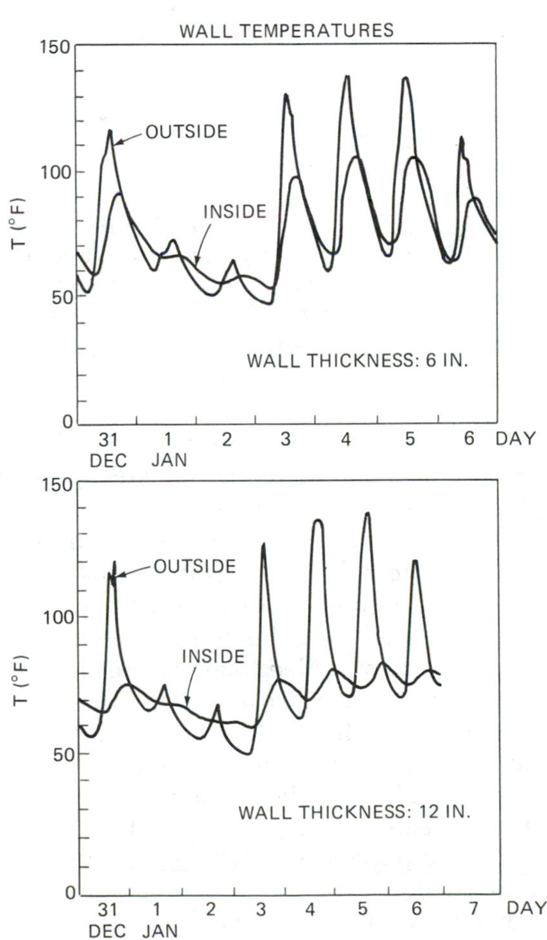

(B) WALL TEMPERATURE VARIATIONS FOR 6 IN., 12 IN., AND 24 IN. THICKNESS

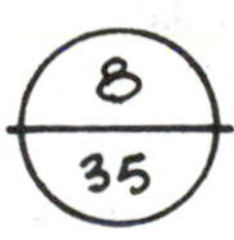

Simulation tests of thermal storage walls. (*From* Energy Conservation through Building Design *edited by Donald Watson. Copyright 1979 by McGraw-Hill, Inc. Used with the permission of the publisher.)*

temperature fluctuations to maintain a comfortable level. Calculations for several geographic locations showed that the optimum 12-in. thickness does not depend on climate.

The relationships between wall thickness and thermal conductivity are shown in *Fig. 8-36*. For each value of conductivity, there is a wall thickness that will give a maximum annual solar energy yield. Optimum thickness decreases as thermal conductivity decreases.

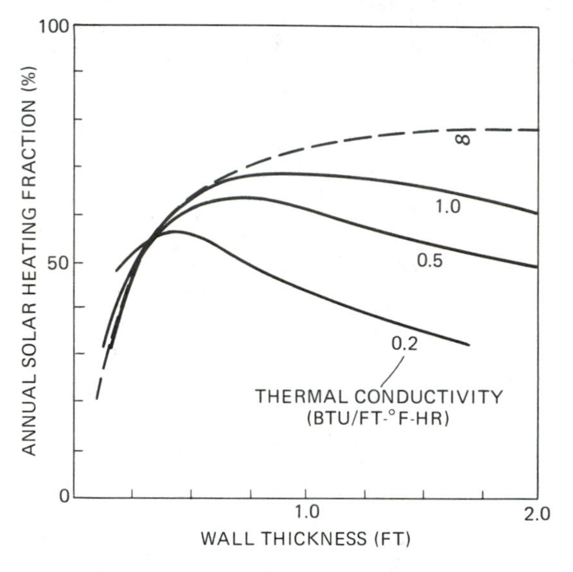

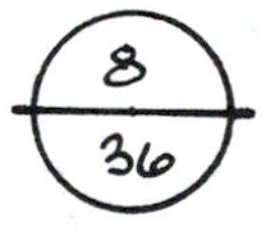

Effect of wall thickness and glass area on the performance of the system. (*From* Energy Conservation through Building Design *edited by Donald Watson. Copyright 1979 by McGraw-Hill, Inc. Used with the permission of the publisher.*)

At the University of Texas Numerical Simulation Laboratory, Fran cisco Arumi conducted tests to identify the design parameters that woulc make a Trombe wall an economical heating supplement during the centra Texas winter without adding to the cooling demand during hot summe months. The results show that for a statistically typical January day (*se Fig. 8-37*), most of the needed heat can be provided by a masonry therma storage wall approximately equal in area to the floor area of the space to b heated if open vents are provided to allow thermocirculation of heated ai (*see Fig. 8-38*). For summer operation, the data show that the wall can b designed with variables which not only prevent additional heat transfe and increased cooling loads, but in fact reduce the required energ consumption. If the wall is shaded in summer and externally vented, it ca curtail daily energy requirements by as much as 50% even without the use o additional insulating panels. This study also identified an optimum wal thickness of about 1 ft.

The best thermal performance can be obtained by combining the direc gain and thermal storage wall systems. A south-facing wall can be designe with sections of brick or solid concrete masonry alternating with window protected from summer radiation by overhangs or shading devices. Thi combination (1) permits some direct sunlight to enter and warm the interic floor and wall elements, (2) achieves higher interior temperatures than th

Location	*Heating degree-days*	*Latitude*	*Solar heating* (Btu/ft²)*	*Solar heating fraction (%)*
Los Alamos, NM	7,350	35.8	60,200	56.5
El Paso, TX	2,496	31.8	50,000	97.5
Ft. Worth, TX	2,467	32.8	38,200	80.8
Madison, WI	7,838	43.0	44,900	41.6
Albuquerque, NM	4,253	35.0	63,600	84.1
Phoenix, AZ	1,278	35.5	38,300	99.0
Lake Charles, LA	1,694	30.1	34,300	90.5
Fresno, CA	2,622	36.8	43,200	83.3
Medford, OR	5,275	42.3	47,400	56.1
Bismarck, ND	8,238	46.8	53,900	46.4
New York, NY	5,254	40.6	48,000	60.2
Tallahassee, FL	1,788	30.3	40,700	97.3
Dodge City, KS	5,199	37.8	58,900	71.8
Nashville, TN	3,805	36.1	39,500	65.2
Santa Maria, CA	3,065	34.8	69,800	97.9
Boston, MA	5,535	42.3	47,100	56.8
Charleston, SC	2,279	32.8	47,900	89.3
Los Angeles, CA	1,700	34.0	53,700	99.9
Seattle, WA	5,204	47.5	42,400	52.2
Lincoln, NE	5,995	40.8	53,500	59.1
Boulder, CO	5,671	40.0	62,500	70.0
Vancouver, BC	5,904	49.1	46,000	52.7
Edmonton, Alb.	11,679	53.5	37,700	24.7
Winnipeg, Man.	11,490	49.8	33,700	22.6
Ottawa, Ont.	8,838	45.3	37,900	31.9
Fredericton, NB	8,834	45.8	40,100	33.9
Hamburg, Germany	6,512	53.2	24,900	27.5
Denmark	6,843	56.0	43,100	43.8
Tokyo, Japan	3,287	34.6	50,300	85.8

*The values in the solar heating column are the net energy flow through the inner face of the wall into the building.

Note: Annual results for 29 locations. Trombe wall system (18 in. thick); thermal conductivity = 1 Btu/hr/ft²/°F; heat capacity = 30 Btu/ft³/°F; vent size = 0.074 ft² per foot of length (each vent); no reverse thermocirculation; load (U_1) = 0.5 Btu/hr/ft²/°F; temperature variation 65 to 75°F.

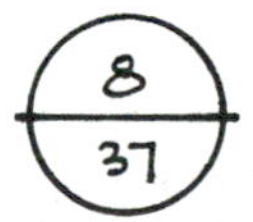

Solar heating contribution of the Trombe wall for various climatic conditions. (*From* Energy Conservation through Building Design *edited by Donald Watson. Copyright 1979 by McGraw-Hill, Inc. Used with the permission of the publisher.*)

thermal storage wall alone, (3) provides less temperature fluctuation than the direct-gain system alone, and (4) provides better distribution of natural light.

The results of geographic studies indicate that passive heating systems using masonry thermal storage walls can work effectively in all U.S. climates. Initial costs are much lower than those for active solar systems and can be recovered in a short period of time through reduced fuel bills and minimum maintenance expenditures. (For detailed passive solar design techniques, see the Los Alamos Scientific Laboratory's *Passive Solar Design Handbook*, Balcomb's *Passive Solar Design Handbook*, and Mazria's *The Passive Solar Energy Book*.)

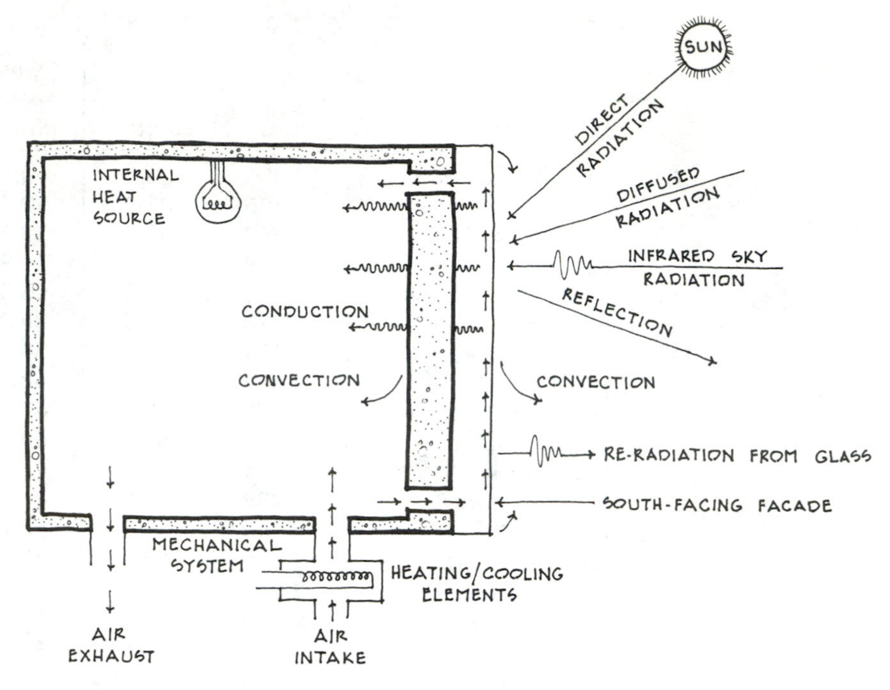

• THIS EXPERIMENTAL SPACE DOES NOT HAVE OPENING TO OUTSIDE
• INTERNAL HEAT SOURCE = 0

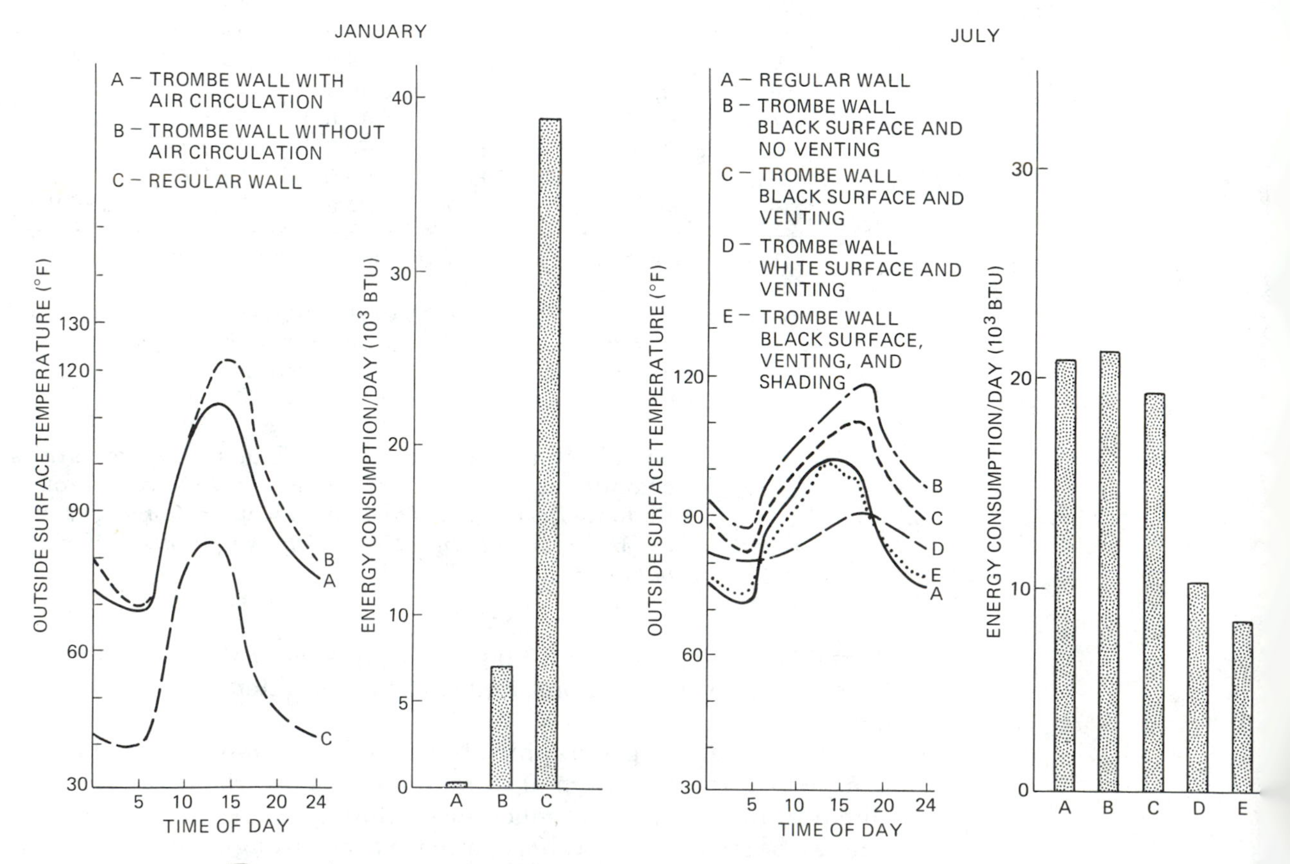

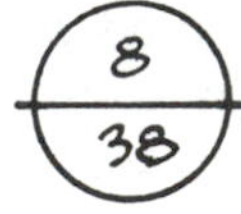

Arumi's tests for optimum summer/winter operation of the Trombe wall system in central Texas. (*From* Energy Conservation through Building Design *edited by Donald Watson. Copyright 1979 by McGraw-Hill, Inc. Used with the permission of the publisher.*)

8.7 ACOUSTICAL PROPERTIES

Environmental comfort in multifamily housing, hotels, office buildings, and private residences can be related as much to acoustical factors as to heating and cooling. Increased technology produces more and more noise sources at the same time when human perception of the need for privacy and quiet has become acute. Interior noise sources such as furnace fans, television sets, vacuum cleaners, video games, and washing machines combine with exterior street traffic, construction equipment, power mowers, and airplanes to create high levels of obtrusive sound. Noise generated by other people is also very aggravating to residents or tenants who can overhear conversation in adjoining rooms or apartments.

For noise that cannot be either eliminated or reduced, steps can be taken to absorb the sound or prevent its transmission through walls, floors, and ceilings. Some building codes now cover acoustical characteristics of construction assemblies. Clay and concrete masonry partitions have been tested and found to provide good sound insulation.

Noise is transmitted in several ways: (1) as airborne sound through open windows or doors, through cracks around doors, windows, water pipes, or conduits, or through ventilating ducts; (2) as airborne sound through walls and partitions; and (3) by vibration of the structure. Acoustical control includes absorbing the sound hitting a wall so that it will not reverberate, and preventing sound transmission through walls into adjoining spaces.

Sound absorption involves reducing the sound emanating from a source within a room by diminishing the sound level and changing its characteristics. Sound is absorbed through dissipation of the sound-wave energy. The extent of control depends on the efficiency of the room surfaces in absorbing rather than reflecting these energy waves. *Sound transmission* deals with sound traveling through barriers from one space into another. To prevent sound transmission, walls must have enough density to stop the energy waves. With insufficient mass, the sound energy will penetrate the wall and be heard beyond it.

8.7.1 Sound Ratings

There are two principal types of sound ratings: absorption and transmission loss. Sound absorption relates to the amount of airborne sound energy absorbed on the wall adjacent to the sound. Sound transmission loss is the total amount of airborne sound lost as it travels through a wall or floor. Each type may be identified at a particular frequency or by class (*see Fig. 8-39*). Sound absorption coefficients (SAC) and noise reduction coefficients (NRC) are measured in sabins, sound transmission loss (STL) in decibels. In both instances, the larger the number, the better the sound insulating quality of the wall.

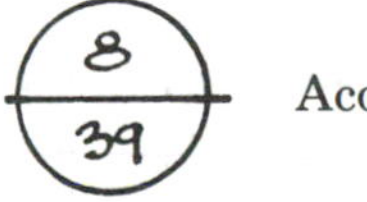

Acoustic ratings.

	Sound absorption	*Sound transmission*
Frequency rating	SAC	STL
Class rating	NRC	STC

8.7.2 Sound Absorption

Sound is absorbed by mechanically converting it to heat. A material that will absorb sound usefully must have a certain "flow resistance"—it must create a frictional drag on the energy of sound. Sound is absorbed by porous, open textured materials, by carpeting, furniture, draperies, or anything else in a room that resists the flow of sound and keeps it from bouncing around. If the room surfaces were capable of absorbing all sound generated within the room, they would have a *sound absorption coefficient* (SAC) of 1.0. If only 50% of it were absorbed, the coefficient would be 0.50.

The percentage of sound absorbed by a material depends not only on its surface characteristics, but also on the frequency of the sound. SAC values for most acoustical materials vary appreciably with sound frequencies. A better measure of sound absorption, which takes frequency variations into account, is the *noise reduction coefficient* (NRC), determined by averaging SAC values at different frequencies. Typical NRC values of various building materials and furnishings are given in *Fig. 8-40.*

Most materials utilized for strength and durability have low sound absorption. Masonry, wood, steel, and concrete all have low absorptions ranging from 2 to 8%. Dense brick and heavy weight concrete block will have 1 to 3%, while lightweight block may be as high as 5%. Painting the surface effectively closes the pores of the material and reduces its absorptive capability even further. Conventional masonry products absorb little sound because of their density and their highly impervious surfaces. Specially designed clay tile units combine relatively high sound absorption with low sound transmission characteristics with little or no sacrifice of strength o fire resistance. Most of these special units have a perforated face shell with the adjacent hollow cores filled at the factory with a fibrous glass pad Perforations may be circular or slotted, uniform or variable in size, and regular or random in pattern (*see Fig. 8-41*). Some proprietary units hav NRC ratings from 0.65 to as high as 0.95, depending on the area an arrangement of the perforations.

Sound absorption and sound reflection are directly related. If at a given frequency a particular material absorbs 75% of the incident sound, it wil reflect the remaining 25%. In acoustical design, sound reflection is just a important as absorption. If too much absorption is provided, or if it i concentrated, the result will tend to "deaden" sound. Too little absorptio will cause reverberation, or the persistence of sound within a room after th source has stopped. In excess, this is the principal defect associated wit poor acoustics. The optimum reverberation time, which varies with roo size and use, can be obtained by controlling the total sound absorptio within a room. Alternating areas of reflective and absorptive materials wi "liven" sound, promote greater diffusion, and provide better acoustic Special sound-absorbing masonry units can be alternated with conve tional units to achieve this effect.

8.7.3 Sound Transmission

Although it is an important element in control of unwanted noise, soun absorption cannot take the place of sound insulation or the prevention noise transmission through building elements. The NRC rating ranks wa systems only by sound absorption characteristics and does not give ar indication of effectiveness in the control of sound transmission.

Sound energy is transmitted to one side of a wall by air. The impact the successive sound waves on the wall sets it in motion like a diaphragı

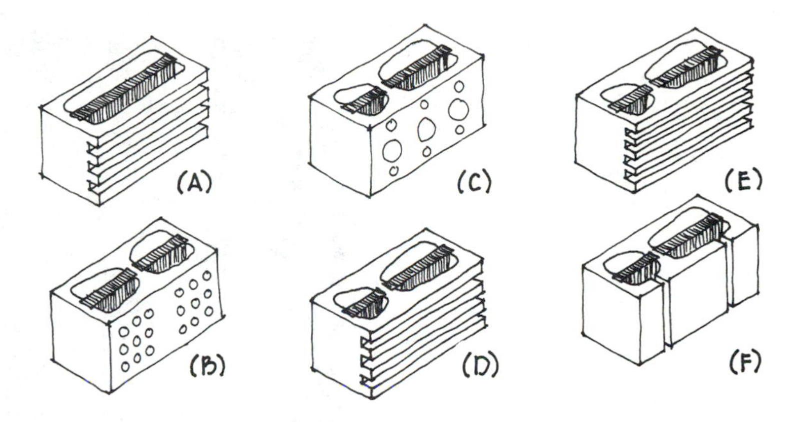

Material	*NRC*	*Material*	*NRC*
Brick, unglazed	0.04	Type C	0.85
Carpet		Type D	0.85
On concrete	0.30	Type E	0.95
On pad	0.55	Type F	0.70
CMU, lightweight		Concrete floor	0.01
Coarse texture	0.40	V.A. tile on concrete	0.03
Medium texture	0.45	Wood floor	0.08
Fine texture	0.50	Marble or glazed tile	0.01
CMU, normal weight		Single strength window glass	0.12
Coarse texture	0.26	Plate glass	0.04
Medium texture	0.27	Gypsum board on 2 × 4 framing	0.07
Fine texture	0.28	Gypsum board on concrete	0.03
Deduct for paint		Plaster on brick or CMU	0.03
All types, spray		Wood paneling on furring	0.13
1 coat	−10%	Draperies	
2 coats	−20%	Lightweight	0.14
Oil, brushed		Medium weight	0.40
1 coat	−20%	Heavy weight	0.55
2 coats	−55%	Furniture (values per sq ft)	
Latex, brushed		Bed	0.80
1 coat	−30%	Sofa	0.85
2 coats	−55%	Wood table, chairs, etc.	0.20
Sound-insulated block		Leather upholstered chair	0.50
Type A (see below)	0.65	Cloth upholstered chair	0.70
Type B	0.70		

Noise reduction coefficients (NRC) of various building materials and furnishings. (*From Brick Institute of America*, Technical Note, *Vol. 9, No. 5, BIA, McLean, Va.*)

Through this motion, energy is transmitted to the air on the opposite side. The amount of energy transmitted depends on the amplitude of vibration of the wall, which in turn depends on four things: (1) the frequency of the sound striking the surface, (2) the mass of the wall, (3) the stiffness of the wall, and (4) the method by which the edges of the wall are anchored. The *sound transmission loss* (STL) of a wall is a measure of its resistance to the passage of noise or sound from one side to the other. If a sound level of 80 dB is generated on one side and 30 dB measured on the other, the reduction in sound intensity is 50 dB. The wall therefore has a 50-dB STL rating. The higher the transmission loss of a wall, the better its performance as a sound barrier.

Some patterns of acoustical tile (back row and right). (*Photo courtesy Stark Ceramics, Inc.*)

8.7.4 STC Ratings

Until the early 1960s, the most common sound rating system was the arithmetic average of STL measurements at nine different frequencies. Heavy walls have a relatively uniform STL curve and are satisfactorily classified by this averaging method. However, lightweight partitions often have "acoustical holes" at critical frequencies (*see Fig. 8-42*). STL averages did not identify these deficiencies, and did not accurately translate acoustical test results into useful design data. *Sound transmission class* (STC) ratings were developed to describe acoustical characteristics more accurately. STC ratings represent the overall ability of an assembly to insulate against airborne noise. They have proven more reliable in classifying the performance of both heavy- and lightweight materials over a wide range of frequencies. The higher STC rating a wall has, the better the wall performs as a sound barrier.

For homogeneous walls, resistance to sound transmission increases with unit weight. When surfaces are impervious, sound may be transmitted only through diaphragm action. The greater a wall's inertia or resistance to vibration, the greater its ability to prevent the transfer of sound. The initial doubling of weight produces the greatest increase in transmission loss.

Porosity, as measured by air permeability, significantly reduces transmission loss through a wall. STC values vary inversely with porosity. Unpainted, open-textured CMU, for instance, will have lower STC values than would be expected on the basis of unit weight. Porosity can be reduced and STC values increased, by sealing the wall surface. The STC value increased by about 8% with one layer of gypsum board, 10% with two coats of paint or plaster, and 15% with two layers of gypsum board (*see Fig. 8-4

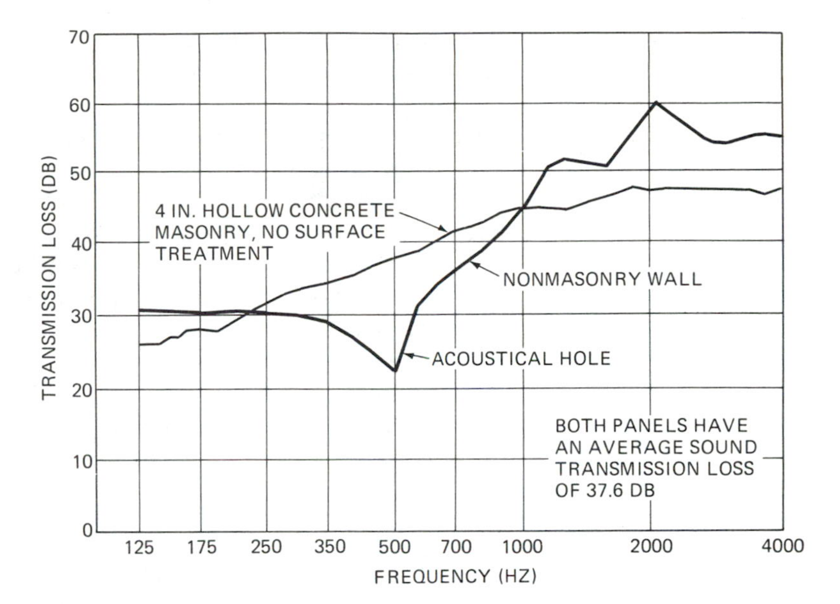

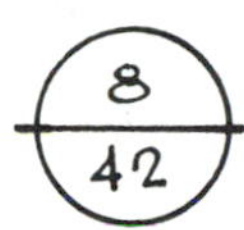

Graphic illustration of an "acoustical hole." (*From National Concrete Masonry Association*, TEK Bulletin 9, *NCMA, Herndon, Va.*)

STC	*Wall Description*
	Unpainted walls
39	4 in. structural clay tile
45	8 in. structural clay tile
45	4 in. face brick
50	8 in. composite face brick and structural clay tile
50	10 in. face brick cavity wall with 2-in. air space
51	6 in. brick wall
52	8 in. solid brick wall, double wythe
59	12 in. solid brick wall
59	10 in. reinforced brick wall
49	8 in. hollow lightweight CMU
48	8 in. hollow lightweight CMU, reinforced and fully grouted
51	8 in. composite brick and hollow lightweight CMU
52	8 in. normal-weight CMU
	Painted walls, two coats latex both sides
43	4 in. hollow lightweight CMU
44	4 in. hollow normal weight CMU
48	6 in. hollow normal weight CMU
55	8 in. lightweight CMU, reinforced and fully grouted
	Plastered walls
50	4 in. brick, $\frac{1}{2}$-in. plaster one side
53	8 in. composite brick and lightweight CMU, $\frac{1}{2}$-in. plaster both sides
56	8 in. grouted, reinforced, lightweight CMU, $\frac{1}{2}$-in. plaster both sides
	Walls with gypsum board on furring strips
53	8 in. solid brick, $\frac{1}{2}$-in. gypsum board on furring strips one side
47	4 in. hollow lightweight CMU, $\frac{1}{2}$-in. gypsum board both sides
48	4 in. normal weight CMU, $\frac{1}{2}$-in. gypsum board both sides
56	8 in. composite brick and lightweight CMU, $\frac{1}{2}$-in. gypsum board both sides
56	8 in. hollow lightweight CMU, $\frac{1}{2}$-in. gypsum board both sides
49	6 in. brick, $\frac{3}{8}$-in. gypsum board over 1-in. styrofoam insulation one side
59	10 in. cavity wall brick and lightweight CMU, $\frac{1}{2}$-in. gypsum board both sides
60	8 in. grouted, reinforced lightweight CMU, $\frac{1}{2}$-in. gypsum board both sides

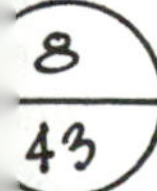

STC ratings of masonry walls.

Sealing both sides of a wall has little more effect than sealing only one side. A sealed surface not only decreases sound transmission, it also reduces sound absorption, which may not be desirable. As a general rule, leave porous surfaces unsealed in noisy areas such as stairwells or corridors, and seal them in living spaces.

Cavity walls have greater resistance to sound transmission than do solid walls of equal weight. Having two wythes separated by an air space interrupts the diaphragm action and improves sound loss. Up to about 24 in., the wider the air space, the more sound efficient the wall. Cavity walls are very effective where a high transmission loss on the order of 70 to 80 dB is required. If the wythes are only an inch or so apart, the transmission loss is less because of the coupling effect of the tightly enclosed air. For maximum benefit, the walls should be further isolated from one another.

8.7.5 Code Requirements

Some building codes now incorporate standards for sound transmission characteristics in buildings of residential occupancy. The standards specify minimum STC ratings for party wall and floor-ceiling separations between dwelling units.

Separating walls generally require an STC of 45 to 50. FHA minimum standards for multifamily housing are shown in *Fig. 8-44*. The required STC values range from a low of 40 to a high of 55. Requirements are lower in the city than in rural areas because higher background noise masks obtrusive sound, effectively raising the threshold of audibility.

Location of partition	*Low background noise*		*High background noise*	
	Bedroom adjacent to partition	*Other rooms adjacent to partition*	*Bedroom adjacent to partition*	*Other rooms adjacent to partition*
Living unit to living unit	50	45	45	40
Living unit to corridor	45	40	40	40
Living unit to public space (average noise)	50	50	45	45
Living unit to public space and service areas (high noise)	55	55	50	50
Bedrooms to other rooms within same living unit	45	—	40	—

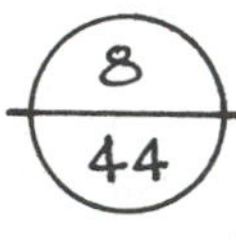

FHA requirements for STC limitations. (*From National Concr[ete] Masonry Association*, TEK Bulletin 39, *NCMA, Herndon, Va.*)

8.8 DIFFERENTIAL MOVEMENT

One of the principal causes of cracking in masonry walls is differenti[al] movement. All materials expand and contract with temperature change[s] but at vastly different rates. All materials change dimension due to stres[s] and some develop permanent deformations when subject to sustained load[s]. Clay masonry expands irreversibly with time, and concrete units shri[nk] and expand through wet-dry cycles. Modern masonry walls are thinner a[nd] more rigid than older construction, often at the expense of flexibility. Usi[ng] masonry with steel and concrete skeleton frames requires very caref[ul] consideration of the movement characteristics of each material. Cracki[ng] can result from restraining movement that originates within the material itself or from differential movement of adjoining materials.

8.8.1 Temperature Movement

Thermal movement can be estimated from the coefficients of thermal expansion which have been determined by laboratory test for most building materials. As shown in *Fig. 8-45*, the potential for expansion varies from the relatively stable characteristics of clay masonry to the highly active movements of metals and woods. Coefficients for masonry units vary with the raw material and type of aggregate used. The stress developed in a restrained element due to temperature change is equal to the *modulus of elasticity* × *coefficient of thermal expansion* × *mean wall temperature change*. For instance, the tensile stress in a fully restrained block wall with a thermal expansion coefficient of 4.5×10^{-6} and modulus of elasticity of 1.8×10^6 for a temperature change of 100°F would be

$$1{,}800{,}000 \times 0.0000045 \times 100°\text{F} = 810 \text{ psi}$$

8.8.2 Moisture Movement

Many building materials expand with increased moisture content and then shrink when drying. In some instances, moisture movement is almost fully reversible, but in others, it causes a permanent dimension change. Moisture shrinkage is of particular concern for concrete masonry units.

ASTM limits the moisture content of concrete masonry depending on the unit's linear shrinkage potential and the annual average relative humidity at the project site. The table in *Fig. 8-46* lists a range of acceptable linear shrinkage values. The object is to limit residual in-the-wall movement. Units with higher shrinkage potential must have a lower moisture content than that of low-shrinkage units. Type I units ensure minimum moisture movement and are recommended to control wall cracking. After units are delivered, it is important to protect on-site stockpiles from rain or other wetting.

Steel reinforcement increases concrete masonry's resistance to the tensile stress of shrinkage. The most common method of shrinkage crack control is the use of horizontal joint reinforcement. Joint reinforcement distributes the stress more evenly through the wall to minimize cracking. Control joints localize cracking so that waterproofing can be applied and moisture penetration prevented. Wall cracking is not usually a problem in masonry that is structurally reinforced to resist externally applied loads.

Clay masonry expands with moisture. The expansion of a wall approximates the sum of the expansions of the individual units. When combined with other types of movement and with the conflicting shrinkage of other materials, this expansion can be significant, and must be anticipated. Although researchers have not yet developed a means to predict accurately the moisture expansion potential of clay masonry, a tentative coefficient of 2×10^{-4} is recommended.

8.8.3 Elastic Deformation

The shortening of axially loaded masonry walls or columns is seldom critical. More often, problems can arise with elastic deformation of horizontal elements such as beams and lintels. Standards limit deflection to $\frac{1}{600}$ of the clear span or 0.3 in. under the combined live and dead load. In veneer

Material	*Average coefficient of lineal thermal expansion* $\times 10^{-6}$ *(in/°F)*	*Thermal expansion, [In. per 100 ft for 100°F temperature increase (to closest* $\frac{1}{16}$ *in.)]*
Clay masonry		
Clay or shale brick	3.6	0.43 ($\frac{7}{16}$)
Fire clay brick or tile	2.5	0.30 ($\frac{5}{16}$)
Clay or shale tile	3.3	0.40 ($\frac{3}{8}$)
Concrete masonry		
Dense aggregate	5.2	0.62 ($\frac{5}{8}$)
Cinder aggregate	3.1	0.37 ($\frac{3}{8}$)
Expanded-shale aggregate	4.3	0.52 ($\frac{1}{2}$)
Expanded-slag aggregate	4.6	0.55 ($\frac{9}{16}$)
Pumice or cinder aggregate	4.1	0.49 ($\frac{1}{2}$)
Stone		
Granite	4.7	0.56 ($\frac{9}{16}$)
Limestone	4.4	0.53 ($\frac{1}{2}$)
Marble	7.3	0.88 ($\frac{7}{8}$)
Concrete		
Gravel aggregate	6.0	0.72 ($\frac{3}{4}$)
Lightweight, structural	4.5	0.54 ($\frac{9}{16}$)
Metal		
Aluminum	12.8	1.54 ($1\frac{9}{16}$)
Bronze	10.1	1.21 ($1\frac{3}{16}$)
Stainless steel	9.6	1.15 ($1\frac{1}{8}$)
Structural steel	6.7	0.80 ($\frac{13}{16}$)
Wood, parallel to fiber		
Fir	2.1	0.25 ($\frac{1}{4}$)
Maple	3.6	0.43 ($\frac{7}{16}$)
Oak	2.7	0.32 ($\frac{5}{16}$)
Pine	3.6	0.43 ($\frac{7}{16}$)
Wood, perpendicular to fiber		
Fir	32.0	3.84 ($3\frac{13}{16}$)
Maple	27.0	3.24 ($3\frac{1}{4}$)
Oak	30.0	3.60 ($3\frac{5}{8}$)
Pine	19.0	2.28 ($2\frac{1}{4}$)
Plaster		
Gypsum aggregate	7.6	0.91 ($\frac{15}{16}$)
Perlite aggregate	5.2	0.62 ($\frac{5}{8}$)
Vermiculite aggregate	5.9	0.71 ($\frac{11}{16}$)

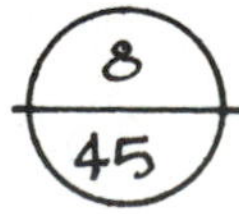

Thermal expansion coefficients of various building materials. (*From Brick Institute of America*, Technical Note 18, *BIA, McLean, Va.*)

Linear shrinkage, (%)	Moisture content, maximum % of total absorption (average of 3 units)		
	Humidity conditions at job site or point of use		
	Humid*	Interme-diate†	Arid‡
0.03 or less	45	40	35
From 0.03 to 0.045	40	35	30
0.045 to 0.065, max	35	30	25

*Average annual relative humidity above 75%.

†Average annual relative humidity 50 to 75%.

‡Average annual relative humidity less than 50%.

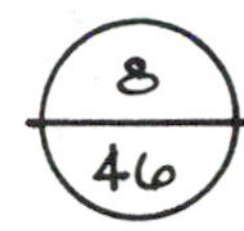

Linear shrinkage values of Type I concrete masonry units, from ASTM C90. (*Copyright, American Society for Testing and Materials, 1916 Race Street, Philadelphia, Pa. 19103. Reprinted, with permission.*)

construction, deformation of the structural frame to which the masonry is attached can also cause distress if unusually high loads are transferred to the veneer.

8.8.4 Plastic Flow

When concrete or steel is continuously stressed, there is a gradual yielding of the material, resulting in permanent deformations equal to or greater than elastic deformation. Under sustained stress this plastic flow, or creep, continues for years, but the rate decreases with time and eventually becomes so small as to be negligible. About one-fourth of the ultimate creep takes place within the first month or so, and one-half of the ultimate creep within the first year. In clay masonry construction, the units themselves are not subject to plastic flow, but the mortar joints are. In concrete block construction, long-term deformations in the mortar and grout are relatively high compared to that of the unit. Joint reinforcement restrains the mortar and grout so that overall deformations of the wall are similar in magnitude to those for cast-in-place concrete. Plastic flow of the mortar in brick walls helps prevent joint separations by compensating for the moisture expansion of the units. The creep deflection of a concrete or steel structural frame to which masonry is rigidly anchored is the most potentially damaging. Steel shelf angles and concrete ledges that sag over a period of time can exert tremendous force on the masonry below. Without pressure-relieving joints, the lower courses can buckle or spall under the load (see Chapters 9 and 14 for construction details).

8.8.5 Effects of Differential Movement

In a cavity wall, the temperature variation in the outer wythe is considerably greater than that of the inner wythe, especially if the wall contains insulation. Cracks can form at an external corner because of the greater relative expansion of the outer wythe. Long walls constructed without pressure-relieving joints develop shearing stresses in areas of minimum cross section. Diagonal cracks often occur between window and door openings, usually extending from the head or sill at the jamb of the opening. When masonry walls are built on concrete foundations that extend above grade, thermal and moisture expansion of the masonry can work against the drying shrinkage of the concrete, causing extension of the masonry wall beyond the corner of the foundation or cracking of the foundation. When the concrete contracts with lowering temperatures, the tensile strength of the masonry may not be sufficient to move the wall back with it, and cracks form in the masonry near the corners. Flashing often serves as a bond break between the foundation and the wall to allow each to move independently.

Brick parapet walls can be particularly troublesome because, with two surfaces exposed, they are subject to temperature and moisture extremes much greater than the building wall below. Differential expansion can cause parapets to bow, to crack horizontally at the roof line, and to overhang corners. Through-wall flashing, although often necessary, creates a plane of weakness at the roof line that may only amplify the problem. If parapets must be a part of the building design, some extra precautions are needed. Extend all expansion joints up through the parapet and space additional joints halfway between those running the full height of the building. Adding steel reinforcement helps to counteract the tensile forces created and prevent excessive movement. The same material should be used for the entire thickness of the parapet.

Where structural steel columns are protected by masonry, excessive temperature movement of the column can be inadvertently transmitted to the masonry and cause cracking. To prevent this problem, provide a positive bond break to isolate the mortar from the steel, and use flexible anchors that can accommodate the differential movement.

Floor and roof slabs poured directly on masonry bearing walls can curl from shrinkage, deflection, and plastic flow of the concrete. If the slab warps, it can rupture the masonry at the building corners and cause horizontal cracks just below the slabs. To permit flexibility, install a horizontal slip plane between the slab and wall running 12 to 15 ft back from the corners and terminating at a control joint. This will relieve the strain at the points where movement is greatest.

Vertical shortening of a steel or concrete structural frame due to shrinkage or contraction can transfer excessive stress to the masonry facing material. Failures are characterized by bowing, by horizontal cracks at shelf angles, by vertical cracks near corners, and by spalling of masonry units at window heads, shelf angles, and other points where stresses are concentrated. Flexible joint material must be intermittently provided in horizontal joints to alleviate these stresses and allow the frame to contract without damage to the masonry.

8.8.6 Movement Joints

Differential movement creates problems in masonry construction only if excessive stress is allowed to develop. Reinforcing steel, flexible anchorage, control joints, and pressure-relieving joints can counteract or relieve the stress. No single recommendation on the positioning and spacing of movement joints can be applicable to all structures. Each building design must be analyzed to determine the potential movements, and provisions made to relieve any excessive stress which might be expected to result from such movement. (Chapters 9 and 14 give some guidelines on joint location and construction.)

9

NON-LOADBEARING CONSTRUCTION

Masonry is used in many non-loadbearing applications as partitions, furring, fireproofing, and veneer. In some instances, the units support their own weight, the dead load of the units above, and wind loads. In other instances, they are adhered to solid backup materials.

Code requirements for nonbearing walls are based on standards and recommended procedures developed by the National Bureau of Standards, the American National Standards Institute (ANSI), the BIA, and the NCMA. Although there are some variations in detail, design for unreinforced masonry is generally governed by empirical lateral support requirements expressed as length or height-to-thickness (h/t) ratios, or by the flexural tensile stress between the mortar and unit. The *Uniform Building Code*, *National Building Code*, and *Basic Building Code* all require that support spacing for interior partitions not exceed 36 times the actual thickness including plaster, and the *Standard Building Code* requires a minimum of 45 times the nominal thickness exclusive of plaster. Minimum thickness requirements for interior partitions and exterior walls vary among the codes depending on certain conditions of construction and lateral support.

Lateral support can be provided by cross walls, columns, pilasters (*see Fig. 9-1*), or buttresses where the limiting span is measured horizontally; and by floors, roofs, or spandrel beams where the limiting span is measured vertically. Anchorage between walls and supports must be able to resist

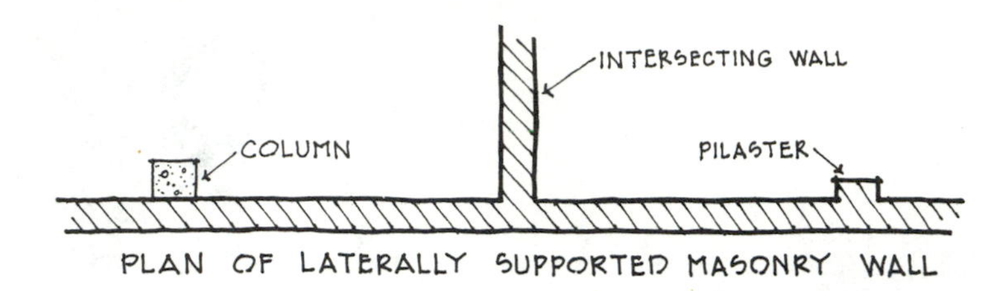

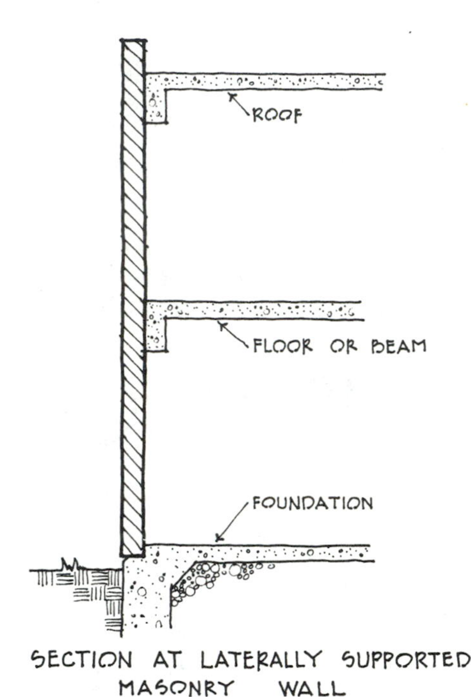

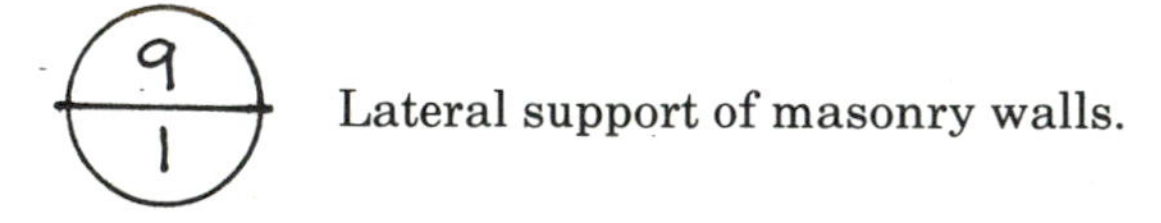

Lateral support of masonry walls.

wind and other horizontal forces acting either inward or outward. All latera support members must have sufficient strength and stability to transfe these forces to adjacent structural members or to the ground. All of the code contain provisions stating that specific limitations may be waived rational engineering analysis is provided to justify additional height width, reduced thickness, and so on. The addition of reinforcing steel relieve tensile stresses will also permit greater freedom in design (s chapters 10 and 11).

9.1 PARTITION WALLS

Partitions are interior, non-loadbearing walls one story or less in heigh that support no vertical load other than their own weight. They may separating elements between spaces, as well as fire, smoke, or soun barriers.

Based on a height/thickness ratio of 36, a single-wythe 4-in. *bri* partition without reinforcement is limited to a 12-ft span, while a 6-in. bri partition can span 18 ft between supports. If the partition is secure anchored at the floor and ceiling, and if the height does not exceed the dimensions, there is no requirement for intermediate walls, piers, pilasters. If additional height is required, an 8-in. through-the-wall brick c be used, or supporting walls, pilasters, etc. can be added at 12- or 18 intervals. Lateral bracing is required in only one direction and can be eith floor and ceiling anchorage or cross-walls, piers, pilasters, and so on, but need not be both.

Structural clay tile is often used for partitioning in schools, hospitals, food-processing plants, and so on, where the imperviousness of a ceramic glazed surface, high durability, and low maintenance are required. Several different types of wall construction may be used depending on the aesthetic requirements for the facing. For the standard 4-, 6-, and 8-in. thicknesses, single units glazed one or both sides are available. Double wythes can be used to get different finishes on each side of the partition (*see Fig. 9-2*). The

9/2 Non-loadbearing structural clay tile partitions. (*From Brick Institute of America*, Technical Note 22, *BIA, McLean, Va.*)

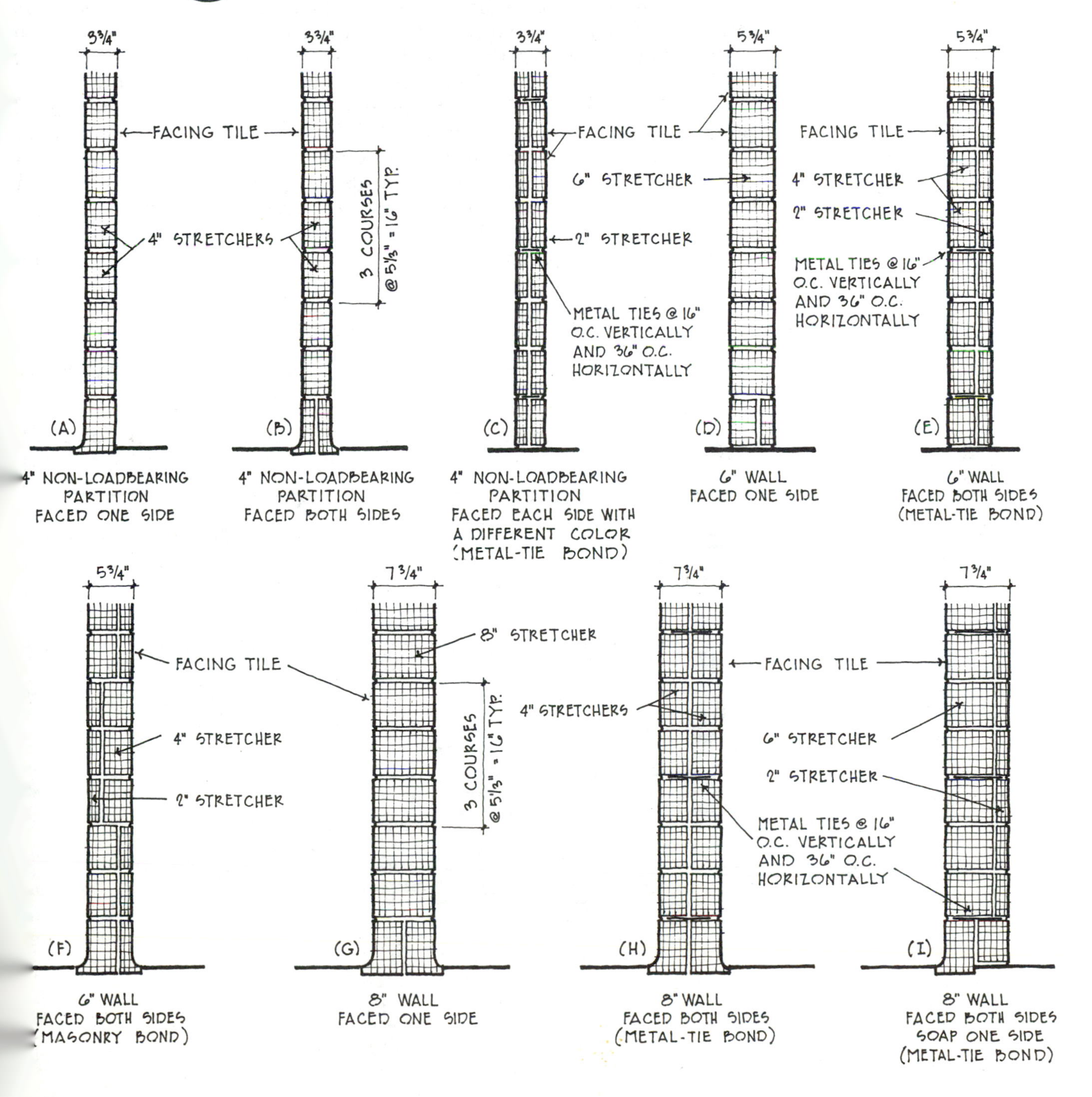

6-and 8-in. walls are actually capable of supporting superimposed structural loads, but the 4-in. partitions are limited to non-loadbearing applications. Lateral support spacing is governed by the same length- or height-to-thickness ratio of 36, giving the same height limitations of 12, 18, and 24 ft without cross-wall bracing.

Concrete block partitions are widely used as interior fire and sound barriers. Decorative units can be left exposed, but standard utility block is usually painted, textured, plastered, or covered with gypsum board. Wood or metal furring strips are attached by mechanical means as described in Chapter 7, or sheet materials may sometimes be laminated directly to the block surface. Code requirements for lateral support are the same as for brick and clay tile.

Hollow CMU and vertical cell tile can be internally reinforced to provide lateral support in lieu of cross walls, pilasters, and so on (*see Fig. 9-3*). A continuous vertical core at the required interval is reinforced with deformed steel bars and then grouted solid. Double-wythe cavity walls of brick may be similarly reinforced without thickening the wall section.

Cavity walls also facilitate the placement of electrical conduit and piping for distribution of utilities within a building. The continuous collar joint easily accommodates horizontal runs. Thickness of cavity walls for computing lateral support requirements is normally taken as the net thickness of the two wythes without the width of the cavity, although requirements vary from code to code.

Gypsum block partitions can be used only as a base for gypsum plaster

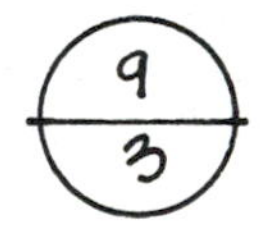

Internal vertical reinforcement for hollow masonry and cavity walls.

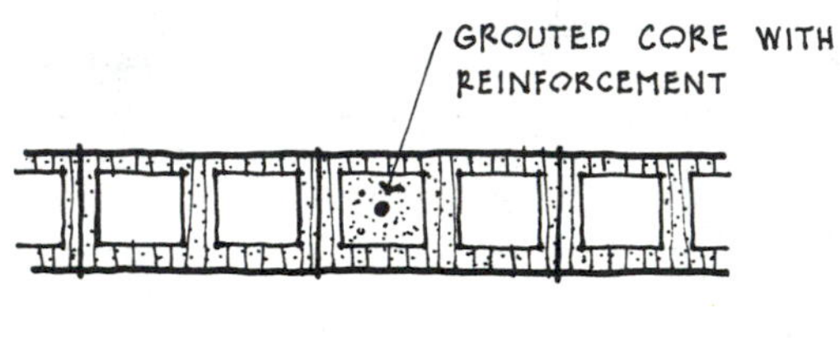

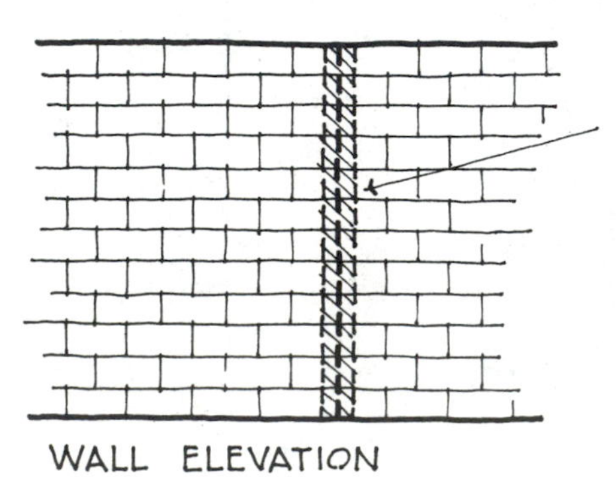

finishes. They provide very thin wall sections with excellent fire ratings. A 2-in.-thick block with plaster directly applied to each side will provide a 4-hour rating. ANSI standards prohibit the application of portland cement plaster, ceramic tile, or marble wainscots unless self-furring metal lath is placed over the gypsum blocks. Some codes also require that gypsum partitions be laid up only with gypsum mortar, and do not permit their use in areas exposed to the weather or in areas where frequent or continuous wetting can be expected.

9.2 EXTERIOR WALLS

Masonry is used in exterior, non-loadbearing applications over structural frames of concrete, steel, wood, or other masonry. The system may be built by either empirical or rational design methods as a *veneer*, a *panel wall*, or a *curtainwall*. Veneers and panel walls are usually of single-wythe construction, but curtainwalls, particularly in high-rise buildings, may be of solid multi-wythe or cavity wall design. In most instances, these non-loadbearing walls are unreinforced, but reinforcement may sometimes be required to increase resistance to flexural tensile stresses.

9.2.1 Veneer Design and Attachment

A veneer is defined as a nonstructural facing attached to a backing for the purpose of ornamentation, protection, or insulation, but not bonded to the backing so as to exert a common reaction under load. There are two basic methods of attaching masonry veneer. *Adhered veneer* is secured by adhesion of a bonding material applied over solid backing. *Anchored veneer* is secured to and supported by mechanical fasteners attached to either a solid backing or structural frame. A veneer supports only its own weight. The backing or frame must be designed to support the vertical and lateral loads imposed by the veneer.

Adhesion attachment is normally limited to relatively thin sections of ceramic veneer, terra cotta, or stone facing. Building codes generally limit the weight of the veneer to 15 lb/sq ft. That weighing more than 3 lb/sq ft may not exceed 36 in. in the greatest dimension or 720 sq in. in area. The bond of an adhered veneer to its backing must be designed to withstand a shearing stress of 50 psi.

A paste of portland cement and water is brushed on the moistened surfaces of the backing and the veneer unit. Type S mortar is then applied to the backing and to the unit, resulting in a mortar thickness of not less than $\frac{1}{2}$ in. or more than $1\frac{1}{4}$ in. If the surfaces are clean and properly moistened, the neat cement paste assures good bond to both surfaces. Adhesion attachment is not common on wood or metal stud framing, but can be successfully accomplished by first applying a scratch coat of cement plaster on metal lath over the studs.

Code requirements do not limit the length or height of adhered veneer except as necessary to control expansion and contraction. Any movement joints that occur in the backing or the frame must be carried through the veneer as well.

Building codes regulate the design of anchored veneers by prescriptive requirements based on empirical data. The system must be designed to resist a horizontal force equal to twice the weight of the veneer. An air space of at least 1 in. should be maintained between the veneer and the backing surface. Noncombustible, noncorrosive lintels must be provided over openings. Deflections are limited to $L/500$ of the span.

Empirical requirements limit the height permitted for anchored veneer. Both the *Uniform Building Code* (*UBC*) and *Basic Building Code* set maximum heights at 25 ft. The *UBC* requires additional support at 12-ft intervals above that level. The *Standard Building Code* and *National Building Code* limit maximum height to 20 ft, but only the National Code requires additional supports above at 12-ft intervals. The National Code also permits an increase to 30 ft at plate height and 38 ft at gable height when, in addition to metal anchors, 16-gauge 2 × 2 in. paper-backed welded-wire mesh is attached to the studs and the 1-in. space between veneer and backing is solidly grouted with portland cement grout. The BIA recommends that brick veneer be self-supporting to heights up to 100 ft with proper detailing to allow differential movement between veneer and frame. There have been some instances of buildings as high as 16 stories with brick veneer supported only at the foundation.

9.2.2 Brick Veneer

Brick veneer is most commonly used over wood and metal stud framing. Flexible metal anchors permit horizontal and vertical movement parallel to the plane of the wall but resist tension and compression perpendicular to it. The veneer must transfer lateral wind loads to the backing, and these meta anchors and their mechanical fasteners are the weakest component of the system.

For securing brick veneer to wood frame construction, corrugated shee metal anchors are used. These should be 22-gauge galvanized steel, at least $\frac{7}{8}$ in. wide × 6 in. long. Corrosion-resistant nails should penetrate the stud a minimum of $1\frac{1}{2}$ in. exclusive of sheathing. The free end of the anchor i placed firmly in the mortar and not on top of the brick, and should extend a least 2 in. into the joint (*see Fig. 9-4*). The anchors are weakest i compression with ultimate load set at 330 lb (*see Fig. 9-5*). The assumed saf compressive load used for empirical code requirements is 80 lb per tie, whic would give a safety factor of 4.1 in compression. Recommended spacing o metal veneer anchors based on these assumed safe loads is given in *Fig. 9-*

Brick veneer is anchored to metal stud frames with 9-gauge corrosion resistant wire looped through a slotted connector for flexibility. The greate deflection of metal stud systems requires more allowance for differenti movement, so corrugated anchors cannot be used. Anchors are attache through the sheathing into the studs with corrosion-resistant, self-tappin screws that can withstand ultimate withdrawal loads of at least 240 lb (s *Fig. 9-7*). Wire anchors with a minimum diameter of $\frac{3}{16}$ in. are used to attac veneer to structural steel. For concrete, wire or flat bar dovetail anchors a recommended. Wire anchors should be at least 6 gauge and 4 in. wide, wi the wire looped and closed (see Chapter 7). Flat-bar dovetail anchors shou be 6 gauge, $\frac{7}{8}$ in. wide, and fabricated so that the end embedded in t masonry is turned up $\frac{1}{4}$ in. Rectangular wire ties and Z-ties are used to atta brick veneer to masonry backup. Z-ties should be used only with so masonry. With concrete masonry backup, continuous horizontal joi reinforcement with wire tabs is recommended. All wire sizes should be least 9 gauge.

The use of brick veneer over metal stud backing is relatively recent the long history of masonry construction. Some questions have been rais regarding the relative rigidity of masonry veneer versus the flexibility of metal frame in resisting lateral loads. However, tests have confirmed t deflection limits of $L/360$ provide adequate stiffness in the studs. Ri

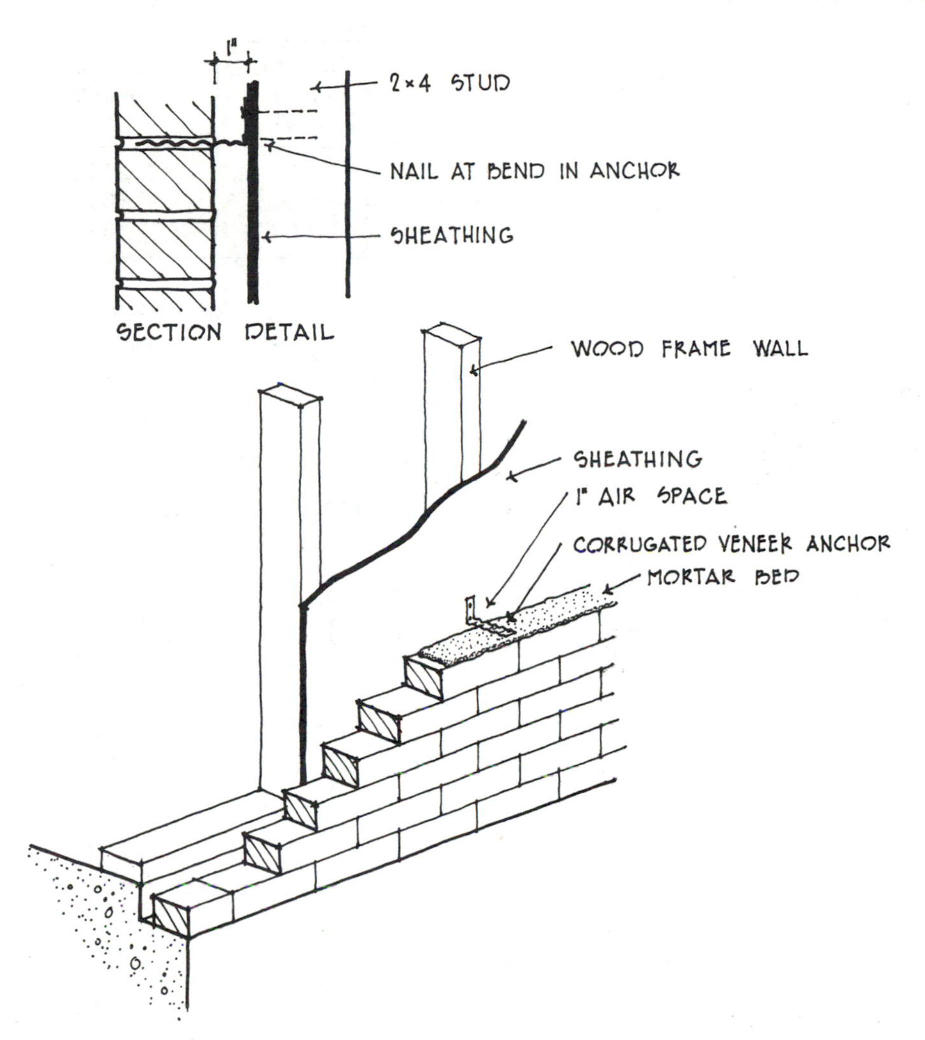

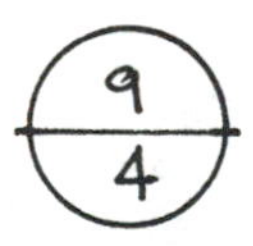

9/4 Brick veneer secured with corrugated metal anchors embedded in mortar.

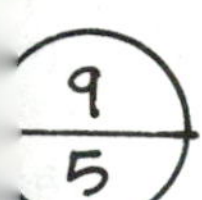

9/5 Ultimate loads on brick veneer anchors. (*From Brick Institute of America*, Technical Note 28, *BIA, McLean, Va.*)

Properties	*Load (lb)*
Pull-out 8d nails, No. 1 Douglas fir studs	500
Pull-out from mortar joint, 22-gauge corrugated ties, $\frac{7}{8}$ in. wide*	875+
Tensile strength, 22-gauge corrugated ties, $\frac{7}{8}$ in. wide*	875+
Compressive strength, 22-gauge corrugated ties, $\frac{7}{8}$ in. wide	330

*Estimated $\frac{7}{8}$ of value for similar anchors 1 in. wide.

sheathing is recommended to help distribute loads and to provide moisture protection for the wall.

Another method of attachment recognized by some building codes and by HUD "Minimum Property Standards" uses galvanized 16-gauge 2 × 2 in. paper-backed welded-wire mesh attached to the studs with galvanized nails

Design wind load (psf)	Spacing horizontal by vertical (in.)	Wall area per anchor (sq ft)
20	24 by 24	4
30	16 by 24	$2\frac{2}{3}$
40	16 by 18	2

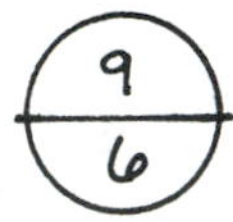

Brick veneer anchor spacing. (*From Brick Institute of America*, Technical Note 28, *BIA, McLean, Va.*)

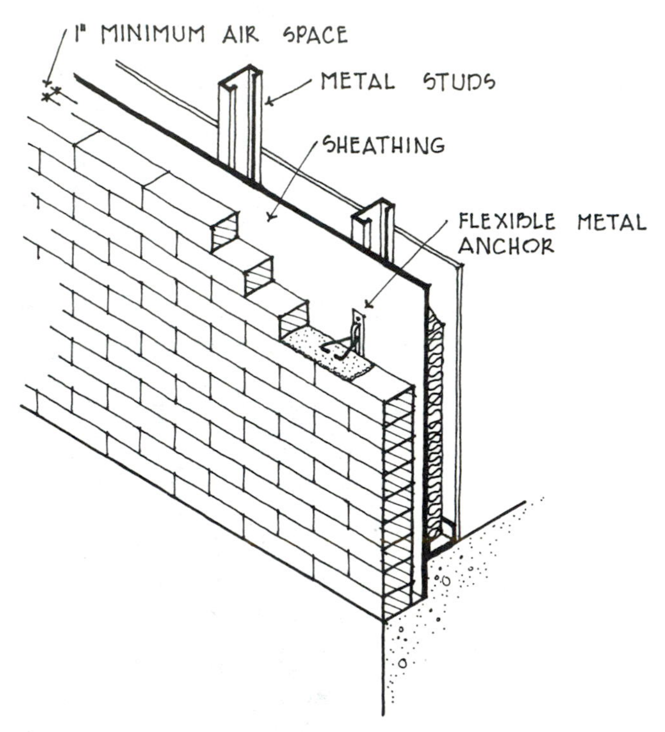

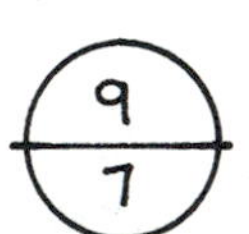

Typical brick veneer installation over a metal stud wall. (*From Brick Institute of America*, Technical Note 28B, *BIA, McLean, Va.*)

or wire ties. The metal veneer anchors are hooked through the mesh, and th 1-in. space between veneer and backing is then grouted solid. This procedu can eliminate the need for sheathing and in some cases, increase th allowable veneer height. It is however, generally limited to masonry units in. or less in thickness (*see Fig. 9-8*).

Grade SW brick is recommended for all exterior veneers in most are of the United States because the facing is isolated from the rest of the wa and therefore exposed to temperature extremes. Type N mortar is suitable i most cases. Type S should be used where a higher degree of flexur resistance is required, and Type M where the brick veneer is in contact wi earth.

Since the overall thickness of a brick veneer wall is approximately in., a foundation wall of the same thickness is required for adequate suppo Many codes permit a nominal 8-in. foundation provided that the top of t wall is corbeled as shown in the detail in *Fig. 9-9*. The total projection of t

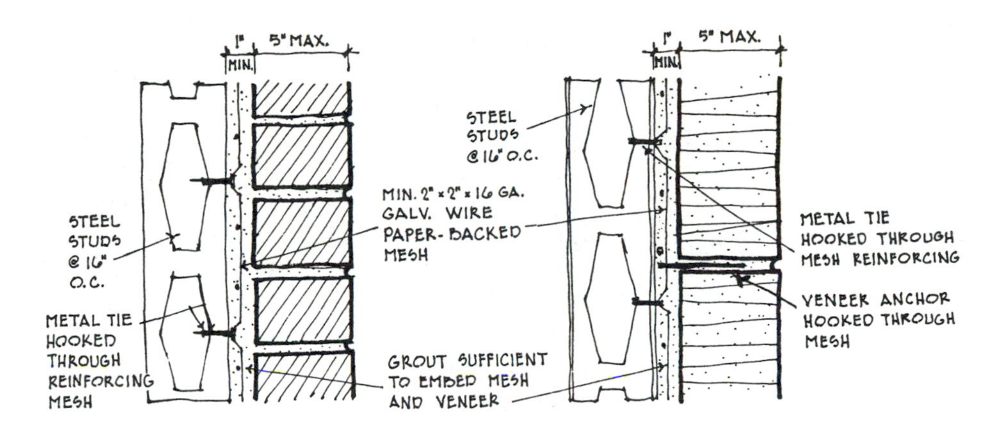

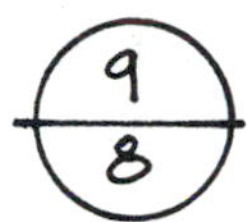

Alternative methods of attaching veneer to metal studs. (*From National Concrete Masonry Association,* TEK Bulletin 79, *NCMA, Herndon, Va.*)

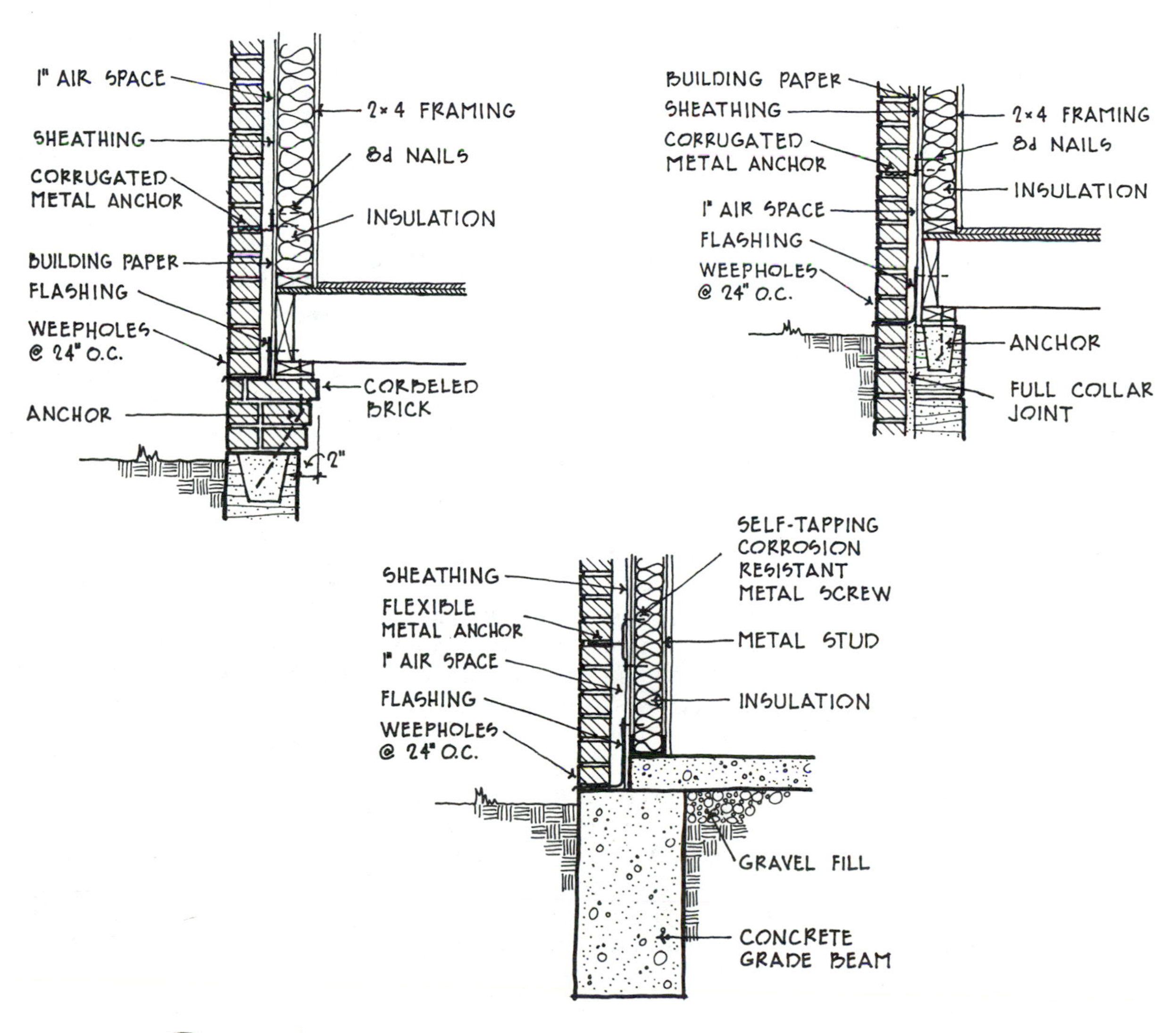

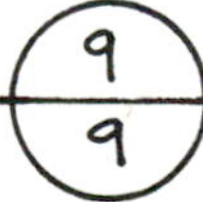

Foundation details for masonry veneer walls. (*From Brick Institute of America*, Technical Note 28, *BIA, McLean, Va.*)

corbel cannot exceed 2 in. with individual corbels projecting not more than one-third the height of the unit. Brick veneer should start below the finish floor line. Moisture entering the wall must be drained to the outside by flashing and weep holes located above grade at the bottom of the wall. Flashing should also be installed at the heads and sills of all openings (*see Fig. 9-10*). The flashing material should be bituminous membrane, sheet metal, or other high-quality material, because replacement in the event of failure is costly if not impossible. Weep holes must be located in the masonry course immediately above all flashing, spaced no more than 24 in. on center horizontally (see Chapter 14 for additional details).

In lieu of steel lintels over openings, brick veneer can be reinforced with $\frac{1}{4}$-in.-diameter steel bars placed horizontally in the first bed joint above the opening. Where spans and loading permit, this method offers a more efficient use of materials (see Chapter 12 for design of masonry lintels).

9.2.3 Concrete Brick and Concrete Block Veneer

This type of construction has increased in use with the variety of colors, textures, and patterns of decorative units made available. Thin veneers of less than 5 in. thickness may be attached in several ways as shown in *Fig. 9-11*. "Spot-bedding" at anchors effectively increases their resistance to compressive lateral loads and strengthens the wall at its weakest point. Hollow CMUs must be anchored in the intermittent mortar beds over the web flanges, or continuous joint reinforcement with wire tabs can be used Solid concrete brick attaches to stud backing with corrugated metal ties in the same manner as clay brick. Anchor spacing based on safe loads of 80 lb per anchor and adjusted for various wind loading is the same as for clay masonry (*see Fig. 9-6*).

Spans across window and door openings are easily accommodate with reinforced hollow concrete masonry. Steel lintels are not required with these units. U-shaped lintel blocks have a deep channel for placement o horizontal reinforcing bars and grout. Depending on the size of the unit an the amount of reinforcing, a masonry lintel can span relatively larg openings (see Chapter 12).

Joint reinforcement is used in concrete masonry to control shrinkag cracks. As the stress increases, it is transferred to and redistributed by th steel. The effectiveness of joint reinforcement depends on the type of morta used and the bond it creates with the wire. Greater bond strength mean greater efficiency in crack control. Only Types M, S, and N portland cement lime mortars are recommended. Minimum mortar cover to the outside face c the block should be $\frac{5}{8}$ in. for the exterior and $\frac{1}{2}$ in. for the interior wall fac Prefabricated corner and T-type reinforcement is recommended for corner and intersecting walls. Splices should be lapped 6 in.

Joint reinforcement is normally located in the first bed joint of a wa and at 8-, 16-, or 24-in. centers above that. In addition to normal placemen joint reinforcement should be located: (1) in the first and second bed join immediately above and below wall openings, extending at least 24 in. pa either side of the opening or to the end of the panel, whichever is less; and (in the first two or three bed joints above floor level, below roof level, and ne the top of a wall. Reinforcement need not be located closer than 24 in. to bond beam.

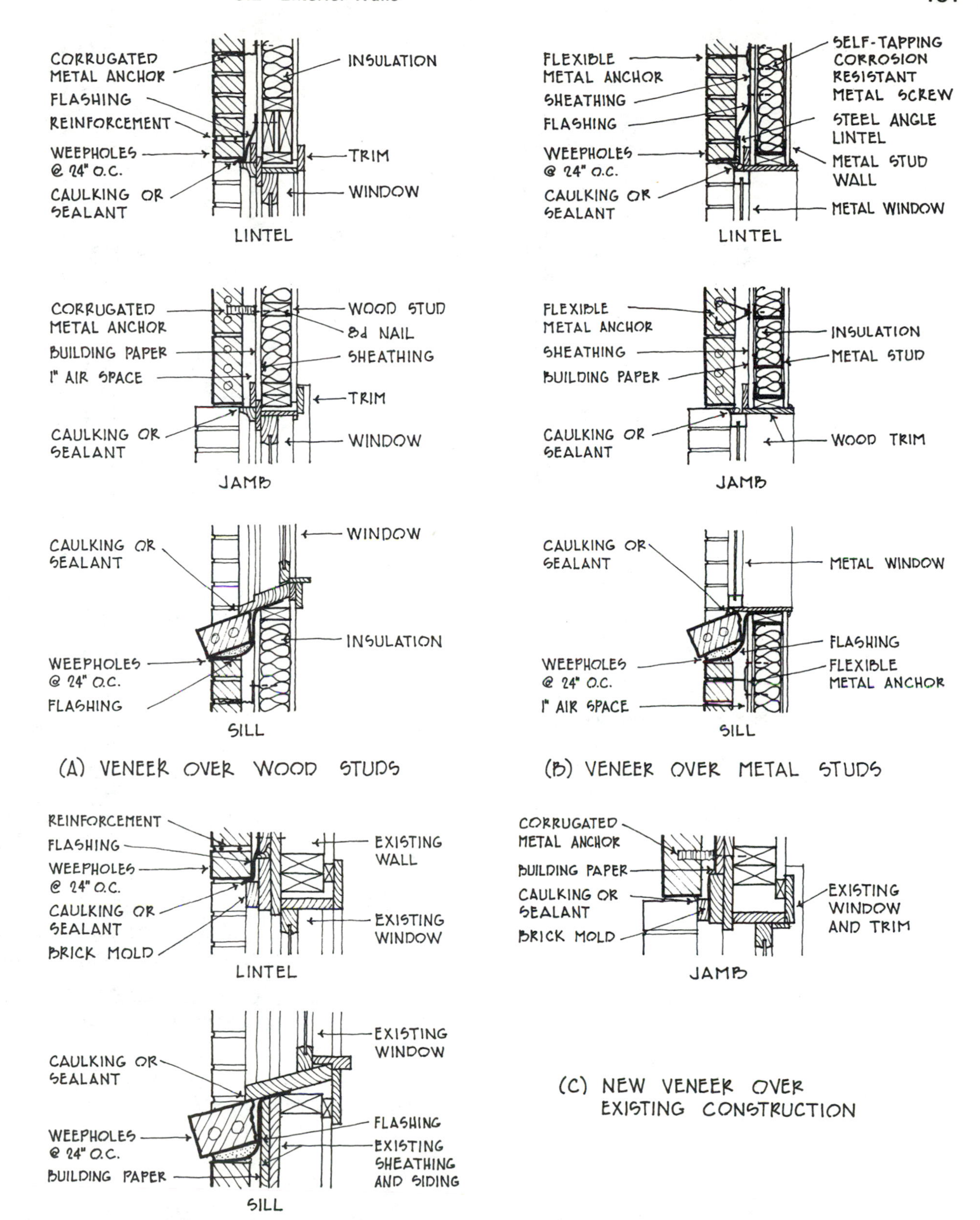

Typical veneer details. (*From Brick Institute of America*, Technical Note 28, *BIA*, *McLean*, *Va.*)

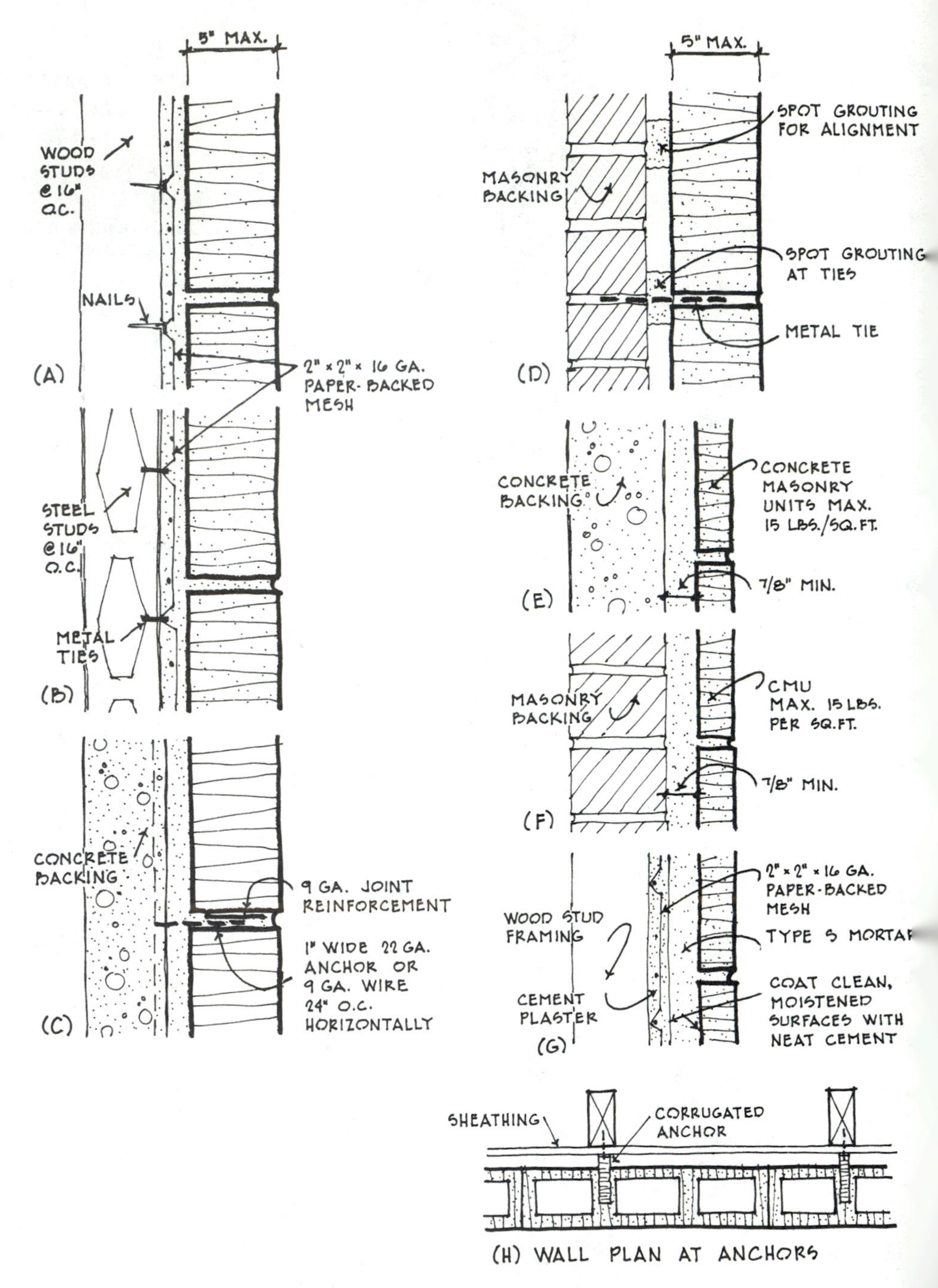

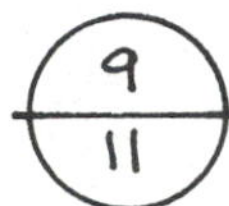

Concrete masonry veneer attachments. (*From National Concre* *Masonry Association*, TEK Bulletin 79, *NCMA, Herndon, Va.*)

9.2.4 Stone Veneer

There are two basic types of stone veneer: (1) rubble or cut stone laid i mortar beds, and (2) thin stone slabs mechanically or adhesively attache Mortar bed construction is generally used in low-rise residential an commercial buildings (*see Fig. 9-12*). The stone may be laid up against backing of concrete, wood or metal studs, or unit masonry. When the ston do not exceed 5 in. in thickness they may be installed without anchors i much the same way as for thin concrete masonry (*see Fig. 9-11*). The bond between stone and mortar or grout must be sufficient to withstand a shearing stress of 50 psi. For larger stones up to 10 in. thick, wire or corrugated sheet metal anchors are required. The connections must be flexible enough to compensate for the irregularities of mortar bed height (*see Fig. 9-13*). Anchors should be spaced 24 in. on center maximum, and support no more than 2 sq ft of wall area. Metal anchors must receive full mortar coverage at the outside face of the wall to prevent rusting and corrosion. All joints and the space between the stone and the backing should be filled solidly with mortar.

Stone slab veneers are most commonly used on commercial buildings

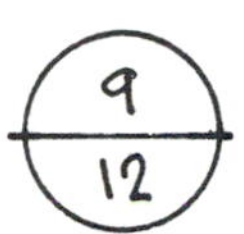

Stone veneer mortar bed installation.

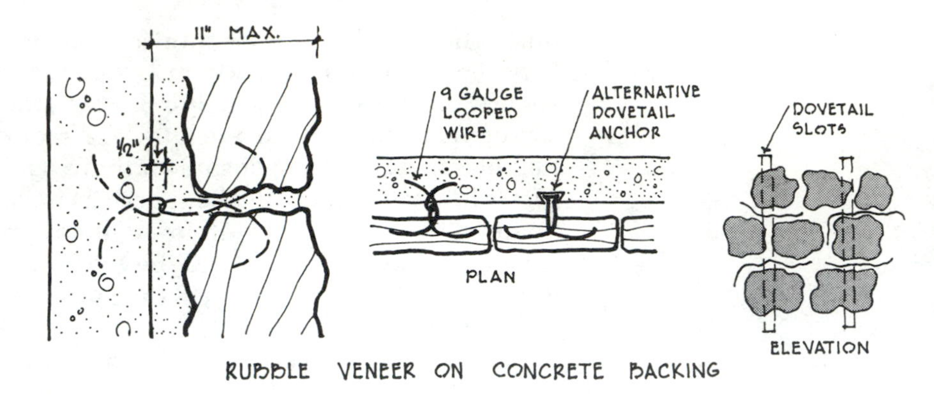

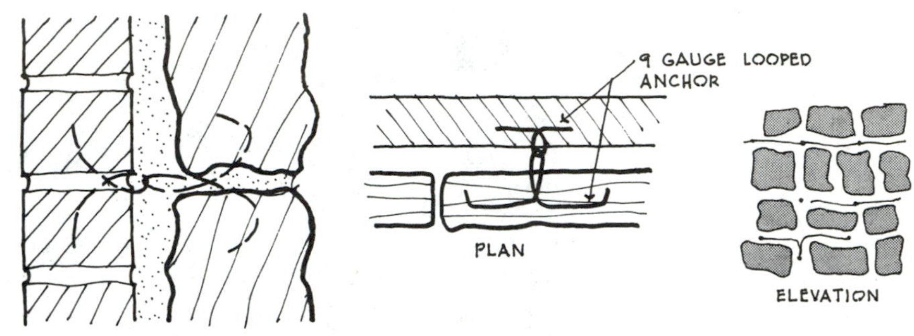

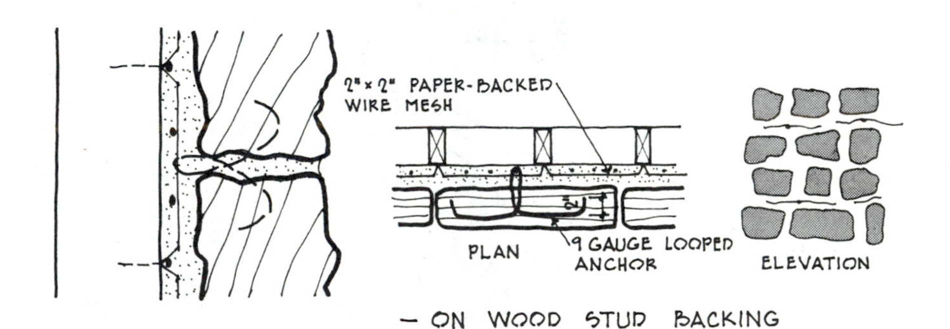

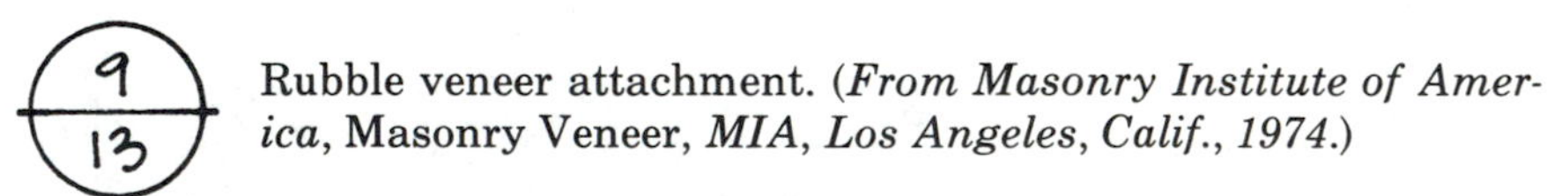

Rubble veneer attachment. (*From Masonry Institute of America*, Masonry Veneer, *MIA, Los Angeles, Calif., 1974.*)

of low-, medium-, and high-rise construction. For adhered attachmen without mechanical anchors, the individual stone units must not exceed 3 in. in the largest dimension, 720 sq in. in area, or 15 lb/sq ft in weight. The may be adhered to a concrete or masonry backing. The surfaces of th backing and the veneer must be cleaned, moistened, and brushed with nea cement paste. Type S mortar is then applied to both surfaces, and the unit are tapped into place forming a collar joint $\frac{1}{2}$ to $1\frac{1}{4}$ in. wide.

There are many ways to attach stone slab veneer with metal anchor All anchoring systems must be designed to resist a horizontal force equal twice the weight of the veneer. Some types of stone are drilled around th perimeter for insertion of corrosion-resistant metal dowels. Dowels a normally spaced 18 in. on center, with a minimum of four for each stone uni The total area of the stone slab usually may not exceed 20 sq ft. Each dow is secured to the backing with wire or sheet metal anchors (*see Fig. 9-14*). In lieu of dowels, wire anchors may be bent into the holes and set with a cementitious mix. Dowel anchoring may not be used on stone slabs less than $1\frac{1}{4}$ in. thick. The space between the veneer and the backing surface may be solidly grouted or spot-bedded at anchor locations and for alignment. Bed joints are usually filled with a mastic caulking rather than a strong mortar which might be subject to shrinkage cracking and subsequent moisture penetration. The mastic provides a weathertight joint and also permits slight movement of the units to relieve stress.

Several other types of mechanical anchorage are shown in *Fig. 9-14*, for attachment to concrete, masonry, or steel framing. Relieving supports, or shelf angles, must be provided at least every 12 ft vertically for veneer installed more than 25 ft above ground level. Most manufacturers prepare

Stone slab veneer details. (*From Masonry Institute of America,* Masonry Veneer, *MIA, Los Angeles, Calif., 1974.*)

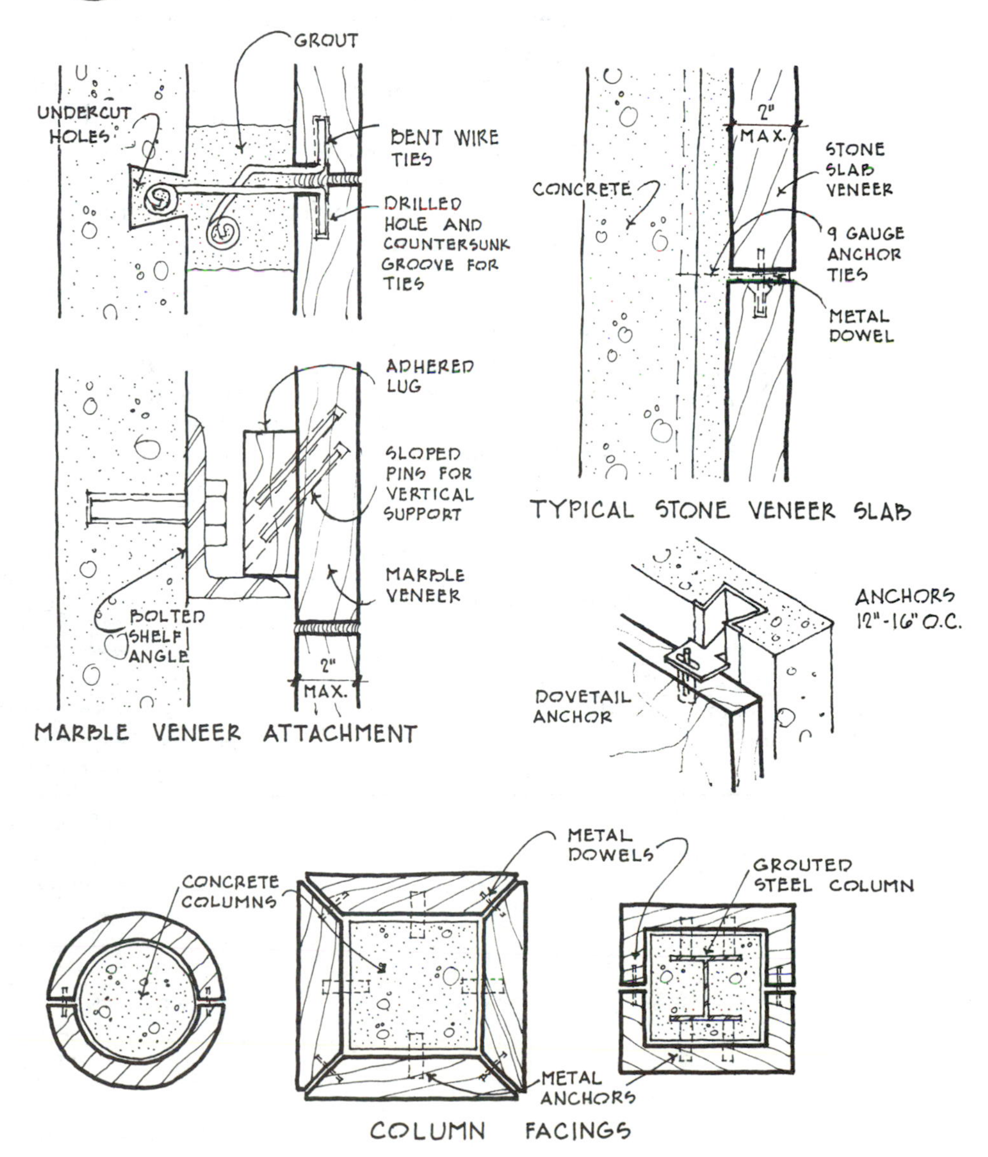

installation drawings using anchorage details that are best suited for their particular type of stone.

9.2.5 Movement Joints

All masonry veneer requires careful consideration of the design and location of movement joints. Brick expansion, CMU shrinkage, and differential movement can create excessive stress, particularly in unreinforced veneer.

Control joints are continuous, vertically weakened wall sections designed to control CMU shrinkage cracking. When thermal and moisture movements create sufficient tensile stress, cracks can be localized at these predetermined points. *Pressure-relieving joints* absorb brick thermal and moisture expansion and relieve the stress of differential movement. (Details of movement joint construction are covered in Chapter 14.)

Guidelines for concrete masonry control joint spacing have been developed empirically, but since wall layouts vary with opening placement, some judgment must be used in determining exact locations. ACI recommendations for Type I, moisture-controlled units are given in *Fig. 9-15.* If Type II units are used, spacing should be reduced by half. Where walls are solidly grouted, spacing should be reduced by one-third. In addition to these general locations, control joints should be placed at points of weakness or high stress concentration such as (1) all abrupt changes in wall height; (2) all changes in wall thickness such as columns and pilasters; (3) coincidentally with movement joints in floors, roofs, and foundations; and (4) at one or both sides of all window and door openings.

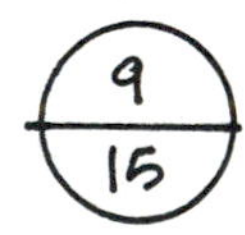

Control joint spacing for unreinforced concrete masonry, Type I units. (*From Frank A. Randall and William C. Panarese,* Concrete Masonry Handbook for Architects, Engineers, and Builders, *Portland Cement Association, Skokie, Ill., 1976: and National Concrete Masonry Association,* TEK Bulletin 3, *NCMA, Herndon, Va.*)

Recommended control joint spacing	*Vertical spacing of joint reinforcement*			
	None	*24 in.*	*16 in.*	*8 in.*
Ratio of panel length to height (*L*/*H*)	2	2.5	3	4
Panel length (ft) not to exceed (*L*) regardless of (*H*)	40	45	50	60

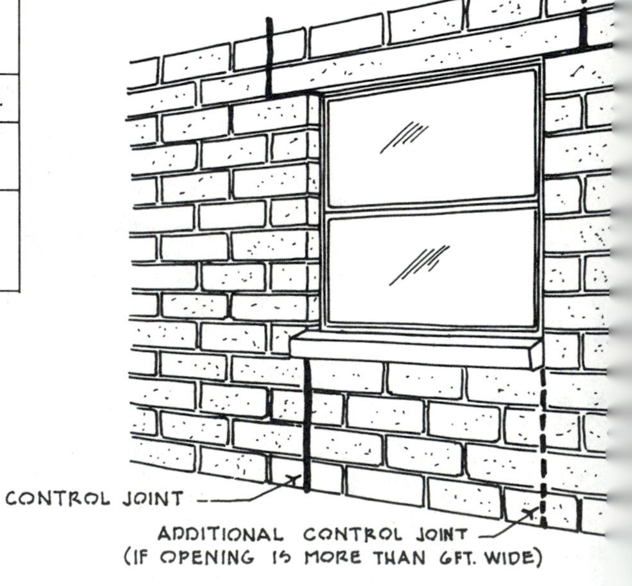

CONTROL JOINTS AT OPENINGS

For brick masonry, the width and spacing of pressure-relieving joints can be estimated from the following formulas:

$$W = [0.0002 + 0.0000045(T_{max} - T_{min})]L$$

and

$$S = \frac{24{,}000}{T_{max} - T_{min}}(p)$$

where W = total expansion of the wall, in.

0.0002 = coefficient of moisture expansion

0.0000045 = coefficient of thermal expansion

L = length of wall, in.

S = maximum spacing of movement joints, ft.

p = ratio of opaque wall area to gross wall area

T_{max} = maximum mean wall temperature, °F

T_{min} = minimum mean wall temperature, °F

The values determined are not precise. Total theoretical expansion will be modified by indeterminate compensating factors such as degree of restraint, plastic flow of mortar, and variations of workmanship. The formulas can be used for general estimations, supplemented by field experience and good judgment. Movement joints should also be located at points of high stress concentrations, as indicated in *Fig. 9-16*.

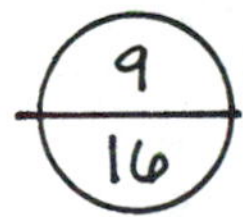

Additional brick expansion joint locations for various building plan shapes. (*From Harry C. Plummer*, Brick and Tile Engineering, *Brick Institute of America, McLean, Va., 1962.*)

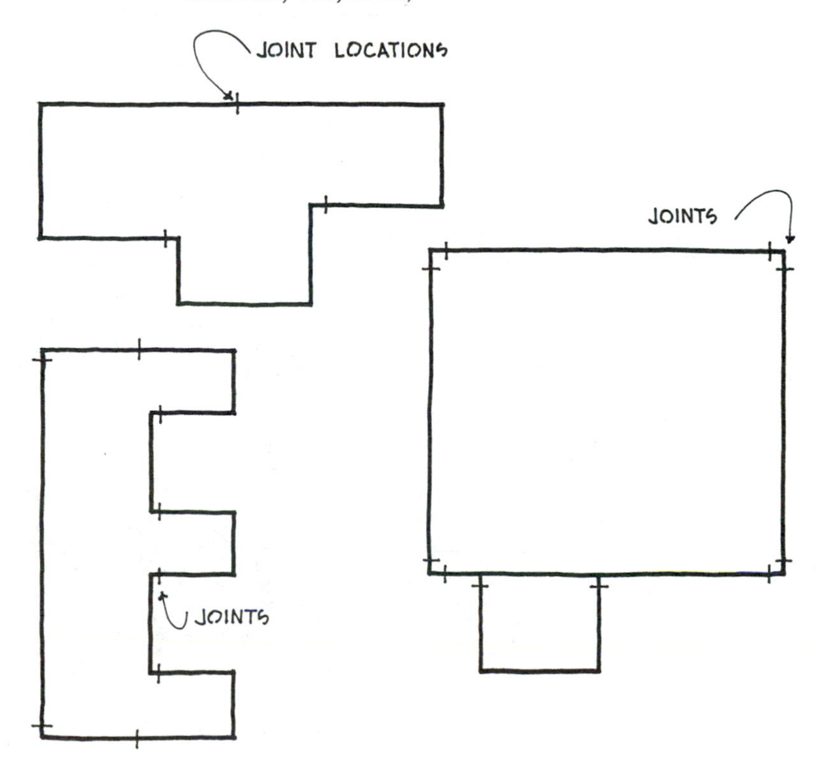

9.2.6 Panel Walls

A masonry panel wall is an exterior, non-loadbearing wall wholly supported at each floor by a concrete slab or beam, or by steel shelf angles. (Veneer walls constructed between relieving angles on high-rise buildings are technically panel wall sections.) Panel walls must be designed to resist lateral forces and transfer them to adjacent structural members. Design is governed by empirical height or length-to-thickness ratios in most codes. Top and bottom anchorage must secure the panel against lateral loads, but must also permit differential movement between the masonry and the steel or concrete framing. To accommodate movement, connections should be flexible, and allowance should be made for expansion and contraction (*see Fig. 9-17*).

Differential movement can be critical in panel wall construction. Steel and concrete are both subject to temperature and/or moisture shrinkage as well as permanent deflections caused by plastic flow. Without proper movement joints below a support, the masonry panel is actually squeezed between the frame members and can fail in bending under an axial load that it is not reinforced to withstand. Clay masonry is particularly vulnerable in these circumstances since it is subject to irreversible moisture expansion. If restrained by the concrete or steel structure, the brick panel can buckle or spall. Pressure-relieving joints will absorb the movement and protect the masonry. Minimum clearance should be $\frac{3}{8}$ in., and the joint should be filled with an elastic material such as neoprene and then pointed with sealant instead of mortar.

Shelf angles that are bolted to framing members are better able to move under stress than are angles welded rigidly in place. Bolted connections also permit adjustments for proper alignment in both the horizontal and vertical directions. Flashing should be installed immediately above all shelf angles or supports, and weep holes must be located just above the flashing to drain accumulated moisture.

Glass block may be used in non-loadbearing veneer or in panel wall construction in much the same way as clay and concrete masonry. The compressive strength of the units is sufficient to carry the dead load of the material weight for a moderate height. Intermediate supports at floor and

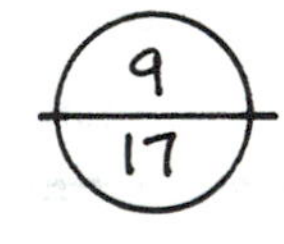

Flexible connections transfer lateral loads to the structural frame, but allow differential movement. (*From Brick Institute of America*, Technical Notes 17L, 28B, *BIA, McLean, Va.; National Concrete Masonry Association*, TEK Bulletin 93, *NCMA, Herndon, Va.*)

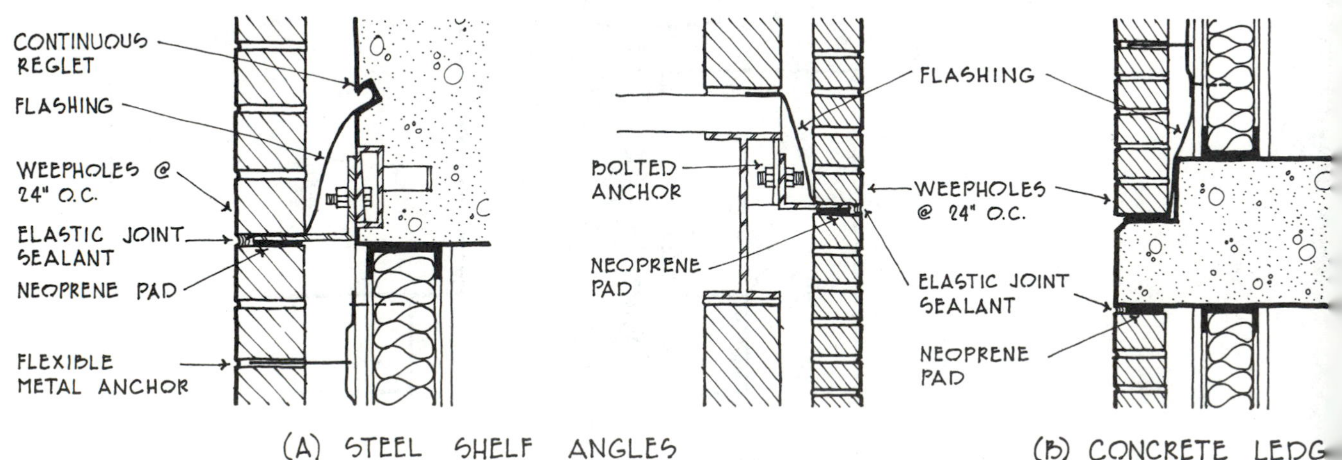

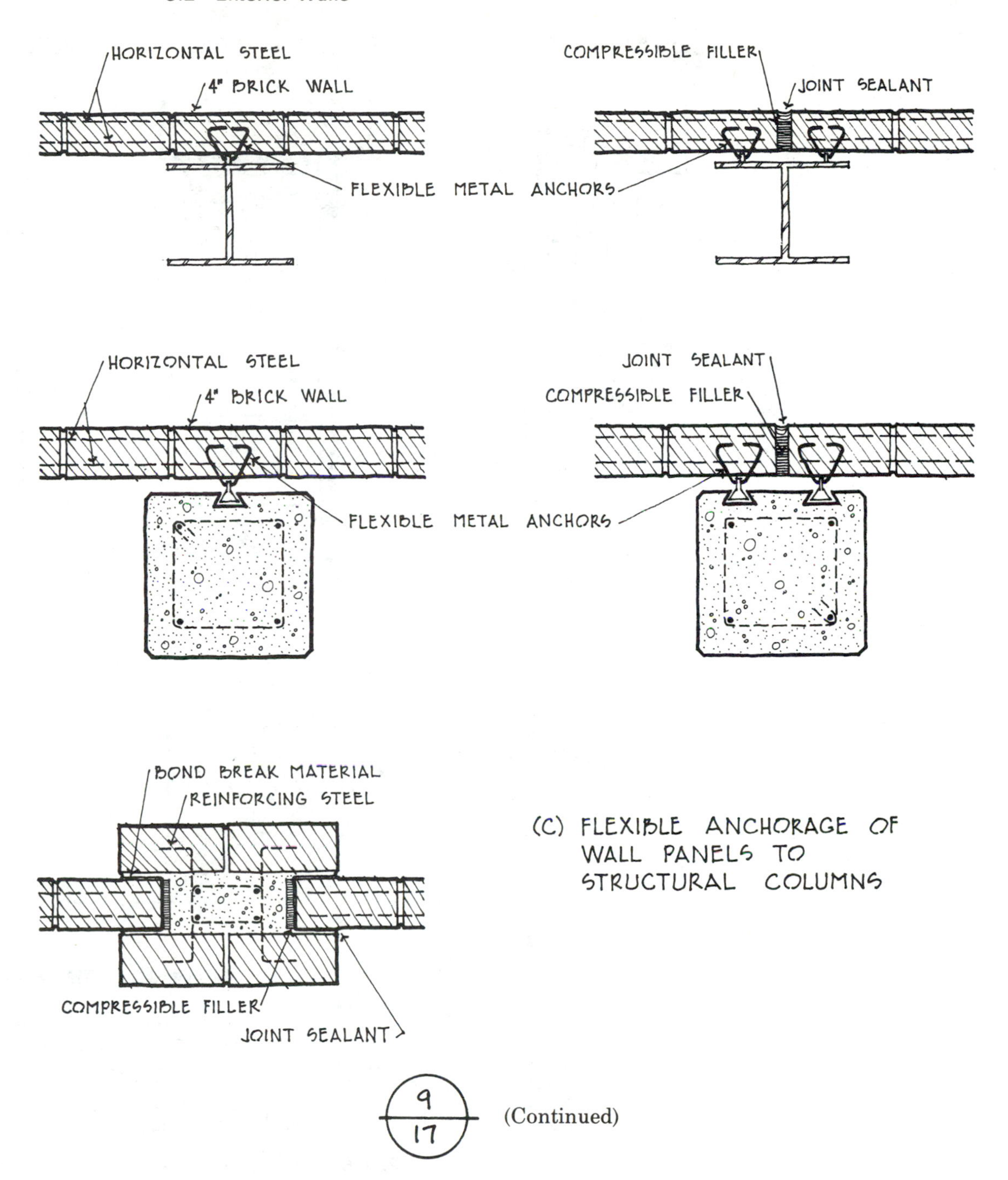

(C) FLEXIBLE ANCHORAGE OF WALL PANELS TO STRUCTURAL COLUMNS

(Continued)

roof slabs require the same care in detailing to allow expansion and contraction of dissimilar materials (*see Fig. 9-18*). Movement joints at the perimeter of the panels should be at least $\frac{1}{2}$ in. Glass blocks are normally laid in Type S or Type N portland cement mortar, and bed joints reinforced with ladder-type horizontal joint reinforcement 24 in. on center vertically. Head and jamb recesses are sometimes required to increase lateral resistance. If no jamb recess is provided in the adjacent wall, jamb anchors may be required for larger panels (*see Fig. 9-19*). Some codes require a minimum thickness of 3 or $3\frac{1}{2}$ in.

Size and area limitations for glass block wall panels vary among the major model building codes. The requirements are as follows:

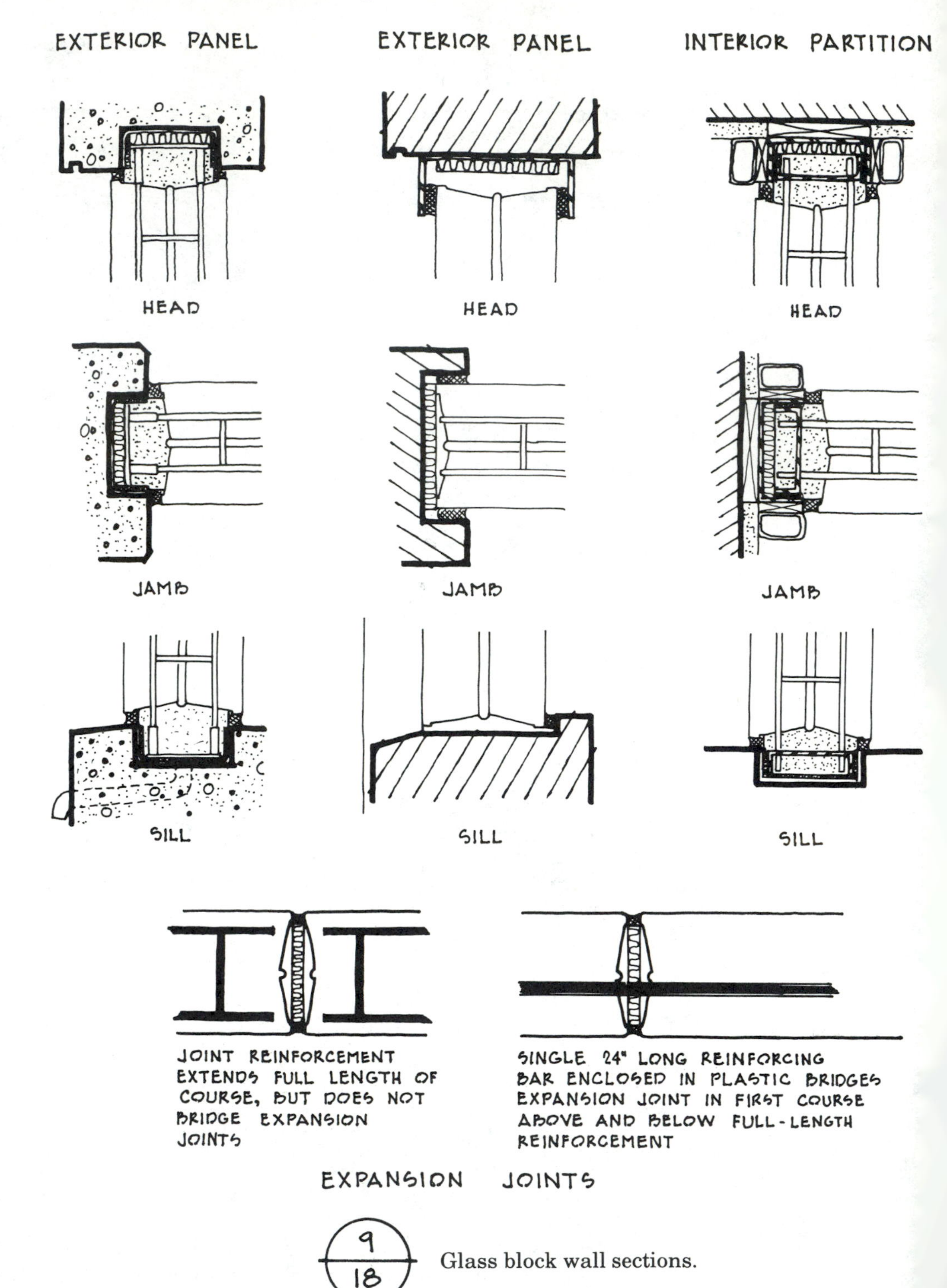

Glass block wall sections.

1. *Uniform Building Code:* Exterior walls are limited to 144 sq ft o unsupported wall surface or 15 ft maximum dimension. Interio walls may not exceed 250 sq ft of unsupported area or 25 ft in an dimension.
2. *Basic Building Code:* Exterior walls are limited to 25 ft in lengt and 20 ft in height between structural supports and/or expansio joints. Area may not exceed 250 sq ft. Any panel over 144 sq ft. i

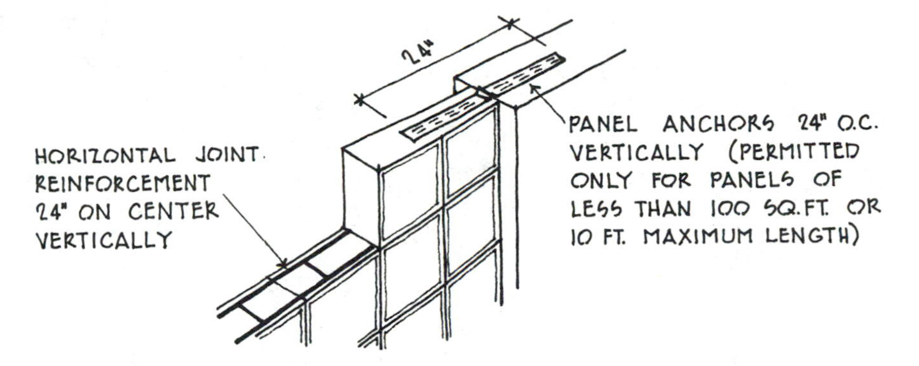

(A) PANEL ANCHOR DETAIL

Glass block size (in.)	Minimum radius for curved panels
5	3′-9″
6	4′-9″
8	6′-3″
10	7′-11″

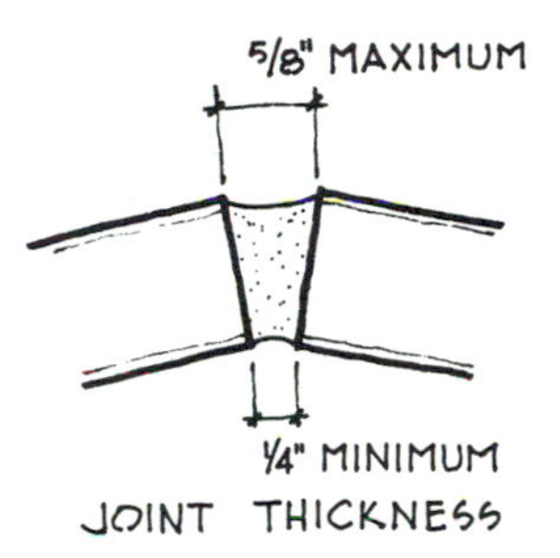

(B) REQUIREMENTS FOR CURVED WALL PANELS

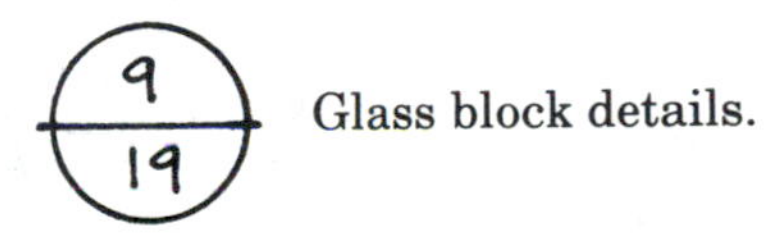

Glass block details.

area must have supplementary anchorage to structural supports for stiffening. (No requirements are listed for the area of interior walls.)

3. *National Building Code* and *Standard Building Code:* Exterior panels may not exceed 144 sq ft of unsupported wall surface, or 25 ft in length, or 20 ft in height between supports. Interior panels are limited to 250 sq ft of unsupported area or 25 ft maximum dimension between supports.

9.2.7 Curtainwalls

Masonry curtainwalls are designed by the rational engineering procedures outlined in Chapters 10, 11, and 12. By using this method, empirical height-to-thickness ratio restrictions are waived, and walls can be built to span multiple structural bays. Curtainwalls may be single- or multi-wythe design, and are "partially reinforced" against lateral loads. Long walls are attached only at columns or bearing walls for transfer of horizontal wind loads. Multistory walls are wholly supported at the foundation, without intermediate shelf angles, and are attached at the floors only for lateral load transfer.

Walls must be designed for both positive and negative wind pressure. The maximum moment in the wall (*see Fig. 9-20*) is used to calculate the required area of reinforcement.

Long walls are considered to span horizontally between columns, and

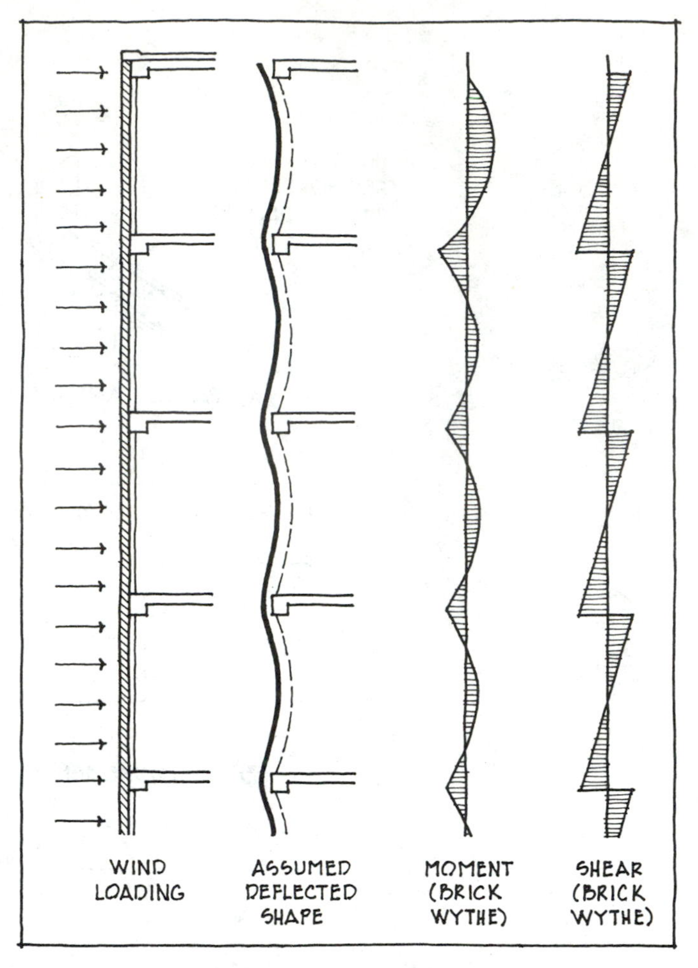

Lateral wind loading. (*From Brick Institute of America*, Technical Note 28B, *BIA*, *McLean*, *Va*.)

must be designed for flexure and bending in this direction. The design i
considred "partially reinforced" because only horizontal steel is used t
resist loads, and the amount is less than that required by code to b
considered "reinforced masonry." For nominal 4-in. brick walls, BI
engineers have prepared the design tables in *Fig. 9-21* based on a maximu
deflection of $L/200$ for simply supported solid walls without openings. Th
area of reinforcement is required in each face of the wall for the variou
support conditions, span lengths, and wind loads shown. After the require
area is determined from Table *A*, the bar size and spacing may be selecte
from Table *B*. Ladder or truss-type joint reinforcement is best since the wire
cannot be misaligned during construction. Special conditions at opening
must be investigated on an individual basis since they reduce the section o
the wall and its stiffness, affecting the required area of reinforcemen
strength, and distribution of lateral load.

Multistory walls that span vertically from floor to floor may requir
vertical reinforcement to resist bending in that direction. It may then be bes
to design the wall as reinforced masonry, or as a loadbearing element fo
additional stiffness (see Chapters 10 and 11). Alternatively, the wall may b
tied to metal stud infill walls for additional lateral load distribution betwee
floors as previously discussed (*see Figs. 9-6 and 9-7*).

TABLE A REQUIRED AREA OF REINFORCEMENT PER VERTICAL FOOT FOR *EACH FACE OF WALL* (SQ IN./FT)

Span (ft)	*Support condition*							
	Simple span or two continuous spans				*Three or more continuous spans*			
	Wind pressure (psf)				*Wind pressure (psf)*			
	15	*20*	*25*	*30*	*15*	*20*	*25*	*30*
10	0.035	0.047	0.059	0.070	0.028	0.038	0.047	0.056
12	0.051	0.068	0.084	0.101	0.041	0.054	0.068	0.081
14	0.069	0.092	0.115	0.138	0.055	0.074	0.092	0.110
16	0.090	0.120	0.150	0.180	0.072	0.096	0.120	0.144
18	0.114	0.152	0.190		0.091	0.122	0.152	0.182
20	0.141	0.188			0.113	0.150	0.188	
22	0.170				0.136	0.182		
24	0.203				0.162	0.216		
26					0.190			

TABLE B AREA OF REINFORCEMENT *PER FACE* OF WALL FOR VARIOUS BAR SIZES AND SPACING (SQ IN./VERTICAL FT) (BASED ON 3 COURSES PER 8-IN. HEIGHT)

Reinforcement per face	*Spacing of reinforcement*							
	Every course	*Every 2nd course*	*Every 3rd course*	*Every 4th course*	*Every 5th course*	*Every 6th course*	*Every 7th course*	*Every 8th course*
one #2 bar	0.225	0.112	0.075	0.056	0.045	0.037	0.032	0.028
one $\frac{3}{16}$-in. wire	0.124	0.062	0.041	0.031				
one #9 gauge wire	0.077	0.039	0.026					

TABLE C MAXIMUM *VERTICAL* SPACING OF WALL ANCHORS (IN.)

Wind pressure (psf)	*Horizontal wall span (ft)*								
	10	*12*	*14*	*16*	*18*	*20*	*22*	*24*	*26*
15	24	24	24	24	24	24	22	20	18
20	24	24	24	22	20	18	16	15	
25	24	24	20	18	16	14			
30	24	20	17	15	13				

Note: No. 6 gauge galvanized wire anchors. Load distribution assumed to be 600 lb per anchor. Maximum vertical spacing limited to 24 in. Spacing is for an interior support of a continuous wall. Spacing at an end support is 24 in.

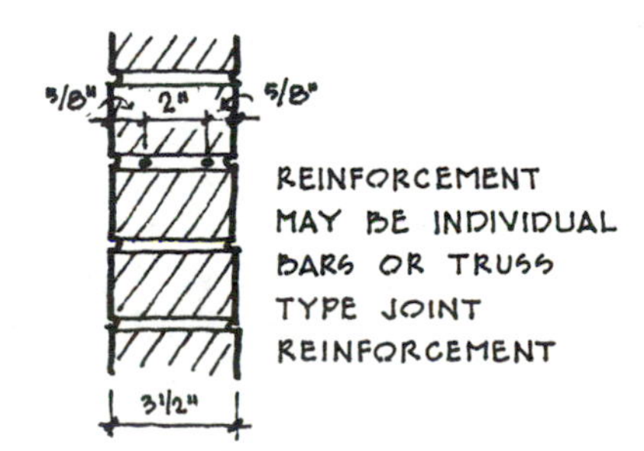

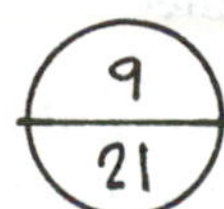

Reinforcing requirements for 4-in. brick curtainwalls spanning horizontally. (*From Brick Institute of America*, Technical Note 17L, *BIA*, *McLean*, *Va.*)

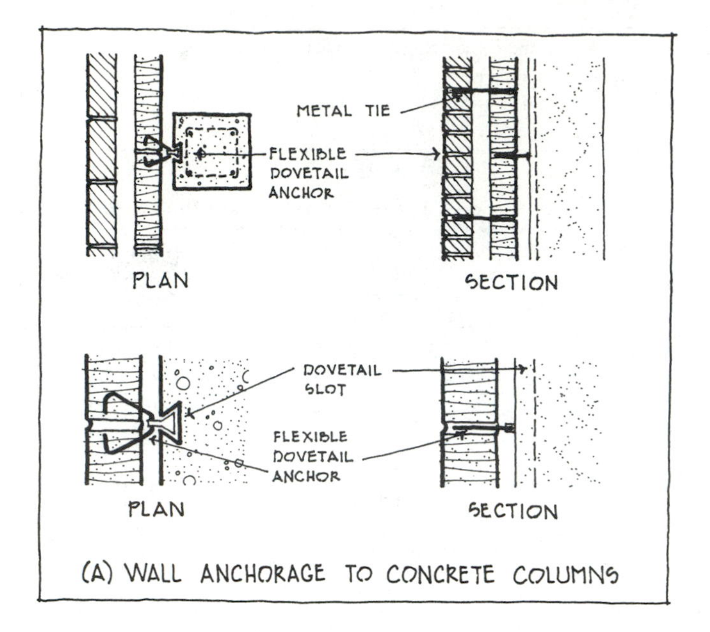

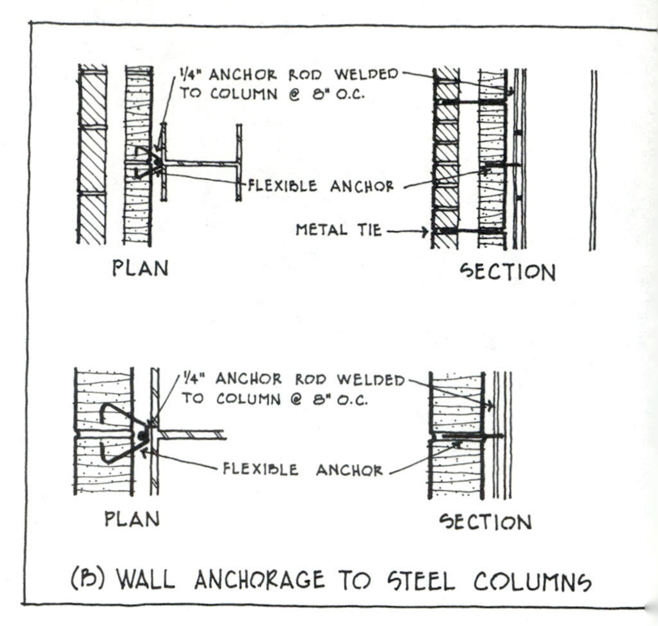

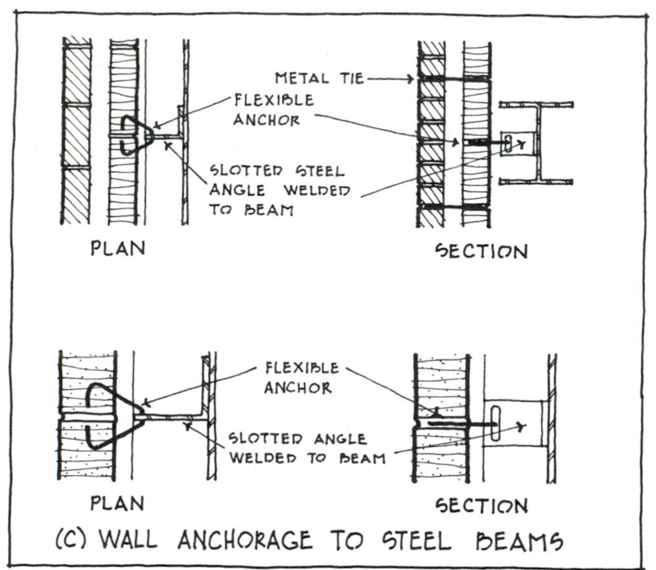

Flexible anchorage to structural columns and beams. (*From Brick Institute of America*, Technical Note 18B, *BIA*, *McLean, Va.*)

All masonry curtainwall design calculations must be verifiable to cod authorities for waiver of the h/t ratios. For both vertical and horizonta curtainwalls, flexible connections must be used to accommodate differentia movement (*see Fig. 9-22*).

9.3 SCREEN WALLS AND GARDEN WALLS

Perforated masonry *screen walls* may be built with specially designed bloc or clay tile units, with standard cored blocks laid on their sides, or with soli units laid in an open pattern (*see Fig. 3-18*). As sun screens, the walls a often built along the outside face of a building to provide shading f windows and glass areas. Screen walls can generally be anchored only the floor line or at vertical structural projections such as columns pilasters. For this type of application, they are designed as panel walls curtainwalls.

Concrete masonry screen wall units should have a minimum compre sive strength of 1000 psi on the gross area when tested with their hollow cel

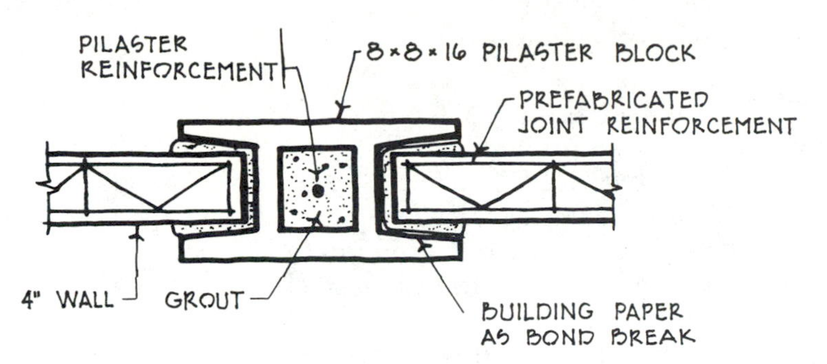

(A) INTERMEDIATE PILASTER

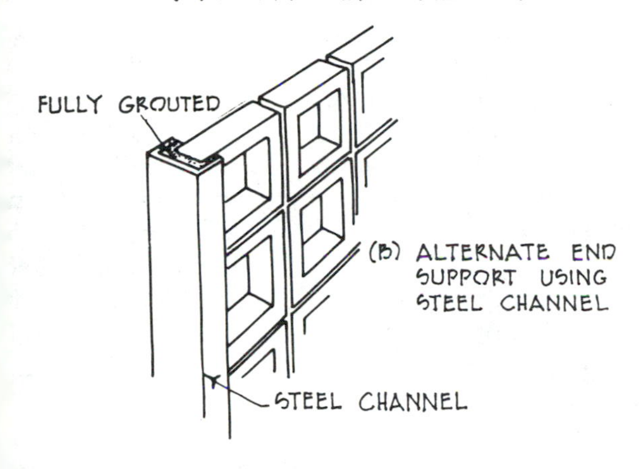

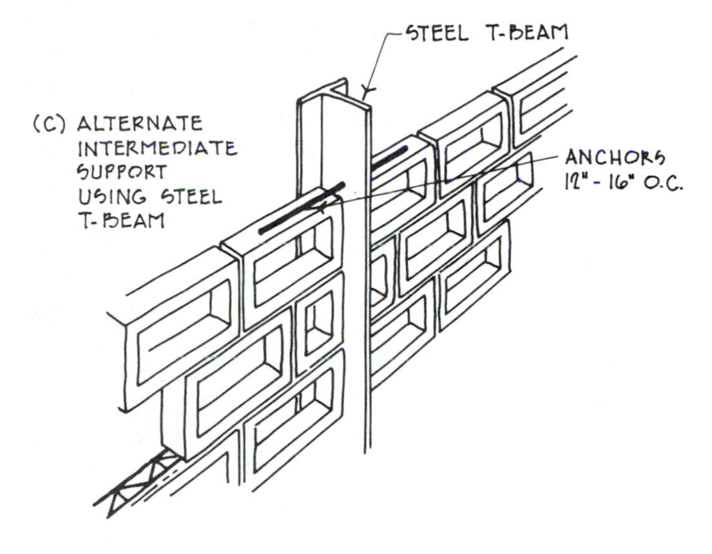

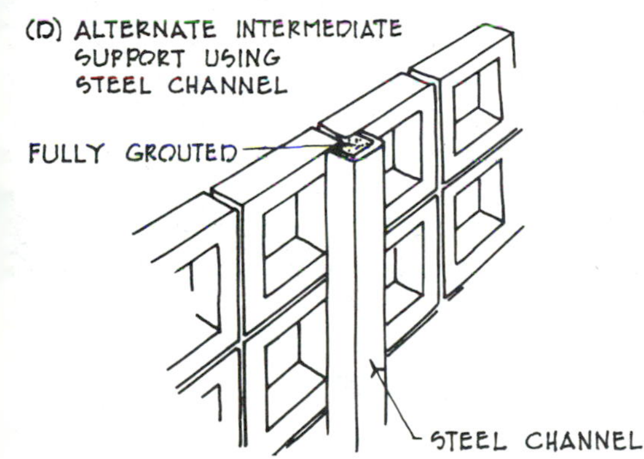

NOTE: PROVIDE WIRE ANCHORS IN MORTAR JOINTS WELDED TO CHANNEL 12" TO 16" O.C.

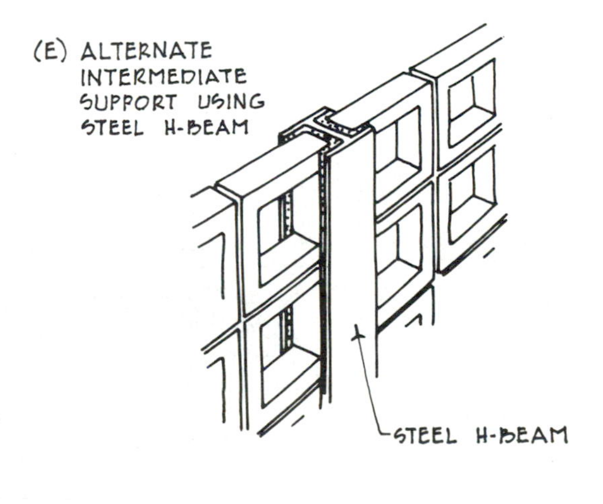

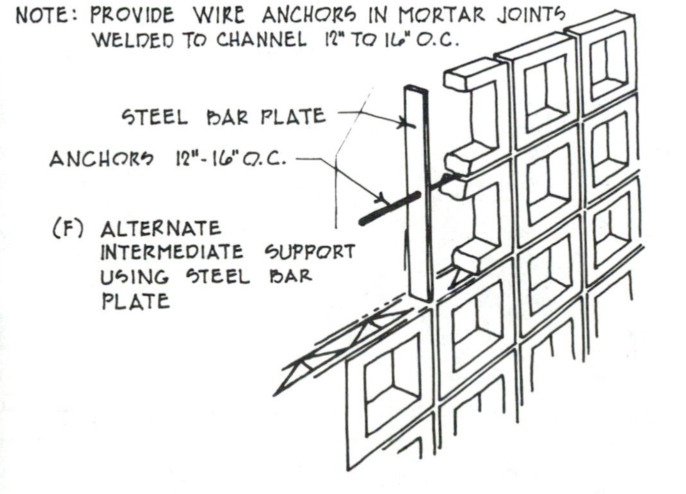

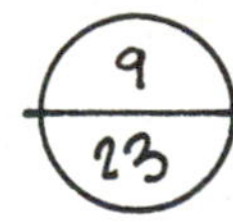

Methods of providing lateral support for masonry screen walls. (*From National Concrete Masonry Association*, TEK Bulletin 5, *NCMA, Herndon, Va.*)

parallel to the direction of the load. They should be Type I, moisture-controlled units in conformance with ASTM C90 or C145. Brick should be Grade SW conforming to ASTM C216, and clay tile units should be Grade NB, ASTM C530. Mortar for exterior screen walls should be Type M or Type S.

Screen walls are governed by h/t ratio lateral support requirements. Lateral support can be provided with either steel or reinforced masonry piers or pilasters (*see Fig. 9-23*).

Non-loadbearing exterior screen walls should have a minimum thickness of 4 in. and maximum clear spans as shown in *Fig. 9-24*. The maximum distance between lateral supports may be measured vertically or horizontally, but need not be limited in both directions. Adequate anchorage to structural supports must be provided so that lateral stresses are transferred to the foundation. Because of the reduced stiffness of perforated screen walls, horizontal joint reinforcement is recommended (particularly for stack bond designs) and should equal two 9-gauge wires placed 12 or 16 in. on center vertically. Steel reinforcement can also be embedded in bond beam courses, or grouted into continuous vertical cells. If screen walls must support vertical loads other than their own weight, the allowable compressive stress on the gross area should not exceed 50 psi. Where units are laid with interrupted bed joints, the allowable stresses should be reduced in proportion to the reduction in mortar bed area.

Lateral support for *concrete masonry garden walls and fences* is usually provided by reinforced pilasters or internal vertical reinforcement. Pilaster spacing and reinforcement requirements are listed in *Fig. 9-25*. Foundations should be placed in undisturbed soil below the frost line. For stable soil conditions where frost heave is not a problem, a shallow continuous footing or pad footing provides adequate stability. Where it is necessary to go deeper to find solid bearing material, where location in relation to property lines restricts footing widths, or where the ground is steeply sloping, a deep pier foundation provides better support. In each

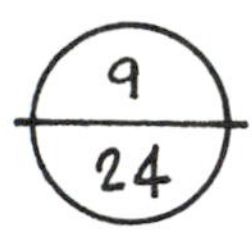

Maximum span for various concrete masonry screen walls. (*From National Concrete Masonry Association*, TEK Bulletin 5, *NCMA, Herndon, Va.*)

Construction	Minimum nominal thickness, t (in.)	Maximum distance between lateral supports (height or length, not both)			
		Nominal thickness of wall (in.)			Other nominal thicknesses
		4	6	8	
Non-loadbearing reinforced*					
Exterior	4	10′–0″	15′–0″	20′–0″	30t
Interior	4	16′–0″	24′–0″	32′–0″	48t
Unreinforced					
Exterior	4	6′–8″	10′–0″	13′–4″	20t
Interior	4	12′–0″	18′–0″	24′–0″	36t
Loadbearing					
Reinforced*	6	Not recommended	12′–6″	16′–8″	25t
Unreinforced	6		9′–0″	12′–0″	18t

*Total steel area, including joint reinforcement, not less than 0.002 times the gross cross-sectional area of the wall, not more than two-thirds of which may be used in either vertical or horizontal direction.

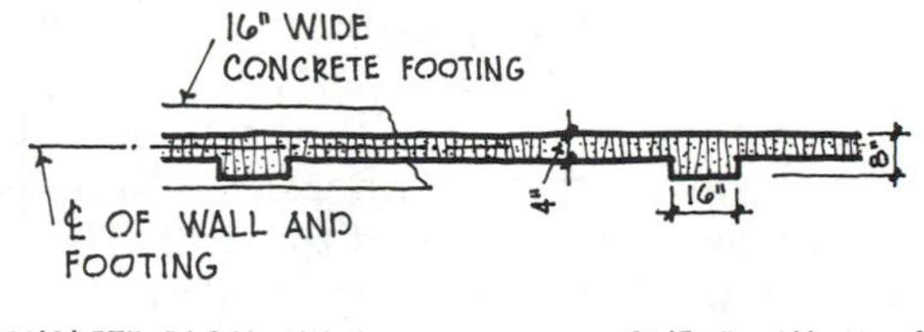

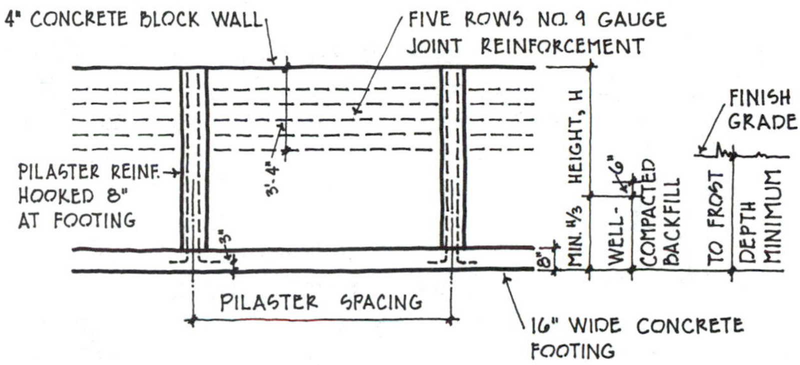

(A) Fence with pilasters

Pilaster spacing for wind pressure					*Reinforcement for wind pressure*			
5 psf	*10 psf*	*15 psf*	*20 psf*	*H*	*5 psf*	*10 psf*	*15 psf*	*20 psf*
19′–4″	14′–0″	11′–4″	10′–0″	4′–0″	1—No. 3	1—No. 4	1—No. 5	2—No. 4
18′–0″	12′–8″	10′–8″	9′–4″	5′–0″	1—No. 3	1—No. 5	2—No. 4	2—No. 5
15′–4″	10′–8″	8′–8″	8′–0″	6′–0″	1—No. 4	1—No. 5	2—No. 5	2—No. 5

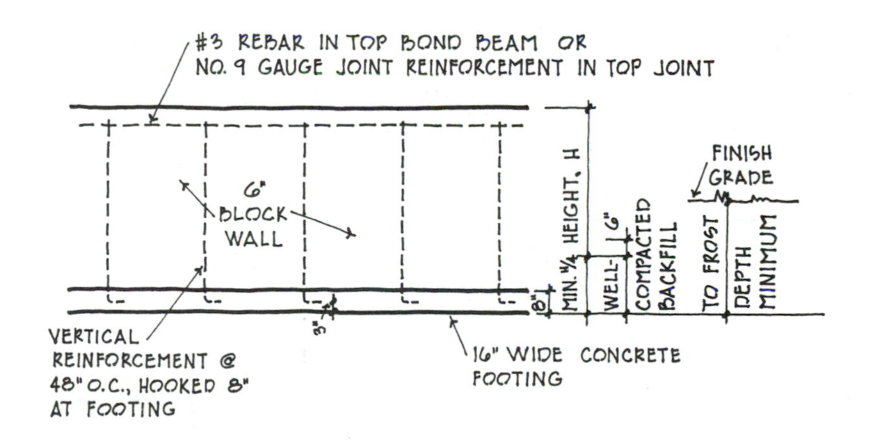

(B) Fence without pilasters

H	*Reinforcement for wind pressure*			
	5 psf	*10 psf*	*15 psf*	*20 psf*
4′–0″	1—No. 3	1—No. 3	1—No. 4	1—No. 4
5′–0″	1—No. 3	1—No. 4	1—No. 5	1—No. 5
6′–0″	1—No. 3	1—No. 4	1—No. 5	2—No. 4

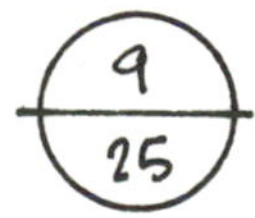

Pilaster spacing and reinforcing requirements for concrete block fences. (*From Frank A. Randall and William C. Panarese,* Concrete Masonry Handbook for Architects, Engineers, and Builders, *Portland Cement Association, Skokie, Ill., 1976.*)

instance, the supporting pilaster is tied to the foundation by continuous reinforcing steel. A vertical control joint for expansion and contraction may be provided on one side of the pilaster support. Reinforcing steel in the panel sections may run vertically or horizontally, depending on the type of units used and the bed joint design of the wall. The designs shown are based on wind loading conditions, but are not intended to resist lateral earth pressure as retaining walls.

Brick garden walls may take a number of different forms. A straight wall without pilasters must be designed with sufficient thickness to provide lateral stability against wind and impact loads. Rule of thumb is that for a 10-lb/sq ft wind load, the height above grade should not exceed three-fourths of the square of the wall thickness ($h \leq \frac{3}{4}t^2$). If lateral loads exceed 10 lb/sq ft, the wall should be designed with reinforcing steel.

Brick "pier-and-panel" walls are composed of a series of thin (nominal 4 in.) panels braced intermittently by reinforced masonry piers (*see Fig. 9-26*). Reinforcing steel and foundation requirements are given in the tables in *Fig. 9-27*. Foundation diameter and embedment are based on a minimum soil bearing pressure of 3000 lb/sq ft. Reinforcing steel requirements vary with wind load, wall height, and span. Horizontal steel may be individual bars or wires, or may be prefabricated joint reinforcement, but must be continuous through the length of the wall with splices lapped 16 in.

Since the panel section is not supported on a continuous footing, it actually spans the clear distance between foundation supports, functioning as a deep wall beam (see Chapter 12). Masons build the sections on temporary 2 × 4 wood forms that are removed after the wall has cured for at least 7 days.

9/26 Brick pier-and-panel garden walls. (*From Brick Institute of America, Technical Note 29A, BIA, McLean, Va.*)

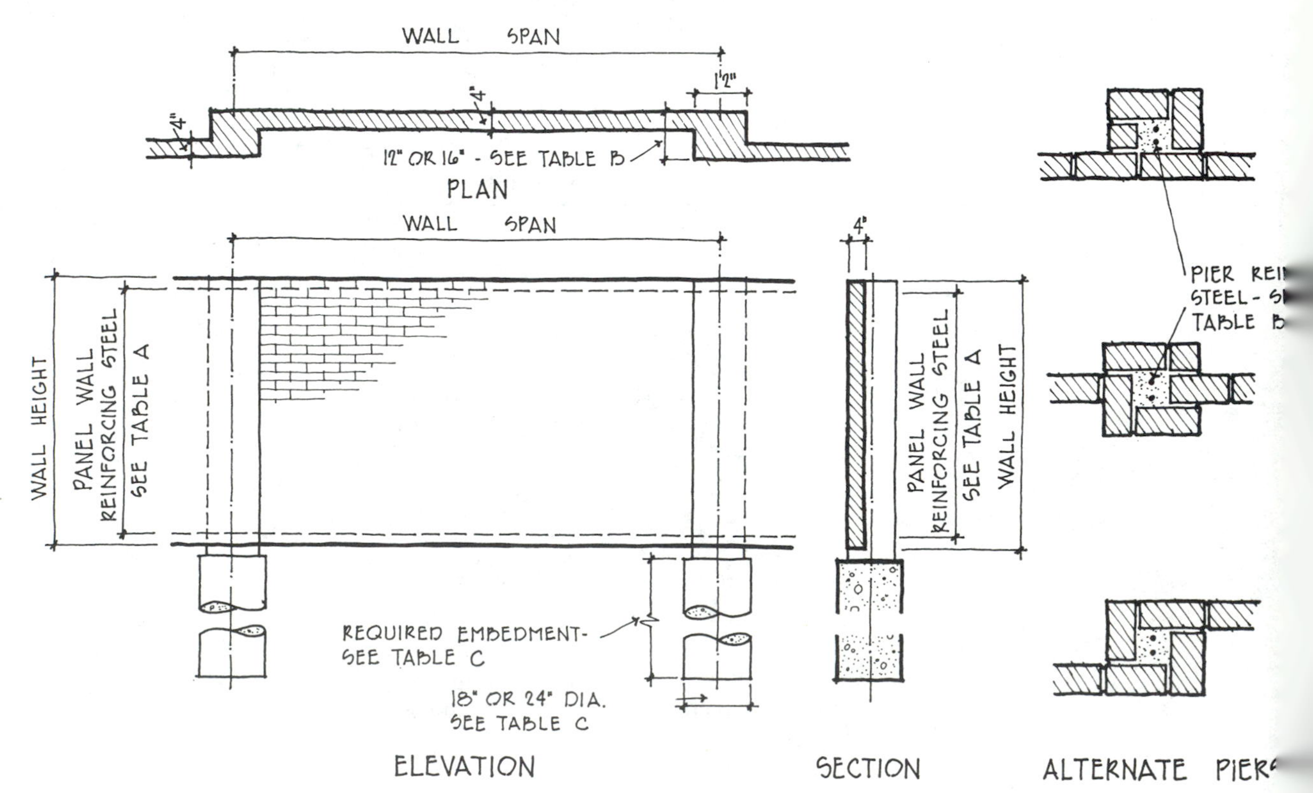

TABLE A PANEL WALL REINFORCING STEEL									
	*Vertical Spacing** (*in.*)								
Wall span (*ft*)	*Wind load 10 psf*			*Wind load 15 psf*			*Wind load 20 psf*		
	A	*B*	*C*	*A*	*B*	*C*	*A*	*B*	*C*
8	45	30	19	30	20	12	23	15	9.5
10	29	19	12	19	13	8.0	14	10	6.0
12	20	13	8.5	13	9.0	5.5	10	7.0	4.0
14	15	10	6.5	10	6.5	4.0	7.5	5.0	3.0
16	11	7.5	5.0	7.5	5.0	3.0	6.0	4.0	2.5

*A, two - No. 2 bars; B, two - $\frac{3}{16}$-in. diam wires; C, two - 9 gauge wires.

TABLE B PIER REINFORCING STEEL*									
Wall span (*ft*)	*Wind load 10 psf*			*Wind load 15 psf*			*Wind load 20 psf*		
	Wall height (*ft*)			*Wall height* (*ft*)			Wall height (*ft*)		
	4	*6*	*8*	*4*	*6*	*8*	*4*	*6*	*8*
8	2#3	2#4	2#5	2#3	2#5	2#6	2#4	2#5	2#5
10	2#3	2#4	2#5	2#4	2#5	2#7	2#4	2#6	2#6
12	2#3	2#5	2#6	2#4	2#6	2#6	2#4	2#6	2#7
14	2#3	2#5	2#6	2#4	2#6	2#6	2#5	2#5	2#7
16	2#4	2#5	2#7	2#4	2#6	2#7	2#5	2#6	2#7

*Within heavy lines 12 by 16-in. pier required. All other values obtained with 12 by 12-in. pier.

TABLE C REQUIRED EMBEDMENT FOR PIER FOUNDATION*									
Wall span (*ft*)	*Wind load 10 psf*			*Wind load 15 psf*			*Wind load 20 psf*		
	Wall height (*ft*)			*Wall height* (*ft*)			*Wall height* (*ft*)		
	4	*6*	*8*	*4*	*6*	*8*	*4*	*6*	*8*
8	2′-0″	2′-3″	2′-9″	2′-3″	2′-6″	3′-0″	2′-3″	2′-9″	3′-0″
10	2′-0″	2′-6″	2′-9″	2′-3″	2′-9″	3′-3″	2′-6″	3′-0″	3′-3″
12	2′-3″	2′-6″	3′-0″	2′-3″	3′-0″	3′-3″	2′-6″	3′-3″	3′-6″
14	2′-3″	2′-9″	3′-0″	2′-6″	3′-0″	3′-3″	2′-9″	3′-3″	3′-9″
16	2′-3″	2′-9″	3′-0″	2′-6″	3′-3″	3′-6″	2′-9″	3′-3″	4′-0″

*Within heavy lines 24-in. diam. foundation required. All other values obtained with 18-in. diam. foundation.

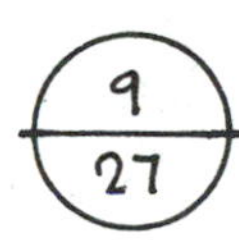

Design tables for brick pier-and-panel garden walls. (*From Brick Institute of America*, Technical Note 29A, *BIA*, *McLean*, *Va.*)

Serpentine walls and "folded plate" designs are laterally stable because of their shape. This permits the use of very thin sections without need for reinforcing steel or other lateral support. For non-loadbearing walls of relatively low height, rule-of-thumb design based on empirically derived geometric relationships is used.

Since the wall depends on its shape for lateral strength, it is important that the degree of curvature be sufficient. Recommendations for brick and CMU walls are illustrated in *Fig. 9-28*. The brick wall is based on a radius of curvature not exceeding twice the height of the wall above finished grade, and a depth of curvature from front to back no less than one-half of the height. A maximum height of 15 times the thickness is recommended for the CMU wall, and depth-to-curvature ratios are slightly different. Free ends of a serpentine wall should be supported by a pilaster or short-radius return for added stability. Thicker sections and taller walls may be built if proper design principles are applied to resist lateral wind loads.

All freestanding masonry walls, regardless of thickness, must be properly capped to prevent excessive moisture infiltration from the top. The appearance and character of a wall is substantially affected by the type of cap or coping selected. Designers may choose from several appropriate materials, including natural stone, cast stone, terra cotta, metal, brick, or

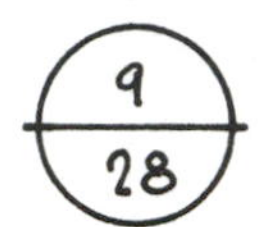

Serpentine walls. (*From Brick Institute of America*, Technical Note 29A, *BIA, McLean, Va.; and Frank A. Randall and William C. Panarese*, Concrete Masonry Handbook for Architects, Engineers, and Builders, *Portland Cement Association, Skokie, Ill., 1976.*)

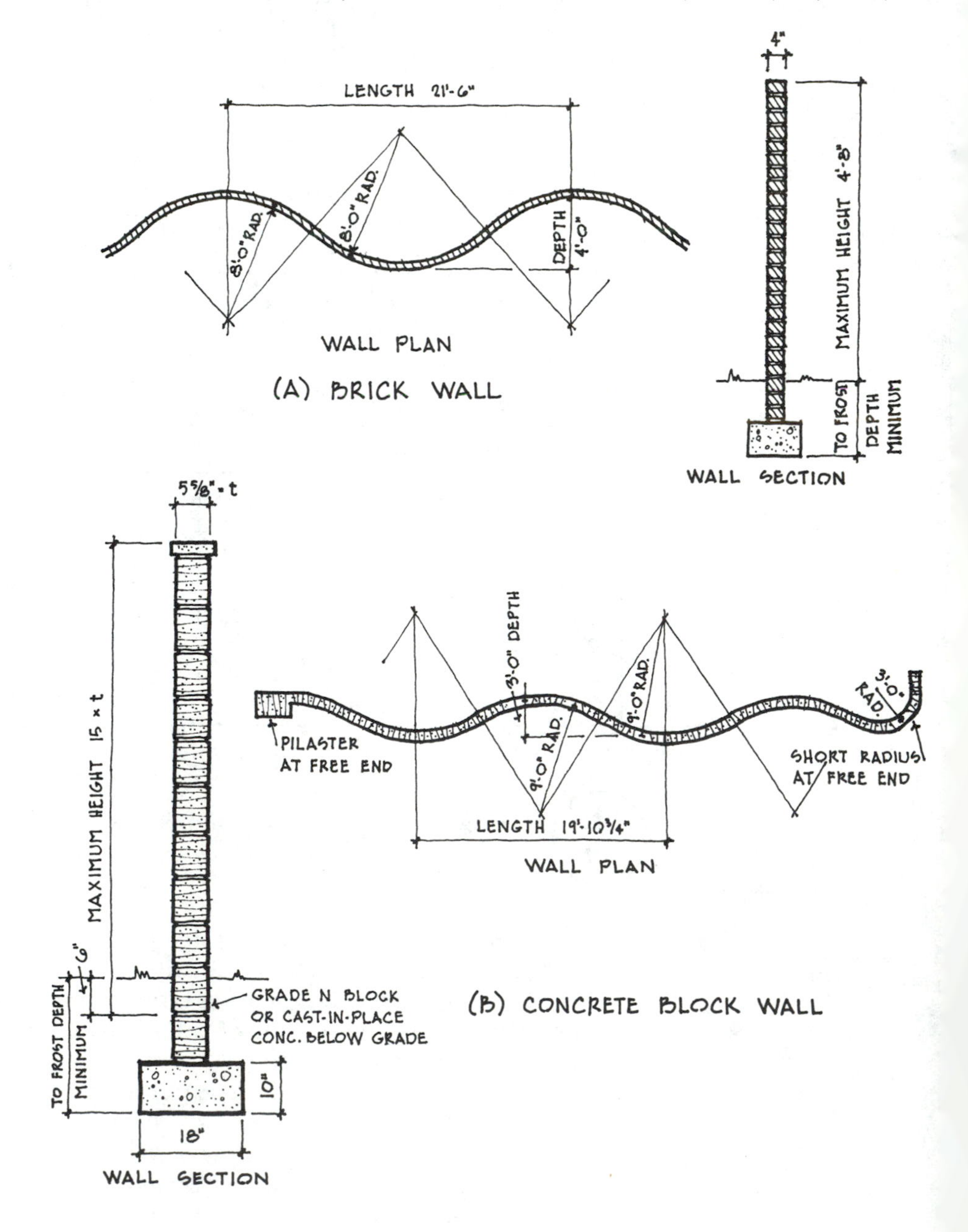

concrete masonry (*see Fig. 9-29*). The coping should project beyond the face of the wall a minimum of $\frac{1}{2}$ in. on both sides, and should provide a positive drip to prevent water from flowing down the face of the wall. Copings should be secured with metal anchors and flexible movement joints should be provided. For wet and/or cold climates, through-wall flashing should be installed immediately below the coping to prevent damage from the freezing and thawing of absorbed water.

Natural stone is used to build freestanding dry-stack and mortared walls. *Dry-stack walls* laid without mortar are generally 18 to 24 in. wide and depend only on gravity for their stability. Trenches are dug to below the frost line, and if the ground slopes, may take the form of a series of flat terraces. A concrete footing may be poured in the trench, but walls are often laid directly on undisturbed soil. Two rows of large stones laid with their top planes slightly canted toward the center will provide a firm base. All stones placed below grade should be well packed with earth in all the crevices. Stones should be well fitting, requiring a minimum number of shims. A bond stone equal to the full wall width should be placed every 3 or 4 ft in each course to secure the inner and outer wythes. All of the stones should be slightly inclined toward the center of the wall so that the weight leans in on itself (*see Fig. 9-30*). Greater wall heights require more incline from base to

9/29 Wall caps and copings. (*From Brick Institute of America*, Technical Note 29A, *BIA, McLean, Va.*)

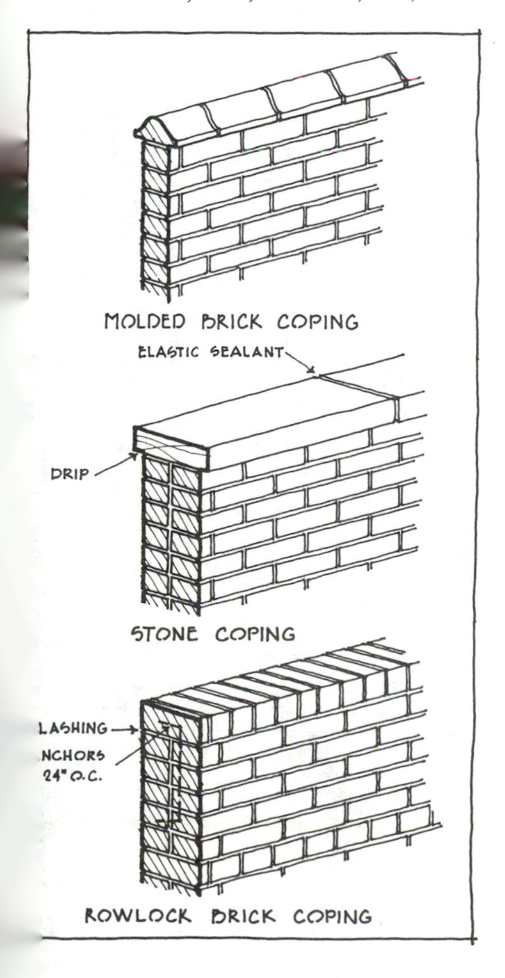

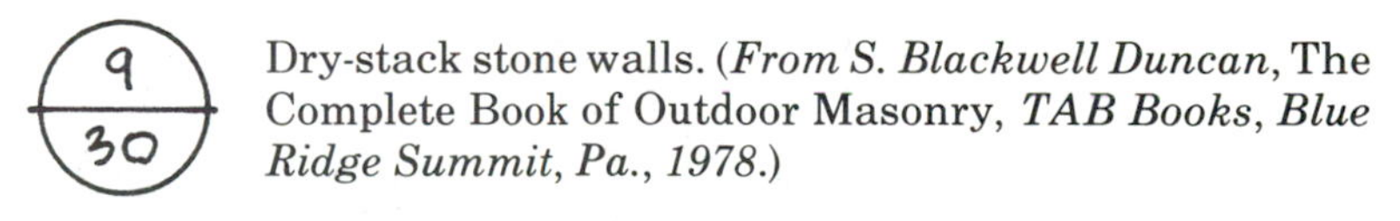

9/30 Dry-stack stone walls. (*From S. Blackwell Duncan*, The Complete Book of Outdoor Masonry, *TAB Books, Blue Ridge Summit, Pa., 1978.*)

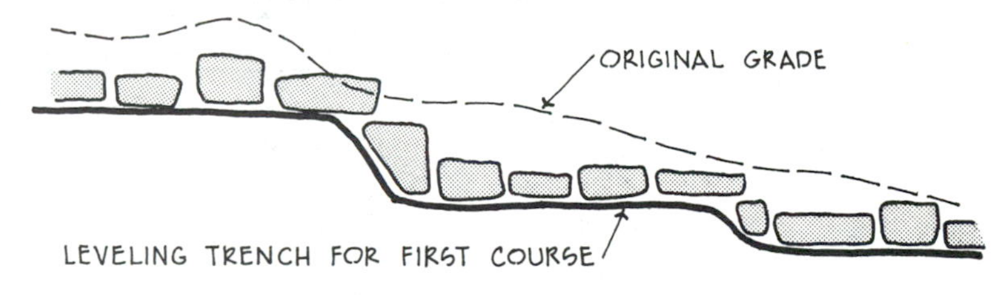

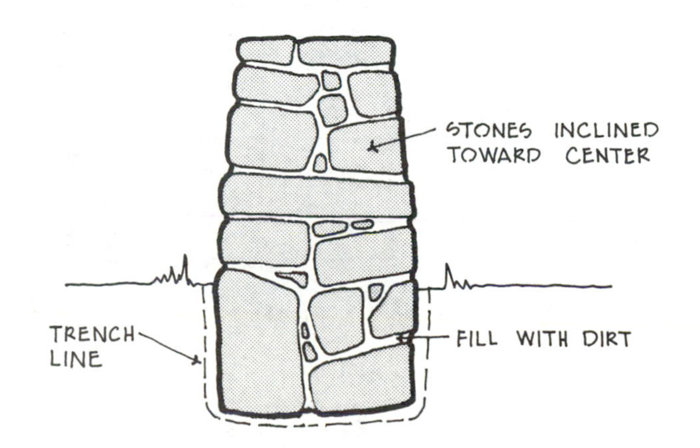

cap. Wall ends and corners are subject to the highest stress and should be built with stones tightly interlocked for stability. Relatively flat slabs of roughly rectangular shape work best for cap stones. The top course should be as level as possible for the full length of the wall.

Mortared stone walls are laid on concrete footings poured below the frost line. Rubble stone or fieldstone walls are laid up in much the same way as dry-stack walls except that the voids and cavities are filled with mortar. Type S, M, or N mortar should be used, and each course should be laid in a full mortar bed for maximum bond and strength. In ashlar construction,

9/31 Masonry fireplace details. (*From Harry C. Plummer*, Brick and Tile Engineering, *Brick Institute of America, McLean, Va., 1962.*)

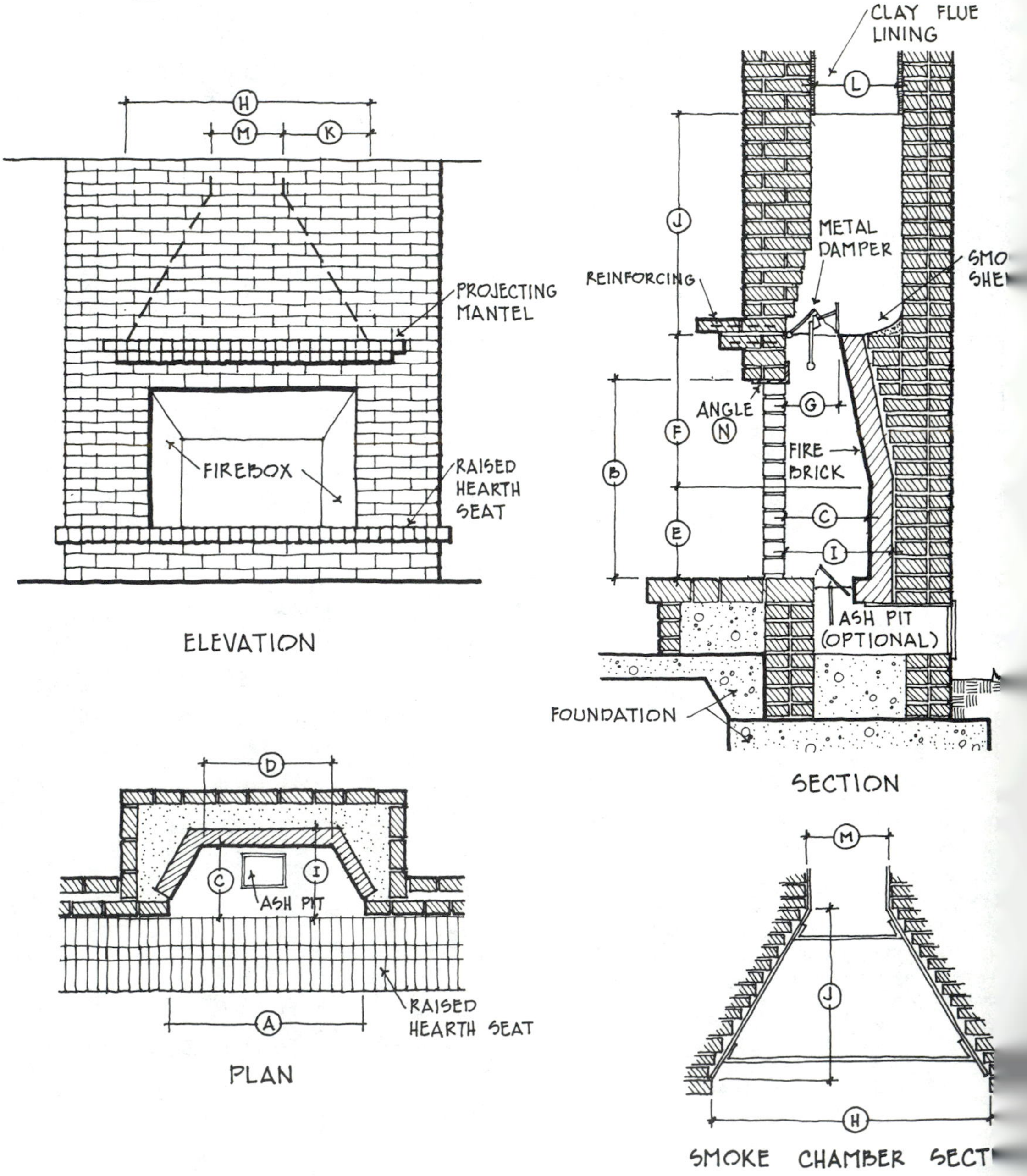

building codes generally require that bond stones be uniformly distributed and account for no less than 10% of the exposed face area. Mortared rubble stone walls less than 24 in. thick must have bond stones at a maximum of 3 ft on center vertically and horizontally. For thicknesses greater than 24 in., provide one bond stone for each 6 sq ft of wall surface. For non-loadbearing conditions, the minimum thickness of the wall must be sufficient to withstand all horizontal forces and vertical dead loads. For relatively low structures, a mortared wall thickness of as little as 8 in. has proven satisfactory.

9.4 FIREPLACES

Residential fireplace design has evolved over the centuries toward standardization of the functional elements that assure successful operation. A fireplace must have proper fuel combustion, good chimney draw, and maximum heat radiation. Design should also provide simplicity of construction and fire safety, particularly when adjacent building elements are of combustible materials.

The proper functioning of a fireplace is related to the shape and relative dimensions of the combustion chamber or fire box, the proper locating of the fireplace throat in relation to the smoke shelf, and the ratio of the flue area to the area of the fireplace opening (*see Figs. 9-31 through 9-33*). The shape of

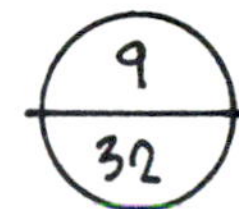

Standard fireplace dimensions (*see Fig. 9-31*). (*From Harry C. Plummer,* Brick and Tile Engineering, *Brick Institute of America, McLean, Va., 1962.*)

Finished fireplace opening							*Rough brick work and flue sizes*									*Steel angles†*
										*New sizes**			*Old sizes*			
A	*B*	*C*	*D*	*E*	*F*	*G*	*H*	*I*	*J*	*K*	*L M*	*R‡*	*K*	*L*	*M*	*N*
24	24	16	11	14	18	$8\frac{3}{4}$	32	20	19	10	8×12	8	$11\frac{3}{4}$	$8\frac{1}{2}$×	$8\frac{1}{2}$	A-36
26	24	16	13	14	18	$8\frac{3}{4}$	34	20	21	11	8×12	8	$12\frac{3}{4}$	$8\frac{1}{2}$×	$8\frac{1}{2}$	A-36
28	24	16	15	14	18	$8\frac{3}{4}$	36	20	21	12	8×12	10	$11\frac{1}{2}$	$8\frac{1}{2}$×	13	A-36
30	29	16	17	14	23	$8\frac{3}{4}$	38	20	24	13	12×12	10	$12\frac{1}{2}$	$8\frac{1}{2}$×	13	A-36
32	29	16	19	14	23	$8\frac{3}{4}$	40	20	24	14	12×12	10	$13\frac{1}{2}$	$8\frac{1}{2}$×	13	A-42
36	29	16	23	14	23	$8\frac{3}{4}$	44	20	27	16	12×12	12	$15\frac{1}{2}$	13 ×	13	A-42
40	29	16	27	14	23	$8\frac{3}{4}$	48	20	29	16	12×16	12	$17\frac{1}{2}$	13 ×	13	A-48
42	32	16	29	14	26	$8\frac{3}{4}$	50	20	32	17	16×16	12	$18\frac{1}{2}$	13 ×	13	A-48
48	32	18	33	14	26	$8\frac{3}{4}$	56	22	37	20	16×16	15	$21\frac{1}{2}$	13 ×	13	B-54
4	37	20	37	16	29	13	68	24	45	26	16×16	15	25	13 ×	18	B-60
0	37	22	42	16	29	13	72	27	45	26	16×20	15	27	13 ×	18	B-66
0	40	22	42	16	31	13	72	27	45	26	16×20	18	27	18 ×	18	B-66
2	40	22	54	16	31	13	84	27	56	32	20×20	18	33	18 ×	18	C-84
4	40	24	64	20	28	13	96	29	61	36	20×24	20	36	20 ×	20	C-96
6	40	24	76	20	28	13	108	29	75	42	20×24	22	42	24 ×	24	C-108

ew flue sizes conform to modular dimensional system. Sizes shown are nominal. Actual size is $\frac{1}{2}$ in. less each dimension.

ngle sizes: A, 3 by 3 by $\frac{3}{16}$ in.; B, $3\frac{1}{2}$ by 3 by $\frac{1}{4}$ in.; C, 5 by $3\frac{1}{2}$ by $\frac{5}{16}$ in.

ound flues.

	Rectangular liners				Round Liners	
	Standard		Modular			
Fireplace width	Outside dimensions	Effective area	Nominal outside dimensions	Effective area	Inside diameter	Effective area
24	$8\frac{1}{2} \times 8\frac{1}{2}$	52.56	8 × 12	57	8	50.26
			8 × 16	74		
30–34	$8\frac{1}{2} \times 13$	80.50			10	78.54
			12 × 12	87		
	$8\frac{1}{2} \times 18$	109.69	12 × 16	120	12	113.00
36–44	13 × 13	126.56	16 × 16	162	15	176.70
46–56	13 × 18	182.84	16 × 20	208		
58–68	18 × 18	248.06	20 × 20	262	18	254.40
70–84			20 × 24	320	20	314.10
			24 × 24	385	22	380.13
					24	452.30

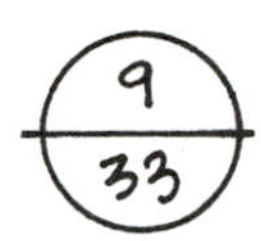

Recommended flue sizes (in inches). (*From Harry C. Plummer,* Brick and Tile Engineering, *Brick Institute of America, McLean, Va., 1962.*)

the combustion chamber influences both the draft and the amount of hea radiated into the room. The dimensions recommended in the tables may b varied slightly to correspond with brick coursing for modular and nonmodu lar unit sizes, but it is inadvisable to make significant changes. A multifac fireplace can be a highly effective unit, but presents certain problems o draft and opening size that must sometimes be solved on an individua basis. The single-face fireplace is the most common and the oldest design and the majority of the standard detail information is based on this type.

The structural design of residential wood-burning fireplaces an chimneys is a fairly simple and well-known empirical procedure. Fc commercial and industrial chimneys however, the sizes are sufficientl larger that a number of different forces are in effect. The principal force acting on a large masonry chimney are (1) wind pressure; (2) the weight (the chimney, including its lining and foundation; (3) the foundatio reaction; and (4) the expansion and contraction of the chimney walls due t the difference in temperature between the combustion gases inside and th surrounding air.

Design wind loads are listed in most building codes by geographic are and height zone. Since industrial chimneys can often be as tall as 100 ft, th potential lateral loads are significant. Chimneys constructed in area subject to earthquakes must be designed as reinforced masonry. A samp design of a reinforced brick chimney 80 ft tall is shown in *Fig. 9-34*. Th chimney shell measures 15 in. at the base and 9 in. in the upper section, wi the profile tapering toward the top. Design for an identical chimney unreinforced construction is shown in *Fig. 9-35*. Shell thickness at the ba of this chimney is almost 2 ft exclusive of the flue lining. Many unreinforc chimneys perform satisfactorily without horizontal reinforcement to resi thermal stresses. However, as an added safety factor for high flue-g temperatures, two No. 2 bars placed in bed joints 2 in. from the interior a exterior faces of the wall and spaced 16 in. on center vertically may be use

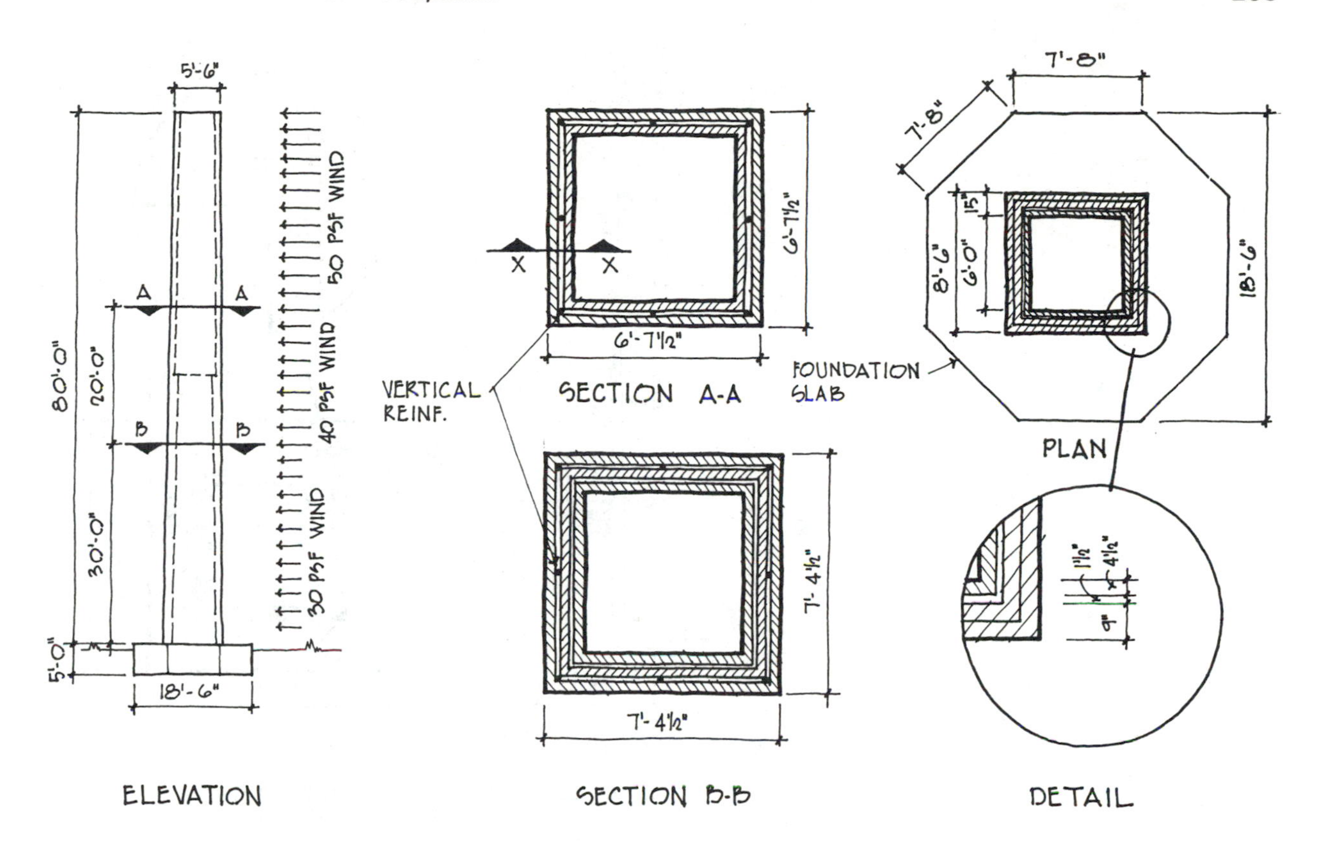

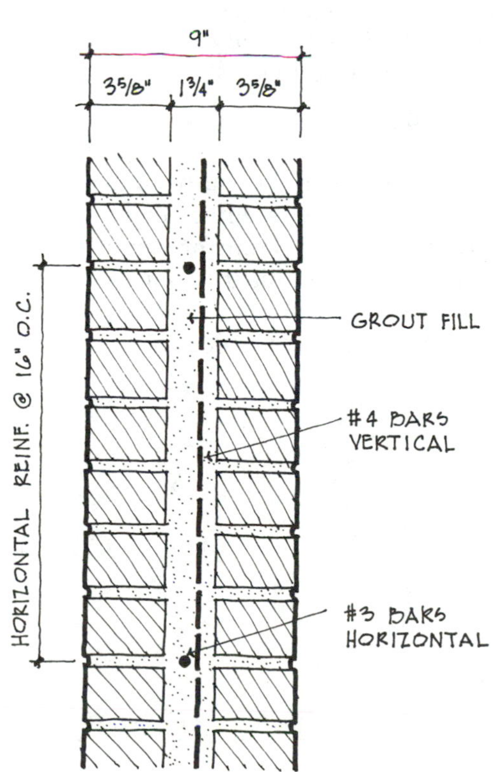

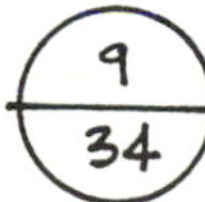

Reinforced brick masonry industrial chimney. (*From Brick Institute of America*, Technical Note 19B, *BIA*, *McLean*, *Va.*)

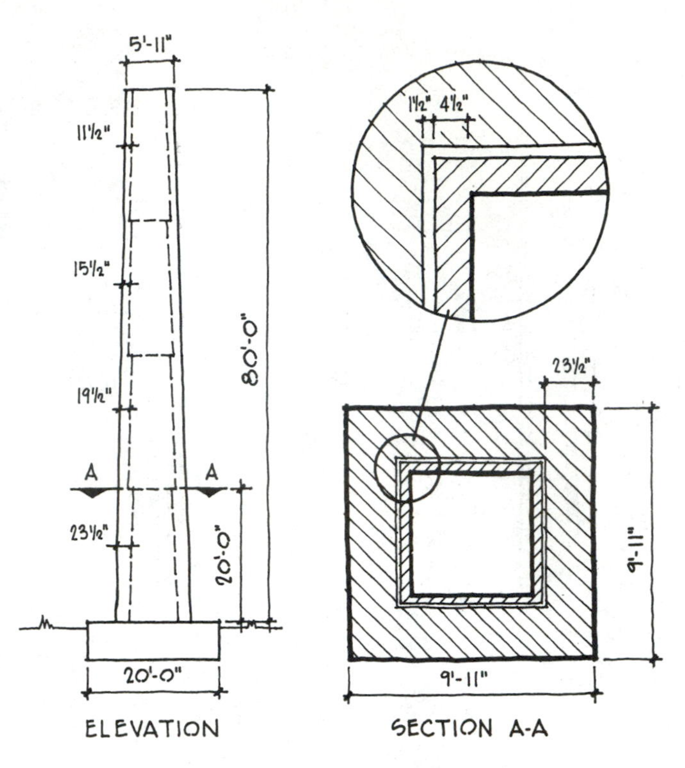

Unreinforced brick masonry industrial chimney. (*From Brick Institute of America*, Technical Note 19B, *BIA*, *McLean*, *Va.*)

Unreinforced masonry chimneys can be built to withstand win pressures exceeding 30 lb/sq ft, but in areas where considerable latera forces may be anticipated, the addition of steel reinforcement usually offer a more economical solution. In areas not subject to seismic activity, an where wind pressures are less severe, unreinforced chimneys may be safel and economically built to a height approaching 100 ft.

10

PLAIN AND PARTIALLY REINFORCED LOADBEARING MASONRY

Extensive structural engineering design is beyond the intended scope of this text. This chapter and Chapters 11 and 12 give only an overview of the structural concepts of masonry bearing wall design. For detailed design analysis methods and formulas, the reader should consult Schneider and Dickey's *Reinforced Masonry Design*, Amrhein's *Masonry Design Manual*, or Plummer's *Brick and Tile Engineering*.

The general concept of a bearing wall structure is combined action of the floor, roof, and walls in resisting applied loads. The bearing walls can be considered as continuous vertical members supported laterally by the floor system. Vertical live loads and dead loads are transferred to the walls by the floor system acting as a horizontal flexural member. The floor system also acts as a diaphragm to carry lateral loads and horizontal reactions to the walls for transfer to the foundation. Gravity loads and lateral loads applied from only one side will induce moment in an exterior wall. The total moment is a result of the combination of vertical and lateral loading. Since gravity loads oppose the tension from this bending moment, the primary stresses controlling loadbearing systems are compression and shear. The diagrams in *Fig. 10-1* illustrate the typical forces acting on bearing walls.

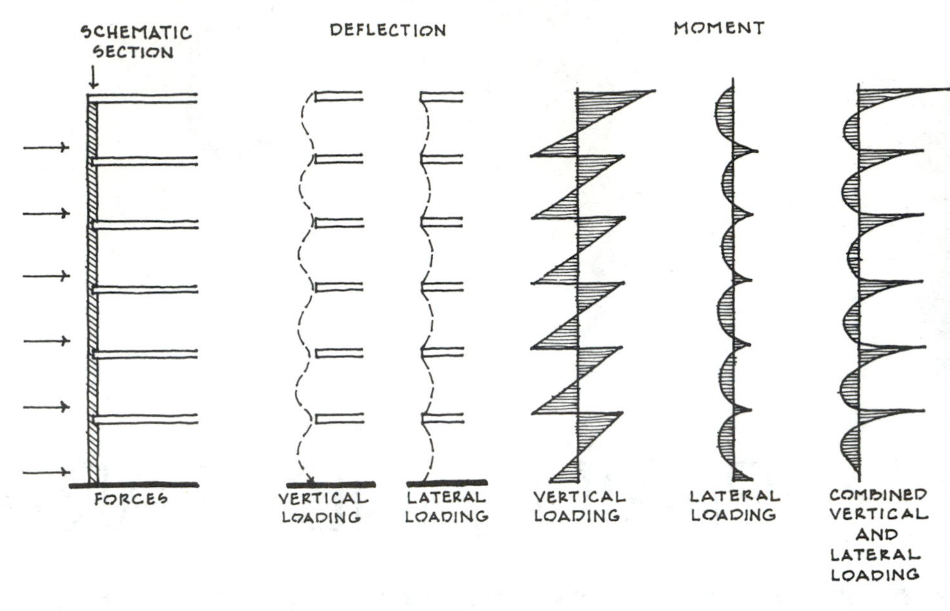

Vertical and horizontal forces acting on a bearing wall.

10.1 BEARING WALL SYSTEMS

In unreinforced masonry, shear and flexural stresses are resisted only by the mass of the wall and by the bond between mortar and units. Shearing stresses in bearing wall buildings, however, will seldom control the wal type and thickness. Although flexural stresses may control the design c shear walls under certain conditions, it is the bearing stresses that wi generally govern in loadbearing structures (*see Fig. 10-2*).

Skeleton frame systems function through the use of rigid connection at the intersection of beams, girders, and columns, or by use of diagona bracing. This results in a concentration of stresses at joints in the frame an at the foundation. In a bearing wall system, or box frame, the structura floors and walls constitute a series of intersecting planes with the resultin forces acting along continuous lines rather than at intermittent point The use of masonry in multistory loadbearing applications is dependent o the cohesive action of the structure as a whole. Overall strength ar successful performance in resisting loads are directly related to the strengt and performance of the floor and roof framing systems and the connection which must be capable of transferring forces horizontally to shear wall The floor system must be sufficiently rigid to function as a horizont diaphragm and the connections between walls and floors must be adequa to transmit lateral forces.

10.1.1 Plain Masonry

Plain masonry contains no reinforcing steel. It is very strong in compr sion, but weak in tension and shear. Small lateral loads and overturni moments are resisted by the mass of the wall, but if lateral loads a significant, plain masonry is limited by the bond between the mortar a the units, and the precompression effects of vertical loads.

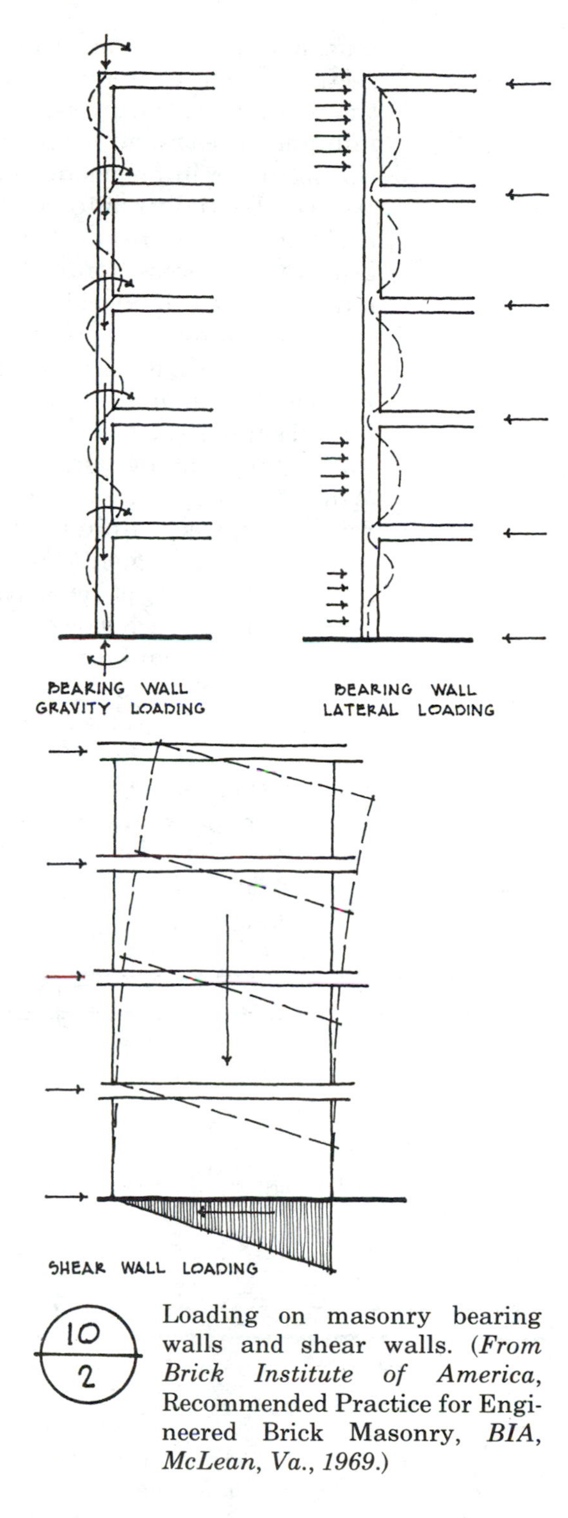

10/2 Loading on masonry bearing walls and shear walls. (*From Brick Institute of America,* Recommended Practice for Engineered Brick Masonry, *BIA, McLean, Va., 1969.*)

10.1.2 Partially Reinforced Masonry

Where lateral loads are a significant factor in the design of loadbearing buildings, flexural strength can be increased by placing steel reinforcing bars in mortar bed joints or in grouted cells or cavities. The hardened grout binds the masonry units and steel together so that they act as a single element in resisting applied forces. Requirements for fully reinforced masonry are based on a minimum area of steel (0.002 times the cross-sectional area of the wall) and maximum spacing of reinforcement (six times the wall thickness, or 48 in.). Partially reinforced walls have no minimum area requirement. They are designed the same as plain masonry except that reinforcement may be added to resist isolated flexural tensile stresses. Partially reinforced masonry uses steel only where design analysis indicates that tensile stress is developed. The amount and location, however, are less than code requirements specify for fully "reinforced" masonry (see Chapter 11).

For example, the effect of wind forces perpendicular to an exterior wall is usually negligible in the lower stories because the compressive stress is sufficient to suppress the tension developed as the wall spans between lateral supports. However, in the upper stories where wind loads are higher and axial loads lighter, the allowable tensile stresses may be exceeded. When this happens, it can be provided for by the selective location of reinforcing steel within the wall to resist the tension.

Code requirements for partially reinforced masonry walls limit the maximum spacing of vertical reinforcement to 8 ft. Reinforcement must also be provided at each side of openings and at each corner of the walls. The *UBC* requires that horizontal reinforcement at least 0.2 sq in. in area (one No. 4 or two No. 3 bars) be provided in bond beams at the top of footings, at the bottom and top of wall openings, at roof and floor levels, and at the top of parapet walls (*see Fig. 10-3*). ANSI and NCMA standards require reinforcement in these same locations but do not stipulate a minimum bar diameter. They do require $\frac{1}{4}$-in.-diameter bars placed 16 in. on center vertically (or their equivalent) for units laid in stack bond.

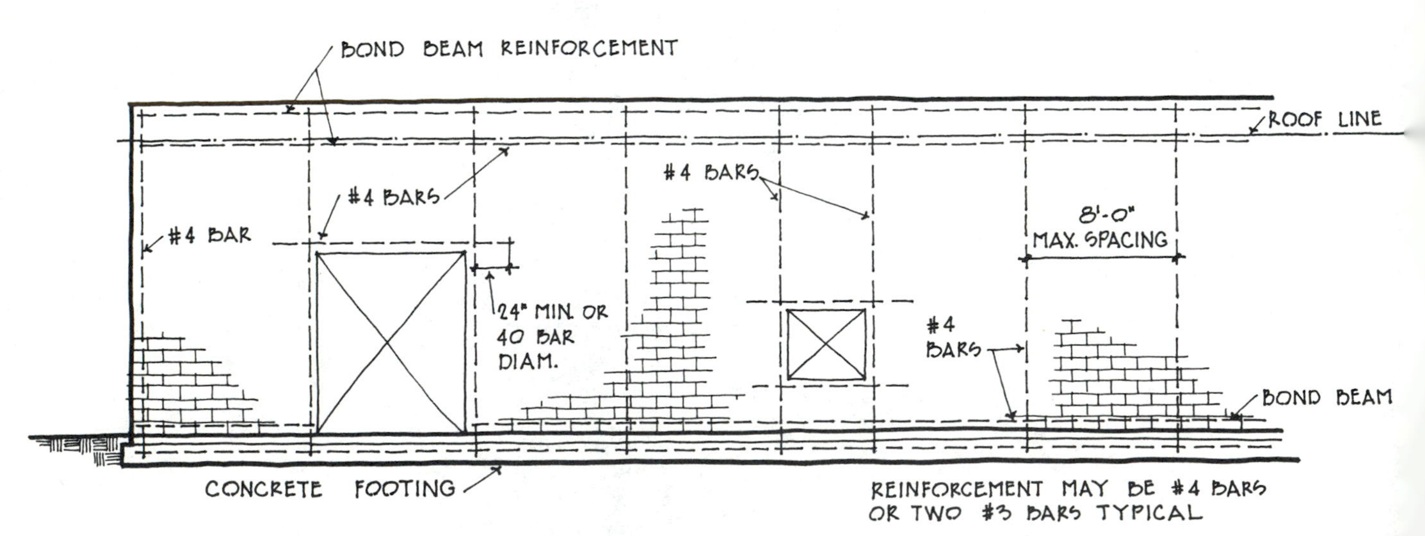

10-3 Partially reinforced masonry wall. (*From Masonry Institute of America,* Reinforcing Steel in Masonry, *MIA, Los Angeles, Calif.*)

10.1.3 Axial Load Distribution

When a superimposed axial load is applied to a masonry wall, it is assum[ed] to be distributed uniformly through a triangular section of the wall (*see Fi[g.] 10-4*). Bearing pads or plates should be used to distribute concentrated lo[ad] stresses and to permit any slight movement that may occur. An allowab[le]

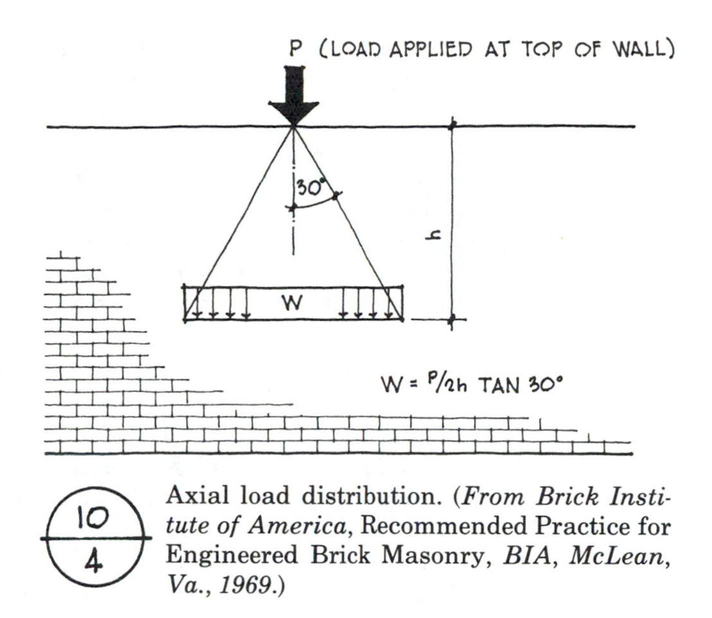

10-4 Axial load distribution. (*From Brick Institute of America*, Recommended Practice for Engineered Brick Masonry, *BIA, McLean, Va., 1969.*)

increase in design stress is permitted for loads of a purely local nature, such as at beam and girder reactions. Both the BIA code and the *UBC* permit an increase of 50% for loads bearing on less than one-third of the area of support, provided that the edge distance surrounding the bearing plate is a minimum of one-fourth of the parallel plate dimension. For walls, the area of support is assumed to equal the thickness of the wall times the distance between bearings. When a concentrated load results from a beam or girder bearing directly on a masonry wall, the reaction will not generally be in the center of the bearing area. Due to deflection of the member, it will move toward the inner face of the support. It may be assumed in such cases that the vertical reaction is one-third of the bearing distance from this inner face. If the vertical load develops an eccentricity that falls outside of the middle two-thirds of the wall, excessive tensile bending stresses may develop requiring the addition of steel reinforcing.

10.1.4 Foundations

Although the weight of a loadbearing structure is greater than that of a similar frame building, the required soil bearing capacity is often less because the bearing walls distribute the weight more evenly. Bearing wall structures are compatible with all of the common types of foundations, including reinforced concrete spread footings, pile footings, caissons, or mat footings. The foundation wall below grade may be of either concrete or masonry, but should be doweled to the footing to assure joint action of the wall and the foundation.

10.1.5 Lateral Load Distribution

The transverse strength of unreinforced masonry walls depends on three factors: (1) the type and design of the masonry unit, (2) the degree of bond between mortar and unit, and (3) the quality of workmanship. Higher transverse strengths are developed with solid masonry units because the wide mortar bed permits development of greater bond. Failure in lateral loading usually occurs as a ruptured bed joint, resulting from bond failure. Factors that will increase mortar bond strength therefore also increase the transverse strength of the wall. Portland cement-lime mortars generally produce greater bond than proprietary masonry cement mortars. Greater transverse strengths are also developed with full, unfurrowed mortar joints.

When a lateral load is applied perpendicular to a wall, it will be transmitted to vertical and horizontal edge supports. The proportion of the load transmitted either vertically or horizontally will depend on the flexural resistance and rigidity of the wall in both spans, the degree of fixity or restraint developed at the edges, the horizontal-to-vertical span ratio, and the distribution of the loads applied to the wall panel.

The need to investigate lateral wind resistance is greater in the case of nonbearing or lightly loaded walls. Lateral stability of masonry walls is increased by applied vertical loads producing compressive stresses which must be overcome by tensile stresses of the transverse loads before failure can occur. In the lower stories of loadbearing buildings, these compressive stresses are generally sufficient to overcome tension, but occasionally in the upper stories of taller buildings where exterior walls are subject to high lateral wind loads and small axial loads, the allowable flexural tensile stresses may be exceeded, thus requiring reinforcement.

10.1.6 Diaphragms

To function satisfactorily as a diaphragm, a roof or floor system must be capable of transmitting lateral forces to the shear walls without exceeding a deflection that would cause damage to any of the vertical structura elements. Deflection must be limited to prevent excessive tensile stresse and bending in the walls perpendicular to the direction of the lateral force

Maximum span-to-width or depth ratios are usually used to contro diaphragm deflection. If the diaphragm system is designed with the prope ratio, deflection generally will not be critical. A typical table of horizonta diaphragm proportions is reproduced in *Fig. 10-5.*

Maximum span-to-width ratios for roof or floor diaphragms. (*From Bric Institute of America,* Recommended Practice for Engineered Bric Masonry, *BIA, McLean, Va., 1969.*)

Diaphragm construction	*Maximum span/width ratio*	
	Masonry and concrete walls	*Wood and light steel walls*
Concrete (cast-in-place or precast)	Limited by deflection	
Steel deck (continuous sheet in a single plane)	4:1	5:1
Steel deck (without continuous sheet)	2:1	$2\frac{1}{2}$:1
Poured reinforced gypsum roofs	3:1	4:1
Plywood (nailed all edges)	3:1	4:1
Plywood (nailed to supports only—blocking may be omitted between joists)	$2\frac{1}{2}$:1*	$3\frac{1}{2}$:1
Diagonal sheathing (special)	3:1*	$3\frac{1}{2}$:1
Diagonal sheathing (conventional construction)	2:1*	$2\frac{1}{2}$:1

*The use of diagonally sheathed or unblocked plywood diaphragms for buildings having masonr reinforced concrete walls shall be limited to one-story buildings or to the roof of a top story.

The stiffness of the floor or roof diaphragm affects the distribution of lateral forces to the shear walls. No diaphragm is infinitely rigid, and no diaphragm capable of carrying loads is infinitely flexible. For the purposes of analysis, diaphragms are classified as *rigid*, *semirigid or semiflexible*, or *flexible*. Cast-in-place concrete or flat precast concrete slabs are considered rigid. Steel joists with structural concrete deck are considered semirigid or semiflexible, and steel or wood joists with wood decking are considered flexible. The distribution of horizontal loads to the vertical resisting elements depends on the relative rigidity of the diaphragm and the shear walls.

Diaphragms may be constructed of concrete, wood, metal, or combinations of materials. Working stresses for materials such as steel and reinforced concrete are well known and easily designed once the loading and reaction conditions are known. Where a diaphragm is made up of units such as plywood, precast concrete planks, or steel deck units, its characteristics are largely dependent on methods of attachment to one another and to supporting members. Such attachments must resist shearing stresses and provide proper ties to the supporting shear walls. The designer must ensure this action by appropriate detailing at the juncture between horizontal and vertical structural elements. There are no standard details which can be used without question. Each juncture of floor and wall must be analyzed as a separate problem and an appropriate detail developed for the specific conditions of the situation.

10.1.7 Shear Walls

The lateral load absorbed by a floor or roof diaphragm is transferred to shear walls. Shear walls are designed to resist lateral forces applied parallel to the plane of the wall (*see Fig. 10-2*). The orientation of the bearing walls in a building can minimize lateral load stresses and take advantage of the compressive and shear resistance of the walls. Shear walls perform best when they are also loadbearing, because the added loading offers greater resistance to overturning stresses. If all the bearing walls in a building are oriented in the same direction, they will resist lateral loads in only one direction. Nonbearing shear walls may then be needed in the other direction.

If analysis indicates that tension will be developed in unreinforced walls, the size, shape, or number of walls must be revised, or the wall must be designed as reinforced or partially reinforced masonry. (For seismic loading, see Chapter 11.)

10.1.8 Floor–Wall Connections

The box frame system of lateral load transfer is critically dependent on proper connection details. Unless the walls and diaphragms can transfer applied forces, the system does not work. Connections may be required to transmit axial loads, shear, bending moments, and so on. These may act separately or in combination. Connections can be formed with bolts, reinforcing dowels, mechanical devices, or welding, and may be either fixed or hinged. Although neither complete restraint nor a completely hinged condition actually exists, these assumptions may be made for purposes of calculation. Each individual condition will dictate the type of connection needed, and the engineer can usually design a variety of solutions for a given problem. The design and detailing of structural connections are covered in depth in the engineering texts listed at the beginning of this chapter. *Figures 10-6* through *10-9* show some typical details for attaching floor and roof systems to loadbearing masonry walls.

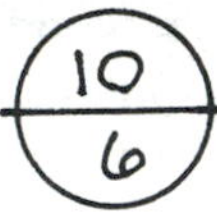

Floor–wall connections, restrained and hinged. (*From Brick Institute of America*, Technical Note 24B, *BIA, McLean, Va.*)

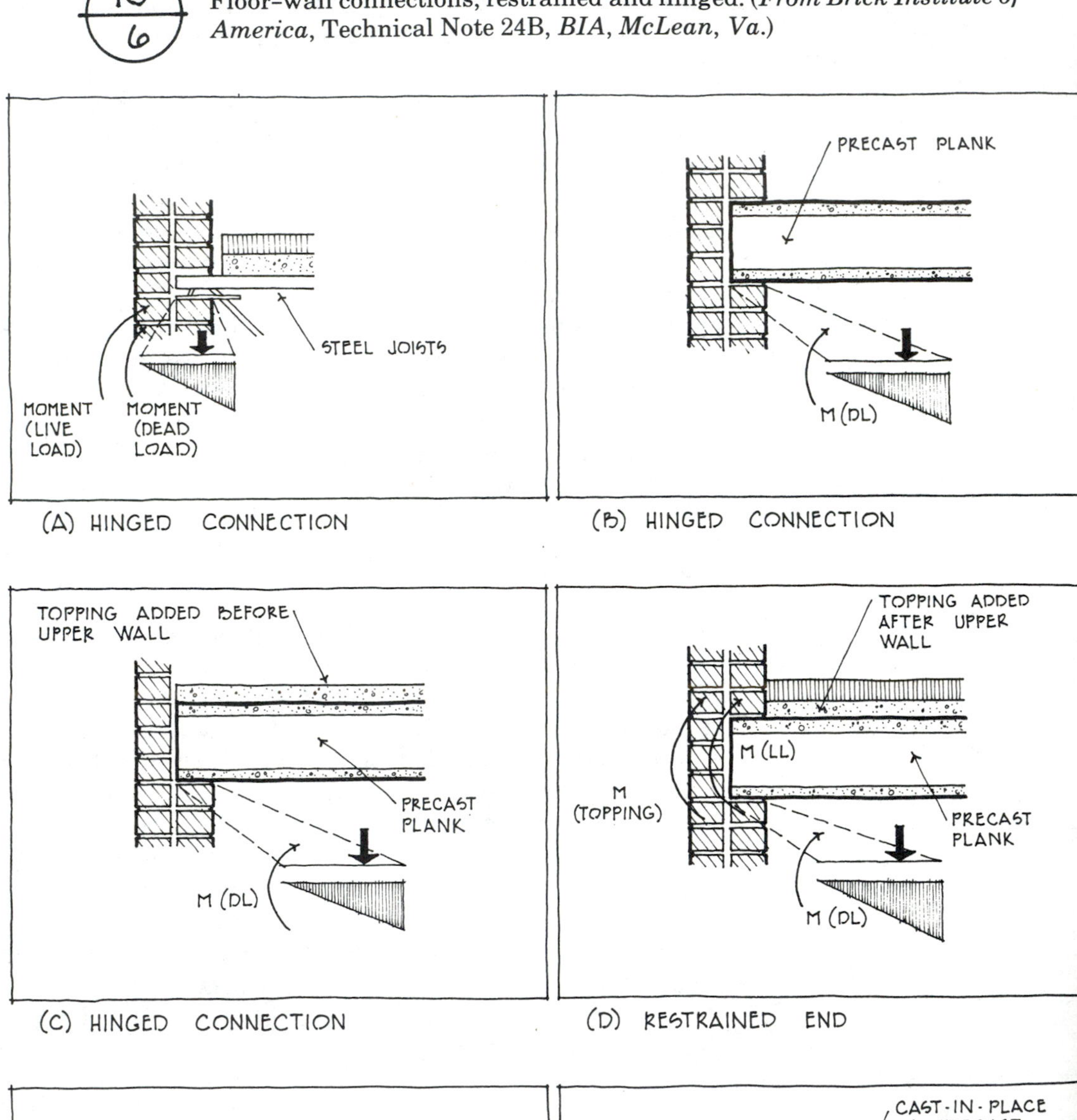

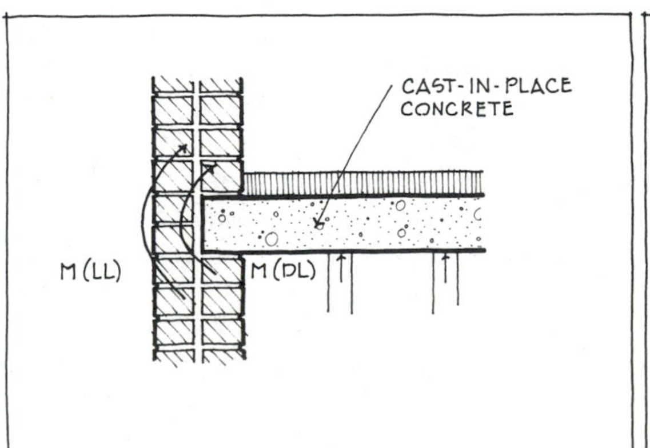

(E) RESTRAINED END

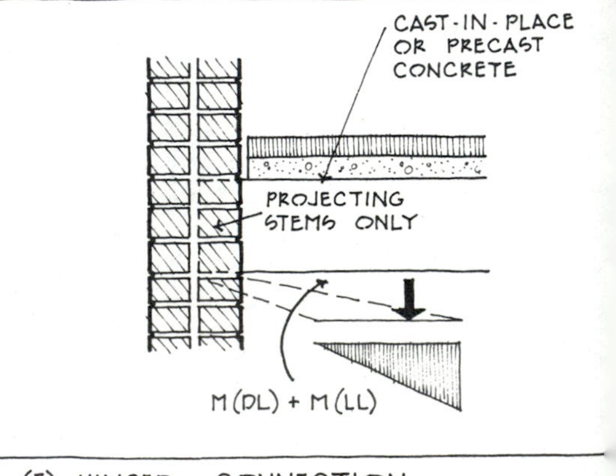

(F) HINGED CONNECTION

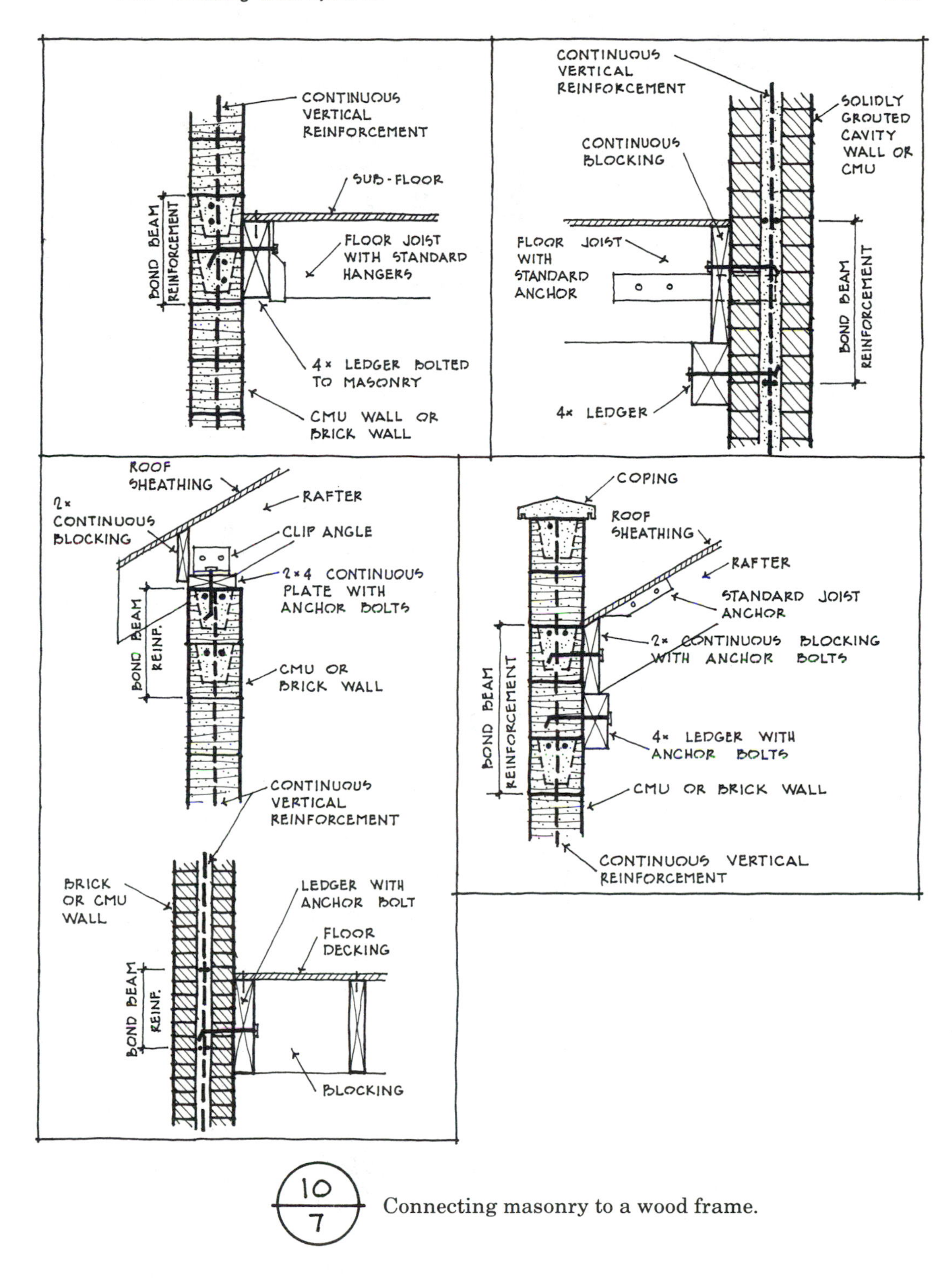

Connecting masonry to a wood frame.

10.1.9 Differential Movement

A building is a dynamic structure, and its successful performance depends on construction which allows movement of different elements to take place without causing distress. This movement can be accommodated by the use of flexible connections or, where the forces are relatively small, by the use of rigid connections properly reinforced to resist the stresses that will be developed. Movement joints are not as critical in loadbearing buildings

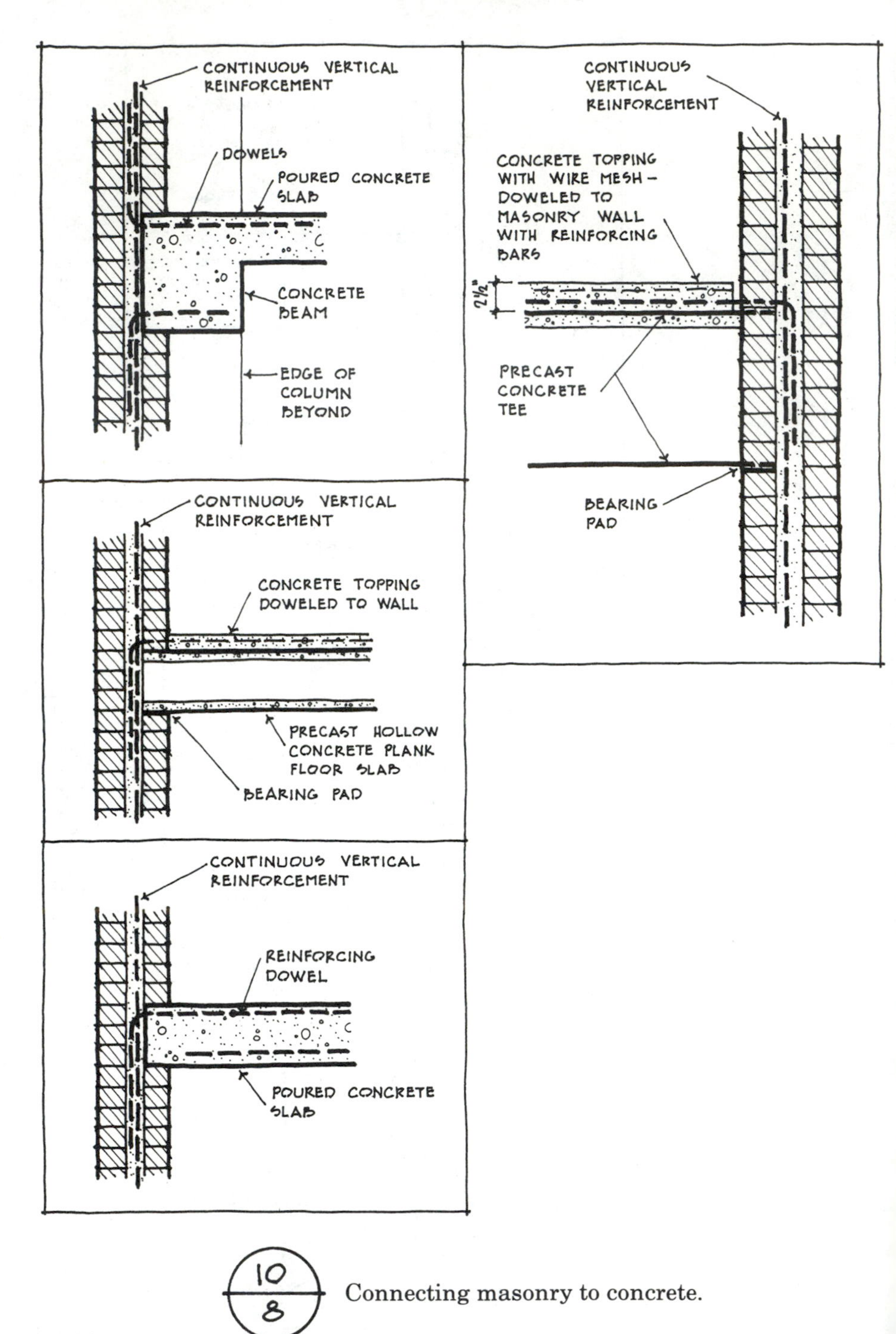

Connecting masonry to concrete.

because the walls are at a higher stress level and thus can better resi movement. One critical location is the joint between interior loadbearin walls and exterior non-loadbearing walls which are designed to act a flanges to resist shear. The exterior wall will undergo more therm movement and less elastic and creep-induced movement than will th interior wall. Ties between these walls must be designed for the effects this movement as well as for shear stress. Each building must be analyze as a unique situation and provisions made to relieve any potential excessive stresses.

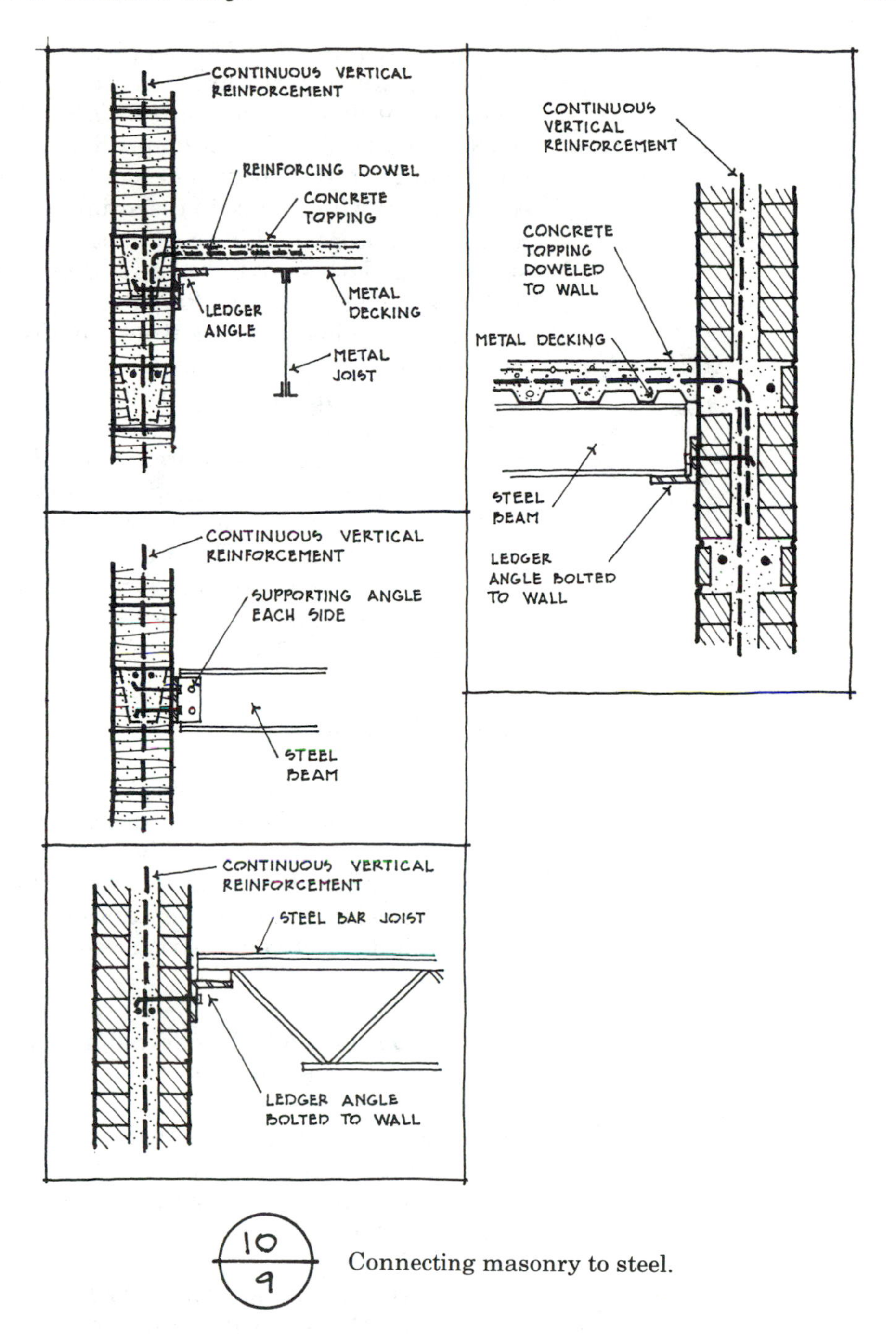

Connecting masonry to steel.

10.2 STRUCTURAL DESIGN

Empirical design methods may still be used for one- and two-story buildings, but rational engineering analysis will give better and more economical results for both plain and partially reinforced masonry. Empirical requirements are very simplistic in their application. Height or length-to-thickness ratios are used in conjunction with minimum wall thicknesses to determine the required section of a given wall. *Rationally designed* masonry does not require new techniques of analysis, merely the application of accepted engineering principles already developed for other structural systems. The rational design method is based on the properties of the component materials rather than on arbitrary empirical limitations, and may be used for high-rise or low-rise construction.

Code requirements for empirically designed unreinforced masonry loadbearing structures are based on the American National Standards Institute's *American Standard Building Code Requirements for Masonry, ANSI A41.1* (1970). The *National Building Code, Uniform Building Code*, and *Standard Building Code* contain sections governing this type of construction which contain essentially the same information as the ANSI Standard. The *Basic Building Code* incorporates *ANSI A41.1* as a reference standard and does not list any additional requirements.

Design methods for rational analysis of solid clay masonry construction are contained in the Brick Institute of America's *Recommended Practice for Engineered Brick Masonry* (1969), and for concrete masonry in the National Concrete Masonry Association's *Specification for the Design and Construction of Load-bearing Concrete Masonry* (1970). These two documents are listed as reference standards in the *Basic Building Code, National Building Code*, and *Standard Building Code*. The *Uniform Building Code* contains detailed requirements which are based on the same engineering principles. The American Concrete Institute (ACI) has issued *Building Code Requirements for Concrete Masonry Structures, ACI-531* (1979), and several other organizations are presently preparing rational design standards.

10.2.1 Empirical Design

ANSI A41.1 was first published in 1944, based on a 1931 Department of Commerce Publication. It contains no mathematical formulas for masonry design because it was written before any comprehensive testing had been performed and such formulas derived. Its requirements are based on very conservative information (the standard is being reviewed at the time of this writing) that effectively restricts economical use to buildings of relatively low height.

Materials. ANSI code requirements for quality of the masonry units, cement, aggregate, mortar, and grout are listed by reference to ASTM Standards. All five types of mortar described in ASTM C270 are recognized in the code, and their permitted use for the several kinds of masonry a listed in *Fig. 10-10* are based on the compressive strength of the mortar Stronger mortars are required for foundations, for cavity walls, and fo walls of hollow units.

Allowable stresses. *ANSI A41.1* lists allowable compressiv stresses which vary with unit and mortar type. Applied loads must b limited so that the maximum compressive stress in the extreme fiber of th wall does not exceed the value given in *Fig. 10-11*. To determine stresses i the masonry, the effects of all loads and conditions of loading, and th influence of all forces affecting design and strength must be considered. I the appendix to *ANSI A41.1*, the authoring committee recognizes th desirability of including allowable stresses for shear and for tension i flexure, but forgoes inclusion of such requirements because of a lack reliable data at the time the code was written—thus the need for conserv tive, empirical guidelines.

Lateral support requirements. In lieu of design by the allowab stress method, arbitrary limits were established to govern the ratio of th unsupported height or length to the nominal thickness of masonry wal

Kind of masonry	Types of mortar permitted
Foundations	
Footings	M or S
Walls of solid units	M, S or N
Walls of hollow units	M or S
Hollow walls	M or S
Masonry other than foundation masonry	
Piers of solid masonry	M, S, or N
Piers of hollow units	M or S
Walls of solid masonry	M, S, N or O
Walls of solid masonry, other than parapet walls or rubble stone walls, not less than 12 in. thick or more than 35 ft in height, supported laterally at intervals not exceeding 12 times the wall thickness	M, S, N, O, or K
Walls of hollow units; loadbearing or exterior, and hollow walls 12 in. or more in thickness	M, S, or N
Hollow walls, less than 12 in. in thickness where assumed design wind pressure:	
Exceeds 20 lb/sq ft	M or S
Does not exceed 20 lb/sq ft	M, S, or N
Glass block masonry	M, S, or N
Nonbearing partitions of fireproofing composed of structural clay tile or concrete masonry units	M, S, N, or O
Gypsum partition tile or block	Gypsum
Fire brick	Refractory air-setting mortar
Linings of existing masonry, either above or below grade	M or S
Masonry other than above	M, S, or N

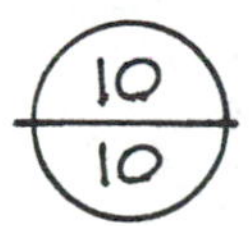

Types of mortar permitted for various kinds of masonry. (*From ANSI A41.1*)

and partitions. Lateral support must be provided in *either* the horizontal *or* the vertical direction within these limits. For solid masonry walls or bearing partitions laid in Type M, S, N, or O mortar, the ratio of unsupported height or length to nominal thickness cannot exceed 20. For walls of hollow masonry units or for cavity walls, regardless of the type of mortar used, the ratio cannot exceed 18. In computing the ratio for cavity walls, the value for thickness is taken as the sum of the nominal thicknesses of the inner and outer wythes only, excluding the width of the cavity itself.

When the limiting distance is to be measured horizontally, lateral support can be provided by cross walls, piers, or pilasters. When measured vertically, support may be obtained from floors, beams, girders, or roofs. Lateral support gives the wall sufficient strength to resist wind loads and other horizontal forces acting either inward or outward. These members must be adequately bonded or anchored to the wall and must be capable of transferring forces to adjacent structural members or directly to the ground. Pilasters may be either bonded into the wall or mechanically fastened to it across a continuous movement joint (*see Fig. 10-12*).

Wall thickness. Empirical standards require that clay and concrete masonry walls be at least 12 in. thick for the uppermost 35 ft of the wall and increase 4 in. for each successive 35 ft or fraction thereof measured

Construction; grade of unit	Allowable compressive stresses gross cross-sectional area (except as noted) (psi)				
	Type M mortar	Type S mortar	Type N mortar	Type O mortar	Type K mortar
Solid masonry of brick and other solid units of clay or shale; sand–lime or concrete brick					
8000 plus (psi)	400	350	300	200	100
4500 to 8000 (psi)	250	225	200	150	100
2500 to 4500 (psi)	175	160	140	110	75
1500 to 2500 (psi)	125	115	100	75	50
Grouted solid masonry of brick and other solid units of clay or shale; sand–lime or concrete brick					
4500 plus (psi)	350	275	200	—	—
2500 to 4500 (psi)	275	215	155	—	—
1500 to 2500 (psi)	225	175	125	—	—
Solid masonry of solid concrete masonry units					
Grade A concrete	175	160	140	100	—
Grade B concrete	125	115	100	75	—
Masonry of hollow units	85	75	70	—	—
Piers of hollow units, cellular spaces filled with mortar or concrete	105	95	90	—	—
Hollow walls (cavity or masonry bonded)					
Solid units					
2500 + psi	140	130	110	—	—
1500 to 2500 psi	100	90	80	—	—
Hollow units	70	60	55	—	—
Stone ashlar masonry					
Granite	800	720	640	500	—
Limestone or marble	500	450	400	325	—
Sandstone or cast stone	400	360	320	250	—
Rubble stone, coursed, rough, or random	140	120	100	80	

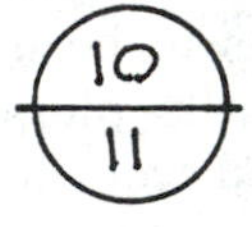

Allowable compressive stresses in unit masonry. (*From ANSI A41.1*)

downward from the top of the wall (*see Fig. 10-13*). There are, however several exceptions to this rule.

1. Where solid masonry bearing walls are stiffened by masonry cross walls or by reinforced concrete floors at distances of 12 ft or less the wall may be 12 in. thick for the uppermost 70 ft of the wall, and increase 4 in. in thickness for each successive 70 ft or fraction thereof.

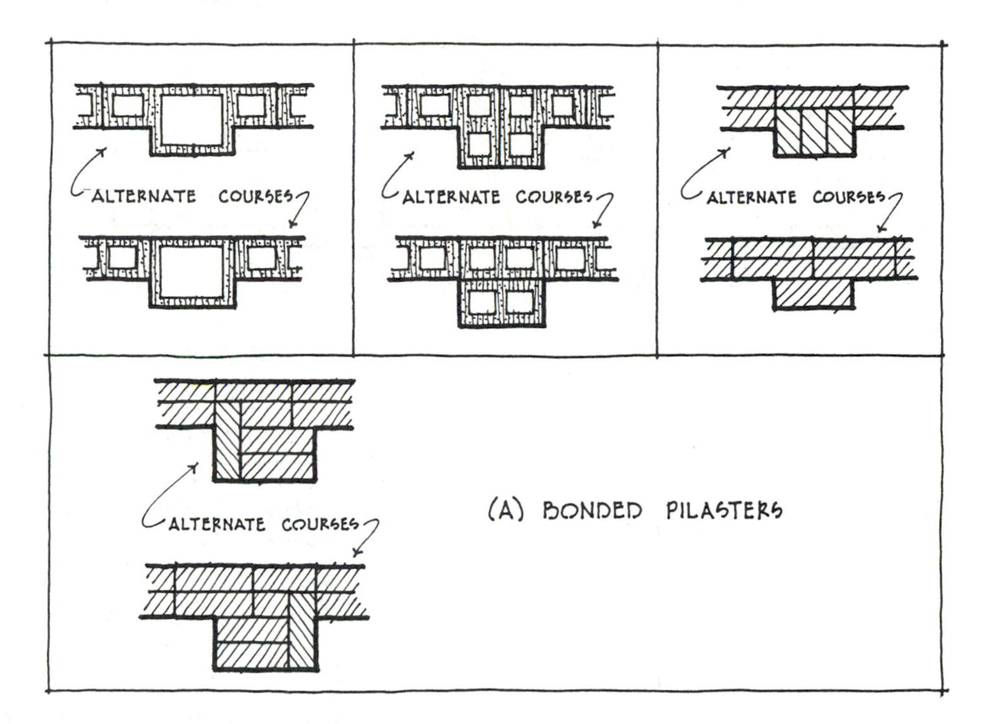

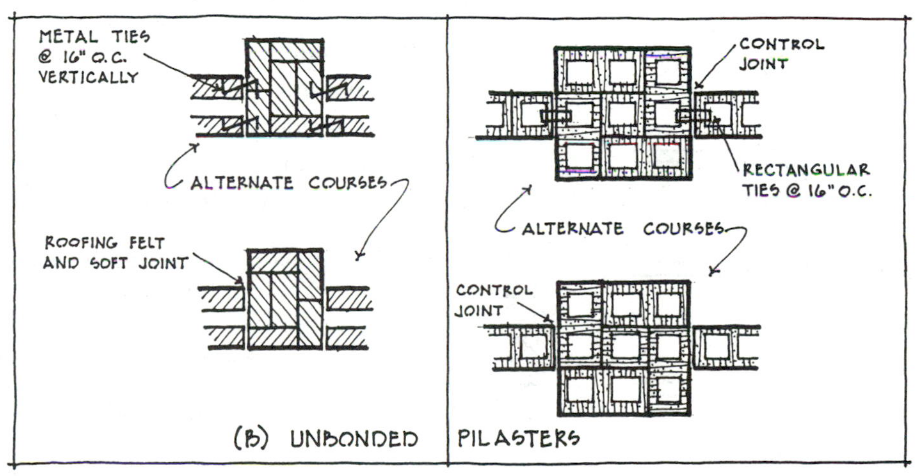

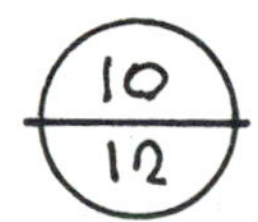

Masonry pilasters. (*From Harry C. Plummer*, Brick and Tile Engineering, *Brick Institute of America, McLean, Va., 1962.*)

2. Top-story bearing walls of a building 35 ft or less in height may be 8 in. thick provided that the wall height itself does not exceed 12 ft and that the roof construction does not impart lateral thrust to the walls.

3. In residence buildings of less than three stories, walls may be 8 in. thick when their height does not exceed 35 ft and the roof does not impart horizontal thrust. In one-story residence buildings and private garages, such walls may be 6 in. thick if their height does not exceed 9 ft generally or 15 ft at the peak of a gable.

4. Masonry walls above roof level which enclose stairways, machine rooms, penthouses, or shafts, and which do not exceed 12 ft in

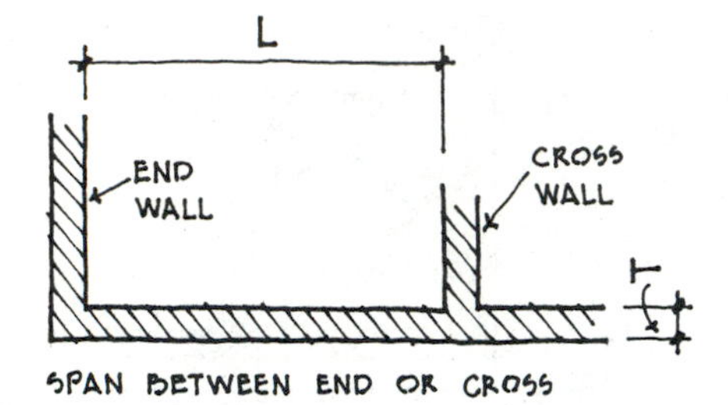

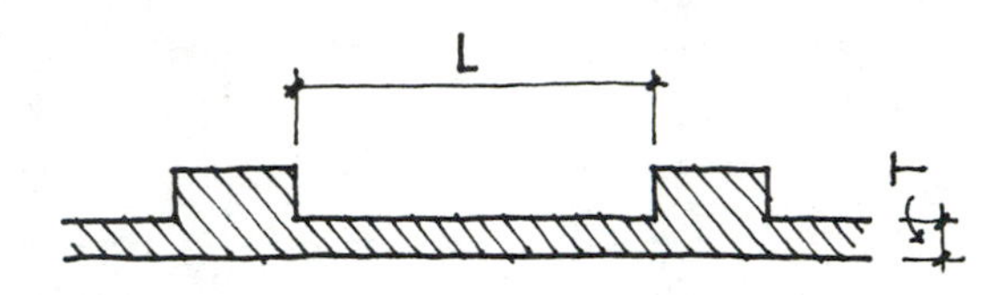

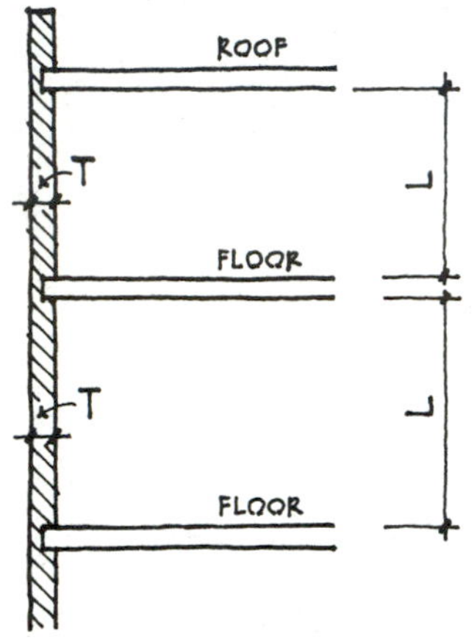

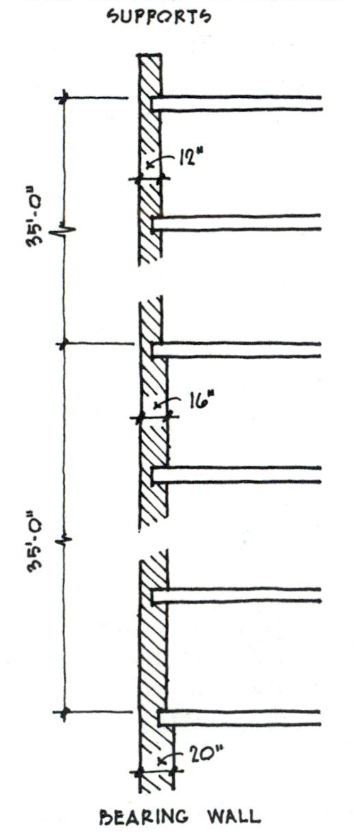

MAXIMUM RATIO OF UNSUPPORTED HEIGHT OR LENGTH TO NOMINAL THICKNESS	
Type of masonry	*Ratio L/T*
Solid masonry walls	20
Hollow-unit masonry walls	18
Cavity walls	18*

*Thickness equal to sum of the nominal thicknesses of the inner and outer wythes.

Minimum Thickness:
Bearing walls: As shown in adjacent sketch, except as follows:

1. Stiffened walls
2. Top-story walls
3. Residence buildings
4. Penthouses, roof structures
5. Hollow walls (cavity or masonry bonded)

Exterior non-bearing walls: Four inches less in thickness than required for bearing walls but not less than 8 in. (except where 6-in. walls are permitted).

Height-to-thickness ratios and minimum thickness requirements based on ANSI A41.1. (*From National Concrete Masonry Association*, TEK Bulletin 73, *NCMA*, *Herndon*, *Va.*)

height, may be 8 in. thick, and are not considered as increasing the height or requiring any increase in the thickness of the wall below

5. Cavity walls are generally limited to a 35-ft height except that 10-in. cavity walls may not exceed 25 ft in height. The inner and outer masonry wythes must be at least 4 in. thick, and the width of the cavity must be between 2 and 3 in.
6. Walls of composite construction cannot exceed the height or lateral support limitations for either of the types of units used.

Bonding. Two different types of bonding are covered by the ANSI Code: bonding with masonry unit headers (*see Fig. 10-14*), and bonding with metal ties. Metal ties are preferable, and should be $\frac{3}{16}$-in.-diameter corrosion-resistant steel rods or wires embedded in the horizontal bed joints. There must be one tie for each $4\frac{1}{2}$ sq ft of wall area with a maximum vertical spacing of 18 in. and a maximum horizontal spacing of 36 in. Ties in alternate courses must be staggered (*see Fig. 10-15*). For use with solid units ties should be Z-shaped with ends bent at 90° angles to form hooks at least

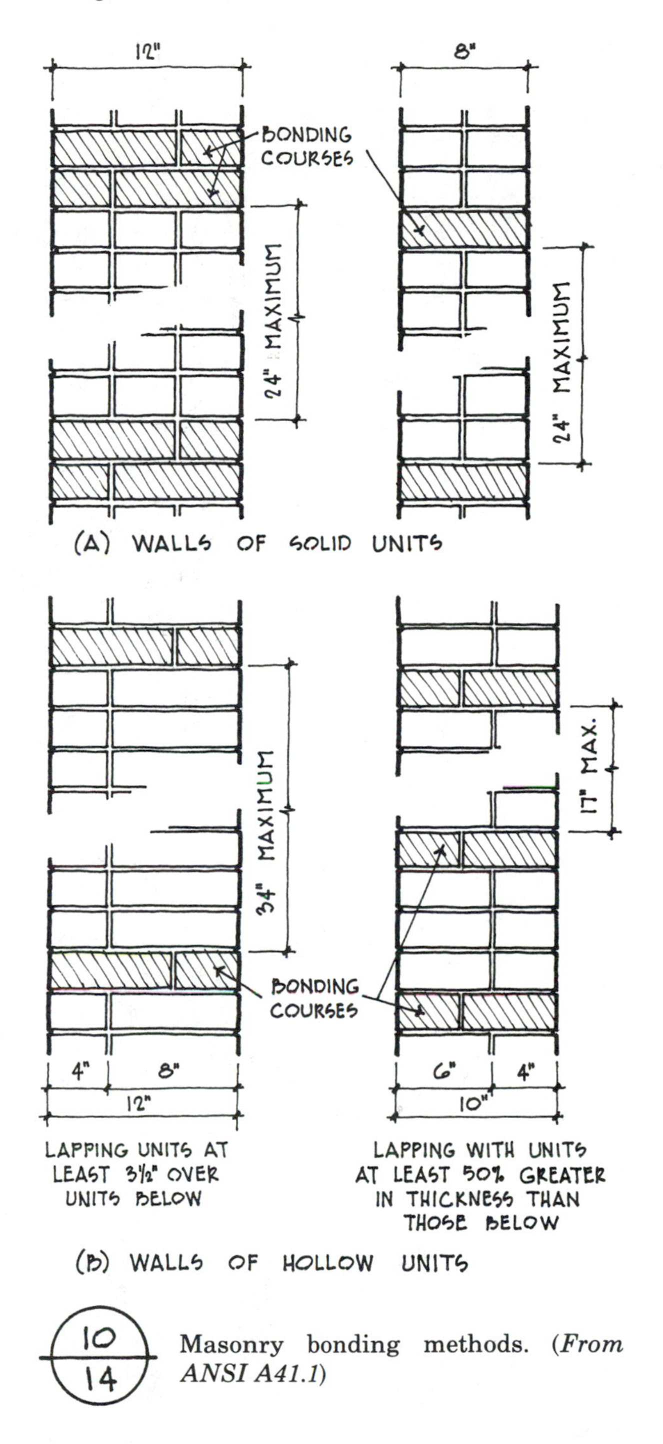

10
14
Masonry bonding methods. (*From ANSI A41.1*)

in. long. For hollow units, rectangular shapes are used. Additional ties must be provided at all openings, spaced not more than 3 ft apart around the perimeter and within 12 in. of the opening. Walls bonded with metal ties must conform to the allowable compressive stress, lateral support, thickness (excluding cavity), height, and mortar requirements for cavity walls.

Intersecting bearing walls that are built up simultaneously must have at least 50% of the units at the intersection laid in a true bond (*see Fig. 10-16*). If the walls are built up separately, the first must be regularly toothed or blocked with 8-in.-maximum offsets. Metal anchors 2 ft long with bent legs and a minimum section of $\frac{1}{4} \times 1\frac{1}{2}$ in. are then used to tie the joints at 4-ft vertical intervals.

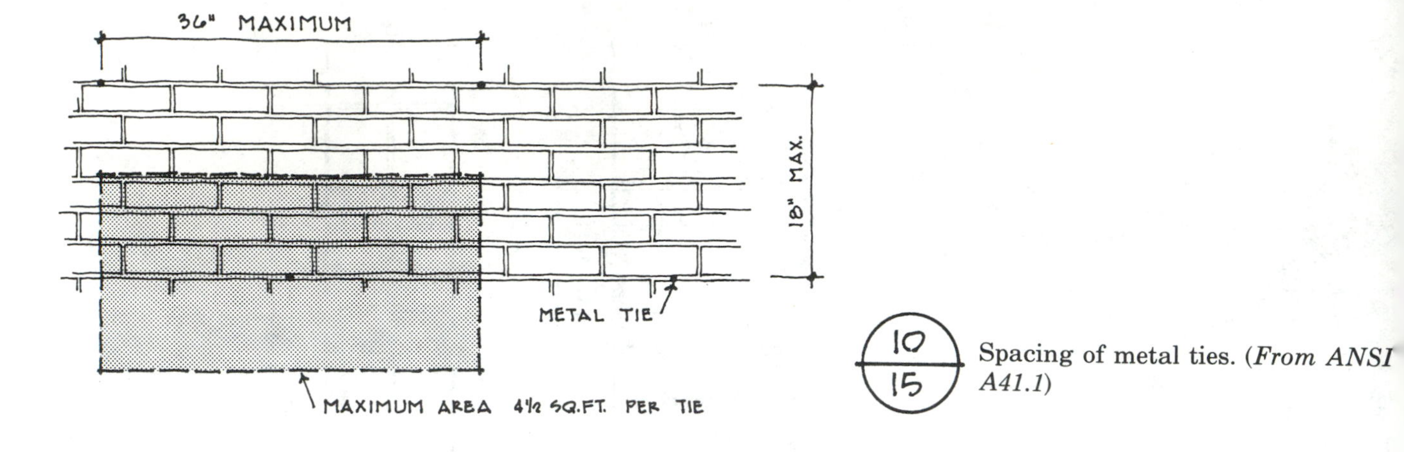

Spacing of metal ties. (*From ANSI A41.1*)

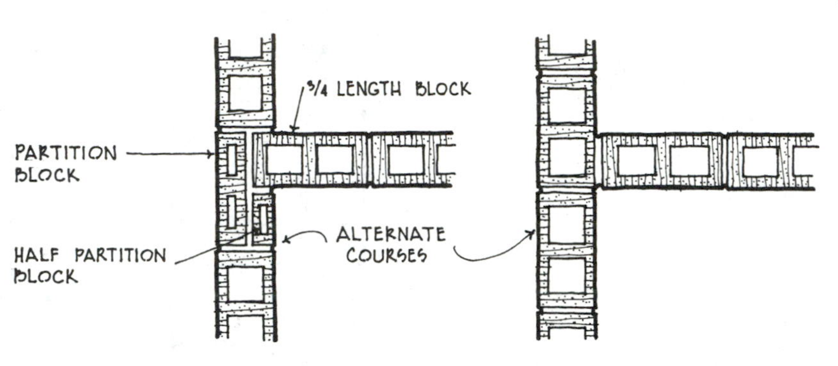

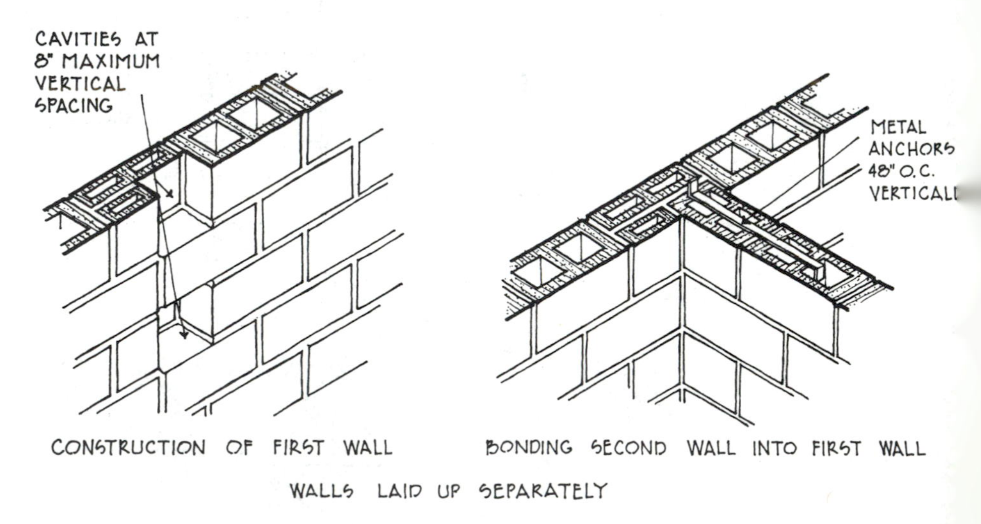

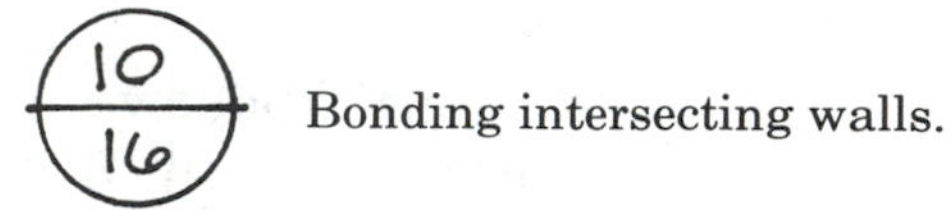

Bonding intersecting walls.

Members bearing on masonry. Lintels supported by masonr walls must have a minimum of 4 in. bearing at each end. Concentrate loads must have at least 3 in. actual bearing on either solid masonry whic is at least 4 in. deep, or on a metal bearing plate designed to safely distribu the loads. The ends of floor joists bearing on the walls must be securel anchored at 4-ft intervals and have at least 3 in. actual bearing on sol masonry which is a minimum of $2\frac{1}{4}$ in. deep.

Empirically designed unreinforced walls do not have adequate strength to resist substantial lateral loads. Buildings subject to seismic conditions or high wind loads must be rationally designed as either reinforced or partially reinforced masonry.

10.2.2 Rational Analysis

The rational design of loadbearing masonry buildings is based on a general analysis of the structure to determine the magnitude, line of action, and direction of all forces acting on the various members. All dead loads, live loads, lateral loads, and other forces, such as those resulting from temperature changes, impact, and unequal settlement, are considered. The combinations of loads that produce the greatest stresses are used to size the members. The performance of loadbearing masonry walls, pilasters, and columns subject to both concentric and eccentric axial loading can be predicted with reasonable accuracy by a rational analysis of the design using well-known principles of mechanics and data obtained from laboratory tests on compressive and transverse strengths of masonry. The method of analysis will depend on the complexity of the building with respect to height, shape, wall location, and openings in the wall.

Whereas empirical codes place arbitrary limits on wall height and thickness, rational analysis determines the thickness and section required to resist the actual stresses in the wall. Empirical codes arbitrarily dictate the spacing of lateral supports, but rational analysis calculates actual lateral loads and provides shear walls or pilasters to resist specific forces and provide stability. Rational design establishes allowable compressive stresses based on the characteristics of the materials, the slenderness of the wall, the eccentricity of applied loads, and the workmanship and control to be exercised. Rational design also takes into account shear and flexural stresses, which are not even considered in the empirical codes.

10.2.3 Codes

Accepted standards for rational design methods used throughout the United States may be found in the BIA's *Recommended Practice for Engineered Brick Masonry* (*BIA-1969*), the NCMA's *Specification for the Design and Construction of Load-bearing Concrete Masonry* (*NCMA-1970*), the ACI's *Building Code Requirements for Concrete Masonry Structures*, (*ACI-531*), and in the *Uniform Building Code* (*UBC*).

To provide the benefits of higher allowable stresses and greater wall heights between lateral supports without an arbitrary minimum wall thickness, the codes impose restrictions and quality standards on materials, design, and construction. They limit the permissible variation in strength and dimension of units, control mortar types, require high-quality workmanship characterized by full head and bed joints, and impose limits on variation in alignment and plumbness. To guarantee a high quality of workmanship, the standards also require architectural or engineering inspection to verify proper performance of the work in compliance with the drawings and specifications. Without such inspection, reduced values of allowable stresses must be used.

It should be emphasized that design based on rational analysis is no better than the assumptions made as to exposure, loading, quality of construction obtained, and particularly in the case of members used to resist bending, the method of support and degree of fixity or continuity at supports.

Structures are no stronger than the foundations on which they rest. The rigidity of masonry assemblages cannot tolerate displacements at the top and bottom of vertical loadbearing elements. Foundation settlements and the action of unstable soil may induce unsightly and sometimes structurally serious cracking. For poor soil conditions, a bearing wall system has some advantage over skeleton frames in that the loads are delivered in lines rather than points, thus keeping the soil bearing stress lower. (Walls may also be designed as thin, deep, reinforced beams which, when working with the footings, will be able to span weak spots or depressions in the subsoil; see Chapter 12.)

10.2.4 Materials

The compressive, shearing, and flexural strengths of masonry are affected by the properties of both the units and the mortar in which they are laid. The compressive strength of the individual unit components has great effect on the compressive strength of the assembly. Although mortar strength is a factor, its greatest effect is on the flexural and shearing strengths of the construction. The ultimate compressive strength of the masonry (upon which allowable stresses are based) may be determined by prism tests or may be "assumed" from the strength of the units and the type of morta used in the assembly. For these reasons, the standards place specifi requirements and limitations on the materials that can be used in rationally designed masonry structures.

Brick units must comply with ASTM C62 or ASTM C216 for buildin brick and face brick, and must be of Grade MW or SW. Concrete masonr must meet the requirements of ASTM C90 for hollow loadbearing units ASTM C145 for solid loadbearing units, or ASTM C55 for concrete brick Mortar and grout for use in unreinforced masonry construction mus comply with ASTM C270, Types M, S, or N. If values for assume compressive strength and allowable stresses are taken from code tables *BIA-1969* and the *UBC* require portland cement–lime mixes. They do no permit the use of proprietary mortar mixes unless actual strength has bee established by laboratory test. All of the codes prohibit the addition o calcium chloride or admixtures containing calcium chloride to any morta or grout in which reinforcement, metal ties, or anchors will be embedde Metal or wire ties used in multi-wythe construction must be of corrosio resistant finish to protect against the destructive action of the cemen When different kinds or grades of units or mortars are used in composi walls, stresses may not exceed the allowable for the weakest of th combinations of units and mortars unless a higher strength is substa tiated by laboratory test.

10.2.5 General Design Requirements

In determining stresses, the effects of all dead and live loads must be tak into account, and stresses must be based on actual rather than nomin dimensions. Consideration must be given to the effects of lateral loa eccentricity of vertical load, nonuniform foundation pressure, deflectio and thermal and moisture movements. All critical loading conditions mu be calculated. Fixity, or end restraint, must also be considered, as it affec resistance to applied loads.

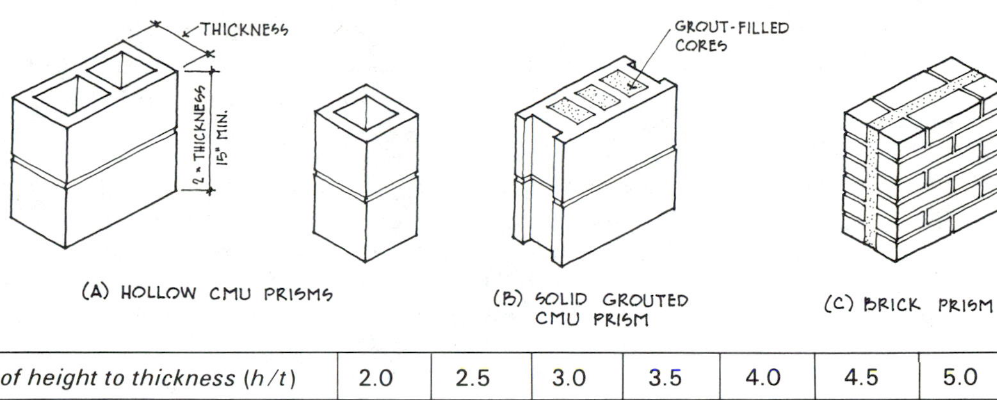

Ratio of height to thickness (h/t)	2.0	2.5	3.0	3.5	4.0	4.5	5.0
*Correction factor**	0.73	0.80	0.86	0.91	0.95	0.98	1.00

*Interpolate to obtain intermediate values.

(D) Correction factors from *BIA-1969*

Ratio of height to thickness (h/t)	1.5	2.0	2.5	3.0
*Correction factor**	0.86	1.00	1.11	1.20

*Interpolate to obtain intermediate values.

(E) Correction factors from *NCMA-1970*

Masonry test prisms. (*From Brick Institute of America,* Recommended Practice for Engineered Brick Masonry, *BIA, McLean, Va., 1969; and National Concrete Masonry Association,* Specification for the Design and Construction of Loadbearing Concrete Masonry, *NCMA, Herndon, Va., 1970.*)

Flexural, shear, and axial stresses resulting from wind, blast, or earthquake forces must be added to the stress of dead and live loads. Connections must be designed to resist such lateral forces acting either inward or outward. For combined stresses due to these temporary or transient lateral forces plus the dead and live loads, allowable stresses may be increased $33\frac{1}{3}\%$ provided that the strength of the resulting section is not less than that required for dead loads and vertical live loads alone. For purposes of stress calculation, wind, blast, and earthquake forces may be assumed never to occur simultaneously.

10.2.6 Masonry Strength

In the rational design of masonry structures, allowable axial, flexural, shear, and bearing stresses are based on the compressive strength of the masonry at 28 days. This ultimate compressive strength (f'_m)* may be determined from laboratory testing, or may be based on an assumed value if the units, mortar, and workmanship all conform to requirements of the code. Prism tests for determination of f'_m will generally yield higher values than those assumed from tables based on unit and mortar strength. A prism is defined as a small assemblage of masonry units and mortar which is used primarily to predict the strength of full scale masonry members (*see Fig. 10-17*).

*Notations and symbols may vary among the several building codes and standards referenced. To avoid confusion, these have been standardized for use throughout this text. See Appendix B for complete list.

When prism tests are not made, assumed values for compressive strength of the masonry may be used if the units, mortar, and workmanship are all in compliance with the code requirements. The tables in *Fig. 10-18*

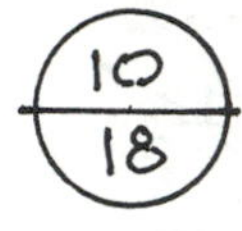

Assumed compressive strength of masonry. (Table A, *from National Concrete Masonry Association*, Specification for the Design and Construction of Load-bearing Concrete Masonry, *NCMA, Herndon, Va., 1970;* Table B, *from Brick Institute of America*, Recommended Practice for Engineered Brick Masonry, *BIA, McLean, Va., 1969;* Table C, *from American Concrete Institute*, Building Code Requirements for Concrete Masonry Structures, 1979.)

TABLE A

Compressive strength of the units (psi)*	*Assumed compressive strength of concrete masonry, f'_m (psi)*	
	Types M and S mortar	*Type N mortar*
Over 1500 to 2500	1,151 to 1,550	875 to 1100
Over 2500 to 4000	1,551 to 2,000	1101 to 1250
Over 4000 to 6000	2,001 to 2,400	1251 to 1350
Over 6000	2,400	1350

*Gross area for masonry of solid units; net area for masonry of hollow units.

TABLE B

Compressive strength of units (psi)	*Assumed compressive strength of brick masonry, f'_m (psi)*					
	Without inspection			*With inspection*		
	Type N mortar	*Type S mortar*	*Type M mortar*	*Type N mortar*	*Type S mortar*	*Type M mortar*
14,000 plus	2140	2600	3070	3200	3900	4600
12,000	1870	2270	2670	2800	3400	4000
10,000	1600	1930	2270	2400	2900	3400
8,000	1340	1600	1870	2000	2400	2800
6,000	1070	1270	1470	1600	1900	2200
4,000	800	930	1070	1200	1400	1600
2,000	530	600	670	800	900	1000

TABLE C

Compressive test strength of masonry units (psi) on the net cross-sectional area	*Assumed compressive strength of concrete masonry, f'_m (psi)*	
	Types M and S mortar	*Type N mortar*
6000 or more	2400	1350
4000	2000	1250
2500	1550	1100
2000	1350	1000
1500	1150	875
1000	900	700

show assumed values of f'_m for brick and concrete masonry. For unit strengths between those listed, values for f'_m should be interpolated. Once the compressive strength of the masonry has been established, allowable stresses may be determined and rational design carried out according to the formulas given in the codes.

10.3 BRICK MASONRY

Allowable axial and flexural compressive stresses, shear stresses, modulus of elasticity, and modulus of rigidity for unreinforced loadbearing brick construction are controlled by the ultimate compressive strength of the masonry (f'_m). These figures are based on an assumption of good bond between the units and the mortar or grout. The table in *Fig. 10-19* lists *BIA-1969* allowable stresses for use in design calculations (note that higher values are permitted for tension and shear when architectural or engineering inspection is provided during construction). Calculated stresses in a wall or column section may not exceed these allowable figures. If the stresses in a proposed member are greater than these values, the designer should consider a larger section or the addition of reinforcing steel to increase resistance to applied loads.

A design formula is given in the codes for the determination of maximum allowable vertical loads on walls and columns. Since most masonry structural members tend to be tall, thin sections, the potential for

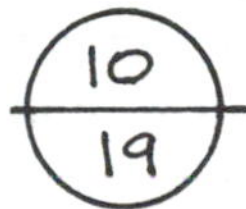

Allowable stresses in unreinforced brick masonry. (*From Brick Institute of America*, Recommended Practice for Engineered Brick Masonry, *BIA, McLean, Va., 1969.*)

Description		*Allowable stresses (psi)*	
		Without inspection	*With inspection*
Compressive, axial			
Walls	(f_m)	$0.20f'_m$	$0.20f'_m$
Columns	(f_m)	$0.16f'_m$	$0.16f'_m$
Compressive, flexural			
Walls	(f_m)	$0.32f'_m$	$0.32f'_m$
Columns	(f_m)	$0.26f'_m$	$0.26f'_m$
Tensile, flexural			
Normal to bed joints			
M or S mortar	(F_t)	24	36
N mortar	(F_t)	19	28
Parallel to bed joints			
M or S mortar	(F_t)	48	72
N mortar	(F_t)	37	56
Shear			
M or S mortar	(V_m)	$0.5\sqrt{f'_m}$, but not to exceed 40	$0.5\sqrt{f'_m}$, but not to exceed 80
N mortar	(V_m)	$0.5\sqrt{f'_m}$, but not to exceed 28	$0.5\sqrt{f'_m}$, but not to exceed 56
Bearing			
On full area	(f_m)	$0.25f'_m$	$0.25f'_m$
On $\frac{1}{3}$ area or less	(f_m)	$0.375f'_m$	$0.375f'_m$
Modulus of elasticity		$1000f'_m$, but not to exceed 2,000,000 psi	$1000f'_m$, but not to exceed 3,000,000 psi
Modulus of rigidity		$400f'_m$, but not to exceed 800,000 psi	$400f'_m$, but not to exceed 1,200,000 psi

bending due to slenderness and eccentricity is relatively great and must be considered in the calculation of these maximum loads. Two stress reduction factors, the *slenderness coefficient* (C_s) and the *eccentricity coefficient* (C_e) were developed to accommodate this tendency and assure adequate performance under axial loading conditions. The eccentricity coefficient is used to reduce the allowable axial load in lieu of performing a separate bending analysis. The slenderness coefficient reduces the allowable to prevent buckling. To determine these two factors, it is first necessary to find the end eccentricity ratio (e_1/e_2), the ratio of the maximum virtual eccentricity to the wall thickness (e/t), and the ratio of unsupported height to wall thickness (h/t).

10.3.1 Eccentricity

Eccentric loading is the application of force on a wall or column at a point other than the centroid. Eccentricity is the distance between the centroidal axis and the applied load. Virtual eccentricity (e) is the eccentricity required to produce axial and bending stresses equivalent to those produced by applied axial and transverse loads. It is calculated by dividing the moment at the section under investigation by the resultant axial load. A virtual eccentricity of some magnitude occurs at both the top and bottom of any wall or column. The *end eccentricity ratio*, e_1/e_2, is the ratio of the smaller to the larger virtual eccentricity. Values of e_1/e_2 for various end conditions are shown in *Fig. 10-20*. The value of the eccentricity coefficient required for load calculation may be taken from Table *B* in *Fig. 10-21*. Linear interpolation is permitted within the table.

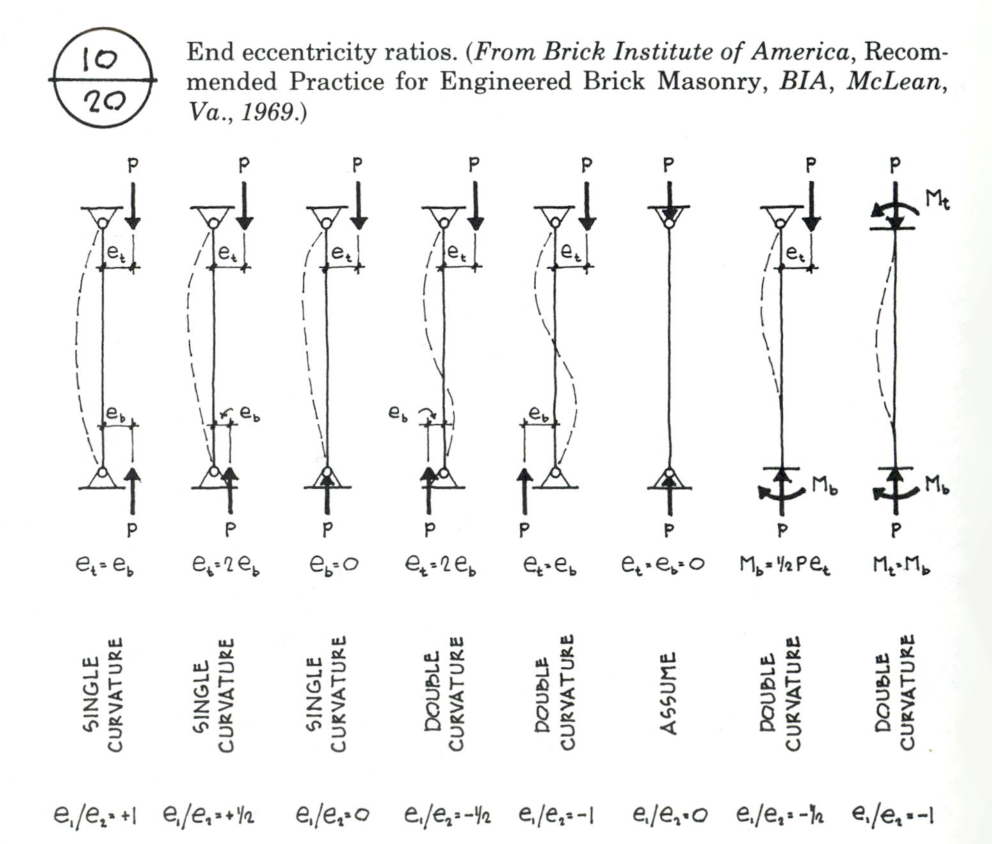

End eccentricity ratios. (*From Brick Institute of America*, Recommended Practice for Engineered Brick Masonry, *BIA*, *McLean*, *Va.*, *1969*.)

TABLE A SLENDERNESS COEFFICIENTS (C_s)

h/t	e_1/e_2								
	-1	$-\frac{3}{4}$	$-\frac{1}{2}$	$-\frac{1}{4}$	0	$+\frac{1}{4}$	$+\frac{1}{2}$	$+\frac{3}{4}$	$+1$
5.0	1.00	1.00	1.00	1.00	1.00	1.00	1.00	1.00	1.00
7.5	1.00	1.00	1.00	1.00	1.00	0.98	0.96	0.93	0.90
10.0	1.00	0.99	0.98	0.96	0.93	0.91	0.88	0.84	0.80
12.5	0.95	0.94	0.92	0.90	0.87	0.83	0.79	0.75	0.70
15.0	0.90	0.88	0.86	0.83	0.80	0.76	0.71	0.66	0.60
17.5	0.85	0.83	0.81	0.77	0.73	0.69	0.63	0.57	0.50
20.0	0.80	0.78	0.75	0.71	0.67	0.61	0.55	0.48	0.40
22.5	0.75	0.73	0.69	0.65	0.60	0.54	0.47	0.39	
25.0	0.70	0.67	0.64	0.59	0.53	0.47	0.39		
27.5	0.65	0.62	0.58	0.53	0.47	0.39			
30.0	0.60	0.57	0.52	0.47	0.40				
32.5	0.55	0.52	0.47	0.41					
35.0	0.50	0.46	0.41						
37.5	0.45	0.41							
40.0	0.40								

TABLE B ECCENTRICITY COEFFICIENTS (C_e)

e/t	e_1/e_2								
	-1	$-\frac{3}{4}$	$-\frac{1}{2}$	$-\frac{1}{4}$	0	$+\frac{1}{4}$	$+\frac{1}{2}$	$+\frac{3}{4}$	$+1$
0 to:									
$\frac{1}{20}$(0.05)	1.00	1.00	1.00	1.00	1.00	1.00	1.00	1.00	1.00
$\frac{1}{12}$(0.083)	0.90	0.90	0.89	0.89	0.88	0.88	0.87	0.87	0.87
$\frac{1}{8}$(0.125)	0.82	0.81	0.80	0.79	0.78	0.77	0.76	0.75	0.74
$\frac{1}{6}$(0.167)	0.77	0.75	0.74	0.72	0.71	0.69	0.68	0.66	0.65
$\frac{5}{24}$(0.208)	0.73	0.71	0.69	0.67	0.65	0.63	0.61	0.59	0.57
$\frac{1}{4}$(0.250)	0.69	0.66	0.64	0.61	0.59	0.56	0.54	0.51	0.49
$\frac{7}{24}$(0.292)	0.65	0.62	0.59	0.56	0.53	0.50	0.47	0.44	0.41
$\frac{1}{3}$(0.333)	0.61	0.57	0.54	0.50	0.47	0.43	0.40	0.36	0.32

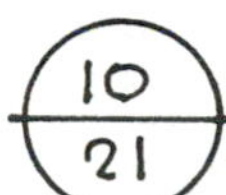

Slenderness and eccentricity coefficients. (*From Brick Institute of America,* Recommended Practice for Engineered Brick Masonry, *BIA, McLean, Va., 1969.*)

The *ratio of the maximum virtual eccentricity to the wall thickness, e/t,* is used in selecting the eccentricity coefficient (C_e) from tables given in the code. The design of an unreinforced member requires that e/t be less than or equal to $\frac{1}{3}$. Where the e/t ratio exceeds $\frac{1}{3}$, a larger section must be considered or reinforcing steel must be added.

10.3.2 Slenderness

The slenderness ratio (h/t) is the ratio of the effective height of a member to its effective thickness. When a wall is laterally supported top and bottom, its *effective height* is the actual height of the wall measured from the top of the lower floor to the bearing area of the upper floor. Where there is no lateral support at the top of a wall, effective height is assumed to be twice the height of the wall above the bottom lateral support. The *effective thickness* of unreinforced solid walls is equal to the actual thickness. For cavity walls loaded on only one wythe, the effective thickness equals the actual thickness of the loaded wythe. For cavity walls loaded on both wythes, each wythe is considered to act independently and the effective thickness of each wythe is its actual thickness. The effective thickness of metal-tied walls is determined the same as for cavity walls unless the collar joints are solidly filled with mortar or grout, in which case the effective thickness is the total actual thickness. Where raked mortar joints are used, the thickness of the member should be reduced by the depth of raking.

Slenderness ratios for unreinforced loadbearing walls cannot exceed the value computed by $h/t \leq 10\ (3 - e_1/e_2)$. For unreinforced loadbearing columns, the slenderness ratio is the greater value obtained by dividing the effective height (h) in any direction by the effective thickness (t) in the corresponding direction and cannot exceed the value computed by $h/t \leq 5$ $(4 - e_1/e_2)$. The slenderness coefficient used in the calculation of allowable vertical loads may be taken from Table *A* in *Fig. 10-21* by using the calculated h/t ratio and the correct e_1/e_2 ratio for the proposed conditions. Linear interpolation is also permitted within this table.

10.3.3 Cross-Sectional Area

The last quantity needed in the formula for calculating allowable vertical loads is the cross-sectional area (A_g) of the member. For solid walls and columns, A_g is the actual gross cross-sectional area of the member. For cavity walls loaded on one wythe, A_g is the actual gross cross-sectional area of the loaded wythe. For cavity walls loaded on both wythes, A_g is the actual gross cross-sectional area of the wythe under consideration. Metal-tied walls are determined the same as cavity walls unless the collar joint is solidly filled with mortar or grout. If raked mortar joints are used, the thickness used to determine A_g must be reduced accordingly.

10.3.4 Allowable Vertical Loads

The basic formula for unreinforced loadbearing brick masonry walls and columns where the virtual eccentricity (e) does not exceed $t/3$ is

$$P = C_e C_s f_m A_g \qquad (10.$$

where P = allowable vertical load
C_e = eccentricity coefficient
C_s = slenderness coefficient
f_m = allowable axial compressive stress (from *Fig. 10-19*)
A_g = gross cross-sectional area

The value of $C_eC_sf_m$ in this formula is the average allowable compressive stress permitted in the member and should not be taken as the maximum compressive stress permitted in the extreme fiber. If the maximum virtual eccentricity (e) exceeds $t/3$, the maximum tensile stress in the masonry cannot exceed the values given in *Fig. 10-19*. If this value is exceeded, the member must be designed with reinforcing steel.

In normal design procedure, a wall thickness is assumed, the loads determined, and resulting stresses checked against the allowables. Using this method, the formula for allowable vertical loads would be written $f_m = P/C_eC_sA_g$, where f_m = maximum average stress on the section being checked. Allowable loads may also be determined for a given wall thickness and masonry strength from the table in *Fig. 10-22*, which was developed as a design aid by the BIA. A design example is given in *Fig. 10-23*.

Actual wall thickness (in.)	A_g (sq. in. per ft)	f'_m								
		1000	1500	2000	2500	3000	3500	4000	4500	5000
2.5	30.0	6.0	9.0	12.0	15.0	18.0	21.0	24.0	27.0	30.0
3.5	42.0	8.4	12.6	16.8	21.0	25.2	29.4	33.6	37.8	42.0
4.5	54.0	10.8	16.2	21.6	27.0	32.4	37.8	43.2	48.6	54.0
5.5	66.0	13.2	19.8	26.4	33.0	39.6	46.2	52.8	59.4	66.0
6.5	78.0	15.6	23.4	31.2	39.0	46.8	54.6	62.4	70.2	78.0
7.5	90.0	18.0	27.0	36.0	45.0	54.0	63.0	72.0	81.0	90.0
8.5	102.0	20.4	30.6	40.8	51.0	61.2	71.4	81.6	91.8	102.0
9.5	114.0	22.8	34.2	45.6	57.0	68.4	79.8	91.2	102.6	114.0
10.5	126.0	25.2	37.8	50.4	63.0	75.6	88.2	100.8	113.4	126.0
11.5	138.0	27.6	41.4	55.2	69.0	82.8	96.6	110.4	124.2	138.0
12.5	150.0	30.0	45.0	60.0	75.0	90.0	105.0	120.0	135.0	150.0
13.5	162.0	32.4	48.6	64.8	81.0	97.2	113.4	129.6	145.8	162.0
14.5	174.0	34.8	52.2	69.6	87.0	104.4	121.8	139.2	156.6	174.0
15.5	186.0	37.2	55.8	74.4	93.0	111.6	130.2	148.8	167.4	186.0

Note: Allowable load $= P/C_eC_s = 0.20\, f'_m A_g/1000$.

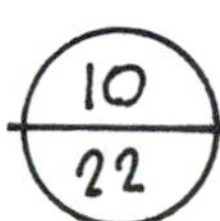

Allowable compressive loads (kips/ft). (*From Brick Institute of America,* Technical Note 24H, *BIA, McLean, Va.*)

10.3.5 Shear Walls

When a lateral force is applied parallel to a bearing wall, it acts as a loadbearing shear wall and must meet the requirements for both shear walls and biaxial bending. Nonbearing shear walls may be stiffened by adding flanges, or by providing positive connection to bearing walls. The effectiveness of the wall in resisting lateral loads is thus increased. Where shear walls are required, design analysis will determine if loads are sufficient to require steel reinforcing. Allowable stresses for unreinforced walls are given in *Fig. 10-19*. Requirements for effective flange width are shown in *Fig. 10-24*.

Example A:

A nominal 8-in. wall (assume $7\frac{1}{2}$-in. actual thickness) is supported laterally by the floor system at 10-ft intervals. Assuming the wall is to be constructed of 4000 psi brick and Type S mortar, determine the allowable concentric load.

From *Fig 10-18B*, for 4000-psi units and Type S mortar, with inspection, a compressive strength (f'_m) of 1400 psi may be assumed. Since the wall is loaded concentrically, the virtual eccentricity (e) and the end eccentricities will be zero.

$e_1/e_2 = 0$

$t = 7\frac{1}{2}$ in.

$e/t = 0$

$h = 120$ in.

$h/t = 120/7.5 = 16$

$C_e = 1$ (from *Fig. 10-21B*)

$C_s = 0.77$ (from *Fig. 10-21A*).

From *Fig. 10-19*,

$f_m = 0.20\, f'_m = (0.20)(1400) = 280$ psi

$A_g = 12(7.5) = 90$ sq in./ft

$P = C_e C_s f_m A_g = (1)(0.77)(280)(90) = 19{,}400$ lb/ft of wall,

or 19.4 K/lin ft from *Fig. 10-22* by linear interpolation

Example B:

Determine the required brick masonry strengths for the wall described above with the load (neglecting the weight of the wall) at the eccentricity shown below.

$e_1 = 1$ in.; $e_2 = 2$ in.

$e_1/e_2 = -\frac{1}{2}$ (double curvature)

$h/t = 16$

$e_{(max)} = 2$ in.

$e/t = 2/7.5 = 0.27$

$C_s = 0.84$ (from *Fig. 10-21A*)

$C_e = 0.62$ (from *Fig. 10-21B*)

$f_m = P/C_e C_s A_g = 19{,}400/(0.62)(0.84)(90) = 415$ psi

$f_m = 0.20 f'_m$ (from *Fig. 10-19*)

$f'_m = f_m/0.20 = 415/0.20 = 2075$ psi required

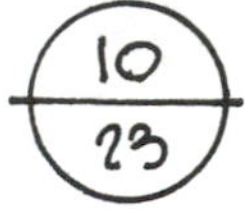

Design example. (*From Brick Institute of America,* Recommended Practice for Engineered Brick Masonry, *BIA, McLean, Va., 1969.*)

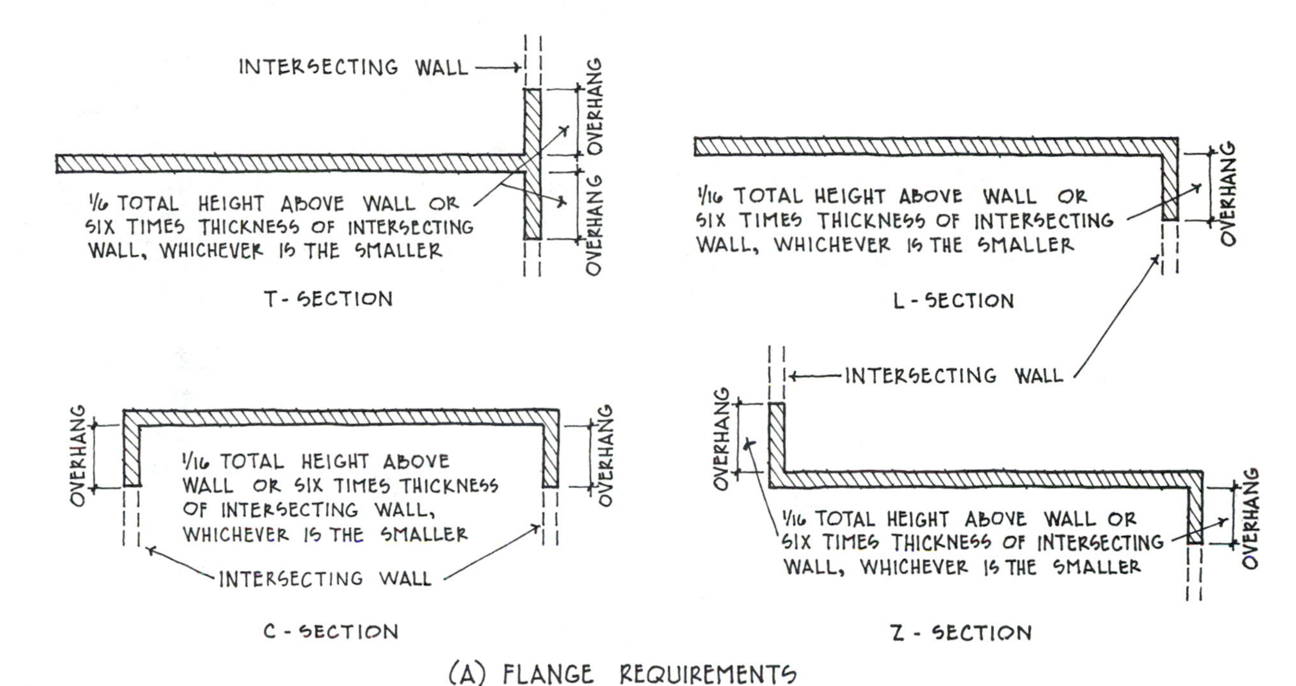

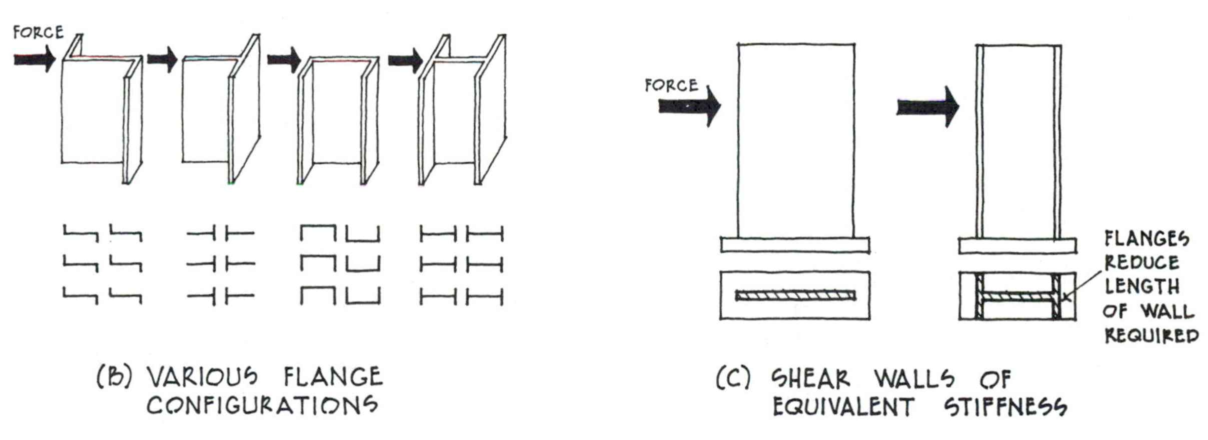

10/24 Shear walls with flanges. (*From Brick Institute of America*, Technical Note 24C, *BIA, McLean, Va.;* and *Theodore Leba, Jr.*, Design Manual—The Application of Non-reinforced Concrete Masonry Load-Bearing Walls in Multistory Structures, *National Concrete Masonry Association, Herndon, Va., 1969.*)

10.4 CONCRETE UNIT MASONRY

Under *NCMA-1970* and *ACI-531*, allowable axial and flexural stresses for concrete masonry construction are based on the strength of the masonry (f'_m), and axial stresses are further modified by a slenderness ratio. Values for shear and for tension in flexure are stated as maximums for various mortar types (*see Fig. 10-25*). Allowable stresses may be increased $33\frac{1}{3}$% for wind, blast, and earthquake forces combined with normal dead and live loads. When there is no engineering or architectural inspection to ensure that standards of workmanship are enforced, allowable stresses shown in the table must be reduced by one-half.

	Type of construction			
	Hollow units		*Grouted or solid units*	
Type of stress	*Type M or S mortar*	*Type N mortar*	*Type M or S mortar*	*Type N mortar*
Shear (psi) V_m	34*	23*	34	23
Tension in flexure F_t				
Normal to bed joints	23*	16*	39	27
Parallel to bed joints	46*	32*	78	54

*Net mortar bedded area.

Note: Allowable *axial* stress in unreinforced masonry shall not exceed the values determined by the following formulas:

$$\text{Walls: } f_m = 0.20f'_m[1\text{-}(h/40t)^3]$$

$$\text{Columns: } f_m = 0.18f'_m[1\text{-}(h/30t)^3]$$

where f_m = allowable axial stress
f'_m = masonry strength
h = effective height
t = effective thickness

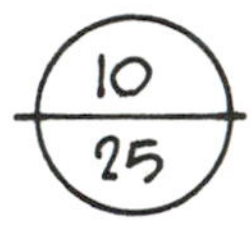

Allowable stresses in unreinforced masonry with inspected workmanship. (*From National Concrete Masonry Association*, Specification for the Design and Construction of Loadbearing Concrete Masonry, *NCMA, Herndon, Va., 1970; National Concrete Masonry Association*, TEK Bulletin 24, *NCMA, Herndon, Va.*)

10.4.1 Slenderness

The effect of slenderness on the performance of concrete masonry walls an columns is an important consideration, just as it is for brick masonry. Fo loadbearing columns and wall sections, the slenderness ratio is taken as th ratio of the effective height to effective thickness, and may not exceed 20

10.4.2 Effective Section

In computing the slenderness of concrete masonry walls and columns, th *effective height* depends on the condition of support at the top. If the wall column is laterally supported at both the bottom and the top, the effecti height is the clear distance between these supports. If the member is free the top, the effective height is taken as twice the height above the low support. The effective height of a wall is considered in only one directio while a column must be considered in both directions if the degree of fixity the top differs in each direction. In such cases, the larger effective heig must be used in determining the slenderness ratio. These requirements a basically the same as for brick masonry.

Requirements for determining the *effective thickness* of concre masonry, however, are quite different than for brick. *NCMA-1970* conside the effect of stiffening pilasters (internal or external) and gives coefficier for "stiffened walls." Except for cavity walls, the effective thickness o wall without pilasters is its nominal thickness (as opposed to actual thickness for brick walls). For walls supported at vertical intervals (i.e., by floors and roof), and stiffened by properly bonded pilasters at regular intervals, the effective thickness is obtained by multiplying the actual thickness between pilasters by the corresponding coefficient shown in *Fig. 10-26*, where

$$\frac{T_p}{T_w} = \frac{\text{pilaster thickness}}{\text{wall thickness}} \tag{10.2}$$

Linear interpolation between values given in the table is permitted, but not extrapolation outside the stated limits.

For cavity walls loaded on both wythes, the effective thickness is determined by the formula

$$T = \tfrac{2}{3}(T_o - W_c) \tag{10.3}$$

where T_o is the overall wall thickness including cavity and W_c is the width of the cavity. For cavity walls loaded on only one wythe, the effective thickness is for the loaded wythe only, and is computed for stiffened walls as indicated above. The effective thickness of a column in any direction is its actual thickness in the direction under consideration.

The *effective area* for stress calculations is based on the net cross-sectional area of the mortar bed. This will vary for solid and hollow unit masonry depending on the mortaring technique used (i.e., full mortar beds or face-shell bedding).

Pilaster spacing (O.C.) / *Pilaster width*	$T_p/T_w = 1$	$T_p/T_w = 2$	$T_p/T_w = 3$
6	1.0	1.4	2.0
8	1.0	1.3	1.7
10	1.0	1.2	1.4
15	1.0	1.1	1.2
20 or more	1.0	1.0	1.0

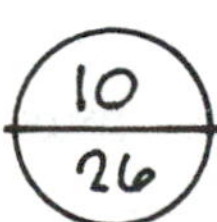

Coefficients for stiffened walls. (*From National Concrete Masonry Association*, Specification for the Design and Construction of Loadbearing Concrete Masonry, *NCMA, Herndon, Va., 1970.*)

10.4.3 Eccentricity

In walls, columns, and cavity walls loaded on both wythes, the eccentricity of the load is considered with respect to the member's center of resistance. In cavity walls loaded on only one wythe, the eccentricity of the load is considered with respect to the centroidal axis of the loaded wythe.

For eccentric loads, *NCMA-1970* requires that (1) axial compressive stresses may not exceed the allowable value f_m, (2) flexural compressive stresses may not exceed $0.30 f'_m$, and (3) with combined loading, the ratio of actual stress (f) to allowable stress (F) for axial and flexural loads may not add up to more than unity—that is, $f_a/F_a + f_b/F_b \leq 1.0$. In most cases, these three factors are all that need be considered as long as the capacity for vertical load (P) is governed by compressive stresses. For walls of hollow concrete masonry units, the limitation on eccentricity (e) is merely $e_{max} = S$ (section modulus)/A_n (net area). As long as (e) is less than the ratio S/A_n, the axial compressive stress f_a is greater than the flexural stress f_b, and the wall is under compression across its entire thickness. Under these conditions, the load-carrying capacity (P) of a wall or column depends on the compressive strength of the masonry, the amount of eccentricity, and the slenderness ratio. In other words, compression governs the design of the member.

Values of eccentricity that are greater than the ratio S/A_n produce flexural stresses which are greater than axial stresses and result in tension at the opposite face of the wall or column. Under these conditions, the maximum load the member can safely carry depends on the allowable tensile stress permitted by code ($f_m - 0.75 f_b \leq F_t$); that is, tension governs. Research has indicated that hollow concrete masonry walls can carry substantial loads even at relatively high eccentricities. For a 8-in. block wall, the tables in *Fig. 10-27* list allowable loads in kips per foot of wall length for various eccentricities as provided in *NCMA-1970*.

10.4.4 Axial Loads

The two formulas given in *Fig. 10-25* for allowable axial stress on unreinforced concrete masonry walls and columns are used to determine whether or not given axial loads are within acceptable limits. Both formulas include a reduction factor for slenderness using a ratio of h/t. Calculations for wall sections are made using the formula

$$f_m = 0.20\, f'_m \left[1 - \left(\frac{h}{40t}\right)^3\right] \qquad (10.$$

where f_m = allowable axial stress, psi
f'_m = masonry strength, psi
h = effective height, in.
t = effective thickness, in.

The modifier for f'_m is a safety factor, and the $h/40t$ is a slenderness reduction factor. In computing the allowable stress on CMU columns, these values are reduced to give the formula

$$f_m = 0.18\, f'_m \left[1 - \left(\frac{h}{30t}\right)^3\right] \qquad (10.$$

The area used for axial stress calculations should be the net cross-sectional area for hollow units, and the gross cross-sectional area for solid fully grouted units. Where only one wythe of a cavity wall supports the vertical load, the area used for calculations is the net cross-sectional area the loaded wythe.

To determine allowable loads, multiply the allowable axial stress (f by the net area (A_n) to obtain the formulas for walls,

$$P = 0.20\, f'_m \left[1 - \left(\frac{h}{40t}\right)^3\right](A_n) \quad \text{or} \quad P = f_m A_n \qquad (10$$

Construction: Single-Wythe 8-in. Hollow Loadbearing Concrete Block

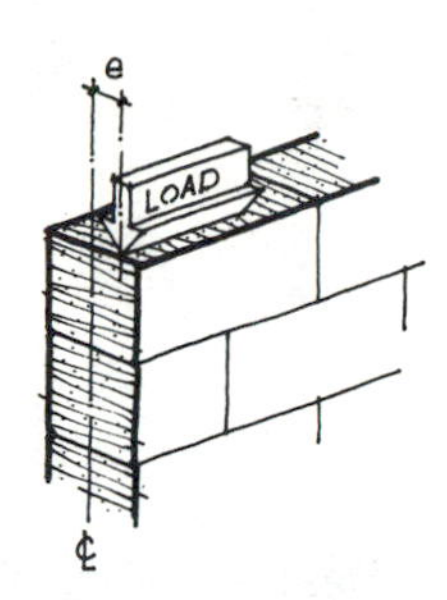

$e \leq S/A_n$
$S/A_n = 80/A_n S = 1.78''$

Prism Strength:
$f'_m = 3500$ psi
$P = F_a F_b A_n S/F_b S + F_a e A_n$

$I = 304.8$ in.^{4}/ft
$A_n = 45.0$ in.^{2}/ft
$S = 80$ in.^{3}/ft

Wall height (ft-in.)	Virtual Eccentricity, e (in.)									
	Axial	0.2	0.4	0.6	0.8	1.0	1.2	1.4	1.6	1.8
4-0	31.39	29.20	27.29	25.62	24.14	22.82	21.64	20.58	19.62	18.74
4-8	31.33	29.14	27.24	25.57	24.10	22.78	21.60	20.54	19.58	18.70
5-4	31.24	29.06	27.17	25.50	24.03	22.72	21.54	20.48	19.52	18.65
6-0	31.14	28.96	27.07	25.41	23.95	22.64	21.47	20.41	19.46	18.59
6-8	31.00	28.84	26.96	25.30	23.85	22.54	21.38	20.33	19.37	18.51
7-4	30.84	28.69	26.82	25.17	23.72	22.43	21.26	20.22	19.27	18.41
8-0	30.64	28.51	26.65	25.01	23.57	22.28	21.13	20.09	19.15	18.29
8-8	30.41	28.29	26.45	24.82	23.39	22.12	20.97	19.94	19.01	18.15
9-4	30.14	28.04	26.21	24.60	23.19	21.92	20.78	19.76	18.84	17.99
10-0	29.83	27.75	25.94	24.35	22.95	21.69	20.57	19.56	18.64	17.81
10-8	29.48	27.42	25.63	24.06	22.67	21.44	20.33	19.33	18.42	17.60
11-4	29.08	27.05	25.28	23.73	22.36	21.14	20.05	19.06	18.17	17.36
12-0	28.62	26.63	24.89	23.36	22.02	20.81	19.74	18.77	17.89	17.09
12-8	28.12	26.16	24.45	22.95	21.63	20.45	19.39	18.43	17.57	16.78
13-4	27.56	25.63	23.96	22.49	21.20	20.04	19.00	18.07	17.22	16.45

(A) Allowable load (kips per foot) of wall when eccentricity of load is less than S/A_n and compression governs.

Construction:
Single-Wythe 8-in.
Hollow Loadbearing
Concrete Block

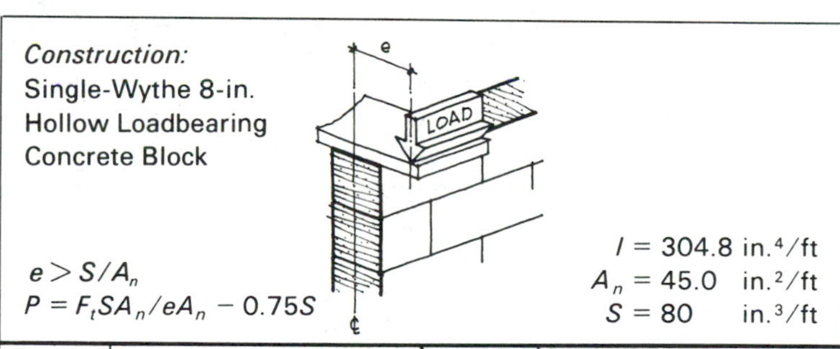

$e > S/A_n$
$P = F_t S A_n / e A_n - 0.75S$

$I = 304.8$ in.^{4}/ft
$A_n = 45.0$ in.^{2}/ft
$S = 80$ in.^{3}/ft

e (in.)	Load (kips/ft) Type M and S mortars	Load (kips/ft) Type N mortar	e (in.)	Load (kips/ft) Type M and S mortars	Load (kips/ft) Type N mortar
2.00	2.74	1.90	4.00	0.69	0.48
2.20	2.11	1.47	4.20	0.64	0.45
2.40	1.72	1.19	4.40	0.60	0.42
2.60	1.45	1.01	4.60	0.56	0.39
2.80	1.25	0.87	4.80	0.53	0.37
3.00	1.10	0.77	5.00	0.50	0.35
3.20	0.98	0.68	5.20	0.46	0.33
3.40	0.89	0.62	5.40	0.45	0.31
3.60	0.81	0.56	5.60	0.43	0.30
3.80	0.74	0.52	5.80	0.41	0.29

(B) Allowable load when eccentricity is greater than S/A_n and tension governs.

0
27

Allowable axial loads for various eccentricities. (*From National Concrete Masonry Association*, TEK Bulletin 31, *NCMA, Herndon, Va.*)

and for columns

$$P = 0.18\,f'_m\left[1-\left(\frac{h}{30t}\right)^3\right](A_n) \qquad \text{or} \qquad P = f_m A_n \tag{10.7}$$

Bearing stresses from concentrated loads on unreinforced masonry may not exceed $0.25\,f'_m$ for bearing on full area, or $0.375f'_m$ for bearing on one-third area or less. The allowable bearing stress on a reasonably concentric area greater than one-third but less than the full area should be interpolated between these two values.

The distribution of compressive stresses from concentrated loads will vary with the style of bonding and the height and length of the masonry units. It is assumed that even with joint reinforcement, stack bond walls will not distribute an axial load to adjacent units. With a standard running bond, recommendations from various industry experts vary from a 60° from horizontal slope distribution pattern to a bell-shaped 2:1 slope. The illustration in *Fig. 10-28* shows the different intensities of stress at the lower floor which result from these various patterns.

10.4.5 Shear Walls and Diaphragms

Concrete masonry shear walls and diaphragms function in exactly the same way as brick masonry. The walls absorb lateral loads applied paralle

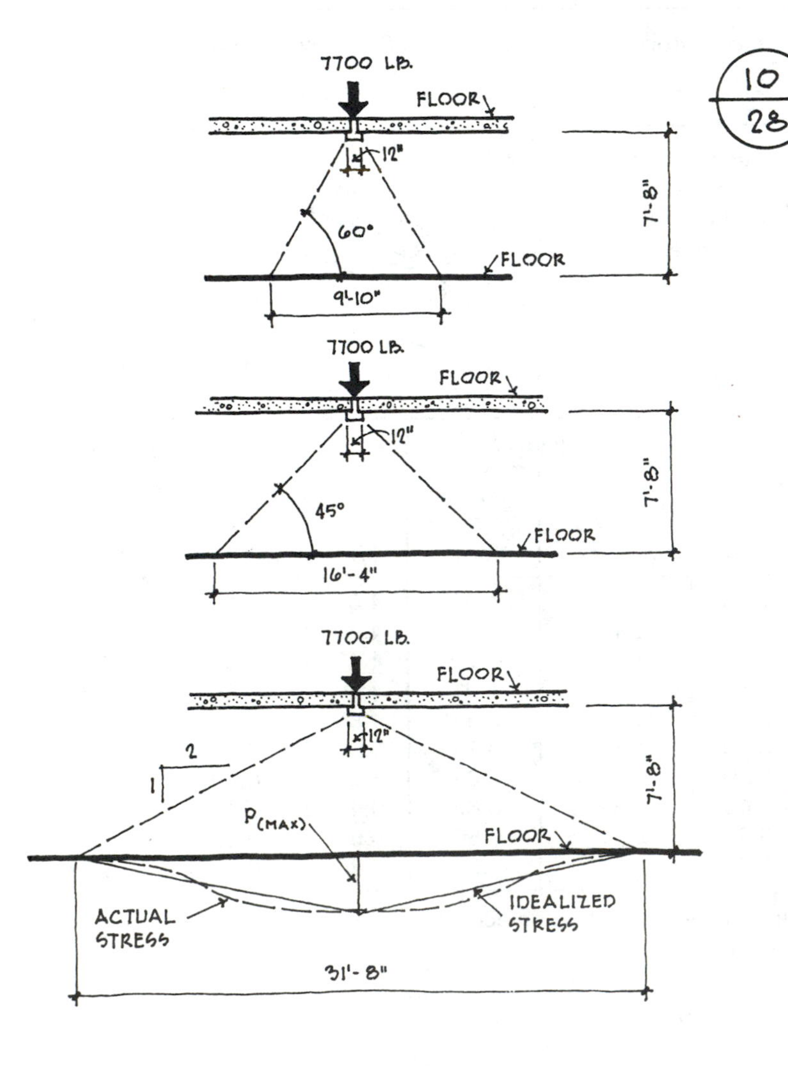

10/28 Axial load distribution in concrete mason (*From Theodore Leba, Jr.*, Design Manua The Application of Non-reinforced Concr Masonry Load-Bearing Walls in Multist Structures, *National Concrete Masonry Ass ation, Herndon, Va., 1969.)*

to their long axis, and must be securely anchored to the floor or roof diaphragms for stress transfer (see Sections 10.1.7 and 10.3.5).

10.4.6 Flexural Strength

Research has indicated that the flexural strength of concrete block walls laid in running bond is substantially greater in the horizontal than in the vertical span. Walls spanning vertically between lateral supports are dependent on a number of variables for flexural strength, all of which are related to the tensile bond developed between the units and the mortar.

The stiffness of vertically spanning bearing walls increases as the vertical load on them increases. Before tensile bending stress can develop in a wall, the horizontal load must overcome the compressive stress due to the vertical load. In effect, the bearing wall is prestressed in compression. The table in *Fig. 10-29* was developed by the NCMA to show the relationship between allowable uniform lateral loads on an 8-in. block wall and vertical

Allowable horizontal uniform load (in lb/sq ft) on vertical walls in a simple span of various heights and with various axial loads. (*From National Concrete Masonry Association*, TEK Bulletin 27, *NCMA, Herndon, Va.*)

CONSTRUCTION:
Unreinforced, Single-wythe 8-in. Hollow Loadbearing Concrete Block Laid Up with Type M or S Mortar

$I = 304.8$ in. 4/ft
$A_n = 45.0$ in. 2/ft

AXIAL LOAD KIPS/FT.
℄
WALL HEIGHT
LATERAL LOAD LB./SQ.FT.

Wall height (ft-in.)	*Axial load (kips/ft)*									
	0.0	*0.25*	*0.50*	*0.75*	*1.0*	*2.0*	*3.0*	*4.0*	*5.0*	*10.0*
	← $f'_m = 1000$ psi →								$f'_m =$ 1350 psi	$f'_m =$ 2000 psi
4-0	76	95	113	132	150	224	298	372	446	816
4-8	56	69	83	96	110	164	218	273	327	598
5-4	42	53	63	74	84	126	167	209	250	458
6-0	33	42	50	58	66	99	132	165	197	361
6-8	27	34	40	47	54	80	107	133	160	292
7-4	22	28	33	39	44	66	88	110	132	241
8-0	19	23	28	32	37	55	74	92	111	203
8-8	16	20	24	27	31	47	63	78	94	173
9-4	13	17	20	24	27	41	54	68	81	149
10-0	12	15	18	21	23	35	47	59	71	129
10-8	10	13	15	18	21	31	41	52	62	114
										$f'_m =$ 2250 psi
11-4	9	11	14	16	18	27	36	46	55	101
12-0	8	10	12	14	16	24	32	41	49	90
12-8	7	9	11	13	14	22	29	36	44	80
13-4	6	8	10	11	13	20	26	33	39	73

axial loads on the same wall. The values shown are based on the allowable stresses listed in *NCMA-1970*, but do not take into account the $33\frac{1}{3}$% increase permitted for intermittent loads such as wind. As the axial load increases, the allowable lateral load also increases dramatically.

The flexural strength of horizontally spanning walls is increased by incorporating horizontal joint reinforcement in the bed joints. Although it will not materially increase the load at which the first cracking will appear, horizontal reinforcement will increase the ultimate strength of the wall. If lateral loads are high, and the tension developed exceeds that permitted by code, the walls must be designed as reinforced or partially reinforced masonry, or the area must be increased.

11

REINFORCED MASONRY

Steel reinforcing bars may be added to brick and concrete masonry structures for increased tensile, shear, and compressive strength. Building codes define "reinforced" masonry by the amount of steel reinforcing used in the design. Unlike partially reinforced masonry, which uses steel only in selective locations to resist specific tensile stresses, fully reinforced masonry must contain the arbitrary minimum amount of steel prescribed by code and located as required in both the horizontal and vertical directions. The arbitrary minimum amount of required steel was developed to resist major seismic stresses. In areas where earthquakes are not a design factor, partially reinforced masonry is more appropriate.

11.1 BASIC DESIGN THEORY

Reinforced masonry is used where compressive, flexural, and shearing stresses are higher than those permitted for plain masonry. The arbitrary amount of reinforcing required by code was developed for areas where high winds or earthquake activity subject buildings to severe lateral dynamic loads. The concept of the box frame or shear wall system utilizes floor diaphragms to distribute these stresses to walls that resist the loads in shear. Reinforcing steel adds ductility and strength to the wall. The masonry, in turn, protects the steel and bears the load with minimum deflection and maximum damping of earthquake energy. A masonry box frame system is much stiffer than a steel or concrete frame.

Reinforced masonry walls may be of double-wythe construction with a grouted cavity to accommodate the steel reinforcing, or of single-wythe hollow units with grouted cores. Steel wire reinforcement may be laid in horizontal mortar joints as long as code requirements for wire size and mortar cover are met. Vertical reinforcement must be held in position at the top and bottom as well as at regular vertical intervals.

Grouting of cores or collar joint cavities may be done in either low lifts of 8 in. maximum height, or high lifts of up to 4 ft. Low-lift grouted walls are filled in 8-in. increments as the wall is built. For high-lift grouting, double-wythe walls must be bonded with rectangular metal ties of at least 9-gauge wire. Tie spacing may not exceed 24 in. on center horizontally and 16 in. on center vertically for running bond or 24 in. horizontally and 12 in. vertically for stack bond. No crimps or water drips are permitted. High-lift grouted walls may be laid to their full height if bottom cleanouts are provided and grout pours are placed in 4-ft-maximum lifts. Mortar fins and protrusions into the cavity or core must be removed as construction progresses to prevent obstructions that would inhibit proper grout placement.

11.1.1 Minimum Amount of Steel

Reinforcing steel contributes to the load-carrying capacity of masonry columns because the lateral ties prevent buckling and confine the masonry within. The vertical steel in walls does not take any of the axial load becaus it is not restrained against buckling. It does, however, increase the overal load-carrying capacity and is an added factor of safety in wall design. A increase in allowable axial compressive and flexural loads usually ac companies the addition of reinforcing to plain masonry. Building code often limit the height of unreinforced masonry because of lateral instabilit and low resistance to tensile forces and bending moments. If reinforcin steel is added, the allowable heights may be increased.

Code requirements for the minimum amount of steel in "reinforced" masonry are arbitrarily derived. J. E. Amrhein and J. L. Noland address th question of how much reinforcing steel is actually needed in masonr structures in an article entitled "Why Reinforce Masonry Walls: Why Thi Amount of Steel?" This article, which appeared in the Feburary, 1976 issu of *Masonry Industry* magazine, briefly outlines the theory of reinforcin steel in masonry construction.

> Upon first glance it would appear that the ratios stated for minimum amount of reinforcement were arbitrary. This is partially true but along with the arbitrary engineering judgment which helped develop these minimum ratios many of the requirements that were used for reinforced concrete were revised and applied to reinforced masonry.
>
> Reinforced concrete walls require twice as much steel as reinforced masonry walls. The area of the horizontal reinforcement of reinforced concrete walls shall be not less than 0.0025, and that of the vertical reinforcement not less than 0.0015 times the area of the wall. This is a total minimum of 0.004 times the area of the wall, $0.004bt$, as compared to 0.002 times the area of the wall, $0.002bt$, for masonry.
>
> Minimum reinforcing steel is placed in concrete to compensate for the shrinkage, moisture change, and temperature stresses that develop not only during the life of the structure, but

also during the initial placing and curing of the concrete. In the case of masonry, the masonry units, either clay or concrete, have already been fabricated and have attained full strength. Concrete units have experienced most of the shrinkage that will occur and clay units have shrunk to their minimum volume during the firing process and, indeed, cannot shrink any more. Mortar and grout which are part of a masonry wall do have slight shrinkage characteristics, but these materials comprise only about one half or much less of the volume of the wall. In keeping with the concepts of reinforced concrete, only half or less of the material that shrinks is used. Thus only half of the steel is required; hence the minimum amount of steel for reinforced masonry is about 50% of that required for reinforced concrete.

Reinforcing steel also helps masonry resist volume changes due to moisture and temperature variations, and substantially reduces the possibility of cracking from these stresses.

Building codes set minimum requirements for spacing of the steel bars. The majority require a maximum spacing of 4 ft on center for principal reinforcement. Some engineers question the universal applicability of such prescriptive requirements, but for the present, most codes do base design on arbitrary ratios and spacings regardless of wall thickness or steel strength. Different combinations of bar sizes and spacing can give the same ratio of steel area to masonry area, and some consideration must be given to economically achieving the minimum standards. Steel spaced too closely will slow construction, can inhibit grout placement, and may be unnecessarily expensive. It is more economical to use larger bars or bundles of bars spaced farther apart as long as the maximum spacing is not exceeded. For instance, No. 3 bars at 8 in. on center give the same area of steel per foot of wall as No. 7 bars at 48-in. centers. The closer spacing requires more time and expense than is necessary to produce the same result. *Figure 11-1* gives steel area and spacing for different bar sizes and wall thicknesses. These tables were developed by the Masonry Institute of America as a design aid in the selection of reinforcement to satisfy code requirements.

11.1.2 Design Codes

The design of reinforced masonry is covered by standards similar to those for unreinforced masonry. The Basic, National, and Standard Codes incorporate by reference the requirements of *BIA-1969* (*Recommended Practice for Engineered Brick Masonry*), *NCMA-1970* (*Specification for the Design and Construction of Loadbearing Concrete Masonry*), and *ANSI A41.2* (*Building Code Requirements for Reinforced Masonry*). The *Uniform Building Code* (*UBC*) outlines specific requirements and procedures that are similar in scope to these three standards, but does not refer specifically to them. Allowable stresses and design formulas vary slightly among the codes, but all of the requirements are based on the same principles of design and analysis.

11.1.3 Materials

The quality and strength of the components substantially affect the strength and performance of reinforced masonry. Because of their im-

Steel reinforcing tables. (*From Masonry Institute of America*, Reinforcing Steel in Masonry, *MIA*, *Los Angeles*, *Calif.*)

TABLE A AREA OF STEEL (SQ IN./FT)

Bar			Bar spacing																
			8″	12″	16″	20″	24″	28″	32″	36″	40″	44″	48″	56″	64″	72″	80″	88″	96″
Size	Diam.	Area	0′-8″	1′-0″	1′-4″	1′-8″	2′-0″	2′-4″	2′-8″	3′-0″	3′-4″	3′-8″	4′-0″	4′-8″	5′-4″	6′-0″	6′-8″	7′-4″	8′-0″
two 9 ga.	0.148	0.0345	0.052	0.034	0.026	0.021	0.017	0.015	0.013	0.102	0.101	0.009	0.0086	—	—	—	—	—	—
two 8 ga.	0.162	0.0412	0.062	0.041	0.031	0.025	0.021	0.018	0.015	0.014	0.012	0.011	0.010	— Joint reinforcing	—	—	—	—	—
two $\frac{3}{16}$	0.1875	0.0552	0.083	0.055	0.041	0.033	0.028	0.024	0.021	0.018	0.017	0.015	0.014	—	—	—	—	—	—
two $\frac{1}{4}$	0.250	0.098	1.47	0.098	0.073	0.059	0.049	0.042	0.037	0.033	0.029	0.027	0.024	—	—	—	—	—	—
two $\frac{5}{16}$	0.312	0.152	0.229	0.15	0.114	0.092	0.076	0.065	0.057	0.051	0.046	0.042	0.038	—	—	—	—	—	—
#2	$\frac{1}{4}$	0.049	0.073	0.05	0.036	0.029	0.024	0.021	0.018	0.016	0.015	0.013	0.012	—	—	—	—	—	—
#3	$\frac{3}{8}$	0.110	0.165	0.11	0.083	0.066	0.055	0.047	0.041	0.037	0.033	0.030	0.027	0.024	0.021	0.018	0.016	0.015	0.014
#4	$\frac{1}{2}$	0.196	0.293	0.20	0.147	0.118	0.098	0.084	0.073	0.065	0.059	0.054	0.049	0.042	0.037	0.033	0.029	0.027	0.024
#5	$\frac{5}{8}$	0.307	0.460	0.31	0.230	0.184	0.154	0.132	0.115	0.102	0.092	0.084	0.077	0.066	0.057	0.051	0.046	0.042	0.038
#6	$\frac{3}{4}$	0.442	0.663	0.44	0.332	0.265	0.221	0.189	0.166	0.147	0.133	0.120	0.110	0.095	0.083	0.074	0.066	0.060	0.055
#7	$\frac{7}{8}$	0.601	0.900	0.60	0.450	0.361	0.300	0.258	0.226	0.200	0.180	0.164	0.150	0.129	0.112	0.110	0.090	0.082	0.075
#8	1.0	0.786	1.180	0.79	0.590	0.471	0.392	0.337	0.295	0.261	0.236	0.214	0.196	0.168	0.147	0.131	0.118	0.107	0.098
#9	1.128	1.000	1.50	1.00	0.750	0.600	0.500	0.428	0.375	0.333	0.300	0.273	0.250	0.214	0.187	0.167	0.150	0.136	0.125
#10	1.270	1.270	1.91	0.127	0.953	0.762	0.635	0.544	0.476	0.423	0.381	0.346	0.318	0.272	0.238	0.212	0.191	0.173	0.159
#11	1.410	1.560	2.34	1.56	1.170	0.936	0.780	0.669	0.585	0.520	0.468	0.425	0.390	0.334	0.293	0.260	0.234	0.213	0.195

TABLE B RATIO OF AREA OF STEEL TO GROSS CROSS-SECTIONAL AREA							
Nominal wall thickness, (in.)	*Bar size*	*Spacing of steel reinforcing (in.)*					
		8	*16*	*24*	*32*	*40*	*48*
6	#3	0.00230	0.00115	0.00076	0.00057		
	#4	0.00417	0.00208	0.00139	0.00104	0.00083	0.00069
	#5	0.00646	0.00324	0.00216	0.00162	0.00129	0.00108
8	#3	0.00172	0.00086	0.00057			
	#4	0.00312	0.00156	0.00104	0.00078	0.00063	
	#5	0.00485	0.00242	0.00161	0.00121	0.00097	0.00081
	#6	0.00688	0.00344	0.00229	0.00172	0.00138	0.00115
	#7	0.00938	0.00469	0.00313	0.00234	0.00188	0.00156
10	#4	0.00250	0.00125	0.00083	0.00063		
	#5	0.00388	0.00194	0.00129	0.00097	0.00078	0.00065
	#6	0.00550	0.00275	0.00183	0.00138	0.00110	0.00092
	#7	0.00750	0.00375	0.00250	0.00188	0.00150	0.00125
	#8	0.00988	0.00494	0.00329	0.00247	0.00198	0.00165
	#9	0.01250	0.00625	0.00417	0.00313	0.00250	0.00208
12	#4	0.00208	0.00104	0.00069	0.00052		
	#5	0.00323	0.00161	0.00109	0.00081	0.00065	
	#6	0.00459	0.00232	0.00153	0.00116	0.00092	0.00077
	#7	0.00625	0.00313	0.00208	0.00156	0.00125	0.00104
	#8	0.00823	0.00411	0.00274	0.00206	0.00165	0.00137
	#9	0.01042	0.00521	0.00347	0.00260	0.00208	0.00174

(Continued)

portance to the overall design, the codes place restrictions on the types of materials that may be used. Brick masonry must comply with ASTM C216 or C62, Grade MW or SW. Concrete units must meet the requirements of ASTM C90, C145, or C55. The steel reinforcing must also meet ASTM standards as follows:

Cold-Drawn Steel Wire for Concrete Reinforcement, ASTM A82
Welded-Steel Wire Fabric for Concrete Reinforcement, ASTM A185
Deformed Billet Steel Bars for Concrete Reinforcement, ASTM A615
Rail-Steel Deformed Bars for Concrete Reinforcement, ASTM A616
Axle-Steel Deformed Bars for Concrete Reinforcement, ASTM A617.

Mortar for reinforced masonry must be ASTM C270, Type M by proportion if masonry cement is used, or Type S by proportion if portland cement-lime is used. Grout must comply with the requirements of ASTM C476.

11.1.4 Allowable Stresses in Reinforced Masonry

Reinforced masonry design is governed by allowable stresses in the masonry and the reinforcement, and allowable shear on bolts and anchors. Masonry stresses are based on values of f'_m, the compressive strength of the masonry at 28 days. This strength may be determined by laboratory prism tests or may be based on an assumed value if the units, mortar, and workmanship all conform to requirements of the code (*see Fig. 10-18*).

The allowable stresses vary slightly among the codes. To facilitate comparison among the standards, the requirements for each are summarized in *Figs. 11-2* through *11-6*. Only one set of standards should be applied in any one design. The allowable stresses and design methods given by a standard must be used throughout the design and should not be substituted interchangeably with the requirements of another standard.

Allowable axial load formulas depend not only on the strength of the masonry (f'_m), but on the height-to-thickness ratio as well. As seen in the equations listed in *Figs. 11-2* through *11-6*, the reduction factors used vary among the codes as they did for unreinforced masonry. *BIA-1969* includes a reduction factor for slenderness (C_s), and for eccentricity (C_e), which may be taken from tables in the code. *NCMA-1970* and the *UBC* use the reduction factors ($h/40t$) and ($h/30t$) with separate checks for eccentricity. *ANSI A41.2* reduces the allowable axial stress in walls for increased slenderness and also requires separate stress checks for eccentric loading. Because of the slight variation among the codes, it is perhaps best to consider each one separately.

11.2 ANSI A41.2

The ANSI code governing reinforced masonry design is the oldest of the four accepted standards, first adopted in 1960. It grew out of the *ANSI A41.1* standard for unreinforced masonry, and there is some carryover of empirical requirements. Generally, however, the *ANSI A41.2* guidelines are based on rational engineering analysis and the design of members to resist calculated forces. Four basic assumptions underlie the theory of this as well as the other three current codes of practice: (1) a section that is plane before bending remains plane after bending; (2) moduli of elasticity of the masonry and of the reinforcement remain constant; (3) tensile forces are resisted only by the tensile reinforcement; and (4) reinforcement is completely surrounded by and bonded to the masonry material.

All dead and live loads acting on a structure must be taken into account, including the moments, shears, and direct stresses resulting from wind, blast, or earthquake forces. These temporary stresses must be added to the maximum dead and live load at any section. An increase in allowable stress of $33\frac{1}{3}\%$ is permitted for wind, blast, or earthquake stresses, as long as the strength of the section is not less than that required for dead and live loads alone. For design purposes, wind, blast, and earthquake loads are assumed never to occur simultaneously.

11.2.1 Design of Columns

Column design is based first of all on minimum dimensions and limitations on height. Minor columns not supporting concentrated loads must have a minimum thickness of 8 in. Major structural columns must have a minimum

TABLE A ALLOWABLE STRESSES IN MASONRY	
Description	*Allowable stress*
Compressive	
Axial, f_m	See notes below
Flexural, f_m	$0.33f'_m$
Shear	
Beams with no web reinforcement, v_m	50 psi
Beams with web reinforcement, v	150 psi
Bond	
Plain bars, u	80 psi
Deformed bars (ASTM A305), u	160 psi
Bearing, f_m	$0.25f'_m$
Modulus of elasticity	$1000f'_m$
Modulus of rigidity	$400f'_m$

Notes:

1. The allowable axial load on columns having the ratio p_v less than 0.006 shall be computed by the following formula:

$$P = A_g(0.16f'_m + 0.52p_gF_s)$$

2. The maximum load P' on axially loaded columns having an unsupported length greater than 10 times the least lateral dimension d, shall not exceed:

$$P' = P(1.3 - 0.03h/d)$$

3. The allowable axial load on columns having the ratio p_v 0.006 or more shall be one and one-quarter times that computed by the formulas above.
4. The allowable stresses in reinforced masonry bearing walls with minimum reinforcement as required by Section 9.9(e), *shall be* $0.20f'_m$ *for walls having a ratio of height to thickness of 10 or less, and shall be reduced proportionally to* $0.15f'_m$ *for walls having a ratio of height to thickness of up to 25.*

TABLE B ALLOWABLE STRESSES IN STEEL
Tensile stress:
18,000 psi for structural-grade steel bars
18,000 psi for structural steel shapes
20,000 psi for intermediate-grade steel bars, and hard-grade bars (billet steel, rail steel, or axle steel)
Compressive stress in column verticals:
16,000 psi for intermediate-grade steel bars
20,000 psi for hard-grade steel bars (billet steel, rail steel, or axle steel)

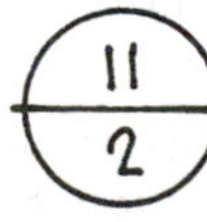

Allowable stresses in reinforced masonry. (*Reproduced with permission from American National Standard Building Code Requirements for Reinforced Masonry, A41.2-1960, copyright 1960, 1970 by the American National Standards Institute. Copies of this standard may be purchased from the American National Standards Institute at 1430 Broadway, New York, N.Y. 10018.*)

TABLE A ALLOWABLE STRESSES IN REINFORCED CONCRETE MASONRY (INSPECTED WORKMANSHIP)*

Description		*Allowable stresses*
Compressive		
Axial	f_m	
Flexural	f_m	$0.33f'_m$ but not to exceed 900 psi
Shear		
No shear reinforcement:		
Flexural members	v_m $1.1\sqrt{f'_m}$	Maximum of 50 psi
Shear walls		
$M/Vd \geq 1$	v_m $0.9\sqrt{f'_m}$	Maximum of 34 psi
$M/Vd < 1$	v_m $2\sqrt{f'_m}$	Maximum of 40 $(1.85 - M/Vd)$
Reinforcement taking entire shear:		
Flexural members	v $3\sqrt{f'_m}$	Maximum of 150 psi
Shear walls		
$M/Vd \geq 1$	v $1.5\sqrt{f'_m}$	Maximum of 75
$M/Vd < 1$	$2\sqrt{f'_m}$	Maximum of 45 $(2.67 - M/Vd)$
Bond		
Deformed bars	u	160 psi
Bearing		
On full area	f_m	$0.25f'_m$
On one-third area or less	f_m	$0.375f'_m$
Modulus of elasticity		$1000f'_m$ but not to exceed 3,000,000 psi
Modulus of rigidity		$400f'_m$ but not to exceed 1,200,000 psi

*See the *NCMA-1970* for notes.

TABLE B ALLOWABLE STEEL STRESSES

	psi
Tensile stress:	
For deformed bars with a yield strength of 60,000 psi or more and in sizes No. 11 and smaller	24,000
For joint reinforcement, 50% of the minimum yield point for the particular kind or grade of steel used, but not to exceed	30,000
For all other reinforcement	20,000
Compressive stress in column verticals:	
40% of the minimum yield strength, but not to exceed	24,000
Compressive stress in flexural members:	
For compression reinforcement in flexural members, the allowable stress shall not be taken as greater than the allowable tensile stress shown above.	
Modulus of elasticity: The modulus of elasticity of steel reinforcement may be taken as 29,000,000 psi.	

Allowable stresses in reinforced concrete masonry. (*From National Concrete Masonry Association,* Specification for the Design and Construction of Load-bearing Concrete Masonry, *NCMA, Herndon, Va., 1970.*)

ALLOWABLE STRESSES IN MASONRY—INSPECTED WORKMANSHIP*			
		Allowable stresses (psi)	
Description		*Related to f'_m*	*Maximum*
Compressive Axial	f_m	See Sections 10.1.3 and 10.1.4	1000
Flexural	f_m	$0.33f'_m$	1200
Bearing On full area On one-third area or less	f_m f_m	$0.25f'_m$ $0.375f'_m$	900 1200
Shear No shear reinforcement Flexural members Shear walls $M/Vd_v \geq 1$ $M/Vd_v < 1$	V_m V_m V_m	$1.1\sqrt{f'_m}$ $0.9\sqrt{f'_m}$ $2.0\sqrt{f'_m}$	50 34 40 $(1.85 - M/Vd_v)$
Reinforcement taking entire shear Flexural members Shear walls $M/Vd_v \geq 1$ $M/Vd_v < 1$	V V V	$3.0\sqrt{f'_m}$ $1.5\sqrt{f'_m}$ $2.0\sqrt{f'_m}$	150 75 45 $(2.67 - M/Vd_v)$
Tension No tension reinforcement Tension normal to bed joints Hollow units Solid and/or grouted units Tension parallel to bed joints in running bond Hollow units Solid and/or grouted units	F_t F_t F_t F_t	$0.5\sqrt{m_o}$ $1.0\sqrt{m_o}$ $1.0\sqrt{m_o}$ $1.5\sqrt{m_o}$	25 40 50 80
Modulus of elasticity Modulus of rigidity		$1000f'_m$ $400f'_m$.	2,500,000 1,000,000

*See *ACI-539* for notes.

Allowable stresses in reinforced concrete masonry. (*From American Concrete Institute*, Building Code Requirements for Concrete Masonry Structures, 1979.)

dimension of 12 in. The maximum unsupported height (h) of columns is limited to 20 times the least lateral dimension (d). Unsupported height is defined as not less than the clear distance between the floor surface and the underside of the deeper beam framing into the column in each direction at the next-higher floor level. For rectangular columns, the combination of vertical and horizontal dimensions which gives the greatest h/d ratio must be used for calculations.

Allowable axial loads on columns are computed in one of several ways depending on the ratio of the volume of lateral reinforcement to the volume

TABLE A MAXIMUM WORKING STRESSES (PSI) FOR REINFORCED SOLID AND HOLLOW UNIT MASONRY*

Type of stress	*Special inspection required*	
	Yes	*No*
Compression		
Axial		
Walls	See Section 2418	One-half of the values permitted under Section 2418
Columns	See Section 2418	One-half of the values permitted under Section 2418
Flexural	$0.33f'_m$ but not to exceed 900	$0.166\ f'_m$ but not to exceed 450
Shear		
No shear reinforcement		
Flexural	$1.1\sqrt{f'_m}$ 50 max.	25
Shear walls		
$M/Vd \geq 1$	$0.9\sqrt{f'_m}$ 34 max.	17
$M/Vd = 0$	$2.0\sqrt{f'_m}$ 50 max.	25
Reinforcing taking all shear		
Flexural	$3.0\sqrt{f'_m}$ 150 max.	75
Shear walls		
$M/Vd \geq 1$	$1.5\sqrt{f'_m}$ 75 max.	35
$M/Vd = 0$	$2.0\sqrt{f'_m}$ 120 max.	60
Bond		
Plain bars	60	30
Deformed bars	140	100
Bearing		
On full area	$0.25f'_m$ but not to exceed 900	$0.125f'_m$ but not to exceed 450
On $\frac{1}{3}$ or less of area	$0.30f'_m$ but not to exceed 1200	$0.15f'_m$ but not to exceed 600
Modulus of elasticity	$1000f'_m$ but not to exceed 3,000,000	$500f'_m$ but not to exceed 1,500,000
Modulus of rigidity	$400f'_m$ but not to exceed 1,200,000	$200f'_m$ but not to exceed 600,000

*See the *UBC* for notes.

Allowable stresses in reinforced masonry. (*Reproduced from the Uniform Building Code, 1982 edition, copyright 1982, with permission of the publisher, the International Conference of Building Officials.*)

of the masonry core (p_v) computed out-to-out of ties. Where the ratio (p_v) is less than 0.006, allowable axial loads are computed by the formula

$$P = A_g(0.16f'_m + 0.52p_g f_s) \tag{11.1}$$

where P = total allowable axial load on column, lb
A_g = overall or gross area of solid masonry units or solidly filled hollow masonry units, or the net cross-sectional area in bearing for hollow masonry, sq in
f'_m = compressive strength of masonry, psi

TABLE B ALLOWABLE STEEL STRESSES	
	psi
Tensile stress:	
For deformed bars with a yield strength of 60,000 psi or more and in sizes No. 11 and smaller	24,000
Joint reinforcement, 50% of the minimum yield point specified in UBC Standards for the particular kind and grade of steel used, but in no case to exceed	30,000
For all other reinforcement	20,000
Compressive stress in column verticals:	
40 percent of the minimum yield strength, but not to exceed	24,000
Compressive stress in flexural members:	
For compression reinforcement in flexural members, the allowable stress shall not be taken as greater than the allowable tensile stress shown above.	

p_g = ratio of effective cross-sectional area of vertical reinforcement to gross area, A_g
f_s = nominal stress in vertical column reinforcement, psi (*see Fig. 11-2*)

For axially loaded columns having an unsupported height greater than 10 times the least lateral dimension, allowable loads are calculated by the formula

$$P' - P\left(1.3 - \frac{0.03h}{d}\right) \tag{11.2}$$

where P' is the total allowable axial load on a column in pounds. The allowable load on columns with a (p_v) ratio of 0.006 or more is taken as 1.25 times that computed by the formulas above.

Slenderness ratios are thus accommodated by a reduction in allowable load for increased height-to-depth relationships. Increased lateral restraint against buckling (represented by increased p_v ratio) is recognized with an increase in the allowable load.

Minimum requirements for vertical reinforcement of columns and pilasters are stated as ratios of the effective cross-sectional area of the reinforcing to the gross area of the masonry. The ratio (p_g) may not be less than 0.005 nor more than 0.04, and the reinforcement must consist of at least four bars, each having a minimum diameter of $\frac{1}{2}$ in. If lapped splices are used, they must be at least 24 bar diameters, and must be capable of transferring the allowable stress by bond. Welded splices must develop the full strength of the bar. All reinforcement must be held firmly in alignment to avoid displacement during grouting operations.

Minimum lateral reinforcement must consist of ties at least $\frac{1}{4}$ in. in diameter, the spacing of which may not exceed any of the following: (1) 16 vertical bar diameters, (2) 48 tie diameters, (3) the least column dimension, or (4) 16 in. This requirement obviously links the need for lateral restraint against bending not only to the size of the steel used, but to the cross-sectional size of the column itself since it is related to slenderness and bending. In columns where the ratio of lateral reinforcement (p_v) is more

TABLE A ALLOWABLE STRESSES IN MASONRY*			
Description		Allowable stresses (psi) Without inspection	With inspection
Compressive			
Axial			
Walls	f_m	$0.25f'_m$	$0.25f'_m$
Columns	f_m	$0.20f'_m$	$0.20f'_m$
Flexural			
Walls and beams	f_m	$0.40f'_m$	$0.40f'_m$
Columns	f_m	$0.32f'_m$	$0.32f'_m$
Shear			
No shear reinforcement			
Flexural members	V_m	$0.7\sqrt{f'_m}$, but not to exceed 25	$0.7\sqrt{f'_m}$, but not to exceed 50
Shear walls	V_m	$0.5\sqrt{f'_m}$, but not to exceed 50	$0.5\sqrt{f'_m}$, but not to exceed 100
With shear reinforcement taking entire shear			
Flexural members	V	$2.0\sqrt{f'_m}$, but not to exceed 60	$2.0\sqrt{f'_m}$, but not to exceed 120
Shear walls	V	$1.5\sqrt{f'_m}$, but not to exceed 75	$1.5\sqrt{f'_m}$, but not to exceed 150
Bond			
Plain bars	u	53	80
Deformed bars	u	107	160
Bearing			
On full area	f_m	$0.25f'_m$	$0.25f'_m$
On one-third area or less	f_m	$0.375f'_m$	$0.375f'_m$
Modulus of elasticity	E_m	$1000f'_m$, but not to exceed 2,000,000 psi	$1000f'_m$, but not to exceed 3,000,000 psi
Modulus of rigidity	E_v	$400f'_m$, but not to exceed 800,000 psi	$400f'_m$, but not to exceed 1,200,000 psi

*See *BIA-1969* for notes.

TABLE B ALLOWABLE STRESSES IN STEEL	
	psi
Tensile stress:	
For deformed bars with a yield strength of 60,000 psi or more and in sizes No. 11 and smaller	24,000
For all other reinforcement	20,000
Compressive stress in column verticals:	
40% of the minimum yield strength, but not to exceed	24,000
Compressive stress in flexural members: For compression reinforcement in flexural members, the allowable stress shall not be taken as greater than the allowable tensile stress given in Section 4.5.1.1.	
Modulus of elasticity: The modulus of elasticity of steel reinforcement may be taken as 29,000,000 psi.	

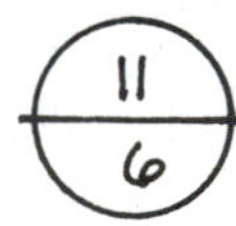

Allowable stresses in reinforced brick masonry. (*From Brick Institute of America*, Recommended Practice for Engineered Brick Masonry, *BIA, McLean, Va., 1969.*)

than 0.006 and increased allowable loads are permitted, the maximum spacing must be reduced to 4 in. Ties must be shaped in a circle or rectangle to surround the vertical steel. Splices may be either welded or lapped 50 diameters.

Eccentricity of axial loading and conditions of end restraint that result in *bending moments* must be considered in the design. Columns must be able to resist axial forces from loads on all floors plus the maximum bending caused by loads on a single adjacent span of the floor under consideration. In other words, eccentricity is not cumulative. All loads from floors above are assumed to be delivered concentrically. Resistance to bending at any floor level must be provided by distributing the moment between the columns immediately above and below the floor in proportion to their relative stiffness and conditions of restraint. For stresses due to combined axial load and bending when the eccentricity (e/t) is $\frac{1}{3}$ or greater, members must be proportioned so that the quantity $f_a/F_a + f_b/F_b$ is less than or equal to 1. If the eccentricity (e/t) is less than $\frac{1}{3}$, design may be based on the uncracked section.

11.2.2 Design of Reinforced Walls

When masonry walls are reinforced with steel that is designed, placed, and laterally anchored as for columns, allowable loads may be determined by the same formulas and under the same conditions as the code provides for column design. The length of the wall considered effective for each element may not exceed the center-to-center distance between loads or the width of bearing plus four times the wall thickness. In other words, if the vertical reinforcing in a wall is restrained against bending, that portion of the wall in which such reinforcing occurs may be considered a column. In areas of concentrated loads, where higher resistance is required, such reinforcing may be desirable to increase the allowable stresses permitted in the wall. A wall section reinforced in this manner functions as an integral column in which the element does not project beyond the face of the wall. Where walls are reinforced with only the minimum required steel in a linear arrangement not restrained by lateral ties, allowable axial compressive stresses are limited to $0.20 f'_m$ when the height to thickness ratio is 10 or less. Allowable stresses must be reduced proportionally to $0.15 f'_m$ as the h/t ratio increases to 25. By either method of reinforcement and stress computation, concentrated loads are not considered as distributed by metal ties across continuous vertical joints. Loads carried on one wythe of a cavity wall are restricted to that wythe, and the other shares neither the load nor the stresses. Similarly, loads are not transferred across head joints when units are laid in stack bond.

All other allowable stresses in reinforced walls such as flexural compressive stresses, shear, bond, and so on, must conform to the limitations set forth in *Fig. 11-2*.

Walls must be designed for eccentric loads and for any lateral forces, pressures, or shears that may be applied. Combined axial and flexural stresses must satisfy the equation $f_a/F_a + f_b/F_b \leq 1$.

Empirical *height-to-thickness ratios* are greater than those for unreinforced walls because of the increased strength and improved performance attributable to the incorporation of steel reinforcing. Bearing walls must have a nominal thickness of at least $\frac{1}{25}$ of the unsupported height or width (whichever is less) but may not be less than 6 in. This is much less restrictive than the minimum 12 in. required for empirically designed

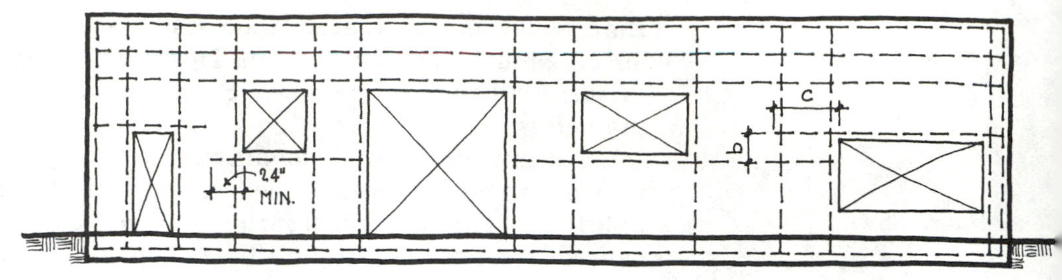

NOTE: WHERE CONTINUITY OF REINFORCEMENT IS DESIRED, C MUST EQUAL 2b OR MORE.

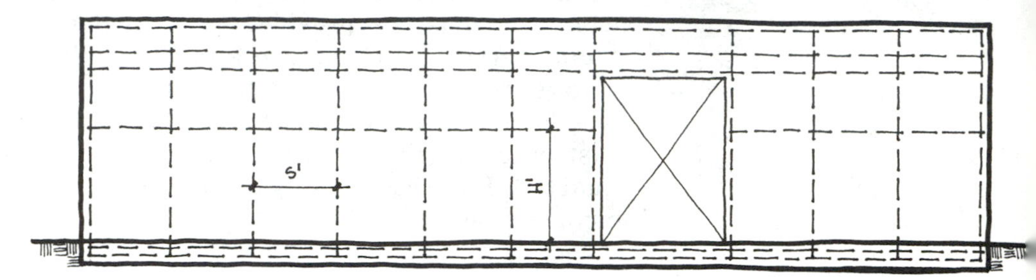

NOTE: IN SEISMIC AREAS, S' = 4 FT. MAX. IN OTHER AREAS, S' = 8FT. MAX. FOR LOW-SHRINKAGE UNITS, H' = 12 FT. MAX. FOR OTHER UNITS, H' = 8 FT. MAX.

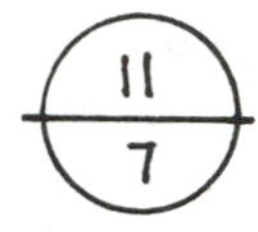

Diagram showing position of vertical and horizontal reinforcement. (*From Albyn Mackintosh*, Design Manual—The Application of Reinforced Concrete Masonry Load-Bearing Walls in Multi-Storied Structures, *National Concrete Masonry Association, Herndon, Va., 1973.*)

unreinforced walls. Panel and enclosing walls must have a thickness at least $\frac{1}{30}$ of the distance between supporting or enclosing members. Both of these restrictions may be waived by the building official with written evidence that the wall meets all other code requirements, including allowable stress limitations.

Walls designed as reinforced masonry must have a minimum area of steel at least 0.002 times the cross-sectional area of the wall, not more than $\frac{2}{3}$ of which (0.0007) may be placed in either direction. The maximum spacing of the principal reinforcement may not exceed six times the wall thickness or 48 in. on center. In addition, horizontal reinforcement must be provided in bond beams at the top of footings, at the bottom and top of wall openings, at roof and floor levels, and at the top of parapet walls. Only reinforcement that is continuous in the wall may be considered in computing the minimum area of steel. There must also be at least one $\frac{1}{2}$-in.-diameter bar around all window and door openings extending at least 24 in. beyond the corner of the openings (*see Fig. 11-7*). This prevents diagonal cracking along the planes of weakness created at these corners.

All reinforced masonry walls must be securely anchored to adjacent structural members such as roofs, floors, columns, pilasters, and intersecting walls. Without such anchorage, forces cannot be properly distributed or transferred to other parts of the structure or to the foundation.

11.3 NCMA-1970

The NCMA design code for reinforced concrete masonry was written 1 years later than *ANSI A41.2*, and some of its restrictions are less conservative, owing to further research and investigation of masonry structures during that decade. It does however, apply only to concrete masonry. The theory of design is essentially the same, but the slenderness reduction facto is expressed differently. Unsupported height restrictions for columns and height-to-thickness ratios for walls are less stringent, and requirements for horizontal reinforcement are different. All dimensions used in CMU design calculations (except h/t ratios) should be actual rather than nominal dimensions.

11.3.1 Design of Columns

Reinforced masonry columns designed under *NCMA-1970* must be at least 12-in. in thickness unless allowable stresses are reduced by one-half, in which case the least lateral dimension may be 8 in. The maximum unsupported height of a column is defined the same as in ANSI requirements, but is increased to 25 times the least lateral dimension. Height-to-thickness ratios of rectangular columns are computed on the basis of the combination of vertical and horizontal dimensions which yields the highest value.

Allowable axial loads on columns are determined by the formula

$$P = A_g(0.20f'_m + 0.65p_g F_s)\left[1 - \left(\frac{h}{30t}\right)^3\right] \qquad (11.3)$$

where P = total allowable axial load on column, lb
A_g = overall gross area of solid masonry units or solidly filled hollow units, or the net cross-sectional area in bearing for hollow masonry, sq in.
f'_m = compressive strength of masonry, psi
p_g = ratio of effective cross-sectional area of vertical reinforcement to gross area, A_g
F_s = allowable stress in vertical column reinforcement, psi
h = unsupported height of column, in.
t = overall depth or thickness of column, in.

The modifier $[1 - (h/30t)^3]$ is the slenderness reduction coefficient, which replaces the formula $P' = P(1.3 - 0.03h/d)$ in the ANSI code. The NCMA coefficient applies to all columns (where the ANSI equation applies only to those columns with an unsupported height greater than 10 times the least lateral dimension). There is no minimum slenderness ratio that must be attained before the height-to-thickness proportions are considered to affect the design. H/t ratios are integral to the calculation of allowable loads in all instances.

Minimum requirements for vertical reinforcement are identical to those in the ANSI standard. That is, the ratio (p_g) of vertical reinforcement may not be less than 0.005 or more than 0.04, and the reinforcing must consist of at least four bars, each having a minimum diameter of $\frac{1}{2}$ in. Lapped splices must be at least 24 bar diameters, and must be capable of transferring allowable stresses by bond. Welded splices must develop the full strength of the bar, and all column reinforcement must be held firmly in its designed position until permanently grouted. *Minimum lateral reinforcement* of columns is slightly different in *NCMA-1970*. All vertical reinforcement must be enclosed by lateral ties no smaller than 9-gauge wire spaced apart no more than 16 bar diameters, 48 tie diameters, or the least column dimension. Ties may be placed either in the mortar joint, or in contact with the vertical steel. There is no differentiation in requirements for various ratios of volume of lateral reinforcement to volume of masonry, nor, in fact, is any such ratio mentioned in this standard. The size and spacing requirement minimums apply equally to all column designs.

Bending moments are considered for eccentric loads and certain conditions of end restraint at the floor under consideration. They are distributed between the columns immediately above and below a given floor in proportion to their relative stiffness and conditions of restraint. For combined axial and bending stresses, requirements are the same as in ANSI. If the eccentricity (e/t) is greater than $\frac{1}{3}$, the quantity $f_a/F_a + f_b/F_b$ may not exceed 1. If e/t is less than $\frac{1}{3}$, design may be based on the uncracked section.

Allowable stresses in flexure, shear, bond, and so on, are given in *Fig. 11-3*. Most of these stresses are based on the compressive strength of the masonry (f'_m) with a maximum value given. Although *NCMA-1970* gives greater detail as to type of loading and reinforcement, the maximum allowables are the same as those in the ANSI standard. All allowable stresses must be reduced by one-half when there is no provision for architectural or engineering inspection to assure compliance with the drawings, specifications, and prescribed construction procedures. For stresses due to wind, blast, or earthquake, allowable stresses may be increased $33\frac{1}{3}$% provided that the strength of the section is not less than that required for dead and live loads alone. Again, wind, blast, and earthquake loads may be assumed never to occur simultaneously.

11.3.2 Design of Reinforced Walls

Walls are required to have a nominal thickness of at least $\frac{1}{30}$ the unsupported height or width of the wall (whichever is less), but may not be less than 6 in Non-loadbearing reinforced panel and enclosing walls must have a thick ness not less than $\frac{1}{36}$ the distance between the supporting or enclosing members. These figures are less restrictive than those permitted by ANSI but here too, all height-to-thickness restrictions may be waived by th building official upon approval of written justification.

Minimum reinforcing requirements call for an area of steel not less than 0.002 times the cross-sectional area of the wall with a maximum o two-thirds in either direction. Spacing of principal reinforcement may no exceed 48 in. or six times the wall thickness. Although these requirement are the same as ANSI, the NCMA code goes on to state that wir reinforcement in horizontal mortar joints (i.e., truss or ladder type used fo shrinkage control) may be calculated as part of the required area of stee Since this type of reinforcing is peculiar to concrete masonry construction, i is logical that this standard makes specific provision for its inclusion. (Th ANSI Code does not exclude horizontal joint reinforcing, but makes n specific reference to it.)

Horizontal reinforcement (other than joint reinforcement) must be pro vided in bond beams at the top of footings, at the bottom and top of wa openings, at roof and floor levels, and at the top of parapet walls. Onl reinforcement that is continuous in the wall may be considered in com puting the minimum area of steel. In addition to the minimum rati required, reinforcing bars of at least $\frac{1}{2}$ in. in diameter must be provide around all window and door openings, and must extend no less than 24 in beyond the corners of the openings.

When bearing walls are reinforced according to these minimu standards, allowable stresses are calculated by the formula

$$f_m = 0.225 f'_m \left[1 - \left(\frac{h}{40t} \right)^3 \right] \qquad (11.$$

where f_m is the allowable axial compressive stress in psi. The modifier for slenderness $[1 - (h/40t)^3]$ reduces the allowable load by a lesser amount than the slenderness coefficient for columns $[1 - (h/30t)^3]$. This takes into account the greater resistance to bending offered by an elongated wall section compared to more symmetrical column shapes, and reflects the assumption that a column is a more critical member in the performance of the building. That is, the loss of a column is more critical than that of a wall since all of the wall would not be likely to fail.

When the reinforcement in bearing walls is designed, placed, and anchored in position as for columns, the allowable stresses may be calculated using the equation for columns, which takes into consideration the actual strength of the steel and the ratio of its area to the area of the masonry. When using this formula, the effective length of wall may not exceed the center-to-center distance between loads nor may it exceed the width of bearing plus four times the wall thickness. This is intended primarily for concentrated loads where wall reinforcing may be increased to create an integral column that does not project beyond the wall surface. Concentrated loads are not considered distributed by metal ties or across continuous vertical joints in multi-wythe walls, or across head joints in stack bond walls.

Walls must be designed for any eccentric loads, lateral forces, pressures, or shears that may occur. Combined axial and bending stresses must satisfy the formula $f_a/F_a + f_b/F_b \leq 1$, as previously outlined. If the eccentricity is less than $\frac{1}{3}$, however, design may be based on the uncracked section. (This requirement is identical to that of *ANSI A41.2.*)

The final provision of *NCMA-1970* is for secure anchorage of reinforced walls to adjacent structural members to assure transfer and distribution of forces. Some typical wall intersections and wall-floor connections are shown in *Fig. 11-8*. These are suggested details only. Anchorage, bonding, and reinforcement requirements may differ for given situations, depending on the amount and type of stress involved. Each problem should be thoroughly investigated to determine the appropriate engineering solution. Proper performance of a loadbearing wall system depends on diaphragm action between the wall and floor systems, and secure anchorage between the two is essential to the structural integrity of the building.

11.4 ACI-539

The most recently adopted code, *ACI-539*, published by the American Concrete Institute, is very similar to *NCMA-1970*, and also applies only to concrete masonry. Allowable axial loads on columns are computed by the formula

$$P = (0.225 f'_m A_n + 0.65 A_s F_s)\left[1 - \left(\frac{h}{40t}\right)^3\right] \qquad (11.5)$$

where A_n is the net area (hollow units) in square inches and A_s is the area of reinforcement in square inches. The slenderness reduction coefficient is less restrictive than NCMA, as indeed most of the ACI requirements are. Minimum vertical reinforcement for columns must be between 0.0025 and 0.04 times the gross cross-sectional area. This minimum is only one-half of that required by NCMA.

Reinforced walls must have a minimum thickness of only $\frac{1}{36}$ the unsupported height or width, and required reinforcing may be spaced 8 ft on center or 12 times the wall thickness. Walls reinforced for concentrated loads have an effective thickness equal to the center-to-center distance

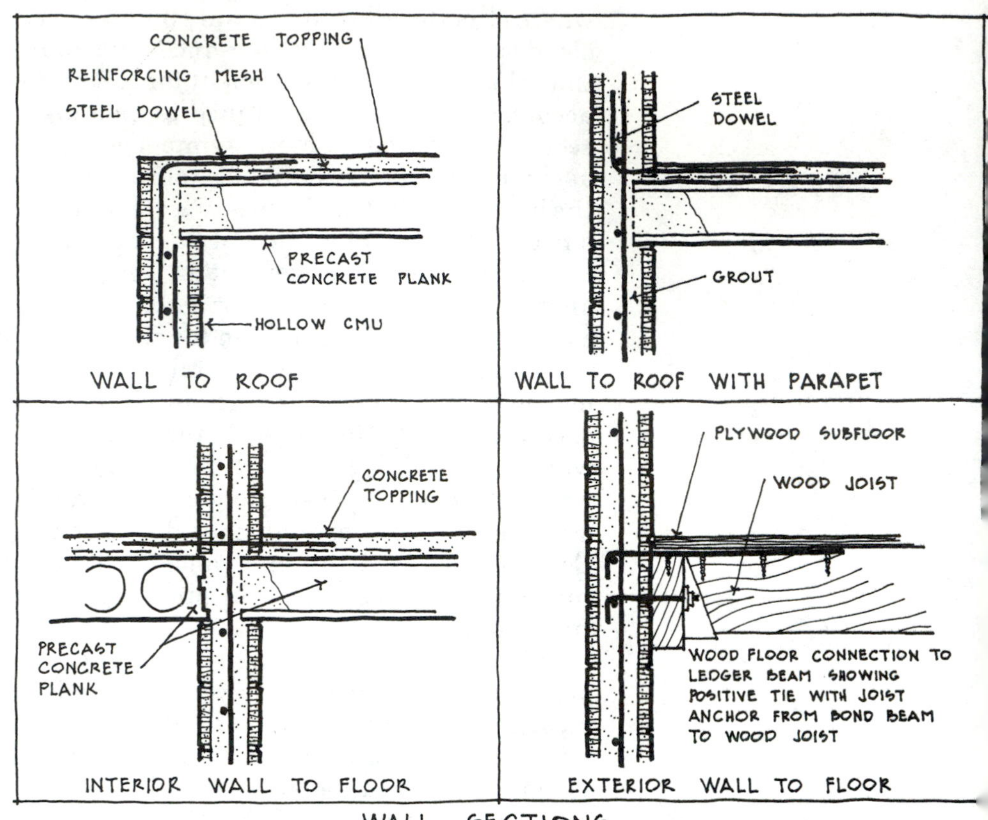

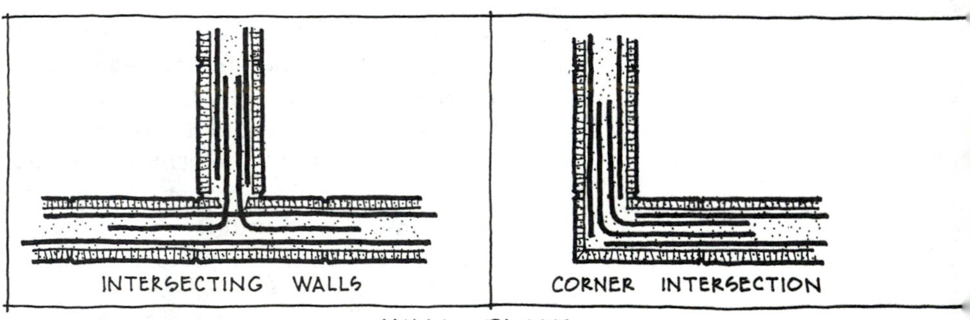

Connection details. (*From Masonry Institute of America,* Reinforcing Steel in Masonry, *MIA, Los Angeles, Calif.*)

between loads, or one-half the wall height. Allowable tension and shear (*se Fig. 11-4*) must be reduced by one-half if there is no architectural o engineering inspection, but compressive stress is reduced by only one-thirc All other major requirements are the same as *NCMA-1970*.

The more liberal design requirements may again be attributed t additional research and greater understanding of the structural behavior c masonry between the 1970 publication of the NCMA requirements and th 1979 adoption of the ACI code.

11.5 UBC-1982

The allowable stresses for reinforced masonry given in the *Unifor Building Code* (*see Fig. 11-5*) are more conservative than in any of the othe design standards (perhaps because the *UBC* is widely used on the We

Coast, where seismic disturbances are of major concern). The code requires that in Seismic Zones 2, 3, and 4, all masonry elements of a structure must be of reinforced design and comply with minimum reinforcing and allowable stress limitations. Although general design considerations are similar to other masonry standards, there are some special requirements related directly to the severity of expected earthquake loading represented by the seismic zone designation. *UBC-1982* applies to both concrete and clay masonry. Minimum thickness requirements are based on nominal unit dimensions, but stresses must be determined for net thickness of the masonry with consideration for reductions such as raked joints.

11.5.1 Design of Columns

The least lateral dimension permitted for primary structural columns designed on the basis of the allowable stresses listed in *Fig. 11-5* is 12 in. Where columns are auxiliary supports stressed to less than one-half the allowable, the minimum dimension may be 8 in. No masonry column is permitted to have an unsupported length greater than 20 times its least lateral dimension. *Allowable axial loads* are computed using a formula similar to the one given in *NCMA-1970* except that the safety factor is greater and the reduction for slenderness is less restrictive:

$$P = A_g(0.18f'_m + 0.65p_gF_s)\left[1 - \left(\frac{h}{40t}\right)^3\right] \tag{11.6}$$

A limitation of 6000 psi maximum is given for ultimate masonry strength. This value is quite high, and would generally require solid clay units with a compressive strength greater than 10,000 psi. Although standard-run bricks are available in compressive strengths as high as 22,500 psi, units most commonly used are in the range 4000 to 8000 psi and yield ultimate strengths of 2500 to 5500 psi when laid with Type M or S mortar.

For *vertical reinforcement* in columns, the ratio (p_g) may not be less than 0.5% or greater than 4%. A minimum of four bars are required, at least $\frac{3}{8}$ in. in diameter, but not greater than No. 10. Where lap splices are used, the lap must be sufficient to transfer the working stress by bond, and may not in any case be less than 30 bar diameters. Welded splices must be full butt-welded. This differs from ANSI and NCMA in minimum bar size ($\frac{1}{2}$ to $\frac{3}{8}$ in.), lap splice length (30 bar diameters to 24), and allowable stress transfer to working stress.

The minimum *lateral reinforcement* required to support the vertical steel in columns varies depending on the seismic zone in which the building is located. Requirements are very detailed, very explicit, and generally very conservative as to bar size, spacing, and placement.

11.5.2 Design of Reinforced Walls

The minimum nominal thickness of reinforced masonry bearing walls under *UBC* requirements is 6 in. The ratio of height or length to thickness may not exceed 25. As in the other codes, however, these limitations may be waived when the building official approves a written justification. (Research reports approved by the International Conference of Building Officials allow 4-in.-thick reinforced bearing walls under certain conditions.) The axial stress in reinforced wall elements may not exceed the value determined by the formula (from Chapter 10)

$$f_m = 0.20f'_m\left[1 - \left(\frac{h}{40t}\right)^3\right] \quad (10.4$$

This formula is only slightly more conservative than NCMA requirements with a safety reduction factor of 0.20 compared to 0.225.

All walls designed by allowable stress must have both *horizontal and vertical reinforcement*. The minimum total area of required steel is 0.002 times the gross cross-sectional area of the wall, with a minimum of 0.0007 in either direction. Maximum spacing is limited to 4 ft on center. Reinforcing bars must be at least $\frac{3}{8}$ in. in diameter except that horizontal join reinforcement may be considered in calculating the minimum required area. Horizontal reinforcement must be provided in bond beams at the top of footings, at the top of wall openings, at structurally connected floor and roof levels, and at the top of parapet walls. Only reinforcement that is continuous in the wall may be considered in calculating the minimum area of reinforcement.

If a wall is constructed of more than two wythes, the required reinforcement must be equally divided into two layers. (Reinforcement over and above the required minimum need not be divided in this manner.) In addition to the minimum requirements, all loadbearing walls must have one No. 4 bar or two No. 3 bars on all sides of, and adjacent to, openings that exceed 24 in. in either dimension. These bars must extend at least 40 bar diameters and at least 24 in. beyond the corners of the openings. The UBC standard is the only one of the four design codes that refers to the size of the opening in its provision for this extra reinforcement.

If the reinforcement in bearing walls is designed, placed, and anchored in position as for columns, allowable axial loads may be computed using the formula given for columns. The length of wall considered effective may not exceed the center-to-center distance between loads, or the width of bearing plus four times the wall thickness. Concentrated loads are not considered distributed by metal ties across continuous vertical joints, or across head joints in stack bond.

As compared to the previous design standards, *UBC-1982* differs primarily in its conservative values for allowable stress, stringent requirements for lateral reinforcing ties in columns, and lower height-to-thickness ratio for walls. Although the differences do not appear to be great, they give an added factor of safety for the structural integrity of bearing wall buildings subjected to seismic disturbance.

11.6 BIA-1969

The design code developed by the Brick Institute of America applies, of course, only to solid brick construction. The requirements for reinforced column and wall design differ from the other standards primarily in the reduction factors for slenderness and eccentricity, and in the method used to determine f'_m.

11.6.1 Design of Columns

The method of calculating *allowable vertical loads* depends on the amou of eccentricity of the applied load. If the maximum virtual eccentricity do not exceed $t/3$, allowable loads are determined by the formula

$$P = C_e C_s (f_m + 0.80 p_g F_s) A_g \quad (11.$$

where C_e is the eccentricity coefficient (*see Fig. 10-21B*) and C_s is the slenderness coefficient (*see Fig. 10-21A*). The derivation of eccentricity and slenderness coefficients (C_e and C_s) are discussed in Chapter 10, and tables are given to determine values for use in design formulas (*see Fig. 10-21*). For relatively small eccentricities, equation (*11.7*) determines permissible loadings. For larger eccentricities exceeding $t/3$, allowable loads are determined on the basis of a transformed section and linear stress distribution. In these cases, the maximum compressive stress in the masonry may not exceed $0.32f'_m$, and the stresses in the reinforcement may not exceed the values given in *Fig. 11-6*. Allowable vertical loads determined by this method must be reduced for slenderness as determined from Table *A* in *Fig. 10-21*.

When columns are subject to bending about both principal axes, these same two methods are used to determine axial loads. When eccentricity exceeds $bt/3$, the second method is used. When eccentricity is less than $bt/3$, allowable loads are determined by equation (*11.7*) above except that the eccentricity coefficient (C_e) must be computed by a formula given in the code.

The area of *vertical reinforcement* for columns must be at least $0.005A_g$, but not more than $0.04A_g$. Columns stressed to less than one-half of their allowable values may have vertical reinforcement reduced to a minimum of $0.002A_g$. Lateral ties must be at least 9-gauge wire, and spacing may not exceed 16 bar diameters, 48 tie diameters, or the least dimension of the column.

11.6.2 Design of Reinforced Walls

When the maximum virtual eccentricity of a reinforced brick loadbearing wall does not exceed $t/3$, *allowable loads* are determined by the formula (from Chapter 10)

$$P = C_e C_s f_m A_g \tag{10.1}$$

The value of $(C_e C_s f_m)$ is the average allowable compressive stress permitted.

If the maximum virtual eccentricity of the load exceeds $t/3$, the allowable load is determined on the basis of a transformed section and linear stress distribution. Reinforcement in compression is neglected unless it is anchored in position with lateral ties as required for columns. The maximum compressive stress in the masonry may not exceed $0.40f'_m$ and the tensile stress in reinforcement must fall within the values given in *Fig. 11-6*. Allowable vertical loads determined in this manner must be reduced for slenderness using the values in *Fig. 10-21*. For walls subject to bending about both principal axes, the requirements and restrictions given above for columns under these conditions should be applied in computing the value of C_e.

Reinforced walls must have a *minimum area of steel* not less than 0.002 times the cross-sectional area of the wall, not more than two-thirds of which may be used in either direction. Maximum spacing of principal reinforcement may not exceed six times the wall thickness or more than 48 in. on center. Horizontal reinforcement must be provided in bond beams at the top of footings, at the bottom and top of wall openings, at roof and floor levels, and at the top of parapet walls. Only reinforcement that is continuous in the wall may be considered in computing the minimum area of steel. In addition to the minimum reinforcement required by the structural design, the equivalent of one No. 4 bar must be provided around all window and door openings and must extend at least 24 in. beyond the corner of the opening.

As in the other design standards, *BIA-1969* provides that when the reinforcement in bearing walls is designed, placed, and anchored in position as for columns, the allowable stresses given for columns may be used. The length of wall to be considered effective may not exceed the center-to-center distance between loads, nor may it exceed the width of bearing plus four times the wall thickness.

11.7 A DESIGN EXAMPLE

Loadbearing masonry systems depend on structural interaction between the walls or bearing elements and the roof or floor acting as a diaphragm to transfer loads. The theory of interaction and performance of the members under loading conditions is basically the same as for unreinforced construction except that the strength of wall and column sections is increased by the steel reinforcing. The sample problem that follows is abbreviated and does not represent the thorough analysis required for a complete loadbearing building. It is intended simply to illustrate the engineering methods used in the design of masonry bearing wall structures.

Loadbearing Brick Wall Design Example:

Based on *BIA-1969*, information from code and physical layout of building

Live load, 50 lb/sq ft

Wind load, 25 lb/sq ft

Minimum wall thickness, 6-in. brick for 2-hour fire rating

Floor system of 8-in.-thick hollow-core concrete plank spanning 28 f
weighs 80 lb/sq ft, including topping

Floor-to-floor height of 9 ft

Information from analysis of higher floors and selection of materials:

Brick compressive strength, 8000 psi

Type S mortar by proportions

Assumed masonry compressive strength (from *Fig. 10-18*) with inspe
tion, $f'_m = 2400$ psi

Allowable stresses (from *Fig. 11-6*):

Axial compression:

$$
\begin{aligned}
f_m &= 0.25f'_m \\
&= 0.25 + 2400 \\
&= 600 \text{ psi}
\end{aligned}
$$

Flexural compression:

$$
\begin{aligned}
F_b &= 0.40f'_m \\
&= 0.40 \times 2400 \\
&= 960 \text{ psi}
\end{aligned}
$$

Allowable tension in reinforcement:

$$f_s = 24{,}000 \text{ lb/sq in.}$$

Cumulative axial loads:

1080 lb/ft: wall weight, dead load
2240 lb/ft: floor dead load
1400 lb/ft: floor live load

Analysis of wall at floor considered. The wall is supported at top and bottom by the floor, so the effective height is the actual height, 8 ft 4 in. With a nominal thickness of 6 in., $t = 5.6$ in.

$$\frac{h}{t} = \frac{8.33 \times 12}{5.6} = 17.9$$

The wall is subject to the following loads:
At the top of the wall:

$$P_T = 1080 + 2240 + \left(\tfrac{1}{2} \times 28 \text{ ft} \times 80 \text{ lb/ft}\right) + 1400 + \left(\tfrac{1}{2} \times 28 \text{ ft} \times 50 \text{ lb/ft}\right)$$

$$= 1080 + 2240 + 1120 + 1400 + 700$$

$$= 6540 \text{ lb}$$

At the bottom of the wall:

$$P_B = 6540 + (8.33 \text{ ft} \times 60 \text{ lb/ft})$$

$$= 6540 + 500$$

$$= 7040 \text{ lb}$$

The moment due to wind is the same at the top and bottom:

$$M = \tfrac{1}{10}\, WL^2 = \tfrac{1}{10}\, (25 \text{ lb/ft})(9.0)^2(12)$$

$$= 2430 \text{ in.-lb/ft}$$

For the moment at the top of the wall due to floor load, assume the following load distribution:

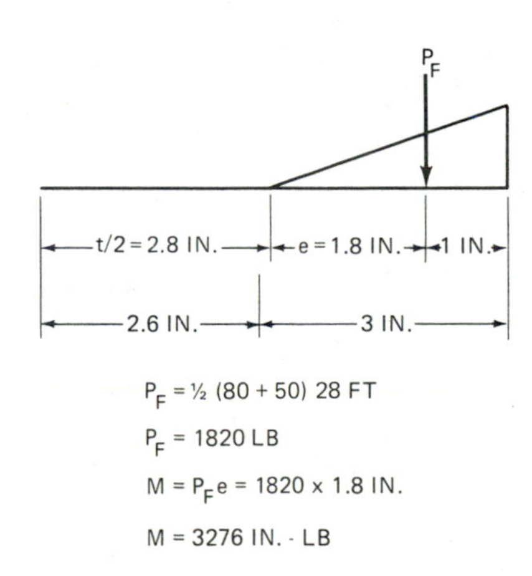

Virtual eccentricity of the wall is a result of the wind load moment and the floor load moment:

At the top of the wall:

$$M = 3276 + 2430$$

$$= 5706 \text{ in.-lb}$$

$$e_t = \frac{M}{P} = \frac{5706}{6540}$$

$$= 0.87 \text{ in.}$$

At the bottom of the wall, $M = 2430$ in.-lb:

$$e_b = \frac{2430}{7040}$$

$$= 0.35 \text{ in.}$$

The wall is single curvature, so e_1/e_2 is positive.
The slenderness ratio must be less than that prescribed in the code:

$$\frac{h}{t} \leq 10\left(3 - \frac{e_1}{e_2}\right) \quad (11.8$$

$$17.9 \leq 10(3 - 0.40)$$

$$\leq 10(2.60)$$

$$\leq 26.0 \qquad \text{OK}$$

Check compressive stress due to axial load:

$C_s = 0.66$ (interpolated from *Fig. 10-21A*)

$C_e = 0.71$ (interpolated from *Fig. 10-21B*)

$$P = C_e C_s f_m A_g \quad \text{(allowable load on wall)} \quad (10.$$

$$= 0.71 \times 0.66 \times 600 \times 5.6 \times 12 \text{ in.}$$

$$= 18{,}894 \text{ lb} > 7040 \text{ lb at bottom of wall therefore, OK}$$

Check flexural tension and compression at location of maximum mome (top of wall). Cumulative dead load and live load induced moments combi for the worst conditions.

Cumulative dead loads:

$$P = 1080 + 2240 + 1120$$

$$= 4440 \text{ lb}$$

$$\frac{P}{A} = \frac{4440}{12 \times 5.6}$$

$$= 66 \text{ psi dead load compression}$$

Live load moment, wind plus eccentric load:

$$M = 5706 \text{ in-lb}$$

Assume reinforcement at center of wall:

$$d = \frac{t}{2} = \frac{5.6}{2} = 2.8 \text{ in.}$$

$$n = \frac{E_s}{E_m} = \frac{29 \times 10^6 \text{ lb/sq in.}}{24 \times 10^5 \text{ lb/sq in.}} = 12$$

$$k = \frac{1}{1 + f_s/nf_m} = \frac{1}{1 + 24{,}000/12 \times 960} = 0.325$$

$$j = 1 - \frac{k}{3} = 1 - \frac{0.325}{3} = 0.892$$

Flexural compressive stress:

$$f_b = \frac{2M}{kjdb} = \frac{2 \times 5706}{0.325 \times 0.892(2.8)^2 \times 12} = 418 \text{ lb/sq in.}$$

The maximum compressive stress is the combination of the axial compression and the flexural compressive stress:

$$f_m = \frac{P_t}{A} + f_b = \frac{6540}{5.6 \times 12} + 418$$

$$= 515 \text{ lb/sq in.} < F_b = 960 \text{ lb/sq in. therefore, OK}$$

The flexural tensile stress in the steel can be determined in several ways. Axial dead load can be used to offset flexural tension, or steel can resist full moment. The latter instance will be used:

$$A_s = \frac{M}{f_s jd} = \frac{5706}{24{,}000 \times 0.892 \times 2.8}$$

$$= 0.95 \text{ sq in. per foot of wall length}$$

Maximum spacing of reinforcement is four times wall thickness:

$$4 \times 5.6 = 22.4 \text{ in.}$$

A #4 bar at 24 in. on center provides an area of steel of 0.098 in/ft of wall (*Fig.* 11-1). Therefore, use #4's at 24 in. on center.

11.8 HIGH-RISK DESIGN

In some areas of the country, seismic disturbances and/or high wind loads add critical temporary loading to building structural systems. The transitory nature of these forces is acknowledged in masonry design codes by a $33\frac{1}{3}\%$ increase in allowable stresses, and by provisions for shear walls and lateral load transfer through diaphragm action. Unreinforced masonry should not be used in areas of high risk if analysis shows that tension develops in the shear walls. The *Uniform Building Code* requires reinforced construction in Seismic Zones 2, 3, and 4.

Seismic forces are caused by a stress buildup within the earth's crust. An earthquake is the sudden relief of this stress and consequent shifting o the earth mass along an existing fault plane. Primary vibration waves create a push–pull effect on the ground surface. Secondary waves traveling at about half the speed set up transverse movements at right angles to the first shock. Structures may experience severe lateral dynamic loading under such conditions, and must be capable of absorbing this energy and withstanding the forces of seismic ground motion. Reinforcing steel in masonry buildings resists shear and tensile stresses and provides th structure with the ductility or flexibility necessary to sustain such loa reversals.

Framing systems designed to withstand seismic and high wind load may be either flexible structures with low damping characteristics or rigi structures with high damping. Steel and concrete skeleton frames ar generally classified as moment-resisting space frames in which the joint resist forces primarily by flexure. This flexibility, although effective i dissipating the energy of the seismic loads, can cause substantial secon dary, nonstructural damage to windows, partitions, and mechanic equipment. Loadbearing masonry buildings are categorized as box frame in which lateral forces are resisted by shear walls. Masonry building designed in compliance with current building codes have been ver successful in withstanding seismic forces. The rigidity inherent in th masonry systems often assures little or no secondary damage.

Damping is the ability of a structure to diminish its amplitude o vibration with time through dissipation of energy by internal friction resistance. It is generally recognized that the response of buildings t earthquake motion is influenced by the natural period of vibration of th building. This is taken into consideration in determining the horizont force factor (*K*). The greater the weight and rigidity of the structure, th greater the forces it must resist. Under the *Uniform Building Code*, bo systems are given the highest *K* factor for lateral forces, 1.33. The capaci of masonry structures to absorb seismic energy is such that unit stress remain extremely low with a very high factor of safety. Damping is n itself a recognized numerical coefficient in code equations. It is th characteristic, however, which contributes most significantly to the hig safety factor. The greater the damping effect, the less the response of th building to an earthquake. There is actually less distress in a loadbearing shear wall building because of its stiffness and lack of deflection. This accounts for the decreased secondary damage compared to flexible buildings, which yield and deflect, breaking glass, plaster, drywall partitions, stairs, mechanical piping, and other costly and potentially dangerous elements.

11.8.1 Building Layout

One of the most important aspects of shear wall and seismic design are qualitative elements regarding symmetry and location of resisting members, relative deflections, anchorage, and discontinuities. Shear walls resist horizontal forces acting parallel to the plane of the wall through resistance to overturning and shearing resistance (*see Fig. 10-2*). The location of these walls in relation to the direction of the applied force is critical. Since ground motion may occur in perpendicular directions, location of resisting elements must coincide with these forces. Shear walls may be either bearing or nonbearing elements. It is best to combine the functions of such members and, whenever possible, design the building with both transverse and longitudinal loadbearing shear walls. Designing all the loadbearing walls to resist lateral forces improves overall performance because an increased number of shear walls distributes the load and lowers unit stresses. Shear walls that are also loadbearing have greater resistance to seismic forces because of the stability provided by increased axial loads.

If the bearing walls function dually as shear walls, then general building layout becomes a very important aspect of seismic design. The building shear wall layout should be symmetrical to eliminate torsional action. Several compartmented floor plans are shown in *Figs. 11-9* and *11-10*. The regular bay spacing lends itself to apartment, hotel, hospital, condominium, and nursing home functions, where large building areas are subdivided into smaller areas. By changing the span direction of the floor elements, loadbearing shear walls can resist forces from two or more directions. The radial walls of the round building can actually absorb seismic shocks from any direction and dissipate the earthquake energy with very low levels of stress.

In skeleton frame buildings, elevator and stair cores of concrete or masonry are often used as shear walls even though they may not have axial loadbearing capacity. Providing a good balance between the amount of shear wall along each of the principal axes will give greatest economy and best performance. It is best to design uniform wall lengths between openings and to provide wall returns wherever possible to minimize variations in relative rigidity and to decrease maximum shear.

11.8.2 Shear Walls and Diaphragms

A box frame structural system must provide a continuous and complete path for all of the assumed loads to follow from the foundation to the roof, and vice versa. This is achieved through the interaction of floors and walls securely connected along their planes of intersection. Lateral forces are carried by the floor diaphragms to vertical shear walls parallel to the

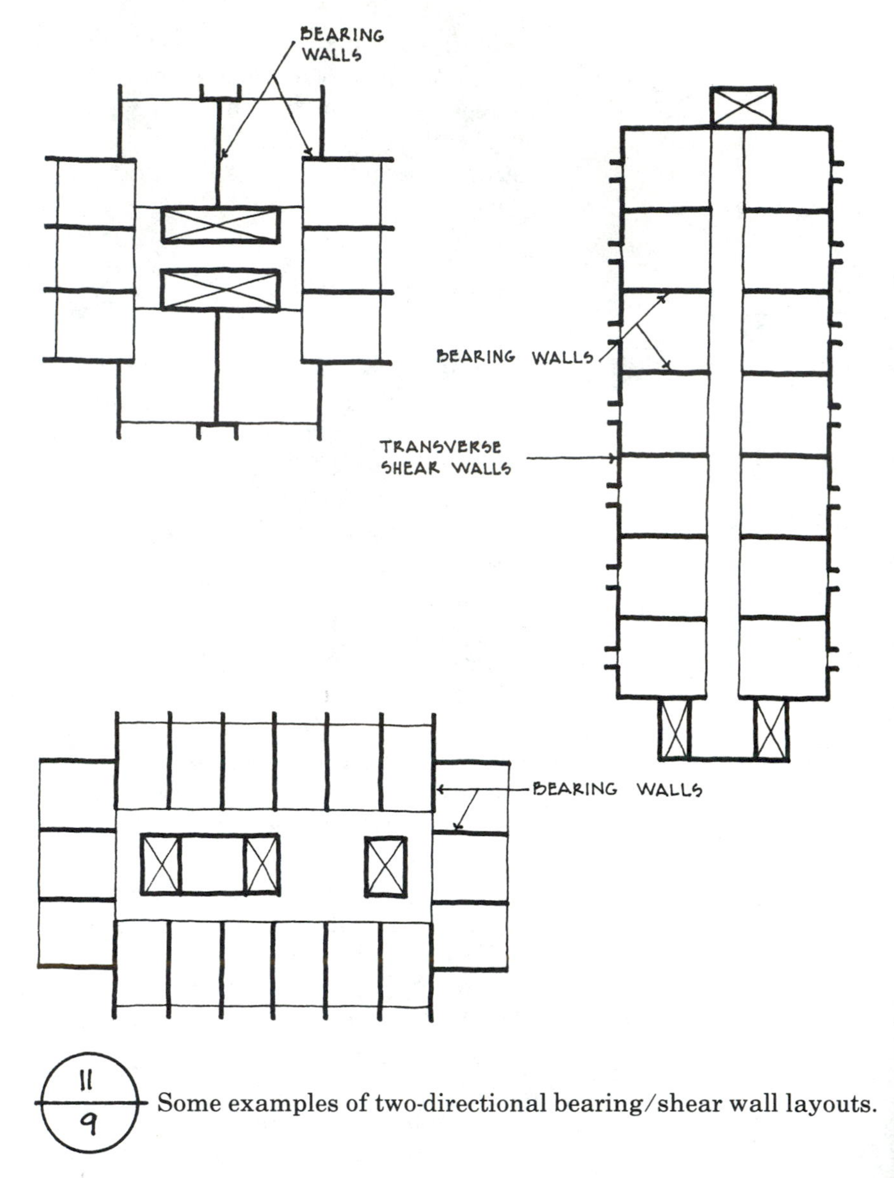

11-9 Some examples of two-directional bearing/shear wall layouts.

direction of the load. The *shear walls* act as vertical cantilevered masonr beams subject to concentrated horizontal forces at floor level, and transfe these lateral forces to the foundation by shear and flexural resistance. Th load transfer induces shear stresses in the wall.

When the various elements of a structure are rigidly connected to on another and one element is deflected by a force, the other elements mus move equally at the point of connection or failure will occur somewhere i the system. The amount of horizontal load carried by a shear wall i proportional to its relative rigidity or stiffness. The rigidity of a shear wa is inversely proportional to its deflection under unit horizontal loa Resistance is a function of the cube of the wall length.

One method of increasing the stiffness of shear walls as well as the resistance to bending is the use of intersecting walls or flanges (refer to *Fi 10-24*). Although codes limit the effective length of flanges, walls with *L*, *I*, or *C* shapes have better flexural resistance for loads applied perpendic lar to their flange surface. Shear stresses at the intersection of the walls a dependent on the type of bonding used to tie the two elements together.

Another method that may be used to increase the stiffness of a bearin

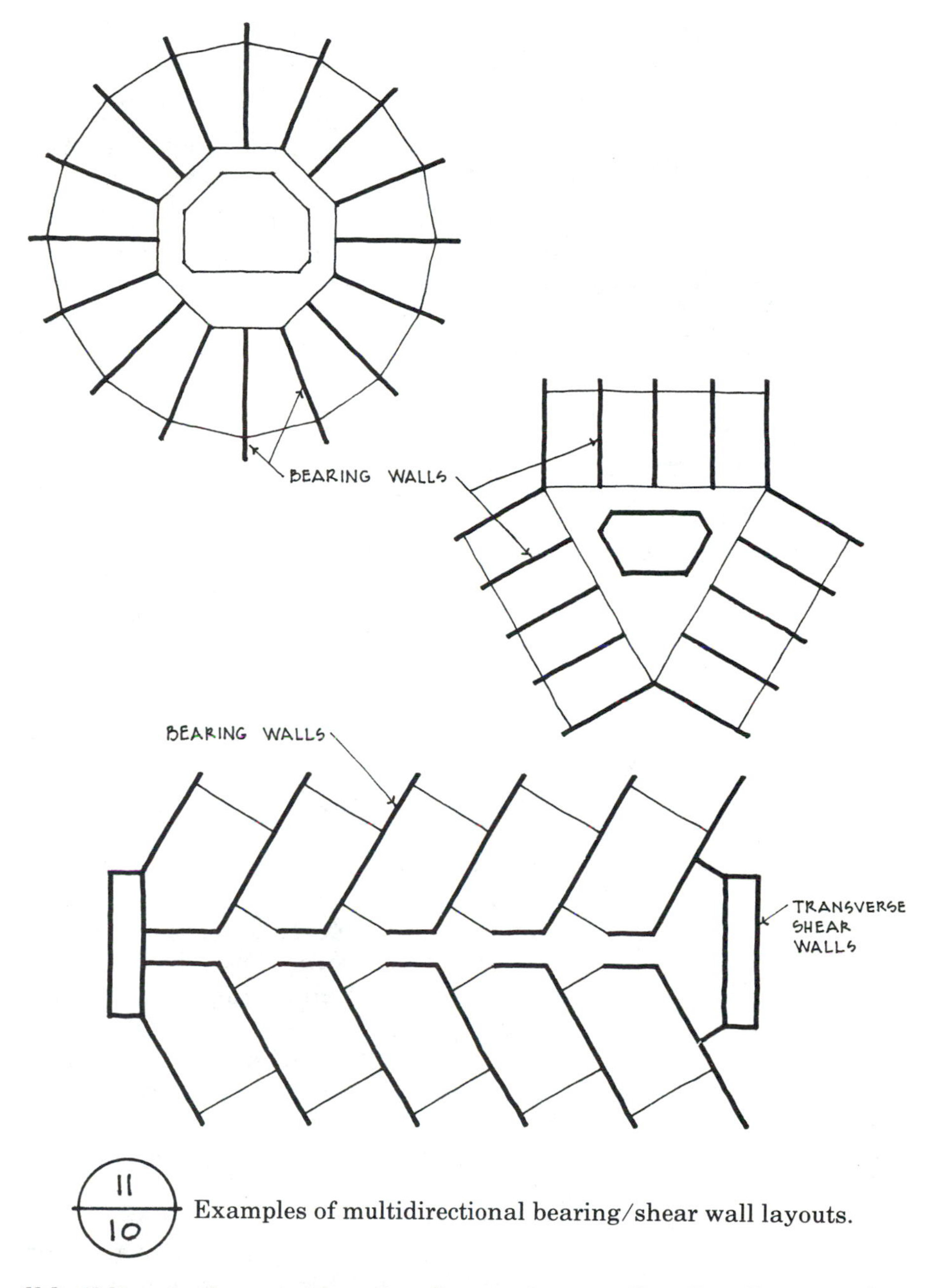

Examples of multidirectional bearing/shear wall layouts.

wall building is the coupling of co-linear shear walls. The illustrations in *Fig. 11-11* indicate the effect of coupling on stress distribution from forces parallel to the wall. In parts (A) and (D), a flexible connection between the walls is assumed so that they act as independent vertical cantilevers in resisting the lateral loads. Walls (B) and (E) assume the elements to be connected with a more rigid member capable of shear and moment transfer so that a frame-type action results. This connection may be made with a steel, reinforced concrete, or reinforced masonry section. The plate action in parts (C) and (F) assumes an extremely rigid connection between walls such as full-story-height sections or deep rigid spandrels.

Lateral forces are distributed horizontally to shear walls by the floor and roof systems acting as *diaphragms*. A diaphragm must be able to transmit lateral forces without excessive deflection that could cause distress to vertical elements, particularly to walls perpendicular to the

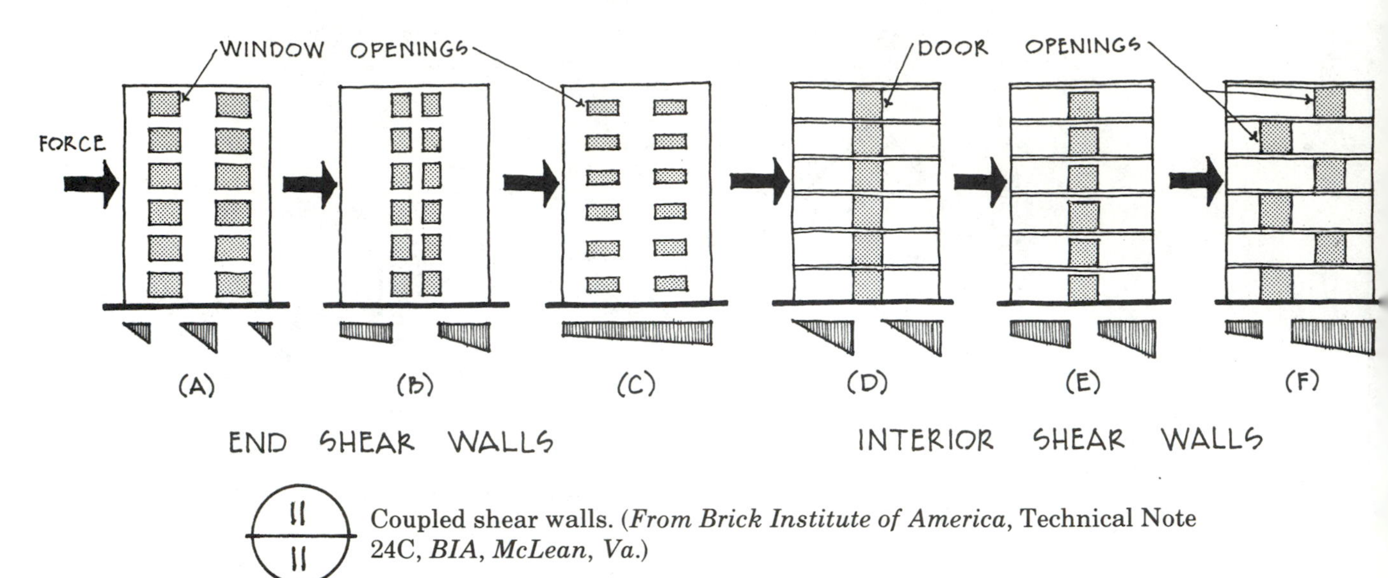

11-11 Coupled shear walls. (*From Brick Institute of America*, Technical Note 24C, *BIA, McLean, Va.*)

shear walls (*see Fig. 11-12*). Rigid and semirigid diaphragms do no experience excessive deflection under load. They distribute lateral forces t the shear walls in proportion to their relative rigidities. The nature of th floor–wall connection also influences the transfer of stress (see Chapter 10)

11.8.3 Diagonal Tension

Analysis of the damage to masonry walls caused by earthquake stres shows that cracks in shear walls typically follow a diagonal path. Th plane of failure extends from near the top corner where the load is applie diagonally downward toward the bottom support (*see Fig. 11-13*). This is th same mode of failure produced by diagonal tension or racking tests, i which 4×4 ft masonry panels are subjected to diagonal loading a opposing corners. Shear strength is independent of unit properties such a initial rate of absorption and compressive strength, but it is affected b mortar type and workmanship. Since the failure is in tension, masonry wi weak bond characteristics (as shown by flexural strength) also shows lo diagonal tensile or shear strength. Failure occurs without explosi popping or spalling of unit faces. The steel reinforcement holds the wa together after a shear failure to prevent the panel from separating aft cracks appear. This stability under maximum stress prevents catastroph structural failure and increases the factor of safety in the aftermath seismic disturbances. It is this performance characteristic that led to t minimum steel areas called for in "reinforced" masonry.

11.8.4 Building Code Requirements

The *Uniform Building Code* is the most widely used for design of structu in seismic areas. Quantitative requirements are based on assumptio made to establish the magnitude of the design forces. The introduction the section on earthquake regulations states: "Stresses shall be calcula as the effect of a force applied horizontally at each floor or roof level ab the base. *The force shall be assumed to come from any horizon direction. . .*" [emphasis added]. It is obvious from this directive why

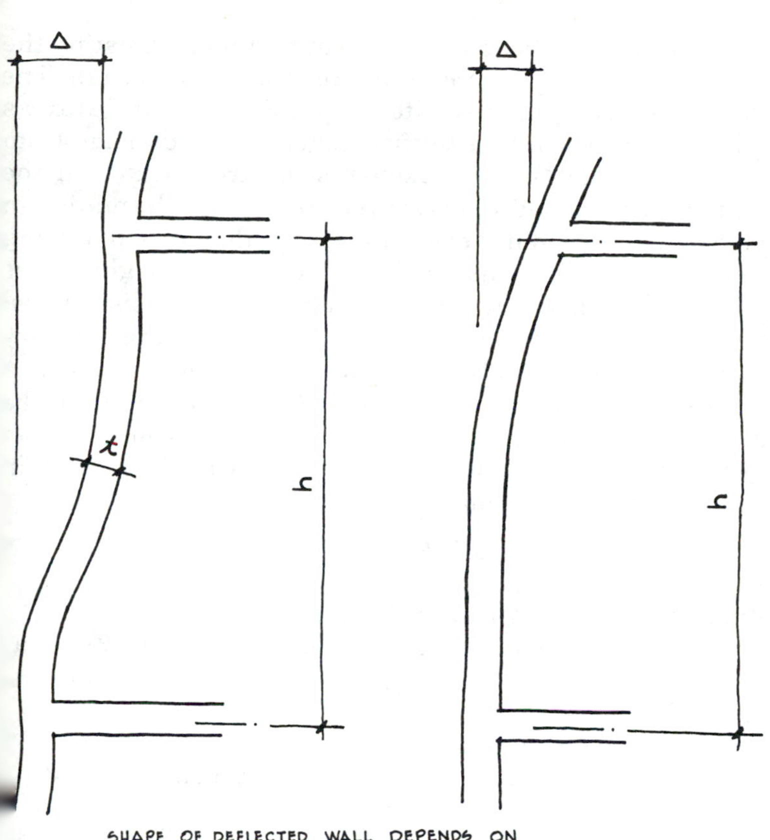

11/12 Diaphragm deflection limitations. (*From Brick Institute of America*, Technical Note 24C, *BIA*, *McLean*, *Va.*)

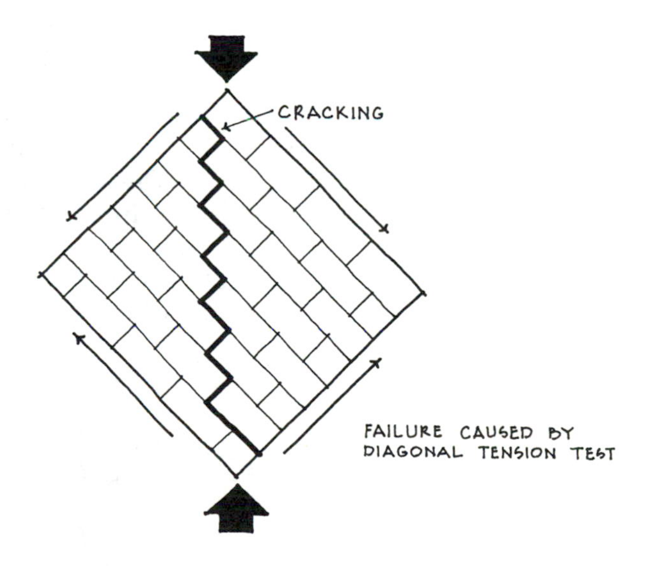

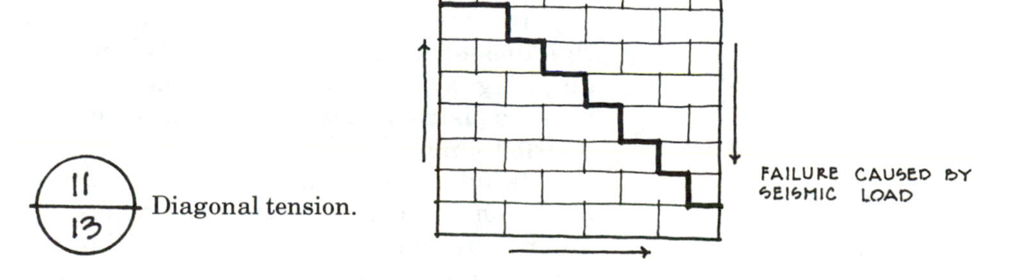

11/13 Diagonal tension.

building layout and location of shear walls is so important in resisting the seismic forces and at the same time providing an economical design. The code includes a seismic risk map of the United States which identifies areas and regions of the country according to their potential for damage from seismic activity (*see Fig. 11-14*). All masonry structures located in the higher-risk Seismic Zones 2, 3, and 4 must be reinforced with the minimum amount of steel required for "reinforced" masonry. Box systems using masonry shear walls and diaphragms are limited to 160 ft in height unless properly substantiated technical data are submitted which establish lateral forces and distribution by dynamic analyses. In fact, loadbearing masonry structures 200 ft tall have been constructed in seismically active areas.

Every building or structure designed under *UBC* requirements must be capable of resisting minimum total lateral seismic forces assumed to act nonconcurrently in the direction of each of the main axes of the structure in accordance with the following formula:

$$V = ZIKCSW \tag{11.9}$$

where V = total lateral force or shear at the base
Z = numerical coefficient dependent on the Seismic Zone as follows: Zone 1, $Z = \frac{3}{16}$; Zone 2, $Z = \frac{3}{8}$; Zone 3, $Z = \frac{3}{4}$; and Zone 4, $Z = 1$
I = occupancy importance factor
K = numerical coefficient for type or arrangement of resisting elements
C = numerical coefficient for the fundamental period of vibration of the structure
S = numerical coefficient for site-structure resonance
W = total dead load, including partitions (plus 25% of live load for storage and warehouse occupancies, and snow load if greater than 30 lb/sq ft)

Once the magnitude of the lateral force has been determined, the code gives three additional formulas for distribution of these loads over the height of the structure and for computing the concentrated force at the top of the building. The total lateral force is distributed in accordance with the formula

$$V = F_t + \sum_{i=1}^{n} F_i \tag{11.10}$$

where V = total lateral force or shear at base
F_t = that portion of V considered concentrated at the top of the structure in addition to F_n
F_i = lateral force applied to level i
$i = 1$ = first level above base
n = that level which is uppermost in the main portion of the structure

Distribution of horizontal shear must be in proportion to the relative rigidities of the resisting elements and must consider the rigidity of the diaphragm. Overturning moments caused by either wind or seismic loads must be distributed to the resisting elements in the same proportion as the distribution of shears.

In addition to the calculation and resolution of total seismic forces acting on a structure, consideration must be given to individual structural and nonstructural elements of the building. Parts or portions of structures, nonstructural components, and their anchorage to the main structure

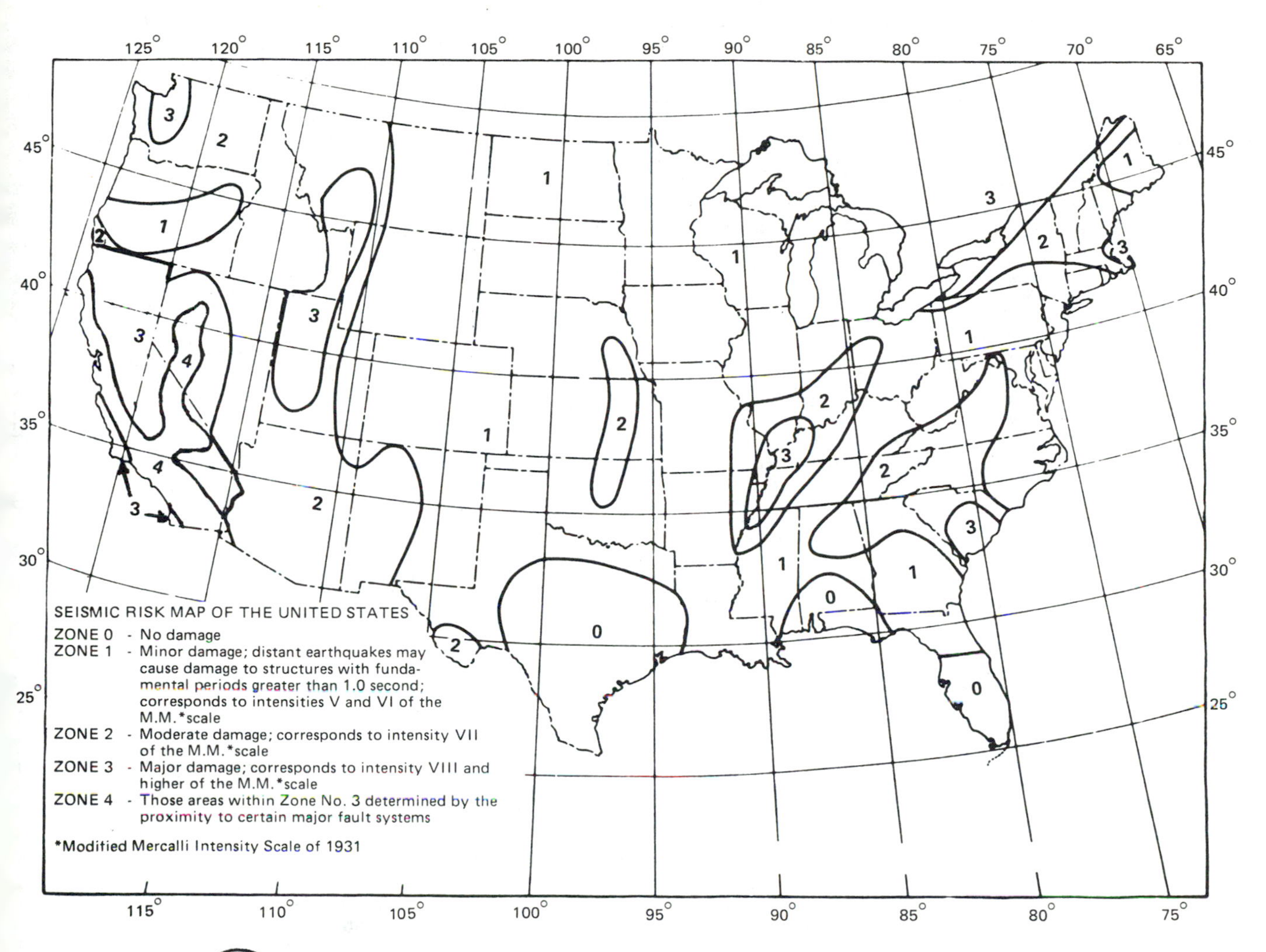
SEISMIC RISK MAP OF THE UNITED STATES
ZONE 0 - No damage
ZONE 1 - Minor damage; distant earthquakes may cause damage to structures with fundamental periods greater than 1.0 second; corresponds to intensities V and VI of the M.M.*scale
ZONE 2 - Moderate damage; corresponds to intensity VII of the M.M.*scale
ZONE 3 - Major damage; corresponds to intensity VIII and higher of the M.M.*scale
ZONE 4 - Those areas within Zone No. 3 determined by the proximity to certain major fault systems
*Modified Mercalli Intensity Scale of 1931

11
14

Seismic zone map of the continental United States. (*Reproduced from the Uniform Building Code, 1982 edition, copyright 1982, with permission of the publisher, the International Conference of Building Officials.*)

system must be designed for lateral forces in accordance with the following formula:

$$F_p = ZIC_pW_p \tag{11.11}$$

where F_p = lateral forces on a part of the structure and in the direction under consideration
Z = numerical coefficient for Seismic Zone designation
I = occupancy importance factor (same as for building)
C_p = numerical coefficient (*Fig. 11-15*)
W_p = weight of a portion of a structure or nonstructural component

The distribution of these forces must be according to the gravity loads pertaining to them.

All of the seismic design requirements of the UBC are based on translating dynamic forces into static lateral forces. This does not, however, preclude the use of more sophisticated dynamic analyses since the code does allow submission of properly substantiated technical data. (Another seismic design method is currently being developed by the Building Seismic Safety Council under the auspices of the National Bureau of Standards.)

Part or portion of buildings	Direction of horizontal force	Value of C_p*
Exterior bearing and nonbearing walls, interior bearing walls and partitions, interior nonbearing walls and partitions—see also Section 2312 (j) 3 C. Masonry or concrete fences over 6 ft high	Normal to flat surface	0.3†
Cantilever elements Parapets	Normal to flat surfaces	0.8
Chimneys or stacks	Any direction	
Exterior and interior ornamentations and appendages	Any direction	0.8
When connected to, part of, or housed within a building		
Penthouses, anchorage and supports for chimneys and stacks and tanks, including contents		
Storage racks with upper storage level at more than 8 ft in height, plus contents	Any direction	0.3‡
All equipment or machinery		
Suspended ceiling framing systems (applies to Seismic Zones Nos. 2, 3, and 4 only)	Any direction	0.3§
Connections for prefabricated structural elements other than walls, with force applied at center of gravity of assembly	Any direction	0.3¶

*C_p for elements laterally self-supported only at the ground level may be two-thirds of value shown.

†See also Section 2309 (b) for minimum load and deflection criteria for interior partitions.

‡W_p for storage racks shall be the weight of the racks plus contents. The value of C_p for racks over tw storage support levels in height shall be 0.24 for the levels below the top two levels. In lieu of th tabulated values, steel storage racks may be designed in accordance with UBC Standard No. 27-11

Where a number of storage rack units are interconnected so that there are a minimum of fou vertical elements in each direction on each column line designed to resist horizontal forces, th design coefficients may be as for a building with K values from Table No. 23-1, $CS = 0.2$ for use in th formula $V = ZIKCSW$ and W equal to the total dead load plus 50% of the rack-rated capacity. Wher the design and rack configurations are in accordance with this paragraph, the design provisions i UBC Standard No. 27-11 do not apply.

For flexible and flexibly mounted equipment and machinery, the appropriate values of C_p sha be determined with consideration given to both the dynamic properties of the equipment an machinery and to the building or structure in which it is placed but shall be not less than the liste values. The design of the equipment and machinery and their anchorage is an integral part of th design and specification of such equipment and machinery.

For essential facilities and life safety systems, the design and detailing of equipment whic must remain in place and be functional following a major earthquake shall consider drifts i accordance with Section 2312 (k).

§Ceiling weight shall include all light fixtures and other equipment which is laterally supported b the ceiling. For purposes of determining the lateral force, a ceiling weight of not less than 4 poun per square foot shall be used.

¶The force shall be resisted by positive anchorage and not by friction.

Horizontal force factor (C_p) for elements of structure and nonstructu components. (*Reproduced from the* Uniform Building Code, *1982 editi copyright 1982, with permission of the publisher, the Internatio Conference of Building Officials.*)

12

BEAMS, LINTELS, AND ARCHES

The use of reinforcing steel in masonry construction permits the design of flexural members such as beams, girders, and lintels to span horizontal openings. This gives a continuity of materials, finishes, and fire ratings by eliminating the introduction of other materials solely for flexural spans.

The design of reinforced masonry beams and girders is based on the straight-line theory of stress distribution. The required steel is determined by actual calculated stress on the member. The reinforcement needed to resist this stress is then provided in the necessary amounts and locations.

Design requirements of the National, Basic, and Standard building codes are based on *ANSI A41.2*, *BIA-1969*, and *NCMA-1970* by reference standard. The *UBC*, although based on the same sources, outlines detailed requirements without specific reference standards. Material standards and allowable stresses are the same as in Chapter 11 for reinforced walls and columns (*see Figs. 11-2 through 11-6*).

12.1 ANSI A41.2

In computations of flexural stress, the ANSI code requires that members be designed to resist at all sections the maximum bending moment and shears produced by dead load, live load, and other forces determined by the principles of continuity and relative rigidity.

The span length of freely supported beams is defined as the clear span plus the depth of the beam (may not exceed the distance between centers of supports). Beam depth is taken as the distance from the centroid of the tensile reinforcement to the compression face. The effective width may not be greater than four times the wall thickness in solid masonry, or more than the width of the solidly filled section plus the length of the unit in hollow masonry (as long as this also does not exceed four times the wall thickness). The clear distance between lateral supports of a beam may not exceed 32 times the least width of the compression flange. The moments and shears shown in *Fig. 12-1* are used for design calculation.

If longer of two adjacent spans does not exceed the shorter by more than 20%, loads are uniformly distributed, and live load is no greater than three times dead load, use the following:	
Positive moment	
Interior spans	$wl^2/16$
End spans	
If discontinuous end is unrestrained	$wl^2/11$
If discontinuous end is integral with support	$wl^2/14$
Negative moment	
At interior face of first interior support	
Two spans	$wl^2/9$
More than two spans	$wl^2/10$
At face of other interior support	$wl^2/11$
Shear	
In end members at first interior support	$1.15wl/2$
At other supports	$wl/2$

Shears and moments for beams and griders.

In applying the principle of continuity, center-to-center distances ma be used to determine the moment. Moments at the faces of a support must b used for the design of beams and girders at such points.

If compression steel is used, it must be anchored by ties or stirrups a least $\frac{1}{4}$ in. in diameter spaced a maximum of 16 bar diameters or 48 t diameters apart. These ties or stirrups must be used throughout the distanc where the compression steel occurs. When members are subject to combine axial and flexural stresses, they must be proportioned so that the quanti $f_a/F_a + f_b/F_b$ does not exceed 1.

12.1.1 Shear and Diagonal Tension

Shearing stress in masonry flexural members is taken as a measure diagonal tension, and is computed by the formula

$$v = \frac{V}{bjd} \quad (12$$

or

$$v = \frac{V}{b'jd} \qquad (12.2)$$

where v = shearing stress, psi
V = total shear, lb
b = width of rectangular flexural member or width of flange for T or I sections, in.
b' = width of web in T and I flexural members, in. (actual width when solid masonry units are used, and the width of the filled core area plus the thickness of adjacent webs when hollow masonry units are used)
j = ratio of distance between centroid of compression and centroid of tension to the depth, d
d = depth from compression face of beam to centroid of longitudinal tensile reinforcement, in.

Formula (*12.2*) should be used for T and I-shaped sections.

When the computed shearing stress exceeds the allowable for masonry in an unreinforced web (50 psi under this code), steel web reinforcing must be added. The reinforcing must be provided in the area theoretically requiring the steel and extend beyond that area an additional distance equal to d. The total allowable shear stress for beams with web reinforcing is 150 psi, three times that permitted for plain, unreinforced members.

12.1.2 Web Reinforcement

Steel web reinforcement in beams and girders may consist of stirrups or bent bars, although bent bars are seldom used in masonry construction. The area of steel required in stirrups which are placed perpendicular to the longitudinal reinforcement is computed by the formula

$$A_v = \frac{V's}{f_v jd} \qquad (12.3)$$

where A_v = total area of web reinforcement in tension within a distance of s (measured in a direction parallel to that of the main reinforcement, or the total area of all bars bent up in any one plane,) sq in.
V' = excess of the total shear over that permitted in the masonry, lb
s = spacing of stirrups or of bent bars in a direction parallel to that of the main reinforcement, in.
f_v = tensile stress in web reinforcement, psi

When web reinforcement is required to resist calculated shearing stresses, it must be spaced so that every 45° line (representing a potential crack) extending from the mid-depth of the beam to the longitudinal tension bars is crossed by at least one line of web reinforcement.

12.1.3. Bond Stress

In masonry flexural members where tensile reinforcement is parallel to the compressive face, bond stress is calculated by the formula

$$u = \frac{V}{\Sigma_o jd} \qquad (12.4)$$

where u = bond stress, psi
V = total external shear at the section, lb
Σ_0 = sum of perimeters of bars in one set, in.

The quality of mortar required to achieve proper bond to beam reinforcement is governed by material standards and volumetric mix proportions (see Chapter 11).

12.1.4 Anchorage Requirements

ANSI outlines specific requirements for anchorage of reinforcing steel for various conditions of design. The code calls for tensile reinforcement to be adequately anchored by bond, hooks, or mechanical anchors in or through the supporting members. In all cases except a lapped splice, every bar must be extended at least 12 bar diameters beyond the point at which it is no longer needed to resist stress. Except for positive reinforcement at interior supports of continuous members, plain bars in tension must terminate in standard hooks or bends (*see Fig. 12-2*). Mechanical devices capable of developing the strength of the bar without damage to the masonry may be used in lieu of hooks if tests are presented to show the adequacy of such devices. Single separate bars used as web reinforcement must be anchored at each end by bond, hooks, bends, or welding. Bars or stirrups forming a simple "U" perpendicular to the longitudinal reinforcement may be hooked or bent around a longitudinal bar of equal or greater diameter. When longitudinal bars are bent to act as web reinforcing, they must be continuous with the longitudinal reinforcement throughout the region of

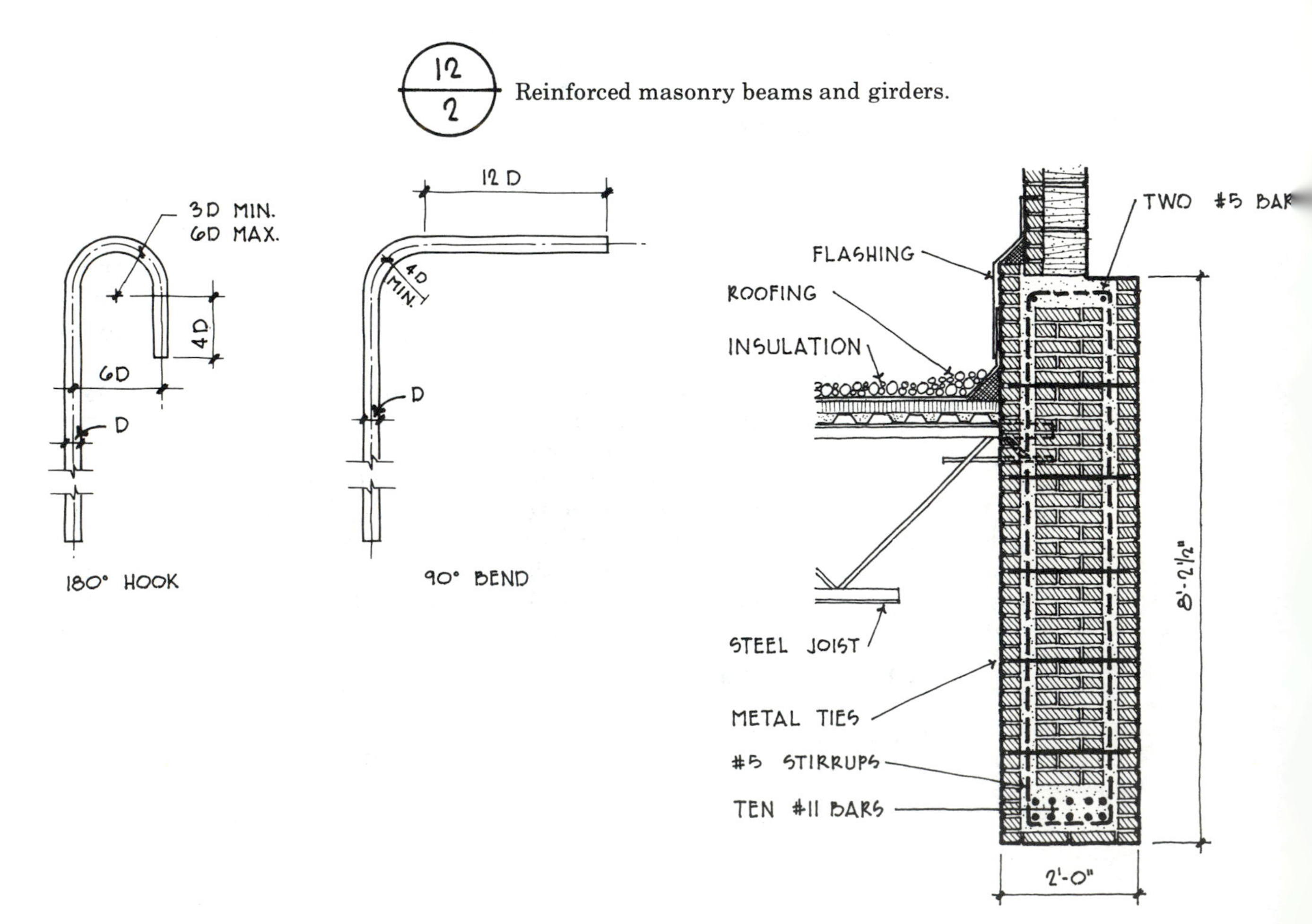

Reinforced masonry beams and girders.

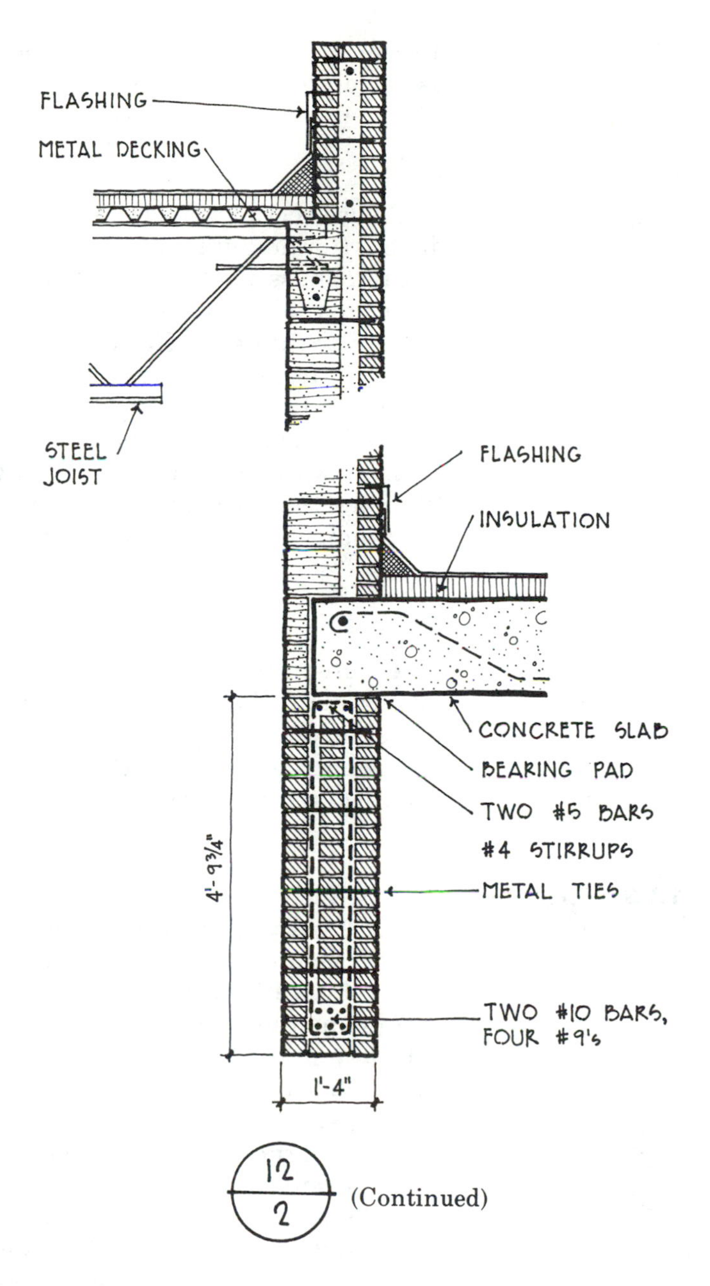

(Continued)

tension. The tensile stress in each bar must be fully developed in both the upper and lower half of the beam by adequate anchorage of bond or hooks.

12.2 NCMA-1970

NCMA code requirements for the design of flexural members are based on the ANSI standard and are therefore identical in many respects. Basic design assumptions are the same except that specific moment formulas are not listed, and span length and depth are not defined. Effective width b is more narrowly defined for concrete masonry construction, and is limited to six times the wall thickness in running bond, and three times the wall thickness in stack bond. *NCMA-1970* also permits a reduction in calculated tensile fiber stress due to lateral forces by the direct stress due to vertical loads. This simply acknowledges the stabilizing effect on bending created by vertical loading (refer to Chapter 11).

Other than these items, the only difference between the two codes is in the shear resistance required of web reinforcing. The nominal shear stress in reinforced concrete masonry members, as a measure of diagonal tension, is calculated by the formula

$$v = \frac{V}{bd} \tag{12.5}$$

For T or I sections, b' is substituted for b, where b' is the width of the web in inches (for hollow masonry, width is equal to the width of the filled cell plus the thickness of adjacent webs). This formula does not take into account the ratio of the distance between centroid of compression and centroid of tension to the depth. However, the major difference between the codes is not this, but the fact that when computed shearing stresses exceed the allowable for unreinforced webs, steel reinforcing must be provided to carry the entire stress instead of just the excess as in *ANSI A41.2*. This, of course, significantly increases the reinforcing requirements (by a factor of 50 psi), and changes the formula used for computation. The area of steel required in perpendicular stirrups is calculated by the formula

$$A_v = \frac{Vs}{f_v d} \tag{12.6}$$

The substitution of V for V' as used in *ANSI A41.2* expresses th requirement that the web reinforcement carry all of the calculated shea stress instead of only that in excess of the allowable for unreinforced beams

The remainder of the NCMA requirements for flexural design ar identical to those previously outlined.

12.3 BIA-1969

The BIA code covers only clay masonry construction, but is also based o *ANSI A41.2* with modifications similar to those made in *NCMA-1970*. Basi design assumptions again are the same, with effective width limited to fou times the wall thickness. There is, however, an additional requirement fo minimum reinforcement. The ratio of required tensile reinforcement t effective masonry area may not be less than $80/f_y$ (where f_y is the yiel strength of the steel) unless the total reinforcement provided at every sectio is at least one-third greater than that required by analysis. *BIA-1969* is th only one of the four codes which links the required amount of reinforcemen to the strength of the steel actually used.

The requirements for shear reinforcement in webs are the same as th NCMA standard. That is, the steel must be designed to carry the entir stress instead of just the excess above that allowed for unreinforce members.

The types of web reinforcement permitted in the BIA code are the sam as in *ANSI A41.2*, but the formulas for calculating required area of steel f stirrups and bent bars are the same as *NCMA-1970*. The only oth variation between this and the two codes previously discussed regar anchorage requirements. *BIA-1969* has an additional provision whic states that no flexural bar may be terminated in a tension zone unle certain limiting conditions are satisfied. All other requirements are iden cal to those of *ANSI A41.2*.

12.4 UBC-1982

The *Uniform Building Code* bases its requirements for the design masonry flexural members on a combination of standards from ANSI, BI and NCMA. Shear reinforcement is required to resist the entire calculat stress, and thus all of the formulas for calculating area of steel use V inste of V'. However, three of the formulas do incorporate the ratio of the distance between centroid of compression and centroid of tension to depth. Formulas for shear and for steel requirements under *UBC* are as follows:

Shearing unit stess:

$$v = \frac{V}{bjd} \tag{12.7}$$

Stirrups:

$$A_v = \frac{Vs}{f_v jd} \tag{12.8}$$

The formula for computation of bond stress is the same as the other three codes, with Σ_o further defined as "the perimeter of all effective bars crossing the section on the tension side. To be effective, the bars must be properly developed by hooks, lap, or embedment on each side of the section."

12.5 DEEP WALL BEAMS

Deep concrete wall beams have been an accepted and widely used construction technique since the early 1930s. It was not until 30 years later, however, that the concept was applied to masonry construction. In the 1960s, a research program was instituted in California to test the capacity of unit masonry construction to carry and transfer applied axial loads in this manner. In 1974, the ICBO sanctioned the use of deep masonry wall beams under the *Uniform Building Code.*

The concept is based on a wall spanning between columns or footings instead of having continuous line support at the bottom as in conventional loadbearing construction (*see Fig. 12-3*). If soil bearing capacities permit this type of concentrated load, the wall may be designed as a flexural member and must resist forces in bending rather than in direct compression. Bearing walls, nonbearing walls, and shear walls may all employ this principle to advantage in some circumstances.

One of the advantages of deep wall beams is the increased height-to-thickness ratio permitted. Where h/t ratios are normally limited to 25 or 30, design of the wall as a deep beam increases the limit to 36 for loadbearing and shear walls and 48 for other uses. In most instances, this results in reduced wall thickness and a more economical design. For instance, a 16-ft-high, reinforced loadbearing wall supported on continuous footings would have a minimum thickness of 8 in. to satisfy an h/t ratio of 25. As a flexural member supported on isolated footings, however, the same wall could be reduced to a 6-in. thickness. For an unsupported height of 16 ft, a nonbearing wall can be reduced from 8 in. to 4 in. in thickness. When the required steel is governed by the code minimum of 0.002 times the cross-sectional area, reduced wall thickness translates directly into a savings in steel. A drop from 8 in. to 6 in. reduces the amount of steel required by about 27%. Even where additional reinforcing must be added for stirrups or concentrated loads, a savings can still be realized. The elimination of continuous footings and the reduced wall weight also requires less concrete in the foundation and lowers overall costs.

Deep wall beams may also be used to advantage to open up the ground floor of a loadbearing structure. The bearing wall on the floor above can be supported on columns to act as a deep wall beam and transfer its load to the supports. This alternative permits the design of larger rooms and open spaces than might be possible with regularly spaced bearing walls.

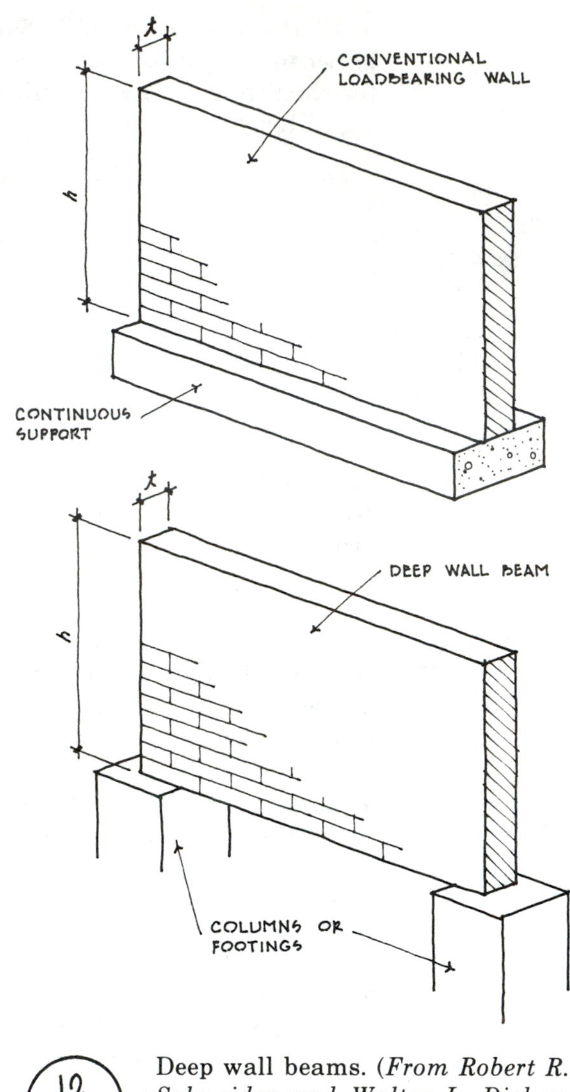

12/3 Deep wall beams. (*From Robert R. Schneider and Walter L. Dickey,* Reinforced Masonry Design, *© 1980, p. 215. Reprinted by permission of Prentice-Hall, Inc., Englewood Cliffs, N.J.*)

Although each design situation must be investigated independentl the application of rational engineering methods can quickly determin whether deep wall beams offer an advantage. If it is determined to be th most economical solution for a particular problem, the flexural desig formulas used for ordinary beams may then be applied to the walls.

The *UBC* also permits prefabrication of masonry beams and deep wa panels. These methods can speed construction and reduce overall job cost particularly when the design calls for a large number of repetitious units

12.6 LINTELS

Lintels are used to span small openings in masonry walls. Materials inclu steel, reinforced masonry, stone, concrete, precast concrete, and cast sto (wood lintels are common only to adobe construction). Lintels should b carefully analyzed to determine loads and stresses. Many of the cracks th appear over door and window openings result from excessive deflection lintels which have been improperly or inadequately designed.

12.6.1 Load Determination

Regardless of the material used to form or fabricate a lintel, one of the most important aspects of design is the determination of applied loads. *Figure 12-4* shows an elevation of an opening with a concrete plank floor and concrete beam bearing on the wall, and a graphic illustration of the distribution of these loads. The triangular area (ABC) immediately above the opening has sides at 45° angles to the base and represents the area of wall weight actually carried by the lintel. Arching action of the masonry will carry other loads outside the triangle provided that the height of the wall above the apex is sufficient to resist arching thrusts. (Arching action may be assumed only when the masonry is laid in running bond, or when sufficient bond beams distribute the loads in stack bond.) For most lintels of ordinary thickness, load, and span, a depth of 8 to 16 in. above the apex is generally sufficient.

In addition to the dead load of wall contained within the triangular area, the lintel will also carry any uniform floor loads occurring above the opening and below the apex of the triangle. In *Fig. 12-4*, the distance D is greater than $L/2$, so the floor load may be ignored. If arching action does occur as described above, loads outside the triangle may be neglected.

Determination of the lintel load. (*From Brick Institute of America*, Technical Note 31B, *BIA, McLean, Va.*)

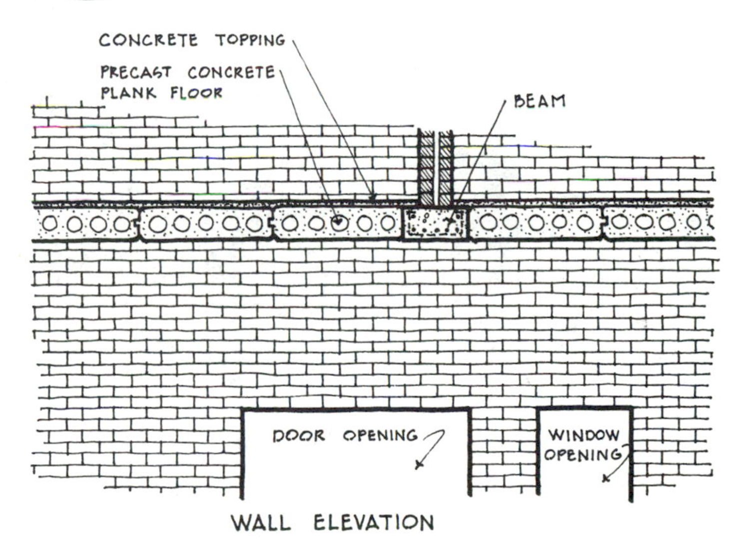

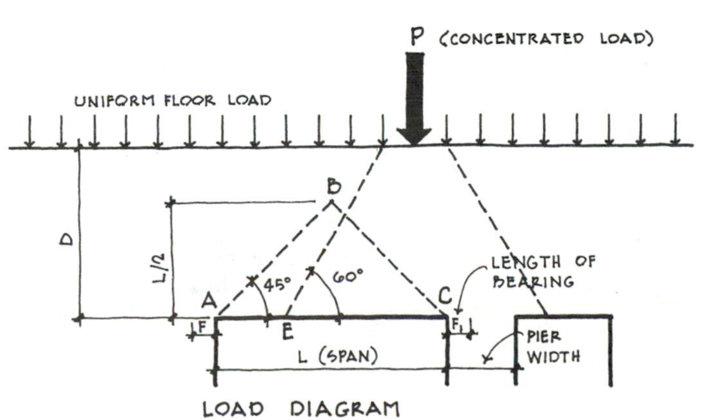

Consideration must also be given to concentrated loads from beams, girders, or trusses which frame into the wall above the opening. These loads are distributed over a length of wall equal to the base of a trapezoid whose summit is at the point of load application and whose sides make an angle of 60° with the horizontal. In *Fig. 12-4*, the portion of concentrated load carried by the lintel is distributed over the length *EC* and is considered as a uniform load partially distributed. The sum of all loads is used to calculate the size of lintel required to span the opening.

12.6.2 Steel Lintels

Structural steel shapes are commonly used to span masonry openings. Steel angles are the simplest shapes and are suitable for openings of moderate width where superimposed loads are not excessive. For wider openings or heavy loads, steel beams with suspended plates may be required (*see Fig. 12-5*). The horizontal leg of a steel angle should be at least $3\frac{1}{2}$ in. wide to adequately support a nominal 4-in. wythe of brick. Generally, angles should be a minimum of $\frac{1}{4}$ in. thick to satisfy code requirements for exterior steel members.

The method of design for steel lintels is basically the same as for steel beams. Although the computations are relatively simple, proper steel lintel

12/5 Steel lintels. (*From Brick Institute of America*, Technical Note 31B, *BIA, McLean, Va.*)

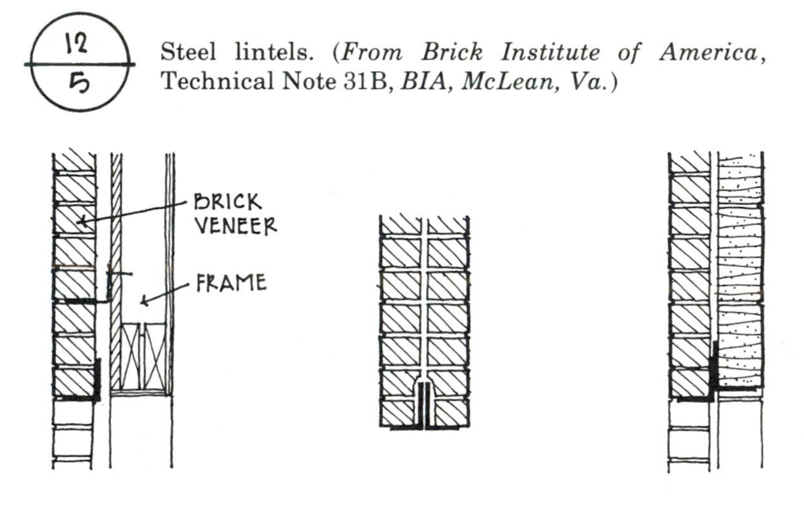

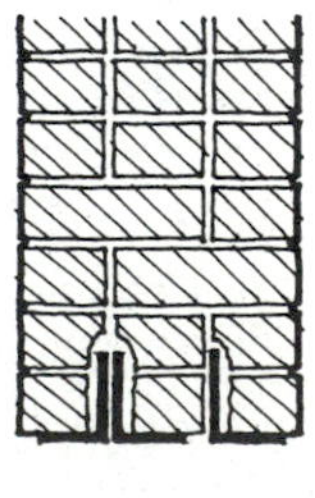

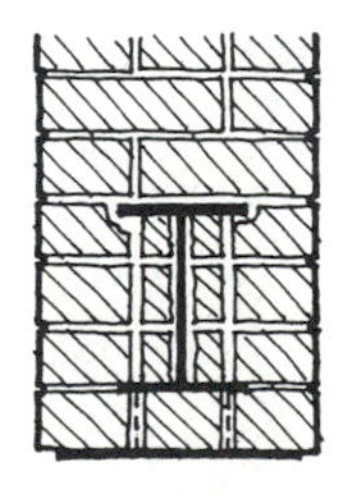

design should take into consideration (1) loading, (2) bending moment, (3) reactions, (4) required section modulus, (5) selection of section, (6) bearing area, (7) deflection, and (8) shear. If steel members are subject to torsional loads, a special analysis should be made.

The maximum *bending moment* caused by uniformly distributed loads above an opening is determined by the formula

$$M = \frac{wL^2}{8} \tag{12.9}$$

where M = bending moment, ft-lb
w = total uniform load, lb/lin foot
L = span of lintel, center to center of end bearing

To this should be added the bending moment caused by concentrated loads. Where such loads are located far enough above the lintel to be distributed as shown in *Fig. 12-4*, the bending moment formula for a uniform load partially distributed may be used. Otherwise, concentrated load bending moments are used [refer to the *Manual of Steel Construction*, published by the American Institute of Steel Construction (AISC)].

A steel lintel should be selected by first determining the required *section modulus* using the formula

$$S = \frac{12M}{F_s} \tag{12.10}$$

where S is the section modulus in cubic inches and F_s is the allowable stress in bending of the steel in psi. The allowable stress (F_s) for ASTM A-36 structural steel is 22,000 psi for members laterally supported. Under most conditions, masonry walls provide sufficient lateral stiffness to permit the use of the full 22,000 psi, particularly for certain wall-floor framing conditions.

Using the tables inthe AISC manual, an angle (or other steel shape) is selected which has an elastic section modulus equal to, or slightly greater than, that required. Within the limitations of minimum steel thickness and length of horizontal leg required, the lightest weight member having the required section modulus should be chosen.

To determine overall lintel length, required *bearing area* must be calculated. Compressive bearing stresses in the supporting masonry should not exceed the allowable unit stress for the type of masonry used (refer to *Figs. 10-11, 10-19, 10-25,* and *11-2* through *11-6*). The reaction at each end of the lintel is one-half the total uniform load plus a portion of any concentrated load or partially distributed uniform load. The required area may be found by the formula

$$A_b = \frac{R}{f_m} \tag{12.11}$$

where A_b = required bearing area, sq in.
R = reaction, lbs
f_m = allowable compressive stress in masonry, psi

The width of the selected steel section divided into the required bearing area A_b will determine the length of bearing required (F and F_1 in *Fig. 12-4*). In no instance should this length be less than 4 in. The table in *Fig. 12-6* lists length of bearing required for angles with $3\frac{1}{2}$- and 4-in. horizontal legs for various end reactions.

f_m (psi)	$3\frac{1}{2}$-in. leg horizontal					f_m (psi)	4-in. leg horizontal				
	Length of bearing, in.						Length of bearing, in.				
	4	5	6	7	8		4	5	6	7	8
400	5,600	7,000	8,400	9,800	11,200	400	6,400	8,000	9,600	11,200	12,800
350	4,900	6,125	7,350	8,575	9,800	350	5,600	7,000	8,400	9,800	11,200
300	4,200	5,250	6,300	7,350	8,400	300	4,800	6,000	7,200	8,400	9,600
275	3,850	4,813	5,775	6,738	7,700	275	4,400	5,500	6,600	7,700	8,800
250	3,500	4,375	5,250	6,125	7,000	250	4,000	5,000	6,000	7,000	8,000
225	3,150	3,938	4,725	5,513	6,300	225	3,600	4,500	5,400	6,300	7,200
215	3,010	3,763	4,515	5,268	6,020	215	3,440	4,300	5,160	6,020	6,880
200	2,800	3,500	4,200	4,900	5,600	200	3,200	4,000	4,800	5,600	6,400
175	2,450	3,063	3,675	4,288	4,900	175	2,800	3,500	4,200	4,900	5,600
160	2,240	2,800	3,360	3,920	4,480	160	2,560	3,200	3,840	4,480	5,120
155	2,170	2,713	3,255	3,798	4,340	155	2,480	3,100	3,720	4,340	4,960
150	2,100	2,625	3,150	3,675	4,200	150	2,400	3,000	3,600	4,200	4,800
140	1,960	2,450	2,940	3,430	3,920	140	2,240	2,800	3,360	3,920	4,480
125	1,750	2,188	2,625	3,063	3,500	125	2,000	2,500	3,000	3,500	4,000
115	1,610	2,013	2,415	2,818	3,220	115	1,840	2,300	2,760	3,220	3,680
110	1,540	1,925	2,310	2,695	3,080	110	1,760	2,200	2,640	3,080	3,520
100	1,400	1,750	2,100	2,450	2,800	100	1,600	2,000	2,400	2,800	3,200
85	1,190	1,488	1,785	2,083	2,380	85	1,360	1,700	2,040	2,380	2,720
75	1,050	1,313	1,575	1,838	2,100	75	1,200	1,500	1,800	2,100	2,400
70	980	1,225	1,470	1,715	1,960	70	1,120	1,400	1,680	1,960	2,240

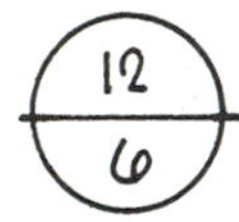

End reaction (in pounds) and required length of bearing (in inches) for steel angle lintels. (*From Brick Institute of America*, Technical Note 31B, *BIA, McLean, Va.*)

Before a selected steel lintel section is incorporated into the fina design, it should be checked for *deflection*. The design data in *Fig. 12-7* i based on a maximum deflection of $\frac{1}{600}$ of the span which is recommended fo masonry flexural members. The table in *Fig. 12-7*, which was developed b BIA engineers, contains tabulated load values to assist in the selection o proper steel lintel sizes.

Using steel lintels to span openings in masonry walls requires carefu attention to flashing details, and to provisions for differential movement o the steel and masonry. Code requirements for fireproofing of steel member should also be thoroughly investigated. If fireproofing is required, it may b simpler to design the lintel as a reinforced masonry section. Steel lintel should be painted or galvanized to minimize corrosion.

12.6.3 Concrete Masonry Lintels

Reinforced concrete masonry lintels are widely used today. Not only do the cost less than structural steel lintels, but the danger of steel corrosion an subsequent masonry cracking is eliminated, as is the painting and mai tenance of exposed steel. Concrete masonry walls do not lend themselve easily to the use of steel angle lintels. Openings are more commonly spanne with U-shaped lintel blocks grouted and reinforced with deformed steel bar

In some instances, cast-in-place or precast concrete sections can l used. *Cast-in-place lintels* are subject to drying shrinkage and have surfa textures that are not always compatible with the adjoining masonr

Horizontal leg	Angle size	Weight per foot (lb)	Span in feet (center to center of required bearing)										Resisting moment (ft-lb)	Elastic section modulus
			3	4	5	6	7	8	9	10	11	12		
$3\frac{1}{2}$	$3 \times 3\frac{1}{2} \times \frac{1}{4}$	5.4	956	517	262	149	91	59					1082	0.59
	$\times \frac{5}{16}$	6.6	1166	637	323	184	113	73					1320	0.72
	$\times \frac{3}{8}$	7.9	1377	756	384	218	134	87	59				1558	0.85
$3\frac{1}{2}$	$3\frac{1}{2} \times 3\frac{1}{2} \times \frac{1}{4}$	5.8	1281	718	406	232	144	94	65				1448	0.79
	$\times \frac{5}{16}$	7.2	1589	891	507	290	179	118	80				1797	0.98
	$\times \frac{3}{8}$	8.5	1947	1091	589	336	208	137	93	66			2200	1.20
$3\frac{1}{2}$	$4 \times 3\frac{1}{2} \times \frac{1}{4}$	6.2	1622	910	580	338	210	139	95	68			1833	1.00
	$\times \frac{5}{16}$	7.7	2110	1184	734	421	262	173	119	85	62		2383	1.30
	$\times \frac{3}{8}$	9.1	2434	1365	855	490	305	201	138	98	71		2750	1.50
	$\times \frac{7}{16}$	10.6	2760	1548	978	561	349	230	158	113	82	60	3117	1.70
4	$4 \times 4 \times \frac{7}{16}$	11.3	2920	1638	1018	584	363	239	164	116	85	62	3299	1.80
	$\times \frac{1}{2}$	12.8	3246	1820	1141	654	407	268	185	131	95	70	3666	2.00
$3\frac{1}{2}$	$5 \times 3\frac{1}{2} \times \frac{1}{4}$	7.0	2600	1460	932	636	398	264	184	132	97	73	2933	1.60
	$\times \frac{5}{16}$	8.7	3087	1733	1106	765	486	323	224	161	119	89	3483	1.90
	$\times \frac{7}{16}$	12.0	4224	2371	1513	1047	655	435	302	217	160	120	4766	2.60
	$\times \frac{1}{2}$	13.6	4875	2736	1746	1177	736	488	339	244	179	134	5500	3.00
$3\frac{1}{2}$	$6 \times 3\frac{1}{2} \times \frac{1}{4}$	7.9	3577	2009	1283	888	650	439	306	221	164	124	4033	2.20
	$\times \frac{5}{16}$	9.8	4390	2465	1574	1090	798	538	375	271	201	151	4950	2.70
	$\times \frac{3}{8}$	11.7	5200	2922	1865	1291	945	636	443	320	237	179	5867	3.20
	$\times \frac{1}{2}$	15.3	6828	3834	2448	1695	1228	818	570	412	305	230	7700	4.20
4	$6 \times 4 \times \frac{1}{4}$	8.3	3739	2099	1340	928	679	458	319	231	171	129	4216	2.30
	$\times \frac{5}{16}$	10.3	4552	2556	1632	1129	827	562	391	283	209	158	5133	2.80
	$\times \frac{3}{8}$	12.3	5365	3012	1923	1331	974	665	463	335	248	187	6050	3.30
	$\times \frac{7}{16}$	14.3	6178	3469	2214	1533	1122	764	532	384	284	215	6967	3.80
	$\times \frac{1}{2}$	16.2	6990	3925	2506	1734	1270	857	597	431	319	241	7883	4.30

ote: Allowable loads to the left of the heavy line are governed by moment, and to the right by deflection.

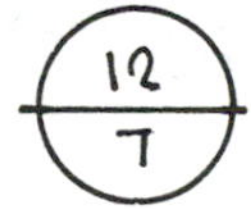

Allowable uniform superimposed load (in pounds) per foot for steel angle lintels (F_s = 22,000 psi). (*From Brick Institute of America*, Technical Note 31B, *BIA, McLean, Va.*)

Precast concrete lintels are better in some respects since they are delivered to the job site ready for use, do not require temporary shoring, and can carry superimposed loads as soon as they are in place. These sections can be produced with surface textures closely matching that of the masonry, and can be scored vertically to simulate mortar joints. Precast lintels may be one-piece, or may be split into two thinner sections. Split lintels are relatively light weight and easily handled. Split lintels, however, are not recommended to support combined wall and floor loads because of the difficulty involved in designing the heavily loaded inner section to match the deflection of the outer section, which may carry only wall loads. Differential deflection could cause critical stress concentrations in the wall. Mortar for bedding precast lintels should be the same quality as that used in laying the wall, and at least equal to ASTM C270, Type N.

Reinforced concrete masonry lintels are constructed with specially formed lintel units, bond beam units, or standard units with depressed, cut-out, or grooved webs to accommodate the steel bars (*see Fig. 12-8*). Individual units are laid end to end to form a channel in which continuous reinforcement and grout are placed. Among the major advantages of CMU lintels over steel are low maintenance and elimination of differential movement between dissimilar materials. Concrete masonry lintels are often designed as part of a continuous bond beam course, which helps to further distribute shrinkage and temperature stresses in the masonry above openings. This type of installation is more satisfactory in areas subject to seismic activity.

Units used for lintel construction should comply with the requirements of ASTM C90, Hollow Loadbearing Units, and should have a minimum

12/8 CMU lintel sections.

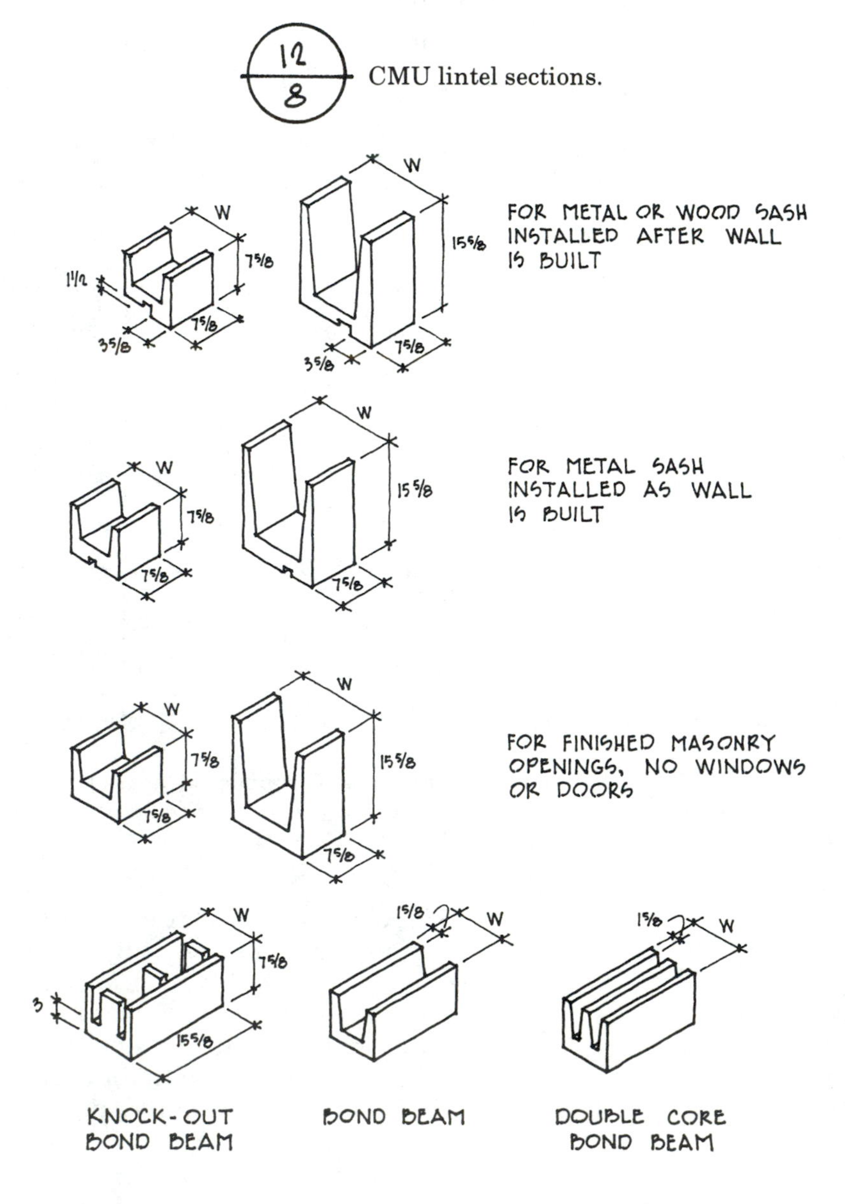

compressive strength adequate to provide the masonry compressive strength (f'_m) used in the design. Mortar should be equal to that used in constructing the wall and should meet the minimum requirements of ASTM C270, Type N. Grout for embedment of reinforcing steel should comply with ASTM C476, and maximum aggregate size should be consistent with job conditions to ensure proper placement. Mortar used for laying the masonry units should not be used to fill the lintel core unless it has a low lime content. The first course of masonry above the lintel should be laid with full mortar bedding so that the cross webs as well as the face shells of the units bear on the lintel and reduce the shear stress between the grout filled core and the face shells.

Determination of the magnitude of imposed loads follows the same method outlined above for steel lintels, with those loads occurring outside the 45° triangle neglected because of natural arching action. For ordinary wall thicknesses, loads, and spans, 8 to 16-in. of masonry above the apex of the triangle is required to resist arching thrusts. Arching action also requires an adequate length of masonry at each end of the lintel to resist lateral thrusts. When the end of a long-span lintel supporting relatively heavy loads is located near a wall corner or another opening, it may be necessary to neglect arching action and design the lintel for all superimposed loads applied above the opening. If walls are laid in stack bond, no loads are distributed beyond a vertical joint unless structural elements such as bond beams are designed within the wall to effect such distribution. Generally, the loads from floor joists or rafters in residential and light commercial buildings may be considered as uniformly distributed when the height of the masonry between the lintel and the bearing plane is more than one-third of the joist or rafter spacing. If members bear directly on the lintel or are relatively heavy, they should be treated as concentrated loads.

A minimum end bearing of 8 in. is recommended for reinforced CMU lintels with relatively modest spans. For longer spans or heavy loads, bearing stresses should be calculated to ensure that the allowable compressive stress of the masonry is not exceeded. High stress concentrations may require the use of solid units or solidly grouted hollow units for one or more courses under the lintel bearing so that loads are distributed over a larger area.

The design table in *Fig. 12-9* developed by NCMA, is based on typical

Required reinforcement for simply supported CMU lintels. (*From National Concrete Masonry Association*, TEK Bulletin 25, *NCMA, Herndon, Va.*)

Type of loading	*Lintel section nominal size (in.)*	*Required reinforcing*							
		Clear span							
		3′-4″	*4′-0″*	*4′-8″*	*5′-4″*	*6′-0″*	*6′-8″*	*7′-4″*	*8′-0″*
Vall loads (200–300 lb/lin ft)	6 × 8 6 × 16	1-No. 3	1-No. 4	1-No. 4	2-No. 4	2-No. 5 1-No. 4	1-No. 4	1-No. 4	1-No. 4
loor and roof loads (700–1000 lb/lin ft)	6 × 16	1-No. 4	1-No. 4	2-No. 3	1-No. 5	2-No. 4	2-No. 4	2-No. 5	2-No. 5
Vall loads (200–300 lb/lin ft)	8 × 8 8 × 16	1-No. 3	2-No. 3	2-No. 3	2-No. 4	2-No. 4	2-No. 5	2-No. 6 2-No. 5	2-No. 5
loor and roof loads (700–1000 lb/lin ft)	8 × 8 8 × 16	2-No. 4 2-No. 3	2-No. 3	2-No. 3	2-No. 4	2-No. 4	2-No. 4	2-No. 5	2-No. 5

equivalent uniform loads of 200 to 300 lb/lin ft for wall loads, and 700 to 1000 lb/lin ft for combined floor and roof loads. The table can be used to determine required lintel size and reinforcing for various spans subject to this type of loading. Where concentrated loads occur, calculation of total shear and moment is required, and lintels may be sized on the basis of the table in *Fig. 12-10*, which lists resisting moments and shears for various unit sizes and prism strengths. A design example is given in *Fig. 12-11*.

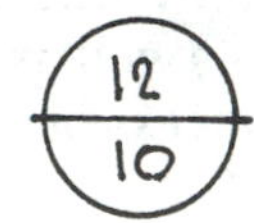

Allowable shear and resisting moments for CMU lintels. (V_v, allowable shear based on shear stress; V_u, allowable shear based on bond stress). (*From National Concrete Masonry Association*, TEK Bulletin 81, *NCMA*, Herndon, Va.)

Steel reinf.	$f'_m = 1500$ psi						$f'_m = 2000$ psi					
	Bottom cover $1\frac{1}{2}$ in.			Bottom cover 3 in.			Bottom cover $1\frac{1}{2}$ in.			Bottom cover 3 in.		
	V_v (kip)	V_u (kip)	M (ft-kip)	V_v (kip)	V_u (kip)	M (ft-kip)	V_v (kip)	V_u (kip)	M (ft-kip)	V_v (kip)	V_u (kip)	M (ft-kip)
TABLE A 6-IN. LINTELS (8 IN. HIGH)												
1-#3	1.422	1.009	0.983*	1.063	0.743	0.685	1.642	1.021	0.995*	1.227	0.754	0.734†
1-#4	1.407	1.291	1.335	1.048	0.945	0.813	1.625	1.310	1.615	1.210	0.960	0.989
1-#5	1.392	1.554	1.501	1.033	1.132	0.899	1.608	1.582	1.834	1.193	1.152	1.106
1-#6	1.377	1.802	1.624	1.018	1.305	0.957	1.590	1.836	2.001	1.175	1.330	1.188
2-#5	1.392	2.965	1.802	1.033	2.154	1.059	1.608	3.024	2.238	1.193	2.201	1.326
2-#6	1.377	3.434	1.906	1.018	2.484	1.102	1.590	3.503	2.391	1.175	2.535	1.393
TABLE B 8-IN. LINTELS (8 IN. HIGH)												
1-#4	1.908	1.311	1.633	1.421	0.961	1.001	2.203	1.329	1.766†	1.640	0.976	1.210
1-#5	1.887	1.581	1.855	1.400	1.152	1.119	2.179	1.607	2.252	1.617	1.173	1.366
1-#6	1.867	1.837	2.025	1.380	1.331	1.202	2.156	1.870	2.477	1.594	1.356	1.481
1-#7	1.847	2.076	2.163	1.360	1.496	1.265	2.132	2.115	2.666	1.570	1.526	1.571
2-#6	1.867	3.505	2.422	1.380	2.536	1.411	2.156	3.576	3.013	1.594	2.588	1.770
2-#7	1.847	3.954	2.537	1.360	2.848	1.455	2.132	4.036	3.183	1.570	2.907	1.840
TABLE C 8-IN. LINTELS (16 IN. HIGH)												
1-#5	4.484	3.926	6.472†	3.997	3.482	5.739†	5.178	3.974	6.549†	4.616	3.526	5.811†
1-#6	4.464	4.614	8.755	3.977	4.087	7.228	5.155	4.677	9.177†	4.592	4.145	8.079†
2-#5	4.484	7.588	9.911	3.997	6.720	8.172	5.178	7.704	11.988	4.616	6.827	9.097
2-#6	4.464	8.881	10.981	3.977	7.856	9.019	5.155	9.032	13.379	4.592	7.993	11.016
2-#8	4.423	11.292	12.719	3.937	9.967	10.368	5.108	11.512	15.714	4.545	10.166	12.847
TABLE D 12-IN. LINTELS (8 IN. HIGH)												
1-#5	2.878	1.618	2.461	2.135	1.182	1.497	3.323	1.641	2.705†	2.466	1.200	1.812
1-#6	2.847	1.884	2.714	2.104	1.368	1.629	3.287	1.914	3.290	2.430	1.391	1.987
2-#5	2.878	3.111	3.073	2.135	2.265	1.841	3.323	3.165	3.751	2.466	2.307	2.263
2-#6	2.847	3.609	3.327	2.104	2.613	1.962	3.287	3.677	4.096	2.430	2.665	2.433
2-#8	2.785	4.513	3.679	2.042	3.237	2.103	3.216	4.606	4.599	2.358	3.304	2.651
TABLE E 12-IN. LINTELS (16 IN. HIGH)												
1-#5	6.838	3.994	4.937†	6.096	3.545	4.382†	7.896	4.035	4.987†	7.038	3.583	4.429
1-#6	6.807	4.705	9.170†	6.065	4.171	8.130†	7.860	4.760	9.278†	7.003	4.222	8.229
2-#5	6.838	7.755	12.782†	6.096	6.874	10.814	7.896	7.858	12.952†	7.038	6.969	11.486
2-#6	6.807	9.099	14.643	6.065	8.055	12.078	7.860	9.235	17.688	7.003	8.180	14.622
2-#8	6.745	11.611	17.299	6.003	10.256	14.164	7.789	11.817	21.159	6.931	10.443	17.370

*Moment reduced by 25% since $p = A_s/bd$ is less than $100/f_y$, where f_y is yield strength of reinforcement.

†Moment governed by allowable tensile stress in steel (f_s = 20,000 psi).

Determine steel reinforcement required for the simply supported reinforced concrete masonry lintel shown.

Assume:

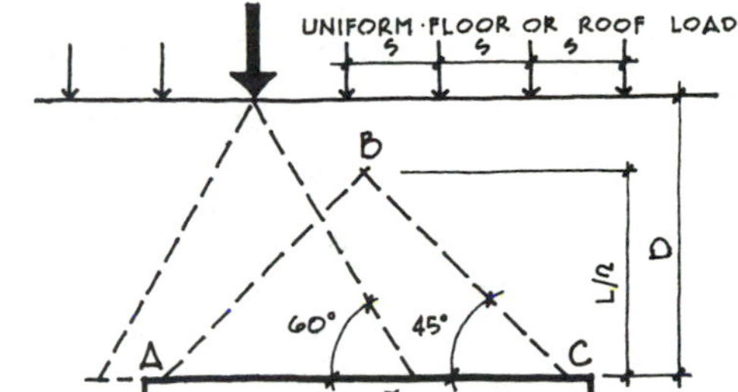

Effective span, $L = 6$ ft
Nominal wall thickness = 8 in.
Wall weight = 34 lb/sq. ft
Lintel weight (8 × 8 in.) = 54 lb/ft
Uniform floor load = 300 lb/ft

(A) Assuming no arching action ($D = 2$ ft)

Total uniform load, $w = 2(34) + 54 + 300 = 422$ lb/ft

$$\text{Moment} = \frac{wl^2}{8} = \frac{(422)(6)^2}{8}$$

$M = 1899$ ft-lb = 1.899 ft-kip
Assuming 8-in. bearing length, shear (at face of support),

$$V = \frac{wl}{2} - \frac{w}{3} = \frac{(422)(6)}{2} - \frac{422}{3}$$

$V = 1125$ lb = 1.125 kip

From Fig. 12-10 ($f'_m = 1500$ psi, cover = 1.5 in):
For 1 − No. 6 bar, $V_u = 1.837$ kips; $M = 2.025$ ft-kip
From Fig. 12-10 ($f'_m = 2000$ psi, cover = 1.5 in.):
For 1 − No. 5 bar, $V_u = 1.607$ kips; $M = 2.252$ ft-kip

(B) Assuming arching action ($D = 4$ ft)
Moment (due to uniform load of lintel)

$$M = \frac{(54)(6)^2}{8} = 243 \text{ ft-lb} = 0.243 \text{ ft-kip}$$

Moment (due to wall weight—triangular area)

$$M = \frac{(34)(6)^3}{24} = 306 \text{ ft-lb} = 0.306 \text{ ft-kip}$$

Total moment = 0.549 ft-kip

Shear (at face of support):

$$V = \frac{(54)(6)}{2} + \frac{(3)(3)(34)}{2} - \frac{54}{3} - \frac{(0.3)(0.15)(34)}{2}$$

$$= 162 + 153 - 18 - 0.8 = 296 \text{ lb} = 0.296 \text{ kip}$$

From Fig. 12-10 for 8 × 8 in. lintel with 1-No. 4 bar, allowable shear and moment for $f'_m = 1500$ psi and $f'_m = 2000$ psi are greater than actual calculated stresses. Therefore, 1-No. 4 bar will satisfy design requirements.

Design example. (*From National Concrete Masonry Association*, TEK Bulletin 81, *NCMA, Herndon, Va.*)

12.6.4 Reinforced Brick Lintels

Standard brick masonry units are also adaptable to reinforced lintel design even though they contain no continuous channels for horizontal steel (*see Fig. 12-12*). Reinforcing may be located in bed joints or in a widened collar joint created by using bats broken to one-half of the normal unit depth. Horizontal cell clay tile are ideally suited for lintels since, like concrete blocks, they contain natural channels for the steel.

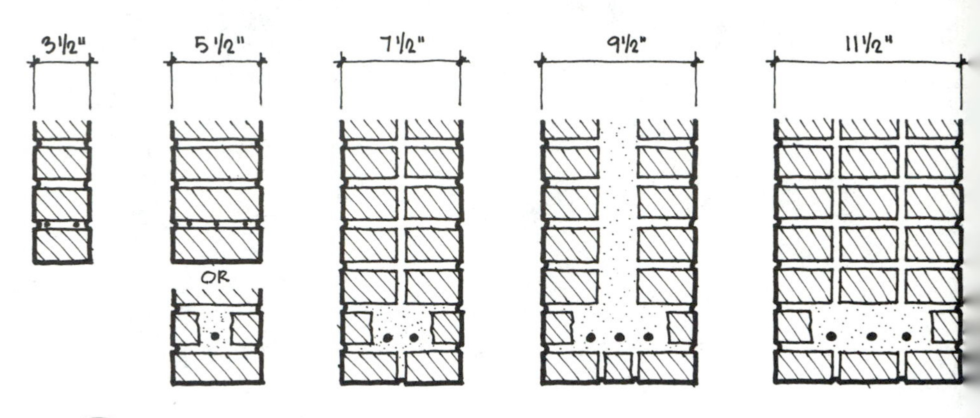

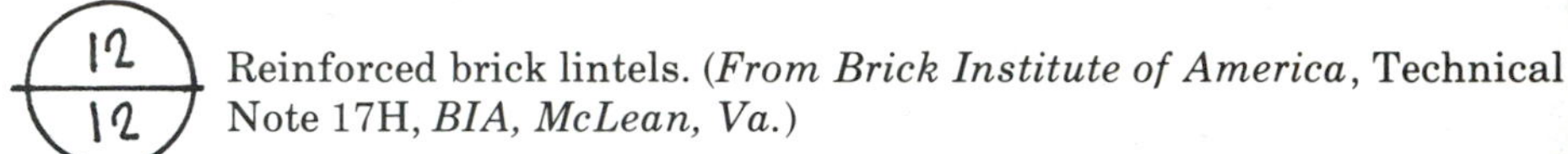

Reinforced brick lintels. (*From Brick Institute of America*, Technical Note 17H, *BIA, McLean, Va.*)

The physical dimensions of clay masonry lintels are partially deter mined by the type of wall that they support and the type of unit used i construction. The design width of brick lintels should be the same as th wall thickness. Design depth is, to a degree, determined by the course heigh of the units. Depth can be limited by the height of masonry above th opening, in which case compression steel may be required to provid adequate resistance to bending. *Figure 12-13*(*B*) shows a reinforced bric lintel capable of carrying the same loads as the steel angles in *Fig. 12-13*(*A* The labor costs of laying the greater number of masonry units is more tha offset by the reduced amount of steel required. The result is a mor economical structural member and a more efficient use of materials.

The BIA has developed a tabulated set of resisting moments an shears as a design aid for clay masonry lintels. The tables in *Figs. 12-14* an

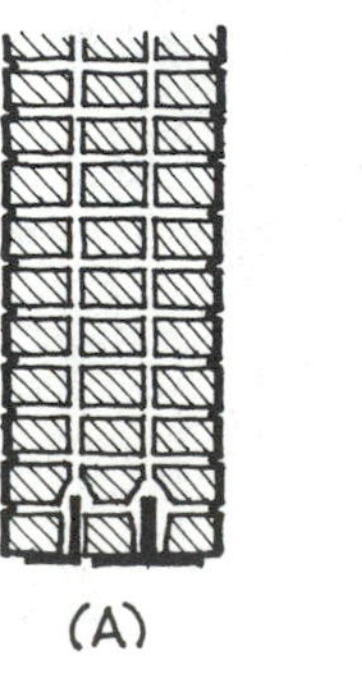

(A) Steel lintel and (B) reinforced bri lintel with the same load-carrying capa ity. (*From Brick Institute of Ameri* Technical Note 17H, *BIA, McLean, V*

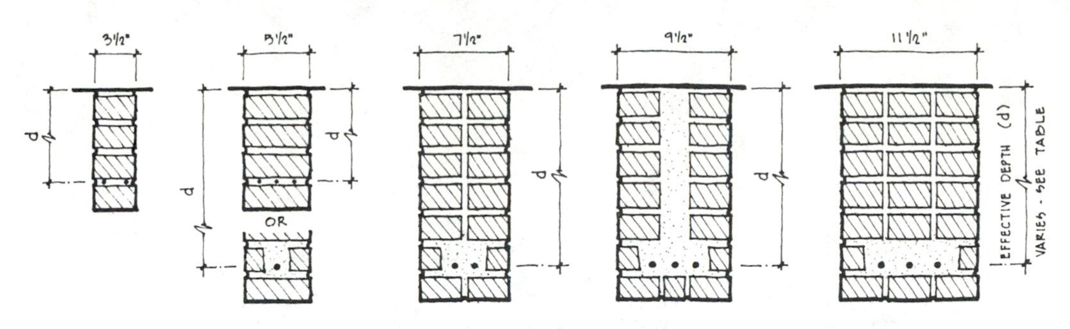

TABLE A NOMINAL 4-IN. LINTELS (REINFORCING IN BED JOINT)

Depth	d = 5.1"	(2 courses)	d = 7.8"	(3 courses)	d = 10.5"	(4 courses)
Reinf.	M	V_o	M	V_o	M	V_o
1-#2	*390*	220	605	455	820	620
2-#2	750	430	1180	890	1600	1210

TABLE B NOMINAL 6-IN. LINTELS (REINFORCING IN BED JOINT)

Depth	d = 5.1"	(2 courses)	d = 7.8"	(3 courses)	d = 10.5"	(4 courses)	d = 13.1"	(5 courses)	d = 15.8"	(6 courses)
Reinf.	M	V_o	M	V_o	M	V_o	M	V_o	M	V_o
2-#2	780	440	*1200*	905	*1630*	1230	*2060*	1550	*2490*	1400
3-#2	1140	645	1770	1340	*2410*	1820	*3040*	2290	*3670*	2080

TABLE C NOMINAL 6-IN. LINTELS (REINFORCING IN GROUT SPACE)

Depth	d = 4.6" (2 courses)			d = 7.3" (3 courses)			d = 10.0" (4 courses)			d = 12.6" (5 courses)			d = 15.3" (6 courses)			d = 18.0" (7 courses)		
Reinf.	M	V_m	V_o	M	V_m	V_o	M	V_m	V_o	M	V_m	V_o	M	V_m	V_o	M	V_m	V_o
1-#3	770	—	790	*1230*	—	1260	*1700*	—	1730	*2170*	—	2230	*2640*	—	2720	*3120*	—	3200
1-#4	1060	—	1010	*2160*	—	1630	*3000*	—	2260	*3840*	—	2890	*4700*	—	3550	*5530*	—	4180
1-#5	1190	1060	1210	2600	1730	1970	4400	2410	2750	*5810*	3090	3540	*7120*	3800	4330	*8410*	4480	5120

TABLE D NOMINAL 8-IN. LINTELS

Depth	d = 4.6" (2 courses)			d = 7.3" (3 courses)			d = 10.0" (4 courses)			d = 12.6" (5 courses)			d = 15.3" (6 courses)			d = 18.0" (7 courses)		
Reinf.	M	V_m	V_o	M	V_m	V_o	M	V_m	V_o	M	V_m	V_o	M	V_m	V_o	M	V_m	V_o
2-#3	1380	1540	1550	*2420*	2480	2500	*3350*	3420	3440	*4280*	4380	4400	*5220*	5340	5360	*6160*	6300	6340
3-#3	1570	1500	2260	3360	2430	3660	*4930*	3360	5070	*6310*	4310	6490	*7690*	5250	7920	*9100*	6210	9360
2-#4	1630	1470	1970	3530	2380	3200	*5890*	3310	4450	*7560*	4250	5700	*9230*	5190	6960	*10900*	6140	8240
2-#5	1800	1410	2350	4000	2300	3860	6820	3210	5380	10170	4140	6930	*13970*	5070	8500	*16600*	6010	10100
2-#6	1910	1350	2720	4350	2230	4480	7510	3120	6290	11300	4040	8110	15600	4950	9950	20500	5880	11800

TABLE E NOMINAL 10-IN. LINTELS

epth	d = 7.3" (3 courses)			d = 10.0" (4 courses)			d = 12.6" (5 courses)			d = 15.3" (6 courses)			d = 18.0" (7 courses)			d = 20.6" (8 courses)		
einf.	M	V_m	V_o	M	V_m	V_o	M	V_m	V_o	M	V_m	V_o	M	V_m	V_o	M	V_m	V_o
2-#3	*2450*	—	2510	*3380*	—	3470	*4310*	—	4430	*5260*	—	5400	*6200*	—	6370	*7150*	—	7350
3-#3	*3600*	3110	3710	*4980*	4300	5120	*6380*	5510	6560	*7780*	6720	8020	*9190*	7930	9440	*10600*	9150	10900
2-#4	4130	3060	3230	*5950*	4240	4480	*7630*	5440	5760	*9320*	6640	7020	*11000*	7850	8300	*12700*	9060	9580
3-#4	4630	2990	4750	7960	4160	6610	*11200*	5340	8480	*13700*	6530	10400	*16300*	7730	12300	*18800*	8930	14200
2-#5	4720	2950	3900	7980	4120	5460	*11500*	5300	7010	*14100*	6490	8590	*16700*	7680	10200	*19300*	8890	11800
2-#6	5170	2870	4540	8850	4020	6380	13300	5180	8220	18300	6350	10100	*23300*	7540	12000	*26900*	8730	13900
3-#5	5320	2880	5710	9110	4030	8000	13700	5190	10300	18800	6360	12600	*24600*	7540	15000	*28500*	8720	17300
2-#7	5530	2780	5150	9600	3910	7250	14500	5060	9380	20200	6220	11500	26500	7400	13700	33400	8560	15900
3-#6	5750	2790	6660	9980	3910	9310	15100	5060	12000	21000	6210	14800	27500	7380	17600	34800	8550	20400

TABLE F NOMINAL 12-IN. LINTELS

pth	d = 10.0" (4 courses)			d = 12.6" (5 courses)			d = 15.3" (6 courses)			d = 18.0" (7 courses)			d = 20.6" (8 courses)			d = 23.3" (9 courses)		
inf.	M	V_m	V_o	M	V_m	V_o	M	V_m	V_o	M	V_m	V_o	M	V_m	V_c	M	V_m	V_o
#4	*5990*	—	4530	*7680*	—	5810	*9380*	—	7070	*11100*	—	8370	*12800*	—	9650	*14500*	—	10900
#4	*8840*	5080	6670	*11300*	6520	8540	*13800*	7970	10400	*16400*	9420	12400	*18900*	10900	14300	*21500*	12400	16200
#5	*9060*	5040	5500	*11600*	6480	7070	*14200*	7920	8660	*16800*	9370	10200	*19500*	10800	11800	*22100*	12300	13400
#4	9800	5010	8750	14800	6430	11200	*18200*	7870	13800	*21600*	9310	16300	*25000*	10800	18800	*28300*	12200	21400
#6	10100	4920	6450	15000	6330	8300	*19800*	7770	10200	*23500*	9210	12100	*27100*	10600	14000	*30800*	12100	15900
#5	10400	4940	8080	15500	6350	10400	*21000*	7780	12800	*24900*	9220	15100	*28700*	10700	17500	*32600*	12100	19800
#7	11000	4800	7340	16500	6190	9470	23000	7610	11600	30100	9030	13800	*36400*	10500	16000	*41400*	11900	18200
#5	11300	4850	10600	17000	6250	13600	23500	7670	16800	30700	9090	19900	*37800*	10500	23000	*43000*	12000	26100
#6	11400	4800	9440	17200	6200	12200	23800	7600	14900	31200	9020	17700	39400	10400	20500	*45500*	11900	23400
#6	12400	4710	12300	18800	6090	16000	26100	7480	19600	34300	8880	23300	43400	10300	27000	53200	11700	30700
#7	12400	4670	10700	18800	6040	13900	26200	7430	17100	34500	8840	20300	43500	10300	23500	53300	11700	26800
#7	13300	4590	14000	20300	5940	18200	28400	7310	22400	37500	8690	26600	47600	10100	30900	58600	11500	35100

te: Resisting moments in italics are controlled by the steel, others are controlled by the masonry. Resisting moments are given in foot-pounds. Resisting shears are given in pounds.

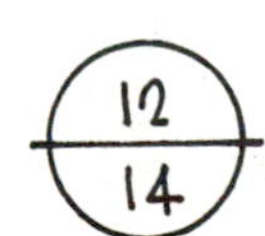

Design tables for reinforced brick lintels. (*From Brick Institute of America, Technical Note 17H, BIA, McLean Va.*)

TABLE G NOMINAL 4 IN. LINTELS

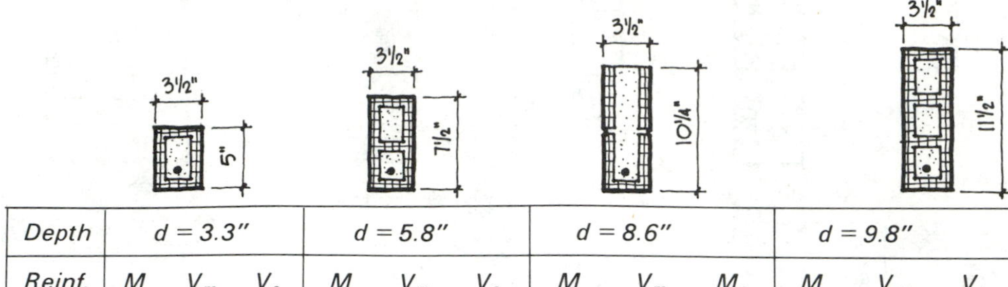

Depth	$d = 3.3''$			$d = 5.8''$			$d = 8.6''$			$d = 9.8''$		
Reinf.	M	V_m	V_o	M	V_m	V_o	M	V_m	M_o	M	V_m	V_o
1-#3	375	500	540	***950***	905	975	***1430***	1370	1470	***1640***	1570	1690
1-#4	430	470	680	1140	865	1250	2220	1320	1890	2760	1510	2170
1-#5	460	450	805	1270	830	1490	2520	1270	2290	3160	1470	2630
1-#6				1370	805	1730	2760	1240	2660	3470	1430	3070
1-#7				1440	780	1950	2960	1200	3020	3750	1390	3490

TABLE H NOMINAL 6-IN. LINTELS

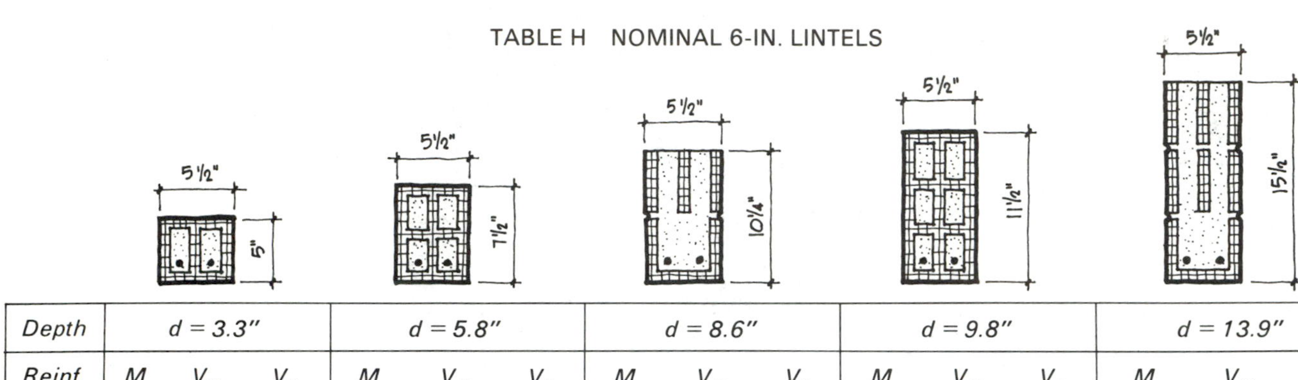

Depth	$d = 3.3''$			$d = 5.8''$			$d = 8.6''$			$d = 9.8''$			$d = 13.9''$		
Reinf.	M	V_m	V_o	M	V_m	V_o	M	V_m	V_o	M	V_m	V_o	M	V_m	V_o
2-#3	635	780	1060	1630	1410	1930	***2830***	2120	2910	***3250***	2430	3340	***4680***	3510	4810
2-#4	720	730	1340	1930	1340	2450	3770	2040	3730	4710	2350	4290	***8260***	3410	6220
2-#5	760	695	1580	2130	1290	2950	4260	1970	4500	5330	2270	5190	9760	3310	7560
2-#6				2270	1240	3410	4620	1910	5240	5840	2210	6050	10800	3230	8840

TABLE I NOMINAL 8-IN. LINTELS

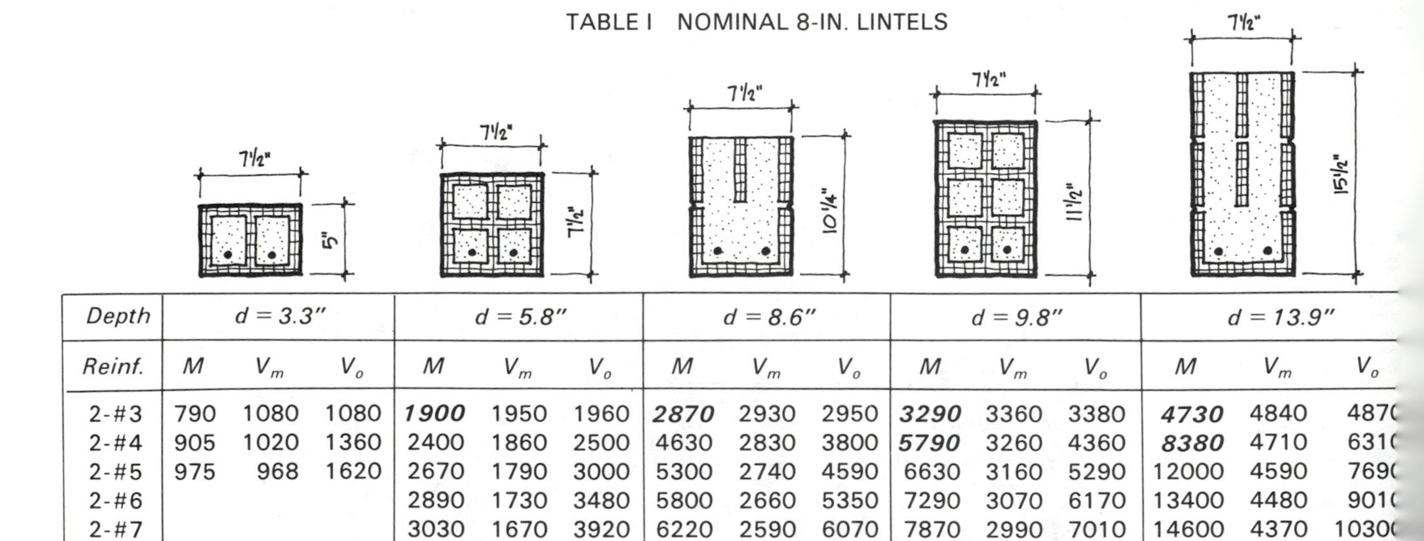

Depth	$d = 3.3''$			$d = 5.8''$			$d = 8.6''$			$d = 9.8''$			$d = 13.9''$		
Reinf.	M	V_m	V_o	M	V_m	V_o	M	V_m	V_o	M	V_m	V_o	M	V_m	V_o
2-#3	790	1080	1080	***1900***	1950	1960	***2870***	2930	2950	***3290***	3360	3380	***4730***	4840	4870
2-#4	905	1020	1360	2400	1860	2500	4630	2830	3800	***5790***	3260	4360	***8380***	4710	6310
2-#5	975	968	1620	2670	1790	3000	5300	2740	4590	6630	3160	5290	12000	4590	7690
2-#6				2890	1730	3480	5800	2660	5350	7290	3070	6170	13400	4480	9010
2-#7				3030	1670	3920	6220	2590	6070	7870	2990	7010	14600	4370	10300
2-#8				3150	1620	4350	6570	2520	6760	8350	2910	7810	15600	4280	11500

Note: Resisting moments in italics are controlled by the steel, others are controlled by the masonry.

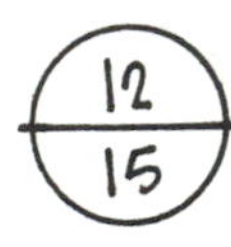

Design tables for reinforced clay tile lintels. (*From Brick Institute of America*, Technical Note 17H, *BIA, McLean, Va.*)

12-15 list resisting moments and shears for various reinforced brick and structural clay tile sections with several areas of reinforcing steel. Resisting moments are determined by either the masonry or the reinforcement, and the value listed is that which governs design. Where moment is governed by steel, the value is shown in italics. The two values of shear are diagonal tension in the masonry (V_m), and bond on the tensile steel (V_o). For lintels in which no stirrups are provided, the smaller resisting shear governs. V_m is based on an allowable working stress in diagonal tension of 50 psi. When web reinforcement is provided, the working stress increases to 150 psi. In other words, resisting shears are tripled if stirrups are provided. However, before such high shearing stresses are reached, bond shear may govern the design. In Tables *A* and *B* in *Fig. 12-14*, bond governs the shear values and V_m is therefore not listed.

Where the resisting moment of the masonry governs design, it is often possible to increase this value by using units with higher compressive strengths than that assumed (8000 psi, which yields f'_m of 2000 psi). Although it is not precise, an estimate of the increase in resisting moment with higher compressive strength masonry may be obtained by increasing M proportionately to the increase in f'_m.

Tables *A* through *F* in *Fig. 12-14* for brick lintels are based on modular units with $\frac{1}{2}$-in. mortar joints. The effective depths in Tables *A* and *B* assume $\frac{1}{4}$-in. reinforcing bars centered within the bed joint. In Tables *C* through *F*, effective depth assumes reinforcement placed in grout with $\frac{1}{2}$ in. clear between the steel and the top of the unit immediately below. True effective depth will therefore vary slightly with the bar diameter, but for the range of sizes shown here, the difference is not significant. Tables *G* through *I* in *Fig. 12-15* list resisting moments and shears for various cross sections of tile, and are also based on modular unit sizes with $\frac{1}{2}$-in. mortar joints. The configurations shown serve as examples only, and the tables apply to all cross-sectional arrangements of grouted tile lintels of the same overall dimensions. The effective depths for the tile lintels assume that the reinforcement is completely embedded in grout with $\frac{3}{4}$ in. clear between the steel and the tile units.

12.6.5 Precasting

Reinforced brick lintels are normally built in place by using temporary shoring which is left in place until the section has cured sufficiently to carry superimposed loads. The soffit brick may be standard units or special shapes, and are laid with mortar in the head and collar joints only.

Reinforced brick lintels may also be precast in the same manner as beam sections. This eliminates the need for shoring and permits field operations to progress at a more rapid rate.

12.7 ARCHES

Brick and stone masonry arches, vaults, and domes have been used since earliest history to span horizontal openings. In the ruins of Babylonia, archaeologists discovered an arch which they estimate was constructed around the year 1400 B.C. In more recent history, brick arches have been used for long spans with heavy loading as in the railway bridge at Maidenhead, England, built in 1835, which spans 128 ft with a rise of 24.3 ft. A railway bridge constructed in Baltimore in 1895 spans 130 ft with a rise of 26 ft.

Arches are still widely used in contemporary construction in forms which vary from *segmental*, *elliptical*, *Tudor*, *Gothic*, *semicircular*, and *parabolic* to *flat or jack arches* (*see Fig. 12-16*). The primary advantage of an arch is that under uniform loading conditions, the induced stress is principally compression rather than tension. For this reason, an arch will frequently provide the most efficient structural span. Since masonry's resistance to compression is greater than to other stresses, it is an ideal material for the construction of arches.

Arches are divided structurally into two categories. *Minor arches* are those whose spans do not exceed 6 ft with a maximum rise/span ratio of 0.15, and equivalent uniform loads on the order of 1000 lb/ft. These are most often used in building walls over door and window openings. *Major arches* are those whose spans or loadings exceed the maximum for minor arches. With larger spans and uniformly distributed loads, the parabolic arch is often the most structurally efficient form.

12.7.1 Minor Arch Design

In a fixed masonry arch, three conditions must be maintained to ensure the integrity of the arch action: (1) the length of span must remain constant, (2) the elevation of the ends must remain unchanged, and (3) the inclination of the skewback must be fixed. If any of these conditions are altered by sliding, settlement, or rotation of the abutments, critical stresses can develop and may result in structural failure. Adequate foundations and high-quality mortar and workmanship are essential to proper arch construction. The compressive and bond strength of the mortar must be high, and only Types M, S, or N are recommended. It is also particularly important that mortar joints be completely filled to assure maximum bond and even distribution of stresses.

Arches are designed by assuming a shape and cross section based on architectural considerations or empirical methods, and then analyzing the shape to determine its adequacy to carry the superimposed loads. Minor arch loading may consist of live and dead loads from floors, roofs, walls, and other structural members. These may be applied as concentrated loads or as uniform loads fully or partially distributed. The dead load on an arch is the weight of the wall area contained within a triangle immediately above the opening. The sides of the triangle are at 45° angles to the base, and its height is therefore one-half of the span. Such triangular loading is equivalent to a uniformly distributed load of 1.33 times the triangular load. Superimposed uniform loads above this triangle are carried beyond the span of the opening by arching action of the masonry wall itself when running bond patterns are used. Uniform live and dead loads below the apex of the triangle are applied directly on the arch for design purposes. Minor concentrated loads directly or nearly directly on the arch may safely be assumed equivalent to a uniformly distributed load twice the magnitude of the concentrated load. Heavy concentrated loads should not be allowed to bear directly on minor arches (especially jack arches).

There are two basic theories for verification of the stability of an assumed arch section. The *elastic theory* considers the arch as a curved beam subject to moment and shear, whose stability depends on internal stresses. For arches subject to nonsymmetrical loading that can cause

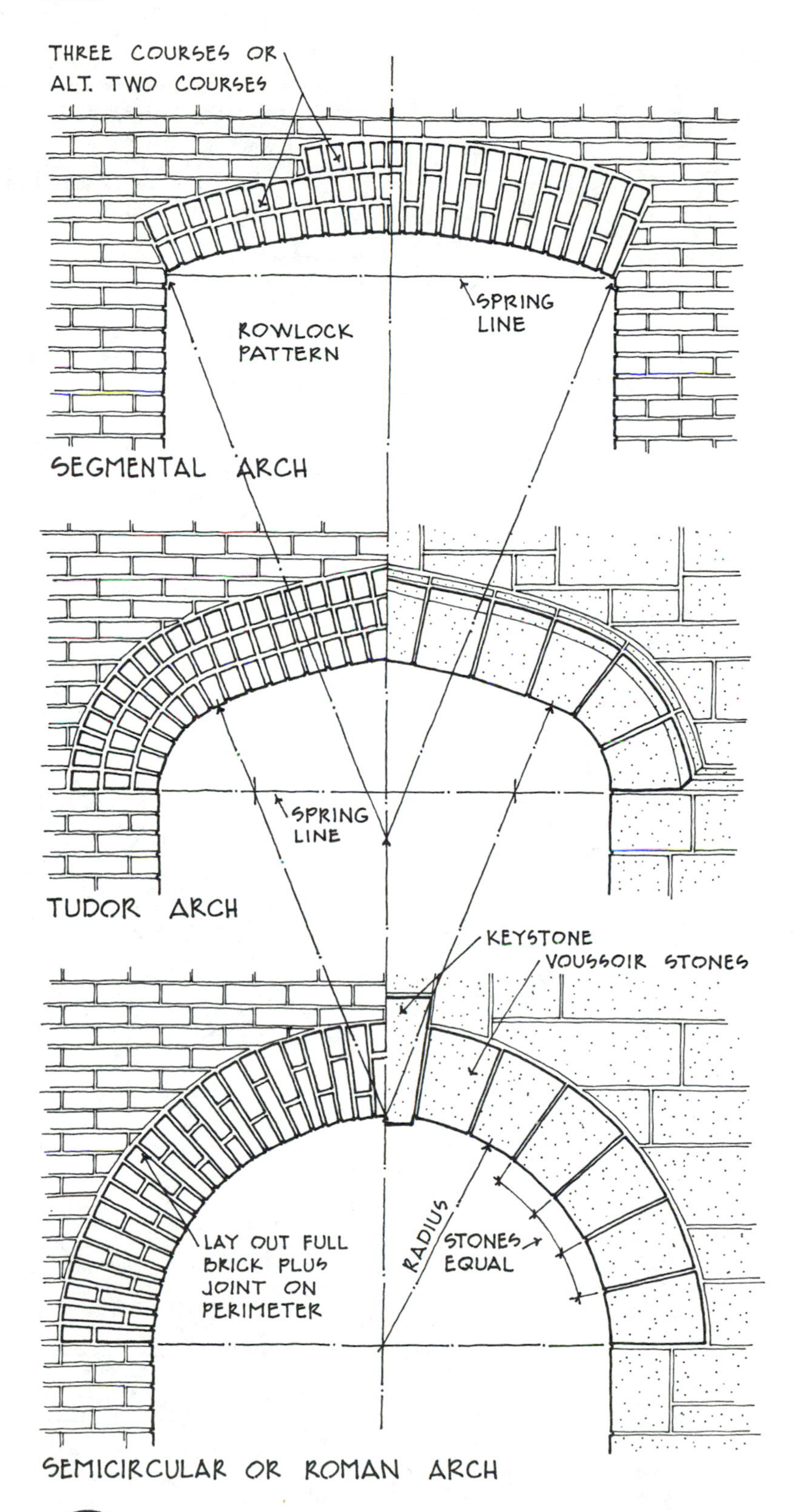

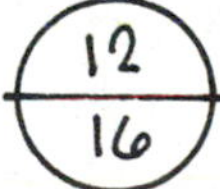

Masonry arch forms. (*From Brick Institute of America*, Technical Note 31, *BIA, McLean, Va.)*

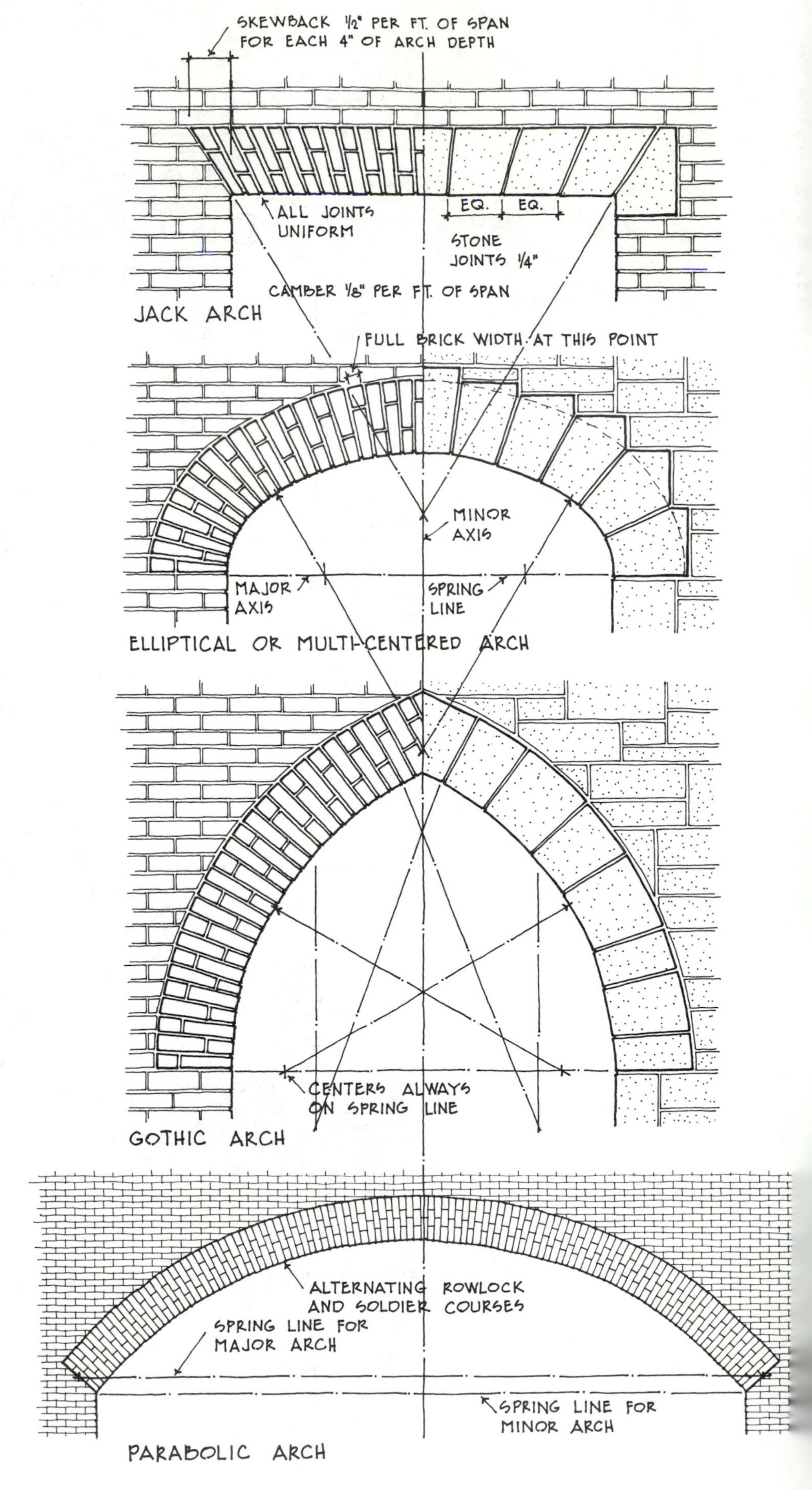
SKEWBACK 1/2" PER FT. OF SPAN
FOR EACH 4" OF ARCH DEPTH
ALL JOINTS
UNIFORM
EQ.
EQ.
STONE
JOINTS 1/4"
CAMBER 1/8" PER FT. OF SPAN
JACK ARCH
FULL BRICK WIDTH AT THIS POINT
MINOR
AXIS
MAJOR
AXIS
SPRING
LINE
ELLIPTICAL OR MULTI-CENTERED ARCH
CENTERS ALWAYS
ON SPRING LINE
GOTHIC ARCH
ALTERNATING ROWLOCK
AND SOLDIER COURSES
SPRING LINE FOR
MAJOR ARCH
SPRING LINE FOR
MINOR ARCH
PARABOLIC ARCH

12/16 (Continued)

tensile stress development, the elastic theory provides the most accurate method of analysis. There are many methods of elastic analysis for arch design, but in most instances, their application is complicated and time consuming. Such detailed engineering discussions are beyond the scope of this book, and the reader is referred to Valerian Leontovich's *Frames and Arches* for further information.

A second theory of analysis is the *line-of-thrust method*, which considers the stability of the arch ring to be dependent on friction and the reactions between the several arch sections or voussoirs. In general, the line-of-thrust method is most applicable to symmetrical arches loaded uniformly over the entire span or subject to symmetrically placed concentrated loads. For such arches, the line of resistance (which is the line connecting the points of application of the resultant forces transmitted to each voussoir) is required to fall within the middle third of the arch section so that neither the intrados nor extrados of the arch will be in tension (*see Fig. 12-17* for arch terminology).

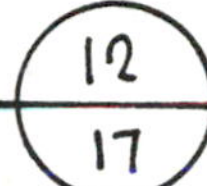

Arch terminology (see the Glossary, Appendix A). (*From Brick Institute of America*, Technical Note 31A, *BIA, McLean, Va.*)

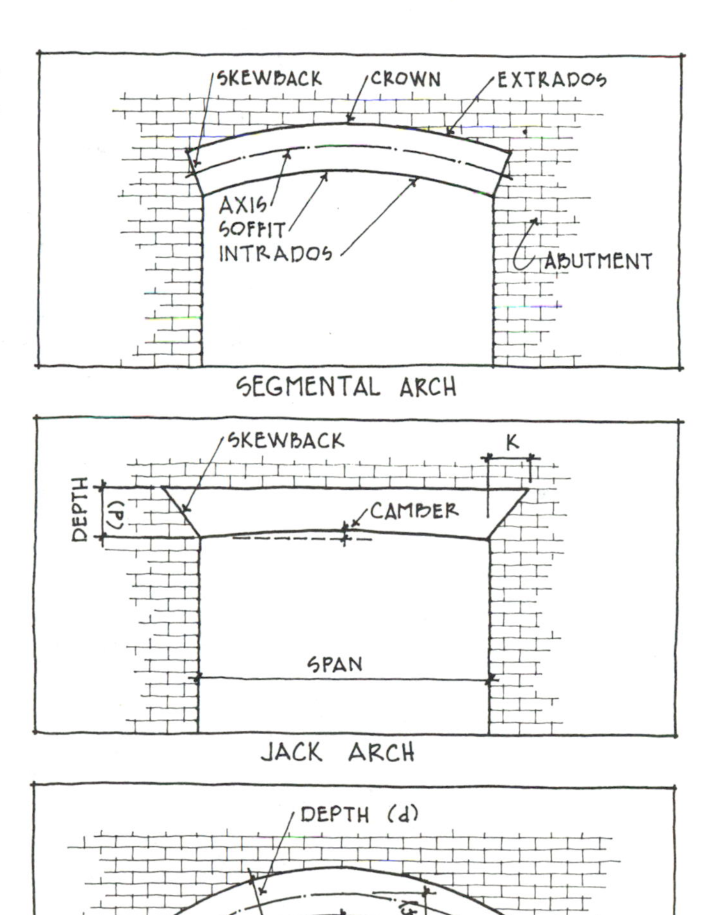

12.7.2 Graphic Analysis

The simplest and most widely used line-of-thrust method is based on th hypothesis of "least crown thrust," which assumes that the true line o resistance of an arch is that for which the thrust at the crown is the leas possible consistent with equilibrium. This principle can be applied by stati methods if the external forces acting on the arch are known and the point o application and direction of crown thrust are assumed. Normally, th direction of the crown thrust is assumed as horizontal and its point o application as the upper extremity of the middle one-third of the section (i.e. two-thirds the arch depth from the intrados). This assumption has bee proven reasonable for symmetrical arches symmetrically loaded, but is no applicable to unsymmetrical or partially distributed uniform loads.

With these assumptions, the forces acting on each section of an arc may be determined by analytical or graphic methods. The first step in th procedure is to determine the joint of rupture. This is the joint for which th tendency of the arch to open at the extrados is the greatest and whic therefore requires the greatest crown thrust applied at the upper extremit of the middle third to prevent the joint from opening. At this joint, the line c resistance of the arch will fall on the lower extremity of the middle third o the section. For minor segmental arches, the joint of rupture is ordinaril assumed to be the skewback of the arch. (For major arches with highe rise/span ratios, this will not be true.) Based on the joint of rupture at th skewback and the hypothesis of least crown thrust, the magnitude an direction of the reaction at the skewback may be determined graphicall (*see Fig. 12-18*).

In this analysis, only one-half of the arch is considered since it i symmetrical and uniformly loaded over the entire span. *Figure 12-18(* shows the external forces acting on the arch section. For equilibrium, th

12/18 Graphic arch analysis. (*From Brick Institute of America,* Technical Note 31, *BIA, McLean, Va.*)

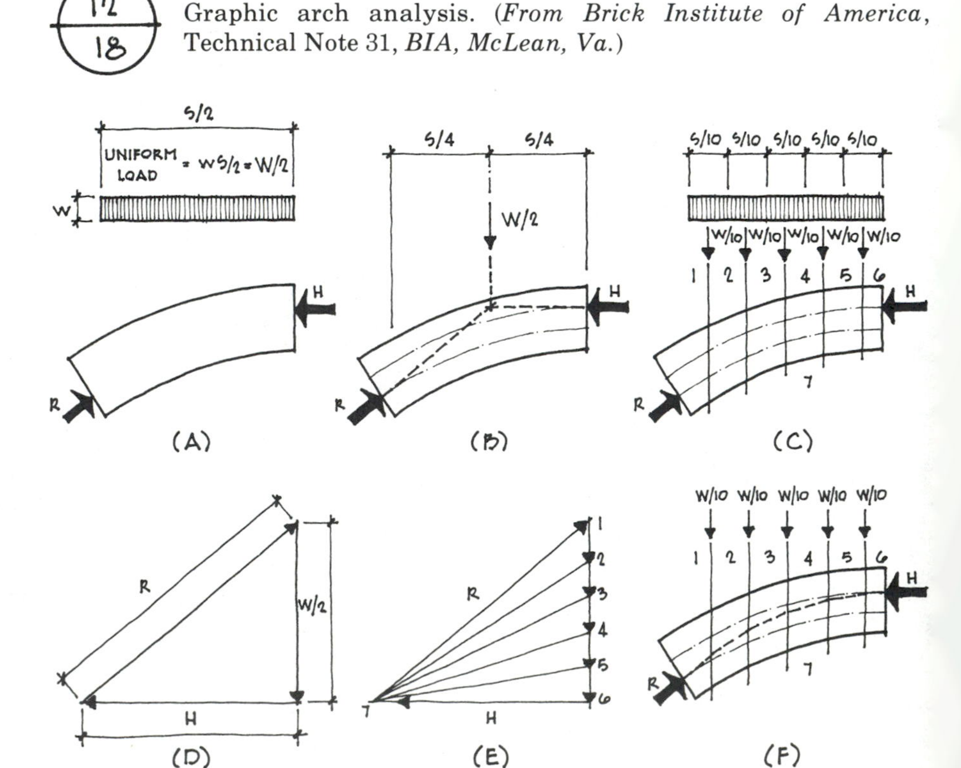

lines of action of those three forces ($W/2$, H, and R) must intersect at one point as shown in *Fig. 12-18(B)*. Since the crown thrust (H) is assumed to act horizontally, this determines the direction of the resisting force (R). The magnitude of the resistance may be determined by constructing a force diagram as indicated in *Fig. 12-18(D)*. The arch is divided into voussoirs and the uniform load transformed into equivalent concentrated loads acting on each section [*see Fig. 12-18(C)*]. Starting at any convenient point (in this example between the reaction and the first load segment past the skewback) numbers are placed between each pair of forces, so that each force can subsequently be identified by a number (i.e., 1–2, 5–6, 7–1, and so on). The side of the force diagram which represents $W/2$ [*Fig. 12-18(D)*] is divided into the same number of equivalent loads, and the same numbers previously used for identification are placed as shown in *Fig. 12-18(E)* to identify the forces in the new force diagram. Thus, the line 7–1 is the skewback reaction, 6–7 the horizontal thrust, and so on. From the intersection of H and R (7–1 and 6–7) a line is drawn to each intermediate point on the leg representing $W/2$.

The equilibrium polygon may now be drawn. First extend the line of reaction until it intersects the line of action of 1–2 [*see Fig. 12-18(F)*]. Through this point, draw a line parallel to the line 7–2 until it intersects the line of action of 2–3. Through this point, a draw a line parallel to the line 7–3, and so on, and complete the polygon in this manner. If the polygon lies completely within the middle third of the arch section, the arch is stable. For a uniformly distributed load, the equilibrium polygon, which coincides with the line of resistance, will normally fall within this section, but for other loading conditions it may not.

The off-center location of the voussoir reactions will produce stresses which differ from the axial stress H/A, where A is the cross-sectional area of the arch ($A = bd$). These stresses are determined by the formula

$$f_m = \frac{H}{bd} \pm \frac{6He}{bd} \tag{12.12}$$

where f_m = maximum compressive stress in the arch, psi
H = horizontal thrust, lb
b = thickness of the arch, in.
d = depth of the arch, in.
e = the perpendicular distance between the arch axis and the line of action of the horizontal thrust

Maximum allowable compressive stresses in an arch are determined on the basis of the compressive strength of the units and mortar, and are governed by the same code requirements as other masonry construction (refer to *Figs. 10-11, 10-19*, and *10-25*).

A number of mathematical formulas have been developed for the design of minor arches. Among the structural considerations are three methods of failure of unreinforced masonry arches: (1) by rotation of one section of the arch about the edge of a joint, (2) by the sliding of one section of the arch on another or on the skewback, and (3) by crushing the masonry.

12.7.3 Rotation

The assumption that the equilibrium polygon lies entirely within the middle third of the arch section precludes the rotation of one section of the arch about the edge of a joint or the development of tensile stresses in either the intrados or extrados. For conditions other than evenly distributed uniform loads where the polygon may fall outside the middle third, however, this method of failure should be considered.

12.7.4 Sliding

The coefficient of friction between the units of a masonry arch is at leas 0.60 without considering the additional resistance to sliding resulting from the bond between the mortar and the masonry units. This corresponds to an angle of friction of approximately 31°. If the angle between the line o resistance and a line perpendicular to the joint between sections is less thar the angle of friction, the arch is stable against sliding. This angle may b determined graphically as shown above. For minor segmental arches, th angle between the line of resistance and a line perpendicular to the joint i greatest at the skewback. This is also true for jack arches if the joints ar radial about a center at the intersection of the planes of the skewbacks However, if the joints are not radial about this center, each joint should b investigated separately for resistance to sliding. This can be most easil accomplished by constructing an equilibrium polygon.

12.7.5 Crushing

A segmental arch is one whose curve is circular but is less than a semicircl The minimum recommended rise for a segmental arch is 1 in. per foot c span. The horizontal thrust developed depends on the span, depth, and ris of the arch.

The graph in *Fig. 12-19* identifies thrust coefficients (H/W) fc segmental arches subject to uniform loads over the entire span. Once th

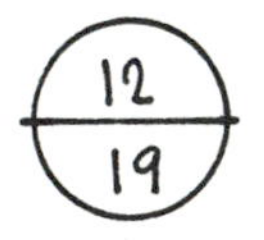

Thrust coefficients for segmental arches. (*From Brick Institute of America*, Technical Note 31A, *BIA, McLean, Va.*)

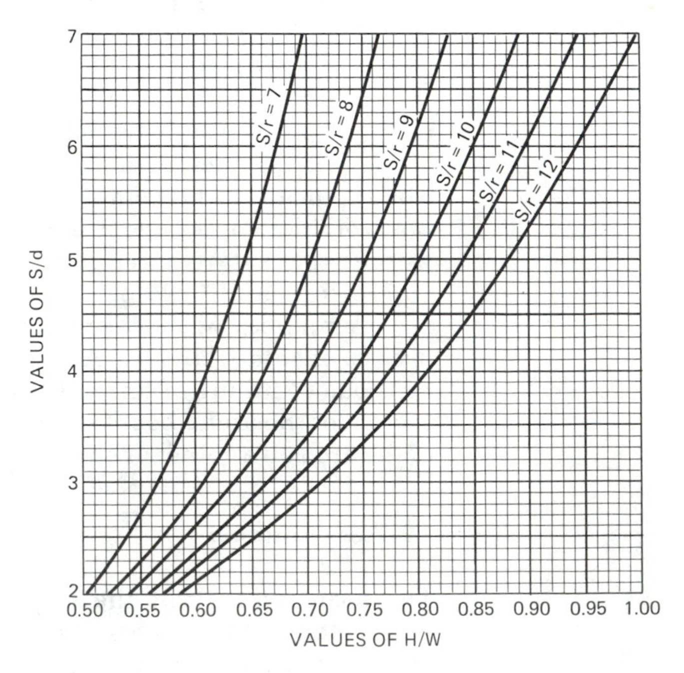

thrust coefficient is determined for a particular arch, the horizontal thrust (H) may be determined as the product of the thrust coefficient and the total load (W). To determine the proper thrust coefficient, first determine the characteristics S/r and S/d of the arch, where (S) is the clear span in feet, (r) is the rise of the soffit in feet, and (d) is the depth of the arch in feet. If the applied load is triangular or concentrated, the same method may be used, but the coefficient (H/W) is increased by one-third for triangular loading and doubled for concentrated loads.

Once the horizontal thrust has been determined, the *maximum compressive stress* in the masonry is determined by the formula

$$f_m = \frac{2H}{bd} \tag{12.13}$$

This value is twice the axial compressive stress on the arch due to the load H because the horizontal thrust is located at the third point of the arch depth.

The common rule for *jack arches* is to provide a skewback (K measured horizontally; *see Fig. 12-17*) of $\frac{1}{2}$ in. per foot of span for each 4 in. of arch depth. Jack arches are commonly constructed in depths of 8 and 12 in. with a camber of $\frac{1}{8}$ in. per foot of span. To determine the *horizontal thrust* at the spring line for jack arches, the following formulas may be used:

For uniform loading over full span:

$$H = \frac{3WS}{8d} \tag{12.14}$$

For triangular loading over full span:

$$H = \frac{WS}{2d} \tag{12.15}$$

Maximum compressive stress may be determined by the formulas

$$f_m = \frac{2H}{6d} \tag{12.16}$$

For uniform loading over full span:

$$f_m = \frac{3WS}{4bd} \tag{12.17}$$

or for triangular loading:

$$f_m = \frac{WS}{6d} \tag{12.18}$$

12.7.6 Thrust Resistance

The horizontal thrust resistance developed by an arch is provided by the adjacent mass of the masonry wall. Where the area of the adjacent wall is substantial, thrust is not generally a problem. However, at corners and openings where the resisting mass is limited, it may be necessary to check the resistance of the wall to this horizontal force. The diagram in *Fig. 12-20* shows how the resistance is calculated. It is assumed that the arch thrust attempts to move a volume of masonry enclosed by the boundary lines *ABCD*. For calculating purposes, the area *CDEF* is equivalent in resistance. The thrust is thus acting against two planes of resistance, *CF* and *DE*. Resistance is determined by the formula

$$H_1 = v_m n x t \tag{12.19}$$

where H_1 = resisting thrust, lb.
v_m = allowable shearing stress in the masonry wall, psi
n = number of resisting shear planes
x = distance from the center of the skewback to the end of the wall, in.
t = wall thickness, in.

By using this principle, the minimum distance from a corner or opening at which an arch may be located is easily determined. To do so, write the formula to solve for x, substituting actual arch thrust for resisting thrust

$$x = \frac{H}{v_m n t}$$

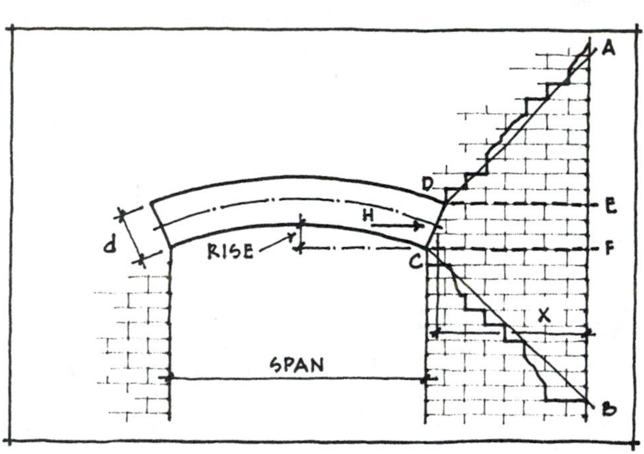

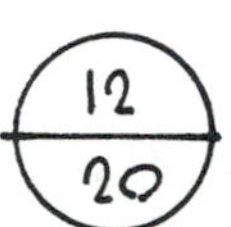

Horizontal thrust resistance. (*From Brick Institute of America*, Technical Note 31A, *BIA, McLean, Va.*)

12.7.7 Design Tables for Semicircular Arches

The semicircular and segmental are perhaps the most popular and widely used arch forms for contemporary design and construction. The following design tables have been developed by the BIA to simplify the analysis of these arches for normal loading conditions.

Since masonry arches are usually integral with the wall rather than free-standing, basic design assumptions can be made which assist in design analysis. For instance, the spring line is assumed to be located on a horizontal line one-fourth of the span length above the horizontal axis. The arches are assumed to be fully restrained at the spring line and that portion of a semicircular arch which is above this line is analyzed in a manner similar to that for parabolic arches. In the determination of the capacity for uniform loads, the limiting factor is the compressive strength of the masonry. In determining the capacity for concentrated loads, the limiting factors are bending at the centerline of span, shear at the spring line, and maximum compressive stress. Tensile stresses are not permitted to develop at midspan. Since axial forces develop in the arch ring from the concentrated and uniform loads, interaction formulas were developed for each loading condition. These formulas combine the axial stresses with the bending stresses.

In all of the formulas given here for use with the BIA design tables, the arch depth and thickness are measured in inches, the span in feet, and the following loading conditions were considered in analyzing a semicircular arch.

For *uniform loads*, the tables in *Figs. 12-23* through *12-26* give the allowable uniform loads occurring over the entire span length for a 1-in.-thick arch ring. *Figure 12-21* shows design requirements and limitations.

The table in *Fig. 12-27* gives allowable *concentrated loads* occurring at the center line of span for a 1-in.-thick arch ring, and *Fig. 12-22* shows design requirements and limitations.

When uniform loads are combined with concentrated loads, the concentrated load capacity of the arch ring increases. This additional capacity is due to the compressive stress from the uniform load equalizing

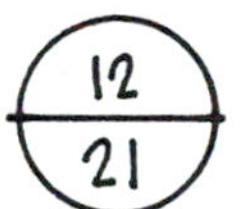

Uniform load on a semicircular arch. (*From Brick Institute of America*, Technical Note 31C, *BIA, McLean, Va.*)

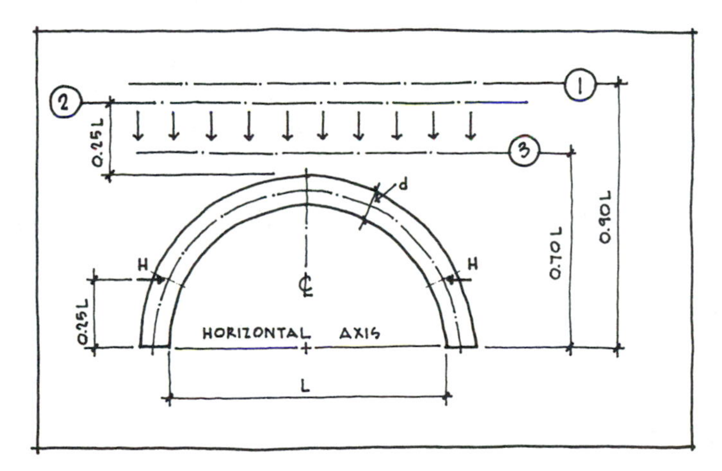

(1) Uniform loads occurring between lines 1 and 3 (0.90L and 0.70L) are those provided for in the load tables.

(2) Uniform loads occurring above line 1 may be ignored (at the discretion of the designer) provided arching action occurs in the masonry above the arch ring.

(3) There must be a minimum height of masonry (line 3) equal to 0.70L, above the horizontal axis. No superimposed loads are permitted below this line.

(4) The maximum design height of masonry is 0.25L above the crown for walls higher than line 2.

(5) In all cases, the horizontal thrust (H) must be checked. For a given arch, the horizontal thrust is directly proportional to the uniform load.

(6) The portion of wall that resists the horizontal thrust is assumed to be nonyielding to any lateral movement.

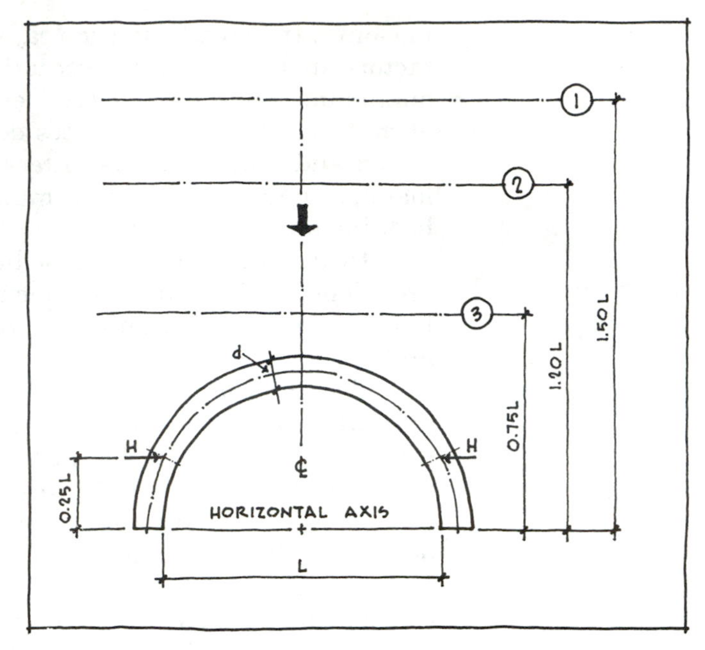

(1) Concentrated loads occurring between lines 2 and 3 (1.20*L* and 0.75*L*) are those provided for in the table.

(2) Concentrated loads occurring between lines 1 and 2 may be divided by the span length (*L*) and considered as equivalent uniform loads.

(3) Concentrated loads occurring above line 1 (1.50*L*) may be ignored (at the discretion of the designer) provided arching action occurs in the masonry above the arch ring.

(4) In all cases, condition (4) for uniform loads (*see Fig. 12-21*) must be used, with the resulting thrusts added to those of the concentrated loads.

(5) There must be a minimum height of masonry (line 3) equal to 0.75*L* above the horizontal axis. No superimposed loads are permitted below this line.

(6) In all cases, the horizontal thrust (*H*) must be checked. For a given arch, the horizontal thrust is directly proportional to the concentrated load.

(7) The proportion of wall which resists the horizontal thrust is assumed to be nonyielding to any lateral movement.

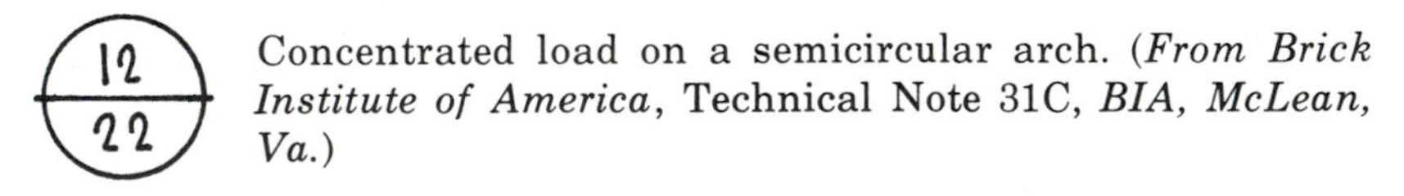

Concentrated load on a semicircular arch. (*From Brick Institute of America*, Technical Note 31C, *BIA, McLean, Va.*)

the tensile bending stress at midspan from the concentrated load (M/S P/A). The additional capacity may be expressed by the formula

$$P^* = \frac{H_{DL}d}{1.34L} \qquad (12.$$

where H_{DL} is the horizontal thrust caused by a uniform dead load. T values of P' and H' in *Fig. 12-28* are the allowable capacitites govern by compression or shear. They should be used only as a check for combin loadings. In all cases, the actual load must be less than P', and less than t allowable load P plus the additional capacity, P^*. The total horizon thrust must be checked and should be less than the maximum allowable a uniform load.

L	d = 3.5 in.		d = 7.5 in.		d = 11.5 in.		d = 15.5 in.	
	W	H	W	H	W	H	W	H
2	810	697	1520	1496	2071	2295	2509	3094
4	424	686	857	1489	1230	2289	1556	3089
6	277	659	591	1474	870	2277	1124	3078
8	193	611	444	1447	669	2257	875	3061
10	134	533	349	1406	538	2227	713	3036
12	86	420	280	1347	445	2185	597	3001
14	—	—	226	1268	374	2128	510	2956
16	—	—	180	1164	317	2055	440	2898
18	—	—	141	1034	269	1964	383	2825
20	—	—	105	873	228	1852	335	2737

Allowable uniform loads on semicircular arches: $f_m = 300$ psi, $t = 1$ in. (*From Brick Institute of America*, Technical Note 31C, *BIA, McLean, Va.*)

L	d = 3.5 in.		d = 7.5 in.		d = 11.5 in.		d = 15.5 in.	
	W	H	W	H	W	H	W	H
2	1082	930	2028	1996	2762	3061	3347	4127
4	569	919	1145	1989	1642	3055	2077	4121
6	376	892	792	1973	1164	3043	1501	4111
8	268	844	599	1947	897	3023	1172	4094
10	195	766	474	1906	724	2993	957	4068
12	137	653	385	1847	602	2951	804	4034
14	87	498	317	1767	510	2894	690	3988
16	—	—	261	1664	437	2821	600	3930
18	—	—	212	1533	377	2730	526	3858
20	—	—	170	1373	325	2618	464	3770

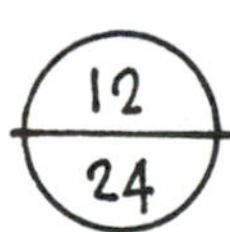

Allowable uniform loads on semicircular arches: $f_m = 400$ psi, $t = 1$ in. (*From Brick Institute of America*, Technical Note 31C, *BIA, McLean, Va.*)

L	d = 3.5 in.		d = 7.5 in.		d = 11.5 in.		d = 15.5 in.	
	W	H	W	H	W	H	W	H
2	1353	1163	2536	2495	3453	3827	4185	5159
4	714	1152	1434	2488	2055	3821	2597	5154
6	475	1125	993	2473	1458	3809	1879	5143
8	343	1077	753	2446	1125	3789	1468	5126
10	255	1000	600	2405	911	3759	1201	5101
12	188	886	491	2347	760	3717	1012	5066
14	131	731	408	2267	647	3660	870	5021
16	80	527	341	2164	558	3587	759	4962
18	—	—	284	2033	485	3496	669	4890
20	—	—	234	1872	423	3384	594	4802

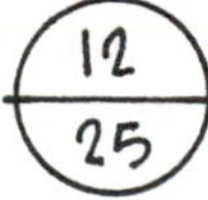

Allowable uniform loads on semicircular arches: $f_m = 500$ psi, $t = 1$ in. (*From Brick Institute of America*, Technical Note 31C, *BIA, McLean, Va.*)

L	d = 3.5 in.		d = 7.5 in.		d = 11.5 in.		d = 15.5 in.	
	W	H	W	H	W	H	W	H
2	1624	1396	3044	2995	4145	4593	5023	6192
4	859	1385	1722	2988	2467	4587	3118	6186
6	573	1359	1194	2973	1752	4575	2257	6175
8	418	1310	908	2946	1353	4555	1765	6158
10	315	1233	725	2905	1097	4525	1445	6133
12	238	1120	596	2846	918	4483	1219	6099
14	174	964	499	2767	783	4426	1050	6053
16	118	760	421	2663	678	4353	918	5995
18	—	—	356	2532	592	4262	812	5922
20	—	—	299	2372	520	4150	723	5834

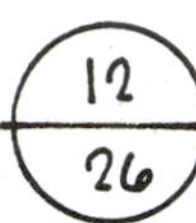

Allowable uniform loads on semicircular arches: $f_m = 600$ psi, $t = 1$ in. (*From Brick Institute of America*, Technical Note 31C, *BIA, McLean, Va.*)

L	f_m = 300 to 600 psi							
	d = 3.5 in.		d = 7.5 in.		d = 11.5 in.		d = 15.5 in.	
	P	H	P	H	P	H	P	H
2	15	12	703	547	1078	841	1451	1131
4	9	9	126	101	1075	841	1449	1131
6	13	16	38	35	418	332	1445	1132
8	20	25	40	41	97	86	967	764
10	28	36	48	52	88	84	199	172
12	38	49	58	67	93	95	162	150
14	48	65	70	83	104	111	160	156
16	61	83	84	102	117	130	168	171
18	75	103	100	124	133	152	181	191
20	90	125	117	148	151	177	198	215

Note: Values may be linearly interpolated except where horizontal lines occur. At these lines, the allowable load is

$$\frac{0.241(L + 0.083d)^3 + 0.134(L + 0.083d)^2 d}{1.34(L + 0.083d) - 0.778d}$$

or the value above the line, whichever is smaller. The horizontal thrust is

$$0.778P + 0.134(L + 0.083d)$$

or the value above the line, whichever is smaller.

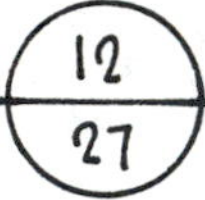

Allowable concentrated loads on semicircular arches ($t = 1$ in.). (*From Brick Institute of America*, Technical Note 31C, *BIA, McLean, Va.*)

Any segmental arch where f/L' is greater than 0.29 but less than 0.50 considered as an equivalent semicircular arch, as shown in *Fig. 12-2* Twice the radius is the equivalent L for use with the tables.

The method of analysis presented here for the design of semicircul and segmental arches is a simplified but conservative approach to complex structural problem. An analysis for all possible assumptions a loading conditions is beyond the scope of this book, but most loadi

	f_m = 300 psi				f_m = 400 psi				f_m = 500 psi				f_m = 600 psi				f_m = 300 to 600 psi			
	d = 3.5		d = 7.5		d = 3.5		d = 7.5		d = 3.5		d = 7.5		d = 3.5		d = 7.5		d = 11.5 in.		d = 15.5 in.	
L	P'	H'	P'	H'	P'	H'	P'	H'	P'	H'	P'	H'	P'	H'	P'	H'	P'	H'	P'	H'
2	328	256	703	547	328	256	703	547	328	256	703	547	328	256	703	547	1078	841	1451	1131
4	326	256	701	547	326	256	701	547	326	256	701	547	326	256	701	547	1075	841	1449	1131
6	324	256	698	548	324	256	698	548	324	256	698	548	324	256	698	548	1072	842	1445	1132
8	273	203	695	549	321	257	695	549	321	257	695	549	321	257	695	549	1065	842	1442	1133
10	235	170	691	550	310	227	691	550	316	257	691	550	316	257	691	550	1060	842	1440	1134
12	211	143	685	551	275	194	685	551	310	258	685	551	310	258	685	551	1055	843	1436	1135
14	195	125	685	504	252	169	676	552	304	258	676	552	304	258	676	552	1050	843	1431	1136
16	187	110	630	453	237	149	670	553	287	188	670	553	297	258	670	553	1045	843	1422	1137
18	183	98	587	410	228	133	661	554	273	169	661	554	289	259	661	554	1040	842	1411	1137
20	184	88	555	374	225	120	723	505	266	151	652	555	280	259	652	555	1025	842	1400	1138

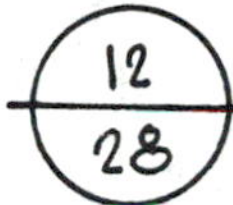

Maximum concentrated load on semicircular arches under combined loading conditions (t = 1 in.). (*From Brick Institute of America*, Technical Note 31C, *BIA, McLean, Va.*)

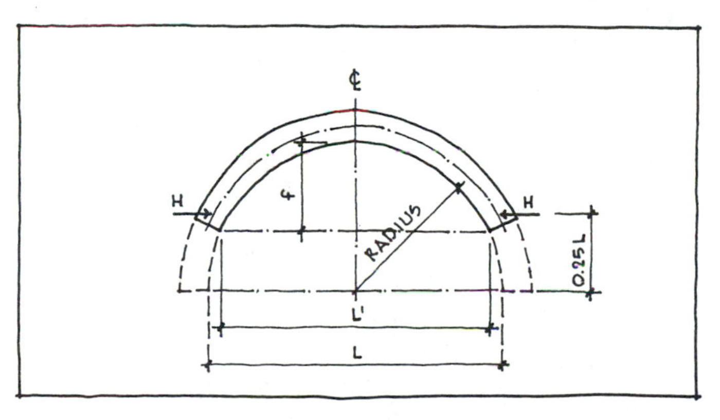

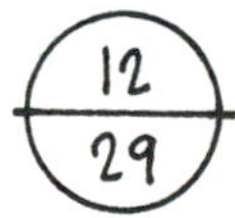

Segmental arch equivalent. (*From Brick Institute of America*, Technical Note 31C, *BIA, McLean, Va.*)

conditions encountered will be similar to those in *Figs. 12-21* and *12-22*. To load an arch asymmetrically defeats the benefits of its natural load-carrying structure and induces bending stresses which may cause failure. If openings in a masonry wall will be loaded asymmetrically, or if design assumptions and limitations do not comply with those given above, consideration should be given to other structural forms such as lintels or beams. If conditions for arched openings are other than those covered here, special engineering analysis should be made.

12.7.8 A Design Example

The following sample problem is taken from BIA *Technical Note 31C*.

Design an arch to meet the requirements shown in *Fig. 12-30*. The arch is semicircular; the horizontal axis is 6 ft above the base; the span (L) is 10 ft; the arch ring depth (d) is 12 in. ($11\frac{1}{2}$ in. actual); and the nominal wall thickness (t) is 8 in. ($7\frac{1}{2}$ in. actual). A beam reaction of 5000 lb is located at the centerline of the span and 17 ft above the base. The uniform load

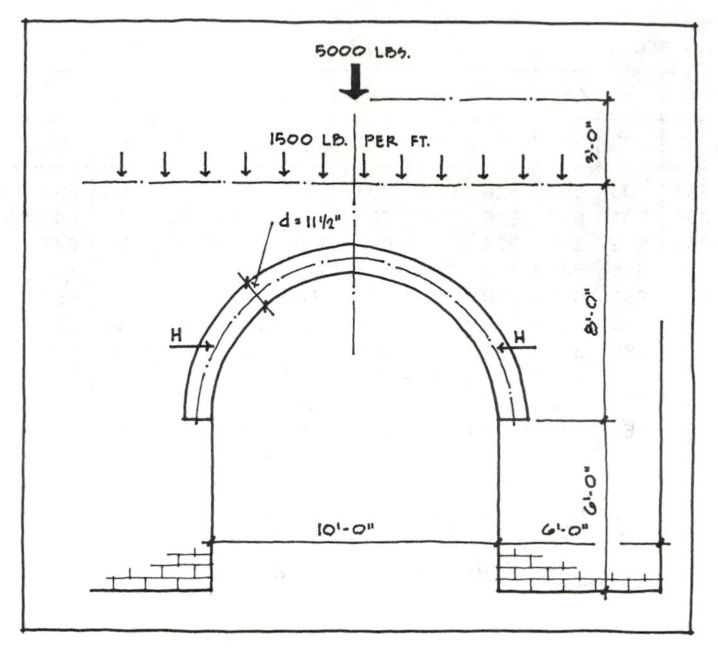

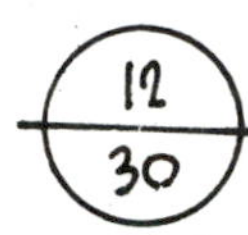

Arch loading for a sample problem. (*From Brick Institute of America*, Technical Note 31C, *BIA*, *McLean, Va.*)

consists of 1000 lb/ft dead load and 500 lb/ft live load occurring 14 ft abov the base. Assume that f_m = 400 psi and the brick masonry weighs 10 lb/s ft per 1 in. of thickness.

Uniform load:

Wall dead load = 0.25(10)(10)(7.5) =	188 lb/ft
Arch dead load = 1(10)(7.5) =	75
Floor dead load =	1000
Floor life load =	500
Total uniform load =	1763 lb/ft

Concentrated load: 5000 lb

$$\frac{DL}{TL} = \frac{1263}{1763} = 0.72$$

All of the following calculations will be with 1 in. of wall thickness; actu t = 7.5 in.; f_m = 400 psi; d = 11.5 in.; and L = 10 ft.

Uniform load. The uniform load occurs at $\frac{8}{10} = 0.8L$. From *F 12-24*,

$$W = 724 \text{ lb/ft}$$

$$H = 2993 \text{ lb}$$

$$W\text{(actual)} = \frac{1763}{7.5} = 235 \text{ lb/ft} < 724 \qquad \text{OK}$$

$$H\text{(actual)} = \frac{235}{724}(2993) = 970 \text{ lb}$$

Concentrated load. The concentrated load occurs at $\frac{8+3}{10} = 1.1L$. From *Fig. 12-27,*

$$P = 88 \text{ lb}$$

$$H = 84 \text{ lb}$$

$$P\text{(actual)} = \frac{5000}{7.5} = 677 \text{ lb} > 88 \qquad \text{NG}$$

However, since there is combined loading, advantage may be taken of the increased capacity due to this condition. Therefore,

$$P^* = \frac{H_{DL}d}{1.34L}$$

$$= \frac{970(0.72)(11.5)}{1.34(10)} = 600 \text{ lb}$$

$$P\text{(allowable)} = 88 + 600 = 688 \text{ lb}$$

$$667 < 688 \qquad \text{OK}$$

From *Fig. 12-28*, $P' = 1060$ lb.

$$667 < 1060 \qquad \text{OK}$$

$$H\text{(actual)} = \frac{667}{88(84)} = 636 \text{ lb}$$

Horizontal thrust:

$$H\text{(total)} = 970 + 636 = 1606 < 2993 \qquad \text{OK}$$

At this point, the wall shear caused by the horizontal thrust at the spring line should be checked. Assume that $v_m = 40$ psi, and $n = 2$.

$$x = \frac{H}{v_m nt} = \frac{1606(7.5)}{40(2)(7.5)} = 20.5 \text{ in.} < 6 \text{ ft} \qquad \text{OK}$$

The overturning moment of the support due to horizontal thrust should be checked next. In this example, the horizontal thrust is calculated 1606(7.5)(8.5) = 102,000 ft.-lb. Resistance to overturning is a function of the overall axial load, wall shape, and reinforcement (if any). This is a separate analysis that should be performed after considering the total loading conditions on the entire structure.

12.7.9 Major Arch Design

Major arches are those with spans greater than 6 ft or rise-to-span ratios of more than 0.15 (*see Figs. 12-31* and *12-32*). The design of these elements is a much more complicated structural problem than minor arches because of increased loading and span conditions. Leontovich's book, *Frames and Arches*, gives formulas for arches with rise-to-span ratios (f/L) ranging from 0.0 to 0.6. These are straightforward equations by which redundant moments and forces in arched members may be determined. The equations are based on a horizontal and vertical grid coordinate system originating at the intersection of the arch axis and the left skewback. Each set of equations depends on the conditions of loading. Moments, shears, and axial thrusts are determined at various increments of the span. No tensile stresses should be permitted in unreinforced masonry arches under static loading conditions. For a detailed analysis of major structural arch design see Leontovich, *Frames and Arches*. For relatively high-rise, constant section arches, his Method A of Section 22 applies.

12.7.10 Arch Construction

Arches are constructed with temporary shoring or centering to carry the dead load of the material and other applied loads until the arch itself has gained sufficient strength. Temporary bracing should never be removed until it is certain that the masonry is capable of carrying all imposed loads. For unreinforced masonry arches, it is generally recommended that centering remain in place for 7 days after the completion of the arch. Wher

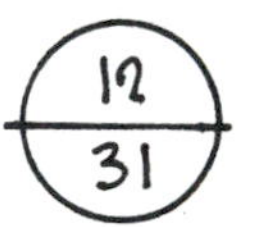

First National Bank of Fayetteville, Arkansas. Polk Stanley Gra architects. (*Photo courtesy BIA.*)

United Bank Tower in Austin, Texas. Zapalac and Associates, architects. (*Photo courtesy BIA.*)

loads are relatively light, or where the majority of the wall load will not be applied until some later date, it may be possible to remove the centering earlier.

Masonry arches may be built of special brick or stone shapes to obtain mortar joints of constant thickness, or of standard brick units with joint thickness varied to obtain the required curvature. The method selected should be determined by the arch dimensions and by the desired appearance.

It is especially important in a structural member such as an arch that all mortar joints be completely filled. Brick arches are usually built so that units at the crown will be laid in soldier bond or rowlock header bond (*see Fig. 12-16*). Under many circumstances, it is difficult to lay units in soldier bond and still obtain full joints. This is especially true where the curvature of the arch is of short radius with mortar joints of varying thickness. In such cases, the use of two or more rings of rowlock headers is recommended. In addition to facilitating better jointing, rowlock headers provide a bond through the wall to strengthen the arch.

13

RETAINING WALLS, BELOW-GRADE WALLS, AND POOLS

Masonry walls may be quite effectively used to retain earth in landscap construction, below-grade building structures, and even in swimmin pools.

13.1 RETAINING WALLS

Retaining walls are built to withstand lateral earth pressure in excavatio and backfill operations. They must be safe against overturning, but th pressure under the toe of the footing should not exceed the bearing strengt of the soil. The friction between the footing and the soil, plus the pressure c any earth in front of the wall, must keep the member from sliding forwar

Because retaining walls are often used in landscape applications an may not enclose habitable space, attention to design and detailing is ofte cursory. However, the walls are exposed to extremes of weather, are i contact with earth and often saturated with water, and must resi considerable forces. Such severity of use and exposure demands caref attention not only to design and details, but to materials and workmansh as well. Some of the primary considerations should be (1) a proper cap c coping to prevent water collecting or standing on top of the wall, (2) waterproof coating on the back of the wall to prevent saturating th masonry, (3) permeable backfill behind the wall to collect water and preve soil saturation and increased hydrostatic pressure, (4) weep holes or drai lines to drain moisture, and (5) expansion joints to permit longitudin thermal and moisture movement.

There are four basic types of masonry retaining walls (*see Fig. 13-1*). Reinforced cantilever walls (B) offer the most economical design, and are most commonly used. The vertical stem is reinforced to resist tensile stresses. The concrete footing anchors the stem and resists overturning and sliding due to both vertical and lateral forces.

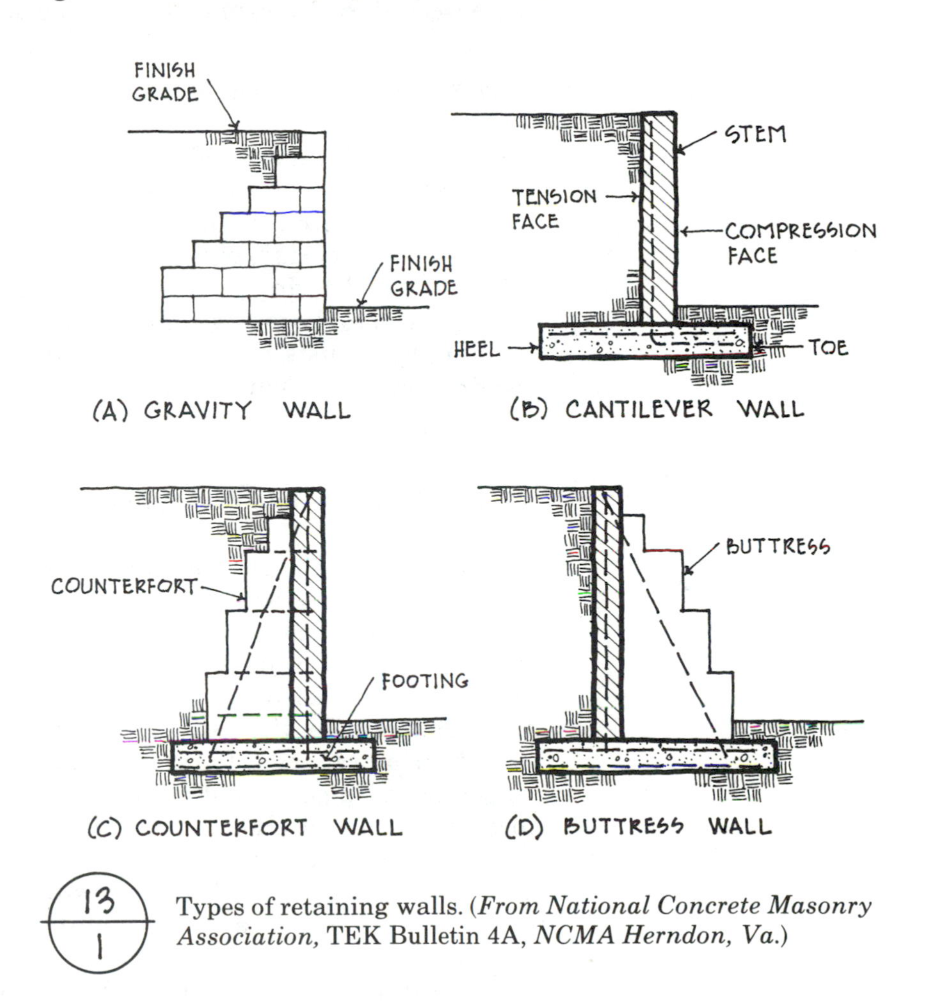

Types of retaining walls. (*From National Concrete Masonry Association,* TEK Bulletin 4A, *NCMA Herndon, Va.*)

13.1.1 Lateral Earth Pressure

The magnitude and direction of this primary force acting on the wall is dependent on the height and shape of the surface, and on the nature and physical properties of the backfill.

The simplest way of determining lateral earth pressure is the *equivalent fluid method.* This method assumes that the retained earth will act as a fluid, and the wall is designed to withstand the pressure of a liquid assumed to exert the same pressure as the actual backfill material. Assumed equivalent-fluid unit weights vary with the nature of the soil in the backfill. The *Uniform Building Code* permits walls retaining drained earth to be designed for a pressure equivalent to that exerted by a fluid weighing not less than 30 lb/cu ft and having a depth equal to that of the retained earth. The *National Building Code* requires equivalent-fluid weights ranging from 25 lb/cu ft for well-drained granular soils of average or high permeability; to 40 lb/cu ft for poorly drained soils and clayey or silty soils of low permeability; and 62.4 lb/cu ft for retained soils that are not drained.

13.1.2 Surcharge

Additional loads are created by operating automobiles, trucks, or construction equipment on the soil surface behind a retaining wall. If activities of this nature are anticipated, the design must make allowance for the increased lateral pressures that will be imposed on the wall.

13.1.3 Overturning and Sliding

Retaining walls must safely resist overturning and sliding forces induced by the retained earth. Unless otherwise required by code, the factor of safety against overturning should not be less than 2.0, and against sliding 1.5. In addition, the bearing pressure under the footing should not exceed the allowable soil bearing pressure. In the absence of controlled tests substantiating the actual bearing capacity of the soil, each of the four major model building codes lists allowable pressures for different types of soil. Local requirements may vary slightly and should be checked to assure design conformance.

13.1.4 Drainage and Waterproofing

Failure to drain the backfill area causes a buildup of hydrostatic pressure which can quickly become critical if rainfall is heavy. In mild climates weep holes at the base of the wall should be provided at 5- to 10-ft intervals In areas where precipitation is heavy or where poor drainage conditions exist, prolonged seepage through weep holes can cause the soil in front o the wall and under the toe of the footing to become saturated and lose som of its bearing capacity. In these instances, a continuous longitudinal drai of perforated pipe should be placed near the base with discharge areas located beyond the ends of the wall (*see Fig. 13-2*).

Industry experts recommend that backfill against brick retainin walls from the top of the footing to within 12 in. of finished grade should b coarse gravel, 2 ft wide, and running the entire length of the wall. T prevent the infiltration of fine fill material or top soil, a layer of 50-lb roofin felt is laid along the top of this course. Weep holes or drain lines at th bottom of the wall to relieve moisture buildup in the gravel fill should exten the full length of the wall.

Waterproofing requirements for the back face of a retaining wall wi depend on the climate, soil conditions, and type of masonry units use Seepage through a brick wall can cause efflorescence if soluble salts ar present, but a parge coat or waterproof membrane will prevent this wate movement. Walls of porous concrete units should invariably receiv waterproof backing because of the excessive expansion and contractio that accompanies variable moisture content. In climates subject to freezin a waterproof membrane can prevent the potentially destructive action c freeze–thaw cycles when moisture is present in the units.

13.1.5 Expansion Joints

The number and location of expansion joints will depend to a certain exte on local conditions of climate and exposure. Joints should always b provided at wall offsets. In straight walls, spacing should not exceed 20 t 30 ft. Joints should be designed with a key for lateral stability, but still allo

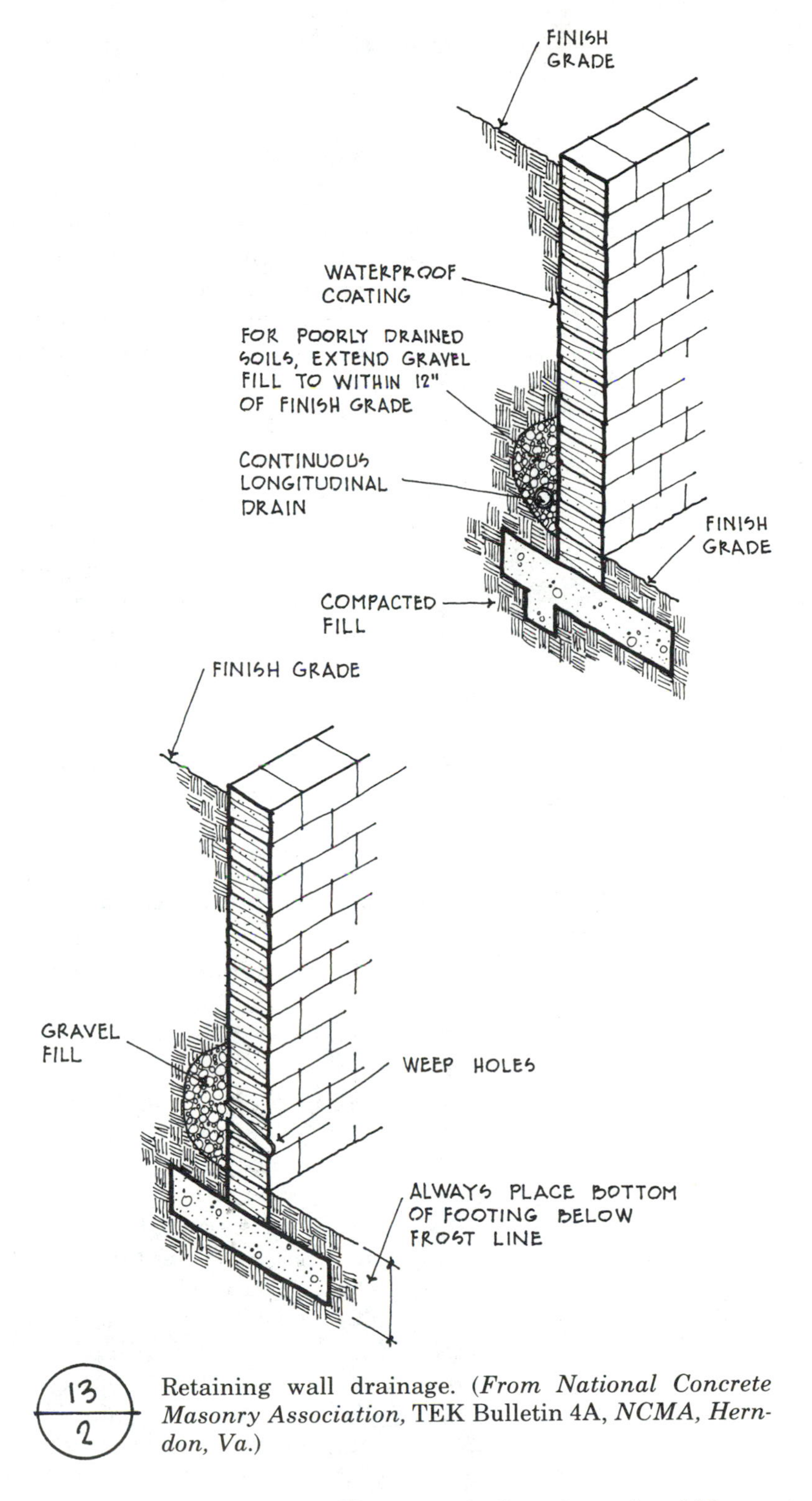

Retaining wall drainage. (*From National Concrete Masonry Association,* TEK Bulletin 4A, *NCMA, Herndon, Va.*)

for longitudinal movement (see Chapter 14). Openings should be protected at the back to prevent clogging with backfill material.

13.1.6 Footings

Concrete footings for retaining walls should be placed on firm undisturbed soil. In areas subject to freezing, they should also be placed below the frost line to avoid heave and possible damage to the wall. If the soil under the footing consists of soft or silty clay, it is advisable to place 4 to 6 in. of well-compacted sand before pouring the concrete.

13.1.7 Materials

Brick masonry for retaining walls should be ASTM C216, Grade SW, with a minimum strength of 5000 psi. Hollow concrete units should be ASTM C90 normal weight, with an oven-dry density of 125 lb/cu ft or more. Mortar should comply with the requirements of ASTM C270, Type M, and grout aggregate should be fine sand. (Mortar and grout should set for 7 days before backfilling.) Concrete for footings should have a minimum compressive strength of 2000 psi.

13.1.8 Concrete Masonry Cantilever Walls

Reinforced CMU cantilever retaining walls are designed to resist overturning and sliding with resultant forces that fall within the middle third of the footing. Many design tables have been developed to simplify selection of wall dimensions and steel reinforcing in CMU retaining walls, including those by the NCMA and the Masonry Institute of America. Graphic examples of design parameters for 4- to 10-ft-high walls are given in Newman's *Standard Structural Details for Building Construction*. The tables and drawings in *Fig. 13-3* represent the most commonly recommended solutions.

13.1.9 Brick Cantilever Walls

Fig. 13-4 shows three different methods of locating vertical reinforcement in a brick masonry cantilever retaining wall. Each offers certain advantage depending on wall thickness, bar spacing, and unit type. Where 8-in. hollow units are available, as shown in part (B), they can often be less expensive than a double-wythe wall. If only standard units are available, grout pockets may be used (A). A double-wythe grouted cavity wall, however offers greatest flexibility because bar spacing is not limited by the fixed dimensions of the units (C).

The BIA recommends horizontal truss or ladder-type reinforcement at a maximum of 16 in. on center vertically. In grouted cavity walls, one No. bar at 20 in. on center vertically may be substituted.

The table in *Fig. 13-5* was developed for preliminary design of brick walls with a maximum height of 6 ft. The following assumptions are used and materials must meet these requirements: (1) brick strength must be in excess of 6000 psi in compression; (2) steel design tensile strength must be 20,000 psi; (3) no surcharge is permitted; and (4) wall thickness is 10 in. (9 in. actual thickness) with reinforcing steel located at the center line of the wall.

For significantly taller walls (6 to 20 ft), the details and tables in *Fig 13-6* and *13-7* are used. Grout pockets for these walls should generally be spaced 4 ft on center except when adjacent to expansion joints. Expansion joints (at wall offsets and every 30 ft maximum) should be centered between pockets spaced no more than 3 ft apart.

The bottom of the footing must be below the frost line and at a

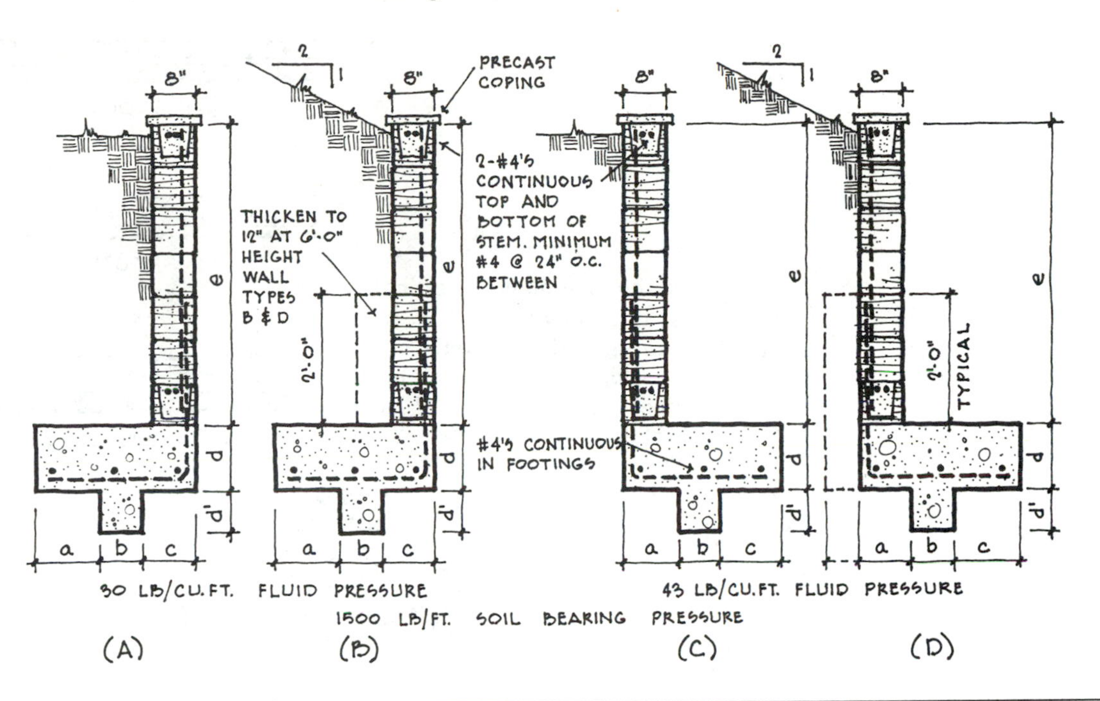

	Wall type			
	A	B	C	D
Maximum stem height, *e*, 4′-0″				
Footing dimensions				
a	1′-4″	1′-2″	1′-6″	1′-0″
b	8″	1′-0″	8″	1′-0″
c	6″	1′-0″	6″	1′-2″
d/d′	12″/12″	12″/12″	12″/8″	12″/12″
Vertical stem reinforcing	No. 4′s at 24″	No. 4′s at 16″	No. 4′s at 16″	No. 4′s at 16″
dowels	No. 4′s at 24″	No 4′s at 16″	No. 4′s at 16″	No. 4′s at 16″
Maximum stem height, *e*, 6′0″				
Footing dimensions				
a	1′-10″	3′-8″	2′-0″	2′-0″
b	12″	12″	8″	1′-0″
c	1′-0″	6″	10″	1′-0″
d/d′	12″/12″	1′-4″/1′-0″	12″/8″	12″/12″
Vertical stem reinforcing	No. 5′s at 24″	No. 5′s at 16″	No. 4′s at 16″	No. 5′s at 16″
dowels	No. 5′s at 16″	No. 5′s at 8″	No. 4′s at 16″	No. 5′s at 16″

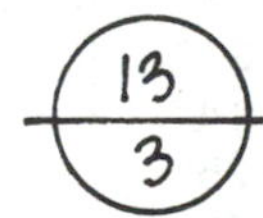

Design of CMU cantilever retaining walls.

elevation where the soil is of sufficient strength to withstand the toe pressure shown in the tables. If the footing elevation must vary, steps must be located adjacent to a grout pocket with the pocket extending to the top of the lower footing. Once the elevation for the bottom of the footing has been established, the $(H + D)$ dimension to finished grade on the high side is known, and the H dimension for entering the design tables may be determined.

All brick retaining walls should be laid in running bond pattern with masonry headers or metal ties every sixth course.

Methods of placing vertical reinforcement in brick retaining walls. (*From Brick Institute of America*, Technical Note 17N, *BIA*, McLean, Va.)

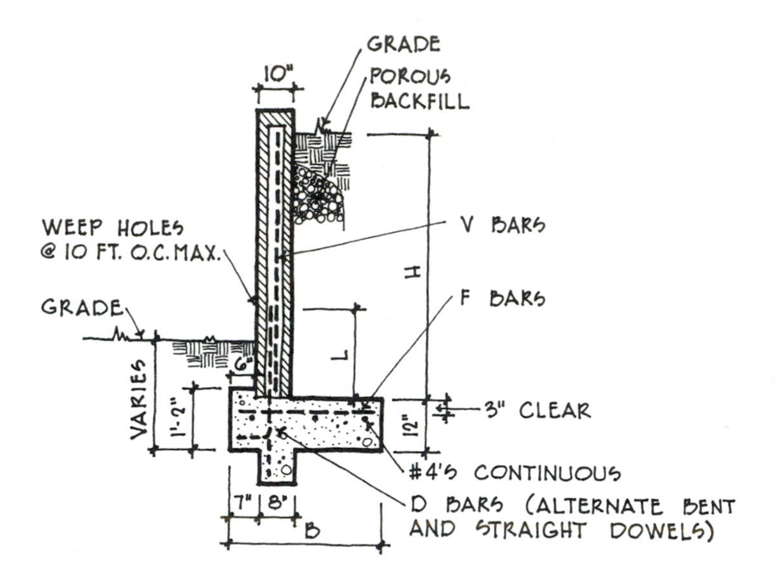

H	*B*	*L*	*D-bars*	*V-bars*	*F-bars*
2′-0″	1′-9″	1′-10″	No. 3 at 40″	—	No. 3 at 40″
2′-6″	1′-9″	2′-4″	No. 3 at 40″	—	No. 3 at 40″
3′-0″	2′-0″	2′-10″	No. 3 at 40″	—	No. 3 at 40″
3′-6″	2′-0″	3′-4″	No. 3 at 40″	—	No. 3 at 40″
4′-0″	2′-4″	1′-4″	No. 3 at 27″	No. 3 at 27″	No. 3 at 27″
			No. 4 at 40″	No. 3 at 40″	No. 3 at 40″
4′-6″	2′-8″	1′-6″	No. 3 at 19″	No. 3 at 38″	No. 3 at 19″
			No. 4 at 35″	No. 3 at 35″	No. 3 at 35″
5′-0″	3′-0″	1′-8″	No. 3 at 14″	No. 3 at 28″	No. 3 at 14″
			No. 4 at 25″	No. 3 at 25″	No. 3 at 25″
			No. 5 at 40″	No. 4 at 40″	No. 4 at 40″
5′-6″	3′-3″	1′-10″	No. 3 at 11″	No. 3 at 22″	No. 3 at 11″
			No. 4 at 20″	No. 4 at 40″	No. 3 at 20″
			No. 5 at 31″	No. 4 at 31″	No. 4 at 31″
6′-0″	3′-6″	2′-0″	No. 3 at 8″	No. 3 at 16″	No. 3 at 8″
			No. 4 at 14″	No. 4 at 28″	No. 3 at 14″
			No. 5 at 20″	No. 5 at 40″	No. 4 at 20″

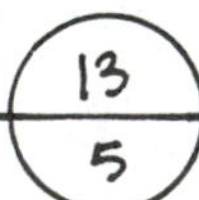

Simplified design of brick cantilever retaining walls and footings. (*From Brick Institute of America,* Technical Note 17N, *BIA, McLean, Va.*)

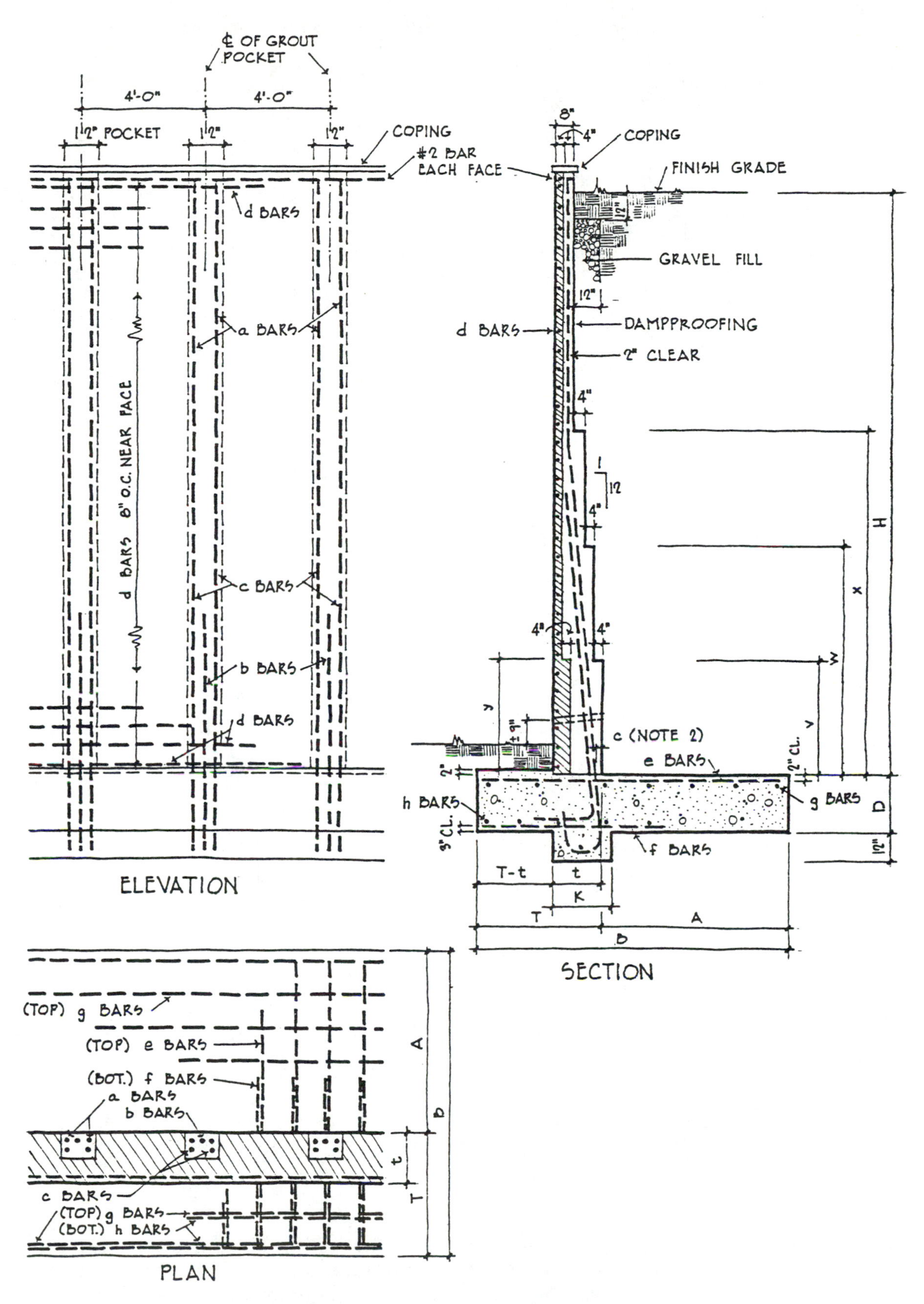

Brick cantilever retaining wall design example. (*From Brick Institute of America,* Technical Note 17G, *BIA, McLean, Va.*)

TABLE A STEM									TABLE B FOOTING							
Dimensions							*Loads*		*Dimensions*					*Loads*		
H (ft)	*Nom. t (in.)*	*c (in.)*	*v (ft)*	*w (ft)*	*x (ft)*	*y (ft)*	*V (kips per 4 ft)*	*M (ft-kips per 4 ft)*	*B*	*T*	*A*	*D*	*K*	*P (kips per ft)*	*M_{OT} (ft-kips per ft)*	*Toe pressure (kips/sq ft*
6	8	2	—	—	—	—	2.06	4.13	3′-6″	1′-3″	2′-3″	1′-0″	—	0.70	1.63	1.33
7	8	2	—	—	—	—	2.80	6.55	4′-0″	1′-3″	2′-9″	1′-0″	—	0.92	2.45	1.54
8	8	2	—	—	—	—	3.66	9.78	4′-6″	1′-3″	3′-3″	1′-0″	—	1.16	3.48	1.74
9	12	5	—	—	1	—	4.63	13.92	5′-0″	1′-6″	3′-6″	1′-0″	—	1.43	4.76	2.00
10	12	4	—	—	2	—	5.72	19.10	5′-6″	1′-9″	3′-9″	1′-0″	—	1.73	6.35	2.11
11	12	3	—	—	3	—	6.92	25.42	6′-0″	2′-0″	4′-0″	1′-3″	1′-3″	2.14	8.75	2.31
12	12	2	—	—	4	—	8.24	33.00	6′-6″	2′-3″	4′-3″	1′-3″	1′-3″	2.52	11.10	2.37
13	16	5	—	1	5	—	9.67	41.96	7′-0″	2′-6″	4′-6″	1′-3″	1′-3″	2.90	13.80	2.62
14	16	4	—	2	6	—	11.21	52.41	7′-6″	2′-9″	4′-9″	1′-6″	1′-6″	3.43	17.70	2.82
15	16	3	—	3	7	—	12.87	64.46	8′-0″	3′-0″	5′-0″	1′-6″	1′-6″	3.89	21.40	2.87
16	16	2	—	4	8	—	14.64	78.23	8′-6″	3′-3″	5′-3″	1′-6″	1′-6″	4.51	26.70	3.04
17	20	5	1	5	9	1	16.53	93.84	9′-0″	3′-6″	5′-6″	1′-9″	1′-9″	5.03	31.40	3.32
18	20	4	2	6	10	2	18.53	111.39	9′-6″	3′-9″	5′-9″	1′-9″	1′-9″	5.72	38.20	3.49
19	20	3	3	7	11	3	20.65	131.01	10′-0″	4′-0″	6′-0″	1′-9″	1′-9″	6.15	42.50	3.52
20	20	2	4	8	12	4	22.88	152.80	10′-6″	4′-3″	6′-3″	2′-0″	2′-0″	6.92	50.80	3.65

TABLE C REINFORCEMENT																
Bars	*a (2 per pocket)*		*b (1 per pocket)*		*c (2 per pocket)*		*d*		*e*		*f*		*g*		*h*	
H (ft)	*Bar size*	*Extension above top of footing*	*Bar size*	*Extension above top of footing*	*Bar size*	*Extension above top of footing*	*Bar size*	*No. of bars*	*Bar size*	*Length of bar*	*Bar size*	*Length of bar*	*Bar size*	*No. of bars*	*Bar size*	*No. of bar*
6	#5	Full height of pocket	—	—	—	—	#2	10	#3	3′-2″	#3	2′-3″	#3	4	#3	3
7	#6		—	—	—	—	#2	11	#4	3′-8″	#3	2′-3″	#3	4	#3	3
8	#6		#6	3′-0″	—	—	#2	13	#4	4′-2″	#3	2′-3″	#3	4	#3	3
9	#7		#6	3′-0″	—	—	#2	14	#5	4′-8″	#3	2′-6″	#3	4	#3	3
10	#7		#7	4′-0″	—	—	#2	16	#5	5′-2″	#3	2′-9″	#3	5	#3	4
11	#8		#7	4′-0″	—	—	#2	17	#6	5′-8″	#3	3′-0″	#3	5	#3	5
12	#8		#8	5′-0″	—	—	#2	19	#6	6′-2″	#3	3′-3″	#3	5	#3	5
13	#8		#8	6′-0″	—	—	#2	20	#7	6′-8″	#4	3′-9″	#4	5	#4	5
14	#9		#10	6′-0″	—	—	#2	22	#7	7′-2″	#4	4′-0″	#4	6	#4	5
15	#10		#9	5′-0″	—	—	#2	23	#7	7′-8″	#4	4′-3″	#4	6	#4	5
16	#10		#11	6′-6″	—	—	#2	25	#8	8′-2″	#5	4′-9″	#4	6	#4	5
17	#10		#8	4′-0″	#8	7′-0″	#2	26	#8	8′-8″	#5	5′-0″	#4	6	#4	5
18	#10		#9	4′-0″	#9	8′-0″	#2	28	#8	9′-2″	#5	5′-3″	#4	7	#4	6
19	#10		#9	4′-0″	#10	9′-0″	#2	29	#9	9′-8″	#6	6′-0″	#4	7	#4	6
20	#10		#11	5′-6″	#10	10′-0″	#2	31	#9	10′-2″	#6	6′-3″	#4	7	#4	6

Design table for grouted, reinforced brick masonry cantilever retaining walls. (*From Brick Institute of America,* Technical Note 17G, *BIA, McLean, Va.*)

13.2 BELOW-GRADE WALLS

Reinforced masonry design through engineering analysis permits th construction of deep basement walls, walls supporting heavy vertical load and walls where unsupported height or length exceeds that recommende for plain masonry. For both residential and commercial buildings, it i often economical to use these low-cost foundation walls to enclose basemer space or underground parking areas. Footings are set below the frost line i undisturbed soil. The masonry walls provide low thermal conductivity and may easily be waterproofed against moisture infiltration and dampness. Below-grade masonry walls also offer excellent enclosures for underground or earth sheltered buildings (see *Earth Sheltered Housing Design* by the University of Minnesota Underground Space Center).

Concrete masonry units are more widely used in below-grade construction than brick, and much research has been done to test their capability and performance. General design considerations must include (1) maximum lateral load from soil pressure, (2) vertical loads from building superstructure, (3) minimum wall thickness required by local codes, and (4) length or height of wall between lateral supports. The vertical and horizontal loads that the walls must resist will vary, of course, as will the manner in which the walls are intended to carry and transmit the loads. Basement walls supporting bearing wall construction must usually support relatively heavy compressive loads in addition to earth pressure or other lateral loads. In skeleton frame construction, columns may extend down to separate footings and carry most of the dead and live loads of the superstructure. In such cases, the basement walls may be subject to appreciable lateral load, but little vertical load. If the columns are closely spaced, or if pilasters are added, the wall may be designed to transmit these lateral loads horizontally and vertically as two-way slabs. If the vertical supports are widely spaced, and the first-floor construction cannot be considered as providing lateral support, a design cantilever action will be required (i.e., retaining wall design).

Lateral earth pressure may be calculated by the equivalent fluid method outlined earlier. It is normally assumed that the stresses created in basement walls by earth pressure against their exterior face are resisted by bending of the walls in the vertical span. This means that the wall behaves like a simple beam supported at top and bottom. Support at the top is provided by the first-floor construction, and bottom support by the footing and basement floor slab. If the first floor is to contribute lateral support, backfilling should be delayed until this construction is in place.

A portion of the lateral earth load is carried by the wall acting as a beam in the horizontal span. The distribution of the total lateral load horizontally and vertically will depend on wall height and length as well as stiffness in both directions. If the length of the wall between supports is no greater than its height, the load is generally divided equally between vertical and horizontal spans.

The overall stability of a below-grade wall may be enhanced by increasing the stiffness in either direction, or by reducing the length of the horizontal span. *Horizontal stiffness* can be increased by incorporating bond beams into the design, or by placing steel joint reinforcement in the mortar joints at intervals of not more than 16 in. Bond beams are most advantageously located at or near the top of the wall, built to extend continuously around the perimeter of the building. When used in this manner, they will also serve to distribute concentrated vertical loads and to prevent or minimize the development of cracks. The increase in flexural strength achieved with horizontal joint reinforcement is limited by the practical amount of steel that can be embedded in the joints, and by the amount of bond strength developed between mortar, reinforcement, and masonry units.

Vertical stiffness may be increased in one of two ways: (1) steel reinforcement may be grouted into hollow cells, or (2) pilasters may be added (*see Fig. 13-8*). Pilasters should project from the wall a distance equal to approximately $\frac{1}{12}$ of the wall height. Pilaster width should be equal to approximately $\frac{1}{10}$ of the horizontal span between supports. The distance between pilasters or between end walls or cross walls and pilasters should not exceed 18 ft for unreinforced walls 10 in. thick, or 15 ft for walls 8 in. thick.

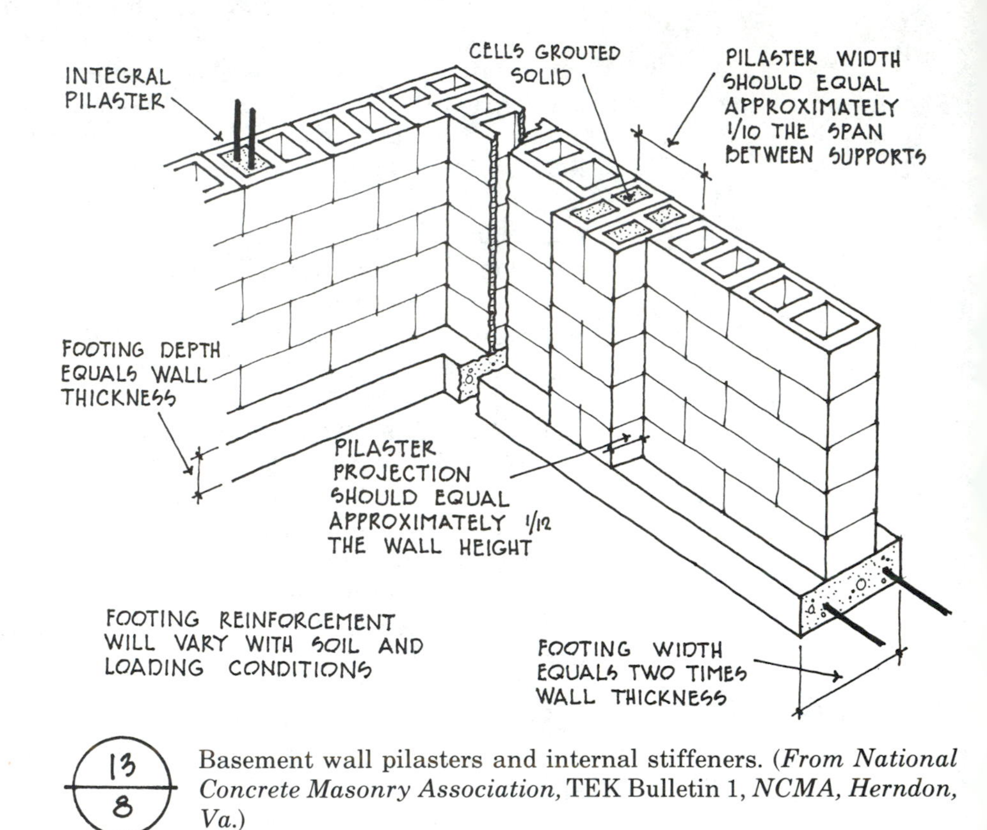

13-8 Basement wall pilasters and internal stiffeners. (*From National Concrete Masonry Association,* TEK Bulletin 1, *NCMA, Herndon, Va.*)

In relying on floors and footings (and roofs in earth-sheltered designs for lateral bracing, proper anchorage of members must be provided t assure transfer of loads. Steel dowels should connect walls securely to th footing. Pilasters, cross walls, and end walls must be bonded wit interlocking masonry units or steel ties. Sill plates should be anchored t the wall at 6-ft maximum intervals (*see Fig. 13-9*). Intersecting walls shoul be anchored with $24 \times \frac{1}{4} \times 1\frac{1}{2}$ in. metal straps spaced not more than 32 ir on center. If a partition does not provide lateral support, strips of metal lat or galvanized hardware cloth may be substituted for the heavier straps Mortar joints at the intersection of cross walls with exterior below-grad walls should be raked out and caulked, to form a control joint to accommo date slight longitudinal movement. Sill plates should be connected wit $\frac{1}{2}$-in.-diameter bolts extended at least 15 in. into the filled cells of th masonry, and spaced to within 12 in. of the end of the plate. Where girder bear on a basement wall, at least two block cores in the top course below th end of the member should be filled with mortar to a depth of 6 in. Pilaste may be bonded to the wall at girder locations to provide additional suppor and should be grouted solid in the top course. Ends of floor joists should b anchored at 6-ft intervals (normally every fourth joist). At least the fir three joists running parallel to a wall should also be anchored to it a intervals not exceeding 8 ft.

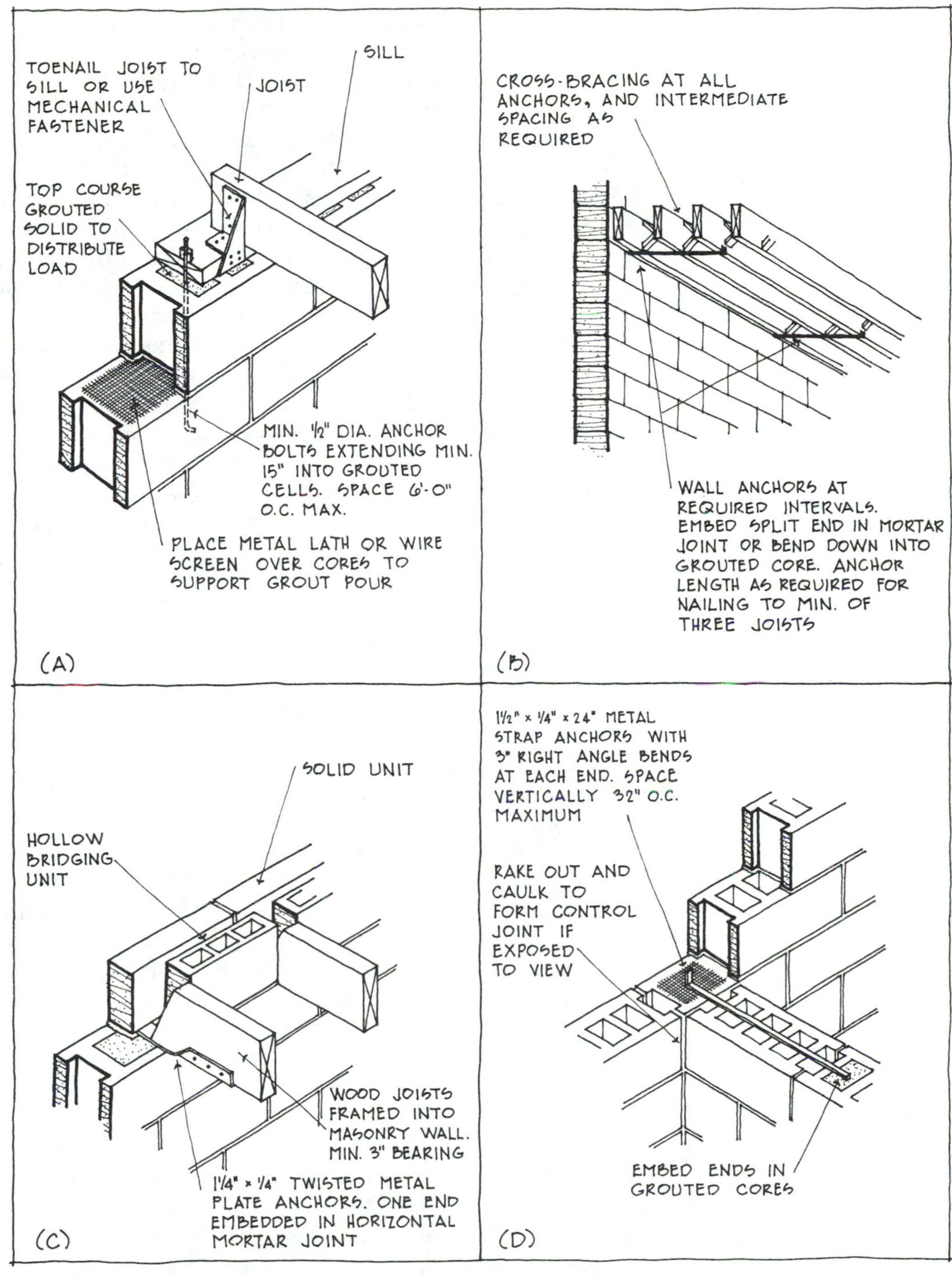

13-9 Basement wall anchorage details. (*From National Concrete Masonry Association,* TEK Bulletin 1, *NCMA, Herndon, Va.*)

Part of the *vertical load* on basement walls is imparted by the weight of the superstructure plus live loads. This is usually transmitted at some point between the centerline of the wall and the inner surface, thus inducing a bending moment. Additional moment is induced at any point where flexural members are restrained by their connection to the wall. These moments tend to counteract the moments from lateral earth pressures at the exterior face. In other words, vertical compressive stresses are effective in reducing the tensile stresses developed in resisting lateral loads. In this regard, it is important to remember that only dead loads may be safely considered as opposing lateral bending stresses, since live loads may be intermittent. Precautions must also be taken in construction scheduling to ensure that the amount of dead load calculated for such resistance is actually present before the lateral load is applied. If early backfill is unavoidable, temporary bracing must be provided to prevent actual stresses from exceeding those assumed in the design.

Other loads applied to below-grade walls may be variable, transient, or moving, such as surcharge, wind, snow, earthquake, or impact forces. The pressures from wind that ordinarily affect basement and foundation walls are those transmitted indirectly through the superstructure: compressive, uplift, shearing, or racking loads. Stresses developed in resisting overturning are not often critical except for lightweight structures subject to high wind loads, or for structures having a high ratio of exposed area to depth in the direction of wind flow. Rational design procedures outlined in Chapters 10 and 11 may be used to analyze and calculate such forces so that the masonry structure will adequately resist all applied loads and induced stresses.

13.2.1 Unreinforced Walls

Code requirements for empirically and rationally designed unreinforced masonry walls are discussed in Chapter 10. The *empirical standards* (*ANSI A41.1*) call for a wall thickness in basement construction equal to that of the wall carried above, with 12 in. as a recommended minimum. Under certain circumstances, however, this may be reduced to 8 or 10 in. provided that height and lateral bracing limitations are not exceeded. *Figure 13-10* provides a summary of these requirements. Generally, unreinforced masonry basement walls are limited in depth to a maximum of 7 ft below finish grade and are applicable to residential and light commercial or industrial buildings. If wall heights must exceed 7 ft to satisfy design requirements, intermittent lateral support in the form of pilasters or cross walls must be provided at least every 15 ft for 8-in. walls and every 18 ft for 10-in. walls, as outlined above.

Rational analysis methods applied to the design of unreinforced basement walls will result in the most economical construction. Exact load determinations will dictate spacing of pilasters or placement of partial reinforcing at critical locations. Compressive and flexural strength, slenderness coefficients, and eccentricity ratios are all considered in the analysis and the wall designed for maximum efficiency and adequate factors of safety within allowable stress limitations. (Refer to Chapter 10 for requirements and methods of calculation.)

13.2.2 Partially Reinforced Walls

In many instances where plain masonry walls are not adequate to resist assumed loads, and fully reinforced walls are not economically justified, a scheme of partial reinforcing may easily satisfy project requirements. Partially reinforced walls are designed in the same manner as plain masonry walls, except that the reinforcement resists flexural stresses. A detailed engineering analysis will determine exact requirements and criteria for placement, but for preliminary planning and estimating, tables developed by NCMA can be used to approximate final conditions. Th

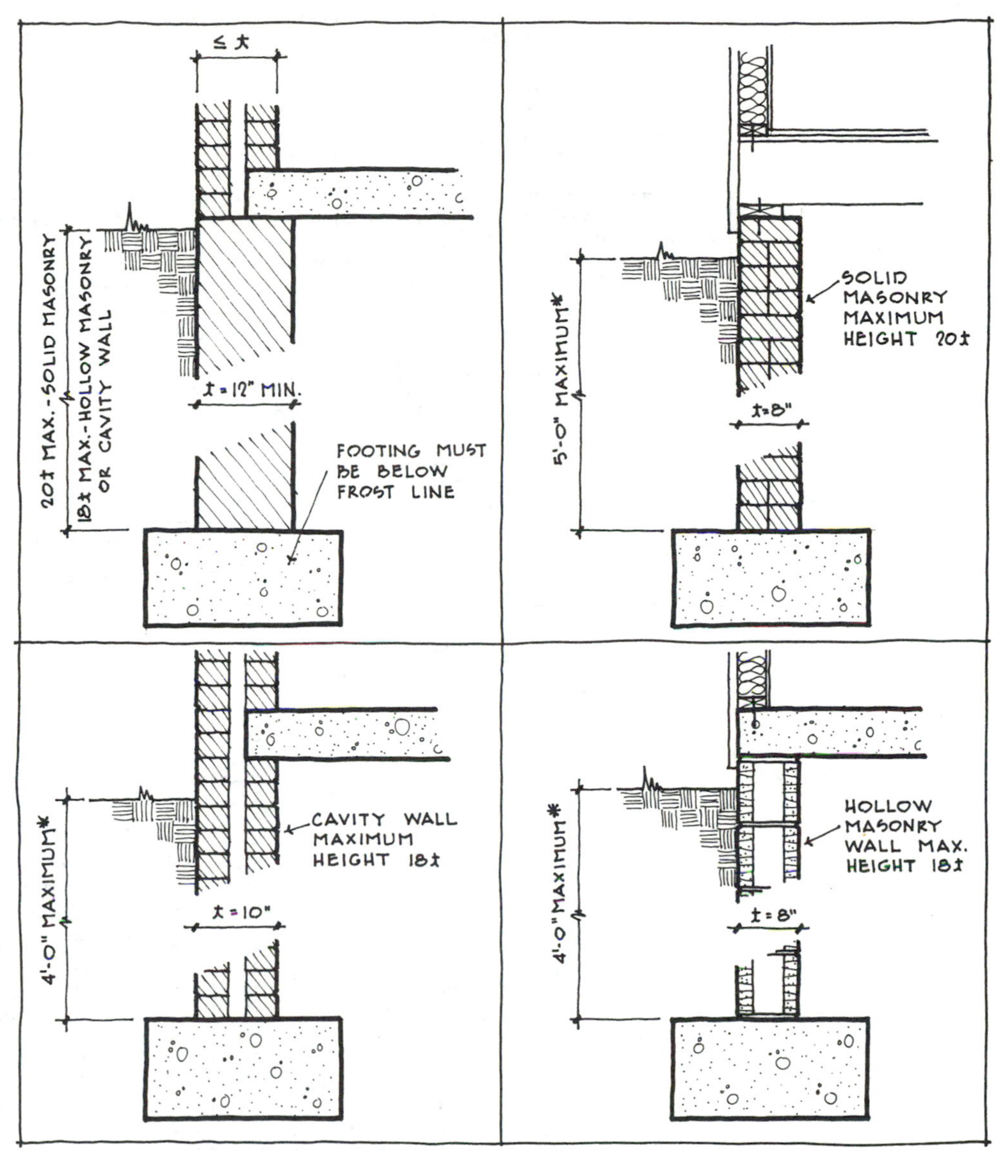

* DEPTH MAY BE INCREASED TO A MAXIMUM OF 7 FT. WITH APPROVAL OF BUILDING OFFICIAL WHEN SOIL CONDITIONS WARRANT SUCH INCREASE. COMBINED HEIGHT OF AN 8" FOUNDATION WALL AND THE MASONRY WALL SUPPORTED MAY NOT EXCEED THE HEIGHT PERMITTED FOR 8" WALLS (18 t CAVITY AND HOLLOW UNIT WALLS, 20 t SOLID WALLS).

Empirical requirements for unreinforced basement walls (based on *ANSI A41.1*).

information in *Fig. 13-11* covers 8-, 10-, and 12-in. walls with compressive strengths (f'_m) of 1200 or 1600 psi. Walls may be as much as 12 ft below grade, but only one soil pressure equivalent is included.

In the preparation of this table, it was assumed that the walls would act as simple beams supported at top and bottom in resisting lateral earth pressures, and that any vertical compressive loads could be considered as axially applied and uniformly distributed over the entire wall length at a maximum of 5000 lb/lin ft. Design assumptions for materials and for allowable stresses are as listed. If the wall under consideration meets the given criteria, the table may be used to determine moment and shear and to select appropriate reinforcement.

Design Assumptions

A. Materials: 1. *Concrete Masonry Units*—Grade N hollow load-bearing units conforming to ASTM C90 except that compressive strength of units should not be less than that required for applicable f'_m; see design standard noted in B below.

2. *Mortar*—Type S conforming to ASTM C270.

3. *Grout*—Fine or coarse grout (ASTM C476) with an ultimate compressive strength (28 days) of at least 2500 psi.

4. *Reinforcement*—Deformed Billet-Steel Bars (ASTM A615); Rail-Steel Deformed Bars (A616); or Axle-Steel Deformed Bars (A617).

B. Allowable Stresses—In accordance with "Specification for the Design and Construction of Load-Bearing Concrete Masonry," NCMA.

Soil equiv.-fluid wt. = 25 pcf

Recommended Vertical Reinforcement with Axial Compressive Load (P) Equal to or Less Than 5000 lb/lin ft

8-in. Walls

Wall Height H (ft)	Wall Depth below Grade, h (ft)	Earth Pressure: Moment (in. lb/ft)	Earth Pressure: Shear (lb/ft)	$f'_m = 1200$ psi, Spacing of Reinforcement (in.): No. 3	No. 4	No. 5	$f'_m = 1600$ psi, Spacing of Reinforcement (in.): No. 3	No. 4	No. 5
10	4	2240	175	24	40	40	24	40	56
	5	3690	260		24	32	16	24	40
	6	6250	360			16		16	24
12	4	2380	180	16	24	40	16	24	40

10-in. Walls

Wall Height H (ft)	Wall Depth below Grade, h (ft)	Moment (in. lb/ft)	Shear (lb/ft)	$f'_m = 1200$ psi: No. 4	No. 5	No. 6	No. 7	$f'_m = 1600$ psi: No. 4	No. 5	No. 6	No. 7
10	6	6250	360	32	40	48	48	40	56	64	72
	7	9000	470	24	32	32	40	24	40	48	56
	8	12200	585	16	24	24	32	16	24	32	40
12	5	4300	270	40	48	56	56	40	56	72	88
	6	6900	375	24	32	40	40	24	40	48	64
	7	10100	495	16	24	32	32	16	24	32	40
14	5	4500	275	32	40	48	56	32	48	56	72
	6	7400	390	16	24	32	40	16	32	40	48

12-in. Walls

Wall Height H (ft)	Wall Depth below Grade, h (ft)	Moment (in. lb/ft)	Shear (lb/ft)	$f'_m = 1200$ psi: No. 4	No. 5	No. 6	No. 7	$f'_m = 1600$ psi: No. 4	No. 5	No. 6	No. 7
10	7	9000	470	40	48	56	56	40	56	80	80
	8	12200	585	32	40	40	40	32	48	56	64
	9	15700	710	24	32	32	32	24	40	48	48
	10	19200	835	16	24	24	32	16	32	40	40
12	6	6900	375	40	56	64	64	40	56	80	96
	7	10100	495	32	40	48	48	32	48	56	72
	8	13800	625	24	32	32	40	24	32	40	56
	9	18100	760	16	24	32	32	16	24	32	40
14	6	7400	390	32	48	56	56	32	48	64	80
	7	11000	510	24	32	40	40	24	32	48	56
	8	15200	650	16	24	32	32	16	24	32	40
16	6	7700	395	24	40	48	56	24	40	56	64
	7	11550	525	16	24	32	40	16	32	40	48
	8	16200	670		16	24	32		24	24	32

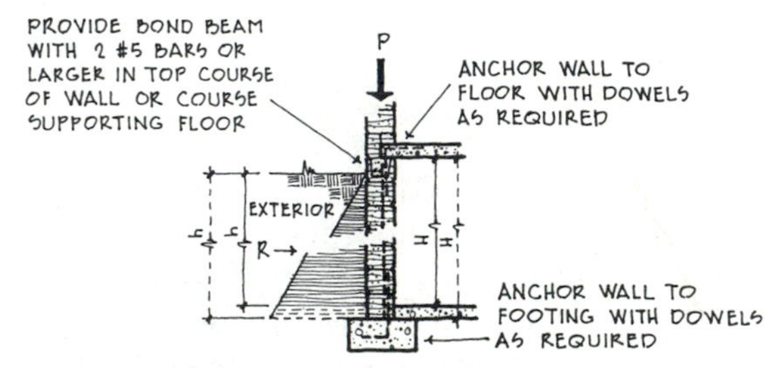

TYPICAL VERTICAL SECTION - PARTIALLY REINFORCED CONCRETE MASONRY BASEMENT WALLS

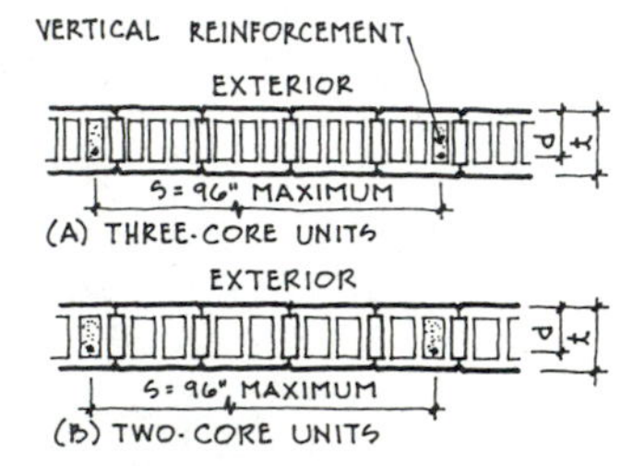

TYPICAL HORIZONTAL SECTION - PARTIALLY REINFORCED CONCRETE MASONRY BASEMENT WALLS

8" WALLS: t = 7 5/8", d = 5"
10" WALLS: t = 9 5/8", d = 7"
12" WALLS: t = 11 5/8", d = 8 3/4"

Design table for partially reinforced CMU basement walls. (*From National Concrete Masonry Association,* TEK Bulletin 56, *NCMA, Herndon, Va.*)

Design Assumptions

A. Materials: 1. *Concrete Masonry Units*—Grade N hollow load-bearing units conforming to ASTM C90 except that compressive strength of units should be not less than that required for applicable f'_m; see design standard noted in B below.

2. *Mortar*—Type S conforming to ASTM C270.

3. *Grout*—Fine or coarse grout (ASTM C476) with an ultimate compressive strength (28 days) of at least 2500 psi.

4. *Reinforcement*—Deformed Billet-Steel Bars (ASTM A615); Rail-Steel Deformed Bars (A616); or Axle-Steel Deformed Bars (A617).

B. Allowable Stresses—In accordance with "Specification for the Design and Construction of Load-Bearing Concrete Masonry," NCMA.

Soil equiv.-fluid wt. = 25 pcf

Recommended Vertical Reinforcement with Axial Compressive Load (P) Equal to or Less Than 5000 lb/lin ft

8-in. Walls

Wall Height *H* (ft)	Wall Depth below Grade, *h* (ft)	Earth Pressure: Moment (in. lb/ft)	Earth Pressure: Shear (lb/ft)	$f'_m = 1200$ psi, Spacing of Reinforcement (in.): No. 4	No. 5	No. 6	No. 7	$f'_m = 1600$ psi, Spacing of Reinforcement (in.): No. 4	No. 5	No. 6	No. 7
10	5	3690	260	16	24	32	40	16	24	40	48
	6	6250	360		16	24	24	16	24	32	40
	7	9000	470			16	16		16	24	24
	8	12200	585								16
12	5	4300	270	16	24	32	32	16	24	32	40
	6	6900	375			16	24		16	24	24

10-in. Walls

Wall Height *H* (ft)	Wall Depth below Grade, *h* (ft)	Moment (in. lb/ft)	Shear (lb/ft)	$f'_m = 1200$ psi, Spacing of Reinforcement (in.): No. 5	No. 6	No. 7	$f'_m = 1600$ psi, Spacing of Reinforcement (in.): No. 5	No. 6	No. 7
10	7	9000	470	24	32	40	24	32	40
	8	12200	585	24	24	32	24	32	40
	9	15700	710	16	24	24	24	24	32
	10	19200	835		16	16	16	24	24
12	7	10100	495	24	32	32	24	32	40
	8	13800	625	16	24	24	16	24	32
	9	18100	760		16	16		16	24
14	6	7400	390	24	32	40	24	32	40
	7	11000	510	16	24	32	16	24	32
	8	15200	650		16	24		16	24

12-in. Walls

Wall Height *H* (ft)	Wall Depth below Grade, *h* (ft)	Moment (in. lb/ft)	Shear (lb/ft)	$f'_m = 1200$ psi, Spacing of Reinforcement (in.): No. 5	No. 6	No. 7	No. 8	$f'_m = 1600$ psi, Spacing of Reinforcement (in.): No. 5	No. 6	No. 7	No. 8
10	8	12200	585	16	24	32	48	16	24	32	48
	9	15700	710	16	24	32	40	16	24	32	48
	10	19200	835	16	24	32	32	16	24	32	48
12	8	13800	625	16	24	32	40	16	24	32	48
	9	18100	760	16	24	32	32	16	24	32	48
	10	22800	905	16	24	24	24	16	24	32	40
	11	28000	1050		16	16	24	16	24	24	32
	12	33100	1200						16	24	24
14	8	15200	650	16	24	32	32	16	24	32	48
	9	20300	800	16	24	24	24	16	24	32	40
	10	26000	955		16	24	24		16	24	32
	11	31800	1115				16			16	24
16	8	16200	670	16	24	32	32	16	24	32	40
	9	21800	825		16	24	24		16	24	32
	10	28200	990			16	16			16	24

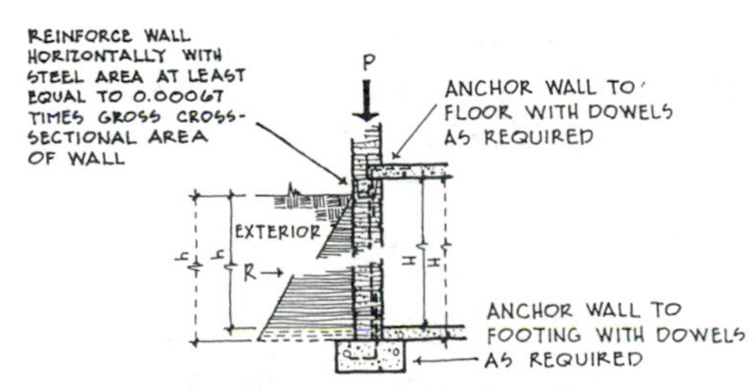

TYPICAL VERTICAL SECTION THROUGH REINFORCED CONCRETE MASONRY BASEMENT WALLS

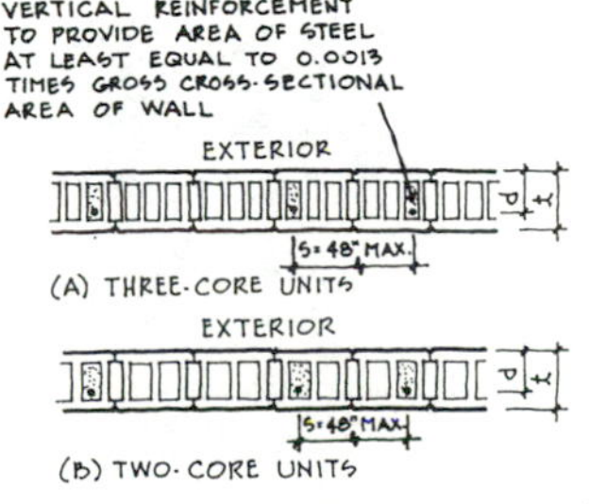

TYPICAL HORIZONTAL SECTION THROUGH REINFORCED CONCRETE MASONRY BASEMENT WALLS

8" WALLS: t = 7 5/8", d = 5"

10" WALLS: t = 9 5/8", d = 7"

12" WALLS: t = 11 5/8", d = 8 3/4"

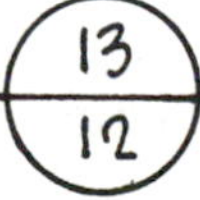

Design table for fully reinforced CMU basement walls. (*From National Concrete Masonry Association,* TEK Bulletin 56, *NCMA, Herndon, Va.*)

13.2.3 Reinforced Walls

All fully reinforced masonry construction is designed and detailed using rational engineering procedures. To be considered fully reinforced, the wall must contain a minimum amount of steel equal to 0.002 times the gross cross-sectional area, with not more than two-thirds of the bars placed in one direction (refer to Chapter 11). Spacing of vertical steel for basement walls may not exceed six times the wall thickness, or 48 in. maximum.

The table in *Fig. 13-12* was developed using the same design assumptions given above for partially reinforced walls, but the combinations of steel bar sizes and spacing given will satisfy code requirements for the minimum area of steel for fully reinforced walls. This table may be used as a design aid whenever the wall being considered meets the basic criteria for size, strength, materials, and so on.

13.2.4 Footings

Some general rules of thumb may be applied to the preliminary design o footings for below-grade concrete masonry walls: (1) cast-in-place concret should have a minimum compressive strength of 2000 psi, (2) footing dept should be equal to the wall thickness, and footing width equal to twice th wall thickness (*see Fig. 13-8*), and (3) the bottom of the footing should b placed in undisturbed soil below the frost line. Allowable soil bearin pressure, of course, must be checked against actual loads in the final desig

13.2.5 Material Requirements

Concrete masonry for below-grade wall construction should be Grade N hollow or solid loadbearing units which meet the minimum standards o ASTM C90 or C145, respectively. Since bond is an important factor in th flexural strength and stiffness of unreinforced walls, either high- o moderate-strength mortars (Type M or S) are required for constructio Type M mortar is recommended for areas in contact with soil. Grout fo filled cores should contain either fine or coarse aggregate, depending on th size and clearances of the grout space; should have a minimum compressiv strength of 2500 psi at 28 days; and should meet the requirements of ASTM C476.

13.2.6 Waterproofing

Proper drainage and waterproofing of below-grade walls is essential, n only to prevent buildup of hydrostatic pressure, but also to preserve d conditions and eliminate dampness in interior spaces.

A 4- to 6-in. bed of crushed stone or gravel should be laid inside th footing covering most of the slab area, with the thickness increased at th edges (*see Fig. 13-13*). Drain tiles should be installed in the thicker perimet areas as shown. These drains should lead to a lower surface outlet or to storm sewer. Weep holes should be installed every 32 in. to drain water in the crushed stone over the drain tile. Floor slabs should be cast at a lev above the weep holes.

The exterior of the walls should be parged or plastered and coated wi bituminous material or other suitable waterproofing. Alternate procedur include the use of a heavy troweled-on coat of fiber-reinforced aspha mastic, built-up membranes of fabric and mastic, liquid-applied polym

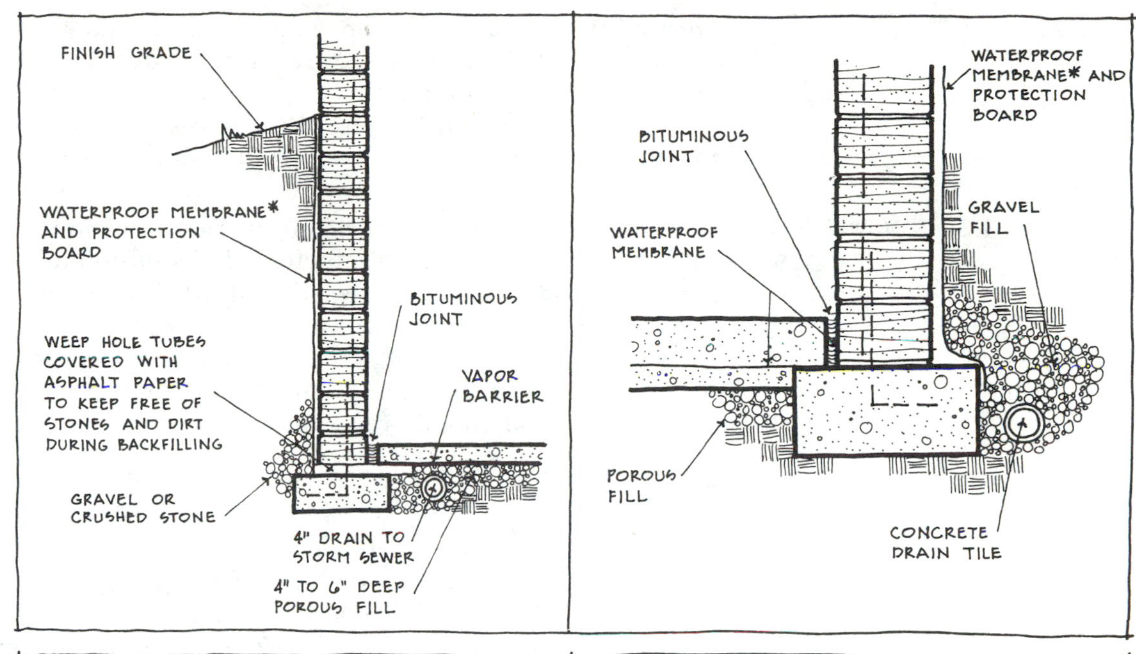

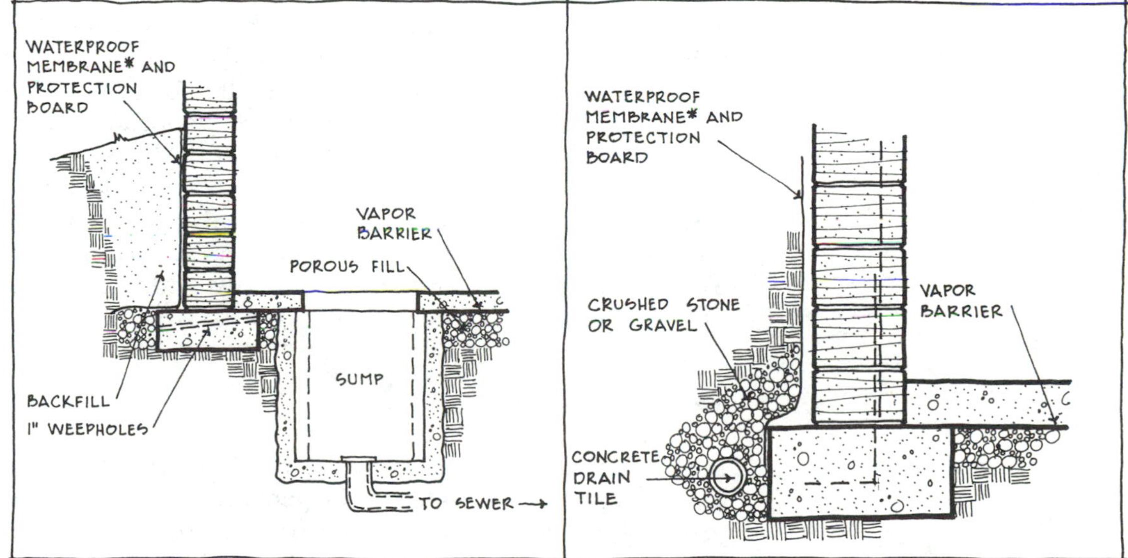

* WATERPROOF MEMBRANE MAY BE PARGE OR PLASTER COAT WITH BITUMINOUS MEMBRANE SURFACE; FIBER-REINFORCED ASPHALT MASTIC; LIQUID-APPLIED POLYMER MEMBRANE; BENTONITE CLAY; EPDM MEMBRANE; OR OTHER SUITABLE MATERIAL. USE PROTECTION BOARD WHERE REQUIRED TO PREVENT TEARS OR PUNCTURES DURING BACKFILLING OPERATIONS.

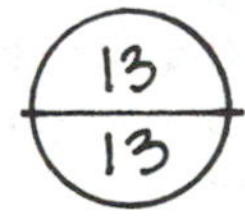

Basement wall drainage. (*From National Concrete Masonry Association,* TEK Bulletins 1, 43, *NCMA, Herndon, Va.*)

membranes, bentonite clay waterproofing, or EPDM membranes. No pinholes, cracks, or open joints should be permitted. The finish grade adjacent to a below-grade wall should slope away from the building to prohibit the accumulation of rainwater or other moisture and prevent its seepage down along the surface of the wall.

Damp or wet conditions in basements are often mistakenly attributed to seepage of moisture through the walls. The trouble is more often due to condensation of moisture from the air inside the space. Condensation is most frequent in warm, humid areas, but can occur wherever the temperature and relative humidity of the inside air are maintained at high levels, either artificially or due to atmospheric conditions. Water vapor will form anytime the surface temperature of the wall is below the dew-point temperature of the interior air.

Condensation can be controlled by regulating ventilation or by the proper location and installation of wall insulation (see Chapters 8 and 14)

13.3 SWIMMING POOLS

Masonry has been used in the construction of swimming pools for many years, and has provided satisfactory, low-maintenance installations fo both public and private facilities. One of the first applications of reinforced brick masonry for outdoor swimming pools was in Springfield, Minnesota in 1936. The pool is 50 × 120 ft in plan, 30 in. deep at the shallow end, and 9 ft-6 in. deep under the diving board. The first major maintenance was performed in 1966 after 30 years of service.

13.3.1 Materials

Because of severe exposure to water and chemicals, and the likelihood o freezing conditions, only high-quality materials should be used in th construction of masonry pools. Brick masonry should comply with th requirements of ACTM C216 and ASTM C62 for facing and building brick respectively, for the exposed and unexposed wythes of the wall. Grade SV should be specified with the additional requirement that maximum cold water absorption should not exceed 8.0% for outdoor pools and 12.0% fo indoor pools. The brick face should be smooth and dense, and minimur strength should comply with structural design requirements. Hollo loadbearing CMUs should conform to ASTM C90, Grade N. Two-core unit are preferable to three-core designs to facilitate the placement of reinforcin steel and grout. Mortar should have a minimum compressive strength c 2500 psi at 28 days, and should comply with the requirements of ASTM C270, Type M. Grout should comply with ASTM C476.

13.3.2 Design

Pool walls are usually designed as reinforced cantilever retaining walls. A a rule, the height of water in the pool and the height of the backfill are th principal factors affecting wall thickness and required reinforcement. Th length and width of the pool do not significantly affect the design. The po walls must be designed to withstand the lateral pressure of the water whe the pool is full, and the pressure of the backfill when the pool is empty. It not advisable to consider the backfill as a restraint against the pressure the water, because careless backfilling, shrinkage of the soil in hot or d weather, and other factors may reduce its supporting value. For the reasons, the reinforcing steel is best placed in the center of the wall.

In selecting dimensions for the pool components, wall heights shou generally not exceed 5 or 6 ft, since lateral pressure increases exponential with water depth. Greater depths needed for diving can be obtained b using a hopper-shaped bottom formed by sloping the pool floor downwa from the base of the wall.

An example of a *brick masonry pool* is shown in *Fig. 13-14*. The desi is based on an ultimate masonry compressive strength (f'_m) of 3000 p (which may be assumed with high-quality workmanship and brick of 90 psi compressive strength). If different wall heights than those shown a required by the design, the table may be used to revise footing sizes a reinforcement.

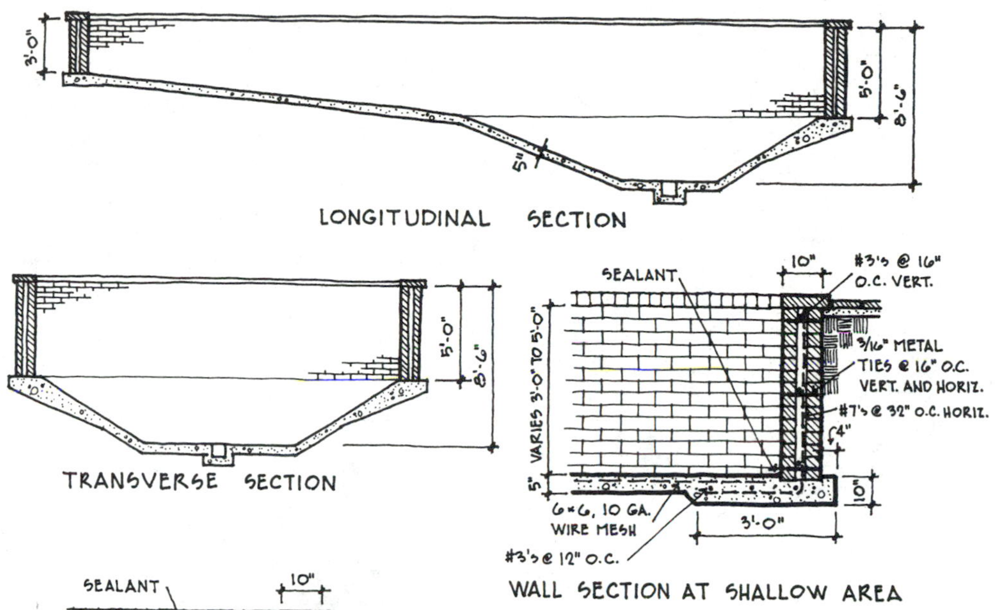

VARIOUS WALL HEIGHT REQUIREMENTS				
			Reinforcing steel§	
*H** *(ft)*	*t†* (in.)	*L‡* *(ft)*	*Bar size*	*Spacing (in.)*
3.0	10	3.0	No. 4	48
3.5	10	3.5	No. 5	48
4.0	10	4.0	No. 5	32
4.5	10	4.5	No. 6	32
5.0	10	5.0	No. 7	32
5.5	10	5.5	No. 5	16
6.0	10	6.0	No. 6	16
6.5	10	6.5	No. 7	16
7.0	10	7.0	No. 8	16
7.5	10	7.5	No. 8	16

*Wall height measured from top of footing to proposed water level.

†Footing thickness measured vertically.

‡Footing length measured horizontally.

§Main reinforcing steel: vertical steel in wall bent into footing.

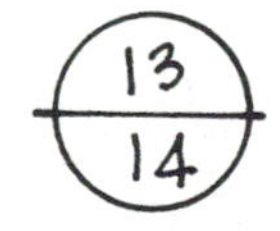

Brick pool design example. (*From Brick Institute of America,* Technical Note 17K, *BIA, McLean, Va.*)

The 8-in. wall section in *Fig. 13-15* gives design data for *concrete masonry pools* with wall heights ranging from 2 ft to 5 ft-4 in. Note that the information given is based on two different types of soil. Type I is a very permeable, coarse-grained material such as clean sand or gravel, and Type II is a fine, silty sand or a granular soil with conspicuous clay content. It was assumed that the principal stresses would be taken in the vertical span, and that the walls would act as cantilever retaining walls. The horizontal reinforcement shown is adequate to resist lateral stresses in small pools in moderate climates. However, pools subject to unusually cold weather while empty should have 25 to 50% more horizontal steel to allow for greater shrinkage stresses from the additive effects of low temperature and drying.

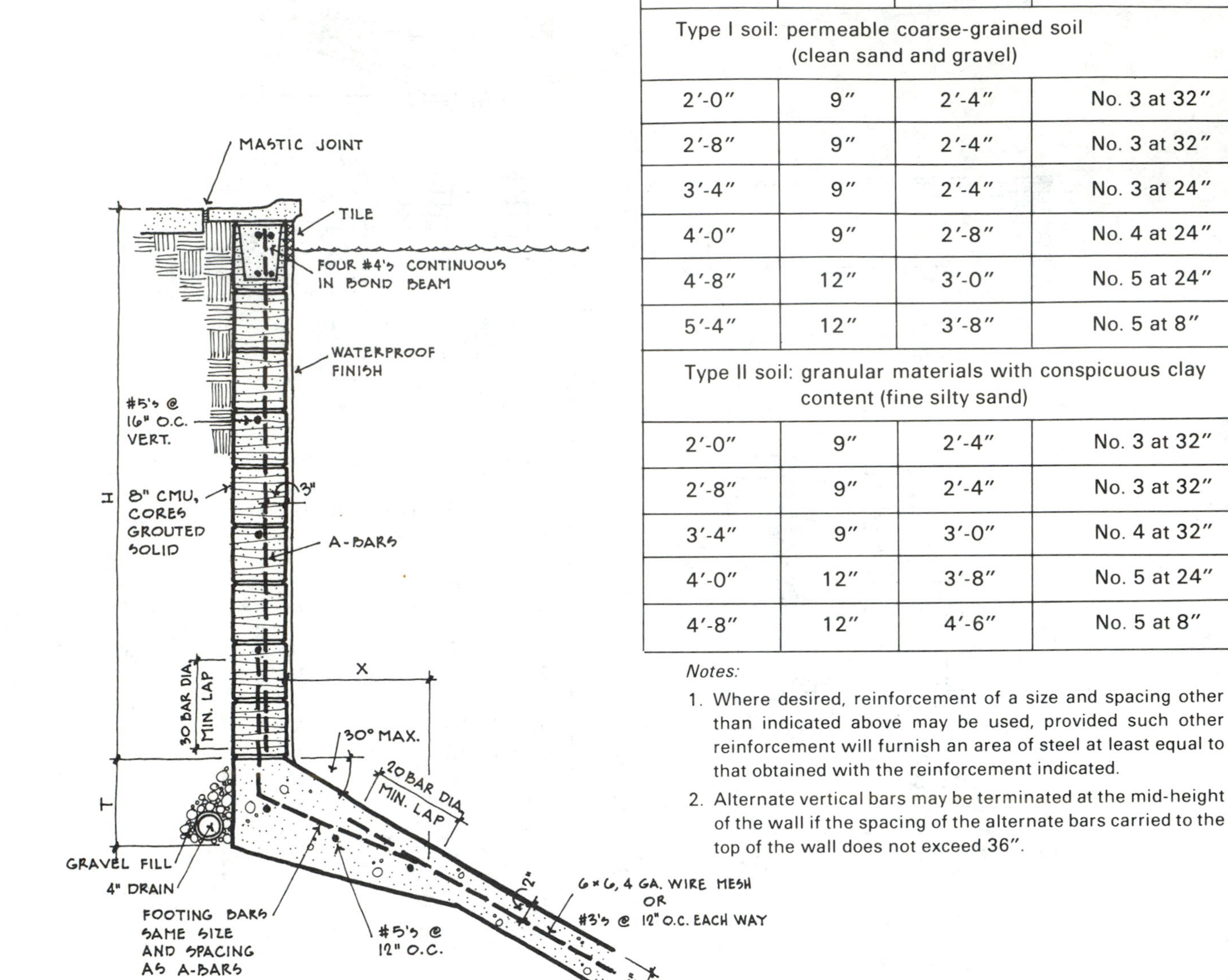

H	*T*	*X*	*Reinforcement A-bars*
Type I soil: permeable coarse-grained soil (clean sand and gravel)			
2′-0″	9″	2′-4″	No. 3 at 32″
2′-8″	9″	2′-4″	No. 3 at 32″
3′-4″	9″	2′-4″	No. 3 at 24″
4′-0″	9″	2′-8″	No. 4 at 24″
4′-8″	12″	3′-0″	No. 5 at 24″
5′-4″	12″	3′-8″	No. 5 at 8″
Type II soil: granular materials with conspicuous clay content (fine silty sand)			
2′-0″	9″	2′-4″	No. 3 at 32″
2′-8″	9″	2′-4″	No. 3 at 32″
3′-4″	9″	3′-0″	No. 4 at 32″
4′-0″	12″	3′-8″	No. 5 at 24″
4′-8″	12″	4′-6″	No. 5 at 8″

Notes:

1. Where desired, reinforcement of a size and spacing other than indicated above may be used, provided such other reinforcement will furnish an area of steel at least equal to that obtained with the reinforcement indicated.
2. Alternate vertical bars may be terminated at the mid-height of the wall if the spacing of the alternate bars carried to the top of the wall does not exceed 36″.

13/15 CMU pool wall design. (*From National Concrete Masonry Association,* TEK Bulletin 50, *NCMA, Herndon, Va.*)

13.3.3 Construction

For brick masonry construction, all vertical and horizontal joints should b completely filled with mortar and tooled concave to produce a smoot uniform, and watertight surface. The vertical collar joint should be ke free of mortar droppings during construction, and may be grouted by eith the low-lift or high-lift method. Drains, inlets, outlets, skimmers, filters a other plumbing, diving boards, ladders, and accessories should be plac and installed as recommended by the manufacturers. In areas whe natural drainage away from the pool is not provided, or where soil does n readily drain, it is advisable to use drain tile and gravel fill around t footing at the low point of the pool.

For concrete masonry pools, the first course of block on the footi

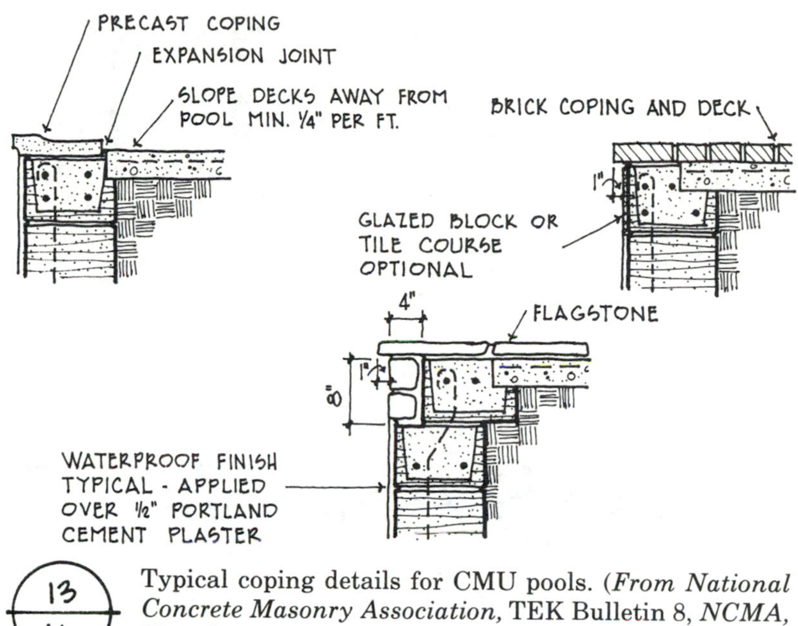

Typical coping details for CMU pools. (*From National Concrete Masonry Association,* TEK Bulletin 8, *NCMA, Herndon, Va.*)

should be laid in a full mortar bed. Subsequent courses may use face-shell bedding. Cleanouts should be provided in the first course of reinforced cores to remove mortar droppings before grouting. The top course of the masonry should receive a coping to protect against water intrusion. Copings may be of concrete, masonry or stone, and flashing should be provided to help prevent leaks (*see Fig. 13-16*). Expansion or contraction joints are not required for walls less than 50 ft in length, and are actually undesirable as possible leakage sources.

A finish must be applied to the interior of concrete masonry swimming pools to provide an impervious, easily cleaned surface which is resistant to the alkali in the block and the chemicals in the water. Plastic liners and organic-base paints have both been used successfully for this purpose. For painted surfaces, the block must first be coated with a $\frac{1}{2}$-in. minimum thickness of portland cement plaster applied as two $\frac{1}{4}$-in. layers. The wall surface should be dampened with water before applying the first coat, and the juncture of wall and footing should be coved. Vinyl or nylon liners are also satisfactory. The liner may be laid over a sand floor without a concrete slab, and should extend up and over the wall. Dimensional tolerance is critical when this method is used, and the pool must not be drained or the fittings will not realign correctly upon refilling. Special epoxy or other synthetic coatings which form a film by chemical reaction may also be used to waterproof concrete block pools. Frost-proof or glazed tile may be used around the waterline for decorative effect.

Walls of brick or CMU pools should be cured for at least 7 days prior to placing and compacting the backfill. Care should be taken not to overstress the walls by operating heavy equipment on the backfill area or by impact loading of large masses of moving earth. As a general rule, heavy equipment should not come closer to the wall than a distance equal to the height of the wall.

Pools should be left filled year-round in most areas to prevent freezing of the subgrade and possible frost heave. For concrete masonry, this will also assure that the moisture content of the units remains essentially constant, eliminating a major source of cracking.

14

DETAILS AND WORKMANSHIP

Masons and bricklayers belong to one of the oldest crafts in history. Th rich architectural heritage of many civilizations attest to the skill an workmanship of the trade, and the advent of modern technological method and sophisticated engineering practices have not diminished the impo tance of this aspect of masonry construction. The best intentions of th architect or engineer will not produce a masterpiece unless the workma ship is of the highest order and the field practices are as exacting an competent as the detailing.

14.1 PREPARATION OF MATERIALS

High-quality construction begins with the proper storage and protection materials. Brick and concrete masonry units are shipped and delivered secured bundles or cubes, sometimes stacked on wooden pallets. If the uni are not delivered on pallets, the contractor must provide a raised platform prevent staining of units from contact with the soil, and absorption soluble salts or other minerals that might cause efflorescence on t finished work. Both the units and the mortar materials must be covered f protection against rain, snow, or ice.

14.1.1 Mortar

The mortar mix required in the project specifications must be carefu controlled at the job site to maintain the intended quality standa Consistent measurement of mortar ingredients for the mix should ensu uniformity of proportions, yields, strengths, workability, and mortar co from batch to batch. Volumetric rather than weight proportioning is most often called for, and most often miscalculated because of variations in the moisture content of the sand. Common field practice is to use a shovel as the standard measuring tool for dry ingredients. However, moisture in the sand causes a "bulking" effect, and the same weight of wet sand occupies more volume than dry sand. Such variables often cause over- or undersanding of the mix, which of course affects both the strength and bonding characteristics of the mortar. Oversanded mortar is harsh and unworkable, provides a weak bond with the masonry units, and performs poorly during freeze–thaw cycles.

Simple field-quality control measures can be instituted through the use of wooden boxes 12 in. square by 12 in. high. The mixer may then determine the exact number of shovels full which equal 1 cu ft. Since the moisture content of the sand will vary constantly because of temperature, humidity, and evaporation, it is good practice to check the volume measurement once or twice a day and make adjustments as necessary.

The other dry ingredients in masonry mortar are normally packaged and labeled only by weight. The standard weights of these ingredients per cubic foot of volume are given in *Fig. 14-1.* The mason may also use the

Material	*Unit weight per cubic foot*
Portland cement	94
Masonry cement	70
Hydrated lime	
Dry	40
Putty	80

Unit weight of mortar ingredients.

wooden measuring boxes to determine equivalent bags or shovels full for volume proportioning of these ingredients.

The amount of mixing water required is not stated as part of the project specifications. The characteristics of flow and workability are the best indicators of a correct mix, and these are best determined by the mason. Experience is the best judge in this case, and authorities generally agree that the mason is always right. A mortar that is easy to work is properly mixed with regard to the amount of water, and this is something the mason decides by the feel of the mortar on the trowel. The water proportion will vary for different conditions of temperature, unit moisture content, unit weight, and so on. It is an empirical judgment call rather than a laboratory test standard. The consistency of grout should be such that at the time of placement, it has a slump of 8 to 11 in. as determined by ASTM C143.

A relatively recent innovation for mortar mixing (particularly on large jobs) is dry-batching of all the ingredients. This eliminates the problem of field adjustments for moisture content of sand by moving the mixing operation to a controlled batching plant where ingredients can be accurately weighed and mixed, then delivered to the job site. A weathertight hopper is provided for on-site storage, and when the contractor is ready for mortar, he simply adds water and mixes. This method virtually guarantees uniformity from batch to batch, and offers greater convenience and efficiency as well.

There are two traditional methods of mixing mortar on the job site. For very small installations, *hand mixing* may be the most economical. It is accomplished using a mason's hoe and a mortar box. Sand, cement, and lime are spread in the box in proper proportions and mixed together until an even color is obtained. Water is then added, and the mixing continues until the consistency and workability are judged to be satisfactory.

More commonly, *machine mixing* is used to combine mortar ingredients. The rotating spiral or paddle-blade mechanical mixers are similar to but of lighter duty than concrete mixers. Normal capacity ranges from 4 to 7 cu ft. About three-fourths of the mixing water, half the sand, and all of the cementitious ingredients are first added and briefly mixed together. The balance of the sand is then added, together with the remaining water. After all batched materials have been combined, they should be mixed from 3 to 5 minutes. Less mixing time may result in nonuniformity, poor workability, low water retention, and less than optimum air content. Overmixing may reduce the strength of the mortar.

The water proportion should always be the *maximum* amount to produce good workability. Water in the mortar is essential to the development of bond, and if the mortar is too dry, the bond will be weak. Both strength and resistance to moisture penetration depend to a great extent on the completeness and strength of this bond between mortar and unit. Mortar batches should be sized according to the rate of usage, to avoid excessive drying and stiffening. Loss of water by absorption and evaporation can be minimized on hot days by wetting the mortar board and covering the mix in the mortar box. Within the first $2\frac{1}{2}$ hours of initial mixing, the mason may add water to restore workability. *Retempering* is accomplished by adding water to the mortar batch and thoroughly remixing. Sprinkling of the mortar is not satisfactory. Mortars containing added color pigment should not be retempered, as the increased water will lighten the color and thus cause variation from batch to batch.

14.1.2 Concrete Masonry Units

Concrete masonry units are cured and dried at the manufacturing plant and when delivered to the job site, their moisture content should be within the specified limits. Concrete units should never be moistened before or during placement because they will shrink as they dry out. If this shrinkage is restrained, as it normally is in a finished wall, stresses can develop that will cause the wall to crack. Proper storage and protection of the units is essential to prevent rain or groundwater absorption from affecting the controlled moisture content.

There may be some instances in which concrete units should ideally be dried beyond the specified limits. Where walls will be exposed to relatively high temperature and low humidity, as in the interior of heated buildings, the units should be dried to the approximate average air-dry condition to which the finished construction will be exposed.

14.1.3 Brick

The moisture content of brick is not controlled in the manufacturing process, and will vary for different units and conditions. The initial rate of absorption (see Chapter 3) will affect the amount of capillary action th takes place between the brick and the mortar. A high-suction brick, if laid dry, will absorb excessive water from the mortar, thus impairing the hydration process and weakening the bond of the finished masonry. The amount of water that will enter a brick by capillary action varies greatly. If the initial rate of absorption is greater than 20 g/min, the units should be moistened before laying. In wetting brick at the job site, sprinkling is not sufficient. The stockpile should be thoroughly sprayed with a hose stream, and the bricks then allowed to surface dry before they are laid. A surface film of water will cause the brick to float on the mortar bed and will prevent proper bonding.

Clay masonry units must also be properly blended for color to avoid unpleasant visual effects. Brick from four different cubes or pallets should be used at the same time. For single-color jobs, this takes advantage of the subtle shade variations produced in the manufacturing process, and on a mingle or mix of colors, will prevent stripes or patchy effects in the finished wall.

14.2 CONSTRUCTION AND WORKMANSHIP

The design of masonry buildings should take into consideration the size of the units involved. The length and height of walls as well as the location of openings and intersections will greatly affect both the speed of construction and the appearance of the finished work. The use of a common module in determining dimensions can easily reduce the amount of field cutting required to fit the building elements together.

14.2.1 Modular Coordination

A number of the common brick sizes available are adaptable to a 4- or 6-in. module, and dimensions based on these standards will generally result in the use of only full or half-size units. Similarly, a standard 16-in. concrete block layout may be based on an 8-in. module with the same reduction in field cutting (*See Fig. 14-2*). In composite construction of brick and concrete

14-2 Planning CMU wall openings based on 8 × 8 × 16 units. (*From National Concrete Masonry Association*, TEK Bulletin 14, *NCMA, Herndon, Va.*)

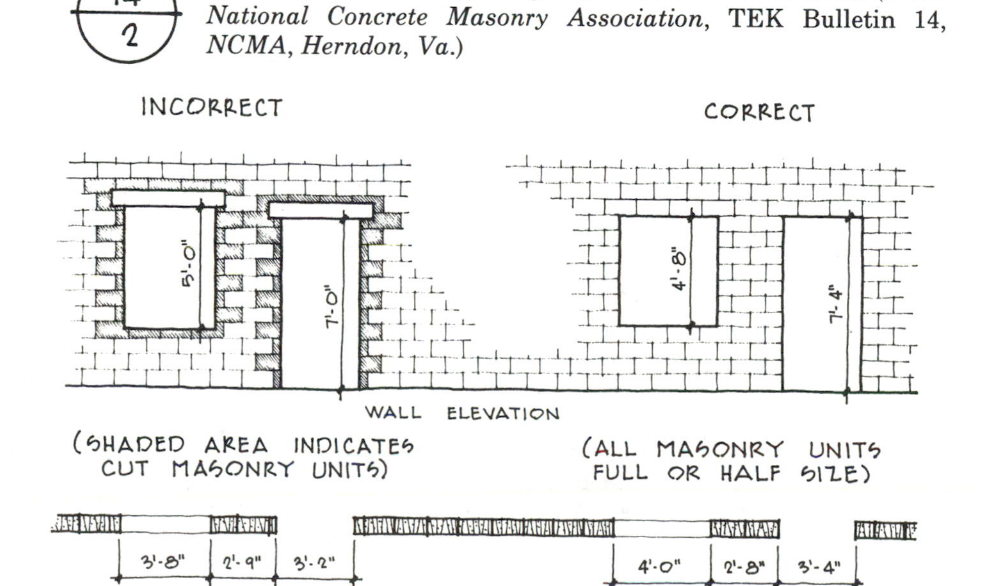

block, units should be selected which will course out the same, thus facilitating the anchorage of the backing to the facing wall as well as the joining and intersecting of the two systems. The tables shown in *Fig. 14-3* indicate the height, width, length, and coursing of various brick sizes, and the relationship between the brick units and standard 8 × 8 × 16 in. concrete blocks. For instance, three courses of Standard Modular brick equal the height of one concrete block course, five courses of Engineer brick equal two courses of concrete block, and so on. As shown in *Fig. 14-4,* the brick and block units work together in both plan and section, thus

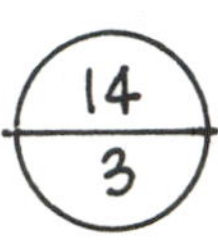

Modular coursing of masonry units. (*From Brick Institute of America,* Principles of Clay Masonry Construction, *BIA, McLean, Va. 1973;* and *Frank A. Randall and William C. Panarese,* Concrete Masonry Handbook for Architects, Engineers, and Builders, *Portland Cement Association, Skokie, Ill., 1976.)*

TABLE A NOMINAL MODULAR SIZES OF BRICK

Unit designation	*Thickness (in.)*	*Face dimensions* Height (in.)	*Face dimensions* Length (in.)	*Number of courses in 16 in.*
Standard	4	$2\frac{2}{3}$	8	6
Engineer	4	$3\frac{1}{5}$	8	5
Economy 8 or Jumbo Closure	4	4	8	4
Double	4	$5\frac{1}{3}$	8	3
Roman	4	2	12	8
Norman	4	$2\frac{2}{3}$	12	6
Norwegian	4	$3\frac{1}{5}$	12	5
Economy 12 or Jumbo Utility	4	4	12	4
Triple	4	$5\frac{1}{3}$	12	3
SCR brick	6	$2\frac{2}{3}$	12	6
6-in. Norwegian	6	$3\frac{1}{5}$	12	5
6-in. Jumbo	6	4	12	4
8-in. Jumbo	8	4	12	4

TABLE B LENGTH OF CMU WALLS WITH STRETCHER UNITS

Number of stretchers	*Wall length**
1	1'4"
$1\frac{1}{2}$	2'0"
2	2'8"
$2\frac{1}{2}$	3'4"
3	4'0"
$3\frac{1}{2}$	4'8"
4	5'4"
$4\frac{1}{2}$	6'0"
5	6'8"
$5\frac{1}{2}$	7'4"
6	8'0"
$6\frac{1}{2}$	8'8"
7	9'4"
$7\frac{1}{2}$	10'0"
8	10'8"
$8\frac{1}{2}$	11'4"
9	12'0"
$9\frac{1}{2}$	12'8"
10	13'4"
$10\frac{1}{2}$	14'0"
11	14'8"
$11\frac{1}{2}$	15'4"
12	16'0"
$12\frac{1}{2}$	16'8"
13	17'4"
$13\frac{1}{2}$	18'0"
14	18'8"
$14\frac{1}{2}$	19'4"
15	20'0"
20	26'8"

*Based on units $15\frac{5}{8}$in. long, and half units $7\frac{5}{8}$in. long, with $\frac{3}{8}$in.-thick head joints.

TABLE C HEIGHT OF CMU WALLS BY COURSES

Number of courses	*Wall height* $\frac{3}{8}$-in. bed joint 8-in. block	$\frac{3}{8}$-in. bed joint 4-in. block	$\frac{7}{16}$-in. bed joint 8-in. block	$\frac{7}{16}$-in. bed joint 4-in. block	$\frac{1}{2}$-in. bed joint 8-in. block	$\frac{1}{2}$-in. bed joint 4-in. block
1	8"	4"	$8\frac{1}{16}$"	$4\frac{1}{16}$"	$8\frac{1}{8}$"	4
2	1'4"	8"	1'$4\frac{1}{8}$"	$8\frac{1}{8}$"	1'$4\frac{1}{4}$"	8
3	2'0"	1'0"	2'$\frac{3}{16}$"	1'$\frac{3}{16}$"	2'$\frac{3}{8}$"	1'
4	2'8"	1'4"	2'$8\frac{1}{4}$"	1'$4\frac{1}{4}$"	2'$8\frac{1}{2}$"	1'4
5	3'4"	1'8"	3'$4\frac{5}{16}$"	1'$8\frac{5}{16}$"	3'$4\frac{5}{8}$"	1'8
6	4'0"	2'0"	4'$\frac{3}{8}$"	2'$\frac{3}{8}$"	4'$\frac{3}{4}$"	2'
7	4'8"	2'4"	4'$8\frac{7}{16}$"	2'$4\frac{7}{16}$"	4'$8\frac{7}{8}$"	2'4
8	5'4"	2'8"	5'$4\frac{1}{2}$"	2'$8\frac{1}{2}$"	5'5"	2'
9	6'0"	3'0"	6'$\frac{9}{16}$"	3'$\frac{9}{16}$"	6'$1\frac{1}{8}$"	3'1
10	6'8"	3'4"	6'$8\frac{5}{8}$"	3'$4\frac{5}{8}$"	6'$9\frac{1}{4}$"	3'5
15	10'0"	5'0"	10'$\frac{15}{16}$"	5'$\frac{15}{16}$"	10'$1\frac{7}{8}$"	5'1
20	13'4"	6'8"	13'$5\frac{1}{4}$"	6'$9\frac{1}{4}$"	13'$6\frac{1}{2}$"	6'10
25	16'8"	8'4"	16'$9\frac{9}{16}$"	8'$5\frac{9}{16}$"	16'$11\frac{1}{8}$"	8'7
30	20'0"	10'0"	20'$1\frac{7}{8}$"	10'$1\frac{7}{8}$"	20'$3\frac{3}{4}$"	10'
35	23'4"	11'8"	23'$6\frac{3}{16}$"	11'$10\frac{3}{16}$"	23'$8\frac{3}{8}$"	12
40	26'8"	13'4"	26'$10\frac{1}{2}$"	13'$6\frac{1}{2}$"	27'1"	13
45	30'0"	15'0"	30'$2\frac{13}{16}$"	15'$2\frac{13}{16}$"	30'$5\frac{5}{8}$"	15'
50	33'4"	16'8"	33'$7\frac{1}{8}$"	16'$11\frac{1}{8}$"	33'$10\frac{1}{4}$"	17'

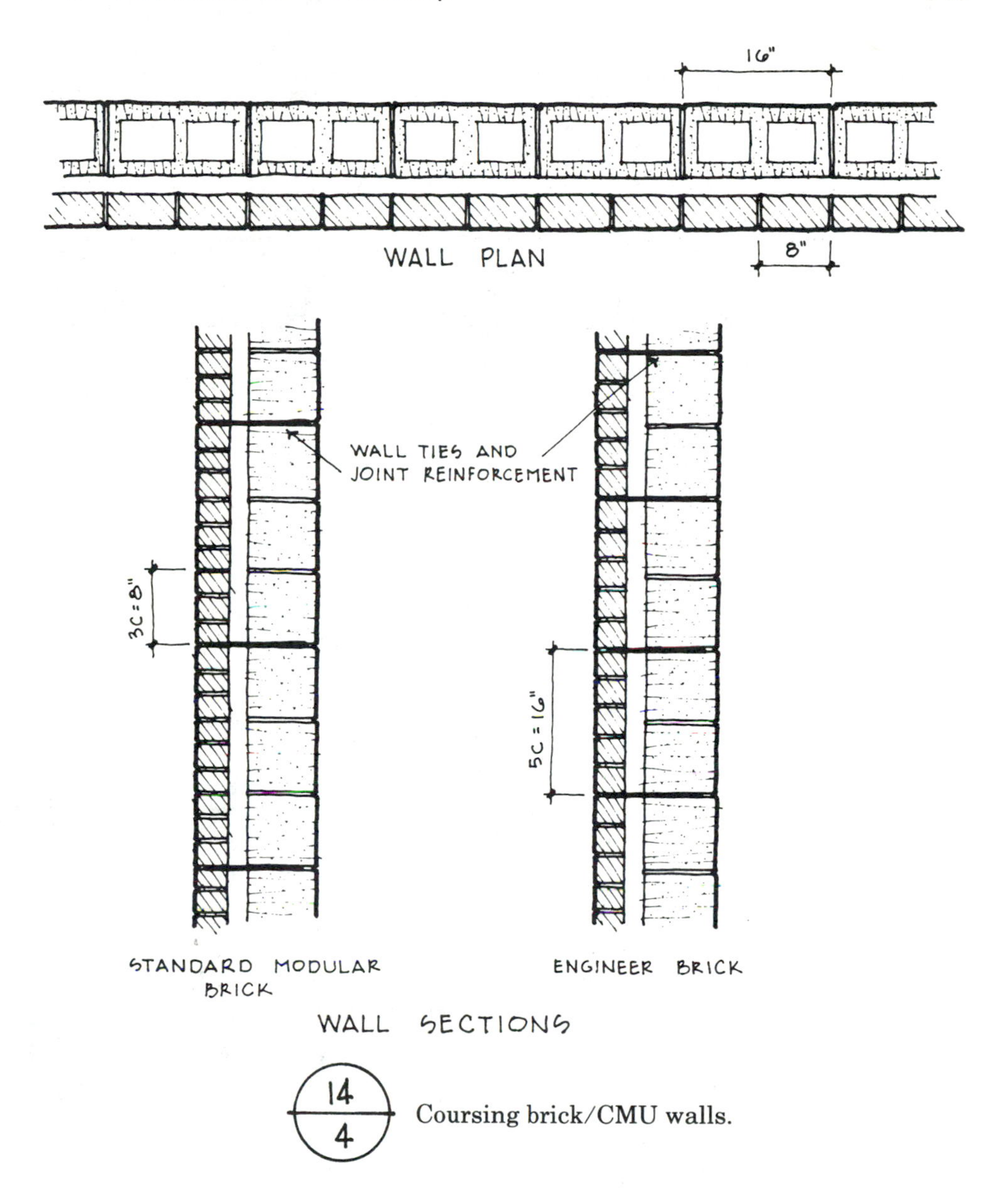

Coursing brick/CMU walls.

increasing the speed with which the mason can lay up a wall and improving the general quality and workmanship of the job.

Corners and intersections in masonry walls can be critical both structurally and aesthetically , and proper planning can greatly facilitate the construction of these elements. When masonry shear walls are used to transfer wind loads and seismic forces, they must be securely anchored to the transverse elements with steel reinforcing, and the coursing and layout of the units can affect the ease with which the steel can be placed. Corner intersections are often points of high stress, and must also be aesthetically pleasing from the exterior if the masonry units are to be left exposed. The use of masonry pilasters as integral stiffening elements in a wall must also be carefully considered in the layout, and dimensions properly set so that the pilaster fits in with the regular coursing. The examples in *Figs. 14-5* through *14-10* illustrate various common methods of detailing.

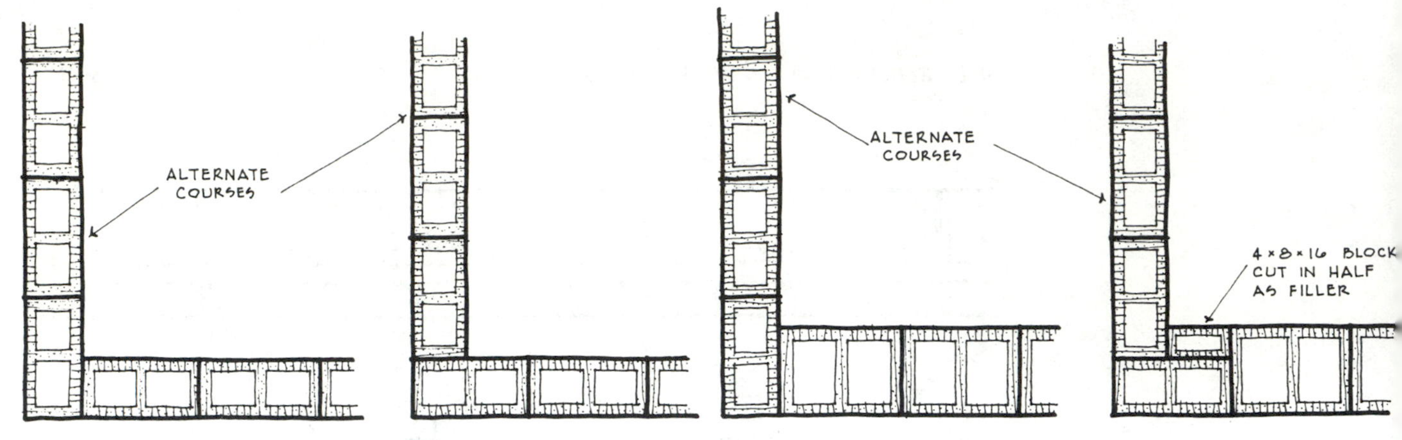

Modular corner layouts.

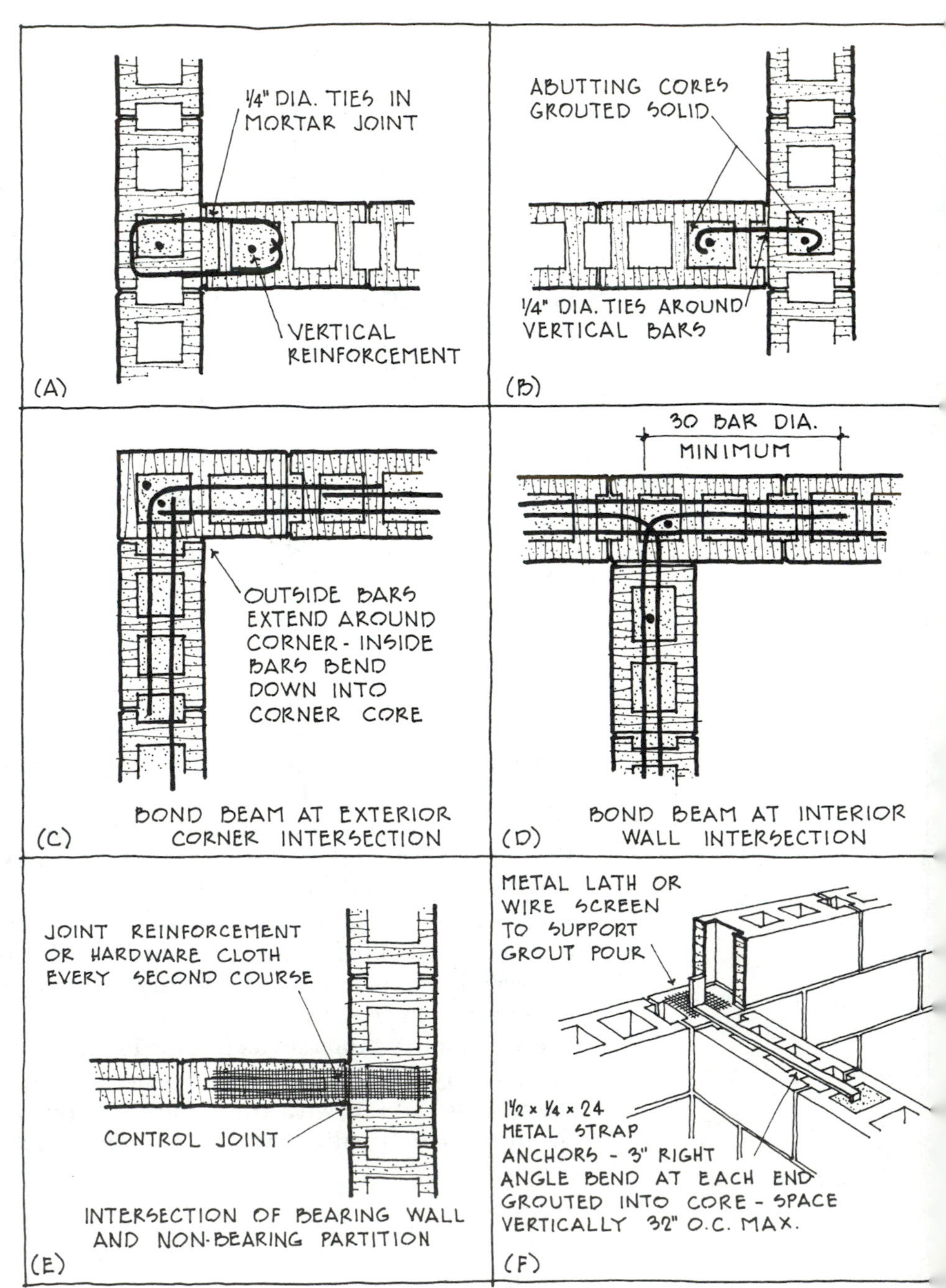

CMU metal-tied and reinforced wall intersections.

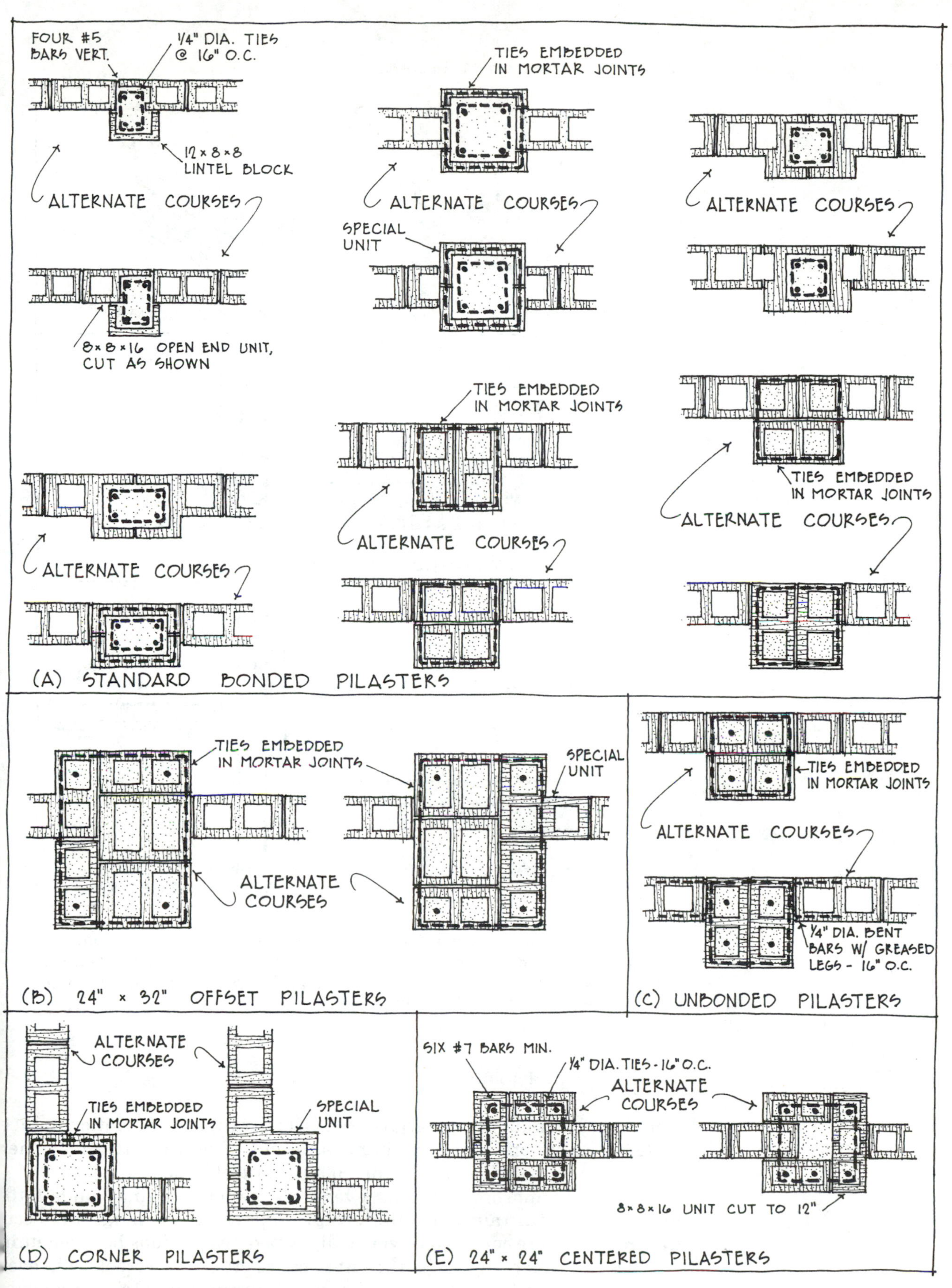

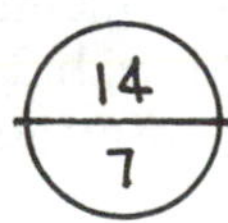

CMU pilaster coursing.

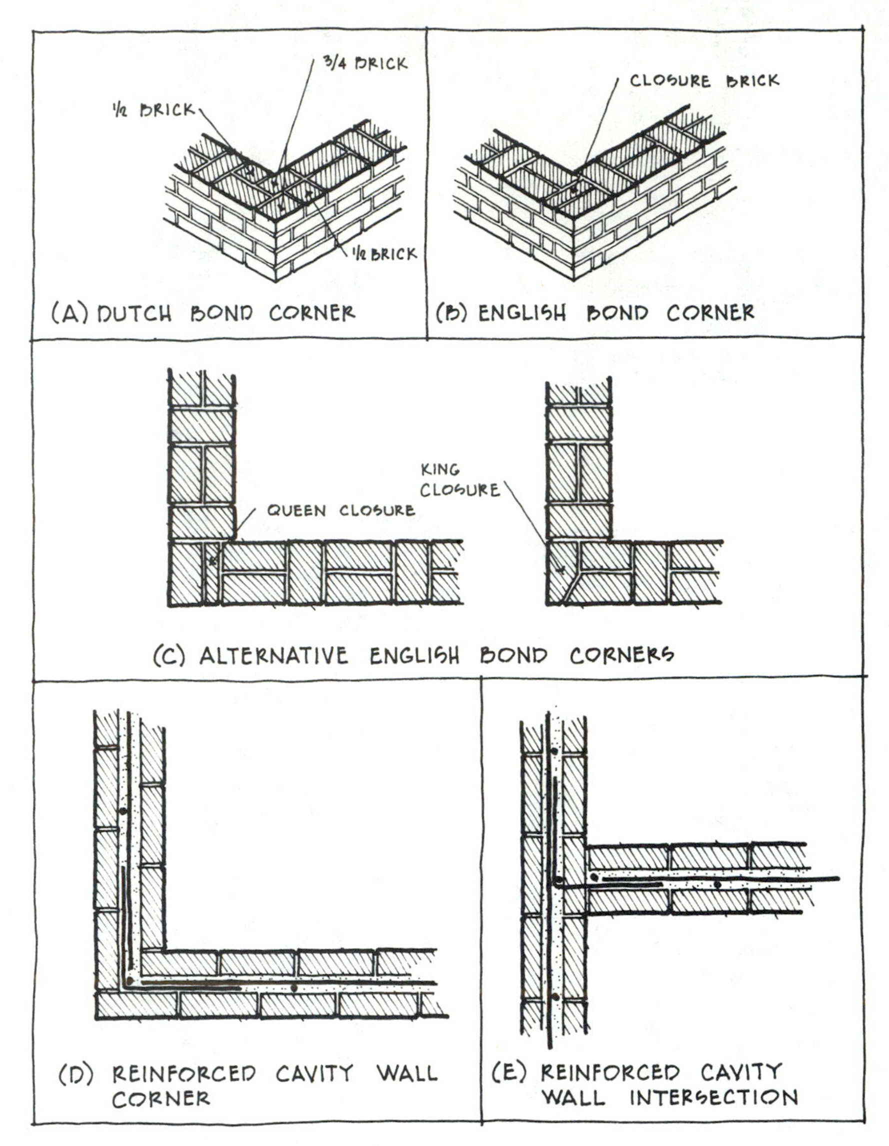

Brick corners and intersecting walls (A, B, and C seldom used today except in restoration work). (*From Harry C. Plummer,* Brick and Tile Engineering, *Brick Institute of America, McLean, Va., 1962*).

14.2.2 Control Joints

Buildings and building materials are in a constant state of motion induce by temperature and moisture changes. In masonry walls, this movemer must be accommodated by expansion and control joints.

Concrete masonry shrinks over long periods of time, and if thi movement is restrained by the design, excessive cracking can occu Control joints are continuous, vertically weakened sections between unit If stresses develop that are sufficient to cause cracks, the cracks will occ at these joints and thus be inconspicuous. Horizontal joint reinforcement used to control shrinkage cracking, but strategically located control join will further assist in the elimination of random cracks, and preve

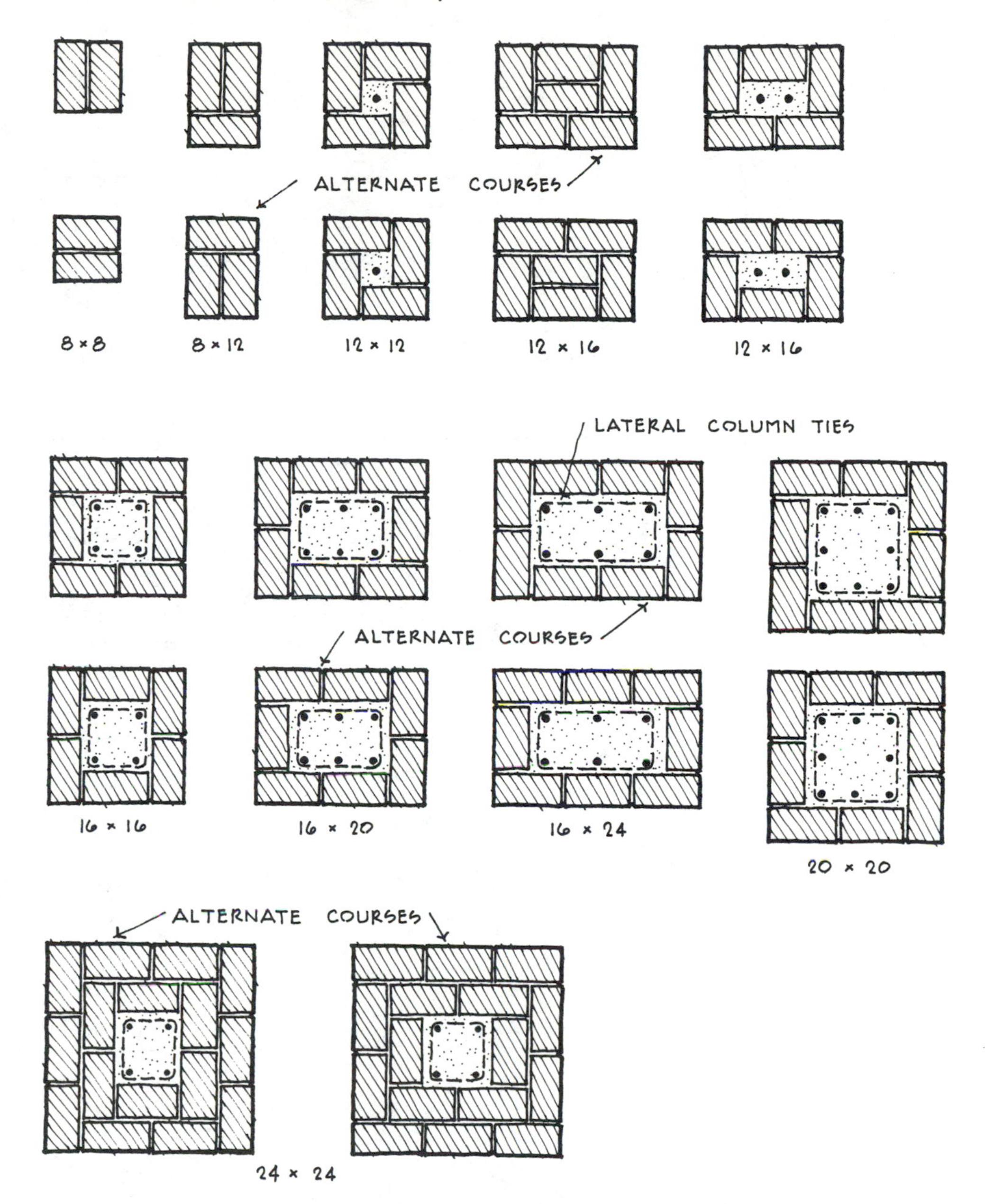

14-9 Brick column coursing. (*From Harry C. Plummer,* Brick and Tile Engineering, *Brick Institute of America, McLean, Va., 1962; and Brick Institute of America,* Principles of Clay Masonry Construction, *BIA, McLean, Va., 1973.*)

moisture penetration which might otherwise occur. Cracking is not as critical in fully reinforced construction since the steel absorbs the tensile stresses.

Control joints must be sealed against weather and may be required to stabilize the wall laterally by means of a shear key. *Figure 14-11* shows several common types of joints. All of these are first laid up in mortar just as any other vertical joint would be. After the mortar has stiffened, the joint is raked out to a depth of about $\frac{3}{4}$ in., with the remaining mortar left to form a backing for caulking compound or other weathertight sealant. The side

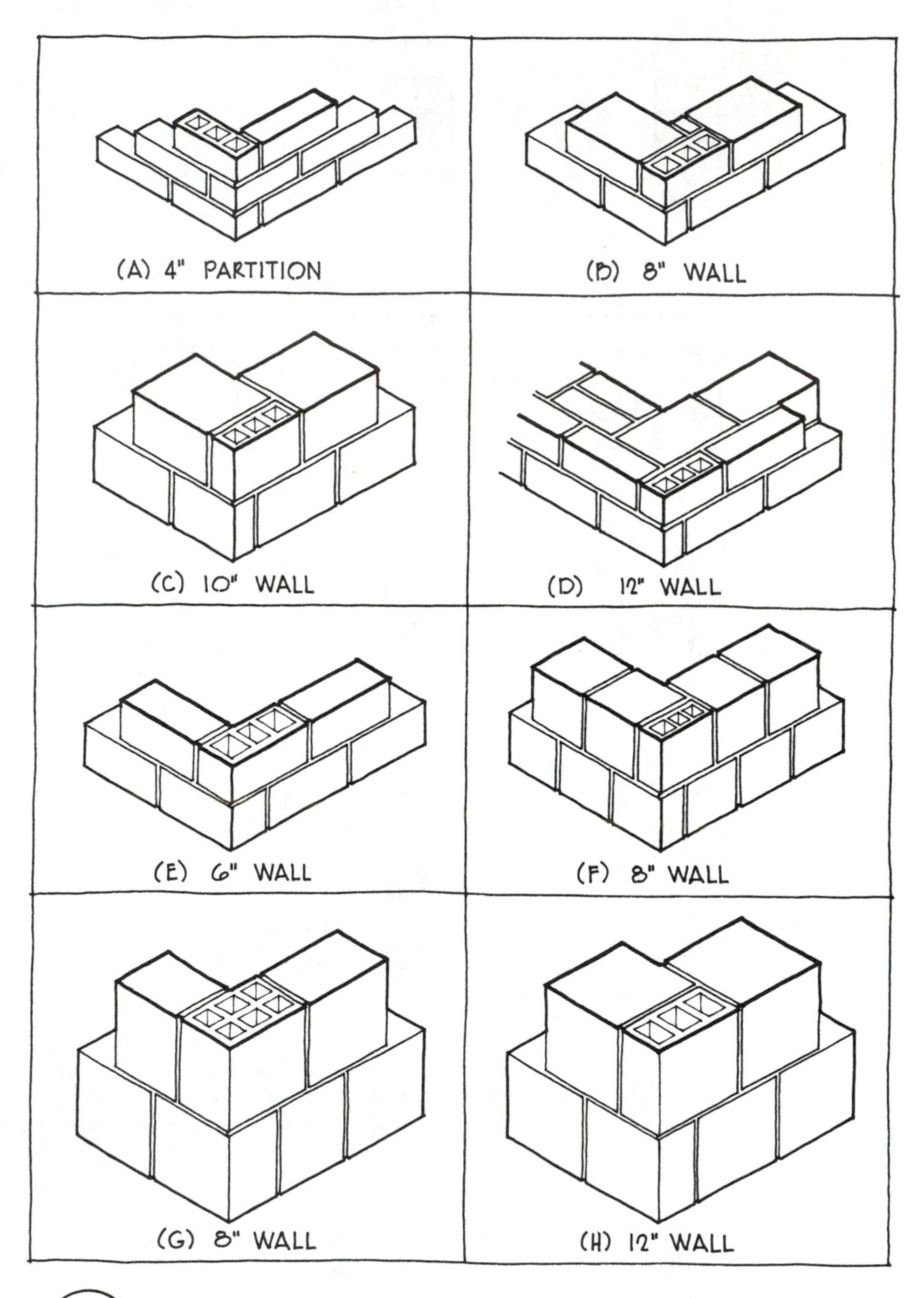

14/10 Structural clay tile corner details. (*From Brick Institute of America,* Technical Note 22, *BIA, McLean, Va.*)

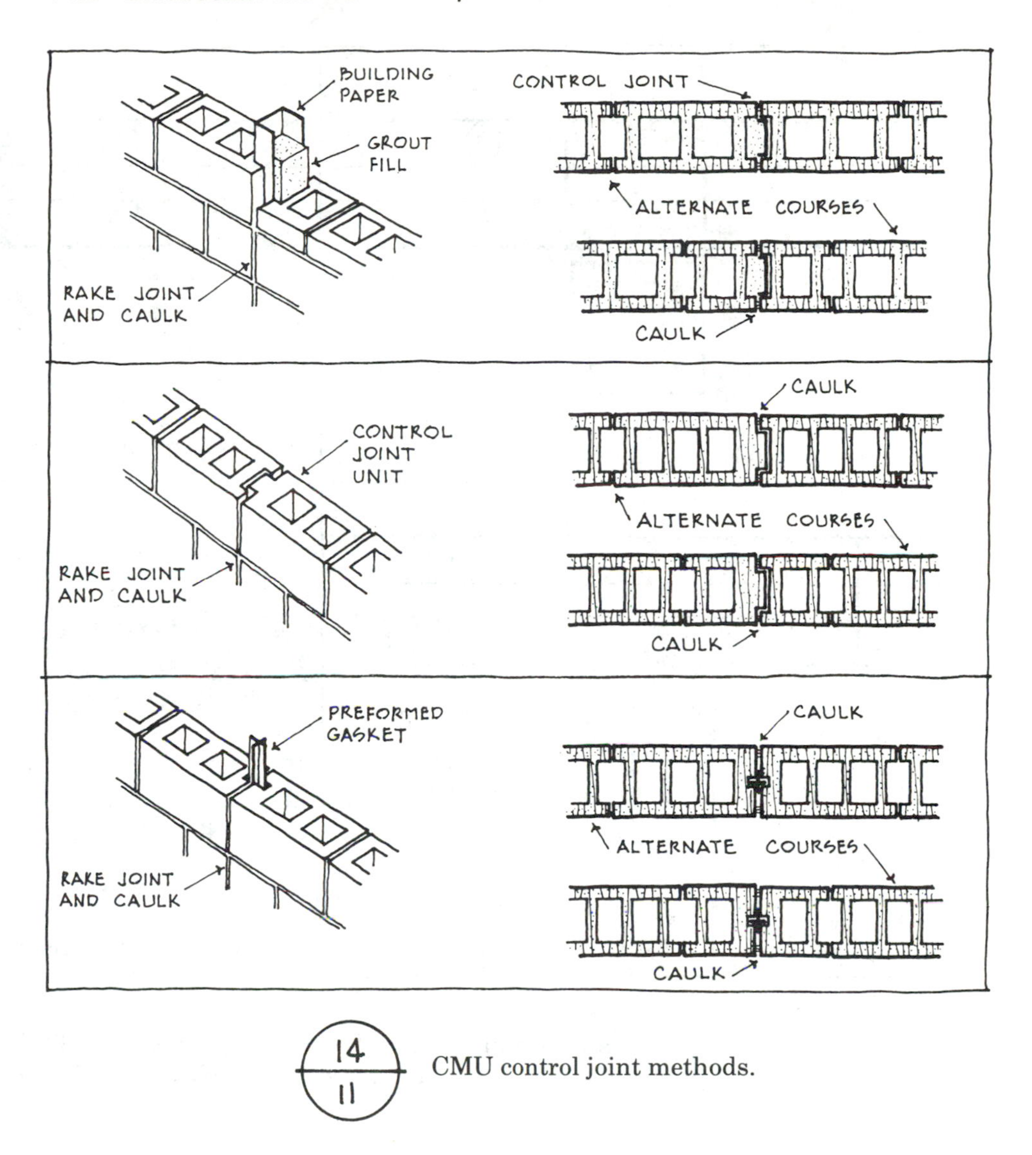

14/11 CMU control joint methods.

faces of the raked joint must be primed with a sealer to prevent absorption of oils from the sealant. The inner face of the joint is greased or treated with some type of bond-break material. Caulking may be applied with a gun or pointing trowel. All of the joints shown provide lateral stability.

14.2.3 Expansion Joints

Brick masonry expands irreversibly with time, and walls must be designed to absorb this growth without inducing stresses that could spall or fracture the unit surfaces. The illustrations in *Fig. 14-12* show several methods of constructing vertical brick expansion joints using roofing felt as a bond-break and premolded compressible filler for expansion. The joints are weatherproofed with an elastic joint sealant.

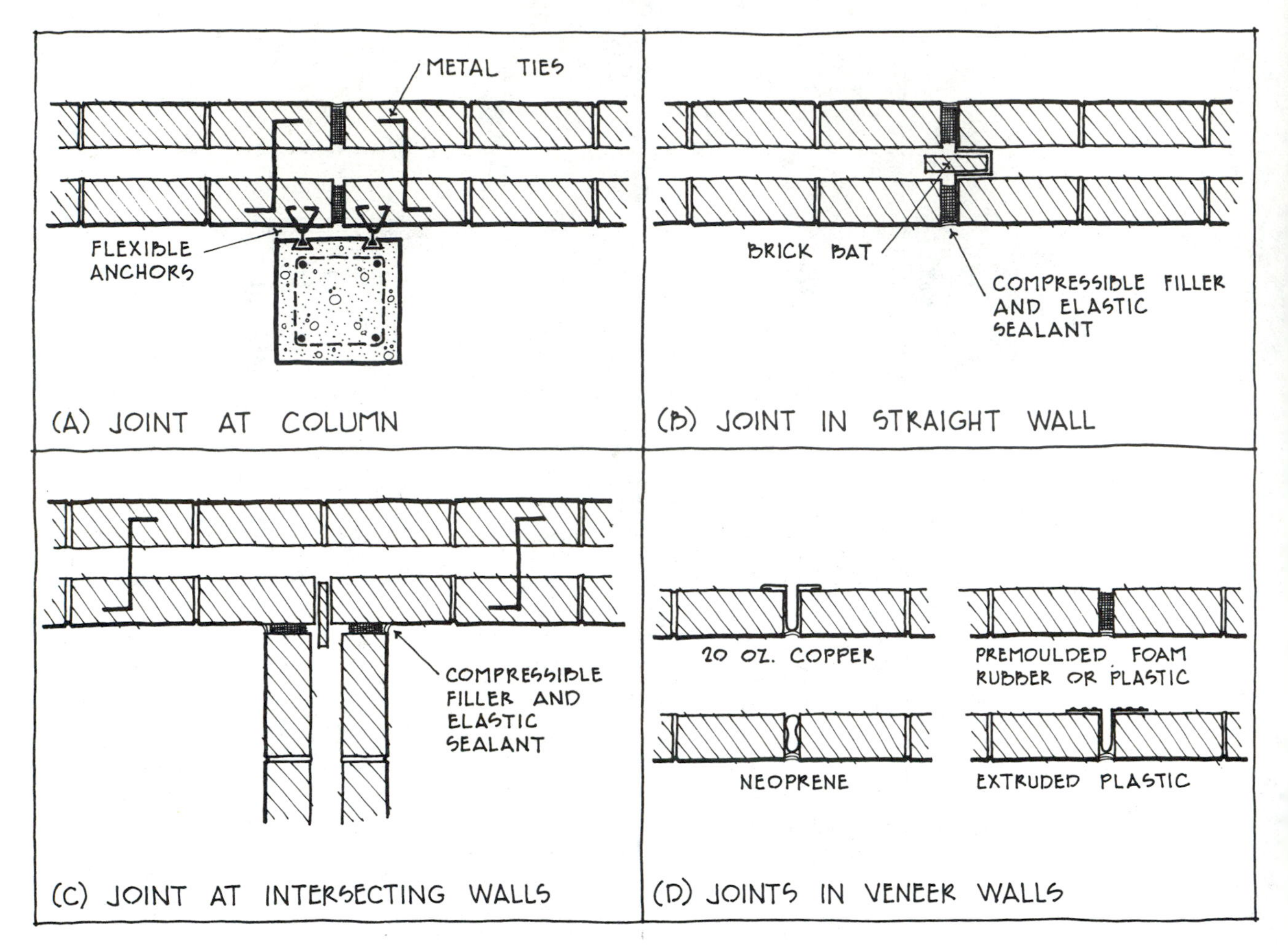

14
12
Brick expansion joint details. (*From Harry C. Plummer,* Brick and Tile Engineering, *Brick Institute of America, McLean, Va., 1962.*)

14.2.4 Movement Joint Locations

Openings, offsets, and intersections are the most effective locations fo movement joints for both brick and concrete masonry construction (*se Figs. 9-15 and 9-16*). In brick walls, expansion joints are often located at o near external corners of buildings, particularly when the masonry i resting on a concrete foundation. The shrinkage of the concrete, combine with the growth of the brick, can actually cause the wall to slip beyond th edge of the foundation (*see Fig. 14-13*). Brick parapet walls can experienc similar slippage from differential movement caused by a variation i temperature exposures. The building enclosure walls are heated and coole from the inside as the seasons change, thus moderating the temperature o the wall itself. The parapet is exposed to higher and lower extremes. As result, cracking, slippage, and separation often occur at this point. On solution to the problem is to anchor the parapet more securely to the wall o roof slab below, to add partial reinforcement to unreinforced walls, and t provide an extra expansion joint in the parapet between those carried u from the building wall below. Brick cavity walls can experience simila

14/13 Cracking at a building corner due to brick expansion restrained by concrete shrinkage. (*Photo courtesy BIA.*)

expansion problems since the outer wythe is exposed to greater temperature fluctuations than the inner wythe. Placement of expansion joints near the external corners and offsets of the building can help alleviate cracking at this point. Added resistance may be obtained by incorporating reinforced bond beams at the base of the wall, at floors, and at roof slabs (*see Fig. 14-14*). When bond beams are used in unreinforced construction, the walls are considered partially reinforced. In addition to resisting the formation of cracks from shrinkage, expansion, or differential movement, the bond beams also help distribute concentrated loads which may be imposed on the wall.

14.2.5 Differential Movement

One of the principal causes of cracking in masonry walls is differential movement between the masonry and other materials. In skeleton frame buildings of steel or concrete, masonry is often used as a veneer or

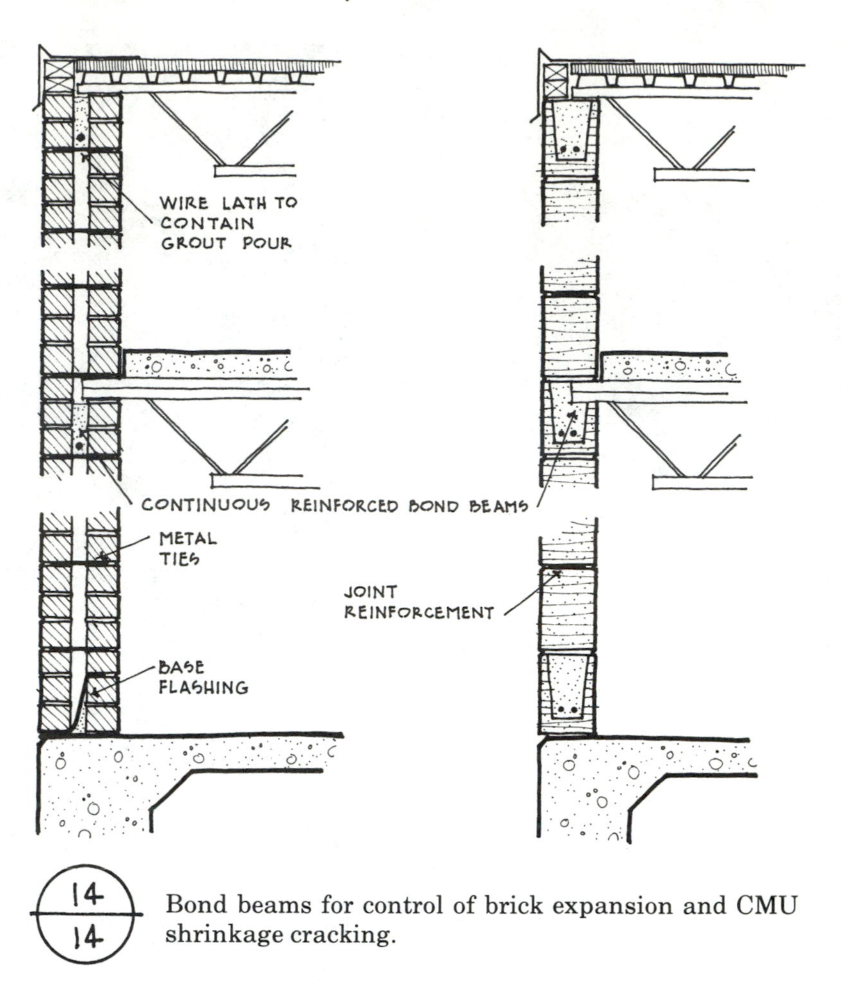

14/14 Bond beams for control of brick expansion and CMU shrinkage cracking.

curtainwall rather than as a structural element. These situations call no only for horizontal expansion joints, but for flexible anchorage as well.

It is normally recommended that masonry walls be supported entirel on the foundation. In some instances, however, the height of the wall or large number of openings may require steel shelf angles or concrete ledge for intermediate support. The difference in thermal and moisture expansio rates of steel, concrete, and masonry are discussed in detail in Chapter 8 The primary means of accommodating this differential movement is wit *horizontal expansion joints.* These pressure-relieving joints are constructe by placing an elastic pad beneath the shelf angle or ledge to absorb th deflection of the frame member (*see Fig. 14-15*). The pad joint is then fille with a permanent elastic sealant matching the mortar color. To make thes horizontal joints even less conspicuous, the bottom course of masonr resting on the framing member may be slightly projected to create a shado line at the joint. In framing, the angle supporting the masonry should b adequately shimmed and anchored to prevent rotation, which could induc high concentrated stresses in the masonry below. Maximum desig deflection of steel angles and concrete ledges should be $\frac{1}{600}$ of the span or 0. in.

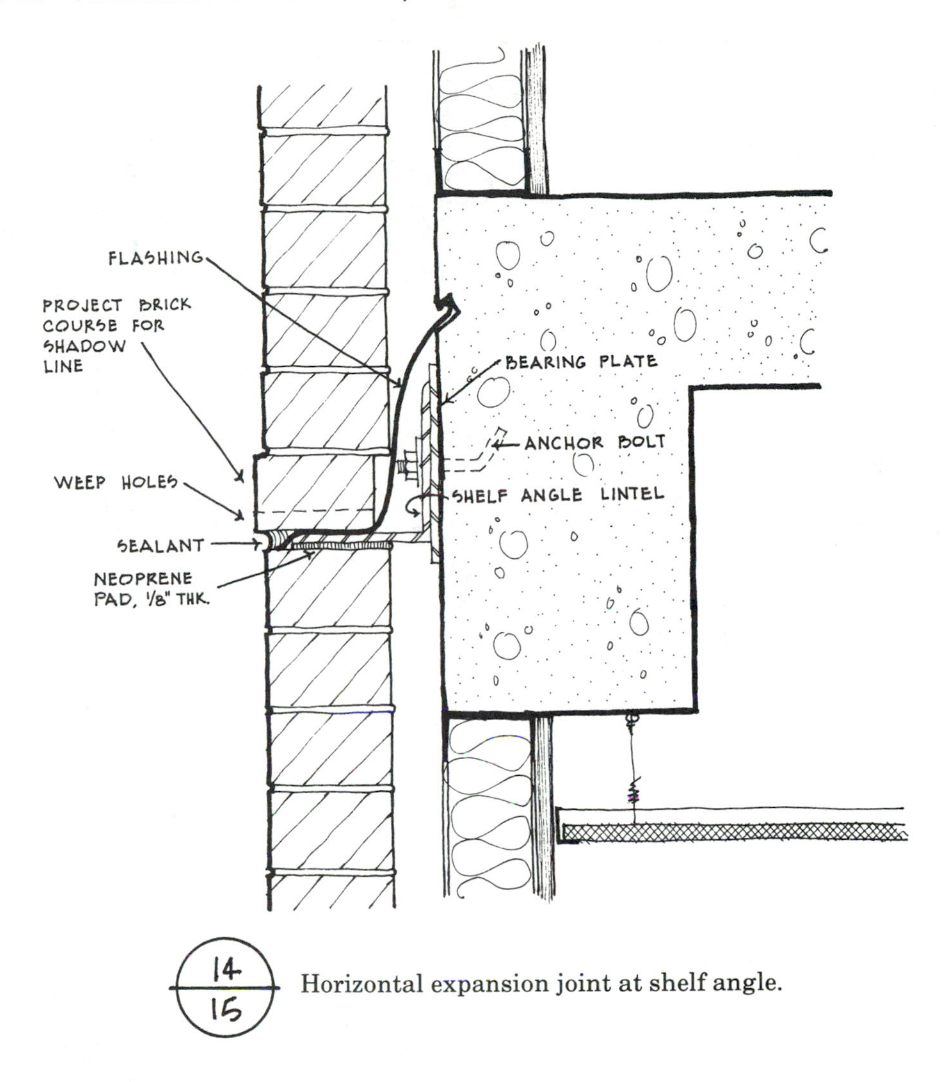

Horizontal expansion joint at shelf angle.

14.2.6 Flexible Anchorage

When masonry walls are used in conjunction with steel or concrete frame buildings, differential movement must also be accommodated in the anchorage of one material to another. Flexible connections will allow relative movement without inducing stress and causing damage. Even if the exterior masonry veneer carries its own weight to the foundation without shelf angles or ledges, concrete or steel columns or floors provide the lateral support which is required by code. Where anchors tie walls to the structural frame, they should be flexible, resisting tension and compression, but not shear. Various types of anchors were discussed in Chapter 7, and the illustrations in *Figs. 14-16* and *14-17* show methods in which these accessories are used.

In loadbearing masonry construction the brick or block walls support concrete floor slabs or steel joist and metal deck floors. The methods of anchorage will vary for different conditions. With concrete block and concrete slabs, there is less concern for differential movement because of the similarity of the material characteristics. Connections may be either rigid or flexible, depending on the particular design situation (*see Fig. 14-18*). In brick masonry, it is more common to provide a bond break or

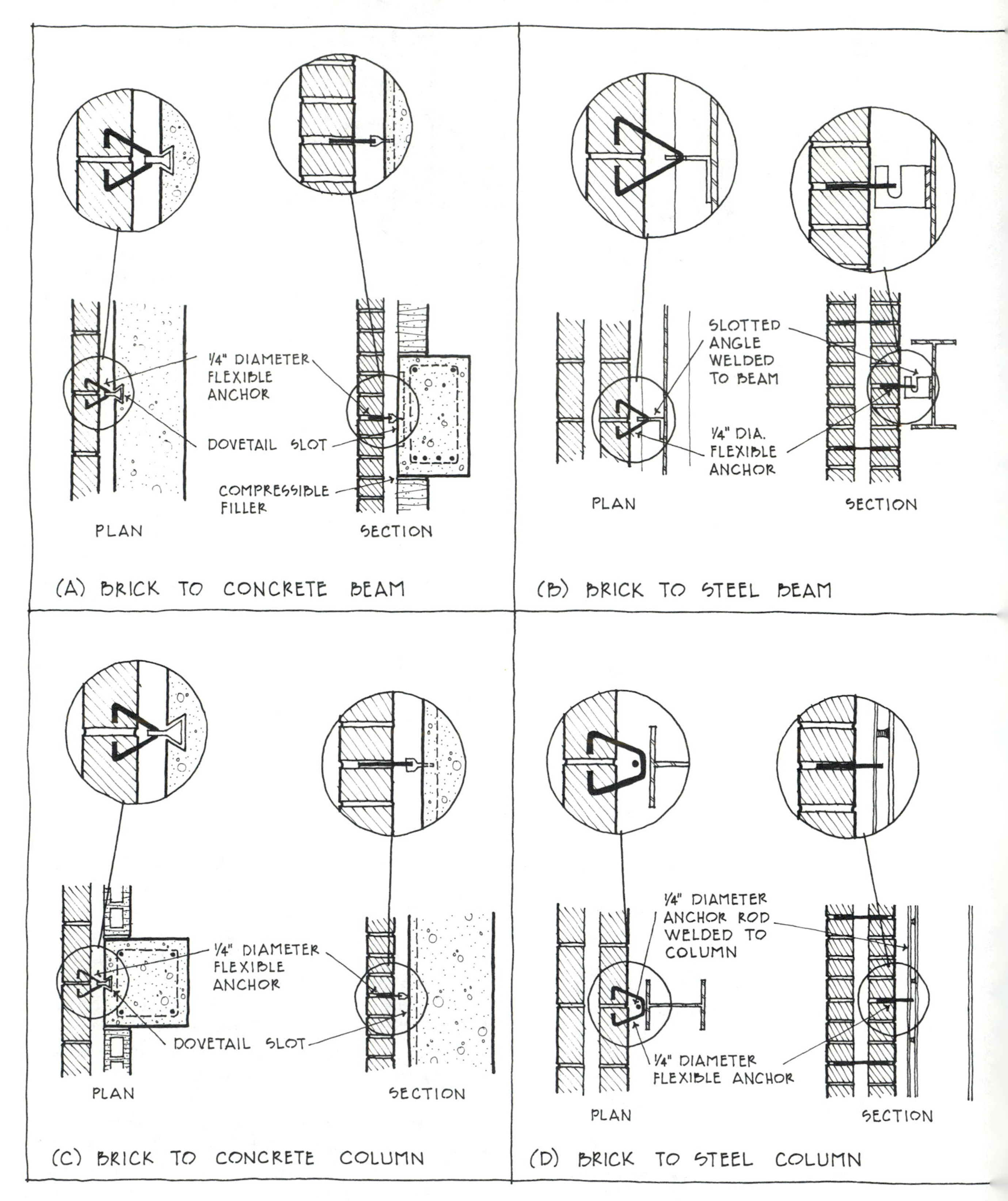

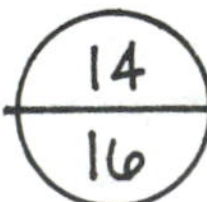

Flexible anchorage of brick masonry. (*From Brick Institute of America*, Technical Note 18, *BIA, McLean, Va.*)

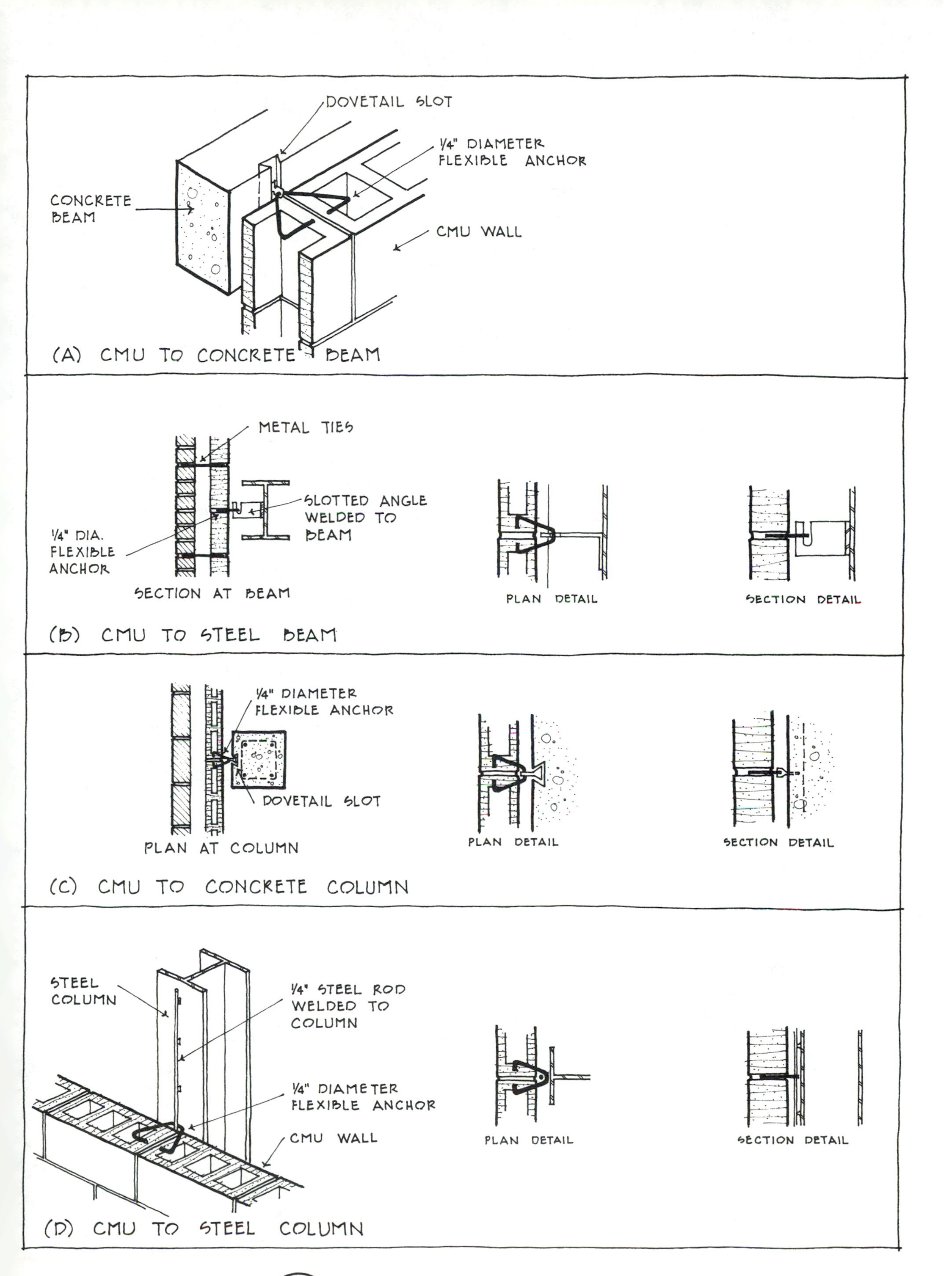
DOVETAIL SLOT
1/4" DIAMETER FLEXIBLE ANCHOR
CONCRETE BEAM
CMU WALL
(A) CMU TO CONCRETE BEAM
METAL TIES
SLOTTED ANGLE WELDED TO BEAM
1/4" DIA. FLEXIBLE ANCHOR
SECTION AT BEAM
PLAN DETAIL
SECTION DETAIL
(B) CMU TO STEEL BEAM
1/4" DIAMETER FLEXIBLE ANCHOR
DOVETAIL SLOT
PLAN AT COLUMN
PLAN DETAIL
SECTION DETAIL
(C) CMU TO CONCRETE COLUMN
STEEL COLUMN
1/4" STEEL ROD WELDED TO COLUMN
1/4" DIAMETER FLEXIBLE ANCHOR
CMU WALL
PLAN DETAIL
SECTION DETAIL
(D) CMU TO STEEL COLUMN

14
17

Flexible anchorage of concrete masonry.

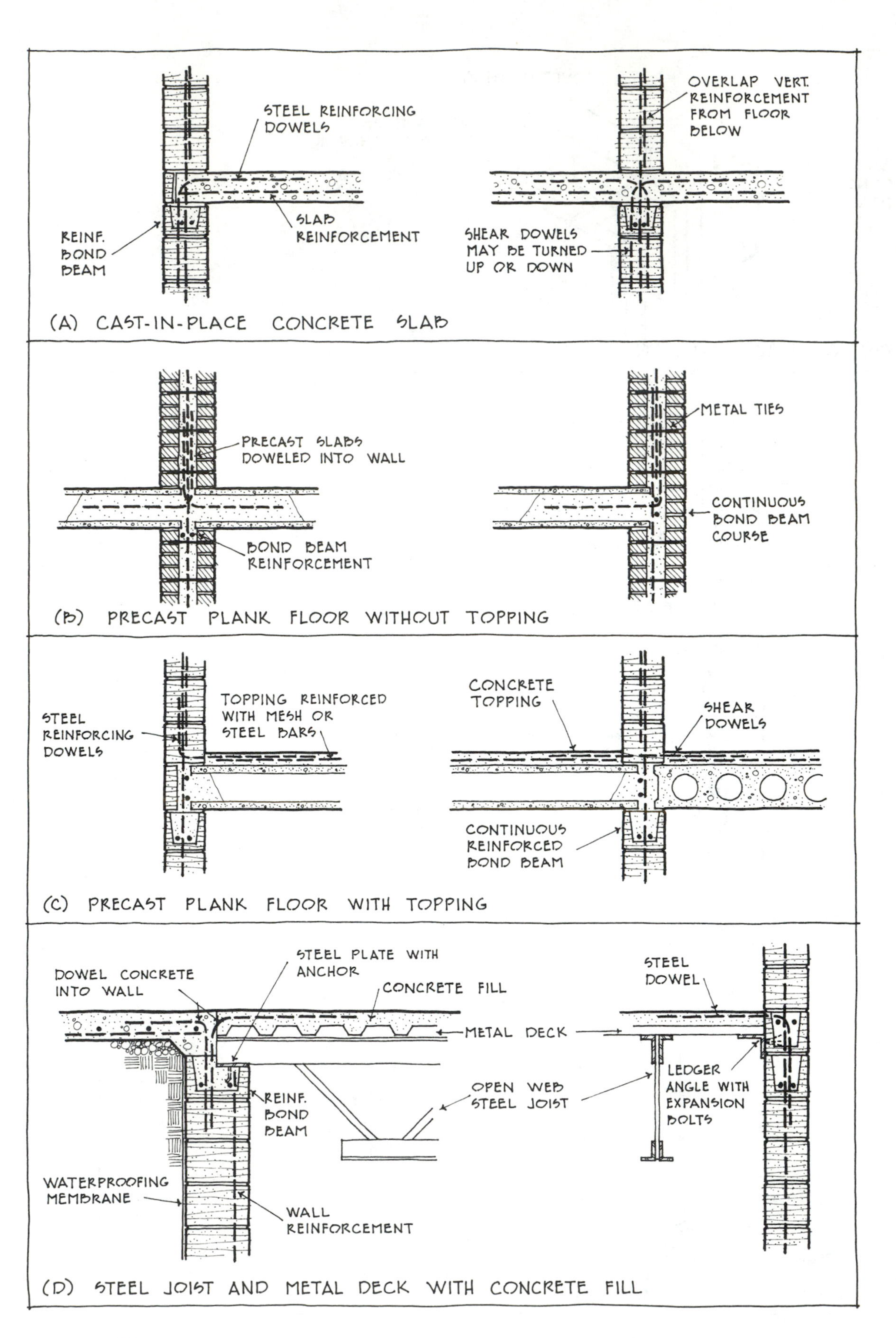

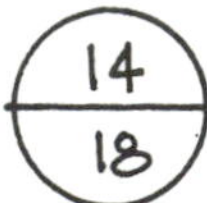

Wall-to-floor connections. (*From Robert R. Schneider and Walter L. Dickey,* Reinforced Masonry Design, *© 1980, p. 345. Reprinted by permission of Prentice-Hall, Inc., Englewood Cliffs, N.J.*)

slippage plane at the point where a concrete slab rests on the wall. Roofing felt or flashing is commonly used for this purpose, and allows each element to move independently while still providing the necessary support. The bond break may be detailed for both conditions where wall-to-slab anchorage is or is not required (*see Fig. 14-19*). Where masonry walls rest on a concrete foundation, anchorage between the two elements may often be eliminated because the weight of the wall and its frictional resistance to sliding are adequate for stability. In shear wall design where floor–wall connections must transfer loads through diaphragm action, anchorage must be designed as part of the engineering analysis.

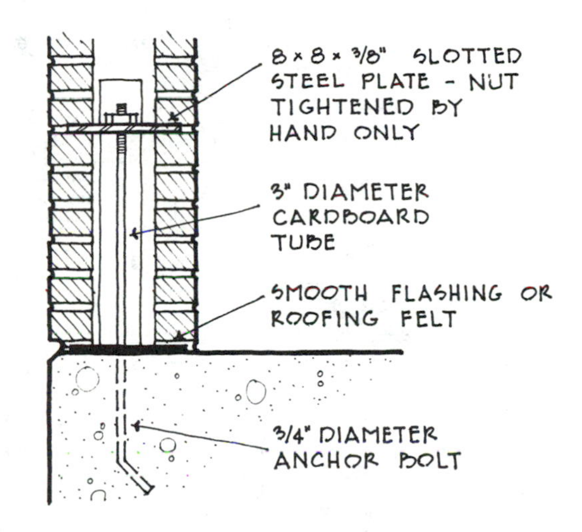

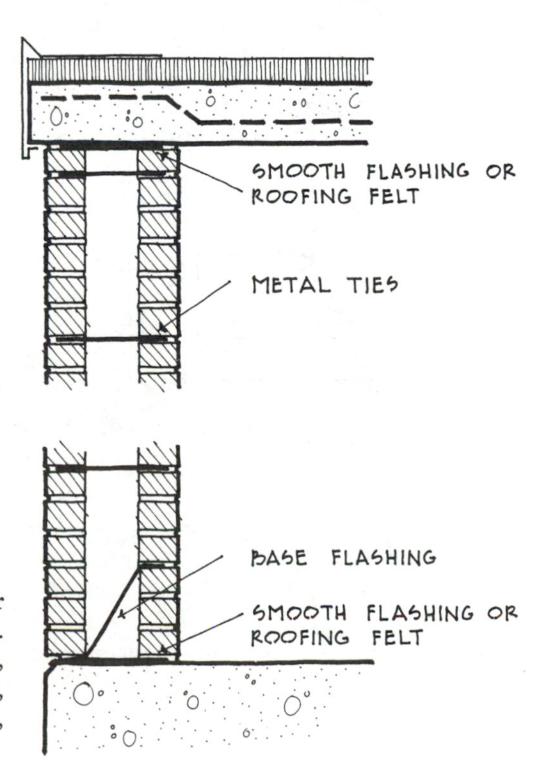

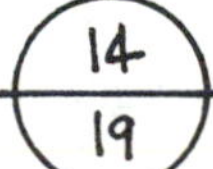

Bond breaks in wall-to-floor and wall-to-roof connections. (*From Harry C. Plummer,* Brick and Tile Engineering, *Brick Institute of America, McLean, Va., 1962.*)

14.2.7 Pattern Bonds

Unit design patterns have been used for centuries to add visual interest to masonry walls. Units may be laid in different positions as shown in *Fig. 14-21*, and arranged in patterns for a variety of effects (*see Figs. 14-20* and *14-22*). The patterns were originally conceived in connection with masonry wall bonding techniques that are not widely used today. In older work constructed without metal ties, rowlock and header courses were used to structurally bond the wythes of a wall together. Most contemporary buildings use the $\frac{1}{3}$ or $\frac{1}{2}$ running bond, or stack bond with very little decorative pattern work. In cavity wall construction, half rowlocks and half

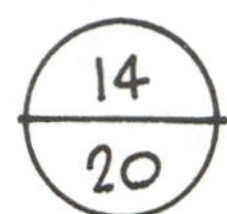

Brick bond patterns. (From *Brick Institute of America*, Technical Note 30, *BIA, McLean, Va.*)

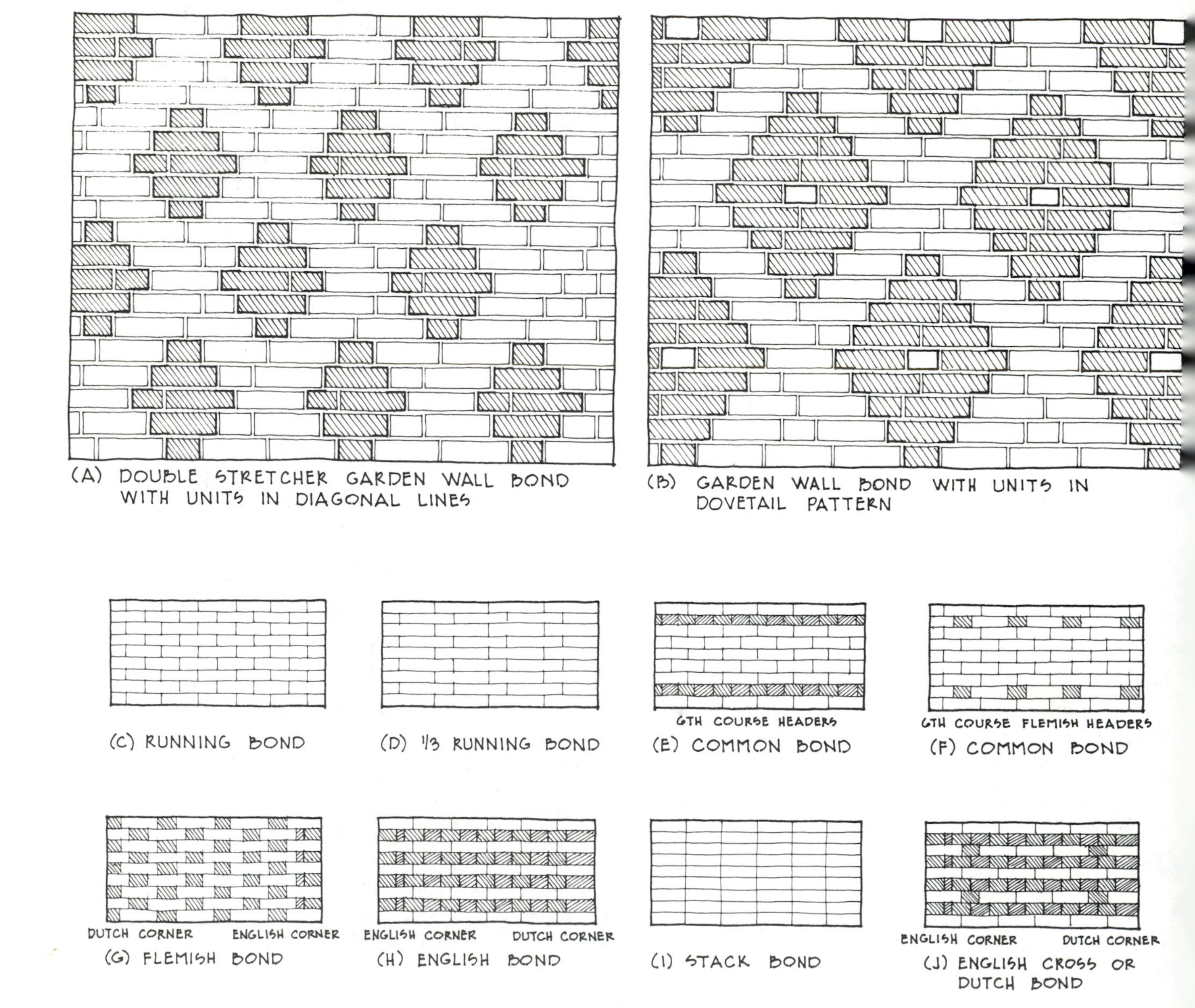

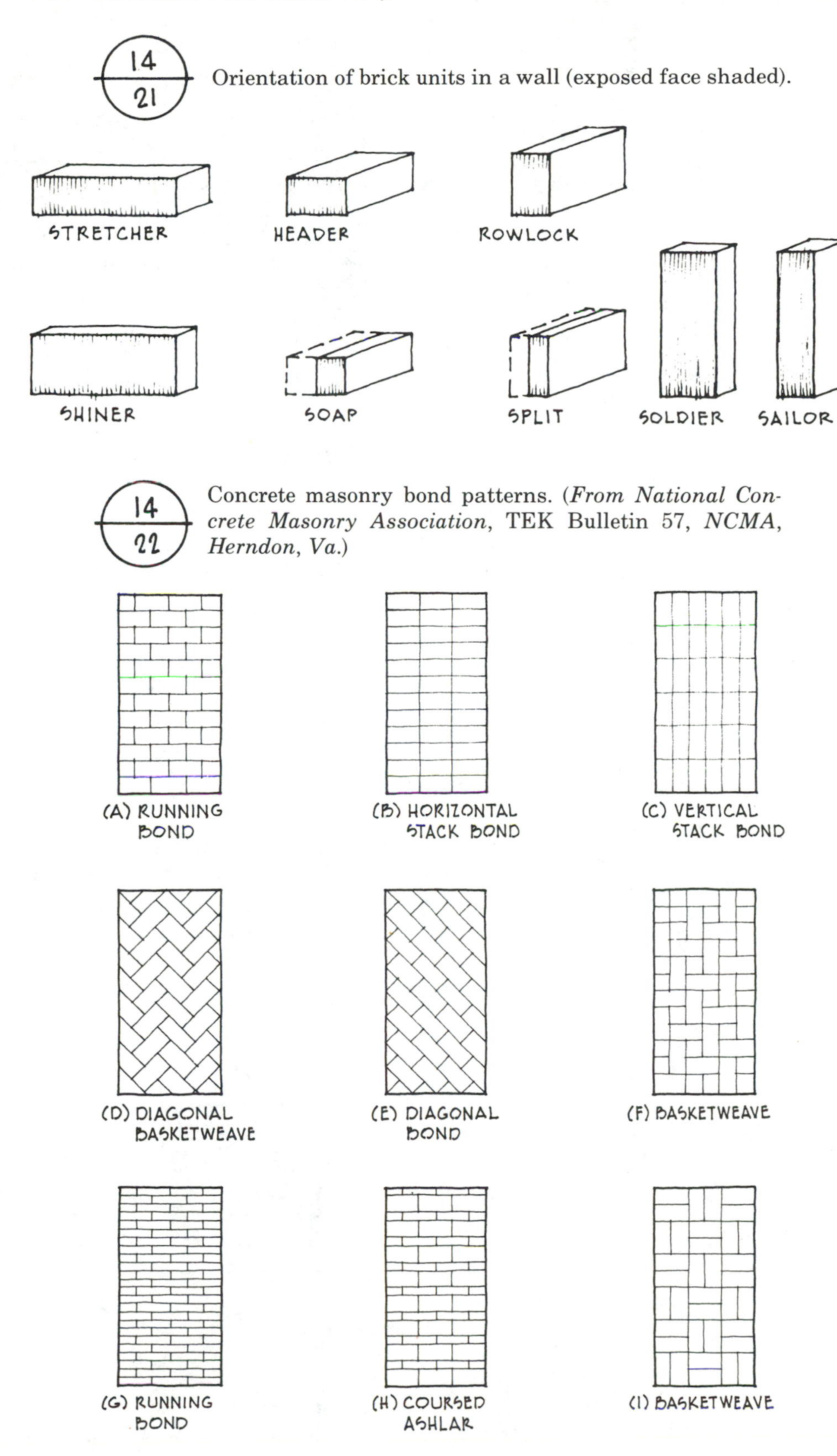

Orientation of brick units in a wall (exposed face shaded).

Concrete masonry bond patterns. (*From National Concrete Masonry Association*, TEK Bulletin 57, *NCMA, Herndon, Va.*)

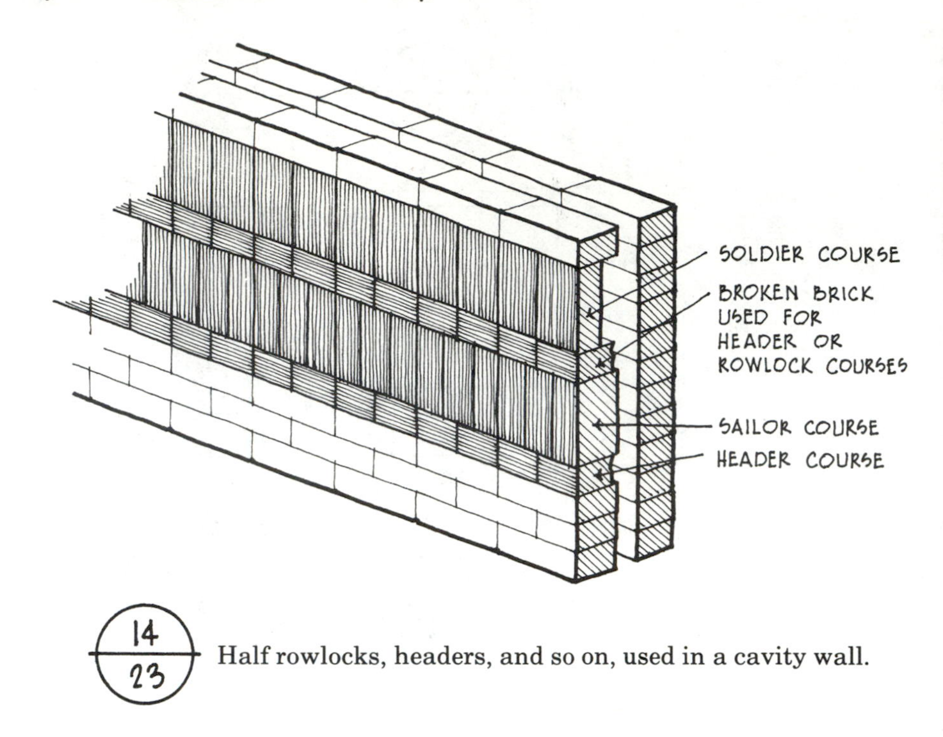

Half rowlocks, headers, and so on, used in a cavity wall.

headers may be used for aesthetic effect on the exterior without the unit actually penetrating the full thickness of the wall (*see Fig. 14-23*).

14.2.8 Mortar Joints

Other variations in aesthetic effect can be achieved by using different types of mortar joints. Two walls with the same brick and the same mortar color can have a completely different appearance depending on the joint treatment used. There are several types of joints common today (*see Fig. 14-24*). The most effective and watertight joints are the *concave* and *V shaped,* tooled joints. Mortar squeezes out of the joints as the masonry units

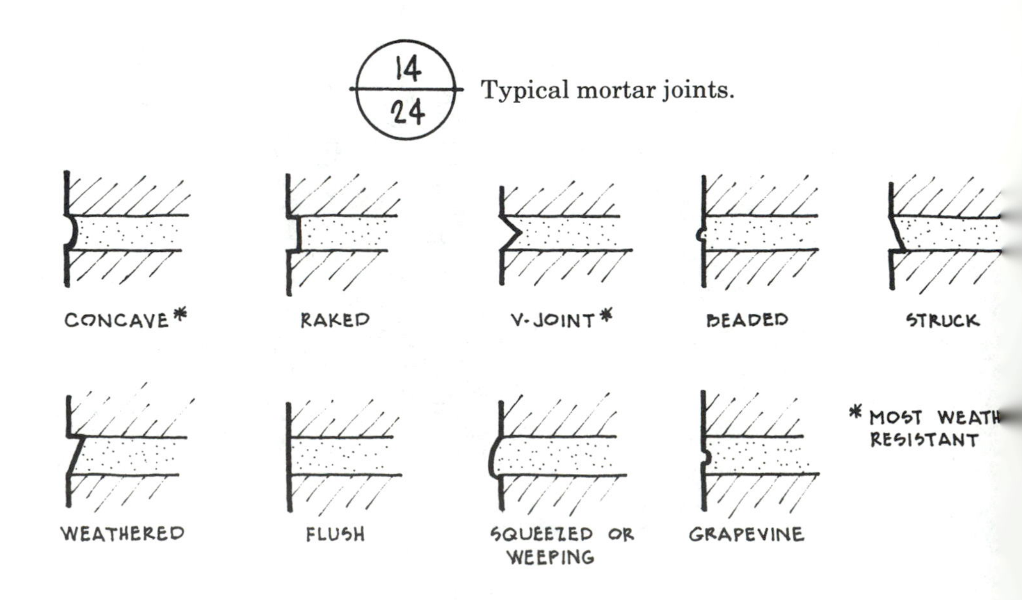

Typical mortar joints.

are set in place, and the excess is struck off immediately with a trowel. After the mortar has become "thumbprint" hard (when a clear thumbprint can be impressed and the cement paste does not stick to the thumb), joints are finished with a steel jointing tool slightly wider than the joint itself. As the mortar hardens, it has a tendency to shrink slightly and separate from the edge of the masonry unit. Proper tooling seals these small cracks and compacts the surface of the mortar making it more dense and impervious.

Horizontal joints should be tooled before vertical joints, using a jointer that is at least 22 in. long and upturned on one end to prevent gouging. Jointers for vertical tooling are small and S-shaped. Although the material most commonly used for these tools is steel, plexiglass jointers are often used to avoid staining of white or light-colored mortars. After the joints have been tooled, mortar burrs or ridges should be trimmed off flush with the face of the unit with a trowel edge, or by rubbing with a burlap bag, a brush, or a piece of carpet.

For exterior walls, proper finishing of the joint is not only for appearance, but also to prevent the penetration of moisture to the interior. The firm pressure applied with a jointing tool compresses the mortar and pushes it tightly against the unit forming an effective seal against moisture penetration. Tooled joints are highly recommended for use in areas subject to heavy rains and high winds. *Rough-cut* or *flush* joints are used when other finish materials, such as stucco, gypsum board, or textured coatings, are to be applied over the masonry. *Weathered* joints are more difficult to form since they must be worked from below, but some compaction does occur, and the joint sheds water easily. *Struck* joints are easily cut with a trowel point, but the small ledge created collects water, which may then penetrate the wall. *Raked* joints are made by scraping out a portion of the mortar while it is still soft, using a square-edged tool. Even though the mortar is compacted by this action, it is difficult to make the joint weathertight, and it is not recommended where rain, high winds, or freezing are likely to occur. The cut of the joint does form a shadow, and tends to give the wall a darker appearance. *Weeping* joints leave excess mortar protruding from the joint to give a rustic appearance, but again are not weathertight. Other, more specialized effects can be achieved with tools to bead or groove the joint.

Mortar color and joint type can be just as important in determining the appearance of a wall as the selection of a unit type or color, and should be carefully considered in the design of the building. Sample panels at the job site can help in evaluating workmanship and appearance of the finished work, and should always be specified on jobs of any mentionable size to assure that the desired effect can be achieved.

Even with high-quality workmanship, some patching or repair of mortar joints may be expected. Mortar in a head joint may have fallen out when the units were being laid, cracks may have formed when the units were aligned, or there may have been insufficient mortar to fill the space left by a broken corner or edge. In addition, any holes left by nails or line pins must be filled with fresh mortar before tooling. The troweling of mortar into joints after the units are laid is known as *pointing*. It is preferable that pointing and patching be done while the mortar is still fresh and plastic, and before final tooling of the joints is performed. If however, the repairs must be made after the mortar has hardened, the joint must be raked or chiseled out to a depth of about $\frac{1}{2}$ in. thoroughly wetted, and repointed with fresh mortar.

14.2.9 Laying Units

The laying up of masonry walls is a very ordered and controlled construction process. Units must remain in both vertical and horizontal alignment throughout the height and length of the structure in order for the coursing to work out with opening locations, slab connections, anchorage to other structural elements, and so on. Laying out of the first course is obviously of critical importance, since mistakes at this point would be difficult, if not impossible, to correct later. The first course must provide a solid base upon which the remainder of the walls can rest.

After locating the corners of the structure, it is a good idea to check dimensions by either measuring or actually laying out a dry course of units. Chalk lines are used to establish initial alignment on the ground, and string lines are used once the walls are up in the air. Base courses must always be laid in a full bed of mortar even if face-shell bedding (for CMU's) is to be used in the rest of the wall. Corner units are laid first and walls worked from outside corners and openings toward the center.

Concrete block should always be laid with the thicker end of the face shell up since this provides a larger mortar bedding area. For vertical joints only the end flanges of the face shells are "buttered" with mortar. Masons often place several blocks on end and apply mortar to the ends of three or four units at one time. Each block is then individually placed in its final position, tapped down into the mortar bed, and shoved against the previously laid block, thus producing well-filled vertical head joints. Once masonry units have been laid and the mortar has begun to set (about 3 minutes), the units cannot be adjusted or realigned by tapping without breaking the bond. If it is necessary to reposition the masonry, remove all the old mortar and replace it with fresh.

Although a slight furrowing of the initial bed joint for concrete block is permissible, furrowing is not considered good workmanship in brick construction. All bed joints are laid full and only slightly beveled away from the cavity to avoid mortar droppings in double-wall construction (*see Fig. 14-25*). Tests conducted by the National Bureau of Standards have proven

Beveled bed joints help prevent mortar droppings from filling the cavity between wythes. (*From Brick Institute of America,* Technical Note 21C, *BIA, McLean, Va.*)

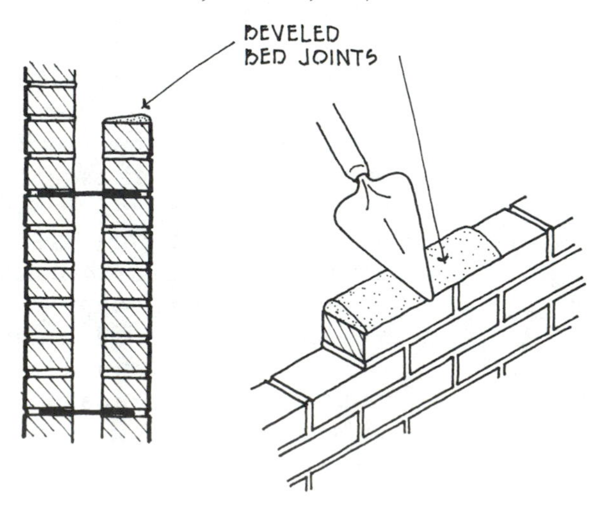

that brick walls with full head and bed joints are stronger and less likely to leak than walls with furrowed bed joints and lightly buttered head joints. Partially filled mortar joints result in leaky walls, reduce the strength of the masonry 50 to 60%, and may contribute to spalling and cracking due to freeze–thaw cycles in the presence of moisture. The mason should use ample mortar on the end of the brick so that when it is shoved into place, mortar will squeeze out. In spreading the bed joint, only a few bricks should be covered at one time so that the mortar will not dry before the units are laid in place.

When installing the last closure unit in a course, all edges of the opening and all vertical edges of the unit should be buttered and the unit carefully lowered into place. If any of the mortar falls out leaving an open joint, the closure unit should be removed and the operation repeated.

For both brick and block masonry, a long mason's level is used as a straightedge to assure correct horizontal alignment. Units are brought to level and made plumb by tapping with the trowel handle.

In cavity wall construction, it is extremely important that the cavity between the outer and inner wythe be kept clean to assure proper drainage of moisture that may enter the wall. If mortar clogs the cavity, it can form bridges for moisture passage, or it may plug weepholes at the bottom of the wall. Some masons use a removable wooden strip to block the cavity during construction and prevent mortar droppings. However, beveling back of the mortar bed and care in placement of the units will allow very little mortar to squeeze toward the cavity. Bricks are rolled into place, keeping most of the squeezed mortar toward the outside of the wall. Any fins that may protrude into the cavity should be cut off or flattened to prevent interference with the placement of reinforcement, grout, or insulation.

Partially completed structures may be subject to loads that exceed their structural capabilities. Wind pressure, for instance, can create four times as much bending stress in a new, freestanding wall as in the wall of a completed building. Fresh masonry with uncured mortar has no tensile strength to resist such lateral forces. Most codes require that new, uncured, unanchored walls be braced against wind pressure. Bracing should be provided until the mortar has cured and the wall has been integrally tied to the structural frame of the building. Bracing is designed on the basis of wall height and expected wind pressures.

14.2.10 Weep Holes

Weep holes are required on cavity wall construction at the base course and at all other flashing levels (such as shelf angles, sills, lintels, and so on). Weep holes should be spaced at approximately 24 in. on center and can be installed by a variety of methods: (1) place oiled rods, rope, or pins in the head joints and remove before final set of the mortar; (2) place sash cord or other suitable wicking material in the head joint; (3) place metal or plastic tubing in the head joint; or (4) eliminate each second or third head joint.

14.2.11 Story Poles

The corners of masonry walls are usually built four or five courses higher than the center of the wall, and as each course is laid, it is checked for level, plumb, and alignment. Story poles are often used to simplify the accurate location of the top of each course or series of courses. A story pole is simply a board with markings 8 in. apart which may be stood at a corner to check proper coursing.

When filling in between the corners of a wall, a string line is stretched from end to end and the top outside edge of each unit is laid to this line. Use of the mason's level between corners is then limited to checking the face of the units to assure that they are in the same plane. This speeds construction time and assures greater accuracy.

14.2.12 SCR Masonry Process

The research division of the BIA undertook a time and motion study a number of years ago to find ways of improving productivity and workman ship. It was found that the bricklaying process itself was not the major time-consuming factor, but rather that the amount of wall area completed per day was in direct relation to the time needed for line adjustment leveling, handling mortar, and reaching for materials. The "SCR masonry process" is the system designed by BIA to improve productivity. The process uses an adjustable scaffold with a materials platform, and the "SCR corner pole" to support the line. The scaffold must be continuously adjustable in height, and must provide a platform for materials about 24 in above that upon which the mason stands. This allows more comfortable working heights by reducing the fatigue of bending and lifting.

The corner pole is a simple tool which eliminates the building of corne leads. They are generally of metal with adjustable coursing scales attached The poles must be rigid enough to resist bending when a string line is pulle from one side; must be easily attachable to the masonry walls at the corners and must be easily plumbed and maintained for the height of the wal Using corner poles assures that the brick coursing is uniform and level, tha the wall is plumb, and that joint thicknesses are more uniform tha attainable by normal methods. The proven results of the SCR process are a increase in mason productivity by eliminating lost time, a reduction i overall cost, and improved, more consistent workmanship.

14.2.13 Reinforcement and Accessories

Metal ties, anchors, horizontal joint reinforcement, and steel reinforcin bars are all placed by the mason as the work progresses. Cavity walls ar tied together with metal ties (see Chapter 7) or with horizontal joir reinforcement, depending on the type of units used and the particula design criteria. Rectangular or Z-ties are generally used when both wythes (the wall are brick. Most codes require a minimum of one $\frac{3}{16}$-in.-diamete steel tie (or its equivalent) for every $4\frac{1}{2}$ sq ft of wall area. Ties must b corrosion resistant, properly spaced, and placed in full mortar bed joints t assure full bond. Crimped ties that form a water drip are not necessary, an can decrease the strength of the tie. When one wythe is concrete block an the other brick, ladder or truss-type horizontal joint reinforcement generally laid in the block courses, with a rectangular extension whic reaches across the collar joint to tie the brick wythe to the block. Both typ of ties may easily be used whether the two wythes are laid up simultan ously or separately.

Vertical reinforcement in a cavity wall is easily placed, and th masonry built up around it. Spacers are required at periodic intervals to hold the reinforcing bars in vertical alignment. If horizontal steel is required in the cavity, it is tied to the vertical members or may rest on the spacers at the proper intervals (*see Fig. 14-26*).

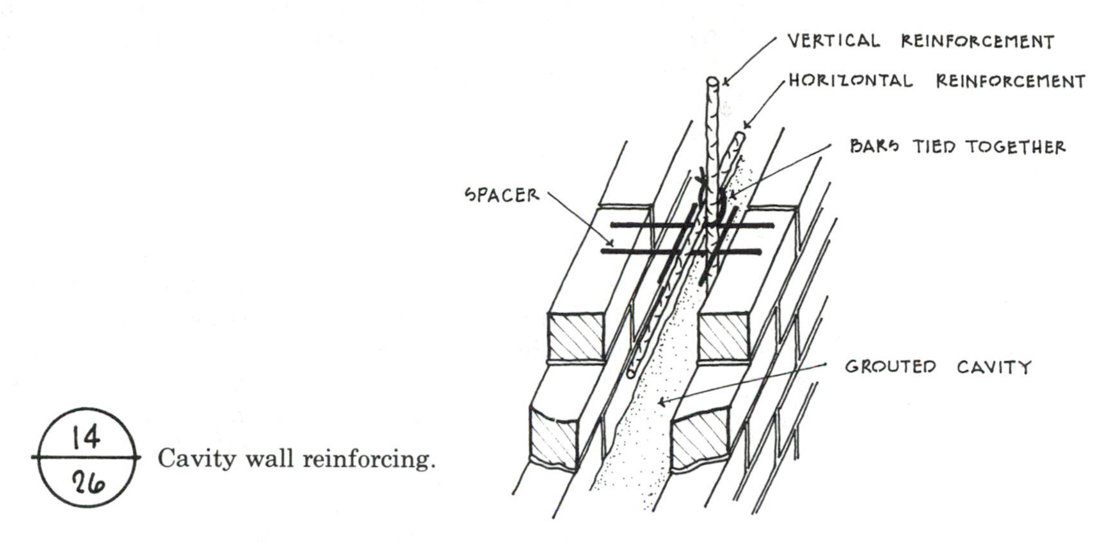

Cavity wall reinforcing.

For single-wythe reinforced CMU walls, special units are made (sometimes called A-blocks) with one end open so that the block may be placed around the vertical steel rather than threaded over the top of the bar (*see Fig. 14-27*). Some specially designed blocks have been produced which can accommodate both vertical and horizontal reinforcing without the need for spacers. The "Ivany Block" shown in *Fig. 14-27* not only has the open ends mentioned above, but incorporates notches in the webs for placement of horizontal bars. This type of unit is very economical for grouted,

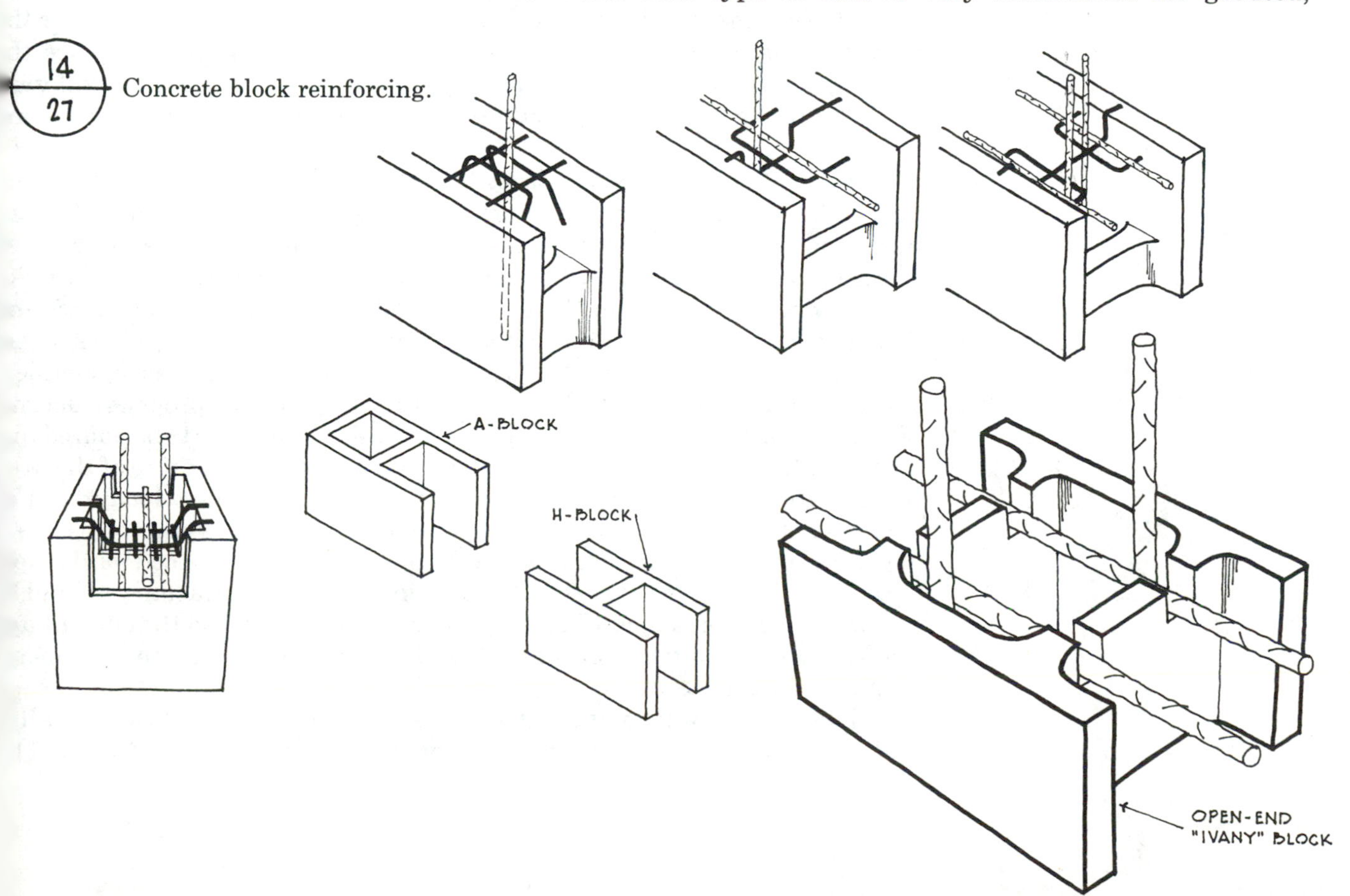

Concrete block reinforcing.

reinforced CMU walls, particularly when the design utilizes wall beams and bond beams requiring large quantities of horizontal steel.

During the course of construction, the mason also places anchorages and cutouts required to fit the work of other trades. These items are furnished and located by others, but incorporated into the wall by the mason. Steel or precast lintels for small openings are also placed by the mason if reinforced masonry lintels are not used in the design.

14.2.14 Insulation

Use of the various types of insulation covered in Chapter 8 will affect the manner in which the masonry walls are laid up. In veneer construction over wood frame, the board or batt insulation and the corrugated metal ties are placed against the frame before the masonry work is begun. If rigid board insulation is used in insulated cavity walls, the backing wall must be laid up higher than the facing wall so that the boards may be attached to it before the facing wythe covers it. If the masons are working overhand from inside the building (as they often do on multistory construction), this makes the insulating process more awkward, and therefore less economical. In these cases, the masons would work better from scaffolding on the outside of th building, but the cost of the installation would increase.

Loose fill insulation does not require that the two wythes of masonr be laid up separately. Both the inner and outer wythes can be laid up simultaneously, and the insulation poured or pumped into place at desig nated vertical intervals.

14.2.15 Grouting

In reinforced masonry construction, the collar joint of a cavity wall, or th vertical cells of hollow units, must be pumped with grout to secure th reinforcing and bond it to the masonry. Methods and procedures will var slightly among contractors, but there are some common practices which a universally applicable.

For hollow masonry construction, the cells that are to be grouted mus be fully bedded in mortar, including the webs and face-shell flanges. In bot brick and CMU work, the importance of keeping the cavity clean has bee stressed before, but should be reemphasized here. Protrusions or fins mortar which project into the cavity will interfere with proper flow an distribution of the grout, and could prevent complete bonding. The space used to maintain alignment of vertical reinforcing will assure comple coverage of the steel and full embedment in the grout for proper structur performance. If bond beams or isolated in-wall columns are to be poured in cavity wall, material must be placed below and/or to either side of the are to prevent the grout from flowing beyond its intended location. F example, if a bond beam is to be poured in a double-wythe brick wa expanded metal lath or metal screen should be placed in the bed joint belo to contain the pour (*see Fig. 14-14*). Grouting of concrete masonry should performed as soon as possible after the units are placed so that shrinka cracking at the joints is minimized, and so that the grout bonds proper with the mortar.

The *low-lift method* of grouting a wall is done in 8 in. lifts a the wall laid up. For cavity wall construction, the first wythe is laid up, followed the second wythe, which is generally left 8 to 12 in. lower. Grout should be well mixed to avoid segregation of materials, and carefully poured to avoid splashing on the top of the brick, since dried grout will prevent proper mortar bond at the succeeding bed joint. At least 15 minutes should elapse between pours to allow the grout to achieve some degree of stiffness before the next layer is added. If grout is poured too quickly, and the mortar joints are fresh, hydrostatic pressure can cause the wall to bulge out of plumb. A displacement of as little as $\frac{1}{8}$ in. will destroy the bed joint bond, and the work must be torn down and rebuilt. The joint rupture will cause a permanent plane of weakness and cannot be repaired by simply realigning the wall.

Bed joints can also be broken by rotation of the brick from uneven suction. To avoid this, the grout level should be kept at or below the center of the top course during construction. If operations are to be suspended for more than 1 hour, however, it is best to build both wythes to the same level, and pour the grout to within $\frac{3}{4}$ in. of the top of the units to form a key with the next pour. Grout that is in contact with brick hardens more rapidly than that in the center of the grout space. It is therefore important that agitation or puddling of the grout take place immediately after the pour and before this hardening begins.

In single-wythe hollow-unit construction, walls are built to a maximum 4 ft height before grout is pumped or poured into the cores. These vertical cores must have the minimum area and dimensions required by the governing building code. Grout may be poured or pumped into the cores, and is then agitated to ensure complete filling and solid embedment of steel.

High-lift grouting of cavity walls requires bonding the two wythes together with metal ties spaced as required by code. One wythe of the wall is built up not more than 16 in. above the other, and vertical grout barriers of solid masonry are placed a maximum of 25 ft apart. Single lifts generally may not exceed 4 ft in height, and must be completed in one operation with no interruptions longer than 1 hour. A period of 30 to 60 minutes must elapse between successive lifts.

Cleanouts must be provided at the base of the wall by leaving out every other unit in the bottom course of the section being poured. A high-pressure air blower is used to remove any debris which may have fallen into the cavity. The cleanouts are filled in after inspection of the cavity, but before the grouting begins. The mortar joints in a wall should be allowed to cure for at least 3 days to gain strength before grouting by this method. In cold, damp weather, or during periods of heavy rain, curing should be extended to 5 days. Temporary lateral bracing may be required to protect fresh masonry against lateral loads until the mortar and grout has fully set (*see Fig. 14-28*).

For single-wythe hollow masonry construction, grouting operations are not performed until the wall is laid up to full story height. Cleanout openings at least 3×4 in. are located at the bottom of every core containing dowels or vertical reinforcement, and in at least every second core that will be grouted, but has no steel. In solidly grouted, unreinforced walls, every other unit in the bottom course should be left out. Building codes generally specify exact cleanout requirements, and should be consulted prior to construction.

The grout should be placed in a continuous operation with no intermediate horizontal construction joints within a story height. Four-foot maximum lifts are recommended, with 30 to 60 minutes between pours to allow for settlement, shrinkage, and absorption of excess water by the units. In each lift, the top 12 to 18 in. are reconsolidated before or during

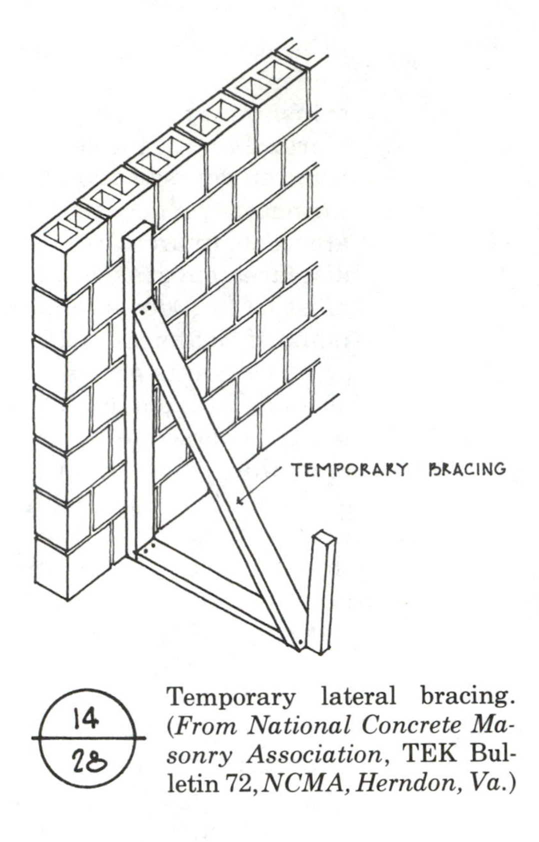

Temporary lateral bracing. (*From National Concrete Masonry Association*, TEK Bulletin 72, *NCMA, Herndon, Va.*)

placement of the next lift. The cross-webs of hollow units are fully em-bedded in mortar to contain the grout within the designated area.

14.2.16 Surface Bonding

In 1967, the U.S. Department of Agriculture introduced surface bonding c concrete block walls as an economical technique for construction of low-cos housing. The system calls for dry-stacking of the units without mortar, the coating both sides of the wall with a thin layer of fiberglass-reinforce cement plaster. Mortar is used in bed joints only at the base and bond bea courses. The $\frac{1}{16}$ to $\frac{1}{8}$-in.-thick surface coating binds the units together in strong composite construction; adds a protective, waterproof shield; an imparts great tensile strength against lateral loads.

Less time and skill are required for building surface-bonded walls, an productivity may be increased by as much as 70%. The waterproofin characteristics of the coating material are excellent, and with the additic of coloring pigments, the walls may be completely finished in one operatic without the added labor involved in painting.

Strength in bending and flexure is equal to and sometimes mu greater than conventionally built walls. Normal units provide slightly le resistance to compressive loads than conventional masonry, but if t bearing surfaces of the blocks are ground, the surface bonded wall is strong in compression as the conventional wall. The compressive streng of any CMU wall is directly related to the strength of the units used in construction. In conventional construction with mortar joints, the wa strength will equal about 50% of the unit strength, compared to 30% f unground surface-bonded units. The lower strength developed in surfa bonded walls is due to the natural roughness of the block, which preclud solid bearing contact between courses. The mortar bed in conventional construction compensates for this roughness and provides uniform bearing, as does precision surface grinding of the dry-stacked units. The table in *Fig. 14-29* reflects these design variables in the allowable stresses for ground and unground units in surface bonded construction. The maximum compressive stress on the gross area of the wall is 45 psi for unground units but increases to 85 psi for ground units, the same as permitted by many codes for conventional construction. Unground units, however, are more than adequate to support the superimposed loads normally encountered in a two- or three-story building.

Compressive:	45 psi based on gross area with unground masonry bearing surfaces 85 psi based on gross area with ground masonry bearing surfaces
Shear:	10 psi based on gross area
Flexural:	*Horizontal Span:* 30 psi based on gross area when units are drystacked in interlocking (running bond) pattern 18 psi based on gross area when units are drystacked in noninterlocking (stack bond) pattern *Vertical Span:* 18 psi based on gross area

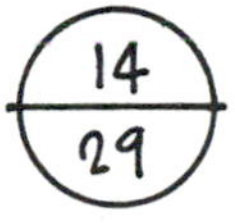

Allowable stresses for surface-bonded masonry. (*From National Concrete Masonry Association,* TEK Bulletin 74, *NCMA, Herndon, Va.*)

The method of constructing a surface-bonded wall is simple. The bottom course is set in mortar to compensate for any unevenness in the footings or slab, and to obtain a good, level base for the remainder of the units. The base course should not include mortared head joints, as this would upset the modular alignment of the succeeding courses. Dry-stacking of the rest of the units begins at the corners, and unit ends should be butted tightly together. Shims may occasionally be required with unground units to maintain the wall level and plumb. Control joints and bond beams are used to control temperature and moisture movements since there are no mortar joints in which to embed horizontal reinforcement (see *Fig. 14-30* for construction details).

Job-site mixing and application of proprietary bonding mortars should be in accordance with the manufacturer's recommendations. Unlike conventional construction, dry-stacked units are lightly sprayed with water just prior to application of the surface bonding mortar. Premixed mortars are smooth textured and easy to apply with a hand trowel or power sprayer. Sprayed-on coatings have slightly less tensile strength than trowel coatings. Troweling tends to orient the short, $\frac{1}{2}$-in. glass fibers in the same plane as the block surface, thus giving greater strength. The surface coating must be a minimum of $\frac{1}{16}$ in. thick, but preferably should average $\frac{1}{8}$ in. In some cases, two-coat spray applications have been made with supplemental troweling of the first coat for strength, and the second coat left as a stipple finish. The finishing techniques obviously lend themselves to regional architectural styles of the Southwest, where stucco is a commonly used material.

Since the basic cementitious material in surface bonding mortars is

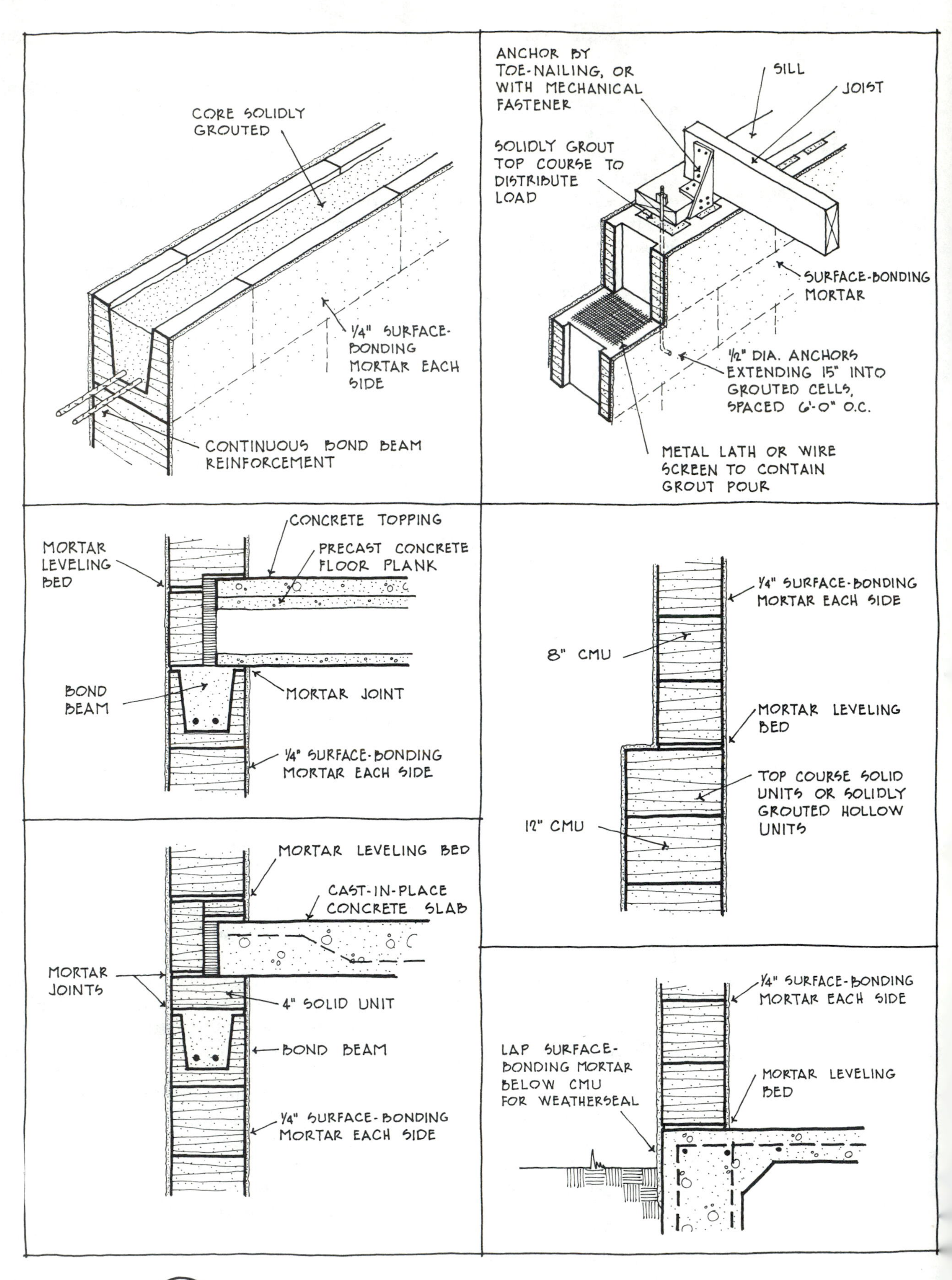

Details for surface-bonded masonry. (*From National Concrete Masonry Association*, TEK Bulletin 88, *NCMA, Herndon, Va.*)

portland cement, the coating should be damp cured for 24 hours after its application to prevent premature evaporation of moisture before the hydration process is complete.

14.3 MASONRY PAVING

Roads made of stone paving blocks were built by the Romans over 2000 years ago, and some are still in use today. Many cities in the United States and Europe have brick streets which continue in service after many years of heavy use, and have proven both durable and easy to maintain. After World War II, economical methods of manufacturing high-strength concrete paving blocks were developed. Since their introduction, concrete pavers have been used extensively in Europe and increasingly in the United States and Canada.

Clay, concrete, and stone paving units may all be used for interior or exterior applications, and may be installed over different subbases suitable for residential and commercial buildings, walkways, patios, streets, and parking areas. Paving assemblies are classified in accordance with the type of base used and the rigidity or flexibility of the paving itself. *Rigid paving* is defined as units laid in a bed of mortar and with mortar joints between the units. *Flexible paving* contains no mortar below or between the units.

Base supports may be rigid, semirigid, flexible, or suspended. A *rigid-base diaphragm* is a reinforced concrete slab on grade, and can accommodate either rigid or flexible paving. A *semirigid continuous base* usually consists of asphalt or other bituminous road pavement, and is suitable for flexible paving. A *flexible base* is compacted gravel or a damp, loose, sand-cement mixture which is tamped in place. Only flexible paving should be laid over this type of base. *Suspended diaphragm* bases are structural floor or roof deck assemblies, the composition of which will vary depending on the type of structural system used. Either flexible or rigid paving may be used on suspended decks.

Selection of the type of paving system to be used will depend to a large extent on the desired aesthetic effect and the intended use. There are a number of design considerations which must be taken into account, particularly for outdoor paving. Heavy vehicular traffic generally requires rigid base diaphragms or semirigid continuous bases. Lighter vehicular traffic may be supported on flexible bases and flexible paving, and pedestrian traffic, of course, can be accommodated by any of these means. Traffic patterns, which dictate the size and shape of a paved area, may also influence the choice of base and cushion material. Successful installations always depend on proper subgrade preparation and removal of all vegetation and organic materials from the area to be paved. Soft spots of poor soil should be removed and refilled with suitable material, then properly compacted.

Site preparation and system selection should also take into consideration the location of underground utilities and storm drainage systems. With rigid concrete bases and rigid masonry paving, access must be provided by means of manholes, cleanout plugs, and so on. If semirigid or flexible bases are used with flexible paving, however, the user may gain unlimited access to underground pipes and cables without incurring the expense of extensive surface repairs. This fact is generally cited as one of the major advantages of flexible masonry paving, which allows utility repairs and alterations by simply removing, stockpiling, and then replacing the paving units and base material. No air hammers or concrete pours are required to complete the

work, and there is reduced danger of damage to utilities by the eliminatio
of such equipment.

14.3.1 Outdoor Paving

Drainage is of major importance in the consideration of all outdoor pavin
systems, and excessive runoff is a growing environmental concern. I
addition to the mortarless paving units which permit a degree of wate
absorption through the joints, masonry paving systems have been de
veloped which lessen the impact of storm drainage even further. Thes
concrete masonry grid pavers (*see Fig. 4-14*) contain open spaces designe
for growth of indigenous grasses and the maintenance of soil permeabilit
while providing a stable base for vehicular traffic. Grid units have also bee
used in a variety of applications for soil stabilization, erosion control, an
aesthetic handling of drainage and access. Installations include shoulde
slopes along highways and under bridges, the lining of canals, constructio
of trailer parks, boat launch ramps, fire lanes adjacent to apartments an
hotels, and erosion control of steep embankments (*see Fig. 14-31*).

14/31 Grid paver installation examples (*From National Concrete Masonry Association*, TEK Bulletin 91, *NCMA, Herndon, Va.*)

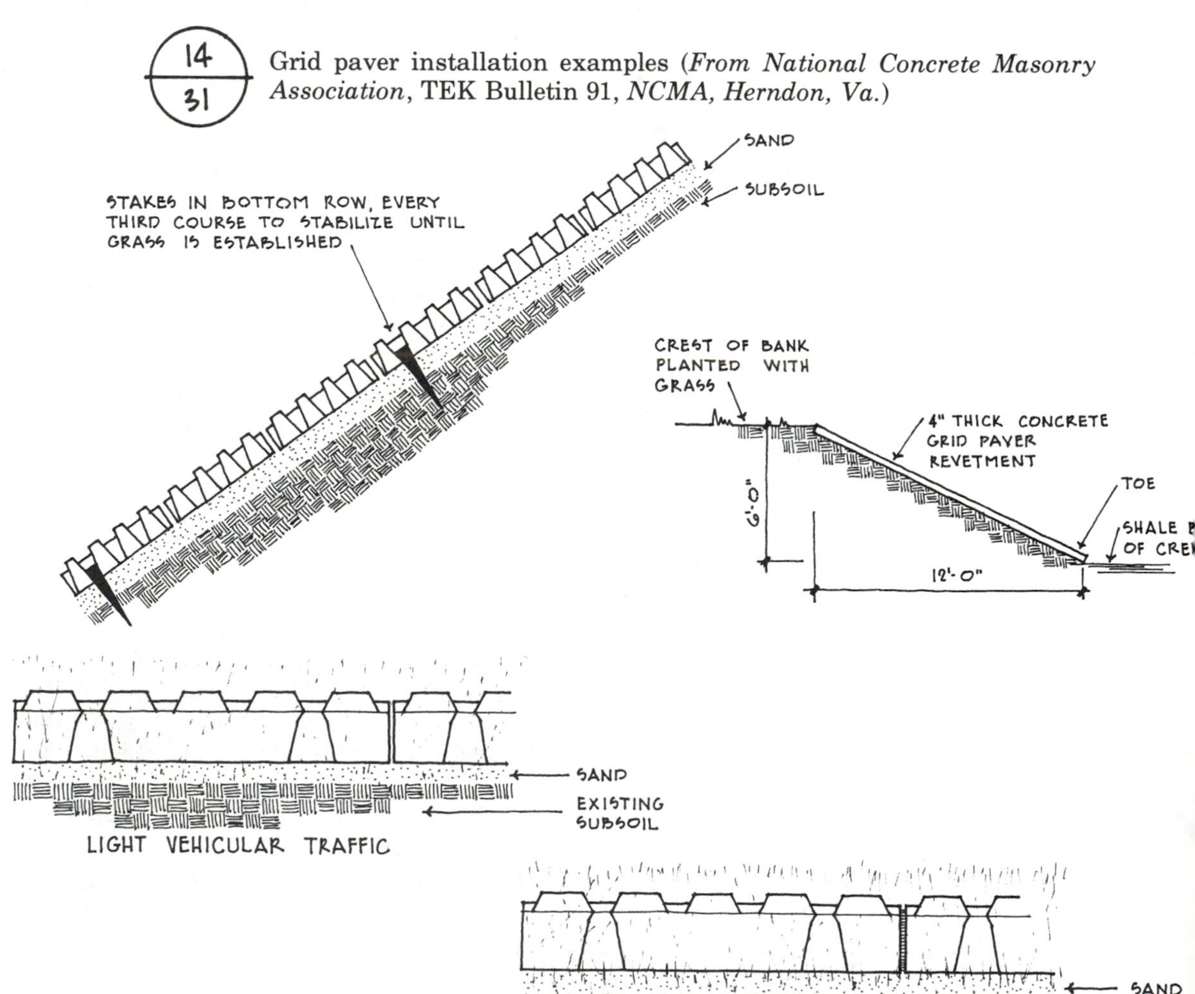

14.3.2 Bases

The successful performance of masonry paving systems depends to a great extent on proper base preparation for the type of pedestrian or vehicular traffic anticipated (*see Fig. 14-32*). *Gravel bases* provide maximum drainage efficiency and prevent the upward flow of moisture by capillary action. Clean, washed gravel should be specified. Bases of unwashed gravel mixed with fine clay and stone dust are popular low-cost systems, but they will cause a loss of porosity and effective drainage due to hardening with the absorption of moisture. Masonry units in direct contact with such contaminated materials may be susceptible to efflorescence as a result of soluble salts leached to the surface (see Section 14.7.1). Gravel bases are

Recommendations for base and paver thickness for concrete masonry pavers. (*From National Concrete Masonry Association*, TEK Bulletin 75, *NCMA, Herndon, Va.*)

Application	*Thickness of sub-base (in.)*		Thickness of concrete masonry paving units (in.)
	Well-drained dry areas	*Low wet areas*	
Light duty Residential Driveways Patios Pool decks Walkways Parking Bicycle path Erosion control Temporary paving	0–3	4–8	$2\frac{1}{2}$–3
Medium duty Sidewalks Shopping malls Residential streets Public parking Bus stops Service roads Cross walks Parking lots Camping areas Mobile home parks Canal lining Safety zones Maintenance areas Farm equipment storage	4–6	10	3–4
Heavy duty City streets Intersections Gas stations Loading docks Loading ramps Industrial floors Stables	8	12	4–6

Note: The sand bed between the sub-base and the concrete paving units is always made 2 in. thick.

suitable only for flexible paving, but can accommodate both pedestrian and light vehicular traffic. Ungraded gravel bases have better interlocking qualities, where graded gravel has a tendency to roll underfoot. Stone screenings compact better than pea gravel, but provide poorer subsurface drainage. Large gravel generally requires the use of heavy road construction equipment for proper preparation, whereas stone screenings of fine gradation lend themselves to compaction with hand tools. If the paving units are turned on edge for greater compressive depth, thin bases of fine screenings can accommodate the same light vehicular loads as thicker beds of coarse gravel. Gravel bases range in thickness from 3 to 12 in., depending on the expected load and the paver thickness.

Concrete bases with either flexible or rigid paving can support heavy vehicular traffic. Existing concrete surfaces may be used, but major cracks must be properly filled and stabilized. If a mortar leveling bed is used over new or existing slabs, the surface should be raked or floated to facilitate good bond. If a noncementitious leveling bed or cushion is used, the surface of new concrete need only be screeded.

Asphalt bases (new or used) can support flexible paving systems for heavy vehicular traffic. Mortar leveling beds can be used, but there will be little or no bond between the mortar and the asphalt. Mortar leveling beds should not be placed on hot or warm asphalt or flash setting of the mortar will occur. Any major defects in an existing asphalt pavement should be repaired before installing masonry pavers.

14.3.3 Setting Beds

Mortar setting beds may be used for rigid paving over concrete bases. A Type M portland cement–lime mortar is generally recommended for outdoc use. Thickness of the bed may vary from $\frac{1}{2}$ to 1 in. Bituminous setting bed composed of aggregate and asphaltic cement may be used over concrete o bituminous bases for flexible paving installations. The mix is generall designed and prepared at an asphalt plant and delivered to the job sit ready for application.

Cushion material is generally placed between mortarless pavers an the base as a leveling layer which compensates for minor irregularities (surface or units. Sand for this purpose should be specified in accordanc with ASTM C144. Under extremely wet conditions, however, sand cushion will provide poor drainage. Sand cushions over gravel bases require membrane to prevent settlement. Dry mixtures of 1 part portland cemer and 3 to 6 parts damp, loose sand may also be used. The higher sand rati mixtures will provide little or no bond between paver and cushion. Roofin felt (15- to 30-lb weight) provides some compensation for minor irregular ties, can be installed rapidly, and adds a degree of resilience for paver installed over concrete bases. Several examples of masonry paving install tions are shown in *Figs. 14-33* through *14-36*.

To prevent horizontal movement of mortarless paving, a method containment must be provided around the perimeter of the paved areas. soldier course set in mortar or concrete, new or existing retaining wall building walls, or concrete curbs will all provide the required stability. Ar new edging that must be installed should be placed prior to the pavir units, and the pavers worked toward the established perimeters. Modul planning in the location of perimeter edging can eliminate or reduce th amount of cutting required to fit the units.

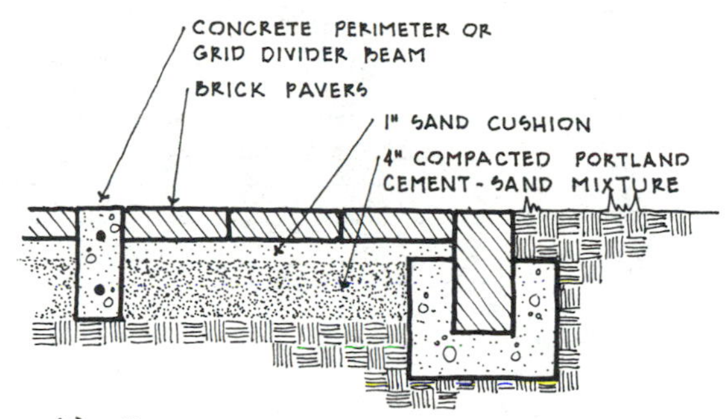

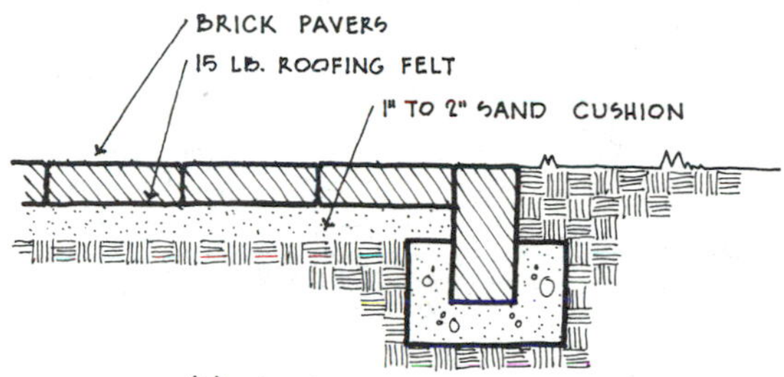

Flexible sand base for mortarless paving. (*From Brick Institute of America*, Technical Note 14A, *BIA, McLean, Va.*)

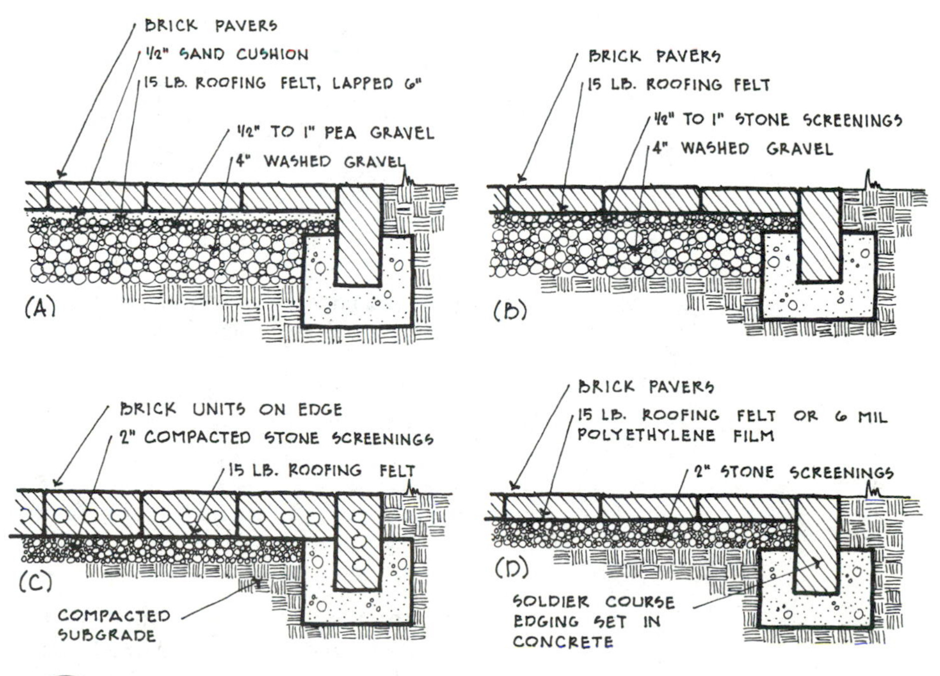

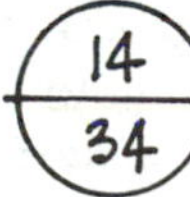

Flexible gravel base for mortarless paving. (*From Brick Institute of America*, Technical Note 14A, *BIA, McLean, Va.*)

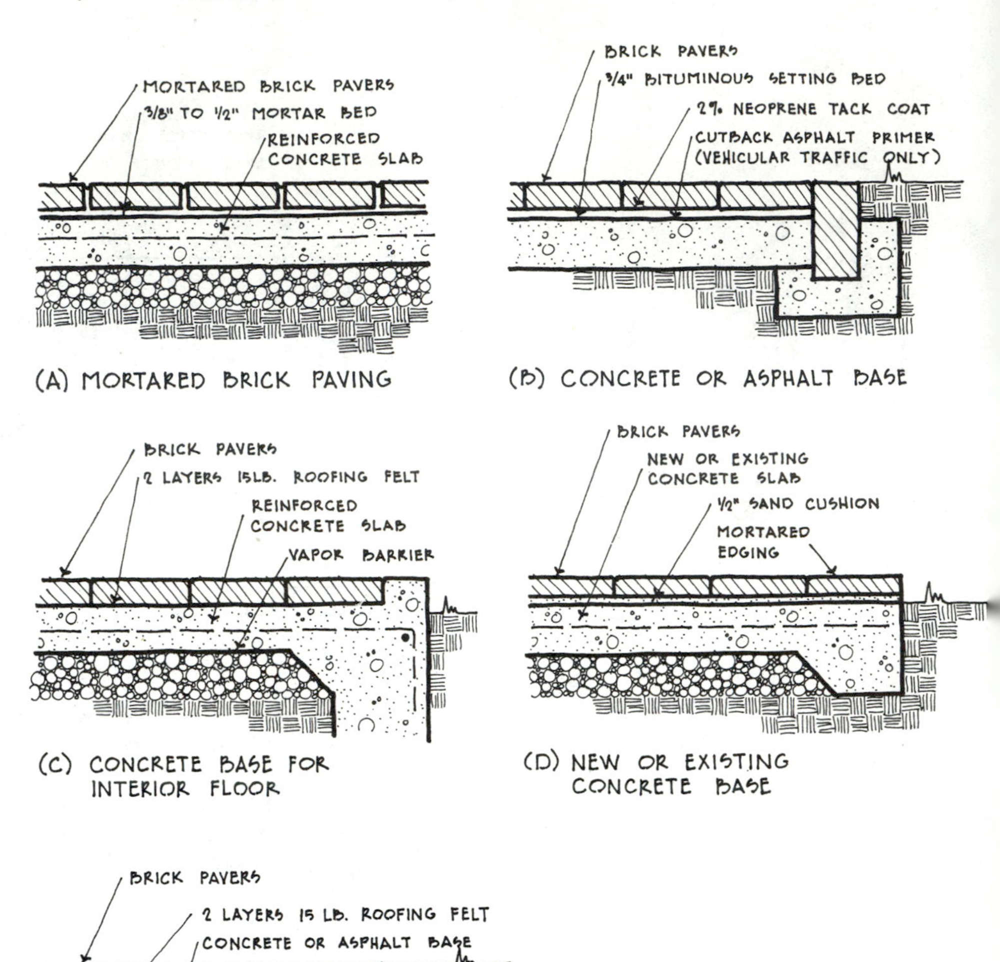

Rigid base for mortared or mortarless paving. (*From Brick Institute of America*, Technical Note 14A, *BIA, McLean, Va.*)

14.3.4 Joints

Installing masonry paving with mortar joints may be done in one of three ways.

1. Using a conventional mason's trowel, the pavers may be buttered and shoved into a leveling bed of mortar.
2. The units may be placed on a mortar leveling bed with $\frac{3}{8}$- to $\frac{1}{2}$-in. open joints into which a grout mixture is then poured. Grout proportions are normally the same as for the mortar, except that the hydrated lime may be omitted. Special care must be taken in pouring this mixture, to protect the unit surfaces from spills and stains that would require special cleaning.

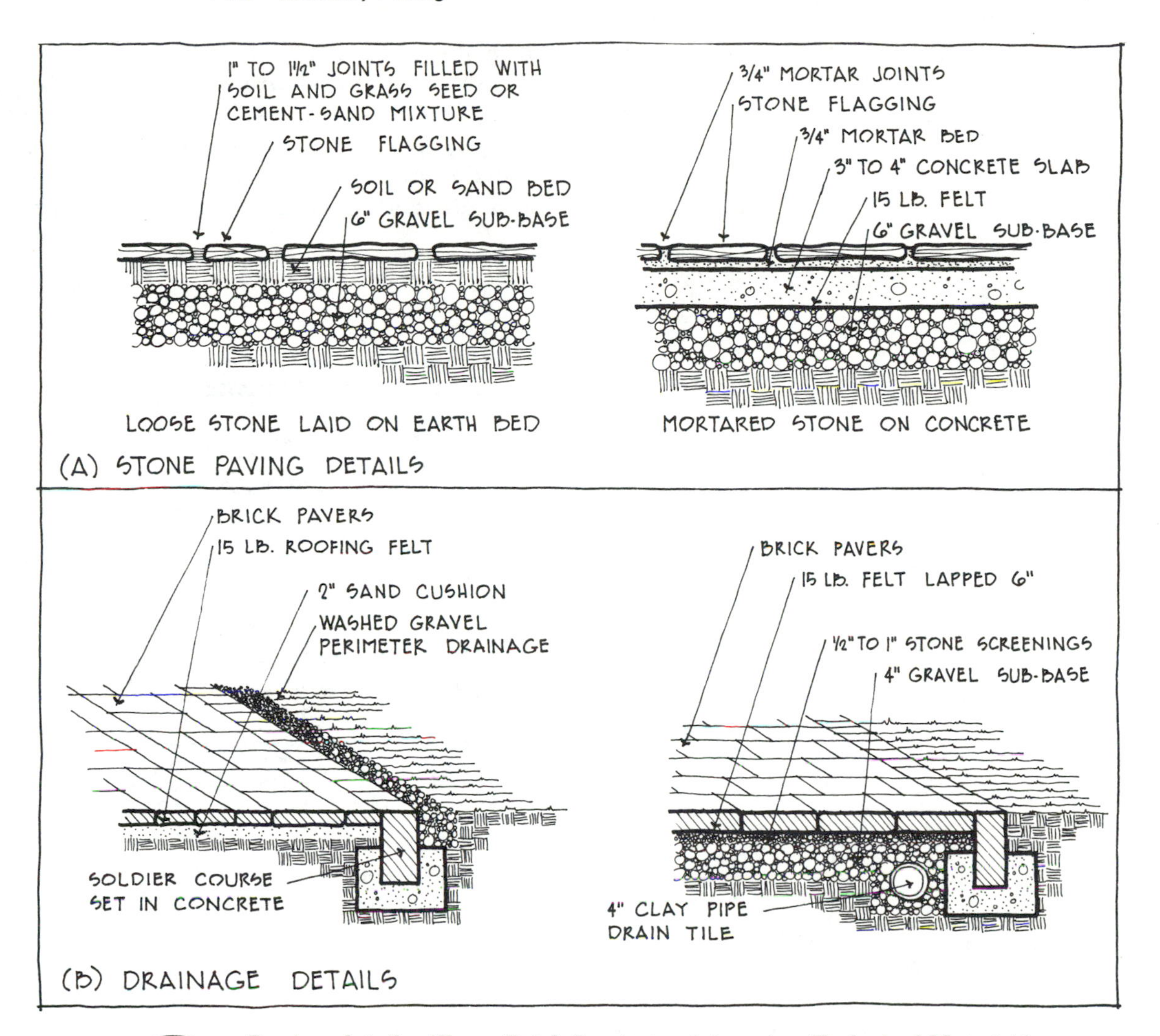

Paving details. (*From Brick Institute of America*, Technical Note 14A, *BIA, McLean Va.; and Charles G. Ramsey and Harold S. Sleeper,* Architectural Graphic Standards, *6th ed., ed. Joseph N. Boaz. Copyright © 1970 by John Wiley & Sons, Inc. Reprinted by permission of John Wiley & Sons, Inc.*)

3. Masonry pavers may also be laid on a cushion of 1 part portland cement and 3 to 6 parts damp, loose sand, and the open joints broomed full of the same mixture. After excess material has been removed from the surface, the paving is sprayed with a fine water mist until the joints are saturated. The installation should then be maintained in a damp condition for 2 or 3 days to facilitate proper curing.

Mortarless masonry paving may be swept with plain dry sand to fill between units, or with a portland cement-sand mixture equivalent to the proportions for Type M mortar. Pavers are generally butted together with only the minimal spacing between adjacent units caused by irregularities of size and shape.

Expansion joints must be provided in rigid masonry paving to accommodate thermal and moisture movements. Standard size and location recommendations to suit all applications are not feasible. Joints should generally be located parallel to curbs and edgings, at 90° turns and angles, and around interruptions in the paving surface. Fillers for these joints must be compressible, and of materials not subject to rot or vermin attack. Solid or preformed materials of polyvinyl chloride, butyl rubber, neoprene, and other elastic compounds are suitable (*see Fig. 14-37*). Even though mortarless masonry paving is flexible and has the ability to move slightly to accommodate expansion and contraction, it is recommended that expansion joints be placed adjacent to fixed objects such as curbs and walls.

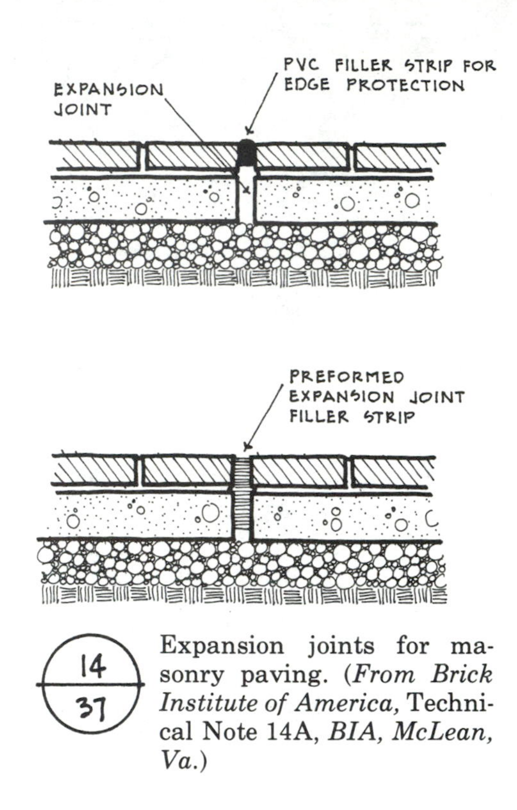

Expansion joints for masonry paving. (*From Brick Institute of America,* Technical Note 14A, *BIA, McLean, Va.*)

14.3.5 Membrane Materials

Membranes of sheet or liquid materials are installed in some pavin applications to reduce or control the passage of moisture, to discourag weed growth, or to form a separating layer or bond break to accommodat differential movement. Roofing felt, polyethylene film, vinyl, neopren rubber, asphaltic liquids, modified urethane, or polyurethane bitumens ar all suitable since they are moisture- and rot-resistant. Liquid types, installed properly, have some advantages over sheet materials because the are seamless and will conform to irregular surfaces. Precautions should b taken during construction to avoid membrane damage, particularly for roo deck installations, where resistance to moisture penetration is of paramount importance.

14.3.6 Masonry Units

The materials available for masonry paving systems have a wide range structural and aesthetic capabilities. Concrete pavers are manufactured i a variety of shapes and sizes (*see Fig. 4-14*). The concrete used in these soli units generally exceeds 5000 psi and often approaches 10,000 psi. Light normal-, or heavyweight aggregates may be used. Dense compactio assures a minimum of voids, and in areas subject to winter freeze–tha cycles, absorption is often less than 5 lb of water per cubic foot of concrete. The interlocking patterns transfer loads and stresses laterally by an arching or bridging action. Distribution of loads over a larger area in this manner reduces point loads, allowing heavier traffic over subbases that would normally require greater strength. Some pavers are manufactured with chamfered edges at the top to reduce stress concentrations and chippage. Chamfers should not, however, reduce the area of the bearing surface by more than 30%.

Solid units are made in thicknesses ranging from $2\frac{1}{2}$ to 6 in. The thicker units are used for heavy service loads, and the thinner ones for light-duty residential areas. The NCMA has developed a table of recommendations for concrete pavers for various uses and base preparations (*see Fig. 14-32*). Individual solid paving units are small in size to facilitate manual installation. The irregular shapes prevent accurate description by linear measure, but units typically will range in size from 15 sq in. to a maximum of about 64 sq in. Some manufacturers are now casting from 6 to 12 units together in clusters which are designed to be laid at the job site by machine or by a team of two workers. After the units are in place, the individual pavers break apart along preformed cleavage lines. This method increases production and can result in a reduction in labor costs.

The concrete paver shapes shown in *Fig. 4-14* are full-size units for interior field areas. Each design series includes edge units and half-length interior units to reduce the amount of job-site cutting required. Units are generally produced in a number of colors, but both color and shape will depend on local availability. Before planning the size and layout of a paved area, check with local manufacturers to verify design-related information.

Clay brick paving units (*see Fig. 3-9*) also come in a number of shapes, sizes, thicknesses, colors, and textures. Coarser-textured, slip-resistant units are recommended for outdoor installations exposed to rain, snow, and ice. This type of exposure also calls for units that are highly resistant to damage from freezing in the presence of moisture. To assure durability under extreme conditions, it is recommended that units have a maximum average cold-water absorption of 8% and a maximum saturation coefficient of 0.78. A minimum average compressive strength of 8000 psi should be required for extruded brick pavers, and 4500 psi for molded units. These dense, hard-burned units generally provide good resistance to abrasion as well. Brick that meets or exceeds the requirements of ASTM C62, Grade SW, will wear well for most nonindustrial purposes.

The BIA does not recommend the use of salvaged or used brick in paving installations. Older manufacturing processes did not assure uniformity in the quality of materials or performance, and units may spall, flake, pit, and crack when exposed to outdoor freeze-thaw cycles. Although used brick may be adequate for small residential jobs, and may provide a pleasing rustic effect, materials of unknown origin and composition should not be used for larger installations unless performance criteria can be tested and verified.

14.3.7 Paving Patterns

Many different effects can be achieved with standard rectangular pavers by varying the bond pattern in which the units are laid (*see Fig. 14-38*). It is important to specify the proper size of unit required for a particular pattern. Any of the patterns shown can be achieved with 4 × 8 in. *actual-dimension*

14/38 Bond patterns for rectangular paving units. (*From Brick Institute of America*, Technical Note 14 Rev., *BIA, McLean, Va.*)

units laid dry and tight, or with *nominal* 4×8 in. units ($3\frac{5}{8} \times 7\frac{5}{8}$ in. actua size) laid with $\frac{3}{8}$-in. mortar or sand joints. Patterns that require the width c the unit to be exactly one-half the length may not be laid dry and tight usin nominal dimension units designed for mortar joints, and vice versa. Th interlocking and herringbone patterns provide greater stability and latera stress transfer. Designs that result in continuous joints (especially longitu dinal joints in the direction of traffic flow) are more subject to shoving displacement, and the formation of ruts.

14.3.8 Brick Floors

Interior brick floors and masonry roof deck plazas can be installed over diaphragm bases in both commercial and residential buildings. Structural design of suspended wood joist, steel frame, or reinforced concrete bases must take into consideration the weight of the masonry and the maximum allowable deflection. The dead weight of mortared or mortarless brick pavers is approximately 10 lb/sq ft per inch of thickness. For the $1\frac{5}{8}$-in. and $2\frac{1}{4}$-in.-thick units, the weight would therefore be 16.25 and 22.5 lb/sq ft, respectively. Diaphragm action is important in maintaining the integrity of mortar joints, and deflections should be limited to 1/600 of the span for mortared paving. Mortarless brick paving may be installed over bases designed for deflections of 1/360 of the span (*see Fig. 14-39*). Corrugated sheet steel or metal decking supported on steel framing can serve as an economical means of constructing brick floors over open spans. The decking serves as combined form and reinforcement. The bricks are placed on a bed of mortar and the vertical joints filled with mortar or grout. For continuous spans, negative steel is placed in the joints (*see Fig. 14-39*).

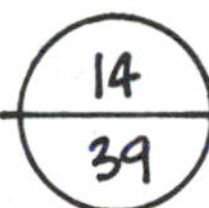

Mortared, mortarless, and reinforced brick flooring. (*From Brick Institute of America*, Technical Note 14B, *BIA, McLean, Va.*)

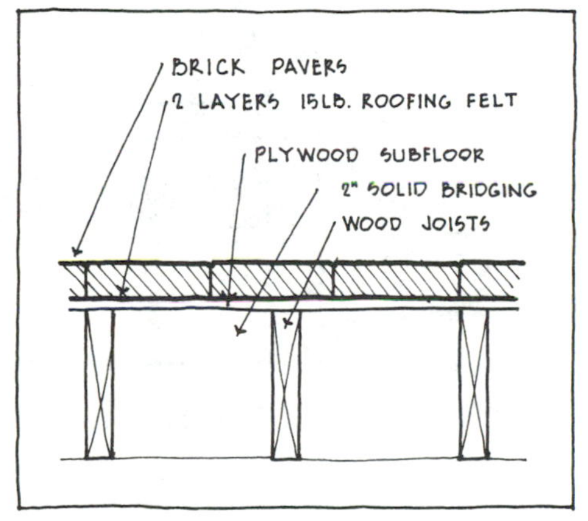

(A) WOOD FRAMING ASSEMBLY

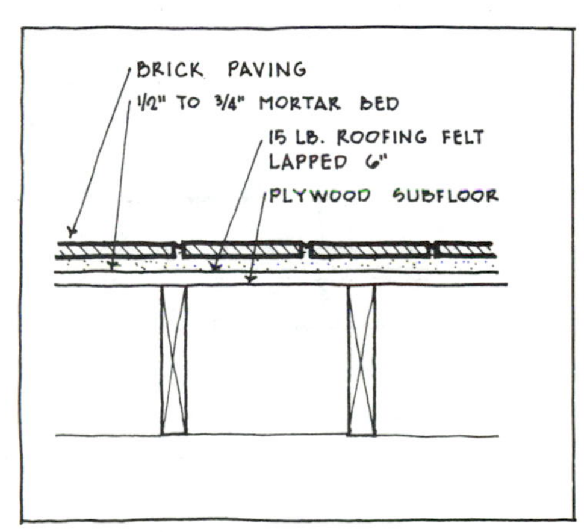

(B) WOOD FRAMING ASSEMBLY

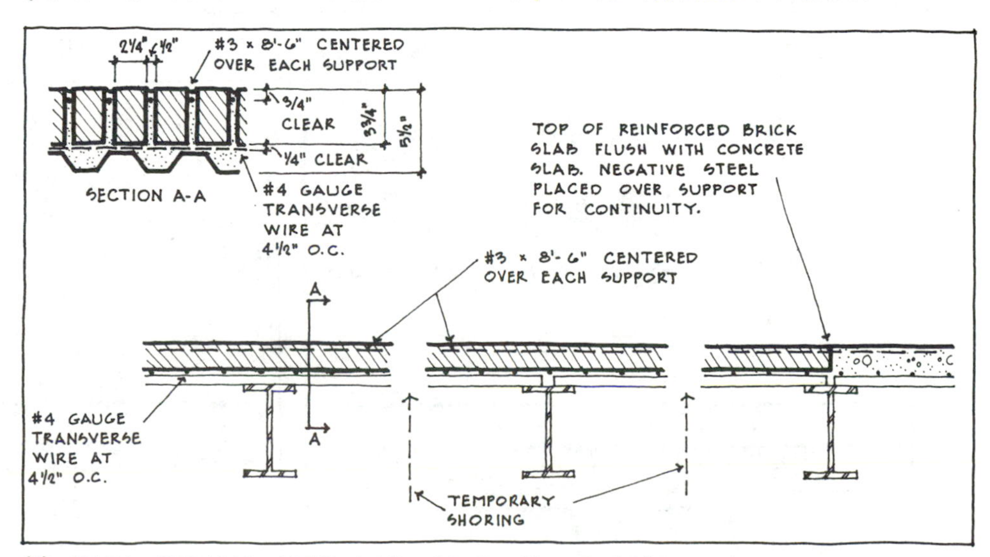

(C) STEEL DECKING-REINFORCED BRICK FLOOR ASSEMBLY

Masonry paving is often used as a decorative wear surface on pedestrian roof deck plazas. For mortarless brick paving, sloping membranes in conjunction with porous base layers will permit rapid drainage and prevent possible damage from alternate freezing and thawing of trapped water. Consideration should be given to horizontal differential movement between supporting base and waterproof membrane. Bituminous membranes are nonelastic, but seamless liquid waterproofing and rubber sheet membranes are usually elastic and capable of adjusting to differential horizontal movement. Since the masonry pavers will also change dimensionally with temperature variations, a slippage plane is recommended between pavers and membrane. Porous gravel cushions and asphalt-impregnated protection boards can withstand both horizontal abrasive movement and vertical traffic loadings (*see Fig. 14-40*).

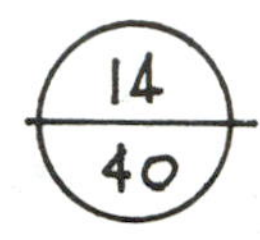

Masonry paver roof decks. (*From Brick Institute of America,* Technical Note 14B, *BIA, McLean, Va.*)

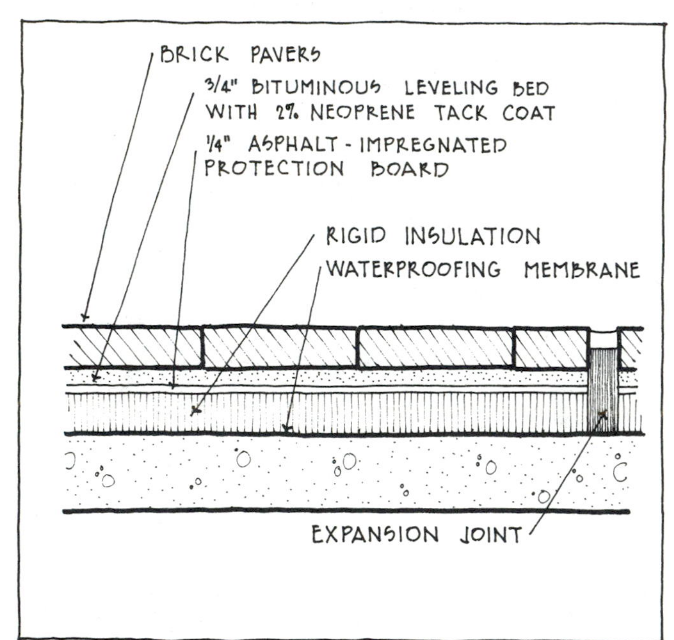

(A) REINFORCED CONCRETE SLAB

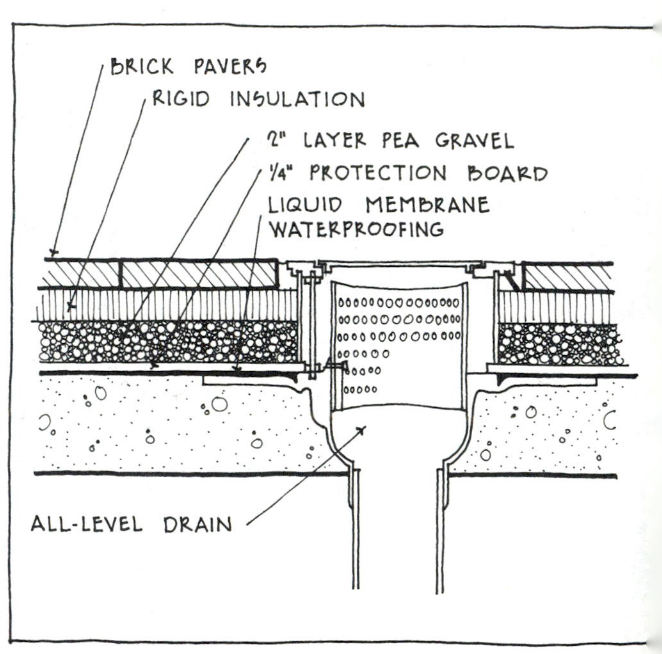

(B) REINFORCED CONCRETE SLAB

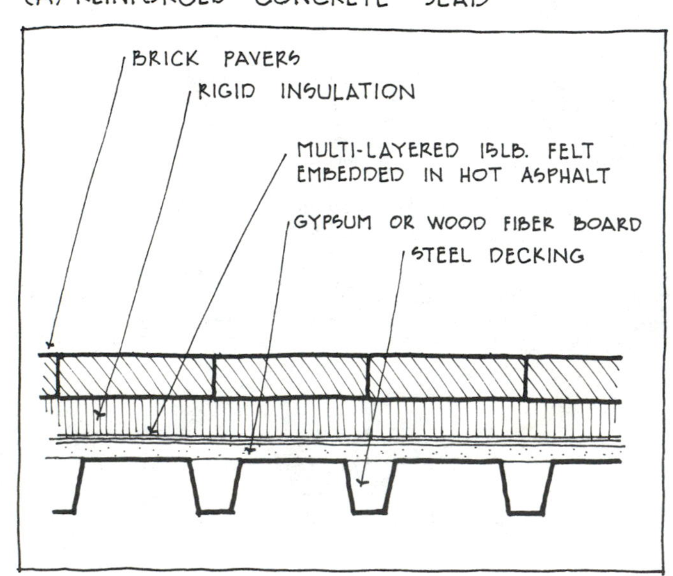

(C) STEEL DECKING BASE

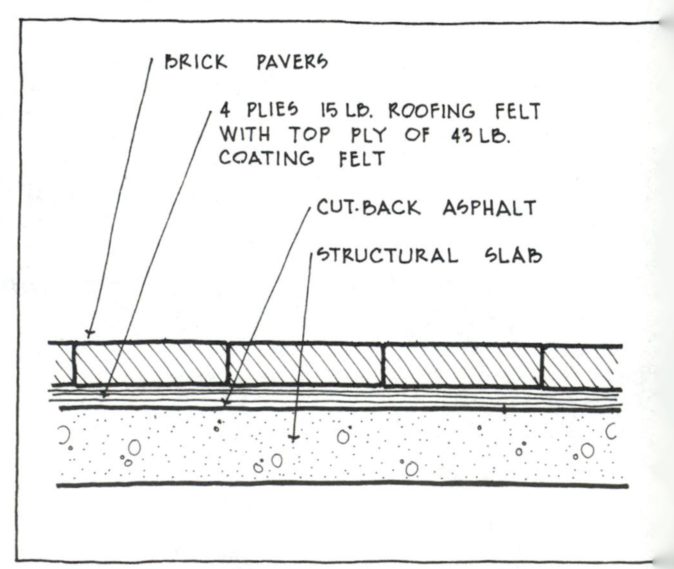

(D) REINFORCED CONCRETE SLAB

Reinforced brick masonry can be used to span open spaces as floor assemblies or to span across fill material that may tend toward uneven settlement. Reinforcement within the masonry section eliminates the need for a separate reinforced concrete slab or other base. These masonry slabs are very practical, especially over short spans, and can satisfy design loadings for both pedestrian and vehicular traffic (*see Fig. 14-41*). Brick floors and roofs are still widely used throughout Mexico and the Middle East, and can be constructed without reinforcing steel if designed as a flat arch (refer to Chapter 12).

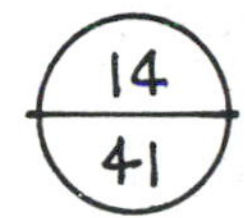

Reinforced brick floor slabs. (*From Brick Institute of America,* Technical Note 14B, *BIA, McLean, Va.*)

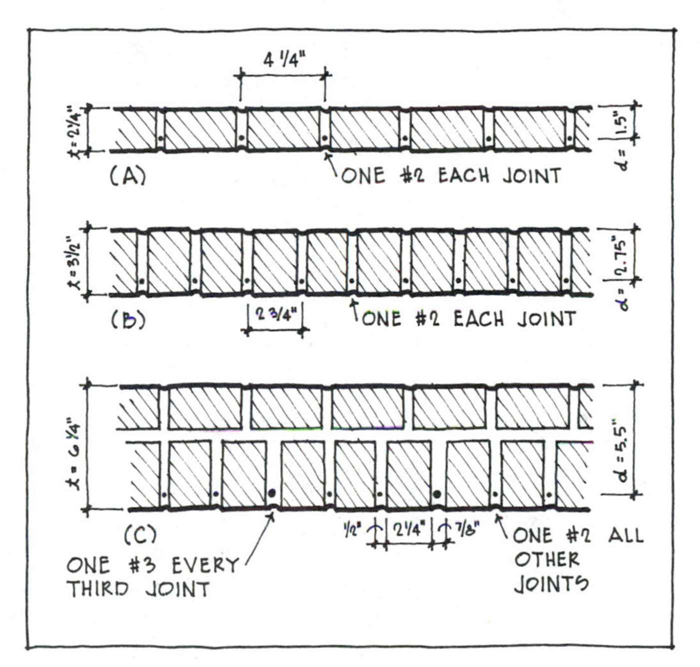

Live load (psf)	Maximum clear span		
	$t = 2\frac{1}{4}$ in. 1 No. 2 each joint [Fig. 14.41(A)]	$t = 3\frac{1}{2}$ in. 1 No. 2 each joint [Fig. 14.41(B)]	$t = 6\frac{1}{4}$ in. 1 No. 3 every 3rd joint 1 No. 2 remaining joints [Fig. 14.41(C)]
30	6′-10″	10′-5″	14′-5″
40	6′-3″	9′-9″	13′-8″
50	5′-10″	9′-2″	13′-1″
100	4′-6″	7′-3″	10′-11″
250	1′-10″	5′-0″	7′-10″

Note: Design parameters are as follows: The brick compressive strength average is 8000 psi. The mortar is type M (1 : $\frac{1}{4}$: 3), portland cement, lime and sand. Reinforcement steel is ASTM A82-66, f_s = 20,000 psi. A simple span loading condition was assumed:

$$M = \frac{wl^2}{8}$$

All mortar joints are $\frac{1}{2}$ in. thick for the slabs shown, except as noted.

14.4 MOISTURE PROTECTION

Water, which may occur as a liquid, solid, or vapor, can be a difficult problem for the masonry designer. As wind-driven rain, it may cause decomposition or staining; as ice, sleet, snow, or hail, it is capable of actual mechanical disintegration; as a vapor, it may penetrate to the interior of a wall, causing decomposition or disintegration; and as surface condensation, it may be ruinous to interior finishes.

Moisture protection for masonry walls is a combination of good design, proper material selection, and high-quality workmanship. The materials used in masonry construction are highly porous, and will absorb rainwater, melting snow, and other atmospheric moisture. Since the materials are known to absorb water, the key to good design is control of moisture flow to prevent damage to the wall or to the building interior. In areas where severe wet weather is to be expected, use of the most weathertight masonry wall type is expedient. Cavity walls with an open air space are known to be most efficient in eliminating moisture from within the wall section. The cavity is designed to collect moisture that enters the wall and redirect it to the exterior through weep holes located at the base of the wall. Even a well-built masonry wall can absorb moisture, especially in coastal areas with heavy precipitation and high winds. The complete separation of cavity wall wythes by an air space is essential to the proper elimination of absorbed moisture. In severe weather areas, masonry bonding of the two wythes must be avoided since the transverse units will act as a bridge carrying water to the interior wythe. Porous, waterproof insulating fill may be used without disrupting the moisture flow in the cavity as long as the material does not block or clog the weep holes.

14.4.1 Flashing

An important consideration in the design of cavity walls is the proper location of flashing and weep holes. It is difficult at best to completely prevent rainwater from entering a wall at parapets, sills, projections, and so on. Without proper flashing, water that does penetrate the wall cannot be diverted back to the exterior. Continuous flashing should be installed at the bottom of the cavity wall and wherever the cavity is interrupted by elements such as shelf angles and lintels. Flashing should be placed over all wall openings not protected by projecting hoods or eaves, at all window sills, spandrels, caps, copings, and parapet walls (*see Fig. 14-42*). Since the purpose of flashing is to collect moisture and divert it to the outside, *weepholes must be provided wherever flashing is used.* The exception to this rule is exterior flashing at parapet walls, where the purpose is not to collect water, but to prevent its entrance altogether. Weepholes are located in the head joints of the outer wythe immediately above the flashing. Spacing of weepholes should generally be 24 in. on center for open joints, and 16 in. on center for those using wick material.

Flashing is generally formed from sheet metals, bituminous membranes, or composite materials selected on the basis of cost and suitability. (Chapter 7 analyzes various materials and their performance in conjunction with masonry.) Both installation and material costs vary widely, and no general recommendation can be made solely on an economic basis. It is critical, however, that only high-quality materials be used since replacement would be both difficult and expensive.

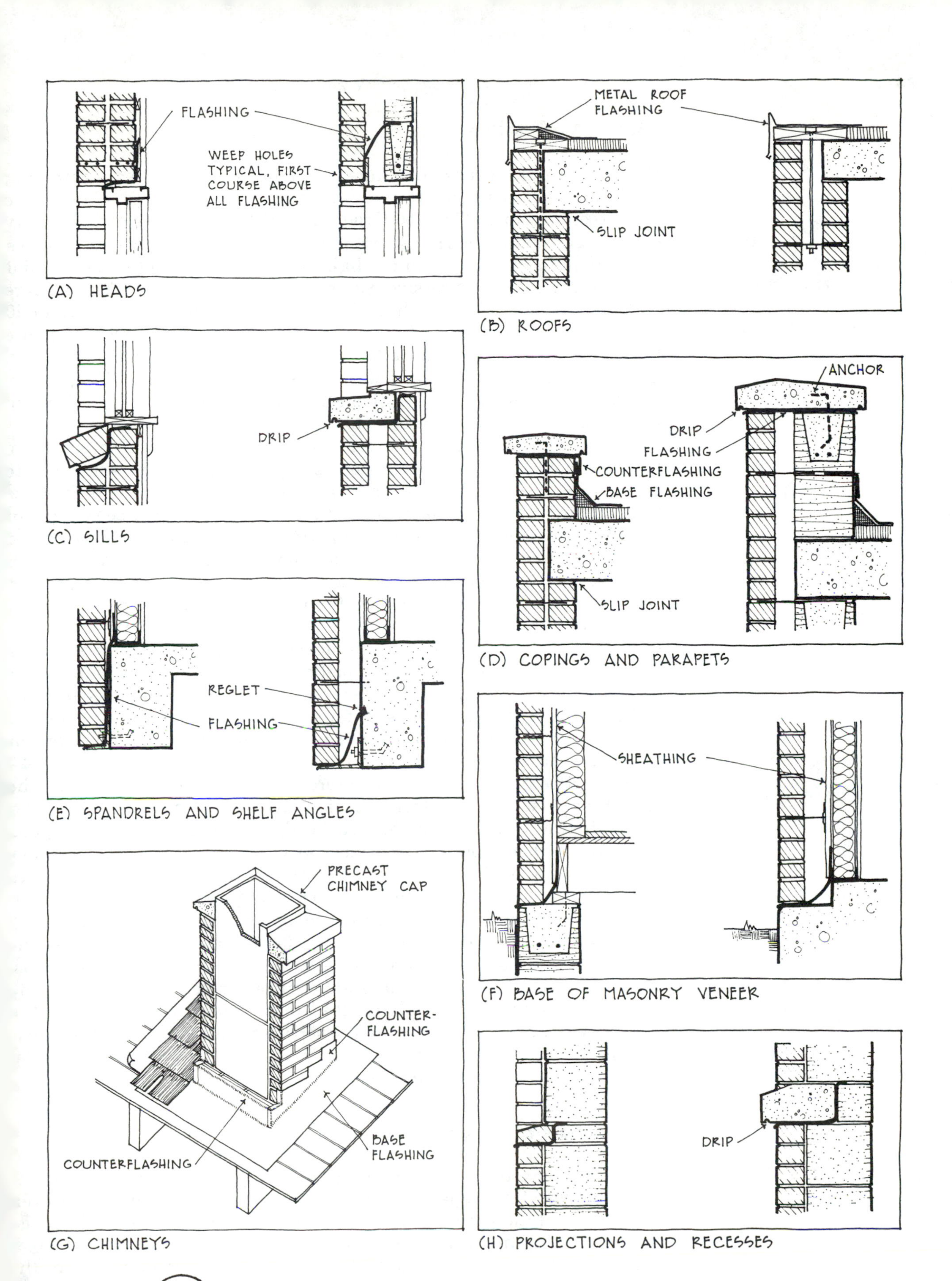

Flashing details. (*From Brick Institute of America*, Technical Note 7A, *BIA, McLean, Va.*)

14.4.2 Material Selection

Proper selection of masonry units and mortar for expected weathering conditions is also an important design factor. Clay brick for exterior use should conform to ASTM C216, Grade MW or SW, depending on the severity of expected conditions. High-suction brick may have as much as 50% less bond strength than moist units. High-suction brick cannot properly bond with mortar because excessive water is absorbed from the mortar before hydration begins. Units with initial rates of absorption greater than 20 g/min should be thoroughly wetted and allowed to surface-dry before laying. This will produce better bond and more weathertight joints. Concrete masonry units should be Grade N, ASTM C90, C129, C145, or C55. Concrete blocks with a low shrinkage potential will exhibit less tendency to form cracks and cause eventual leakage.

Tight mortar joints are essential in maintaining weatherproof walls. Mortars should be selected on the basis of performance, and tests indicate that portland cement–lime mortars are more resistant to water permeance than are proprietary masonry cement mortars. Cracking or separating of bond between mortar and masonry unit invites the intrusion of water, and good bond must be maintained at all contact surfaces to prevent such leakage. Portland cement–lime mortars exhibit greatest bonding characteristics and should be used exclusively in severe weather regions. Type N mortar is generally adequate for above-grade work with moderate exposure but Type S has greater bond strength to withstand severe lateral pressures from wind-driven rain at high velocities. Type M mortar is suitable for below-grade construction in direct contact with the soil (such as basement walls), where hydrostatic pressure forces groundwater against the face of the masonry. All head and bed joints must be fully mortared and tooled for effective weather resistance. The concave and V-joints shown in *Fig. 14-24* are most effective in excluding moisture. Steel jointing tools compress the mortar against the unit face, forming a complete seal against water penetration. Mortar must also be mixed with sufficient water to assure good bond. Although no laboratory measure can be made, empirical studies show that a mortar consistency which is workable to the mason also proves sufficiently moist. There is little danger of the mason adding too much water, since the mortar would become soupy and difficult to handle.

Caulking between masonry and adjacent materials completes the exterior envelope of the building (*see Fig. 14-43*). Door and window jambs, intersections with dissimilar materials, control joints, and expansion joints must all be fully and properly caulked to maintain the integrity of the system. Workmanship in this and other areas must be of high quality. Masons must also assure that cavities of double-wythe walls are kept clear and weep holes free of mortar plugs.

14.4.3 Waterproofing and Dampproofing

Below-grade masonry waterproofing generally consists of a bituminous membrane or other impervious film resistant to water penetration even under hydrostatic pressure. In areas where soil exhibits good drainage characteristics, the membrane may actually be only a dampproof layer designed to retard moisture until the water has drained away from th

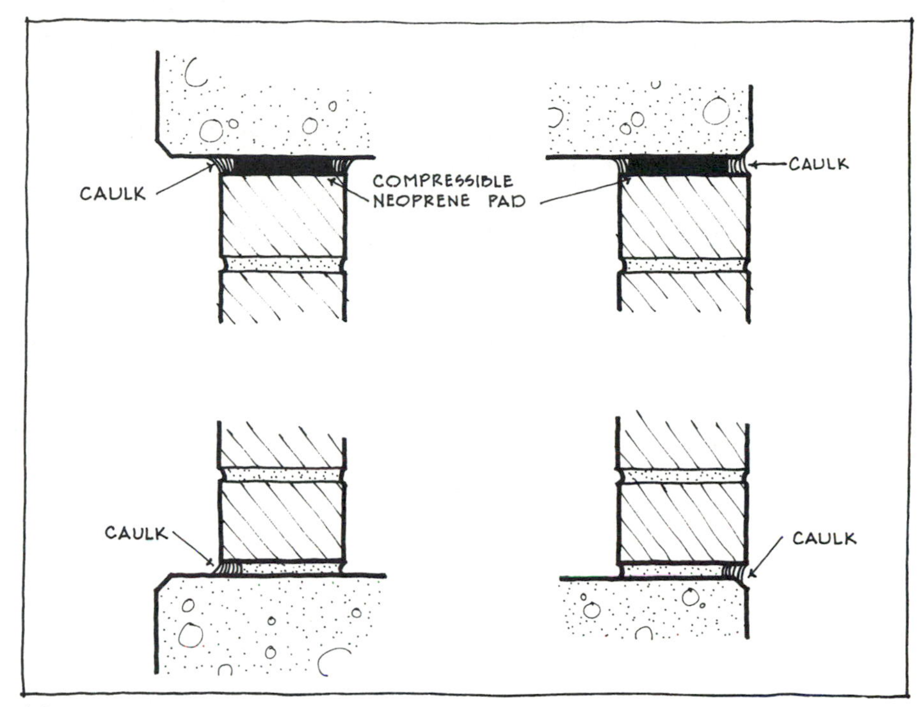

(A) MASONRY ERECTED AFTER CONCRETE (RAKE OUT JOINTS ±1/2" AND CAULK)

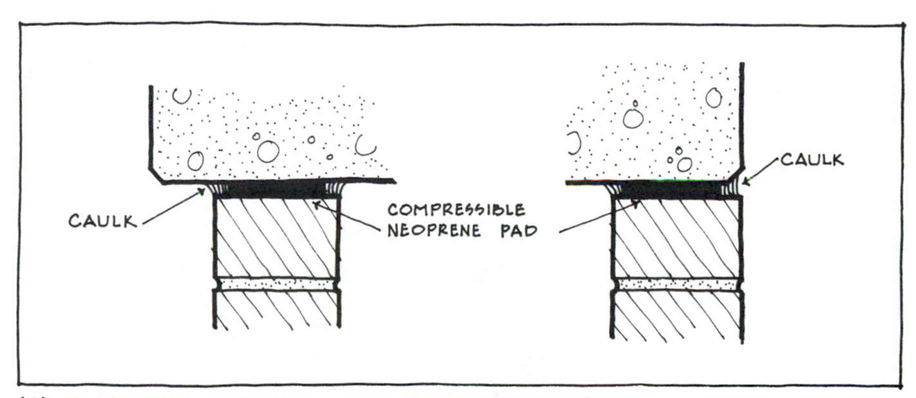

(B) MASONRY ERECTED BEFORE CONCRETE

Caulking details.

building by natural gravity flow (*see Fig. 14-44*). A commonly used protective measure consists of one, or preferably two, coats of cement mortar. This method is known as parging. Although parge coats will retard leakage, wall movements may cause cracks and permit moisture penetration. Impervious membranes with some elasticity offer better assurance against leaks. These may be fluid-applied bituminous or elastomeric products, bentonite clay, or any tested and approved waterproofing system (see Section 13.2.6).

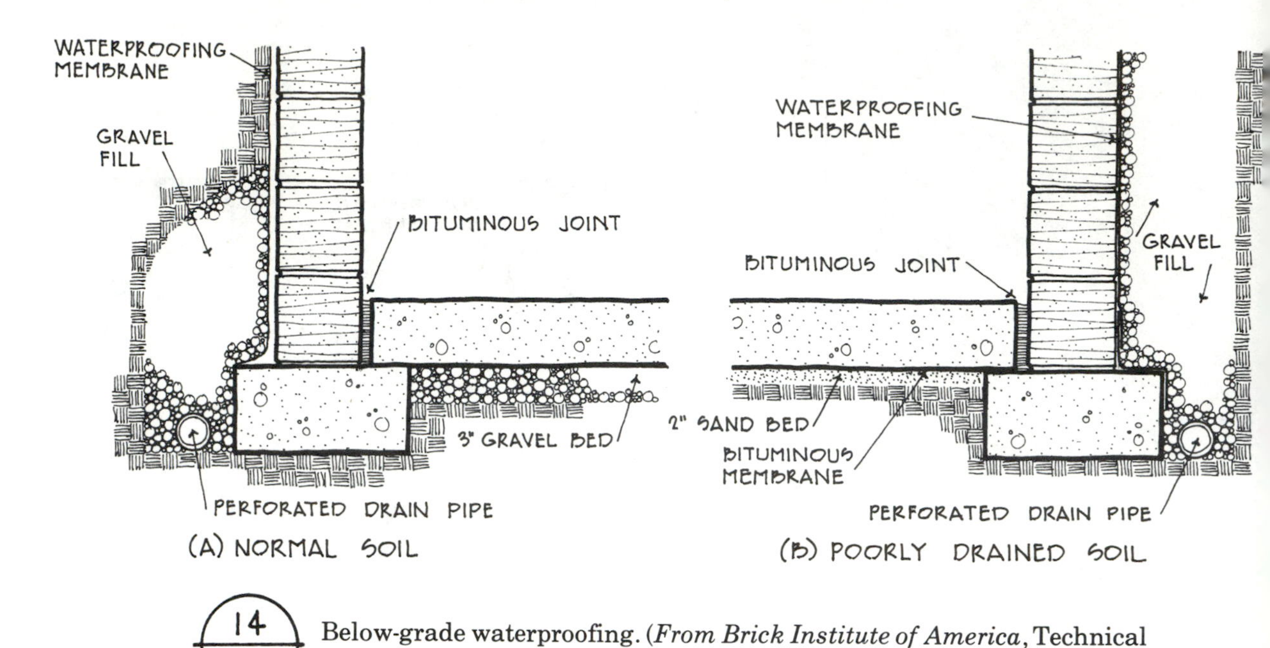

14/44 Below-grade waterproofing. (*From Brick Institute of America*, Technical Note 7, *BIA, McLean, Va.*)

14.4.4 Condensation

Vapor barriers on the cavity face of backup wythes in above-grade construction are often recommended in climates with high humidity. Differences in humidity between inside and outside air cause vapor flow within a wall. Unless controlled by properly placed vapor barriers or by ventilation, the vapor may condense within the wall under certain temperature conditions.

Condensation results from saturated air. The higher the air temperature, the more water vapor the atmosphere can contain before reaching its saturation, or dew point. If warm humid air is sufficiently cooled, the water vapor it contains will condense. Condensation problems are most frequent in insulated buildings of tight construction with occupancies or heating systems that produce humidity. The relative humidity of the air within a building is increased by cooking, bathing, washing, or other activities using water or steam. This rise in the moisture content of the air increases the vapor pressure substantially above that of the outdoor atmosphere, and tends to drive vapor outward from the building through any porous materials that may comprise the enclosure. When wall surface temperatures are substantially below air temperatures, condensation may occur on the wall surface. The greater the humidity level, the less the temperature differential need be to form the condensation. The table in *Fig. 14-45* lists differences in temperature between inside air and inside wall surface for different wall types and temperature variations from inside to outside. If, for instance, an exposed 10-in. brick cavity wall separates outside air that is 50°F warmer than the inside air, the inside wall surface temperature is 11°F lower than the inside air temperature. The graph in *Fig. 14-45* shows that an 11°F temperature drop at the wall will cause condensation for relative humidities of 68% or more. The same wall assembly with vermiculite insulation shows a wall surface temperature differential for the same conditions of only 5°F, which would cause condensation only at humidities greater than 85.5%.

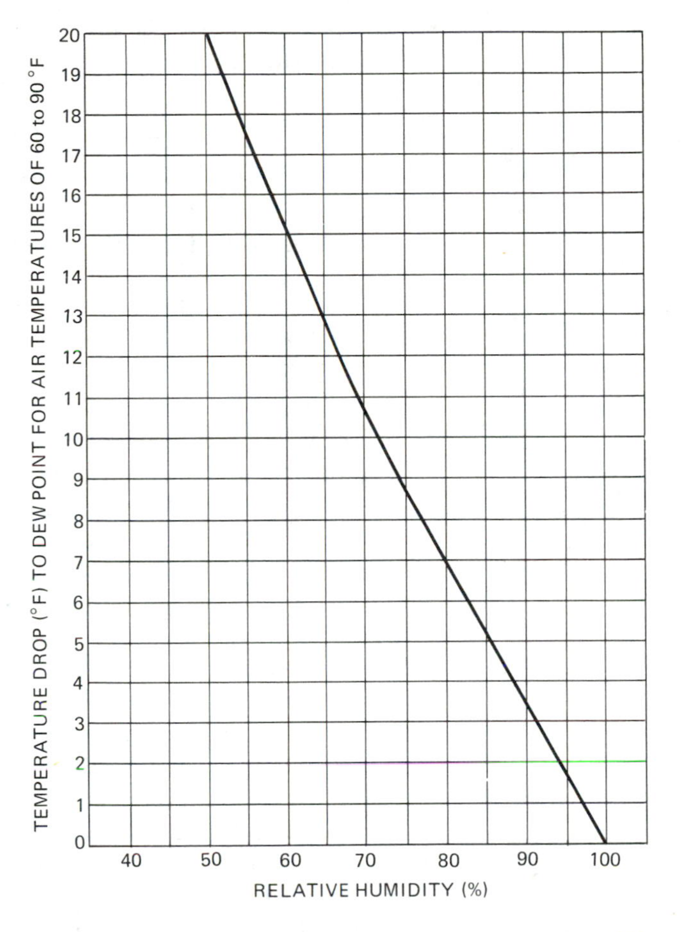

Temperature Drop Curve for Relative Humidities from 50 to 100%

DIFFERENCE IN TEMPERATURE BETWEEN INSIDE AIR AND INSIDE WALL SURFACE FOR VARIOUS TOTAL DIFFERENCES IN TEMPERATURE BETWEEN INSIDE AIR AND OUTSIDE AIR							
	Difference in temperature between inside and outside air (°F)						
Wall construction	*10*	*20*	*30*	*40*	*50*	*60*	*70*
4-in. brick and 4-in. tile, furred and plastered	2	$3\frac{1}{2}$	5	7	9	10	12
"SCR brick," furred and plastered	2	5	7	10	12	14	17
8-in. tile (3-cell), furred and plastered	$1\frac{1}{2}$	3	$4\frac{1}{2}$	6	7	$8\frac{1}{2}$	10
10-in. brick and brick cavity, exposed	$2\frac{1}{2}$	$4\frac{1}{2}$	7	9	11	$13\frac{1}{2}$	16
10-in. brick and tile cavity, plastered	$1\frac{1}{2}$	3	$4\frac{1}{2}$	6	7	$8\frac{1}{2}$	10
10-in. brick and brick cavity, vermiculite insulated, exposed	1	2	3	4	5	6	$6\frac{1}{2}$
10-in. brick and tile cavity, vermiculite insulated, plastered	1	2	$2\frac{1}{2}$	$3\frac{1}{2}$	4	5	6

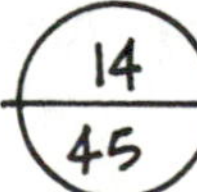

Conditions affecting surface condensation. (*From Brick Institute of America*, Technical Note 7C, *BIA, McLean, Va.*)

If wall surface condensation occurs, it may be eliminated by one of three methods:

1. Reducing the humidity of the inside air by ventilation;
2. Increasing the surface temperature of the interior wall by air movement or other means; or
3. Increasing the heat resistance of the wall by adding an air space or insulation (as in the example above). The table and chart can help in the selection of suitable wall components for various design temperatures and occupancy conditions.

Vapor condensation *within* walls can cause extensive damage to many types of building materials. Wood framing members can warp or decay in the presence of moisture; metal can corrode; insulation can lose its effectiveness; concrete and concrete masonry products can undergo volu metric changes; and freeze–thaw cycles under moist conditions can deterio rate both clay and concrete masonry as well as stone.

Warmer air has higher saturated vapor pressures. If separated by a wall, the higher-pressure vapor will migrate through the wall toward the lower-pressure atmosphere. During the winter this flow is from inside the building toward the outside. In warm humid climates, this flow may reverse during the summer, with vapor traveling from the outside in. When vapor passes through a wall that is warm on one side and cool on the other, it may reach its dew point and condense into water within the wall. The tempera ture drop through a composite wall is directly proportional to the resistance of the various elements. The drop in vapor pressure through the wall is in proportion to the vapor resistance of the constituent parts. The table in *Fig. 14-46* lists vapor resistances for some common building materials, and the table in *Fig. 14-47* shows the saturated vapor pressure for various temperatures.

14.4.5 A Design Example

The average Washington, D.C., 7:30 A.M. January temperature is 30°F a 73% humidity. An interior space is being designed for a 7:30 A.M. condition o 66°F at 50% humidity. The exterior wall is a brick cavity wall with vermiculite insulation. The saturated vapor pressure at 66°F is 0.644 (from *Fig. 14-47*), and at 50% relative humidity, the actual vapor pressure would b 0.322. The saturated vapor pressure at 30°F is 0.165, and at 73% humidity would be 0.120. The indoor vapor pressure is therefore 0.202 greater than th outdoor vapor pressure, indicating that vapor flow will be toward th outside. The actual and saturated vapor pressures passing through the wal may be plotted graphically using the two tables given. At any point wher the actual vapor pressure is greater than the saturated vapor pressure condensation will occur. *Figure 14-48* shows calculations for the pressur values and a graphic illustration of their paths. At the inside face of th outer brick wythe, actual vapor pressure exceeds saturated vapor pressure and condensation will occur at this point. Since the cavity wall is designe for drainage, the moisture will collect at the base of the cavity and b directed out through the weep holes. If the outside temperature drops belo 30°F, the condensation area is greater and may extend through the bric section. To alleviate the possibility of moisture or frost developing withi

Material	Thickness (in.)	Vapor resistance sq ft/hr (in. Hg)
Air		
Still air	1.0	0.0079
Finishes		
Celotex	0.75	0.0798
Fiberboard	0.492	0.0342
Fiberboard, 1 surface asphalt rolled	0.492	0.2545
Fiberboard compressed	0.1875	0.1975
Masonite Presdwood	0.13	0.0938
Masonite Presdwood, tempered	0.13	0.2077
Paint film	—	0.285
Paint, 2 coats with linseed oil	—	0.5843
Paint, 1 coat alum. on wood with gum varnish	—	0.906
Plaster	0.75	0.0672
Plaster, fiberboard or gypsum lath	—	0.0507–0.0487
Plaster, 3 coats flat wall paint	—	0.2341
Plaster, 3 coats oil paint	—	0.2708–0.2586
Plasterboard (drywall)	0.37	0.0289
Insulating materials		
Blanket insulation between coated paper	0.5–1.0	0.5212–0.501
Fiberglas 6-lb PF	1.0	0.639
Foil-surfaced reflective insulation, double-faced	—	11.85–7.74
Insulating lath and sheathing, board type	—	0.0389–0.0291
Insulating sheathing, surface coated	—	0.330–0.229
Mineral wool, unprotected	4.0	0.0344
Vermiculite, loose fill type	2.5	0.0163
Vapor barriers (also see Paint)		
Asphalt felt, 15 lb, soft dull appearance	0.032	0.1509
Asphalt felt, 35 lb/500-sq ft roll	—	0.9793
Asphalt felt, 50 lb/500-sq ft roll	—	1.9586
Duplex Scutan 6-6, asphalt between 2 sheets of kraft paper	0.007	2.158
Kraft paper	—	0.1818
Vapor barrier (Kimberly Clark Corp. data)	—	1.242
Wood		
Fir sheathing	$\frac{3}{4}$	0.340
Pine lap siding	—	0.2036
Pine, Western yellow	$\frac{1}{4}$	1.132
Plywood, Douglas fir 3-ply plain	$\frac{1}{4}$	0.2341–0.1555
Plywood, Douglas fir 5-ply plain	$\frac{1}{2}$	0.3746–0.2973
Masonry		
Brick masonry	4	0.9243
Brick masonry	4	0.9773
Concrete, $5\frac{1}{4}$ gal water/bag cement	1	0.7044
Concrete, $9\frac{1}{4}$ gal water/bag cement	1	0.5008
Concrete, $9\frac{1}{4}$ gal water/bag (with pozzuolanic admixture)	1	0.4011
Hollow tile masonry	4	0.678
Roofing material		
Roll roofing 40 to 65 lb/108-sq ft roll	—	7.737–5.843

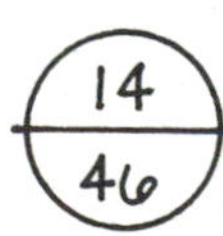

Vapor resistance of various materials. (*From Brick Institute of America,* Technical Note 7D, *BIA, McLean, Va.*)

Temp. (°F)	Vapor pressure (in. Hg)	Temp. (°F)	Vapor pressure (in. Hg)
−20	0.01259	41	0.25748
−19	0.01333	42	0.26763
−18	0.01411	43	0.27813
−17	0.01493	44	0.28899
−16	0.01579	45	0.30023
−15	0.01671	46	0.31185
−14	0.01767	47	0.32386
−13	0.01868	48	0.33629
−12	0.01974	49	0.34913
−11	0.02086	50	0.36240
−10	0.02203	51	0.37611
−9	0.02327	52	0.39028
−8	0.02457	53	0.40492
−7	0.02594	54	0.42004
−6	0.02738	55	0.43565
−5	0.02889	56	0.45176
−4	0.03047	57	0.46840
−3	0.03214	58	0.48558
−2	0.03388	59	0.50330
−1	0.03572	60	0.52159
0	0.03764	61	0.54047
1	0.03967	62	0.55994
2	0.04178	63	0.58002
3	0.04401	64	0.60073
4	0.04634	65	0.62209
5	0.04878	66	0.64411
6	0.05134	67	0.66681
7	0.05402	68	0.69019
8	0.05683	69	0.71430
9	0.05978	70	0.73915
10	0.06286	71	0.76475
11	0.06608	72	0.79112
12	0.06946	73	0.81828
13	0.07300	74	0.84624
14	0.07670	75	0.87504
15	0.08056	76	0.90470
16	0.08461	77	0.93523
17	0.08884	78	0.96665
18	0.09327	79	0.99899
19	0.09789	80	1.0323
20	0.10272	81	1.0665
21	0.10777	82	1.1017
22	0.11305	83	1.1379
23	0.11856	84	1.1752
24	0.12431	85	1.2135
25	0.13032	86	1.2529
26	0.13659	87	1.2934
27	0.14313	88	1.3351
28	0.14966	89	1.3779
29	0.15709	90	1.4219
30	0.16452	91	1.4671
31	0.17227	92	1.5135
32	0.18035	93	1.5612
33	0.18778	94	1.6102
34	0.19456	95	1.6606
35	0.20342	96	1.7123
36	0.21166	97	1.7654
37	0.22020	98	1.8199
38	0.22904	99	1.8759
39	0.23819	100	1.9333
40	0.24767		

14/47 Saturated vapor pressure at various temperatures. (*From Brick Institute of America*, Technical Note 7D, *BIA, McLean, Va.*)

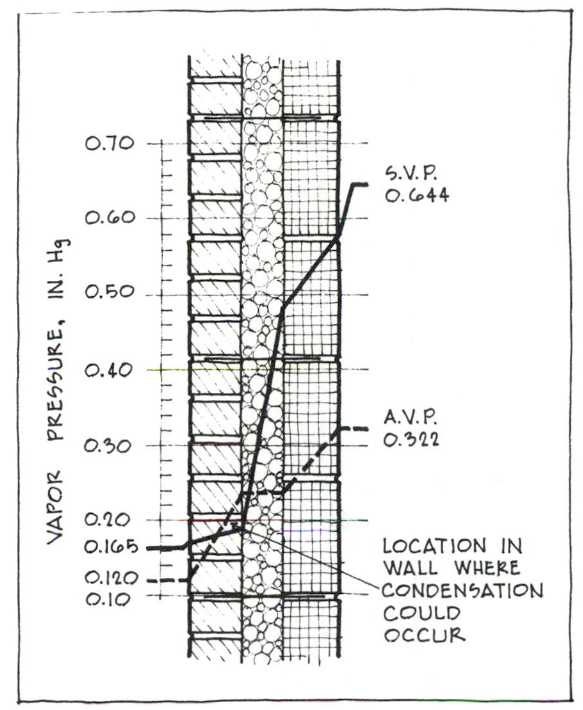

(A) VAPOR PRESSURE GRADIENT CURVES

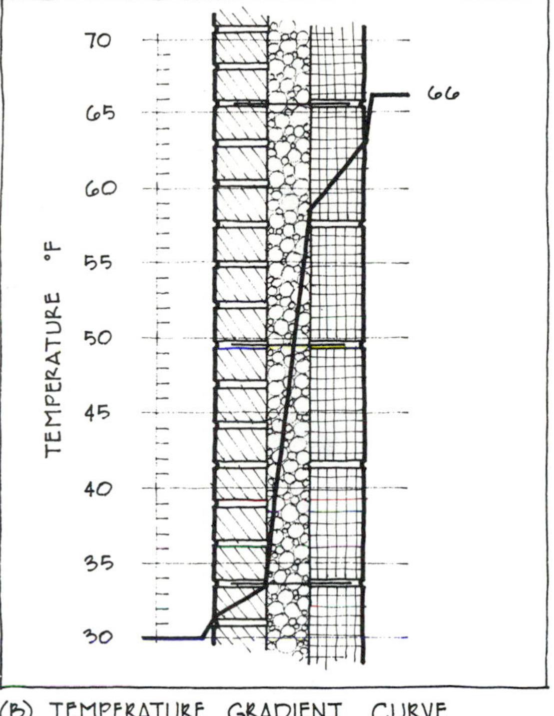

(B) TEMPERATURE GRADIENT CURVE

Material	Thermal analysis				Vapor pressure analysis				
	T.R.	%	T.D.	°F	S.V.P.	V.R.	%	V.P.D.	A.V.P.
Inside air				66	0.644				0.322
Inside film of still air	0.68	9	3	63	0.580				0.322
4 in. unglazed structural clay facing tile	1.11	14	5	58	0.486	0.678	40.5	0.082	0.240
$2\frac{1}{2}$ in. vermiculite insulation	5.50	70	25	33	0.188	0.016	1.0	0.002	0.238
4 in. high-density facing brick	0.44	5	2	31	0.172	0.977	58.5	0.118	0.120
Outside air at 15 mph	0.17	2	1	30	0.165				0.120
Total	7.90	100	36			1.671	100	0.202	

Note: T.R., thermal resistance; T.D., temperature difference in °F; S.V.P., saturated vapor pressure in inches mercury; V.R., vapor resistance in inches of mercury; V.P.D., vapor pressure difference in inches of mercury; A.V.P., actual vapor pressure in inches of mercury.

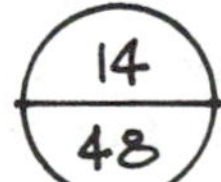

Condensation analysis for the design example. (*From Brick Institute of America*, Technical Note 7D, *BIA, McLean, Va.*)

the brick and deteriorating the material through repeated freeze–thaw cycles, the insulation may be changed to a type with greater vapor resistance than vermiculite, or the collar joint may be filled with grout. This will decrease the actual vapor pressure more rapidly and drop it below the saturated vapor level.

It is obvious that the introduction of vapor barriers within a wall assembly must be studied carefully to avoid trapping moisture in an undesirable location. Regional climatic conditions and the resulting direction of vapor flow must be analyzed and condensation points determined for both summer and winter conditions. If the flow of vapor is impeded by a

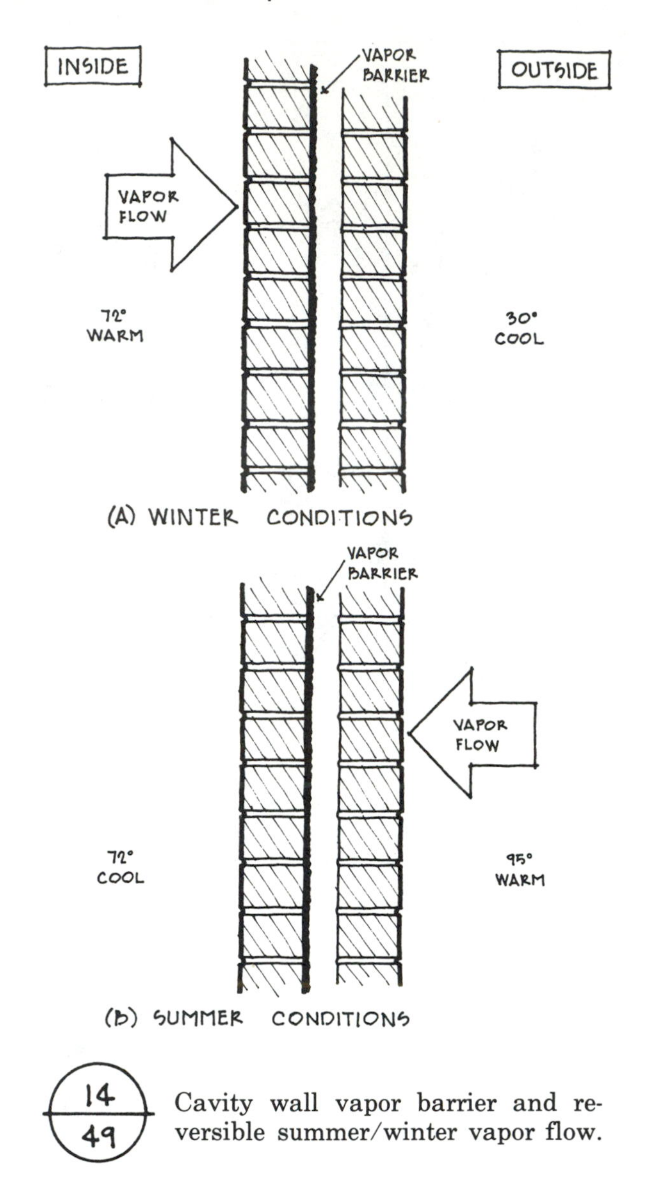

Cavity wall vapor barrier and reversible summer/winter vapor flow.

highly vapor-resistant material on the warm side of the wall, the vapo cannot reach that point in the wall at which the temperature is low enough to cause condensation. Masonry cavity walls are unique in their construc tion and can accommodate vapor flow in either direction without retarding natural moisture drainage. A vapor barrier on the cavity face of the interio wythe will prevent the warm, moist air inside the building from reaching a lower saturation temperature farther out in the wall (*see Fig. 14-49*). Conversely, the path of hot, humid air moving toward air-conditioned interior spaces is blocked at the cavity and condenses within the drainag space. Each design condition must, of course, be studied individually to determine the need and location for a vapor barrier within the wal assembly.

14.4.6 Applied Finishes

In walls of solid masonry or single wythe construction, greater care must b taken to avoid trapping moisture. ASTM specifications for concrete bric and block do not specify water permeability, and protective coatings ar often required to prevent water penetration. Some paint films and various other coatings are impervious to vapor flow and, if placed on the wrong side of the wall, can trap moisture inside the unit. Again, regional climatic conditions must be evaluated in determining the direction of vapor flow. The vapor pressure–temperature differentials discussed above also apply to concrete block walls.

NCMA recommendations for placement of vapor barriers are illustrated in *Fig. 14-50*. The barrier is placed on the moist side of the wall (the side with the highest vapor pressure). If the conditions shown for above-grade walls are revised and the highest vapor pressure is on the exterior, moisture will tend to flow toward the interior spaces, so the barrier should be placed on the outside. In areas of warm summer temperatures and high humidity, such as Houston, an unprotected concrete block wall with gypsum board and an impervious vinyl wall covering on the inside could mildew behind the vinyl. The moisture is able to enter the wall on the warm side, condenses on the cool side, but cannot escape and so is trapped behind the vinyl. The barrier must be on the moist side to prevent such damage.

In areas where summer/winter conditions cause a reverse in the direction of moisture flow, insulation must be used to prevent vapor condensation, or a cavity wall should be used in lieu of solid construction. Placing a vapor barrier and an insulation layer on the outside of the wall moves the saturated vapor pressure temperature sufficiently to avoid coincidence with the actual vapor pressure, and stop the moisture flow on the warm side of the wall (*see Fig. 14-51*). This condition will hold true for both summer and winter conditions and for inward and outward flow of vapor. For various climatic differences, only the thickness and type of insulation need be checked to assure adequate thermal and vapor resistance.

The temperature differentials and temperature gradient curves shown are based only on ASHRAE calculation principles, and do not take into account the time lag created by the thermal mass of the masonry (refer to

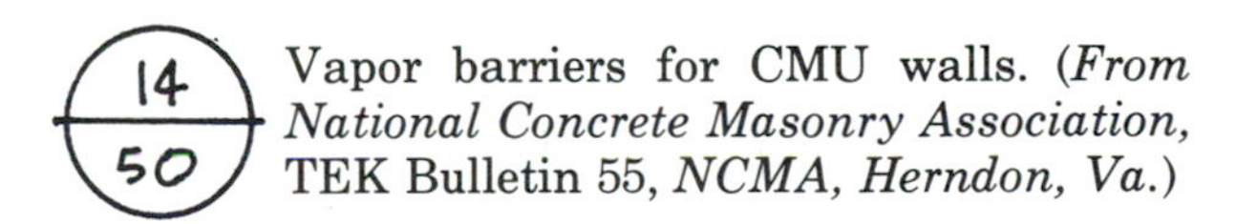

Vapor barriers for CMU walls. (*From National Concrete Masonry Association,* TEK Bulletin 55, *NCMA, Herndon, Va.*)

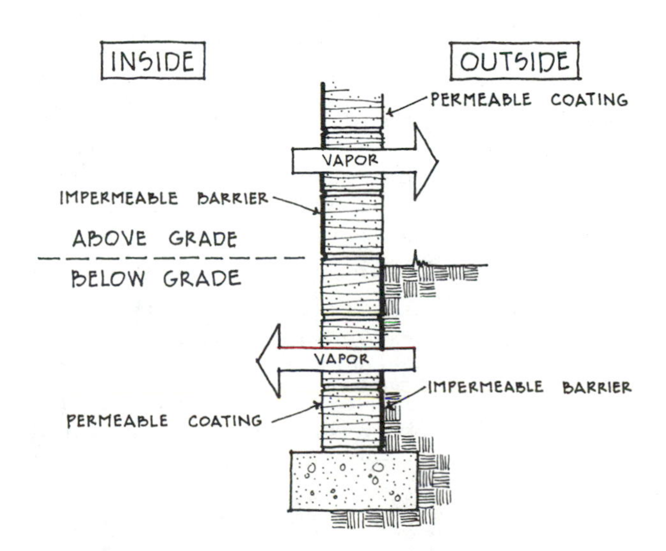

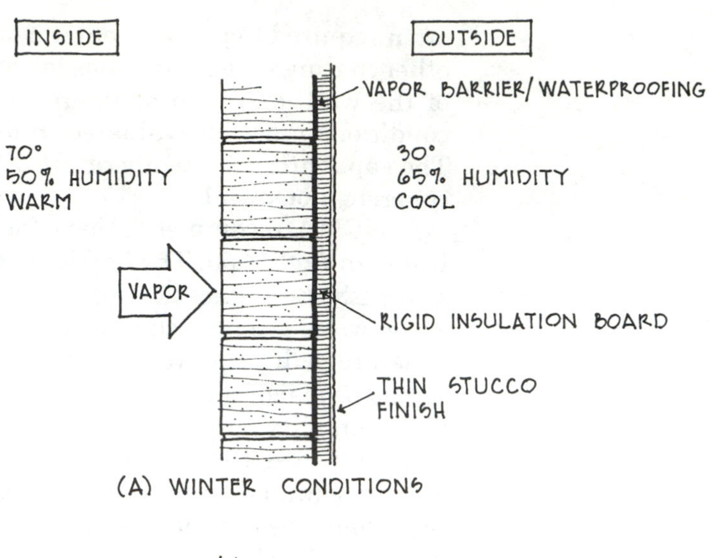

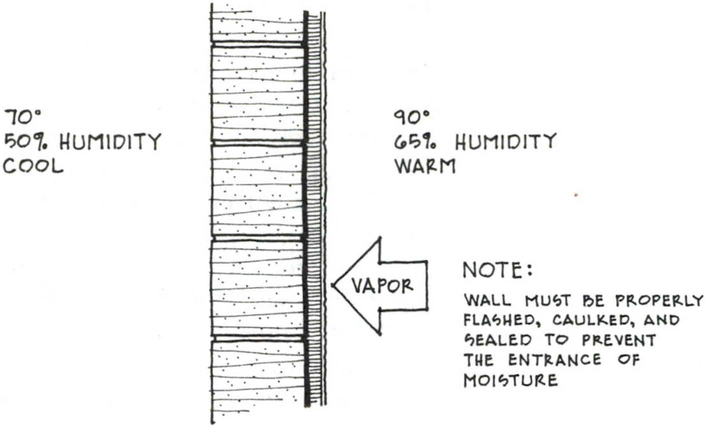

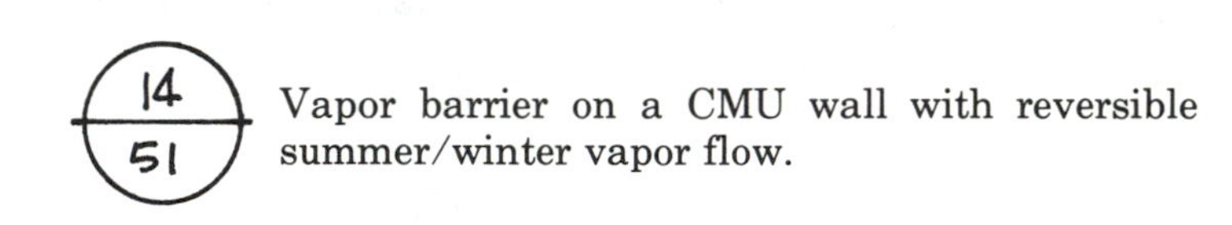

Vapor barrier on a CMU wall with reversible summer/winter vapor flow.

Chapter 8). The ability of masonry to store heat over relatively long period of time can be advantageous in climates where the indoor and outdoo temperature differential would normally reverse vapor flow one or mor times a day. Dense wall materials such as masonry will keep the wal temperature at a more constant level and reduce the temperature fluctua tion which occurs in lightweight thin-skinned construction.

This thermal inertia protects the wall from constantly shifting vapo movement and condensation points. Applied finishes still must be carefull selected on the basis of their permeability. Inadvertent use of an imperme able surface finish on the cold side of a wall can create problems that ar difficult and expensive to correct. A nonporous paint film will prevent som rainwater from entering a wall but, more significantly, it will impede th escape of moisture entering the wall from other sources. Water may ente through pores in materials, partially filled mortar joints, improperl flashed copings, sills, parapet walls, and so on, through capillary contac with the ground, or any number of other means. Moisture escapes from wall in only two ways: (1) through continuous cavities with adequate wee holes, and (2) by evaporation at the wall face (breathing). In single-wythe construction, where there are no cavities or weep holes, the exterior wall face must breathe in order to dry out. If a vapor barrier is used as in *Fig. 14-51*, it must form a complete waterproofing seal which will exclude moisture from any source.

There are numerous types of paint suitable for masonry walls, including cement-based paints, water-thinned emulsions, fill coats, solvent-thinned paints, and high-build coatings. In selecting a paint finish system, there are several things to consider. Paint products that are based on drying oils may be attacked by free alkali from the units or mortar. Alkaline-resistant paints and primers are recommended to prevent this, or the masonry must fully cure before painting. Surface conditions must also be considered, and preparations suitable to the selected finish made. Efflorescence must be removed from a masonry surface and observed for recurrence prior to painting. New masonry must not be washed with acid cleaning solutions. If low-alkali portland cement is not used in the mortar, it may be necessary to neutralize the wall with a 2- to $3\frac{1}{2}$ lb/gal solution of zinc chloride or zinc sulfate and water. Existing masonry must be cleaned of mold, moss, and mildew. Walls must be wetted before any cleaning solution is applied, and thoroughly rinsed afterward to prevent unfavorable paint reactions or chemical contribution to efflorescence.

Previously painted masonry surfaces normally require extensive preparation prior to repainting. All loose material, oil, and dirt must be removed. Penetrating primers may be required to improve adhesion to old paint. Emulsion paints and primers, which are nonpenetrating, generally require cleaner surfaces than do solvent-based paints. Existing cement-based paints may not be repainted with other types (although fresh cement-base paints may be used as primers for another paint when coverage will be within a relatively short time).

New brick masonry is seldom painted, as it is difficult to justify the present and future expense. Once painted, an exterior brick wall will require costly, periodic repainting every 3 to 5 years. The benefits of the desired aesthetic effect must be carefully weighed against maintenance costs before a decision is made to paint over new brick.

Industry experts make the following recommendations concerning the application of paint on masonry walls.

Cement-based paints consist of portland cement, lime, and pigments. They are characterized by good adhesion and the ability to make a wall less permeable to free water while still allowing it to breathe. Although they are not complete waterproofers, cement-based paints help fill and seal porous areas to exclude large amounts of rainwater. The product is available only in powder form since the cementitious components begin to hydrate upon contact with water. White and light colors tend to be most satisfactory, as it is difficult to obtain a uniform color with darker tints. Portland cement paints are generally low in cost but require considerable labor to apply and damp cure. The standard type contains a minimum of 65% portland cement by weight, and is suitable for general use. The heavy-duty type contains 80% portland cement and is used where there is excessive and continuous contact with moisture, such as the walls of a swimming pool.

Fill coats, also called fillers or primer-sealers, are similar in composition, application, and use to cement-based paints, but contain an emulsion paint in place of some of the water. They have improved adhesion and a tougher film than unmodified cement paints, and have greater water retention, allowing the cement a better chance to cure. This can be particularly important in arid climates, where it is difficult to keep the painted surface moist during curing. Fill coats are generally used as base coats on exterior masonry.

Water-thinned emulsion paints, commonly known as *latex paints*, are inexpensive and relatively easy to apply by brush, roller, or spray. Emulsion paints may be applied to damp surfaces, permitting painting shortly after a rain or on walls damp with condensation. As a group, these paints are alkali-resistant and walls do not require neutralizing washes and curing periods before coating. Latex paints have high water vapor permeability, permitting the painted surface to breathe easily. They will not adhere well to moderately chalky surfaces unless special formulations containing emulsified oils or alkyds are used. Vinyl and acrylic latex paints are available in clear coatings for colorless applications.

1. *Butadiene-styrene paint* is a rubber-based latex that develops water resistance more slowly than do vinyl or acrylic emulsions. Light tints are most satisfactory, as deep colors will chalk very rapidly.

2. *Polyvinyl acetate emulsion paint*, or vinyl paint, dries faster, has improved color retention, and has a more uniform, lower sheen than does rubber-based latex.

3. *Acrylic emulsion paints* have excellent color retention, good alkali resistance, and permit recoating in 30 minutes or less. Acrylics have high water-spotting resistance, and may be scrubbed easily. They have demonstrated superior resistance to rain penetration and show excellent overall durability.

4. *Alkyd emulsion paints* are related to solvent-thinned alkyds, but have the general characteristics of a latex. Compared to other emulsion paints, alkyds are slower to dry, have more odor, are not as alkali-resistant and have poorer color retention. They do penetrate better than most other latex paints, and can achieve better adhesion on chalky surfaces. Under normal exposure conditions, alkyd emulsions can serve as a finish coat over a suitable primer.

5. *Multicolored lacquers* are applied only by spray. The finish film appears as a base color with separate dots or particles of contrasting color. These paints can cover many surface defects and irregularities, but must be applied over a base coat of another type, such as vinyl or acrylic emulsion paints.

Solvent-thinned paints are relatively nonporous, and should be applied only to completely dry, clean, interior surfaces. Solvent-thinned paints are not recommended for exterior masonry. Most solvent-thinned paints have good penetration on relatively chalky surfaces.

1. *Oil-based paints* adhere well to porous masonry, but may require several coats to achieve uniform color. These paints are highly susceptible to alkalis, and new masonry must be thoroughly neutralized before application. Oil-based paints are moderately easy to apply, but require several days of drying between coats.

2. *Alkyd paints* have slightly less penetration than oil-based products resulting in better color uniformity at the expense of adhesion. They are more difficult to brush, dry faster, and produce a harder film. New masonry requires neutralization before painting with alkyds.

3. *Synthetic rubber and chlorinated rubber paints* have excellent penetration and good adhesion to previously painted, moderately chalky surfaces as well as to new masonry. They have good alkali resistance, and may be applied directly to alkaline masonry surfaces. Rubber-based paint

are difficult to brush, and darker hues lack color uniformity. They do, however, offer high resistance to corrosive fumes and chemicals, and perform well in industrial applications. Both types of rubber paint require very strong, volatile solvents that have high fire hazard.

4. *Epoxy paints* are of synthetic resins composed of a resin base and a liquid activator. They must be used within a relatively short time after mixing. Epoxy paints can be applied over alkaline surfaces, have good adhesion, and good fume and corrosion resistance. However, they are relatively expensive and somewhat difficult to apply.

Textured coatings of various proprietary mixes are applied thickly to hide minor surface imperfections. They may be of water or solvent base, and are generally classified as breathable coatings. Latex types can be applied over damp surfaces, but solvent types lend themselves to lower application temperatures (35 to 50°F). Since adhesion is mostly mechanical, some types will require primers. These coatings allow passage of water vapor, but cannot accommodate large quantities of water from leaks. Failure can occur as a result of accumulated water freezing behind the film.

Stucco can be used to reduce the permeability of old and new masonry walls, and is a popular finish on concrete block. A three-coat application of portland cement stucco may prove to be the most economical and satisfactory method for treatment of leaky walls where repointing and exterior wall treatment costs would be excessive. Good bond between the stucco and masonry depends on mechanical key and suction, and the texture of concrete masonry provides an excellent subsurface. Lime may be added to the cement-sand mixture for plasticity, but should not exceed 10% by weight or 25% by volume of the cement. Total thickness of the stucco application should be a minimum of $\frac{5}{8}$ in. Walls that are not reinforced against shrinkage and movement cracks can transmit excessive tensile stresses to monolithic stucco coatings and cause cracking of the finish surface as well.

Clear *silicone coatings* do not contain enough solids for waterproofing, and are classified as water repellants. They change the capillary angle of pores in the face of the masonry to repel rather than absorb water. Silicones will not bridge or fill hairline cracks or bond separations at mortar joints, which are usually the principal sources of moisture penetration. Silicones can be used effectively to prevent dirt and soot particles from entering the masonry surface, causing deep stains.

Silicones are available in both water- and solvent-based solutions. Water-based solutions are generally less expensive, but the molecular structure of the solvent types is smaller, allowing better penetration and performance. Silicone sealants do not normally cause perceptible discoloration, but particularly on walls with colored mortar, test panels or areas should be treated and then observed for bleaching or other color change. There are some inherent dangers in the indiscriminate use of silicone sealers. Since water may still penetrate the wall through cracks, faulty flashings, and so on, the moisture taken in must be able to escape by drainage through weep holes or evaporation through the wall surface. Silicones do not seal pores, and water vapor can penetrate the film. However, if capillary water containing soluble salts is traveling toward the treated surface, the salts will be deposited behind the silicone layer at depths of $\frac{1}{8}$ to $\frac{1}{4}$ in. This crystalline formation within the masonry unit can contribute to spalling and/or disintegration of the masonry. So although silicone coatings may appear to control efflorescence, they may actually cause greater damage by preventing escape of the effloresing salts and forcing the masonry to absorb the expansive pressures of internal crystallization. There should be no evidence of efflorescence on a wall that is to receive silicone treatment.

14.5 COLD-WEATHER CONSTRUCTION

Suspension or slow-down of building activity during winter months is costly not only to the construction industry, but to owners, material suppliers, and the general economy. European and Canadian builders, whose techniques are based on colder climates and longer winter seasons, have given U.S. contractors an excellent source of information on cold weather construction. The extensive use of masonry in Europe and the common practice of winter building have established standards and procedures for use in masonry construction in this country. Studies have shown that although the cost of winterizing may run from 0.75 to 1.5% of job costs, these expenditures are more than offset by the economic benefits o early project completion.

Masonry materials and construction exhibit quite different perfor mance characteristics in cold weather than under normal temperature conditions. Hydration and strength development of mortar will procee only at temperatures above freezing, and only when sufficient water i available. Construction may continue during freezing weather, however, i the mortar ingredients are heated and the masonry units and structure ar maintained above freezing during the initial hours after placement.

Mortar mixed using cold but unfrozen ingredients has different plasti properties from mortar mixed under normal conditions. The mix wil exhibit a lower water content, longer setting and hardening time, higher ai content, and lower early strength. Heating of the ingredients prior t mixing, however, will produce mortar with performance characteristic identical to those in the normal temperature range. Frozen mortar assume the outward appearance of being hardened, but suffers a reduction i strength. The water content of mortar during freezing significantly affect its freezing characteristics. Mortars with more than 6 to 8% water conten will expand on freezing, and expansion increases as the water conten increases. If freezing conditions are present, efforts should be made t reduce the normal 11 to 16% moisture content to some volume below 6% t avoid the disruptive expansion forces.

Cold *masonry units* exhibit all the performance characteristics of heated unit except that the volume is the smallest it will be within th completed masonry. Wet, frozen units will show decreased moistu absorption. Preheated units, on the other hand, will withdraw more wat from the mortar because of the absorptive characteristics of a cooling bod Highly absorptive units, by withdrawing water from the mortar, wi decrease the possible disruptive expansion that might occur with initi freezing. Units that are dry, but excessively cold, will withdraw heat fro the mortar and increase the rate of freezing.

During cold-weather construction, it may be best to use a Type II high-early-strength portland cement because of the greater protection will provide the mortar. Antifreeze additives are not recommended. If use in quantities that will significantly lower the freezing point of the morta these additives will rapidly decrease compressive and bond strengt Accelerators that hasten the hydration process are more widely used, b may also have damaging side effects. Calcium chloride is the maj ingredient in proprietary cold-weather admixtures. Although it is a effective accelerator, it has a highly corrosive effect on metal reinforceme and embedded items. High salt contents of accelerating admixtures ma also contribute to efflorescence and cause spalling of the units. In general, the use of such accelerators is not recommended.

Masonry *materials should be stored and protected* at the job site to prevent damage from wet, cold, or freezing weather. Bagged materials and masonry units should be stored elevated to prevent moisture migration from the ground, and covered to protect the sides and tops. Consideration should be given to the method of stockpiling sand to permit heating of the materials if required.

As the temperature falls, the number of different materials requiring heat will increase. Mixing water is the most logical first step since it is easily heated. If none of the other materials are frozen, mixing water may be the only ingredient requiring artificial heat. It should be warmed sufficiently to produce mortar temperatures between 40 and 120°F. Masonry sand, which contains a certain amount of moisture, should be thawed if frozen to remove ice. Sand should be warmed slowly to avoid scorching, and care should be taken to avoid contamination of the material from the fuel source. Dry masonry units should be heated if necessary to a temperature above 20°F at the time of use. Wet, frozen masonry units must be thawed without overheating.

The degree of protection against severe weather which is provided for the work area is an economic balance between mason productivity and cost of the production. Protective apparatus may range from a simple windbreak to an elaborate enclosure. Each job must be evaluated individually to determine needs and cost benefits, but some general rules do apply.

Characteristics such as strength, durability, flexibility, transparency, fire resistance, and ease of installation should be considered when selecting protective materials. Canvas, vinyl, and polyethylene coverings are often used. In most instances, a windbreak or unheated enclosure will reduce the chill factor sufficiently to provide the degree of protection required. Precautions must also be taken to safeguard workers against injury, and enclosures must be adequate to resist wind, snow, and uplift loads. A detailed outline of the recommended specification requirements for cold-weather construction is contained in Chapter 16. The chart in *Fig. 14-52* summarizes heating and protection requirements for various work temperatures.

Cold-weather masonry construction requirements. (*From International Masonry Industry All-Weather Council,* Recommended Practices and Guide Specifications for Cold Weather Masonry Construction, *International Masonry Institute, Washington, D.C., 1977.*)

Workday temperature	*Construction requirement*	*Protection requirement*
above 40°F	**Normal masonry procedures**	*Cover walls with plastic or canvas at end of workday to prevent water entering masonry*
0–32°F	**Heat mixing water to produce mortar temperatures between 40–120°F**	*Cover walls and materials to prevent wetting and freezing; covers should be plastic or canvas*
2–25°F	**Heat mixing water and sand to produce mortar temperatures between 40–120°F**	*With wind velocities over 15 mph provide windbreaks during the workday and cover walls and materials at the end of the workday to prevent wetting and freezing; maintain masonry above freezing for 16 hours using auxiliary heat or insulated blankets*
5–20°F	**Mortar on boards should be maintained above 40°F**	
0–0°F and below	**Heat mixing water and sand to produce mortar temperatures between 40 and 120°F**	*Provide enclosures and supply sufficient heat to maintain masonry enclosure above 32°F for 24 hours*

14.6 PREFABRICATED MASONRY

The technique of prefabricating brick masonry sections was developed in France, Switzerland, and Denmark during the 1950s, and adopted in the United States in the early 1960s. Reducing on-site labor led the construction industry to use of prefabricated building components, but the masonry industry was a late entrant into the field. The evolution and acceptance of a rational design method for masonry, together with improved units and mortars, made masonry prefabrication feasible, and the development of high-bond mortar additives contributed to rapid progress of the work.

Two methods of manufacturing are presently used in masonry prefabrication. *Manual construction* uses conventional labor techniques and preassembles masonry sections at a point other than the final in-place location of the masonry. This may be in a plant, at a remote work yard, or on the job site. Sheltered work areas at the job site allow continuous work operations despite inclement weather, and thus increase productivity. *Casting methods* lend themselves to the use of automated systems, including forming, mechanical unit placing, and machine grouting.

Prefabrication methods have been used successfully on wall panels, columns, and beam sections (*see Fig. 14-53*). A major requirement for the

14-53 Lifting a prefabricated masonry panel into place. (*Photo courtesy BIA.*)

economic feasibility of preassembly is the repetition of design elements in the structure. Large numbers of identical sections may be mass-produced in environmentally favorable locations, then hoisted into place at the job site for field connections and/or grouting and reinforcing. The need for on-site scaffolding can be eliminated, and panelization allows the construction of complicated shapes without the need for expensive falsework and shoring. Shapes with offsets, returns, soffits, or arches are built using jigs and forms at the plant. Reuse of the jigs and forms for repetitive shapes lowers the unit cost considerably. Quality control is more easily maintained under factory conditions by automating mortar batching systems and standardizing curing conditions. Prefabrication may also shorten the total construction time, allow earlier occupancy, and benefit the owner by increased income and lower interim financing costs.

Panel connections for facing materials generally combine the use of shelf angles and welded, bolted, or masonry tie anchors, depending on the type of structural frame used. Allowance must still be made for differential movement between masonry and concrete or steel by expansion/control joints, flexible anchorage, and shelf angle pads. Prefabricated bearing walls and columns within or at the perimeter of a structure use a spliced reinforcement connection like those employed in conventional cast-in-place concrete construction. All connections must be designed to withstand anticipated vertical and lateral loads, including wind and earthquake.

The size of prefabricated masonry panels is limited primarily by transportation and erection requirements. Architectural plan layout may in some cases, preclude the use of preassembled sections. Prefabrication does not have field adjustment capabilities as in normal masonry installation. Panelization may require that other trades build within tolerances that exceed standard construction practice. Although it may not be universally suitable for all construction projects, there are some situations where masonry prefabrication can offer distinct advantages over standard methods.

14.7 MASONRY RESTORATION

Masonry restoration usually involves cleaning, replacing damaged or deteriorated units, and tuckpointing defective mortar joints. Structural failures may be manifested as cracking, uneven settlement, and deterioration of the mortar, the bricks, or the stone. Damage and deterioration are most often caused by moisture in one form or another. Moisture can be absorbed from rain, groundwater, condensation, or even a damp atmosphere. Stone and brick absorb moisture according to their porosity. Eighteenth- and nineteenth-century bricks were soft and porous by today's standards, primarily because firing temperatures were much lower. These bricks absorbed as much as 25% of their weight of water compared to only 4 to 10% for most bricks manufactured in the United States today.

The presence of moisture in a masonry wall can do two things. It can leach soluble salts and other impurities to the surface causing stains; and it can break off the surface of the stone or the units when freezing causes expanding ice crystals to form in the pores. The treatment of efflorescence and some of the cleaning procedures outlined are applicable to new construction as well.

14.7.1 Efflorescence

One problem that clay, concrete, and stone masonry have in common is efflorescence. Efflorescence is the leaching of soluble salts onto the face of the masonry wall, causing stains and streaks (*see Fig. 14-54*). Three types of stains commonly occur, but two of these affect only clay masonry. *White efflorescence,* however, can occur on stone and concrete masonry as well but is most noticeable on dark-colored clay brick. Some forms of efflorescence are temporary. Since the salts are water soluble, the stain will often disappear with washing, or of its own accord with normal rain and weathering. Heavy accumulations or stubborn deposits may be removed with a solution of 1 part muriatic acid to 12 parts water, but the wall must be saturated with water both before and after the scrubbing.

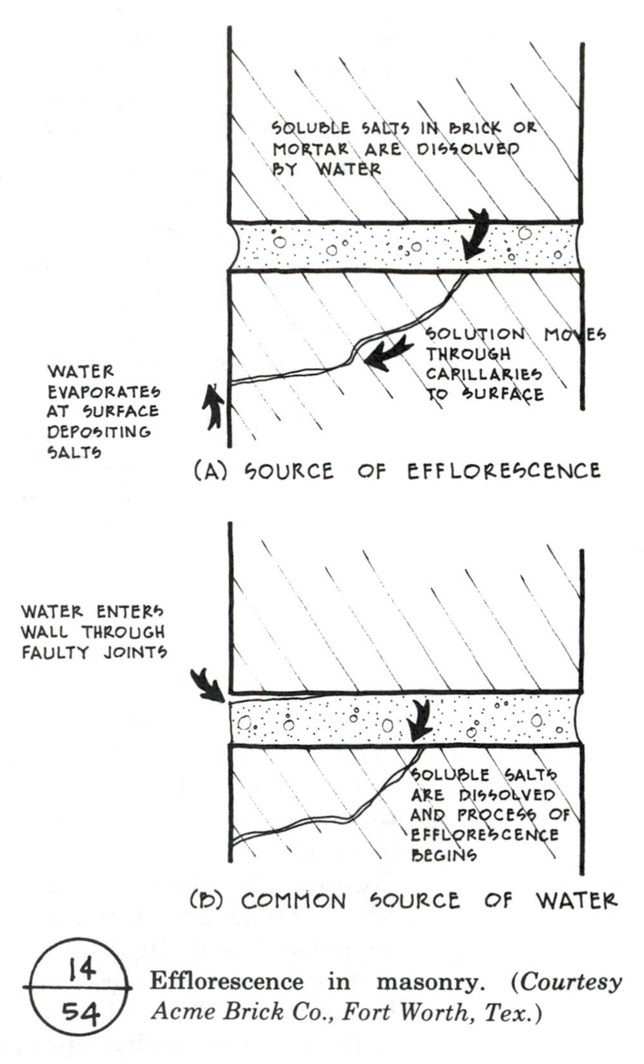

Efflorescence in masonry. (*Courtesy Acme Brick Co., Fort Worth, Tex.*)

There is a permanent form of efflorescence which may occur, subsid or even disappear after heavy rains, and then recur repeatedly over a peri of years. In time, disintegration of the mortar and spalling of the mason units can take place, which then cause the wall to leak. This permanent recurring form of efflorescence is caused primarily by soluble alkali salt usually sodium and potassium sulfates.

Salts are brought to the surface of the masonry in solutions of wat and are deposited there when the water evaporates. The occurrence may be sudden, and is generally brought on by climatic and environmental changes, as in a dry period during cool weather following a rainy period. Hot summer months are not as conducive to this type of efflorescence because the wetting and drying of the wall is generally quite rapid. In late fall, winter, and early spring, however, after rainy periods, when evaporation is much slower and temperatures are cooler, white efflorescent stains are more likely to appear.

Three simultaneous conditions must exist in order for efflorescence to occur: (1) soluble salts must be present within the masonry assembly, (2) there must be a source of water sufficiently in contact with the salts to form a solution, and (3) the wall construction must be such that paths exist for the migration of the salt solution to a surface where evaporation can take place. In conventional masonry construction exposed to weather, it is virtually impossible to ensure that no salts are present, no water penetrates the masonry, and no paths exist for migration. The most practical approach to the prevention and control of efflorescence is to reduce all of the contributing factors to a minimum.

Soluble salts may be present in either the masonry or the mortar, or may be absorbed into the wall through rain or groundwater. Since white efflorescence appears on the face of the masonry units, they are generally assumed to be at fault. This, however, is not usually the case. Virtually all clay brick contains at least some salts, but their efflorescing potential is small. The degree of probability may be easily determined by the wick test included in ASTM C67. Brick units relatively free from impurities are readily available throughout the United States. Dense to moderately absorptive units are least troublesome. Researchers differ in their opinions on concrete masonry, some saying that they have even less efflorescing potential than clay products, and others recording two to seven times as much soluble material.

Mortars also vary in the amounts of soluble salts they contain depending on the type of cement used. Cements are generally the greatest source of soluble materials that contribute to efflorescence. Those with a high alkali content and limestone impurities are most likely to cause problems. Some companies have developed special "low-alkali" and "non-staining" cements for use in masonry mortars. Hydrated limes are relatively pure and generally have 4 to 10 times less efflorescing potential than cements. Therefore, lime is one of the lesser sources, along with well-washed sand and clean, potable water. Soluble salts from the soil may be absorbed into masonry in contact with the ground through the capillary action of groundwater migrating upward into the units. Sulfurous gases in the atmosphere in highly industrialized areas may also contaminate the masonry with soluble salts through soaking with "acid rain."

The source of moisture necessary to produce efflorescence may be either rainwater or the condensation of water vapor within the assembly. Water may also enter the wall because of improper protection from rain and snow during construction.

The greatest cause of efflorescence is faulty design and construction practices. Regardless of the impurity of the materials used, it is unlikely that efflorescence will occur if proper precautions and high-quality workmanship are employed. Some of the more common malpractices are (1) failure to store masonry units off the ground and protect with waterproof covers; (2) failure to cover and protect unfinished walls; (3) inadequately flashed and unprotected parapet walls; (4) absence of drips on cornices o projecting members; (5) poorly filled mortar joints; (6) failure to us dampproof courses at ground level; (7) failure to repair or patch cracked o broken mortar joints and (8) use of dense units and mortar which absor moisture through unrepaired cracks and are then slow to dry out.

To minimize the possibility of efflorescence, the following measure are of greatest importance: (1) use only units of low to moderate absorptio or test the brick for efflorescing potential in accordance with ASTM C67; (2 use only low-alkali, nonstaining, or white cements in the mortar; (3) us mortars of high lime/cement ratio; (4) properly protect materials before an during construction; (5) use flashing materials, weep holes, caulking, an sealants at strategic locations to expedite the removal of moisture that ha entered the wall; and most important of all, (6) construct full, tight weatherproof joints. These precautions are particularly important i regions with high annual rainfall.

Silicone applications are often recommended as a solution to effloresc ing problems. However, if the silicone is applied to a wall that still contain both moisture and salts, the resulting problems may be even mor damaging than the stain. The water in the wall will still take the salts int solution, and as it migrates toward the outer face, most of it will stop at th inner depth of the water repellant. The water will then evaporate throug the silicone as a vapor and deposit the salts inside the masonry unit. Thi interior crystalline buildup can exert tremendous pressure capable o spalling the unit face (*see Fig. 14-55*). Silicone applications are generally no recommended as a treatment for efflorescence unless the chain of contribu tory conditions (moisture, salts, and migration paths) is also broken.

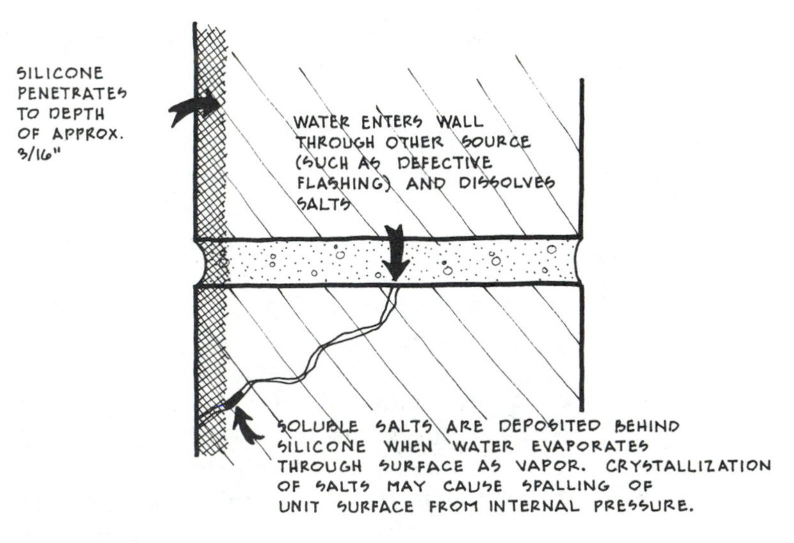
SILICONE PENETRATES TO DEPTH OF APPROX. 3/16"
WATER ENTERS WALL THROUGH OTHER SOURCE (SUCH AS DEFECTIVE FLASHING) AND DISSOLVES SALTS
SOLUBLE SALTS ARE DEPOSITED BEHIND SILICONE WHEN WATER EVAPORATES THROUGH SURFACE AS VAPOR. CRYSTALLIZATION OF SALTS MAY CAUSE SPALLING OF UNIT SURFACE FROM INTERNAL PRESSURE.

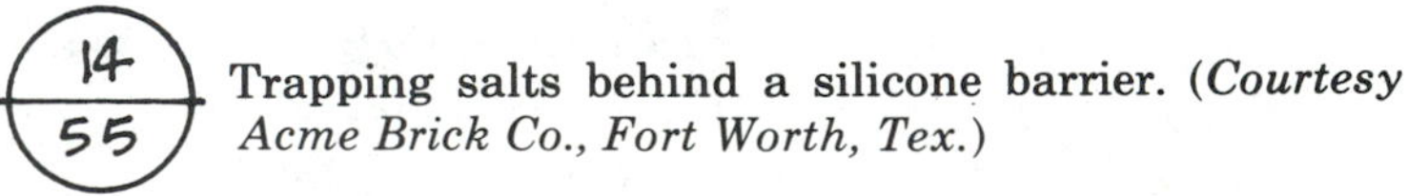
14
55

Trapping salts behind a silicone barrier. (*Courtesy Acme Brick Co., Fort Worth, Tex.*)

In addition to white efflorescence, other stains that are peculiar to cl products are *green vanadium stains* and *brown manganese stains*. Var dium salts originate in the raw materials used to manufacture brick. T resulting stains may occur on red, buff, or white units, but are mo noticeable on the lighter-colored brick. The chloride salts of vanadiu require highly acidic leaching solutions, and the problem of green stain often does not occur unless the walls are washed down with a muriatic acid solution. To minimize the occurrence of green stain, do not use acid solutions to clean light-colored brick, and follow the recommendations of the brick manufacturer for the proper cleaning compounds and solutions. If green stains do appear as a result of acid washing, flush the wall with clean water and then wash or spray with a solution of 2 lb of potassium or sodium hydroxide to 1 gal of water to neutralize the acid. After the solution has been on the wall for 2 or 3 days, the white residue may be hosed off with clean water.

Manganese stain may occur on the surface of mortar or bricks containing manganese coloring agents. The stain may be tan, brown, or gray. The brown stain is oily in appearance, and may streak down over the face of the wall. The manufacturing process chemically changes the manganese into compounds that are soluble in weak acid solutions. The staining may occur because of acid cleaning procedures, or even because of acid rain in some industrial areas. Muriatic acid solutions should never be used to clean tan, brown, black, or gray brick or mortar unless the acids are neutralized during the rinsing operation. The permanent removal of manganese stain may be difficult since it often returns after initial removal. The following procedure, recommended by BIA, has proven effective in removing "brown stain" and preventing its return. Mix by volume 1 part 80% acetic acid, 1 part 30 to 35% hydrogen peroxide, and 6 parts water. Wet the wall thoroughly, and brush or spray on the solution, but do not scrub. The reaction is generally very quick and the stain rapidly disappears. After the reaction is complete, rinse the wall thoroughly with water. Although this solution is very effective, it is dangerous to mix and use, and proper precaution should be taken to protect workers and adjacent surfaces. The permanence of the removal may still be in doubt, and therefore the initial prevention of maganese stains is of paramount importance. Always request and follow the recommendations of the manufacturer in cleaning brick that contains manganese coloring agents.

14.7.2 Cleaning Existing Masonry

For existing masonry buildings, BIA recommends several alternative cleaning methods. *High-pressure steam cleaning* lends itself most readily to a wide variety of unit types. Smooth, hard brick, or glazed brick should always be cleaned by this method. The more impervious the surface, the easier it will clean. *Chemical* or *detergent solutions* are generally not required except for unusually stubborn stains. *Dry sandblasting* should be used only when less destructive methods cannot be successfully used. To lessen the damage caused by sandblasting, softer abrasives other than sand can sometimes be substituted. *Hand washing* is slower and more tedious, must be repeated more often, and is not economical except on small buildings. *High-pressure cold-water cleaning* is a satisfactory method as long as an ample water supply is available, and the disposal of large volumes of water is not a problem. Combined *chemical and steam cleaning* is used primarily to remove applied coatings such as paint. It is a highly specialized field and often requires extensive preliminary analysis. *Wet sand cleaning* is used on hard-surface brick to remove paint or other coatings, but some abrasion will be evident. The water used in this procedure cushions the abrasive action to some degree, and eliminates the problem of dust. Wet aggregate cleaning was specially developed by Western Waterproofing Company for use on soft brick and stone, and is particularly effective on surfaces with flutings, carvings, or other ornamentation. A mixture of water and a friable aggregate free from silica is applied at low pressure with a gentle, but thorough scouring action which cleans effectively without damaging the surface.

The cleaning of historic buildings should be carefully evaluated by a restoration specialist, and may be controlled by government regulations. The decision to clean should be based on condition of the structure, type of staining present, and the effect the cleaning method will have on the surface. In many instances, cleaning may not be a desirable part of the restoration process.

14.7.3 Cleaning New Masonry

The final appearance of exposed brick or concrete block masonry walls depends to a great extent on the attention given to the surfaces during construction and during the cleaning process. Care should always be taken to prevent mortar smears or splatters on the face of the wall, but if such stains do occur, daily cleaning of the finished work can help prevent permanent discoloration. Excess mortar and dust can be brushed from the surface easily when the work is still fresh. For brick walls, a brush of medium-soft bristle is preferable. Any motions that would result in rubbing or pressing mortar particles into the unit face should be avoided. Fo concrete block walls, mortar droppings can be removed with a chisel afte drying and hardening, and the remaining mortar removed with a stiff fibe brush.

Other precautions that may be taken during construction include (1 protecting the base of the wall from rain-splashed mud or mortar dropping by using straw, sand, sawdust, or plastic sheeting spread out on the groun and up the wall surface; (2) turning scaffold boards on edge at the end of th day to prevent possible rainfall from splashing mortar or dirt directly on th wall; (3) covering the tops of unfinished walls at the end of the workday t prevent mortar joint washout or the penetration of water into the wall; an (4) protecting site-stored masonry units from mud or water stains by storin off the ground and covering with waterproof material.

The cleaning process itself can be a source of staining if chemical o detergent cleansing solutions are improperly used, or if windows, doors, an trim are not properly protected from possible corrosive effects. Ne masonry may be cleaned by bucket-and-brush hand scrubbing with wate detergent, muriatic acid solution, or proprietary cleaning compound Cleaning should be scheduled as late as possible in the construction, an the mortar must be thoroughly set and cured. However, long periods of tim should not elapse between completion of the masonry and the actu cleaning, because mortar smears and splatters will cure on the wall an become very difficult to remove. Most surfaces should be thoroughl saturated with water before beginning (saturated masonry will not absor dissolved mortar particles). Confine work to small areas that can be rinse before they dry. Environmental conditions will affect the drying time an reaction rate of acid solutions, and ideally the cleaning crew should be jus ahead of the sunshine to avoid rapid evaporation. Walls should be cleane only on dry days.

Detergent solutions will remove mud, dirt, and soil accumulations. One-half cup dry measure of trisodium phosphate and $\frac{1}{2}$ cup dry measure of laundry detergent dissolved in 1 gal of water is recommended. *Acid cleaners* must be carefully selected and controlled to avoid both injury and damage. Hydrochloric acid dissolves mortar particles, and should be used carefully in a diluted state. Muriatic acid should be mixed with at least 9 parts clean water in a nonmetallic container, and metal tools or brushes should not be used. Acid solutions can cause green or brown stains on some clay masonry, and should not be used on light-colored, brown, black, or gray brick which contains manganese coloring agents. *Proprietary compounds* should be carefully selected for compatibility with the masonry material, and the manufacturer's recommended procedures and dilution instructions should be followed.

Some contractors use *high-pressure water* cleaning combined with high- or low-pressure applied cleaning compounds. If the wall is not thoroughly saturated before beginning, high-pressure application can drive the cleaning solutions into the masonry, where they may become the source of future staining problems. *Dry sandblasting* is still employed as a cleaning method, but is more often used on existing rather than new masonry. It eliminates the dangers of mortar smear, acid burn, and efflorescence inherent in acid cleaning but, if improperly executed, may scar the mortar joints and the masonry surfaces.

All of the cleaning methods outlined should be performed on a small, inconspicuously located test area to determine the effect and the effectiveness of the process. For cleaning of new masonry, the BIA has established guidelines for the selection of methods depending on the type of brick and mortar that has been used (*see Fig. 14-56*).

The method of removal for externally caused stains will depend entirely on the type of material that has been splattered on, or absorbed by the masonry. Many stains can be removed by scrubbing with ordinary kitchen cleansers. Others require the use of a *poultice* or paste made with a solvent or reagent and an inert material. The stain is dissolved, and the solution leached into the poultice. After drying, the powdery substance remaining is simply brushed off. Although repeated applications may be required, the poultice will prevent the stain from spreading during treatment, by actually pulling it from the pores of the masonry. Some of the more common stains and the methods for their removal are listed below.

1. *Paint stains* on both brick and concrete masonry may be removed by a commercial paint remover, or by a solution of 2 lb of trisodium phosphate in 1 gal of water. Apply the liquid with a brush, allow it to remain and soften the paint, then remove with a scraper and wire brush. Wash the surface afterward with clear water.

2. *Iron stains* or welding splatter are removed from clay and from concrete masonry in different manners. For clay brick, spray or brush the area with a solution of 1 lb of oxalic acid crystals, 1 gal of water, and $\frac{1}{2}$ lb of ammonium bifluoride to speed the reaction. This solution shoud be used with caution because it generates hydrofluoric acid, which will etch the brick surface. The etching will be more noticeable on smooth masonry. An alternative method, which may also be used on concrete masonry, uses 7 parts lime-free glycerine with a solution of 1 part sodium citrate in 6 parts lukewarm water mixed with whiting to form a poultice. Apply a thick paste and scrape off when dry. Repeat the process until the stain has disappeared, then wash the area thoroughly with clear water.

Brick category	*Cleaning method*	*Remarks*
Red and red flashed	Bucket and brush hand cleaning High-pressure water Sandblasting	Hydrochloric acid solutions, proprietary compounds, and emulsifying agents may be used. *Smooth texture:* Mortar stains and smears are generally easier to remove; less surface area exposed; easier to presoak and rinse; unbroken surface, thus more likely to display poor rinsing, acid staining, poor removal of mortar smears. *Rough texture:* Mortar and dirt tend to penetrate deep into textures; additional area for water and acid absorption; essential to use pressurized water during rinsing.
Red, heavy sand finish	Bucket and brush hand cleaning High-pressure water	Clean with plain water and scrub brush, or *lightly* applied high pressur and plain water. Excessive mortar stains may require use of cleanin solutions. *Sandblasting is not recommended.*
Light-colored units, white, tan, buff, gray, specks, pink, brown, and black	Bucket and brush hand cleaning High-pressure water Sandblasting	*Do not use muriatic acid!!* Clean with plain water, detergents, emulsifying agents, or suitable proprietary compounds. Manganese colored brick units tend to react to muriatic acid solutions and stair Light-colored brick are more susceptible than darker units to "acid burn" and stains.
Same as light-colored units, etc., plus sand finish	Bucket and brush hand cleaning High-pressure water	Lightly apply either method. (See notes for light-colored units, etc.) *Sandblasting is not recommended.*
Glazed brick	Bucket and brush hand cleaning	Wipe glazed surface with soft cloth within a few minutes of laying units. Use soft sponge or brush plus ample water supply for final washing. Use detergents where necessary and acid solutions only for *very difficult* mortar stain. Do not use acid on salt glazed or metallic glazed brick. Do not use abrasive powders.
Colored mortars	Method is generally controlled by the brick unit	Many manufacturers of colored mortars do not recommend chemica cleaning solutions. Most acids tend to bleach colored mortars. Mi detergent solutions are generally recommended.

Cleaning guide for new brick masonry. (*From Brick Institute of America,* Technical Note 20 Rev., *BIA, McLean, Va.*)

3. *Copper or bronze stains* are removed from both clay and concret masonry by a mixture in dry form of 1 part ammonium chloride and 4 part powdered talc, with ammonia water added to make a thick paste. Apply th paste over the stain and remove when it is dry using a scraper or, if workin on glazed masonry, a wooden paddle.

4. *Smoke stains* are difficult to remove. Scrubbing with a scourin powder that contains bleach using a stiff-bristle brush will generally wor well. Small, stubborn stains are better dealt with using a poultice o trichloroethylene and talc, but the area should be well ventilated to avoid buildup of harmful fumes. In some instances where large areas have bee stained, alkali detergents and commercial emulsifying agents may b brush or spray applied or used in steam cleaners. If given sufficient time t work, this method will work well.

It should be reemphasized that any proposed stain-removal method b tested on a small area before general application is made to a wall c surface.

14.7.4 Tuckpointing

Mortar joints in existing masonry are repaired by cutting out the defective area and replacing it with fresh mortar. This procedure is called tuckpointing or repointing. Generally, it is best to repoint an entire area rather than just the visibly damaged or deteriorated joints because it is usually impossible to identify all of the defective joints by visual inspection alone. Joints should be raked out by hand to a depth of at least $\frac{1}{2}$ in. and all loose particles and dirt removed with a brush or hose stream. Immediately before beginning repointing operations, thoroughly wet the joint with water, then allow it to surface dry. The new mortar should be packed in thin layers until the joint is full, being careful not to stain the surface of the masonry. When the mortar is "thumbprint" hard, tool the surface to match existing joints, making a tight, weather-resistant closure.

For brick or stone that has weathered to a rounded profile, the new mortar joint should be slightly recessed from the unit surface and tooled concave to avoid "feathered" edges that break off easily, leaving voids through which moisture can penetrate. Stippling joints with the end of a stiff brush while the mortar is still soft will give it a worn appearance.

Repointing mortar should match the existing material as closely as possible in color, texture, strength, and hardness. Old mortars were generally soft and may have been mixed from clay, gypsum, lime and sand, natural cement, or portland cement (laboratory analysis can determine exact ingredients). Modern portland cement–lime mortars are much harder than these older mixtures, and in some cases, harder than the brick or stone.

Ideally, the new mortar should have the same density and absorbency as the stones or bricks in the wall. A hard mortar used with soft brick or stone can cause deterioration of the masonry because the two components do not respond to temperature and moisture changes at the same rate, or to the same degree. The softer material will absorb more movement stress and more moisture, thus suffering greater damage from freeze–thaw cycles. Many buildings have been irreparably damaged in this manner. Strong, high-cement-ratio mortars also have greater shrinkage potential, which could leave openings at the mortar-to-unit interface.

Some authorities recommend lime-sand mortars mixed with portland cement in a ratio of 1 part cement to 2 or 3 parts lime. Others prefer 2 parts hydraulic quicklime to 5 parts sand, with the optional addition of portland cement not to exceed a ratio of 1 part to 12 with the lime–sand mixture. For brick masonry restoration, the BIA recommends Type N portland cement–lime–sand mortar (1: 1: 6) if the ingredients of the original mortar are unknown. To compensate for shrinkage, they recommend a prehydration process where the dry ingredients are mixed with only enough water to produce a damp, unworkable mix (will retain its form when pressed into a ball). The mortar is kept in this damp condition for 1 to 2 hours and then sufficient water is added to bring it to a working consistency somewhat drier than conventional mortar. The drier mix facilitates the method of placement used in tuckpointing.

14.7.5 Replacing Masonry

Once the brick or stone in a wall begins to crumble, the deterioration will continue, often at an accelerated rate. The condition can be remedied only by replacement of the affected unit. Where damage or deterioration is extensive, replacement of entire sections of masonry may be required. The brick, stone, terra cotta, or clay tile used for such repairs should match the original material characteristics as closely as possible. Where damage has been caused by excessive moisture penetration from groundwater migration, the installation of a waterproof membrane as a dampproof course may be possible as sections of the original masonry are removed.

15

ECONOMICS OF CONSTRUCTION

Both masonry veneers and reinforced loadbearing systems are highly competitive with other construction materials and techniques. Changing architectural styles have brought more and more masonry to commercial designs, and the material lends itself particularly well to certain building types and applications. Repetitive compartmented plans (hospitals, apartments, hotels, etc.) are ideally suited to bearing wall designs, and masonry veneers are scaling the facades of high-rise buildings across the country. Many factors influence the cost effectiveness of building systems, including initial cost, maintenance, insurance rates, and speed of construction.

15.1 GENERAL COSTS

Most construction systems involve the correlation of several major elements to complete a total structure. First the framing is erected, then the exterior walls are enclosed, and finally the interior partitions are built. Loadbearing masonry construction combines these several elements into a single operation. The bearing walls *are* the structure, the exterior envelope, and the partitions. In addition, masonry walls offer fire resistance, sound attenuation, and thermal massing to control heat loss and heat gain. Some types of masonry, such as brick and architectural concrete block, also give a finished, low-maintenance surface. In order to accurately assess the cost of masonry, all of its functions must be considered. Comparing masonry only as an alternative structural system, for instance, does not give a complete picture. Total building costs must be compared in order to do this.

15.2 COMPARATIVE COSTS

In 1974, the Brick Institute of America retained the American Appraisal Company of Milwaukee, Wisconsin, to conduct just such a comparative study. They investigated the initial construction costs for loadbearing brick, steel frame, and concrete frame buildings. The results were tallied both in dollars and relative cost percentages for over 200 cities in the United States and Canada. The results showed that the loadbearing brick structure was 11 to 30% less expensive than the other systems. The building used for the comparison was Woodlake Towers III, a nine-story luxury apartment building in Fairfax County, Virginia, designed by Collins & Kronstadt, Leahy, Hogan, Collins–Architects (*see Fig. 15-1*). The parameters of the study as stated by American Appraisal Company are as follows:

Woodlake Towers Apartments, Fairfax County, Virginia. Collins & Kronstadt, Leahy, Hogan, Collins, architects. [*From Brick Institute of America*, American Appraisal Company (Milwaukee, Wis.) Cost Comparison Study, *BIA*, *McLean*, *Va.*]

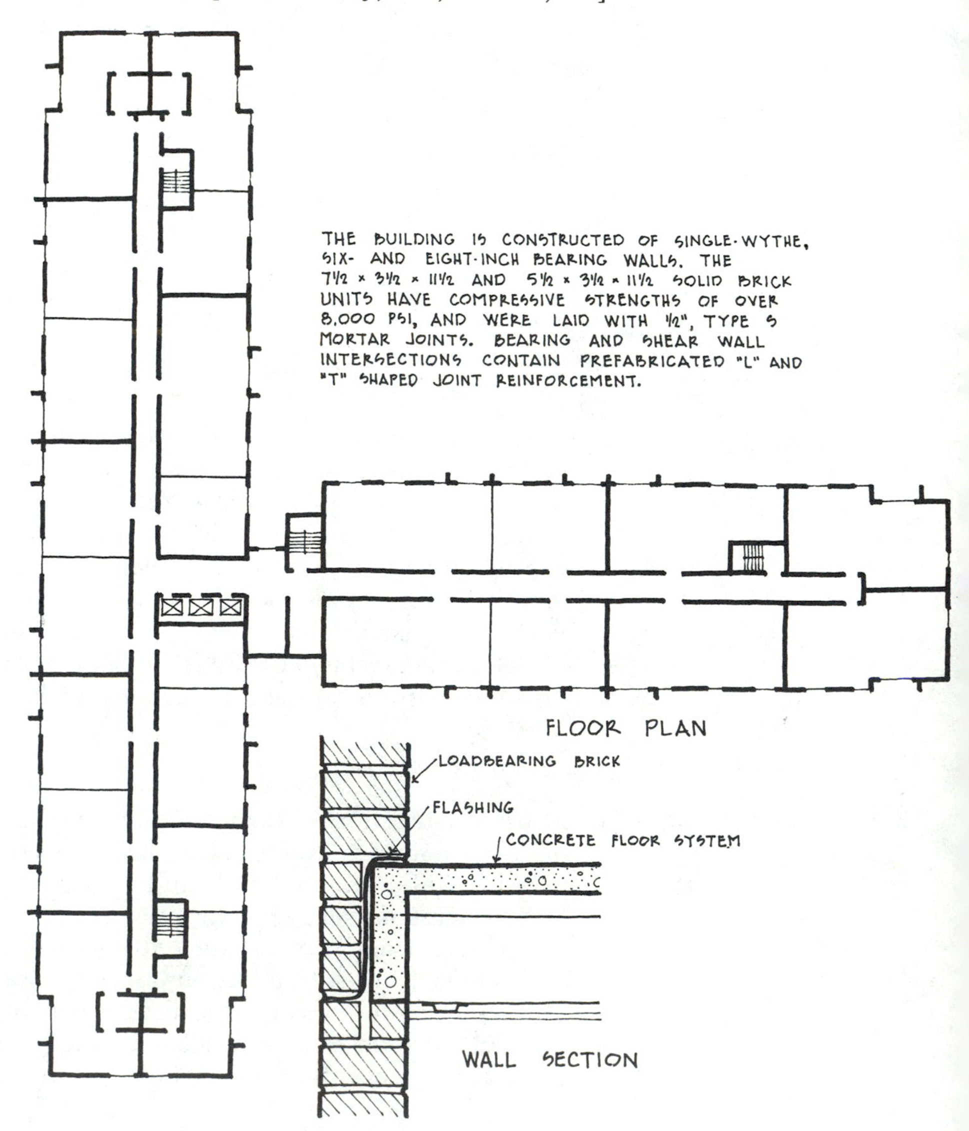

THE BUILDING IS CONSTRUCTED OF SINGLE-WYTHE, SIX- AND EIGHT-INCH BEARING WALLS. THE 7½ × 3½ × 11½ AND 5½ × 3½ × 11½ SOLID BRICK UNITS HAVE COMPRESSIVE STRENGTHS OF OVER 8,000 PSI, AND WERE LAID WITH ½", TYPE S MORTAR JOINTS. BEARING AND SHEAR WALL INTERSECTIONS CONTAIN PREFABRICATED "L" AND "T" SHAPED JOINT REINFORCEMENT.
FLOOR PLAN
LOADBEARING BRICK
FLASHING
CONCRETE FLOOR SYSTEM
WALL SECTION

All cost estimates were made as if the building were to be built in Seismic Zone 0. Then, adjustments were made to each cost for proper design of the structures in Zones 1, 2, and 3 according to the *Uniform Building Code*. We then applied material costs and labor rate modifiers from our computer data bank to the three separate structures . . . for locations in over 200 cities in the United States and Canada.

American Appraisal's results are shown in *Fig. 15-2*. The figures include contractor's fees, profit and overhead, and engineering and architectural fees. Although the 1974 costs do not reflect current rates, the comparative analysis is valid for present markets. Because the building plan lends itself so well to bearing wall design, the initial cost of the loadbearing brick building was significantly lower than for either concrete or steel framing.

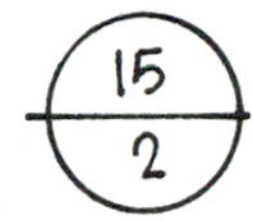

Cost comparison for Woodlake Towers Apartments. [*From Brick Institute of America*, American Appraisal Company (Milwaukee, Wis.) Cost Comparison Study, *BIA, McLean, Va.*]

Location	Total building cost (thousands of dollars)			Relative cost (%)		
	Loadbearing brick	Steel frame	Concrete frame	Loadbearing brick	Steel frame	Concrete frame
labama						
Birmingham	3931	4910	4537	100	125	115
Mobile	3925	4825	4476	100	123	114
Montgomery	3464	4449	3951	100	128	114
laska						
Anchorage	5835	6674	6469	100	114	111
Fairbanks	6432	7201	7184	100	112	112
Juneau	6099	6894	6853	100	113	112
rizona						
Phoenix	4445	5495	5062	100	124	114
Tucson	4201	5398	4784	100	128	114
rkansas						
Fort Smith	3437	4385	3951	100	128	115
Little Rock	3681	4579	4229	100	124	115
alifornia						
Bakersfield	4695	5753	5412	100	123	115
Fresno	4613	5688	5257	100	123	114
Los Angeles	4749	5820	5443	100	123	115
Modesto	4749	5820	5412	100	123	114
Sacramento	4776	5853	5443	100	123	114
San Bernardino	4613	5688	5257	100	123	114
San Diego	4640	5721	5318	100	123	115
San Francisco	4885	5951	5598	100	122	115
olorado						
Colorado Springs	4006	4900	4537	100	122	113
Denver	4222	5094	4815	100	121	114
onnecticut						
Hartford	4520	5481	5155	100	121	113
New Haven	4547	5514	5217	100	121	115
Stamford	4574	5514	5217	100	121	114

Location	*Total building cost (thousands of dollars)*			*Relative cost (%)*		
	Loadbearing brick	*Steel frame*	*Concrete frame*	*Loadbearing brick*	*Steel frame*	*Concrete frame*
Delaware						
Dover	4358	5223	5000	100	120	115
Wilmington	4358	5225	4970	100	120	114
District of Columbia						
Washington	4439	5320	5062	100	120	114
Florida						
Jacksonville	3898	4836	4445	100	124	114
Miami	4493	5436	5093	100	121	113
Orlando	3871	4761	4445	100	123	115
Palm Beach Area	4439	5372	5031	100	121	113
Pensacola	3762	4696	4260	100	125	114
Tallahassee	3627	4578	4167	100	126	115
Tampa	3979	4922	4537	100	124	114
Georgia						
Atlanta	4093	5008	4692	100	122	115
Macon	3358	4320	3858	100	129	115
Savannah	3605	4552	4105	100	126	114
Hawaii						
Honolulu	4613	5655	5381	100	123	117
Idaho						
Boise	3931	4910	4476	100	125	114
Lewiston	4174	5105	4753	100	122	114
Illinois						
Belleville	4499	5431	5124	100	121	114
Chicago	4574	5578	5217	100	122	114
Danville	4087	4965	4723	100	121	116
Galesburg	4114	4997	4692	100	121	114
Peoria	4331	5159	4939	100	119	114
Quincy	3762	4643	4321	100	123	115
Rockford	4060	5030	4599	100	124	113
Rock Island	4141	5062	4753	100	122	115
Springfield	4249	5159	4877	100	121	115
Indiana						
Evansville	4125	5096	4728	100	124	115
Fort Wayne	4114	4997	4661	100	121	113
Gary	4331	5287	4877	100	122	113
Indianapolis	4283	5203	4877	100	121	114
South Bend	4249	5127	4815	100	121	113
Iowa						
Cedar Rapids	4141	4933	4692	100	119	113
Davenport	4249	5127	4846	100	121	114
Des Moines	4222	5030	4815	100	119	114
Dubuque	4006	4836	4599	100	121	115
Sioux City	4006	4804	4537	100	120	113
Kansas						
Topeka	4141	5094	4723	100	123	114
Wichita	4039	5008	4568	100	124	113
Kentucky						
Bowling Green	3686	4584	4198	100	124	114

(Continued)

Location	*Total building cost (thousands of dollars)*			*Relative cost (%)*		
	Loadbearing brick	*Steel frame*	*Concrete frame*	*Loadbearing brick*	*Steel frame*	*Concrete frame*
Kentucky (cont.)						
Frankfort	3952	4965	4476	100	126	113
Lexington	3871	4868	4383	100	126	113
Louisville	4337	5300	4939	100	122	114
Owensboro	3903	4910	4476	100	126	115
Louisiana						
Baton Rouge	3816	4836	4414	100	127	116
Lake Charles	3898	4804	4507	100	123	116
New Orleans	4114	5030	4723	100	122	115
Shreveport	3898	4772	4445	100	122	114
Maine						
Augusta	3686	4618	4198	100	125	114
Portland	3821	4780	4383	100	125	115
Maryland						
Annapolis	4412	5255	5000	100	119	113
Baltimore	4385	5287	5000	100	121	114
Cumberland	3979	4836	4568	100	122	115
Massachusetts						
Boston	4722	5688	5349	100	120	113
Springfield	4601	5481	5217	100	119	113
Worchester	4580	5431	5186	100	119	113
Michigan						
Detroit	4872	5803	5494	100	119	113
Flint	4547	5514	5155	100	121	114
Grand Rapids	4114	5030	4661	100	122	113
Kalamazoo	4331	5191	4877	100	120	113
Lansing	4439	5416	5031	100	122	113
Saginaw	4520	5448	5093	100	121	113
Minnesota						
Duluth	4277	5159	4848	100	121	113
Minneapolis	4331	5223	4908	100	121	113
Rochester	3843	4739	4352	100	123	113
St. Paul	4304	5191	4877	100	121	113
Mississippi						
Biloxi	3573	4535	4136	100	127	116
Jackson	3519	4417	4074	100	126	116
Missouri						
Jefferson City	3681	4545	4198	100	123	114
Kansas City	4466	5352	5062	100	120	113
St. Louis	4553	5463	5217	100	120	115
Springfield	3871	4804	4445	100	124	115
Montana						
Billings	3816	4802	4352	100	126	114
Butte	3935	4997	4541	100	127	115
Great Falls	3903	4910	4476	100	126	115
Helena	3609	4702	4199	100	130	116
Nebraska						
Lincoln	3794	4682	4352	100	123	115
Omaha	4114	4965	4692	100	121	114

(Continued)

Location	Total building cost (thousands of dollars)			Relative cost (%)		
	Loadbearing brick	Steel frame	Concrete frame	Loadbearing brick	Steel frame	Concrete frame
Nevada						
Carson City	4369	5458	4976	100	125	114
Las Vegas	4580	5593	5186	100	122	113
New Hampshire						
Concord	3930	4845	4507	100	123	115
Portsmouth	3767	4682	4260	100	124	113
New Jersey						
Atlantic City	4493	5416	5062	100	121	113
Newark	4845	5835	5525	100	120	114
Trenton	4656	5610	5247	100	120	113
New Mexico						
Albuquerque	3903	4845	4445	100	124	114
Santa Fe	3871	4772	4414	100	123	114
New York						
Albany	4445	5332	5062	100	120	114
Binghamton	4385	5223	5031	100	119	115
Buffalo	4966	5918	5692	100	119	115
Elmira	4304	5159	4877	100	120	113
Jamestown	4364	5332	4970	100	122	114
New York City	5305	6222	5988	100	117	113
Plattsburgh	4310	5203	4877	100	121	113
Rochester	4695	5655	5349	100	120	114
Syracuse	4690	5561	5278	100	119	113
Watertown	4423	5359	5069	100	121	115
Nassau County	4872	5901	5494	100	121	113
Putnam County	4331	5127	4908	100	118	113
Suffolk County	4899	5933	5525	100	121	113
Westchester County	4520	5578	5155	100	123	114
North Carolina						
Charlotte	3388	4357	3920	100	129	116
Raleigh	3437	4320	3982	100	126	116
Winston-Salem	3442	4357	3982	100	127	116
North Dakota						
Bismarck	3789	4707	4321	100	124	114
Fargo	3898	4804	4445	100	123	114
Ohio						
Akron	4818	5707	5494	100	118	114
Cincinnati	4634	5593	5309	100	121	115
Cleveland	4818	5707	5494	100	118	114
Columbus	4493	5416	5093	100	121	113
Dayton	4472	5831	5062	100	121	113
Toledo	4845	5707	5525	100	118	114
Youngstown	4493	5352	5093	100	119	113
Oklahoma						
Oklahoma City	3848	4878	4383	100	127	114
Tulsa	3871	4836	4445	100	125	115
Oregon						
Portland	4310	5235	4908	100	121	114
Salem	4147	5073	4692	100	122	113

(Continued)

Location	Total building cost (thousands of dollars)			Relative cost (%)		
	Loadbearing brick	Steel frame	Concrete frame	Loadbearing brick	Steel frame	Concrete frame
Pennsylvania						
Erie	4391	5332	4970	100	121	113
Harrisburg	4114	4997	4661	100	121	113
Philadelphia	4520	5448	5093	100	121	113
Pittsburgh	4628	5514	5247	100	119	113
Reading	4249	5127	4815	100	121	113
Rhode Island						
Providence	4499	5431	5093	100	121	113
South Carolina						
Charleston	3310	4274	3826	100	129	116
Columbia	3229	4208	3732	100	130	116
South Dakota						
Pierre	3898	4804	4507	100	123	116
Sioux Falls	3952	4836	4507	100	122	114
Tennessee						
Chattanooga	3848	4715	4383	100	123	114
Knoxville	3713	4618	4260	100	124	115
Memphis	3881	4866	4448	100	125	115
Nashville	3735	4611	4321	100	123	116
Texas						
Austin	3735	4664	4260	100	125	114
Corpus Christi	3600	4439	4136	100	123	115
Dallas	4033	4922	4599	100	122	114
El Paso	3492	4417	4013	100	126	115
Fort Worth	3979	4889	4568	100	123	115
Houston	4087	4986	4692	100	122	115
Odessa	3410	4320	3889	100	127	114
San Antonio	3762	4696	4260	100	125	113
Utah						
Salt Lake City	4070	5063	4634	100	124	114
Vermont						
Burlington	3985	4910	4537	100	123	114
Montpelier	3876	4780	4383	100	123	113
Virginia						
Norfolk	3546	4449	4074	100	125	115
Richmond	3600	4481	4136	100	124	115
Roanoke	3469	4423	4013	100	128	116
Washington						
Bellingham	4396	5425	5008	100	123	114
Olympia	4342	5359	4945	100	123	114
Seattle	4532	5589	5163	100	123	114
Spokane	4445	5398	5000	100	121	112
Tacoma	4179	5228	4820	100	125	115
Vancouver	4364	5332	4970	100	122	114
Walla Walla	4228	5235	4815	100	124	114
West Virginia						
Bluefield	3686	4618	4260	100	125	116
Charleston	4331	5223	4939	100	121	114

(Continued)

Location	*Total building cost (thousands of dollars)*			*Relative cost (%)*		
	Loadbearing brick	*Steel frame*	*Concrete frame*	*Loadbearing brick*	*Steel frame*	*Concrete frame*
Wisconsin						
Beloit	4222	5094	4753	100	121	113
Eau Claire	4141	4965	4723	100	120	114
Green Bay	4114	5030	4661	100	122	113
Kenosha	4304	5191	4877	100	121	113
La Crosse	4006	4868	4568	100	122	114
Madison	4277	5127	4846	100	120	113
Milwaukee	4520	5384	5124	100	119	113
Oshkosh	4060	4997	4630	100	123	114
Racine	4249	5159	4784	100	121	113
Wisconsin Rapids	4006	4868	4537	100	122	113
Wyoming						
Cheyenne	3816	4675	4321	100	123	113
CANADA						
Note: Canadian cities have not been adjusted for seismic conditions						
Alberta						
Calgary	4060	4922	4630	100	121	114
Edmonton	4060	4922	4661	100	121	115
British Columbia						
Vancouver	4304	5115	4846	100	119	113
Victoria	4277	5115	4846	100	120	113
Manitoba						
Winnipeg	3952	4696	4476	100	119	113
New Brunswick						
St. John	3762	4471	4321	100	119	115
Newfoundland						
St. Johns	3871	4600	4537	100	119	117
Nova Scotia						
Halifax	3898	4664	4476	100	120	115
Ontario						
Hamilton	4466	5243	5062	100	117	113
Kitchener	4141	4986	4692	100	120	113
Ottawa	4114	4954	4661	100	120	113
Sault St. Marie	4249	4954	4784	100	117	113
Toronto	4385	5179	5000	100	118	114
Windsor	4304	5147	4908	100	120	114
Prince Edward Island						
Charlottetown	3464	4085	4013	100	118	116
Quebec						
Montreal	3816	4535	4352	100	119	114
Quebec	3789	4535	4352	100	120	115
Saskatchewan						
Regina	3952	4761	4507	100	120	114
Saskatoon	3871	4664	4445	100	120	115

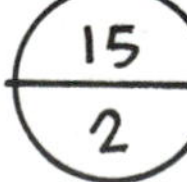

(Continued)

15.3 VALUE ENGINEERING

In estimating the total cost of a system or product, present as well as future costs must be considered before a rational determination can be made. *Value engineering and life-cycle costing methods* look at the overall expenditures throughout the life of a building to arrive at realistic comparisons among various systems, components, or building types. For instance, the fire-resistive properties of masonry structures mean lower insurance rates and lower repair costs if interior spaces do sustain damage from fire (refer to Chapter 8). Masonry thermal characteristics reduce energy consumption for heating and air conditioning. The durability and finish of the surfaces virtually eliminates maintenance and lowers building upkeep costs.

In the mid-1970s, the BIA prepared a long-term cost study of several wall systems using the present-worth method of analysis to determine ultimate life-cycle costs. The present-worth method discounts all future expenditures and credits to their present value. This study used a 10% annual rate of return for calculations. For example, if a building will require painting 5 years from now at a cost of $1000, the present worth of this expenditure is $620.90. Or, more simply stated, $620.90 invested today at 10% interest will be worth $1000 five years from now, and can be used to pay the cost of the painting.

For private investors, income tax rates influence construction decisions. For the purposes of this study, an effective rate of 57% was assumed. Tax-exempt organizations such as churches, schools, governmental bodies, and nonprofit companies are not affected by tax rates or depreciation schedules, so separate calculations were made for these operations.

Initial costs of wall assemblies are estimated from cost index references. Costs of supporting the wall, and the amount of floor space occupied by the wall, are included in this analysis. The cumulative value of these three items determines the total initial cost of construction. *Recovered costs* are reflected as the sum of the value gained by early occupancy and depreciation. *Future costs* include financing, maintenance, air conditioning, heating, and real estate taxes. The study used an expected useful life of 50 years for calculation of all future expenditures.

The four wall assemblies used in the study were:

1. *Masonry cavity wall:* exterior of standard face brick, interior of 4 in. concrete block, 2 in. cavity filled with insulation, and interior finish of fire-and water-resistant drywall
2. *Precast concrete panel:* 6-in.-thick flat panel with 2-in. insulated core, exposed aggregate exterior, and fire- and water-resistant drywall interior
3. *Metal sandwich panel:* factory made, fire-rated sandwich panel with porcelain enamel both sides, and a 2-in. insulated core
4. *Double plate glass:* double $\frac{3}{8}$-in. plate glass wall with vertical venetian blinds inside to control heat load

The tables in *Fig. 15-3* compare the various factors to show what the total cost per square foot of each wall type means in terms of present money. The ultimate cost advantage of the masonry wall can be attributed to lower initial costs and a corresponding savings in finance charges and real estate taxes, and in the significantly lower heating, air-conditioning, and maintenance expenses.

Although the ultimate costs will vary for different buildings, locations, and conditions, the relative costs are more important for comparison, and will remain fairly constant.

TABLE A FOR TAXABLE OPERATIONS				
	Masonry cavity wall	*Precast concrete panel wall*	*Metal panel wall*	*Double-plate glass wall*
Initial cost				
Wall construction	$4.35	$6.80	$8.25	$8.50
Supporting the wall	0.48	0.26	0.12	0.05
Floor space occupancy	2.02	1.44	0.46	0.55
Total initial cost	6.85	8.50	8.83	9.10
Relative initial cost. (%)	100	124	129	133
Recovered cost (credits)				
Early occupancy	—	0.42	0.42	0.42
Depreciation	0.75	0.93	0.97	1.00
Total recovered cost (present worth)	0.75	1.35	1.39	1.42
Relative recovered cost (%)	100	180	185	189
Future cost (charges)				
Financing	0.43	0.53	0.56	0.58
Maintenance	0.28	0.20	0.22	7.34
Air conditioning	0.23	0.30	0.33	7.60
Heating	0.30	0.30	0.30	1.38
Real estate taxes	0.69	0.85	0.88	0.91
Total future cost (present worth)	1.93	2.18	2.29	17.81
Relative future cost (%)	100	113	119	923
Ultimate cost				
Ultimate cost (present worth)	$8.03	$9.33	$9.73	$25.49
Relative ultimate cost (%)	100	116	121	317

TABLE B FOR TAX-EXEMPT OPERATIONS				
	Masonry cavity wall	*Precast concrete panel wall*	*Metal panel wall*	*Double-plate glass wall*
Initial cost				
Wall construction	$4.35	$6.80	$8.25	$8.50
Supporting the wall	0.48	0.26	0.12	0.05
Floor space occupancy	2.02	1.44	0.46	0.55
Total initial cost	6.85	8.50	8.83	9.10
Relative initial cost (%)	100	124	129	133
Future cost (charges)				
Financing	1.01	1.25	1.30	1.34
Maintenance	0.66	0.46	0.52	17.05
Air conditioning	0.38	0.48	0.54	12.35
Heating	0.51	0.51	0.51	2.30
Total future cost (present worth)	2.56	2.70	2.87	33.04
Relative future cost (%)	100	105	112	1,291
Ultimate cost				
Ultimate cost (present worth)	$9.41	$11.20	$11.70	$42.14
Relative ultimate cost (%)	100	119	124	448

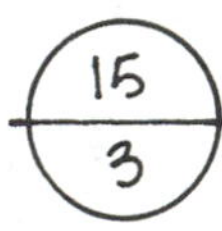

Present worth of the ultimate cost for various wall types (per square foot of wall area). (*From Brick Institute of America, "Walls to Save Dollars," BIA, McLean, Va.*)

Still another study, conducted by the Texas State Building Materials and Systems Testing Laboratory, examined the economic performance and energy usage of a reflective glass facade versus a combination masonry–glass wall for a hypothetical 15-story office building in Dallas, Texas, in mid-1975 (*see Fig. 15-4*). The study also compared the rate of return on investment costs, and the annual rate of return on equity.

The glass curtainwall used 1-in.-thick insulating panels (one pane of $\frac{1}{4}$-in. clear plate, one pane of $\frac{1}{4}$-in. tinted metallic-reflective plate, and a $\frac{1}{2}$-in. air space); $\frac{1}{4}$-in. standard insulated spandrel glass; and an anodized aluminum frame with sealant. The masonry building was of 10-in. brick cavity walls (4-in. face brick, 4-in. lightweight concrete block, and a 2-in.

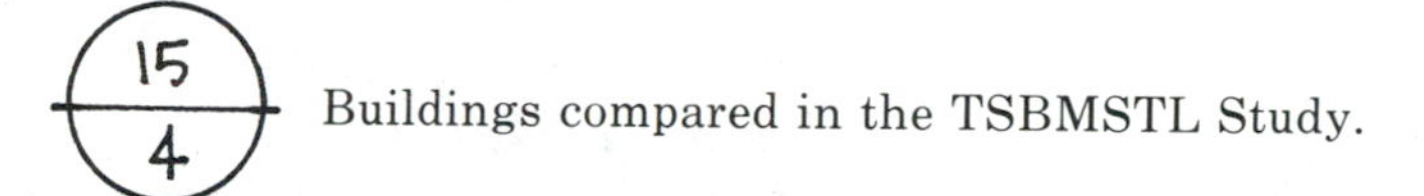

Buildings compared in the TSBMSTL Study.

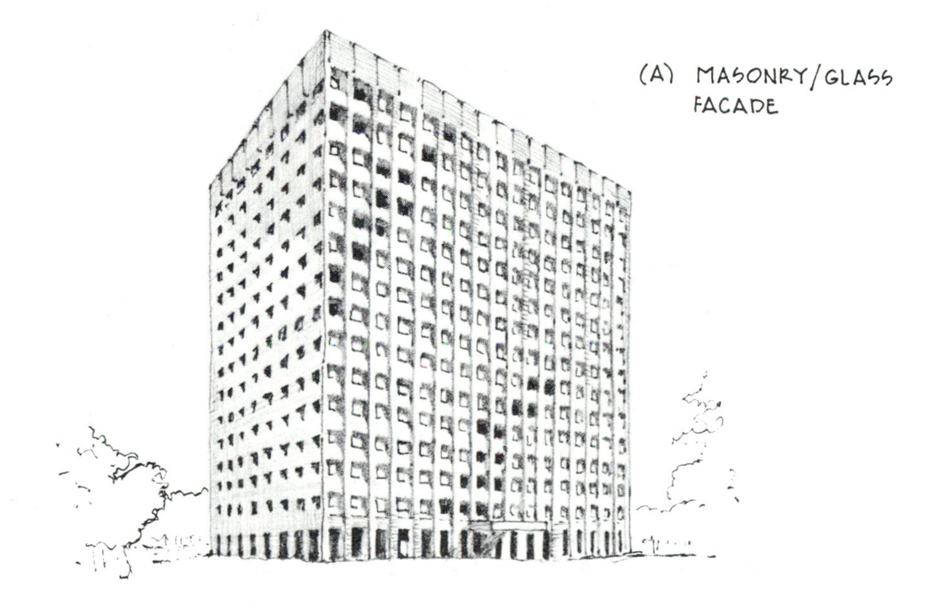

cavity filled with vermiculite insulation); interior finish of $\frac{5}{8}$-in. gypsum plaster painted two coats; and windows of $\frac{1}{4}$-in. tinted glass with venetian blinds. Energy requirements for both buildings were determined using a National Bureau of Standards computer program based on ASHRAE requirements. Window cleaning, painting, glass breakage insurance, and other maintenance items were included to determine operating costs. Real estate investment analysis was based on a technique developed by the Center for Building Research at the University of Texas at Austin.

A summary of the report findings shows that (1) heating and air conditioning costs were higher for the glass wall; (2) the initial cost of the glass facade was higher; (3) the higher foundation and structural framing costs for the masonry were offset by lower wall costs; and (4) higher initial costs for the glass wall produced higher development costs and lower rates of return (*see Fig. 15-5*). The abstract of the study itself concludes that the all-glass facade requires more than 9% higher initial building cost, more than 8% higher investment cost, nearly 34% more cash equity, and more than 4% higher operating costs. The glass facade also reduces the maximum rate of return on investment by more than 22%, and has no effect on rental income.

The three studies examined here each use different cost input and

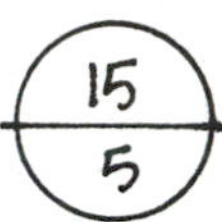

Cost analysis from the TSBMSTL Study. (*Courtesy Acme Brick Co., Fort Worth, Tex.*)

TABLE A INITIAL COST

Cost variable comparison	*Glass building (glass facade)*	*Brick building (brick facade)*
Facade		
Cost per square foot	14.12	7.19
Total cost of facade	1,693,341	862,261
Foundation and framing	145,220	232,765
Heating, venting, and air-conditioning		
Total cost of HVAC	1,020,350	915,150
Cooling capacity (tons/h)	590	510
Heating capacity (10^6 Btu/h)	4.40	4.00
Other building costs	6,929,402	6,929,402
Building cost		
Total building cost	9,788,313	8,939,578
Relative building cost	1.09	1.00
Development cost		
Depreciable dev. cost	1,639,842	1,524,414
Nondepreciable dev. cost	971,722	895,336
Total development cost	2,611,564	2,419,750
Relative development cost	1.08	1.00
Total building and development cost	12,399,877	11,359,328
Land	1,452,384	1,452,384
Investment cost		
Total investment cost	13,852,261	12,811,712
Relative investment cost	1.08	1.00
Mortgage	9,735,510	9,735,510
Equity		
Equity requirement	4,116,751	3,076,202
Relative equity requirement	1.338	1.00

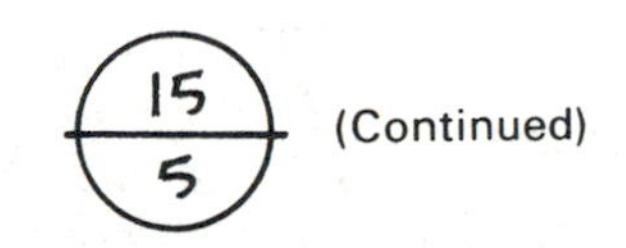

(Continued)

TABLE B FIRST YEAR OPERATING COST		
	Glass building (glass facade)	*Brick building (brick facade)*
Heating, venting, and air-conditioning plant operating cost	$ 45,060	$ 40,614
Window cleaning	17,348	7,864
Blind cleaning*	0	581
Real estate taxes	227,080	210,022
Glass breakage insurance	2,819	480
Wall painting	0	3,310
Other operating costs	477,099	477,099
Total first-year operating costs	769,406	739,970
Relative first-year operating cost	1.039	1.00
(Actual difference in annual operating cost: $29,436)		

Blinds not provided in insulating reflective glass windows but provided for tinted glass windows.

TABLE C RATE OF RETURN ON REAL ESTATE INVESTMENT		
Rate of return	*Glass building (glass facade)*	*Brick building (brick facade)*
Maximum internal rate of return on investment		
Percent	12.6	16.2
Relative return	1.00	1.29
Maximum rate of return on equity including tax shelter		
Percent	8.8	12.6
Relative return	1.00	1.43

different methods of analysis to compare masonry construction with other systems. The results, however, are the same for various geographical areas and various analytical techniques. For certain building types and applications, masonry walls initially and ultimately cost less than other types of construction.

15.4 DESIGN DECISIONS

The cost of masonry construction can be significantly affected by decisions made in the planning and design stages of a project. Basic costs are related, first, to mason productivity. Typical union labor production rates for various types of masonry with concave mortar joints are listed in *Fig. 15-6*. Production figures and costs can be increased or decreased by exercising some of the various options listed below.

1. Larger face-size units can increase the square footage of wall area produced in a day even though the mason may lay fewer units because of greater weight. Note, for instance, the increase of 28% shown in the table when standard brick are replaced by Norman

Unit			Wall		
Description (size and type)	*Weight (lb)*	*Surface area (sq ft)*	*Nominal thickness (in.)*	*Productivity (sq ft/8 hr)*	*Relative productivity*
4" × $2\frac{2}{3}$" × 8" Standard brick	—	0.148	4	160*	1.00
4" × $2\frac{2}{3}$" × 12" Norman brick	—	0.222	4	207	1.28
4" × 4" × 8" Economy brick	—	0.222	4	218	1.36
4" × 4" × 12" Economy brick	—	0.333	4	242	1.51
8" × 5" × 12" Hollow tile	13.7	0.444	8	257	1.61
4" × 5" × 12" Hollow tile	8.6	0.444	4	266	1.66
6" × 5" × 12" Hollow tile	12.4	0.444	6	287	1.79
4" × 8" × 12" Hollow tile	15.2	0.667	4	308	1.93
8" × 8" × 12" Hollow tile	25.3	0.667	8	360	2.25
12" × 8" × 16" Hollow block	31.9	0.889	12	376	2.35
8" × 8" × 16" Hollow block	26.2	0.889	8	467	2.92
4" × 8" × 16" Hollow block	16.4	0.889	4	494	3.09
6" × 8" × 16" Hollow block	21.1	0.889	6	526†	3.29
4" × 12" × 12" Hollow tile	18.9	1.000	4	413	2.58
8" × 8" × 24" Hollow block	34.9	1.333	8	245‡	1.53

*Lowest value.

†Highest value.

‡Because of local union regulations concerning size of unit, two masons to handle each unit were used even though the weight of the unit di not require two masons. The result was a drastic drop in productivity for this size. This size regulation is not a union rule applied nationall

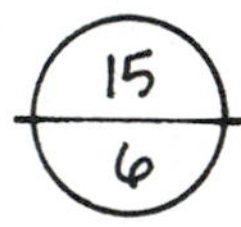

Mason productivity with various unit sizes. (*From National Concrete Masonry Association*, TEK Bulletin 54, *NCMA, Herndon, Va.*)

units. This option is one of the simplest and most cost effective an architect can make. The higher cost of larger units can be offset by lower labor bids and by earlier completion of the work. For some designs, larger masonry units may actually give better proportional scale with the size of the building as well.

2. Rougher, "antiqued" units require normal workmanship and are less expensive to install than are precision units requiring precision workmanship. Type FBA brick or concrete slump block will always generate lower labor costs than precision, ground face units or Type FBX brick.
3. Raked or recessed mortar joints (in addition to being less weather resistant) are more expensive than concave, tooled joints.
4. Colored mortar, although it can be quite attractive, costs more than regular portland cement–lime mortar without added pigments.
5. Field cutting of masonry units can be unnecessarily expensive i the number of cuts required is significantly increased by poor modular or dimensional planning.
6. Stacked bond is sometimes less expensive than $\frac{1}{2}$ or $\frac{1}{3}$ running bond patterns because of reduced cutting required at openings and corners.
7. Designs that permit simultaneous laying of inner and outer cavity wythes from inside the building generate savings in scaffolding and setup expenses.

8. Special unit shapes can be less expensive than field cutting if sufficient quantities are required to justify production expenses.
9. Round corners with a radius of as little as 2 ft can be turned using standard headers instead of specially curved face units.
10. Door jambs covered by bucks and trim eliminate the need for saw-cut or bullnose units and reduce costs.
11. The SCR Masonry Process using calibrated corner poles can reduce costs by as much as 20%.
12. Rational engineering methods show that shelf angles on high-rise buildings are not necessary, and veneers supported only at the foundation yield reduced costs and improved quality.
13. Mechanical and electrical lines and conduit are easily placed in double-wythe cavity walls, but must be cut in to single-wythe sections, thus increasing the labor required for installation.
14. Openings spanned with masonry arches or reinforced masonry lintels eliminate the need for steel angles and the maintenance costs they include.

Careful detailing and thoughtful design practices can enhance the cost economy of any building system. Conscientious planning and material selection, attention to detail, thorough specifications, and on-site construction inspection can all contribute to lower costs in masonry.

16

SPECIFYING AND INSPECTING MASONRY CONSTRUCTION

Specifications are an extremely important part of quality control i masonry construction. To achieve high levels of workmanship and prope performance of the completed work, the architect must carefully outline th products and methods of construction required, and inspect the progress o the job to ensure compliance.

16.1 METHODS OF SPECIFYING MASONRY

It is highly recommended that *reference standards* be used to govern th quality of products included in masonry construction. Standards develope by ASTM cover all of the mortars, unit types, and varieties of stone (se Appendix C), and are widely accepted throughout the industry. It is als recommended that if trade names are used, the special conditions of th contract should include a clause providing for substitutions to be accepte on the basis of quality and price. Lump-sum or unit-price allowances may b included for materials, but must be accompanied by sufficient informatio (including unit size and texture) to indicate to the contractor the amount c labor that will be required for installation.

Most ASTM specifications for masonry products cover two or mor grades and types of units, and the project specifications must identif which ones are required. Omission of this information makes it difficult fc the bidders to accurately estimate the project, and can result in a deman for extras after the contract is awarded.

The size of unit required should be included in the specifications, and it is recommended that actual rather than nominal dimensions be used to avoid ambiguity. In some industries, "nominal" means approximate, whereas in modular masonry units, it means the manufactured dimension plus the thickness of the mortar joint. A nominal 8-in. modular brick can be manufactured at $7\frac{1}{2}$ in. for use with $\frac{1}{2}$-in. joints, or $7\frac{5}{8}$ in. for use with $\frac{3}{8}$-in. joints. Dimensions should be listed with thickness first, followed by the face dimensions of height and then length.

Color and texture are not included in ASTM standards, and requirements must be established by the specifications. If an allowance method is used, the final selection may be made from samples submitted by the contractor or supplier. If trade names are used to identify a color range and finish, or if descriptions are not given in the project specifications, samples of acceptable materials should be available to the contractors for inspection prior to bidding.

16.2 GUIDE SPECIFICATIONS

The guide specifications included here are based on the CSI three-part format. Standard CSI section numbers provide for a breakdown of masonry specifications into several sections covering mortar, masonry accessories, brick, concrete unit masonry, natural stone, and so on. The sample specifications that follow combine several of these sections into a single specification for unit masonry, and a specification for stone. The first sample includes all of the major types of unit masonry construction in order to fully cover the required information. Most projects will use fewer types, and the additional products and related instructions would then simply be deleted.

SECTION 04200 UNIT MASONRY

PART 1 GENERAL

1.1 Work Included

Description of work:

Provide all labor, materials, equipment, and services necessary for and incidental to the installation of all masonry construction s indicated on the drawings and specified herein.

1.2 Related Work

Related work specified elsewhere:

Shop Drawings, Product Data, and Samples, Section 01340

Testing Laboratory Services, Section 01410

Cast-in-Place Concrete, Section 03300

Stone Veneer, Section 04400

Waterproofing and Dampproofing, Section 07100

Insulation, Section 07200

Caulking and Sealants, Section 07920

Lath and Plaster, Section 09200

Stone and Brick Flooring, Section 09600

Painting, Section 09900

Items furnished by others:

The masonry contractor shall install all accessory items that are required in th work and supplied by others, including: bolts, nailing blocks, inserts, anchors flashing, lintels, expansion joints, conduits, etc.

1.3 Quality Assurance

Tests:

Manufacturer or supplier of masonry units shall submit to Architect prior t delivery, certification of compliance of units with specified standards, as deter mined by an acceptable testing agency conforming to the applicable requirement of ASTM E329. Brick shall be tested in accordance with ASTM C67, and concret masonry units in accordance with ASTM C140. If tests made after delivery indicat that units do not conform to specified requirements, costs of such tests shall be born by the supplier.

Masonry cements:

Proprietary masonry cement mixes used in engineered masonry construction sha be subject to laboratory testing to assure compliance with minimum requirement for strength and bond.

Sample panels:

Erect sample panels for each type of masonry required, approximately 4 ft long by ft high, showing the proposed color range, texture, bond, mortar, and quality o work. The sample panel, when accepted, shall become the project standard for bond mortar, quality of work, and appearance. Do not begin work until panel is accepte by Architect.

1.4 Submittals

Samples:

Submit samples of each type of masonry unit and each accessory item require Provide certification of pull-out strength of all masonry ties and anchors. Subm certification of compliance with required standards for all masonry units.

1.5 Product Delivery, Storage, and Handling

Delivery:

Deliver masonry units to job site in undamaged condition. Deliver and handle unit to prevent chipping, breaking, or other damage.

Storage:

Store masonry units off ground and protected from wetting by capillary action, rai or snow, and protected from mud, dust, or other materials and contaminants likel to cause staining or defects.

1.6 Job Conditions

Cold weather construction:

Masonry construction performed when ambient temperature falls below 40°F sha conform to the *Recommended Practices and Guide Specifications for Cold Weath Masonry Construction* published by the International Masonry Industry A Weather Council.

Coverings:

The Contractor shall construct and maintain temporary protection as required permit continuous progress of the work. During construction, partially complete walls which are not enclosed or sheltered shall be kept dry by covering at the end of each day and when work is not in progress with strong, weather-resistant material extended a minimum of 2 ft down each side, and held securely in place.

Protections:

Do not apply uniform floor or roof loads for at least 12 hours or concentrated loads for at least 3 days after building masonry walls or columns.

Staining:

Prevent grout or mortar from staining the face of masonry to be left exposed or to be painted. Remove immediately any grout or mortar in contact with face of such masonry. Protect all sills, ledges, and projections from droppings of mortar. Protect door jambs and corners from damage during construction.

PART 2 PRODUCTS

2.1 Materials

Portland cement:

ASTM C150, Type_____(I, II, or III).

Lime:

Hydrated lime, ASTM C207, Type S.

Sand:

ASTM C144.

Aggregates:

ASTM C404.

Water:

Mixing water must be clean and free of harmful amounts of acids, alkalis, organic materials, or other substances that would adversely affect the quality or appearance of the mortar or the masonry units.

Facing brick:

ASTM C216, Grade _____ (SW or MW), Type _____ (FBX, FBS, or FBA); dimensions _____ (thickness) × _____ (height) × _____ (length); minimum compressive strength ______ psi. Provide brick similar in color, texture, and physical properties to those available for inspection at the Architect's office. Do not exceed variations in color and texture of samples accepted by the Architect.

Glazed brick:

ASTM C126, Grade _____ (S, or SS), Type _____ (I or II); dimensions _____ (thickness) × _____ (height) × _____ (length); minimum compressive strength ______ psi.

Building brick:

ASTM C62, Grade _____ (SW, MW, NW); dimensions _____ (thickness) × _____ (height) × _____ (length); minimum compressive strength ______ psi.

Hollow brick:

ASTM C652, Grade _____ (SW or MW), Type _____ (HBS, HBX, HBA, or HBB); dimensions _____ (thickness) × _____ (height) × _____ (length); minimum compressive strength ______ psi. Provide brick similar in color, texture, and physical properties to those available for inspection in Architect's office. Do not exceed variation in color and texture of samples accepted by Architect.

Hollow loadbearing CMU:

ASTM C90, Grade _____ (N or S), Type _____ (I or II), _______ -weight (normal, medium, or light).
(Designate any special characteristics required such as split-face, ribbed, fluted, fire rating, etc. Stipulate that any decorative architectural concrete masonry units required in the work shall closely match samples available for inspection at the Architect's office.)

Solid loadbearing CMU:

ASTM C145, Grade _____ (N or S), Type _____ (I or II), _______ -weight (normal, medium or light).

Hollow non-loadbearing CMU:

ASTM C129, Type _____ (I or II), _______ -weight (normal, medium, or light).

Concrete building brick:

ASTM C55, Grade _____ (N or S), Type _____ (I or II), _______ -weight (normal or light).

Structural glazed facing tile:

ASTM C126, Grade SS, ground ends, _____ Series (6T or 8W), nominal face size _____ × _____, with coved base course. Color and texture to match samples available for inspection at Architect's office.

Glass block:

Provide _____ in. × _____ in. × _____ in. thick light- _______ (diffusing, reflecting, reducing), sealed glass block with minimum *U*-value of _____, _____ percent light transmission, _______ pattern, as manufactured by _______.

Reinforcement:

Steel reinforcing shall conform to the following ASTM Specifications:

1. Cold-drawn steel wire, ASTM A82
2. Welded steel wire fabric, ASTM A185
3. Billet steel deformed bars, ASTM A615, Grade _____ (40, 50, or 60)
4. Rail steel deformed bars, ASTM A616, Grade _____ (50 or 60)
5. Axle steel deformed bars, ASTM A617, Grade _____ (40 or 60)

Material for anchors and ties:

All anchors and ties shall be coated or corrosion-resistant metal meeting or exceeding the following ASTM Specifications:

1. Zinc coating of flat metal, ASTM A153, Class _____ (B-1, B-2, or B-3)
2. Zinc coating of wire, ASTM A116, Class 3
3. Copper-coated wire, ASTM B227, grade 30HS
4. Stainless steel, ASTM A167, Type 304

Types of anchors and ties:

Provide the following types of anchors and ties for masonry construction:

1. *Wire mesh:* Minimum 20 gauge, $\frac{1}{2}$-in. mesh, galvanized wire, 1 in. less in width than width of masonry.
2. *Corrugated veneer anchors:* Minimum 22 gauge, minimum $\frac{7}{8}$ in. wide × 6 in. long.
3. *Cavity wall ties:* Minimum $\frac{3}{16}$-in. wire diameter, rectangular, at least 2 in. wide or Z-shaped with 2-in. legs, length sufficient to allow 1 in. minimum mortar coverage of ends or legs.

4. *Joint reinforcement:* Prefabricated welded joint reinforcement, longitudinal cross tie wire minimum 9-gauge spaced 16 in. on center; ladder or truss-type design.
5. *Dovetail flat bar anchors:* Minimum 16 gauge, $\frac{7}{8}$ in. minimum width, corrugated, turned up $\frac{1}{4}$ in. at end, or with $\frac{1}{2}$-in. hole within $\frac{1}{2}$ in. of end of bar.
6. *Wire anchors:* Wire anchors shall be minimum $\frac{3}{16}$-in. diameter.
7. *Rigid anchors for intersecting bearing walls:* $1\frac{1}{2}$ in. wide × $\frac{1}{4}$ in. thick × 24 in. minimum length; turn up ends minimum 2 in. or provide cross pins.
8. *Wire ties for grouted reinforced masonry:* Minimum 9-gauge wire bent into rectangular stirrups 4 in. wide and 2 in. shorter than overall wall thickness; form so that tie ends meet in center of one embedded end of stirrup.

Cleaning agents:

Do not use cleaning agents other than water on brick or concrete masonry units without written approval of Architect.

Flashing:

Provide __________ material for base flashing and wall flashing.

2.2 Mixes

Mortar mixes:

Mortar shall comply with the minimum requirements of ASTM C270, Type ______ (M, S, or N).

Grout mixes:

Grout for reinforced masonry shall meet the minimum requirements for Type ______ (PM or PL), ASTM C476.

Admixtures:

No air-entraining admixtures or materials containing air-entraining admixtures shall be used. Air content of mortar and grout shall be limited to 12%. No antifreeze compounds or other substances shall be added to mortar or grout. No calcium chloride shall be included in mortar or grout in which metal reinforcing or accessories will be embedded. Mortar colors shall consist of inorganic compounds not to exceed 15% of the weight of the cement except that carbon black shall not exceed 3% of the weight of the cement. If mortar colors are used in reinforced masonry, the ultimate compressive strength of the masonry shall be determined by prism tests.

PART 3 EXECUTION

3.1 Preparation

Inspection:

Inspect surfaces that are to support masonry work to assure completion to proper lines and grades free of all dirt and other deleterious material. Do not begin work until surfaces not properly prepared have been satisfactorily corrected.

3.2 Field Quality Control

Mortar and grout:

Mix mortar and grout in accordance with the proportion requirements of ASTM C270, and ASTM C476 as applicable. Control batching procedure to ensure proper proportions by measuring materials by volume. Amount of mixing water and mortar consistency shall be controlled by mason. Retempering will be permitted only within the first $2\frac{1}{2}$ hours of initial mix. Any mortar or grout that has partially set shall be discarded.

Allowable tolerances:

1. Maximum variation from plumb in lines and surfaces of columns, walls, and arrises shall not exceed $\frac{1}{4}$ in. 10 ft; $\frac{3}{8}$ in. in any story or 20 ft maximum; or $\frac{1}{2}$ in. in 40 ft.
2. Maximum variation from plumb for external corners, expansion joints, and other conspicuous lines shall not exceed $\frac{1}{4}$ in. in any story or 20 ft maximum; or $\frac{1}{2}$ in. in 40 ft.
3. Maximum variation from level or grades for exposed lintels, sills, parapets, horizontal grooves, and other conspicuous lines shall not exceed $\frac{1}{4}$ in. in any bay or 20 ft maximum; or $\frac{1}{2}$ in. in 40 ft.
4. Maximum variation of linear building line from an established position in plan and related portions of columns, walls, and partitions shall not exceed $\frac{1}{2}$ in. in any bay or 20 ft maximum; or $\frac{3}{4}$ in. in 40 ft.
5. Maximum variation in cross-sectional dimensions of columns and thickness of walls shall be not more than $\frac{1}{4}$ in. smaller or $\frac{1}{2}$ in. larger than shown on plans.

Anchors and ties:

Remove all dirt, ice, loose rust, and scale prior to installation.

Protection of work:

Protect sills, ledges, and offsets from mortar droppings or other damage during construction. Remove misplaced mortar or grout immediately. Protect face materials against staining. Protect door jambs and corners from damage during construction

3.3 Installing Masonry

Preparation:

Verify that initial absorption rate of clay brick is less than 0.025 oz/sq in. per minute. Brick with absorption rates in excess of this amount shall be wetted with clean water 24 hours prior to placement until unit is nearly saturated, and shall be surface dry when laid. During freezing weather, units that require wetting shall be sprinkled with warm or hot water just before laying.
No wetting of concrete unit masonry is permitted.

Installation:

Do not install cracked, broken, or chipped masonry units exceeding ASTM allowances. Use masonry saws to cut and fit exposed units. Lay brick plumb, true to line, and with level courses accurately spaced within allowable tolerances. Unless otherwise shown on the drawings, install masonry work using ½ running bond. Stop horizontal runs at end of workday by racking back in each course; toothing will not be permitted. Adjust units to final position while mortar is soft and plastic. If units are displaced after mortar has stiffened, remove, clean joints and units of mortar, and re-lay with fresh mortar. Adjust shelf angles to keep work level and at proper elevation. When joining fresh masonry to set or partially set masonry, remove loose unit and mortar, and clean and lightly wet exposed surface of set masonry prior to laying fresh masonry. The mason shall place all accessories and reinforcement in the masonry as the job progresses. Place horizontal joint reinforcement in first bed joint and each successive third joint of concrete masonry walls to prevent cracking. Cooperate with other trades to assure proper location of anchors, inserts, penetrations, etc.

Built-in items:

Install bolts, anchors, nailing blocks, inserts, frames, vent flashings, conduit, and other built-in items as masonry work progresses. Avoid cutting and patching. Solidly grout spaces around built-in items. Provide joints around exterior frame openings $\frac{1}{4}$ to $\frac{3}{8}$ in. wide, raked and tooled smooth to a uniform depth of $\frac{3}{4}$ in., read for caulking by others. Build chases in, do not cut. Install chases minimum of one masonry unit length from jambs.

Joints:

Provide nominal joint thickness of $\frac{3}{8}$ in. for concrete unit masonry, ______ in. for brick masonry, and $\frac{1}{4}$ in. for glass block. Do not furrow bed joints for solid masonry units. Provide face-shell bedding for concrete unit masonry except at grouted cells and base course, where full mortar bedding is required. Construct uniform joints. Provide full head and bed joints, shoved tight to prevent penetration of moisture. Provide weather-proof, concave, tooled joints in exposed surfaces when mortar is thumbprint hard, using round jointing tool. Strike joints flush in surfaces to be plastered, stuccoed, or covered with other material or surface-applied finish other than paint. Concave tool exterior joints below grade. Remove mortar protruding into cells or cavities to be grouted. Do not permit mortar droppings to fall into cavities of multi-wythe walls or to block weep holes. Fill with mortar all horizontal joints between top of masonry partitions and underside of concrete beams. Keep movement joints clean of all mortar and debris. For tuckpointing, rake mortar joints to a depth of $\frac{1}{2}$ to $\frac{3}{4}$ in., saturate with clean water, fill solidly with pointing mortar, and tool to match existing joints.

Flashing:

Provide through-wall flashing at base of all cavity walls; at shelf angles; at lintels, heads, and sills of openings in exterior walls; at all locations shown on the drawings; and at any other locations as required to complete the integrity of waterproofed or dampproofed surfaces.

Clean surface of masonry smooth and free from projections that might puncture or otherwise damage flashing membrane. Carefully fit flashing around projections and where dampproof membrane abuts columns, walls, etc. Neatly fold and bed in mastic or mortar so as to direct moisture to the outside. Form membrane to required profiles and install in such a manner as to force any moisture entering the wall to the outside. Hold outer edge of membrane to surface with mastic or mortar. Lap joints 4 in. and seal with mastic or embed in mortar. Form membrane to correct profile without wrinkles or buckles, and protect from punctures and tears during installation.

Weep holes:

Provide weep holes in head joints in first course immediately above all flashing. Leave head joint free and clean of mortar or install weep hole tube in head joint. Space weep holes 24 in. on center maximum for brick masonry, and 32 in. on center maximum for concrete unit masonry. Keep weep holes and area above flashing free of mortar droppings. For backfill material behind retaining walls, and for loose fill insulation in walls, screen cavity side of weep hole against clogging before fill material is placed.

Masonry bonding:

Bond facing and backing of multi-wythe walls as shown on the drawings with masonry headers extended a minimum of 3 in. into backing. If single header does not extend through wall, overlap headers from opposite sides of wall at least 3 in. Provide minimum number of wall headers equal to 4% of wall surface, spaced maximum distance of 24 in. on center either vertically or horizontally.

For multi-wythe walls of hollow concrete masonry units, bond inner and outer wythes by transverse lapping of stretcher unit at least 3 in. over units below, spaced maximum 32 in. on center vertically; or lap with stretcher units at least 50% wider than unit below, spaced maximum 16 in. on center vertically. Bond abutting or intersecting walls and partitions with at least 50% of units at the intersection laid in masonry bond. Provide a minimum of 3 in. of bearing of alternate units on unit below. Masonry bonding is not permitted for grouted or reinforced construction.

Metal-tie bonding:

Provide metal ties for bonding of multi-wythe walls as shown on the drawings. Stagger ties in alternate courses, and provide minimum of one tie for each 4.5 sq ft of wall surface. Maximum distance between adjacent ties not to exceed 18 in. vertically or 36 in. horizontally. Embed ties in horizontal joints of facing and backing. Provide additional ties within 12 in. of openings, spaced maximum 36 in. around perimeter.

In lieu of metal ties, contractor may use continuous prefabricated metal joint reinforcement as specified, spaced not more than 16 in. on center vertically.

For corner intersections of walls carried up separately, provide rigid steel anchors at maximum vertical spacing of 32 in. When intersecting bearing or shear walls are carried up separately, provide rigid steel anchors at a maximum vertical spacing of 2 ft.

Anchor nonbearing partitions abutting or intersecting other walls or partitions with cavity wall ties at vertical spacing not to exceed 4 ft.

Anchoring brick veneer:

Attach brick veneer to backing with metal veneer ties spaced maximum 24 in. on center vertically and horizontally with a minimum of one tie for each 4 sq ft of wal area (based on 20 lb/sq ft wind load). Embed ties at least 2 in. in horizontal joint o facing. Provide additional ties within 12 in. of openings, spaced maximum 36 in around perimeter.

Expansion and contraction:

Provide vertical movement joints where called for on the drawings at intervals of not more than 20 ft on centers, and at all offsets, returns, openings, and intersections with dissimilar materials. Provide continuous bond break at steel columns and members. Provide pressure-relieving joints by placing a continuous $\frac{1}{8}$ in. neoprene pad below shelf angles.

3.4 Reinforced Masonry

Masonry strength:

Provide minimum ultimate compressive strength of ______ psi.

Reinforcement:

Hold vertical reinforcement firmly in place by means of frames or other suitabl devices. Place horizontal reinforcement as masonry work progresses. Provid minimum clear distance between longitudinal bars equal to nominal diameter o bar. Provide minimum clear distance between bars in columns equal to $1\frac{1}{2}$ times ba diameter. Minimum thickness of mortar or grout between masonry and reinforcement shall be $\frac{1}{4}$-in., except than $\frac{1}{4}$-in bars may be laid in $\frac{1}{2}$-in. horizonta mortar joints, and 6-gauge or smaller wires may be laid in $\frac{3}{8}$-in. mortar joints. Colla joints containing both horizontal and vertical reinforcement shall have a minimum width $\frac{1}{2}$ in. larger than the sum of the diameters of the horizontal and vertica reinforcement.

Low-lift grouting:

For grout spaces less than 2 in. in width, place grout at maximum 24-in. intervals i lifts of 6 to 8 in. as the wall is built. Assure that grout core is clean of mortar, morta droppings, and debris. Agitate grout during and after placement to assure complet filling and coverage of reinforcement. If work is to be stopped for 1 hour or more, hol grout $1\frac{1}{2}$ in. below top of masonry. Continue grouting to top of finished wall.

High-lift grouting:

For grout spaces 2 in. or more in width, grout may be placed in lifts not to exceed 4 f For running bond, provide one metal tie for each 3 sq ft of wall with maximu spacing of 16 in. vertically and 24 or 32 in. horizontally for brick and concrete bloc respectively. For stack bond, provide one metal tie for each 2 sq ft of wall with maximum spacing 12 in. vertically and 24 in. horizontally for brick, or 16 in. vertically and horizontally for concrete block.

Keep grout core clean. Provide cleanout holes in bottom course as required for inspection and cleaning. Replace cleanout plugs only after area to be grouted has been accepted. Do not place grout until the entire wall has been in place a minimum of 3 days. Place horizontal grout barriers at convenient intervals. If work is to be stopped for 1 hour or more, hold grout $1\frac{1}{2}$ in. below top of masonry. Continue grouting to top of finished wall.

Forms and shoring:

Provide substantial and tight forms to prevent leakage of mortar or grout. Brace or shore forms to maintain position and shape. Do not remove forms or shoring until masonry has hardened sufficiently to carry its own weight and any other temporary loads that may be placed on it during construction (10 days for girders and beams, 7 days for masonry slabs).

3.5 Cold-Weather Masonry Construction

Surface conditions:

Ice or snow that has formed on the masonry bed shall be thawed by application of heat. Apply heat carefully until top surface is dry to the touch. Any section of completed masonry work that is deemed frozen and damaged shall be removed before continuing construction of that section.

Condition of masonry units:

Use only dry masonry units, except as permitted below. Wet or frozen masonry units shall not be laid. No wetting of concrete masonry units will be permitted.

For brick masonry units used in cold-weather construction, initial rates of absorption may range to a maximum of $1\frac{1}{2}$oz. When sprinkling is required to achieve proper rates, heated water shall be used. Water shall be above 70°F when temperature of units is above freezing, and above 130°F when temperature of units is below freezing.

Construction requirements:

1. *Air temperature 32 to 40°F:* Sand or mixing water shall be heated to produce mortar temperatures ranging from 40 to 120°F.
2. *Air temperature 25 to 32°F:* Sand *and* mixing water shall be heated to produce mortar temperatures ranging from 40 to 120°F. Maintain temperature of mortar on boards above freezing.
3. *Air temperature 20 to 25°F:* Sand *and* mixing water shall be heated to produce mortar temperatures ranging from 40 to 120°F. Maintain mortar temperatures on boards above freezing. Provide sources of heat on both sides of walls under construction. Windbreaks shall be employed when wind is in excess of 15 mph.
4. *Air temperature 20°F and below:* Sand *and* mixing water shall be heated to provide mortar temperatures ranging from 40 to 120°F. Enclosure and auxiliary heat shall be provided to maintain air temperature above freezing. Temperature of units when laid shall be not less than 20°F.

Protection of completed work:

1. *Mean ambient temperature 32 to 40°F:* Masonry completed or not being worked on shall be protected from rain or snow for 24 hours by covering with weather-resistive membrane.
2. *Mean ambient temperature 25 to 32°F:* Masonry shall be completely covered with weather-resistive membrane for 24 hours.
3. *Mean ambient temperature 20 to 25°F:* Masonry shall be completely covered with insulating blankets, or equally protected for 24 hours.

4. *Mean ambient temperature 20°F and below:* Masonry temperature shall be maintained above freezing for 24 hours by enclosure and supplementary heat such as electric heating blankets, infrared heat lamps, or other approved methods.

3.6 Pointing and Cleaning

Pointing:

At final completion of masonry work, cut out any defective joints or holes in exposed masonry and repoint with mortar, tooling to match adjacent joints.

Cleaning:

Dry brush masonry surface after mortar has set at end of each workday and after final pointing. Clean exposed, unglazed masonry with stiff brush and clean water. Cleaning agents may be used only with written approval of Architect. Cleaning agents must be tested on sample wall area of 20 sq ft. Protect adjacent materials from damage due to cleaning operations. Remove efflorescence in accordance with brick manufacturer's recommendations.

Leave work area and surrounding surfaces clean and free of mortar spots, droppings, and broken masonry.

SECTION 04400 STONE VENEER

PART 1 GENERAL

1.1 Work Included

Description of work:

Provide all labor, materials, equipment, and services necessary for and incidental to the installation of all cut stone veneer as indicated on the drawings and specified herein.

1.2 Related Work

Related work specified elsewhere:

Shop Drawings, Product Data, and Samples, Section 01340

Cast-in-Place Concrete, Section 03300

Unit Masonry, Section 04200

Waterproofing and Dampproofing, Section 07100

Insulation, Section 07200

Caulking and Sealants, Section 07920

Lath and Plaster, Section 09200

Stone and Brick Flooring, Section 09600

Items furnished by others:

The masonry contractor shall install all accessory items that are required in the work and supplied by others, including: bolts, nailing blocks, inserts, anchors, flashing, lintels, expansion joints, conduits, etc.

1.3 Quality Assurance

Samples:

Submit three samples 12 × 12 × 1 in. for approval by the Architect. Samples shall be typical of minimum and maximum average color range, texture, and finish, and shall match samples on file in the Architect's office.

Shop drawings:

Prepare and submit to the Architect for approval complete cutting and setting drawings for all stone work. Show in detail the sizes, sections, and dimensions of stone, the arrangement of joints, bonding, anchoring, and other necessary details. Each stone indicated on these drawings shall bear the corresponding number marked on the back or bed with a nonstaining paint. Projecting courses shall have beds in the wall at least 1 in. greater in depth than the projection, or be specially anchored to the structure as shown on the setting drawings.

Clearly indicate provisions for proper anchoring, dowelling, and cramping of the work in keeping with standard practice, and support of stone by ledges, shelf angles, loose steel, etc.

Sample panels:

Erect sample panels for each type of masonry required, approximately 4 ft long by 3 ft high, showing the proposed color range, texture, bond, mortar, and quality of work. The sample panel, when accepted, shall become the project standard for bond, mortar, quality of work, and appearance. Sample panels may be installed in the wall and remain as part of the permanent installation. Do not begin work until panel is accepted by Architect.

1.4 Product Delivery, Storage, and Handling

Loading and shipment:

Carefully pack the fabricated stone for shipment free from stains, saw mud, and other deleterious material. Exercise careful precautions against damage in transit.

Delivery and storage:

Receive and unload stone at job site with necessary care in handling to avoid damaging or soiling. Store clear of the ground on nonstaining skids and adequately protect by covering with waterproof paper or polyethylene.

PART 2 PRODUCTS

2.1 Materials

Stone:

Stone shall be of the highest quality, type ______ as produced by ______. The color, texture, and pattern shall be similar to approved samples selected by the Architect.

Stone thickness shall range from ______ in. to ______ in. as shown on the drawings. No stone shall exceed 10 in. in thickness at any point measured from back to face as laid.

Setting mortars:

All stone shall be set in nonstaining cement-lime mortar composed of one part portland cement ASTM C150, Type I or II (low-alkali), 1 part Type S hydrated lime, and 6 parts sand complying with ASTM C144, except that not less than 5% shall pass a No. 100 sieve. Water shall be clean, nonstaining, and free of deleterious substances. No retempering of mortar will be permitted.

Joint materials:

Joints shall be as follows:

1. Mortar joints as specified above.
2. Sealant joints of an approved system including primer, gasket, and sealant with nonstaining properties conforming to ANSI standard A116.1. Follow manufacturer's recommendations regarding handling, storing, applying, and curing.

3. Expansion joints shall be adequate to allow for thermal and structural differential movement. Filler material for these joints shall be nonstaining and compatible with the sealing compound. Provide neoprene pad at shelf angles.
4. Grout shall be composed of 1 part nonstaining, low-alkali portland cement and $1\frac{1}{2}$ parts fine sand mixed in small quantities in thickest consistency pourable into joints. Stir grout vigorously until used.
5. Weeps shall be composed of mildew- and fungus-proofed felt wicks placed in joints where moisture may accumulate, such as base of cavity walls, continuous angles, flashing, etc.

Anchors and dowels:

Furnish and set all anchors shown on approved shop drawings. All anchors, dowels, etc., shall be of hot-galvanized steel, eraydo alloy zinc, yellow brass, commercial bronze, or stainless steel Type 302 or 304.

2.2 Fabrication and Manufacture

Cutting:

Cut all stone accurately to shape and dimensions with jointing as shown on approved shop drawings. Cut all exposed faces true with beds and joints dressed straight and square unless otherwise shown. Cut all arrises sharp and true. Except where required and noted on the shop drawings, joints shall have a uniform thickness of $\frac{1}{4}$ in.

Washes shall be as deep as practicable, and drips of sufficient width and depth shall be provided on all projecting stones and courses. Cut reglets for flashing, etc. as indicated on the shop drawings. Repair of defects or damaged stone will be permitted only with Architect's approval.

Back-checking and fitting:

Back-check all stone coming in contact with structural work as indicated on the shop drawings. Shape beds to fit supports for all stone resting on structural work. Where stone facing adjoins columns and spandrel beams, cut depth of stone to allow not less than 1 in. clearance between stone and structural work. General Contractor shall furnish necessary field dimensions as required.

Anchors and lewis holes:

Cut holes and sinkages in stone for all anchors as indicated on the shop drawings. Provide lewis holes for all stones that cannot be handled with belts, clamps, or other lifting devices. Do not cut lewis holes in stones less than 4 in. thick without Architect's approval. Lift stone with lewis pins in same plane as holes are drilled.

Provision for other trades:

All cutting and drilling of stone required by other trades will be done in supplier's plant when approved details are furnished prior to fabrication and shipment. Field cutting and fitting shall be the responsibility of the General Contractor.

PART 3 EXECUTION

3.1 Preparation

Inspection:

Verify that all supporting structures and backup walls are completed true to line and grade. Do not begin work until unsatisfactory conditions have been corrected.

Stain prevention:

Waterproof all concrete and masonry surfaces, shelf angles, etc., against which stone is to be applied using a nonstaining asphalt emulsion, vinyl lacquer, or waterproof cement base. Waterproof all unexposed surfaces of the stone to within 1 in. of the exposed faces.

3.2 Installation

Workmanship:

Lay stone with not less than 1 in. of space between the stone and the backing wall. Anchor stones with mechanical devices as shown and/or grout solidly in place with mortar as shown on the shop drawings. The bond between the stone and the mortar and/or the mechanical anchor, as detailed, must be sufficient to withstand a shearing stress of 50 psi after curing for 28 days. Where shown on the drawings, all joints and the space between the stone and the backing shall be filled solidly with mortar, and the joints raked.

Setting:

When setting with mortar, joints not thoroughly wet should be sponged or drenched with clear, nonstaining water. Joints may be tooled when initial set takes place or raked out $\frac{3}{4}$ in. and pointed later with mortar or caulking. Sponge the face along all joints and remove all droppings and splashed mortar immediately. All top joints, wash joints, and reglets shall be pointed flush with an approved elastic caulking compound. All heavy stones shall be set on neoprene pads. All cornices, copings, projecting belt courses, steps, platforms, etc., should be set with unfilled vertical joints. The exterior profile of these joints shall be caulked with rope yarn or picked oakum. After wetting the ends of the stone, thoroughly fill the joint full from above with grout to within $\frac{3}{4}$ in. of the top. After grout has set, remove caulking for pointing.

Cold weather:

In cold weather, follow recommendations of Portland Cement Association, except that no additives will be permitted in the setting mortar. Provide temporary heat as required.

Protection:

Receipt, storage, and protection of cut stone work prior to, during, and subsequent to installation shall be the responsibility of the General Contractor. During construction, carefully cover tops of walls at end of each workday and during any precipitation or other inclement weather. Protect walls at all times from mortar droppings.

Wherever necessary, provide substantial wooden coverings to protect stone work. Provide nonstaining building paper or membrane under the wood. Maintain all coverings until removed to permit final cleaning of stone.

Cleaning:

Wash stone with fiber brushes, soap powder, and clean, nonstaining water.

Waterproofing:

After all stone has been set and cleaned down, apply one coat of approved silicone waterproofing in strict accordance with manufacturer's written instructions.

SECTION 04520 MASONRY RESTORATION

PART 1 GENERAL

1.1 Work Included

Description of work:

Provide all labor, materials, equipment, and services required to clean, repair, tuckpoint, and seal existing brick, stone, and terra cotta where shown on the drawings or specified herein.

1.2 Related Work

Related work specified elsewhere:

Shop Drawings, Product Data, and Samples, Section 01340
Unit Masonry, Section 04200
Waterproofing and Dampproofing, Section 07100
Caulking and Sealants, Section 07920

1.3 Quality Assurance

Samples:

Submit samples of brick, stone, and terra cotta for replacement of deteriorated work.

1.4 Product Delivery, Storage, and Handling

Delivery and storage:

Deliver and handle materials to prevent damage. Store packaged materials above ground, protected from weather.

PART 2 PRODUCTS

2.1 Materials

Brick:

Face brick, ASTM C216, grade, type, size, color, and texture to match existing.

Stone:

Color, texture, and type to match existing.

Terra Cotta:

Color and texture to match existing.

Pointing mortar:

ASTM C270, Type N, prehydrated, or ______________________________

Epoxy adhesive:

Provide moisture-insensitive, low-modulus, two-part epoxy adhesive.

Liquid latex:

Provide H. B. Fuller Co. "TEC Tile Bond."

Aggregate:

Use 30-to 50-mesh silica sand as aggregate in epoxy mortar.

Surface sealant:

Provide Sonneborn "White Roc M-6-50-8 H/S" or Standard Dry Wall "Thoroglaze."

2.2 Mixes

Epoxy mortar:

Mix mortar by volume, 1 part epoxy adhesive, 1 to $1\frac{1}{2}$ parts aggregate. Color of cured mortar shall match existing terra cotta.

Stone setting mortar:

ASTM C270, Type N, non-staining cement and clean, sharp, washed sand.

Stone pointing mortar:

Mix with clean, white sand, and liquid latex in lieu of water to produce a workable mix. Add color pigments as required to match existing joints.

2.3 Fabrication

Stone:

Cut stone to conform with existing shapes and dimensions. Beds and joints shall be dressed straight and at right angles to faces unless otherwise required by design. Exposed arrises shall be sharp and true.

PART 3 EXECUTION

3.1 Preparation

Brick, stone, and terra cotta:

Remove mortar in all defective mortar joints to minimum depth of $\frac{1}{2}$ in. Brush or otherwise clean joints before applying new mortar.

Stone coping:

Remove mortar in stone coping joints to minimum depth of 1 in. Brush or otherwise clean joints before applying new mortar.

3.2 Execution

Brick and stone repair:

All new work shall be toothed into existing so that coursing and bonding will run continuously on all sides of patched areas.

Terra cotta repair:

Repair broken or chipped terra cotta with epoxy mortar to match existing configurations. Do not apply epoxy mortar more than $1\frac{1}{2}$ in. thick. Small cracks that cannot be filled with mortar shall be filled with epoxy adhesive.

Installing stonework:

Set all stones level and plumb, with uniform joints matching existing work. Fill all joints with setting mortar and rake back to minimum 1 in. depth. After setting mortar has set, clean, wet, and fill with pointing mortar. Tool joints flush with stone and rub smooth to match existing work. Remove excess mortar from stone face, and hand clean with soap solution and fiber brushes followed immediately by clear water wash.

Tuckpointing:

Dampen joints just prior to tuckpointing. Apply prehydrated pointing mortar in thin layers until joints are completely filled. Tool joints to smooth, watertight finish matching existing work.

3.3 Cleaning and Sealing

Cleaning:

Clean all repaired and tuckpointed surfaces of dirt, mortar smears, and stains using stiff brush and clean water. Do not use cleaning agents except with written approval of Architect.

Sealing:

Seal all exterior brick, stone, and terra cotta in strict accordance with sealant manufacturer's printed instructions. Apply to clean, dry surfaces only when ambient temperature is above 40°F.

16.3 INSPECTING MASONRY CONSTRUCTION

The construction inspection phase of building projects is becoming increasingly important with the rise in professional liability claims against architects, engineers, and contractors. The purpose of inspection is to assure that the quality of the finished product is in compliance with the requirements of the drawings and specifications, and that the workmanship meets the standards outlined by the design professional.

Good workmanship affects masonry performance, and is essential to high-quality construction. The masonry trade requires highly skilled craftsmen and artisans working cooperatively with the architect and engineer to execute design. The goal of quality workmanship is common to all concerned parties for various reasons of aesthetics, performance, and liability.

Generally, the drawings are used to indicate the size and shape of the building, and give graphic illustrations of unit types, sizes, patterns, and configurations. Large-scale details for joints, intersections, reinforcement, anchorage, and flashing should accompany the plans as further guidelines. The specifications supplement the drawings, and establish a quality standard for both materials and workmanship. By the use of reference standards, performance criteria, and prescriptive requirements, the specifier outlines the required products and their method of installation.

Without field inspection, the architect or engineer has no control over the execution of the design. With the increasing complexity of structural designs used in masonry buildings, the need for inspection to assure design integrity extends beyond aesthetics to concern for the public safety.

Responsibility for construction of a building project lies ultimately with the contractor. The architect or engineer is not a party to the construction contract, but acts solely as the owner's representative in the field. As part of the team, the architect can guide the contractor and offer advice and expertise in solving or avoiding potential problems. The architect must also act as interpreter of design intent, and safeguard the project quality by assuring proper execution of the work according to the requirements of the drawings and specifications. The inspector's authority does not extend to supervision of the work, or to revision of details or methods without the written approval of the parties concerned.

Independent inspection agencies or testing laboratories serve a different function. If required by the specifications, it is their responsibility to test various materials and assemblies to verify compliance with reference standards, design strengths, and performance criteria. Building codes recognize the value of construction inspection by a 50% increase in allowable compressive and shear stresses on masonry sections (refer to Chapter 10). Each building project or contract will vary in its allocation of responsibility during the construction phase, but inspection procedures are vital in assuring the successful translation of the design, drawings, and specifications into a completed structure that functions as intended.

16.3.1 Materials

The inspector must be familiar with the project specifications and must verify compliance of materials at the job site with the written requirements. Manufacturers must supply test certificates showing that the various material properties meet or exceed the referenced standards as to ingredients, strength, dimensional tolerances, durability, and so on.

Unit masonry may be visually inspected for color, texture, and size, and compared to approved samples in the architect's office. Units delivered to the job site should be inspected for physical damage, and storage-protection provisions checked. Stone, brick, or concrete masonry that has become soiled, cracked, chipped, or broken in transit should be rejected. If the manufacturer does not supply test certificates, random samples should be selected and sent to the testing agency for laboratory verification of minimum standards. The inspector should also check the moisture condition of clay masonry at the time of laying since initial suction drastically affects the bond between unit and mortar, and the strength of the mortar itself. Visual inspection of a broken unit can indicate whether field tests of absorption rates should be performed (*see Fig. 16-1*). Concrete masonry units must be protected from wetting at the job site. Units should be rejected if there is a question of excessive shrinkage potential or decreased bond.

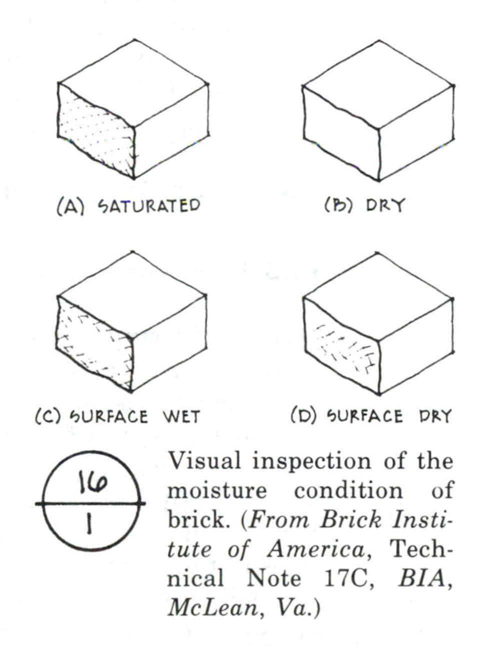

Visual inspection of the moisture condition of brick. (*From Brick Institute of America*, Technical Note 17C, *BIA, McLean, Va.*)

Mortar and grout ingredients should be checked on delivery to assure compliance with the specified design mix, and to guard against damage or contamination. Containers should be sealed with the manufacturer's identifying labels legible and intact. Bagged ingredients that show signs of water absorption should be rejected as unsuitable. If material test certificates are required, check compliance with the specifications.

Acceptable mixing and batching procedures should be established at the start of construction to assure quality and consistency throughout the job. If field testing of mortar prisms is required, laboratory samples should be prepared and tested sufficiently in advance of construction to allow changes or modifications of materials or procedures as indicated. Retempering time should be monitored to preclude the use of mortar or grout that has begun to set.

In some instances, field sample panels are required. These samples may be free-standing, or may be a section of in-place work. The panels can be effectively used to establish material and workmanship standards, and to control the quality of the work throughout construction.

Accessories must also be checked for design compliance. The inspector must assure use of proper anchoring devices, ties, inserts, and reinforcement. Steel bars, shelf angles, lintels, and so on, should carry certification of yield strength and be properly bundled and identified for location within the structure.

16.3.2 Construction

Foundations, beams, floors, and other structural elements that will support the masonry should be checked for completion to proper line and grade before the work begins. Adequate structural support must be assured, and areas cleaned of dirt, grease, oil, laitance, or other materials that might impair bond of the mortar or grout. Overall dimensions and layout must be verified against the drawings and field adjustments made to correct discrepancies. Steel reinforcing dowels must be checked for proper location in relation to cores, joints, or cavities. The inspector should also keep a log of weather conditions affecting the progress of performance of the work. Inspectors should not interfere with the workers or attempt to direct their activities. If methods or procedures are observed that appear to conflict with the specifications or which might jeopardize the quality or pefor-mance of the work, they should be called to the attention of the contractor or foreman, and adjustments made as necessary.

16.3.3 Workmanship

Perhaps the single most important element in obtaining strong, weather-tight masonry walls is full mortar joints. Partially filled head joints or fur-rowed mortar beds will weaken the assembly and offer minimal resis-tance to moisture infiltration. The first course of masonry must be carefully aligned vertically and horizontally, and fully bedded to assure that the remainder of the wall above will be plumb and level. Even if hollow CMU construction requires only face-shell bedding, this critical base course must have full mortar under face and web as well. Head joints must be fully buttered with mortar and shoved tight against the adjacent unit to avoid leaks (*see Fig. 16-2*). Units must not be moved, tapped, or realigned after initial placement, or the bond will be destroyed. If a unit is displaced, all head and bed mortar must be removed and replaced with fresh material. Spot checks for proper bond can be made by lifting a fresh unit out of place to see if both faces are fully covered with adhered mortar.

The inspector should check for proper embedment and coverage of anchors, ties, and joint reinforcement, and should monitor vertical coursing and joint uniformity. Differential widths or thicknesses of mortar joints can misalign the modular coursing and interfere with proper location of openings, lintels, and embedded items. Storypoles, string lines, and tapes or templates should be used to check coursing between corner leads. Nail and line pinholes must be filled with mortar when string lines are removed to

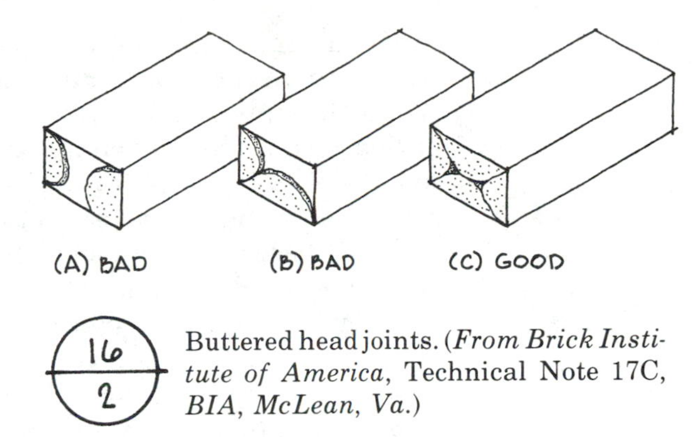

Buttered head joints. (*From Brick Institute of America*, Technical Note 17C, *BIA, McLean, Va.*)

avoid water penetration through these voids. Work of other trades that passes through the masonry should be incorporated during construction of the wall and not cut in later. Collar joints and cavities must be kept free of mortar fins and droppings to avoid plugging weep holes, damaging flashing, or interfering with grout pours. When they become thumbprint hard, joints should be tooled to compress the mortar against the units and seal out water.

The mason should place all vertical and horizontal reinforcement as the work progresses, holding the bars in correct alignment with spacers or wire. Minimum clearances should be maintained, and bar splices lapped and securely tied. The inspector should check to see that reinforcement is free of rust, loose scale, or other materials that could impair bond to the mortar. Care should be taken to avoid moving or jarring vertical steel that is already embedded in lower grouted courses.

Inspection should also include proper installation of flashing, control joints, expansion joints, lintels, sills, caps, copings, and frames. Door frames must be adequately braced until the mortar has set and the masonry work surrounding them is self-supporting.

Grouting is an important factor in the structural integrity of reinforced masonry walls. Cavities and cores should be inspected before the grout is placed, and any remaining dirt, debris, mortar droppings, or protrusions removed before the work proceeds. Cleanout plugs left for high-lift pours allow visual inspection from below by use of a mirror inserted through the opening. Cleanout units should be fully mortared and shoved into place after inspection, then braced against blowout from the fluid pressure of the grout against the uncured mortar. The consistency of the grout should allow for easy pouring or pumping, and complete filling of the space. Rodding to remove air bubbles and pockets also assures that the grout covers fully around and between ties and reinforcement. Grout in contact with the masonry will harden more rapidly than grout in the center of the cavity. It is important that rodding or agitating be done immediately after the pour to avoid disturbance of any hardened grout. Timing of grout pours should be monitored to avoid excessive lateral pressure on uncured joints.

16.3.4 Protection and Cleaning

Throughout the construction period, both the masonry materials and the work must be protected from the weather. Materials must be stored off the ground under waterproof coverings to prevent wetting or staining. Exposed tops of unfinished walls must be covered each night to keep moisture out by draping waterproof paper or canvas 2 ft down each side. Cold weather may require heating of materials and possibly the application of heat during the curing period. Hot, dry climates cause rapid evaporation, and mortar mixes may have to be wetter to compensate for premature drying. The inspector must assure that these precautions are taken to avoid harmful effects, and must also see that completed work is protected from damage during other construction operations.

Suitable cleaning methods must be selected on the basis of the type of stain involved and the type of material to be cleaned. Improper use of cleaning agents can create more problems than are solved by their application. Mortar smears on the face of the masonry should be removed daily before they are fully hardened, and dry-brushed when powdery to prevent stains. Paints, textured coatings, or clear water repellents, if specified, should be applied carefully over clean, dry walls, and adjacent work protected against splatters and drips.

It is the inspector's job to see that the instructions and requirements of the drawings and specifications are carried out in the field. Safeguarding the quality of the work without impeding its progress is best achieved through cooperation with the contractor and workers. Good design and good intentions are not sufficient in themselves to assure quality of the finished product. The inspector can serve as interpreter and mediator in the proper execution of the work, ensuring masonry structures that are as durable and lasting as the materials of which they are made.

* * * * *

Each of these surfaces was once new, flawlessly shaped by the stonecutter for the purpose it was to serve. Now, usage and time have marked and mellowed their form, polished or gouged their surface. Life is imprinted on the stone as on a human face.

Andreas Feininger*

Taken from Man and Stone *by Andreas Feininger. Copyright© 1961 by Andreas Feininger. Used by permission of Crown Publishers, Inc.*

appendix A

GLOSSARY

A

A-Block A hollow unit with one end closed and the opposite end open which forms two cells when laid in the wall.

Absorbed Moisture Moisture held mechanically in the material and having physical properties not substantially different from ordinary water at the same temperature and pressure.

Absorption The weight of water a brick or tile unit absorbs when immersed in either cold or boiling water for a stated length of time, expressed as a percentage of the weight of the dry unit for brick, and as pounds of water per cubic foot of concrete for CMU.

Absorption Rate The weight of water absorbed when a brick is partially immersed for 1 minute, usually expressed in either grams or ounces per minute. Also called suction or **initial rate of absorption**.

Abutment (1) That part of a pier or wall from which an arch springs, specifically the support at either end of an arch, beam, or bridge. (2) A **skewback** and the masonry that supports it.

Accelerated Drying The process of drying block or brick with relatively warm, dry air, or other means than natural **air drying.**

ACI American Concrete Institute.

Note: Boldface type within entries denotes terms for which there are main glossary entries.

Acid-Resistant Brick Brick suitable for use in contact with chemicals, usually in conjunction with acid-resistant mortars.

Actual Dimensions Exact size of masonry units, usually width of mortar joint less than nominal dimensions.

Adhered Attached by adhesion rather than mechanical anchorage, as in adhered veneer.

Adherence The properties of bodies for sticking together.

Adhesion-Type Ceramic Veneer Thin sections of ceramic veneer, held in place by adhesion of mortar to unit and to backing. No metal anchors are required.

Admixture Any material other than water, aggregate, portland cement, or lime that is used as an ingredient in mortar and grout to improve quality or alter properties. May be water repellent, coloring pigment, or agents to retard or hasten setting.

Adobe Soil of diatomaceous content mixed with sufficient water so that plasticity can be developed for molding into masonry units.

Adobe Brick Large roughly molded, unfired, sun-dried clay brick of varying size.

Air Drying The process of drying block or brick without any special equipment simply by exposure to ambient air.

AISC American Institute of Steel Construction.

Alumina The oxide of aluminum; an important constituent of the clays used in brick, tile, and refractories.

American Bond In masonry, a form of bond pattern in which every sixth course is a header course and the intervening courses are stretcher courses.

Anchor In masonry, the unit used to attach a veneer wall to its structural support.

Anchor Block A block of wood replacing a brick in a wall to provide a nailing or fastening surface.

Anchored-Type Ceramic Veneer Thick sections of ceramic veneer held in place by grout and wire anchors connected to backing wall.

Angle Blasting Sandblasting stone at an angle of less than 90°.

Angle Brick Any brick having an oblique shape to fit an oblique, salient corner.

Angle Closer A special-shaped brick used to close the bond at the corner of a wall.

ANSI American National Standards Institute (formerly USAS).

Apron Wall That part of a panel wall between window sill and wall support.

Arch (1) A form of construction in which a number of units span an opening by transferring vertical loads laterally to adjacent units and thus to the supports. An arch is normally classified by the curve of its intrados. (2) Curved compressive structural member spanning openings or recesses; also built flat.

Abutment The skewback of an arch and the masonry that supports it.

Arch Axis The median line of the arch ring.

Back Arch A concealed arch carrying the back of a wall where the exterior facing is carried by a lintel.

Camber The relatively small rise of a jack arch.

Constant-Cross-Section Arch An arch whose depth and thickness remain constant throughout the span.

Crown The apex of the arch ring. In symmetrical arches, the crown is at midspan.

Depth The depth (d) of any arch is the dimension that is perpendicular to the tangent of the axis. The depth of a jack arch is taken to be its greatest vertical dimension.

Extrados The exterior curve that bounds the upper extremities of the arch.

Fixed Arch An arch whose skewback is fixed in position and inclination. Plain masonry arches are, by nature of their construction, fixed arches.

Gothic or Pointed Arch An arch, with relatively high rise, whose sides consist of arcs or circles, the centers of which are at the level of the spring line. The Gothic arch is often referred to as a drop, equilateral, or lacent arch, depending on whether the spacings of the center are respectively less than, equal to, or more than the clear span.

Intrados The interior curve that bounds the lower extremities of the arch (see **Soffit**). The distinction between soffit and intrados is that the intrados is linear, whereas the soffit is a surface.

Jack Arch A flat arch.

Major Arch Arch with span greater than 6 ft and having equivalent uniform loads greater than 1000 lb/ft. Typically, a Tudor arch, semicircular arch, Gothic arch, or parabolic arch. Rise-to-span ratio greater than 0.15.

Minor Arch Arch with maximum span of 6 ft and loads not exceeding 1000 lb/ft. Typically, a jack arch, segmental arch, or multicentered arch. Rise-to-span ratio less than or equal to 0.15.

Multicentered Arch An arch whose curve consists of several arcs of circles which are normally tangent at their intersections.

Relieving Arch One built over a lintel, flat arch, or smaller arch to divert loads, thus relieving the lower member from excessive loading. Also known as discharging or safety arch.

Rise The rise (r) of a minor arch is the maximum height of arch soffit above the level of its spring line. The rise (f) of a major parabolic arch is the maximum height of arch axis above its spring line.

Segmental Arch An arch whose curve is circular but less than a semicircle.

Semicircular Arch An arch whose curve is a semicircle.

Skewback The inclined surface on which the arch joins the supporting wall. For jack arches the skewback is indicated by a horizontal dimension (K).

Soffit The undersurface of the arch.

Span The horizontal distance between abutments. For minor arch calculations the clear span (S) of the opening is used. For a major parabolic arch the span (L) is the distance between the ends of the arch axis at the skewback.

Spring Line For minor arches the line where the skewback cuts the soffit. For major parabolic arches, the term commonly refers to the intersection of the arch axis with the skewback.

Trimmer Arch An arch (usually a low-rise arch of brick) used for supporting a fireplace hearth.

Tudor Arch A pointed four-centered arch of medium rise-to-span ratio.

Voussoir One of the wedge-shaped masonry units which forms an arch ring.

Arch Brick (1) A wedge-shaped brick used in arch or circular construction; has two large faces inclined toward each other. (2) An extremely hard-burned brick from an arch of a scove kiln. Also called **voussoir** brick.

Arch Stone Voussoir stone.

Architectural Terra Cotta Hard-burned, glazed or unglazed clay building units, plain or ornamental, machine extruded or hand molded, and generally larger in size than brick or facing tile. See **Ceramic Veneer.**

Arris A sharp edge of brick, stone, or other building element.

Ashlar Brick A brick whose face has been hacked or broken to resemble roughly dressed stone.

Ashlar Masonry Masonry composed of rectangular units of burned clay or shale, or of stone, generally larger in size than brick and properly bonded, having sawed, dressed, or squared beds, and joints laid in mortar. Ashlar construction may be coursed, random, or patterned.

ASHRAE American Society for Heating, Refrigeration, and Air Conditioning Engineers.

ASTM American Society for Testing and Materials.

Axhammer An axe for spalling or dressing rough stone; has either one cutting edge and one hammer face, or two cutting edges. Also called **mason's hammer.**

B

Back Arch A concealed arch carrying the backing of a wall where the exterior facing is carried by a lintel.

Back Filling (1) Rough masonry built behind a facing or between two faces. (2) Filling over the extrados of an arch. (3) Brickwork in spaces between structural timbers, sometimes called brick nogging.

Backing Brick A utility brick used behind face brick or other masonry.

Backup That part of a masonry wall behind the exterior facing.

Basketweave A checkerboard pattern of bricks or pavers, flat or on edge. Bricks or modular groups of bricks laid at right angles to those adjacent.

Bat A broken brick or piece of brick with one undamaged end. Also called a brickbat. Usually about one-half brick.

Batter Recessing or sloping masonry back in successive courses; the opposite of corbel.

Bearing Plate A plate placed under a truss, beam, or girder to distribute the load.

Bearing Wall A wall that supports a vertical load in addition to its own weight.

Bed (1) In masonry and bricklaying, the side of a masonry unit on which it lies in the course of the wall; the underside when placed horizontally. (2) The layer of mortar on which the masonry unit is set.

Bed Joint In masonry, the horizontal mortar joint or layer of mortar upon which the masonry units are laid.

Bedding Course The first layer of mortar at the bottom of a masonry wall.

Belt Course A narrow horizontal course of masonry, sometimes slightly projected, such as window sills which are made continuous. Sometimes called **string course** or sill course.

BIA Brick Institute of America (formerly **SCPI**).

Blind Header A concealed header in the interior of a wall, not showing on the faces of the wall.

Block (1) A masonry unit; a concrete block. (2) In quarrying, the large piece of stone, usually squared, that is taken from the quarry to the mill for sawing, slabbing, and further working.

Block Bond Same as **flemish bond**.

Blocking A method of bonding two adjoining or intersecting walls, not built at the same time, by means of offsets whose vertical dimensions are not less than 8 in.

Blockwork Masonry of concrete block and mortar.

Bolster A blocking chisel for masonry work. A broad-edged chisel made in a number of sizes, shapes, and weights.

Bond (1) Tying various parts of a masonry wall together by lapping units one over another or by connecting with metal ties. (2) Patterns formed by the exposed faces of the masonry units. (3) Adhesion between mortar or grout and masonry units or reinforcement.

Bond Beam A course or courses of a masonry wall grouted and usually reinforced in the horizontal direction serving as an integral beam in the wall. May serve as a horizontal tie, bearing course for structural members, or as a flexural member itself.

Bond Beam Block A hollow concrete masonry unit with depressed sections forming a continuous channel in which reinforcing steel can be placed for embedment in grout. **Lintel blocks** are sometimes used as bond beam blocks.

Bond Course The course consisting of units that overlap more than one wythe of masonry.

Bonder A bonding unit. Also called a **header.**

Breaking joints Any arrangement of masonry units that prevents continuous vertical joints from occurring in adjacent courses.

Brick A solid masonry unit of clay or shale, formed into a rectangular prism while plastic and burned or fired in a kiln. Similarly shaped units made of portland cement mixes are called concrete brick.

Acid-Resistant Brick Brick suitable for use in contact with chemicals, usually in conjunction with acid-resistant mortars.

Adobe Brick Large roughly molded, sun-dried clay brick of varying sizes.

Angle Brick Any brick shaped to an oblique angle to fit a salient corner.

Arch Brick (1) Wedge-shaped brick for special use in an arch. (2) Extremely hard burned brick from an arch of a scove kiln.

Building Brick Brick for building purposes not specially treated for texture or color. Formerly called common brick. (Covered by ASTM C62.)

Bullnose Brick A brick having one or more rounded corners.

Clinker Brick A very hard burned brick whose shape is distorted or bloated due to nearly complete vitrification.

Common Brick See **Building Brick**.

Concrete Brick Concrete masonry unit of similar size and shape to clay brick.

Cored Brick A brick in which the holes consist of less than 25% of the section.

Dry-Press Brick Brick formed in molds under high pressures from relatively dry clay (5 to 7% moisture content).

Economy Brick A brick whose nominal dimensions are $4 \times 4 \times 8$ in.

Face Brick Brick made especially for exposed use, often treated to produce surface texture. May be made of selected clays, or treated to produce desired color. (Covered by ASTM C216.)

Fire Brick Brick made of refractory ceramic material which will resist high temperatures.

Floor Brick Smooth, dense brick, highly resistant to abrasion, used as finished floor surfaces. (Covered by ASTM C410.)

Gauged Brick (1) Brick that has been ground or otherwise produced to accurate dimensions. (2) A tapered arch brick.

Glazed Brick A brick prepared by fusing on the surface a ceramic glazing material; brick having a glassy surface.

Hollow Brick A masonry unit of clay or shale whose net cross-sectional area in any plane parallel to the bearing surface is between 60 and 75% of its gross cross-sectional area measured in the same plane. (Covered by ASTM C652.)

Jumbo Brick A generic term indicating a brick larger in size than the "standard." Some producers use this term to describe oversize brick of specific dimensions which they manufacture.

Norman Brick A brick whose nominal dimensions are $4 \times 2\frac{2}{3} \times 12$ in.

Paving Brick Vitrified brick especially suitable for use in pavements where resistance to abrasion is important. (Covered by ASTM C902.)

Roman Brick Brick whose nominal dimensions are $4 \times 2 \times 12$ in.

Salmon Brick Generic term for relatively soft underburned brick which are more porous, slightly larger, and lighter colored than hard-burned brick. Usually pinkish orange (salmon) in color.

Sand-Struck Brick See **Soft-Mud Brick**.

SCR Brick See **SCR**.

Sewer Brick Low absorption, abrasive-resistant brick intended for use in drainage structures. (Covered by ASTM C32.)

Soft-Mud Brick Brick produced by molding relatively wet clay (20 to 30% moisture). Often a hand process. When insides of molds are sanded to prevent sticking of clay, the product is sand-struck brick. When molds are wetted to prevent sticking, the product is water-struck brick.

Stiff-Mud Brick Brick produced by extruding a stiff but plastic clay (12 to 15% moisture) through a die.

Wire-Cut Brick Brick formed by forcing plastic clay through a rectangular opening designed to shape the clay into bars. Before burning, wires pressed through the plastic mass cut the bars into brick units.

Brick Ax Same as **Brick Hammer**.

Brick Facing See **Brick Veneer**.

Brick Grade Designation for durability of the unit expressed as SW for severe weathering, MW for moderate weathering, or NW for negligible weathering. (See ASTM C216, C62, and C652.)

Brick Hammer A Steel tool, one end of which has a flat, square surface used as a hammer, for breaking bricks, driving nails, and so on. The other end forms a chisel peen used for dressing bricks. (Also called brick ax or bricklayers hammer.)

Brick Ledge A ledge on a footing or wall which supports a course of masonry.

Brick Trowel A trowel having a flat, trianglar steel blade in an offset handle used to pick up and spread mortar. The narrow end of the blade is called the "point"; the wide end, the "heel."

Brick Type Designation for facing brick that controls tolerance, chippage, and distortion. Expressed as FBS, FBX, and FBA for solid brick, and HBS, HBX, HBA, and HBB for hollow brick. (See ASTM C216 and C652.)

Brick Veneer A facing of brick laid against a wall and not structurally bonded to the wall.

Brickbat See **Bat**.

Building Brick Brick for building purposes not specially treated for texture or color. Formerly called common brick. (Covered by ASTM C62.)

Bullnose Unit A brick or concrete masonry unit having one or more rounded exterior corners.

Buttering Placing mortar on a masonry unit with a trowel.

Buttress A projecting structure built against a wall or building to give it greater strength and stability.

C

Calcium Silicate Brick A sand-lime brick. (Covered by ASTM C73.)

Camber A slight arching or upward curve of a beam or soffit.

Cap Block A solid, flat unit usually $2\frac{1}{4}$ in. thick used as cap or coping for parapet and garden walls, etc.

Capacity Insulation The ability of masonry to store heat as a result of its mass, density, and specific heat. Also called **thermal inertia**.

Cavity Flashing A continuous sheet of waterproofing material which is installed across the gap of a cavity wall.

Cavity Wall A wall built of two or more wythes of masonry units so arranged as to provide a continuous air space within the wall. The facing and backing (inner and outer wythes) are tied together with noncorrosive metal ties.

C/B Ratio The ratio of the weight of water absorbed by a masonry unit during immersion in cold water for 24 hours to weight absorbed during immersion in boiling water for 5 hours. An indication of the probable resistance of brick to freezing and thawing. Also called a **saturation coefficient**. (See ASTM C67.)

Cell See **Core**.

Cement A burned mixture of clay and limestone pulverized for making mortar, grout, or concrete (e.g., portland, pozzolan, masonry, etc.).

Centering Temporary formwork for the support of masonry arches or lintels during construction. Also a temporary structure upon which the materials of a vault or arch are supported in position until the work becomes self-supporting.

Ceramic Any of a class of products, made of clay or a similar material, which are subjected to a high temperature during manufacture, causing the silicates in the clay to melt and fuse together.

Ceramic Color Glaze An opaque, color glaze of satin or gloss finish obtained by spraying the clay body with a compound of metallic oxides, chemicals, and clays. It is burned at high temperatures, fusing glaze to body, making them inseparable. (See ASTM C126).

Ceramic Veneer A type of architectural terra cotta, characterized by large face dimensions and thin sections ranging from $1\frac{1}{8}$ to $2\frac{1}{2}$ in. in thickness.

Chamfer A groove or channel; a bevel edge; an oblique surface formed by cutting away an edge or corner.

Channel Block A hollow unit with web portions depressed less than $1\frac{1}{4}$ in. to form a continuous channel for reinforcing steel and grout.

Chimney Lining Fire clay or terra cotta material, or refractory cement, made to be built inside a chimney.

Chimney Throat That part of a chimney directly above the fireplace where the walls are brought close together.

Cinder Block A lightweight concrete masonry unit made of cinder concrete. Often refers to any lightweight concrete block.

Clay A natural mineral aggregate consisting essentially of hydrous aluminum silicates; it is plastic when sufficiently wetted, rigid when dried, and vitrified when fired to a sufficiently high temperature.

Cleanout Holes Openings in first course of one wythe of brick wall, or cutouts in face shell of first course of block wall, for cleaning out cavity or cores before grouting.

Clear Ceramic Glaze Same as ceramic color glaze except that it is translucent or only slightly tinted, with a gloss finish.

Clinker Brick A very hard burned brick whose shape is distorted or bloated due to nearly complete vitrification.

Closer The last masonry unit laid in a course. It may be whole or a portion of a unit. Supplementary or short-length units are used at corners or jambs to maintain bond patterns.

CMU Concrete masonry unit.

Collar Joint The vertical, longitudinal joint between wythes of masonry.

Common Bond A bond in which every fifth or sixth course consists of headers, the other courses being stretchers.

Common Brick See **Building Brick.**

Compass Brick An arch brick.

Composite Wall A multiple-wythe wall in which at least one of the wythes is dissimilar to the other(s) with respect to type or grade of masonry unit or mortar.

Concave Joint A recessed masonry joint formed in mortar by the use of a curved steel jointing tool. Highly resistant to rain penetration.

Concrete Block A hollow concrete masonry unit made from portland cement and suitable aggregates such as sand, crushed stone, cinders, burned clay or shale, pumice, scoria, air-cooled or expanded blast furnace slag, with or without the inclusion of other materials.

Concrete Brick Concrete masonry unit of similar size and shape to clay brick.

Concrete Masonry Unit Hollow or solid block or solid brick of portland cement, water, and aggregates.

A-Block A hollow unit with one end closed and the opposite end open forming two cells when laid in the wall.

Bond Beam Block A hollow unit with web portions depressed $1\frac{1}{4}$ in. or more to form a continuous channel or channels for reinforcing steel and grout. Lintel blocks are often used to form bond beams.

Bullnose Block A concrete masonry unit which has one or more rounded exterior corners.

Cap Block A solid, flat unit usually $2\frac{1}{4}$ in. thick used as cap or coping for parapets and garden walls, etc.

Channel Block A hollow unit with web portions depressed less than $1\frac{1}{4}$ in. to form a continuous channel for reinforcing steel and grout.

Concrete Block A hollow concrete masonry unit made from portland cement and suitable aggregates such as sand, crushed stone, cinders, burned clay or shale, pumice, scoria, air-cooled or expanded blast furnace slag, with or without the inclusion of other materials.

Concrete Brick A solid concrete masonry unit made from portland cement and suitable aggregates.

Coping Block A solid concrete masonry unit for use as the top and finishing course in wall construction.

Corner Block Concrete masonry unit with one flat end for use in construction of the end or corner of a wall.

Faced Block Concrete masonry units having a special ceramic, glazed, plastic, polished, or ground face surface.

Filler Block Concrete masonry unit for use in conjunction with concrete joists for concrete floor or roof construction.

H-Block A hollow unit with both ends open, and sometimes a continuous bond beam recess at the intersecting web.

Header Block Concrete masonry units which have a portion of one side of the face shell removed to facilitate bonding with adjacent masonry such as brick facing.

Hollow Block A concrete masonry unit whose net cross-sectional area in every plane parallel to the bearing surface is less than 75% of its gross cross-sectional area measured in the same plane.

Jamb Block A block specially formed for jambs of windows or doors, generally with a vertical slot to receive window frames, etc.

Lintel Block A masonry unit consisting of one core with one side open. Usually placed with the open side up like a trough to form a continuous beam, as across a window or door opening.

Offset Block A unit not rectangular in plan and made as a corner block to keep the construction modular in turning the corner.

Paving Block A solid, flat unit used for road and walkway paving.

Pilaster Block Concrete masonry units designed for use in construction of plain or reinforced concrete masonry pilasters and columns.

Sash Block Concrete masonry unit which has an end slot for use in openings to receive jambs of doors or windows.

Scored Block Block with grooves to provide patterns, as, for example, to simulate raked joints.

Shadow Block Block with face formed in planes to develop surface patters.

Sill Block A solid concrete masonry unit used for sills of openings.

Slump Block Concrete masonry units produced in such a way that they slump or sag in irregular form before they harden.

Solid Block A concrete masonry unit whose net cross-sectional area in every plane parallel to the bearing surface is 75% or more of its gross cross-sectional area measured in the same plane.

Split-Face Block Concrete masonry units with one or more faces having a fractured surface for use in masonry wall construction.

U-Block See **Lintel Block**.

Control Joint Continuous, vertically weakened joint designed to accommodate longitudinal movement while providing lateral stability.

Coping The material or masonry units forming a cap or finish on top of a wall, pier, pilaster, chimney, etc. It protects the masonry below from penetration of water from above. Should be projected out from the wall to provide a decorative as well as protective feature.

Coping Brick A brick unit specially formed for use as the top course of an exposed wall.

Corbel A shelf or ledge formed by projecting successive courses of masonry out from the face of the wall.

Corbel Arch Masonry built over a wall opening by uniformly projecting from each side until the units meet at midpoint. No arch action is effected—not a true arch.

Corbel Course A masonry course acting as a corbel or as an ornament of similar appearance.

Corbeled Vault A masonry roof constructed from opposite walls or from a circular base, by shifting courses slightly and regularly inward until they meet. The resulting stepped surface may be smoothed or curved, but no true arch action is involved.

Core A hollow space within a clay or concrete masonry unit formed by the face shells and webs. The holes in clay units.

Cored Brick A brick in which the holes consist of less than 25% of the section.

Corner Block Concrete masonry unit with one flat end for use in construction of the end or corner of a wall.

Course One of the continuous horizontal layers of units bonded with mortar in masonry.

Creeper A brick in the wall adjacent to an arch, cut to conform to the curvature of the extrados.

Cross Joint See **Head Joint.**

CSI Construction Specifications Institute.

Culling Sorting masonry units for size, color, and/or quality.

Culls Masonry units that do not meet the standards or specifications and have been rejected.

Curing The maintenance of proper conditions of moisture and temperature during initial set to develop required strength and reduce shrinkage in concrete products and mortar.

Curtainwall An exterior, nonbearing wall not wholly supported at each floor. May be anchored to columns, spandrel beams, floors, or bearing walls, but not necessarily built between structural elements.

D

Damp course A course or layer of impervious material that prevents capillary entrance of moisture from the ground or from a lower course. Often called a damp check.

Damping Reduction of amplitude of vibrations due to energy loss (as in damping of vibrations from seismic shock).

Dampproofing The preparation of a wall to prevent moisture from penetrating through it.

Dimensions of Masonry Units For masonry units, dimensions are normally given thickness first, height second, and length third.

Drip A projecting piece of material so shaped as to throw off water and prevent its running down the face of the wall or other surface.

Dry-Press Brick Brick formed in molds under high pressures from relatively dry clay (5 to 7% moisture content).

Dutch Bond A pattern bond having the courses made up alternately of headers and stretchers. Same as English cross bond.

E

Eccentricity The distance between the centroidal axis of a member and the applied load.

Economy Brick A brick whose nominal dimensions are $4 \times 4 \times 8$ in.

e_1/e_2 Ratio of virtual eccentricities occurring at the ends of a column or wall under design. The absolute value is always less than or equal to 1.0.

Effective *b* The width of wall assumed to work with reinforcing bars in flexural computations.

Effective Height The height of a member to be assumed for calculating the slenderness ratio.

Effective Thickness The thickness of a member to be assumed for calculating the slenderness ratio.

Efflorescence A powder or stain sometimes found on the surface of masonry, resulting from the deposit of water-soluble salts by the evaporation of water.

End-Construction Tile Structural clay tile units designed to be laid with the axis of the cells vertical.

Engineer Brick A brick whose nominal dimensions are $4 \times 3\frac{1}{5} \times 8$ in.

Engineered Masonry Masonry in which design is based on a rational structural analysis.

Equivalent Thickness The solid thickness to which a hollow unit would be reduced if there were no cores or voids. The gross cross-sectional area divided into the volume per unit of area.

Expansion Joint A joint or space to allow for movement due to volume changes.

Extrados The exterior curve in an arch or vault.

F

Face (1) The exposed surface of a wall or masonry unit. (2) The surface of a unit designed to be exposed in the finished masonry.

Face Shell The side wall of a hollow concrete masonry or clay masonry unit.

Faced Block Concrete masonry units having special ceramic, glazed, plastic, polished, or ground face or surface.

Faced Wall A composite wall in which the masonry facing and backing are bonded together so as to exert a common reaction under load.

Facing Any material used as a finished surface which forms an integral part of a wall (as opposed to a veneer).

Facing Brick Brick made especially for exposed use, often treated to produce surface textures. May be made of selected clays, or treated to produce desired color. (Covered by ASTM C216.)

Facing Tile Structural clay tile for exterior and interior masonry with exposed faces. (Covered by ASTM C212 and ASTM C126.)

Fat Mortar Mortar containing a high percentage of cementitious components. It is a sticky mortar which adheres to a trowel.

Field The expanse of wall between openings, corners, etc., principally composed of stretcher units.

Filler Block Concrete masonry unit for use in conjunction with concrete joists for concrete floor construction.

Filter Block A hollow, vitrified clay masonry unit, sometimes salt-glazed designed for trickling filter floors in sewerage disposal plants.

Fire Box The interior of a fireplace furnace, serving as combustion space.

Fire Brick Brick made of refractory ceramic material which will resist high temperatures.

Fire Clay Clay that is capable of being subjected to high in-use temperatures without further fusing or perceptible softening. It is used extensively for firebrick in the construction of kilns, ovens, and tanks for molten metals.

Fire Division Wall Any wall that subdivides a building so as to resist the spread of fire, but is not necessarily continuous through all stories to and above the roof.

Fire Wall Any wall that subdivides a building to resist the spread of fire and which extends continuously from the foundation through the roof.

Fireproofing Tile Structural clay tile designed for protecting steel structural members from fire.

Flagstone A type of stone that splits easily into flags or slabs; also a term applied to irregular pieces of such stone split into slabs from 1 to 3 in. thick, and used for walks, patios, etc.

Flashing (1) A thin, impervious material placed in mortar joints and through air spaces in masonry to prevent water penetration and/or provide water drainage. (2) Manufacturing method used to produce specific color tones in brick and clay masonry.

Flemish Bond A pattern bond consisting of headers and stretchers alternating in every course, so laid as to always break joints, each header being placed in the middle of the stretchers in courses above and below.

Floor Brick Smooth, dense brick, highly resistant to abrasion, used as finished floor surfaces. (Covered by ASTM C410.)

Flow Measure of mortar consistency, sometimes called initial flow.

Flow after Suction Flow of mortar measured after subjecting it to a vacuum to simulate suction of dry masonry units.

Flue Lining A smooth hollow tile unit used for the inner lining of masonry chimneys.

Flying Buttress A type in which a detached buttress or pier at a distance from a wall is connected to the wall by an arch or portion of arch.

Foundation Wall That portion of a loadbearing wall below the level of the adjacent grade, or below the first-floor beams or joists.

Fretwork In masonry, any ornamental openwork or work in relief.

Frog A depression in the bed surface of a brick. Sometimes called a panel.

Furring Tile Structural clay tile designed for lining the inside of exterior walls to provide air space. Carries no superimposed load.

Furrowing The practice of striking a V-shaped trough in a bed of mortar. (This is not accepted as good practice in masonry construction.)

G

Gauged Brick (1) Bricks that have been ground or otherwise produced to accurate dimensions. (2) A tapered arch brick.

Glazed Brick A brick prepared by fusing on the surface a ceramic glazing material; brick having a glassy surface.

Green Masonry A molded clay masonry unit before it has been fired in a kiln; an uncured concrete masonry unit.

Gross Cross-Sectional Area Area measured by overall dimensions, including voids or core area.

Grounds Nailing strips placed in masonry walls as a means of attaching trim or furring.

Grout A mixture of cementitious material and aggregate to which sufficient water is added to produce pouring consistency without segregation of the constituents. (See ASTM C476.)

High-Lift Grouting Technique of grouting masonry for the full height of the wall, in lifts up to 4 ft.

Low-Lift Grouting Technique of grouting as the wall is constructed.

Grout Bond The adhesion to, and/or the interlocking of, grout with the masonry units and with the reinforcement.

Grouted Masonry Hollow-unit masonry construction in which the empty cores are filled solidly with grout. Also, double-wythe wall construction in which the cavity or collar joint is filled solidly with grout.

H

Hard-Burned Brick Nearly vitrified clay products which have been fired at high temperatures; characterized by relatively low absorption rates and high compressive strengths.

Harsh Mortar A mortar that is difficult to spread; not workable.

H-Block Hollow unit with both ends open, and sometimes a continuous bond beam recess at the intersecting web.

Head Joint The vertical mortar joint between ends of masonry units. Sometimes called cross joint.

Header A masonry unit that overlaps two or more adjacent wythes of masonry to tie them together. Often called **bonder**.

Blind Header A concealed brick header in the interior of a wall, not showing on the faces.

Clipped Header A brick bat placed to look like a header for purposes of establishing a pattern. Also called a false header.

Flare Header A header of darker color than the field of the wall.

Header Block Concrete masonry units which have a portion of one side of the height removed to facilitate bonding with adjacent masonry such as brick facing.

Header Course A continuous bonding course of header brick.

Header Tile Clay tile containing recesses for brick headers in brick faced wall.

Hearth The floor of a fireplace, and the portion of the floor immediately in front of the fireplace.

Height The dimension of a masonry unit measured at right angles to the thickness and length as typically used in a wall. The vertical dimension.

Herringbone Bond The arrangement of bricks in a course in a zigzag fashion, with the end of one brick laid at right angles against the side of an adjacent brick.

High-Bond Mortar Mortar that develops higher bond strengths with masonry units than those normally developed with conventional mortar.

High-Lift Grouting The technique of grouting masonry for the full height of a wall in lifts up to 4 ft.

High-Pressure Steam Curing Any process of curing in saturated steam under pressures usually ranging from 125 to 150 psi; also called autoclave curing. Widely used to produce concrete masonry units.

Hollow Block A concrete masonry unit whose net cross-sectional area in every plane parallel to the bearing surface is less than 75% of its gross cross-sectional area measured in the same plane.

Hollow Brick A masonry unit of clay or shale whose net cross-sectional area in any plane parallel to the bearing surface is 60 to 75% of its gross cross-sectional area measured in the same plane. (Covered by ASTM C652.)

Hollow Masonry Unit A unit whose net cross-sectional area in any plane parallel to the bearing surface is less than 75% of its gross cross-sectional area measured in the same plane.

Hollow Wall See **Cavity Wall.**

I

IMI International Masonry Institute.

Initial Rate of Absorption The weight of water absorbed (expressed in grams per minute) when a brick is partially immersed for 1 minute. Also called suction. (See ASTM C67.)

Intrados The interior curved line of an arch.

J

Jack Arch One having horizontal or nearly horizontal upper and lower surfaces. Also called flat or straight arch.

Jamb Block A block specially formed for jambs of windows or doors, generally with a vertical slot to receive window frames, etc.

Joint Reinforcement Steel bar, wire, or prefabricated reinforcement (ladder or truss type) which is placed in mortar bed joints.

Jointer In masonry, a tool used by masons to form the various types of mortar joints between the courses of masonry.

Joist Anchors Anchors, usually steel straps, used to secure wood joists to masonry walls.

Jumbo Brick A generic term indicating a brick larger in size than the "standard." Some producers use this term to describe oversize brick of specific dimensions which they manufacture.

K

Keystone The wedge-shaped piece at the top of an arch which binds, or locks, all the other members in place.

Kiln A furnace, oven, or heated enclosure used for burning or firing brick or other clay material.

Kiln Run Brick or structural clay tile from one kiln which have not yet been sorted or graded for size or color variation.

King Closer A closer used to fill an opening in a course larger than half a brick; usually about three-fourths the size of a regular brick.

L

Ladder Reinforcement Prefabricated steel reinforcement designed for embedment in the horizontal mortar joints of masonry. Parallel side wires are connected in a single plane by perpendicular cross wires, thus forming a ladder design.

Lap The distance one masonry unit or reinforcing bar extends beyond or over another.

Lateral Support Bracing of walls either vertically or horizontally by columns, pilasters, cross walls, beams, floors, roofs, etc.

Lead The end or corner section of a wall built up and racked back on successive courses. A line is attached to leads as a guide for constructing a wall between them.

Lean Mortar Mortar that is deficient in cementitious components. It is usually harsh and difficult to spread.

Length The dimension of a masonry unit measured between ends. Typically, the dimension that is parallel to the face or length of the wall.

Lime A critical ingredient of masonry mortars and grouts.

Hydrated Lime Quicklime to which sufficient water has been added to satisfy its chemical affinity and convert the oxides to hydroxides.

Lime Putty Hydrated lime in plastic form ready for addition to mortar.

Quicklime A hot, unslaked lime. A calcined material, a major part of which is calcium oxide (or calcium oxide in natural association with lesser amounts of magnesium oxide) capable of slaking with water.

Slaked Lime Formed when quicklime is treated with water; same as hydrated lime.

Lintel A beam or supporting member placed over an opening in a wall.

Lintel Block A masonry unit with one side open, usually placed with the open side up like a trough to form a continuous channel for reinforcing, as across a window or door opening.

Loadbearing Wall Any wall which, in addition to supporting its own weight, supports the building or structure above it.

M

M-Factor Standard modifier used to adjust heat loss calculations for masonry walls.

Major Arch Arch with span greater than 6 ft and equivalent uniform loads greater than 1000 lb/ft. Typically, a Tudor arch, semicircular arch, Gothic arch, or parabolic arch. Rise to span ratio greater than 0.15.

Mason A worker skilled in laying brick, tile, stone, or block.

Masonry Brick, structural clay tile, stone, concrete masonry units, terra cotta, etc., or combinations thereof, bonded with mortar.

Masonry Cement A proprietary, mill-mixed, cementitious material to which sand and water must be added for use in mortar. (See ASTM C91.)

Masonry Nail A hardened-steel nail with a knurled or fluted shank, used for fastening to masonry.

Masonry Unit Natural or manufactured building units of burned clay, concrete, stone, glass, or gypsum, which are used in construction by bonding together with a cementitious agent.

Hollow Masonry Unit One whose net cross-sectional area in any plane parallel to the bearing surface is less than 75% of its gross cross-sectional area.

Modular Masonry Unit One whose nominal dimensions are based on the 4 in. module.

Solid Masonry Unit One whose net cross-sectional area in every plane parallel to the bearing surface is 75% or more of the gross cross-sectional area.

Mason's Hammer A hammer with a heavy steel head, one face of which is shaped like a chisel for trimming brick or stone.

Mason's Level Similar to a carpenter's level, but longer.

Mechanical Bond Tying masonry units together with metal ties, reinforcing steel, or keys.

Minor Arch Arch with maximum span of 6 ft and loads not exceeding 1000 lb/ft. Typically, a jack arch, semicircular arch, segmental arch, or multicentered arch. Rise-to-span ratio less than or equal to 0.15.

Moist-Air Curing Curing with moist air at atmospheric pressure and at a temperature of about 70°F. Used for curing masonry units.

Mortar A plastic mixture of cementitious materials, fine aggregate, and water used to bind masonry and other structural units. (See ASTM C270.)

Fat Mortar Mortar containing a high percentage of cementitious components. It is a sticky mortar which adheres to a trowel.

Harsh Mortar A mortar that is difficult to spread; not workable.

High-Bond Mortar Mortar that develops higher bond strengths with masonry units than those normally developed with conventional mortar.

Lean Mortar Mortar that is deficient in cementitious components. It is usually harsh and difficult to spread.

Mortar Bond The adhesion of mortar to masonry units.

Multi-Wythe Wall A wall composed of two or more wythes or rows of masonry.

N

NBS National Bureau of Standards.

NCMA National Concrete Masonry Association.

Neat Cement In masonry, a pure cement undiluted by sand aggregate or admixtures.

Net Cross-Sectional Area The gross cross-sectional area of a unit minus the area of cores or cellular spaces.

NFPA National Fire Protection Association.

Nominal Dimension A dimension greater than a specified masonry unit's actual dimensions by the thickness of a mortar joint, but not more than $\frac{1}{2}$ in.

Noncombustible Material Any material that will neither ignite nor actively support combustion in air at a temperature of 1200°F when exposed to fire.

Non-Loadbearing Wall A wall that supports no vertical load other than its own weight.

Norman Brick A brick whose nominal dimensions are $4 \times 2\frac{2}{3} \times 12$ in.

O

Offset Block A unit not rectangular in plan and made as a corner block to keep the construction modular on turning the corner.

Ornamental Facing In masonry, a design formed by the laying of stone, brick, tile, or other masonry units so as to produce a decorative effect.

P

Panel Wall A non-loadbearing wall in skeleton frame construction built between columns and wholly supported at each story.

Parging The application of a coat of cement mortar to the back of the facing or the face of the backing in multi-wythe construction.

Partition An interior wall one story or less in height. The term is generally applied to non-loadbearing walls.

Partition Tile Clay tile designed for use in interior partitions.

Paving Block A solid, flat unit used for road and walkway paving.

Paving Brick Vitrified brick especially suitable for use in pavements where resistance to abrasion is important. (Covered by ASTM C902.)

PCA Portland Cement Association

Pendentive A triangular segment of vaulting used to effect a transition at the angles from a square or polygon base to a dome above.

Permeability Property of allowing passage of fluids.

Pier See **Pilaster**.

Pilaster A portion of a wall serving as an integral vertical column, and usually projecting from one or both wall faces. Sometimes called a pier.

Pilaster Block Concrete masonry units designed for use in construction of plain or reinforced concrete masonry pilasters and columns.

Pointing Troweling mortar into a joint after masonry units are laid.

Pointing Trowel A small hand instrument used by masons and bricklayers for pointing up joints.

Prefabricated Masonry Masonry construction fabricated in a location other than its final position in the structure. Also known as preassembled or panelized construction.

Prism A small assemblage of masonry units and mortar used for compressive tests to predict the strength of full-scale masonry members.

Q

Queen Closer A cut brick having a nominal 2 in. horizontal face dimension.

Quoin Projecting courses of brick or stone at the corners and angles of buildings as ornamental features.

R

Racking A method entailing stepping back successive courses of masonry.

Racking Test Laboratory test for shear strength of masonry wall panels measured as diagonal tension.

Raked Joint A type of joint used in brick masonry which has the mortar raked out to a specified depth while the mortar is still green.

Random Rubble Masonry wall built of unsquared or rudely squared stones irregular in size and shape.

R.B.M. Reinforced brick masonry.

Reglet A groove or channel in a mortar joint, or in a special masonry unit, to receive roof flashing or other material which is to be sealed in the masonry.

Reinforced Masonry Masonry units and reinforcing steel bonded with mortar and/or grout in such a manner that the components act together in resisting forces.

Relieving Arch One built over a lintel, flat arch, or smaller arch to divert loads thus relieving the lower member from excessive loading.

Repointing See **Tuckpointing**.

Retempering Moistening and remixing of mortar to a proper consistency for use. Must be done with first $2\frac{1}{2}$ hours after initial mixing.

Reveal Portion of a jamb or recess which is visible from the face of a wall back to the frame.

Rib A plain or molded arch member which forms a support for an arch or vault. Also, a decorative projecting molding on the interior of a vault which is meant to resemble a structural rib.

Roman Brick Brick whose nominal dimensions are $4 \times 2 \times 12$ in.

Rowlock A brick laid on its face edge so that the end is visible in the wall face.

Rubble Rough, broken stones or bricks used to fill in cores of cavity walls or columns. Also rough, broken stone direct from the quarry.

Running Bond Lapping of units in successive courses so that the vertical head joints lap. Placing vertical mortar joints centered over the unit below is called half bond, while lapping $\frac{1}{3}$ to $\frac{1}{4}$ is called third bond or quarter bond.

Rustication In building and masonry, the use of squared or hewn stone blocks with roughened surfaces and edges deeply beveled or grooved to make the joints conspicuous.

S

Salmon Brick Generic term for relatively soft underburned brick which are more porous, slightly larger, and lighter colored than hard-burned brick. Usually, pinkish orange (salmon) in color.

Salt Glaze A gloss finish obtained by thermochemical reaction between silicates of clay and vapors of salt or chemicals.

Sand-Struck Brick See **Soft-Mud Brick**.

Sash Block Concrete masonry unit which has an end slot for use in openings to receive jambs of doors or windows.

Saturation Coefficient The ratio of the weight of water absorbed by a masonry unit during immersion in cold water to weight absorbed during immersion in boiling water. An indication of the probable resistance of brick to freezing and thawing.

Scored Block Block with grooved patterns, as, for example, to simulate raked joints.

SCPI Structural Clay Products Institute, now called Brick Institute of America (**BIA**).

SCR Trademark of the Brick Institute of America.

SCR Acoustile A side-construction, two-celled facing tile, with a perforated face backed with glass wool for acoustical purposes.

SCR Brick Brick whose nominal dimensions are $6 \times 2\frac{2}{3} \times 12$ in.

SCR Masonry Process A process providing greater efficiency, better workmanship, and increased production in masonry construction using story poles, marked lines, and adjustable scaffolding.

Screen Block Concrete masonry units for use in masonry screen walls.

Screen Tile Clay tile manufactured for masonry screen wall construction.

Set A change from a plastic to a hardened state.

Sewer Brick Low absorption, abrasion-resistant brick intended for use in drainage structures. (Covered by ASTM C32.)

Shadow Block Block with face formed in planes to develop surface patterns.

Shale Clay that has been geologically subjected to high pressures until it has hardened.

Shear Wall A wall that resists horizontal forces applied in the plane of the wall.

Shoring Temporary bracing for support.

Shoved Joints Vertical joints filled by shoving a mortared brick against the adjacent unit when it is being laid in a bed of mortar.

Side-Construction Tile Structural clay tile intended for placement with the axis of the cells horizontal.

Sill Block A solid concrete unit used for sills and openings.

Single-Wythe Wall A wall containing only one masonry unit in wall thickness.

Skewback The inclined surface on which an arch joins the supporting wall.

Slenderness Ratio Ratio of the effective height of a member to its effective thickness.

Slump Block Concrete masonry units produced in such a way that they slump or sag in irregular form before they harden.

Slurry A thin, watery mixture of neat cement, or cement and sand.

Slurry Coat A brushed application of slurry generally applied to the back of adhered ceramic veneer units and to the backing.

Slushed Joints Vertical joints filled after the units are laid by "throwing" mortar in with the edge of a trowel. (This method is not accepted as good practice in masonry construction.)

Smoke Chamber The space in a fireplace immediately above the throat where the smoke gathers before passing into the flue; narrowed by corbelling to the size of the flue lining above.

Soap A masonry unit of normal face dimensions, but having only nominal 2 in. thickness.

Soffit The underside of a beam, lintel, or arch.

Soft-Burned Brick Bricks that have been fired at low temperature ranges, producing relatively high absorptions and low compressive strengths.

Soft-Mud Brick Brick produced by molding relatively wet clay (20 to 30% moisture). Often a hand process. When insides of molds are sanded to prevent sticking of clay, the product is sand-struck brick. When molds are wetted to prevent sticking, the product is water-struck brick.

Solar Screen A perforated wall used as a sunshade.

Soldier A stretcher set on end with face showing on the wall surface.

Solid Block A concrete masonry unit whose net cross-sectional area in every plane parallel to the bearing surface is 75% or more of its gross cross-sectional area measured in the same plane.

Solid Masonry Unit A unit whose net cross-sectional area in every plane parallel to the bearing surface is 75% or more of its gross cross-sectional area measured in the same plane.

Solid Masonry Wall A wall built of solid masonry units or solidly grouted hollow units, laid contiguously with joints between units completely filled with mortar or grout.

Spall A small fragment removed from the face of a masonry unit by a blow or by action of the elements.

Split-Face Block Concrete masonry units with one or more faces having a fractured surface for use in masonry wall construction.

Spring Line For minor arches the line where the skewback cuts the soffit. For major parabolic arches the term commonly refers to the intersection of the arch axis with the skewback.

Squinch An arch or corbelling at the upper interior corners of a square tower for support of a circular or octagonal superstructure.

Stack Bond A bonding pattern where no unit overlaps either the one above or below; all head joints form a continuous vertical line.

Stiff-Mud Brick Brick produced by extruding a stiff but plastic clay (12 to 15% moisture) through a die.

Story Pole A marked pole for masonry coursing during construction.

Stretcher A masonry unit laid with its greatest dimension horizontal and its face parallel to the wall face.

String Course A horizontal band of projecting brick or stone for decorative effect in a masonry wall.

Struck Joint A mortar joint that is recessed further at the bottom than at the top. Not suitable for exterior construction because it permits moisture to penetrate the wall.

Structural Clay Tile Hollow masonry building units composed of burned clay, shale, fire clay, or combinations of these materials.

End-Construction Tile Tile designed to be laid with the axis of its cells vertical.

Facing Tile Tile for exterior and interior masonry with exposed faces.

Fireproofing Tile Tile designed for protecting steel structural members from fire.

Furring Tile Tile designed for lining the inside of exterior walls and carrying no superimposed loads.

Header Tile Tile with recesses for brick headers in masonry faced walls.

Loadbearing Tile Tile used in masonry walls carrying superimposed structural loads.

Non-Loadbearing Tile Tile designed for use in masonry walls carrying no superimposed loads.

Partition Tile Tile designed for use in interior partitions.

Screen Tile Tile manufactured for masonry screen wall construction.

Side-Construction Tile Tile intended for placement with the axis of the cells horizontal.

Suction See **Initial Rate of Absorption**.

T

Temper To moisten and mix clay, plaster, or mortar to a proper consistency.

Terra Cotta In construction, a fired clay product used for ornamental work on the exterior of buidings. Now commonly called **ceramic veneer**.

Thermal Inertia The ability of masonry to store heat as a result of its mass, density, and specific heat. Also called **Capacity Insulation**.

Thickness The dimension of a masonry unit at right angles to the face of the wall, floor, or other assembly in which the units are used.

Throat The part of a chimney between the firebox and the smoke chamber through which smoke rises from the combustion area.

Tie Any unit of material that connects the wythes of a multi-wythe wall.

Tooling Compressing and shaping the face of a mortar joint with a special tool other than a trowel.

Toothing Constructing the temporary end of a wall with the end stretcher of every alternate course projecting.

Tuckpointing The filling in with fresh mortar of cut-out or defective mortar joints in masonry.

U

UBC Uniform Building Code.

U-Block See **Lintel Block**.

V

Veneer A single wythe of masonry for facing purposes, not structurally bonded to the backup wall or frame.

Veneered Wall A wall having a facing of masonry units or other weather-resisting, noncombustible materials securely attached to the backing, but not bonded so as to exert a common reaction under load.

Virtual Eccentricity The eccentricity of a resultant axial load required to produce axial and bending stresses equivalent to those produced by applied axial loads and moments. It is normally found by dividing the moment at a section by the summation of axial loads occurring at the section.

Vitrification The condition resulting when kiln temperatures are sufficient to melt the silicates in the body of a clay masonry unit, fusing the grains together and closing the pores.

Voussoir One of the truncated, wedge-shaped masonry units which forms an arch ring.

W

Wall A vertical member of a structure whose horizontal dimension measured at right angles to the thickness exceeds three times its thickness.

Wall Tie A bonder or metal piece that connects wythes of masonry to each other.

Water Retentivity That property of a mortar which prevents the rapid loss of water to masonry units of high suction. It prevents bleeding or water gain when mortar is in contact with relatively impervious units.

Water Table A projection of lower masonry on the outside of a wall slightly above the ground. Often, a damp course is placed at this level to prevent upward penetration of groundwater.

Weep Holes Openings placed in mortar joints of facing material at the level of the flashing to permit the escape of moisture.

Wire-Cut Brick Brick formed by forcing plastic clay through a rectangular opening designed to shape the clay into bars. Before burning, wires pressed through the plastic mass cut the bars into brick units.

Wythe (1) Each continuous vertical section of masonry one unit in thickness. (2) The thickness of masonry separating flues in a chimney.

appendix B

NOTATIONS AND SYMBOLS

Notations and symbols may vary slightly among the several building codes and standards referenced. To avoid confusion, the following standardized list is used throughout this text (except for arches). Notations and symbols are defined as follows:

A_b = required bearing area
A_g = gross area of masonry section
A_n = net area of masonry section
A_s = total area of longitudinal tensile reinforcement
A_s' = area of compressive reinforcement in a flexural member
A_v = area of web reinforcement
b = width of rectangular beam, width of flange for T and I sections, or total width of reinforced masonry column
b' = width of web in T and I members
C_e = eccentricity coefficient
C_s = slenderness coefficient
d = effective depth of flexural member
e = virtual eccentricity
f_a = computed axial compressive stress due to design axial load
f_b = computed flexural stress in the extreme fiber due to design bending loads only

f_s = computed stress in reinforcement due to design loads
f_m = allowable axial compressive stress
f'_m = ultimate compressive strength of masonry
f_v = tensile stress in web reinforcement
F_a = allowable axial compressive stress if members were carrying centroidally applied axial load only
F_b = allowable flexural compressive stress due to bending only
F_s = allowable stress in reinforcement
F_t = allowable flexural tensile stress in masonry
h = effective height of a wall or column
I = moment of inertia about the neutral axis of the cross-sectional area
j = ratio of distance between centroid of compression and centroid of tension to depth (d)
k = ratio of distance between extreme compressive fiber and neutral axis to effective depth (d)
L = length of the wall or segment
M = design moment of a section
n = ratio of modulus of elasticity of steel to that of masonry
p = ratio of the area of flexural tensile reinforcement (A_s) to the area (bd)
p_g = ratio of vertical reinforcement to gross area
p_v = ratio of lateral reinforcement to gross area
P = allowable vertical load
P_a = allowable centroidal axial load for reinforced masonry columns
R = reaction
s = spacing of stirrups or of bent bars in a direction parallel to that of the main reinforcement
S = section modulus
t = effective thickness of a wythe, wall, or column
u = bond stress per unit of surface area of bar
v = unit shearing stress
v_m = computed shear stress due to design load
V = total design shear force
V' = excess of total shear over that permitted in masonry
V_m = allowable shear stress in masonry not including shear reinforcement, if any
w = total uniform load
Σ_o = sum of the perimeters of all the longitudinal reinforcement

appendix C

ASTM REFERENCE STANDARDS

The following reference standards of the American Society for Testing and Materials are related to masonry products, testing, and construction.

Clay Masonry Units

ASTM C27, Fire Clay and High Alumina Refractory Brick
ASTM C32, Sewer Brick
ASTM C34, Structural Clay Loadbearing Wall Tile
ASTM C43, Terms Relating to Structural Clay Products
ASTM C56, Structural Clay Non-Loadbearing Tile
ASTM C62, Building Brick (solid units)
ASTM C106, Fire Brick Flue Lining for Refractories and Incinerators
ASTM C126, Ceramic Glazed Structural Clay Facing Tile, Facing Brick, and Solid Masonry Units
ASTM C155, Insulating Fire Brick
ASTM C212, Structural Clay Facing Tile
ASTM C216, Facing Brick (solid units)
ASTM C279, Chemical Resistant Brick

ASTM C315, Clay Flue Lining
ASTM C410, Industrial Floor Brick
ASTM C416, Silica Refractory Brick
ASTM C530, Structural Clay Non-Loadbearing Screen Tile
ASTM C652, Hollow Brick
ASTM C902, Pedestrian and Light Traffic Paving Brick

Cementitious Masonry Units

ASTM C52, Gypsum Partition Tile or Block
ASTM C55, Concrete Building Brick
ASTM C73, Calcium Silicate Face Brick (sand–lime brick)
ASTM C90, Hollow Loadbearing Concrete Masonry Units
ASTM C129, Hollow Non-Loadbearing Concrete Masonry Units
ASTM C139, Concrete Masonry Units for Construction of Catch Basins and Manholes
ASTM C145, Solid Loadbearing Concrete Masonry Units
ASTM C744, Prefaced Concrete and Calcium Silicate Masonry Units

Natural Stone

ASTM C119, Terms Relating to Building Stone
ASTM C503, Exterior Marble
ASTM C568, Dimension Limestone
ASTM C615, Structural Granite
ASTM C616, Building Sandstone
ASTM C629, Structural Slate

Mortar and Grout

ASTM C5, Quicklime for Structural Purposes
ASTM C33, Aggregates for Concrete
ASTM C91, Masonry Cement
ASTM C105, Ground Fire Clay as a Mortar for Laying up Fire Brick
ASTM C144, Aggregate for Masonry Mortar
ASTM C150, Portland Cement
ASTM C207, Hydrated Lime for Masonry Purposes
ASTM C270, Mortar for Unit Masonry
ASTM C330, Lightweight Aggregates for Structural Concrete
ASTM C331, Lightweight Aggregates for Concrete Masonry Units
ASTM C404, Aggregate for Masonry Grout
ASTM C476, Grout for Reinforced and Nonreinforced Masonry
ASTM C887, Packaged, Dry, Combined Materials for Surface Bonding Mortar

Reinforcement and Accessories

ASTM A82, Cold Drawn Steel Wire for Concrete Reinforcement

ASTM A116, Zinc-Coated Iron or Steel (Class 3)

ASTM A153, Zinc Coating on Iron or Steel Hardware, Class B-1, B-2, or B-3 (hot dip)

ASTM A165, Electro-deposited Coatings of Cadmium on Steel

ASTM A185, Welded Steel Wire Fabric for Concrete Reinforcement

ASTM A496, Deformed Steel Wire for Concrete Reinforcement

ASTM A615, Deformed and Plain Billet-Steel Bars for Concrete Reinforcement

ASTM A616, Rail-Steel Deformed and Plain Bars for Concrete Reinforcement

ASTM A617, Axle-Steel Deformed and Plain Bars for Concrete Reinforcement

ASTM B227, Hard-Drawn Copper-Covered Steel Wire, Grade 30HS

Sampling and Testing

ASTM C67, Sampling and Testing Brick and Structural Clay Tile

ASTM C109, Method of Test for Compressive Strength of Hydraulic Cement Mortars

ASTM C136, Test for Sieve or Screen Analysis of Fine and Coarse Aggregate

ASTM C140, Sampling and Testing Concrete Masonry Units

ASTM C241, Method of Test for Abrasion Resistance of Stone

ASTM C780, Pre-Construction and Construction Evaluation of Mortars for Plain and Reinforced Unit Masonry

ASTM C952, Standard Method of Test for Bond Strength of Mortar to Masonry Units

ASTM D75, Sampling Aggregates

ASTM E72, Conducting Strength Tests of Panels for Building Construction

ASTM E447, Test for Compressive Strength of Masonry Prisms

ASTM E488, Strength of Anchors in Concrete and Masonry Elements

ASTM E518, Test for Flexural Bond Strength of Masonry

ASTM E519, Test for Diagonal Tension in Masonry Assemblages (shear)

Assemblages

ASTM C901, Prefabricated Masonry Panels

ASTM C946, Construction of Dry Stacked, Surface Bonded Walls

appendix D

MASONRY ORGANIZATIONS

Technical guidance, design assistance, cost studies, and general information about masonry materials and construction may be obtained through the following national and regional organizations.

National

BRICK INSTITUTE OF AMERICA
1750 Old Meadow Road
McLean, Virginia 22102

INTERNATIONAL MASONRY INSTITUTE
823 Fifteenth Street, N.W.
Washington, D.C. 20005

NATIONAL CONCRETE MASONRY ASSOCIATION
P.O. Box 781
Herndon, Virginia 22070

PORTLAND CEMENT ASSOCIATION
5420 Old Orchard Road
Skokie, Illinois 60076

Regional

Masonry Industry Program of Arizona
2046 North 16th Street
Phoenix, ARIZONA 85006

Arkansas Masonry Development Trust
2918 West Roosevelt Road
Little Rock, ARKANSAS 72204

Masonry Institute of Kern County
1510 East 19th Street
Bakersfield, CALIFORNIA 93305

Masonry Advisory and Technical Institute
22300 Foothill Blvd.
Hayward, CALIFORNIA 94541

Masonry Institute of America
2550 Beverly Blvd.
Los Angeles, CALIFORNIA 90057

Masonry Trade Promotion of Orange County
4050 Metropolitan Drive, Suite 300
Orange, CALIFORNIA 92668

Masonry Institute of the Inland Empire
164 West Hospitality Lane, Suite 3
San Bernardino, CALIFORNIA 92410

Unit Masonry Association of San Diego
6150 Mission Gorge Road, Suite 115
San Diego, CALIFORNIA 92120

Colorado Masonry Institute
3003 East Third Ave., Suite 301
Denver, COLORADO 80206

Masonry Institute of Connecticut
6 Lunar Drive
Woodbridge, CONNECTICUT 06525

The Trowel Guild
2200 North Florida Mango Road
West Palm Beach, FLORIDA 33409

Masonry Institute of America
59 Chaumont Square, N.W.
Atlanta, GEORGIA 30327

Masonry Institute of Hawaii
905 Umi Street, Suite 206
Honolulu, HAWAII 96819

Masonry Advancement Council of Dupage County
P.O. Box 129
Mount Prospect, ILLINOIS 60056

Illinois Masonry Institute
1550 Northwest Highway, Suite 201
Park Ridge, ILLINOIS 60068

Masonry Institute of Indiana
8123 Castleton Road, Suite B
Indianapolis, INDIANA 46250

Masonry Institute of Iowa
817 S. W. 9th Street, Suite A
Des Moines, IOWA 50309

Kentuckiana Masonry Institute, Inc.
3418 Frankfort Ave., Suite 210
Louisville, KENTUCKY 40207

Masonry Institute of Maryland
2313 St. Paul Street
Baltimore, MARYLAND 21218

Masonry Institute, Inc.
4853 Cordell Ave., Penthouse 1
Bethesda, MARYLAND 20814

Masonry Institute of Massachusetts and New Hampshire
550 Medford Street
Charlestown, MASSACHUSETTS 02129

Masonry Institute of Michigan
24155 Drake Road, Suite 202
Farmington, MICHIGAN 28204

Minnesota Masonry Institute
7851 Metro Parkway, Suite 103
Minneapolis, MINNESOTA 55420

Masonry Institute of St. Louis
1429 S. Big Bend Blvd.
St. Louis, MISSOURI 63117

Nebraska Masonry Institute
11414 W. Center Road, Suite 211
Omaha, NEBRASKA 68144

Masonry Institute of New York City and Long Island
445 Northern Blvd.
Great Neck, NEW YORK 11021

Empire State Masonry Institute
P.O. Box 396
Syracuse, NEW YORK 13206-0396

Unit Masonry Association of Greater Cincinnati
8075 Reading Road, Suite 306
Cincinnati, OHIO 45237

N. E. Ohio Masonry Institute
1737 Euclid Ave., Suite 210
Cleveland, OHIO 44115

Masonry Institute of Dayton
2002 Richard Street
Dayton, OHIO 45403

Oklahoma Masonry Institute
3601 Classen Blvd., Suite 108
Oklahoma City, OKLAHOMA 73118

Masonry Institute of Oregon
3609 S. W. Corbett, Suite 4
Portland, OREGON 97201

Delaware Valley Masonry Institute
134 North Narberth Ave.
Narberth, PENNSYLVANIA 19072

Masonry Institute of Western Pennsylvania
2270 Novelstown Road
Pittsburgh, PENNSYLVANIA 15205

Masonry Institute of Memphis
5575 Poplar, 422
Memphis, TENNESSEE 38114

Masonry Institute of Houston-Galveston
5100 Westheimer, Suite 200
Houston, TEXAS 77056

Masonry Institute of Washington
925 116th Street N.E., Suite 209
Bellevue, WASHINGTON 98004

Spokane Masonry Research and Promotion
8614 East Whitman
Spokane, WASHINGTON 99206

Masonry Institute of Wisconsin
4300 W. Brown Deer Road
Milwaukee, WISCONSIN 53223

BIBLIOGRAPHY

Primary References

AMRHEIN, JAMES E., ET AL. *Masonry Design Manual.* Los Angeles: Masonry Industry Advancement Committee, 1979.

BRICK INSTITUTE OF AMERICA. *Technical Notes on Brick Construction,* Nos. 1–43. McLean, Va.: BIA.

NATIONAL CONCRETE MASONRY ASSOCIATION. *TEK Bulletins,* Nos. 1–97. Herndon, Va.: NCMA.

PLUMMER, HARRY C. *Brick and Tile Engineering.* McLean, Va.: Brick Institute of America, 1962.

SCHNEIDER. ROBERT R., AND WALTER L. DICKEY. *Reinforced Masonry Design.* Englewood Cliffs, N.J.: Prentice-Hall, Inc., 1980

Secondary References

AMERICAN CONCRETE INSTITUTE. *Building Code Requirements for Concrete Masonry Structures,* ACI 531–79. Detroit, Mich.: ACI, 1979.

AMERICAN INSURANCE ASSOCIATION. *National Building Code.* New York: AIA, 1976.

AMERICAN NATIONAL STANDARDS INSTITUTE. *American Standard Building Code Requirements for Masonry,* A41.1. New York: ANSI, 1970.

AMERICAN NATIONAL STANDARDS INSTITUTE. *Building Code Requirements for Reinforced Masonry*, A41.2. New York: ANSI, 1970.

AMERICAN SOCIETY OF HEATING, REFRIGERATING, AND AIR-CONDITIONING ENGINEERS, INC. *ASHRAE Standard 90-75*. New York: ASHRAE, 1975.

AMRHEIN, JAMES E. *Steel in Masonry*. Los Angeles: Masonry Institute of America, 1977.

ARUMI, FRANCISCO N. *Thermal Inertia in Architectural Walls*. Herndon, Va.: National Concrete Masonry Association, 1977.

BALCOMB, J. DOUGLAS, ET AL. *Passive Solar Design Handbook*. Sacramento, Calif.: Solar Energy Information Services, 1981.

BELL, JOSEPTH, ET AL. *From the Carriage Age . . . to the Space Age. . . .* Herdon, Va.: National Concrete Masonry Association, 1970.

BOUDREAU, EUGENE H. *Making the Adobe Brick*. Berkeley, Calif.: Fifth Street Press, 1971.

BRICK INSTITUTE OF AMERICA. *Principles of Clay Masonry Construction*. McLean, Va.: BIA, 1973.

BRICK INSTITUTE OF AMERICA. *Recommended Practice for Engineered Brick Masonry*. McLean, Va.: BIA, 1969.

BUILDING OFFICIALS AND CODE ADMINISTRATORS INTERNATIONAL, INC. *Basic Building Code*. Homewood, Ill.: BOCA, 1978.

CATANI, MARIO J. AND STANLEY E. GOODWIN. "Heavy Building Envelopes and Dynamic Thermal Response." *Journal of the American Concrete Institute*, February, 1976.

COX, WARREN J. *Brick Architectural Details*. McLean, Va.: Brick Institute of America, 1973.

DALZELL, J. RALPH. *Simplified Concrete Masonry Planning and Building*. New York: McGraw-Hill Book Company, 1972.

DICKEY, WALTER L., AND R. W. HARRINGTON. *The Shear Truth About Brick Walls*. San Francisco: Western States Clay Products Association, Inc., 1970.

DUNCAN, S. BLACKWELL. *The Complete Book of Outdoor Masonry*. Blue Ridge Summit, Pa.: TAB Books, Inc., 1978.

EGAN, M. DAVID. *Concepts in Building Fire Safety*. New York: John Wiley & Sons, Inc., 1978.

ELMINGER, A. *Architectural and Engineering Concrete Masonry Details for Building Construction*. Herndon, Va.: National Concrete Masonry Association, 1976.

GOODWIN, STANLEY E. AND MARIO J. CATANI. *Simplified Thermal Design of Building Envelopes for Use with ASHRAE Standard 90-75*. Skokie, Ill.: Portland Cement Association, 1976.

GRIMM, CLAYFORD T. ET AL. *Relative Thermal and Economic Performance of Masonry and Glass Building Enclosures*. Austin, Tex.: Texas Department of Community Affairs, 1975.

HENDRY, ARNOLD W. *Structural Brickwork*. New York: John Wiley & Sons, Inc.

HEYMAN, JACQUES. *The Masonry Arch*. New York: John Wiley & Sons, Inc., 1982.

HUFF, DARREL. *How to Work with Concrete and Masonry*. New York: Popular Science Publishing Co., 1968.

INDIANA LIMESTONE INSTITUTE OF AMERICA, INC. *Indiana Limestone Handbook*. Bedford, Ind.: ILI, 1977.

INTERNATIONAL CONFERENCE OF BUILDING OFFICIALS. *Uniform Building Code*. Whittier, Calif., ICBO, 1982.

INTERNATIONAL MASONRY INDUSTRY ALL-WEATHER COUNCIL. *Recommended Practices and Guide Specifications for Cold Weather Masonry Construction*. Washington, D.C.: International Masonry Institute, 1977.

KREH, R.T., SR. *Masonry Skills*. New York: Van Nostrand Reinhold, Company, 1976.

LEBA, THEODORE, JR. *Design Manual—The Application of Non-reinforced Concrete Masonry Load-Bearing Walls in Multistory Structures*. Herndon, Va.: National Concrete Masonry Association, 1969.

LEONTOVICH, VALERIAN. *Frames and Arches*. New York: McGraw-Hill Book Company, 1959.

LOS ALAMOS SCIENTIFIC LABORATORY, ET AL. *Passive Solar Design Handbook*. Washington, D.C.: U.S. Department of Energy, 1980.

MACKINTOSH, ALBYN. *Design Manual—The Application of Reinforced Concrete Masonry Load-Bearing Walls in Multi-storied Structures*. Herndon, Va.: National Concrete Masonry Association, 1973.

MASONRY INSTITUTE OF AMERICA. *Masonry Veneer*. Los Angeles: MIA, 1974.

MASONRY INSTITUTE OF AMERICA. *Reinforcing Steel in Masonry*. Los Angeles: MIA.

THE MASONRY SOCIETY. *Standard Building Code Requirements for Masonry Construction*. Denver, Colo.: TMS, 1980.

MAZRIA, EDWARD. *The Passive Solar Energy Book*. Emmaus, Pa.: Rodale Press, Inc., 1979.

MCKEE, HARLEY J. *Introduction to Early American Masonry—Stone, Brick, Mortar, and Plaster*. National Trust for Historic Preservation and Columbia University. Washington, D.C.: The Preservation Press, 1973.

NATIONAL CONCRETE MASONRY ASSOCIATION. *Specification for the Design and Construction of Load-Bearing Concrete Masonry*. Herndon, Va.: NCMA, 1970.

NATIONAL FIRE PROTECTION ASSOCIATION. *Fire Protection Handbook*, 14th ed. Cambridge, Mass.: Riverside Press, 1964.

NATIONAL FIRE PROTECTION ASSOCIATION. *Life Safety Code, NFPA No. 101*. Boston: NFPA, 1979.

NATIONAL LIME ASSOCIATION. *Masonry Mortar Technical Notes Series*. Washington, D.C. NLA.

O'BRIEN, JAMES J. *Construction Inspection Handbook*. New York: Van Nostrand Reinhold Company, 1974.

O'NEILL, HUGH. *Stone for Building*. London: William Heinemann Ltd., 1965.

OLGYAY, ALADAR AND VICTOR. *Solar Control and Shading Devices*. Princeton, N.J.: Princeton University Press, 1957.

OLGYAY, VICTOR. *Design with Climate*. Princeton, N.J.: Princeton University Press, 1963.

PRZETAK, LOUIS. *Standard Details for Fire-Resistive Building Construction*. New York: McGraw-Hill Book Company, 1977.

RANDALL, FRANK A. AND WILLIAM C. PANARESE. *Concrete Masonry Handbook for Architects, Engineers and Builders*. Skokie, Ill.: Portland Cement Association, 1976.

SAHLIN, SVEN. *Structural Masonry*. Englewood Cliffs, N.J. Prentice-Hall, Inc., 1970.

SOUTHERN BUILDING CODE CONGRESS. *Standard Building Code*. Birmingham, Ala.: SBCC, 1982.

SZOKOLAY, S.V. *Solar Energy and Building*. London: The Architectural Press Ltd., 1975.

THE UNDERGROUND SPACE CENTER, UNIVERSITY OF MINNESOTA. *Earth Sheltered Housing Design*. New York: Van Nostrand Reinhold Company, 1978.

U.S. DEPARTMENT OF COMMERCE, NATIONAL BUREAU OF STANDARDS. *Fire-Resistance Classifications of Building Construction, Report BMS 92*. Washington, D.C.: U.S. Government Printing Office, 1942.

WATSON, DONALD, ED. *Energy Conservation through Building Design*. New York: McGraw-Hill Book Company, 1979.

Related Material

ALLEN, EDWARD. *Stone Shelters*. Cambridge, Mass.: The MIT Press, 1969.

CALLENDER, JOHN HANCOCK, ED. *Time-Saver Standards for Architectural Design*. New York: McGraw-Hill Book Company, 1974.

COLLINS GEORGE R. *Antonio Gaudi*. New York: George Braziller, Inc., 1960.

CONDIT, CARL W. *American Building*. Chicago: University of Chicago Press, 1968.

CONDIT, CARL W. *The Rise of the Skyscraper*. Chicago: University of Chicago Press, 1952.

DAGOSTINO, FRANK R. *Methods and Materials of Commercial Construction*. Reston, Va.: Reston Publishing Co., Inc., 1974.

DAVEY, NORMAN. *A History of Building Materials*. London: Phoenix House, 1961.

FEININGER, ANDREAS. *Man and Stone*. New York: Crown Publishers, Inc., 1961.

FLETCHER, GORDON A., AND VERNON A. SMOOTS. *Construction Guide for Soils and Foundations*. New York: John Wiley & Sons, Inc., 1974.

GLOAG, JOHN. *The Architectural Interpretation of History*. New York; St. Martin's Press, Inc., 1975.

GUEDES, PEDRO, ED. *Encyclopedia of Architectural Technology*. New York: McGraw-Hill Book Company, 1979.

MARS, G.C., ED. *Brickwork in Italy*. Chicago: American Face Brick Association, 1925.

MERRITT FREDERICK, S. *Building Construction Handbook*. New York: McGraw-Hill Book Company, 1965.

NEWMAN, MORTON. *Standard Structural Details for Building Construction*. New York: McGraw-Hill Book Company, 1968.

OLIN, HAROLD B., ET AL. *Construction Principles, Materials and Methods*. Chicago: U.S. League of Savings Association, 1975.

PARKER, HARRY S., ET AL. *Materials and Methods of Architectral Construction*. New York: John Wiley & Sons, Inc., 1958.

RAMSEY, CHARLES G., AND HAROLD S. SLEEPER. *Architectural Graphic Standards*, 6th ed. Ed. Joseph N. Boaz. New York: John Wiley & Sons, Inc., 1970.

SCHROEDER, W.L. *Soils in Construction*. New York: John Wiley & Sons, Inc., 1975.

SMITH, R. C. *Materials of Construction*. New York: McGraw-Hill Book Company, 1973.

WOODFORDE, JOHN. *Bricks to Build a House*. London: Routledge & Kegan Paul Ltd., 1976.

INDEX

A

Absorption:
brick, 31, 53-55
concrete masonry, 59, 64-65, 73, 74
Accessories:
materials, 105-6, 111, 113-16,
types, 106-15
Acoustical characteristics:
brick, 56-57, 162, 165
clay tile, 162, 164, 165
concrete masonry, 74, 75, 162, 165
sound absorption, 56-57, 75, 161, 162
sound transmission, 56-57, 75, 161, 162-63
STC ratings, 75, 161, 164-66
Actual dimensions:
brick, 35-38
concrete masonry, 66
Adhered veneer, 52, 100, 171, 175, 184
Admixtures:
concrete masonry, 18, 20
mortar, 29, 101, 104, 226
Adobe, 35
Aggregates:
concrete masonry, 18-20, 60-61, 125, 128
grout, 29, 103
mortar, 29
Air entrainment, 20, 26-27, 92
Allowable load formulas, 232-33, 238-40, 252-53, 257, 259, 261, 262, 263
Allowable stresses:
masonry:
reinforced, 248, 249-54
surface bonded, 359
unreinforced, 218-19, 220, 229, 235-36
steel, 249, 250, 253, 254
Aluminum accessories, 106, 116
Anchorage, 171-72
flexible, 56, 75, 123, 170, 176, 188-89, 351-57
rigid, 123, 169, 256, 259, 260, 353, 356-57
Anchored veneer, 52, 175-86
Arches, 297-315
construction, 2, 314-15
design, 298, 301-14
forms, 5, 299-300
graphic analysis, 302-3
major, 298, 314
minor, 298-314
parabolic, 298, 300
semicircular, 299, 306-14
terminology, 301
Arching action, 285

ASTM Standards, 473-75
  assemblages, 475
  brick, 30-33, 40, 41, 42, 43-44, 473-74
  clay tile, 45, 47, 49, 473-74
  concrete masonry, 20, 58-59, 60, 61, 71, 474
  mortar and grout, 95, 97, 101, 103, 474
  reinforcement, 475
  sampling and testing, 475
  stone, 474
Axial loads, 207, 210-11, 225, 232-34, 238-40

B

Basic Building Code, 125, 132, 171, 176, 190, 218, 245, 277
Beams, 277-83
  bond, 180, 196, 256, 258, 262, 263, 290, 291, 351, 352, 366
  deep wall, 198, 226, 283-84
  lintels, 175, 180, 284-97
Bearing plates, 210-11, 224
Bearing wall systems, theory of, 207-16, 225
Below-grade walls, 324-34
Bending moment, 207-8, 213, 244, 255, 258, 277, 278, 281, 287
Block (*see* Concrete masonry units)
Bond:
  mortar, 26, 27, 28, 92, 93, 94-95, 100, 180, 183, 208, 211-12, 241
  patterns, 358-60
  strength, 208, 211-12, 241
  stresses, 279-80
Bond beam:
  brick, 256, 262, 352, 366
  concrete masonry, 180, 196, 256, 258, 262, 290, 291, 352
Bonding methods, 106, 107, 120-21, 123, 222-23, 358
Box frame system, 208, 213, 243, 268, 269, 274
Brick, 30-44, 52-57
  adobe, 35
  ASTM Standards, 30-33, 40, 41, 42, 43-44, 473-74
  building (common), 30-32, 35
  calcium silicate, 60
  chemical-resistant, 42
  cleaning, 397, 407-10, 447-48
  color, 10, 13, 16, 18, 39, 41, 57
  concrete, 58-59, 66
  coursing, 36-38, 342-43, 347
  facing, 30, 32-33, 35
  firebrick, 35, 39
  glazed, 16, 35, 40
  grading, 30-32, 40, 43
  hollow, 43-44
  methods of manufacture, 10-18
  modular coordination, 35-38, 341-46
  pavers, 40-42
  position in wall, 38, 359
  properties:
    abrasion resistance, 10, 55-56
    absorption, 10, 31, 44, 53, 54-55, 92, 340-41
    acoustical characteristics, 56-57, 162-63
    compressive strength, 6, 31, 44, 52, 53
    coring, 13, 35-36, 43-44
    durability, 10, 30-32, 55-56
    expansion, 56, 167-68
    fire resistance, 39, 56, 125-27
    grading, 30-32, 40, 43, 53
    modulus of elasticity, 54
    modulus of rupture, 54
    moisture content, 54-55, 340-41, 445
    thermal characteristics, 56
    types, 32-33, 43
  refractory, 11, 39
  sand-lime, 60
  sculptured brick, 39
  sizes and shapes, 35-40
  solid, 35
  special purpose, 41-42
  terminology, 38-39, 359
  textures, 13, 14, 15, 41, 57
  types, 32-33
Buckling, 230, 244
Buttering, 447
Buttresses, 3, 171

C

Calcium chloride, 20, 104, 105, 106, 226
Calcium silicate brick, 60
Capacity insulation, 138
Caulking, 115, 347, 386-87
Cavity walls, 7-8, 122-23, 148-49, 166, 169, 174, 223, 363, 384
C/B ratios, 31, 53-55
Cement, 3, 18, 20, 26-27, 28
Ceramic veneer (terra cotta), 52
Chimneys:
  industrial, 204-6
  residential fireplace, 202-4
Clay, 10-11, 16-18
Clay tile, 45-47
  ASTM Standards, 45, 47, 49, 473-74
  coursing, 36, 348
  facing, 47-49, 57
  fireproofing, 45-47
  glazed, 49, 50-51
  loadbearing, 45-46
  modular coordination, 36, 49, 341-46, 348
  non-loadbearing, 46-47
  partitions, 173-74
  properties:
    absorption, 45, 47, 53, 54-55
    acoustical characteristics, 162, 164
    compressive strength, 45, 53
    expansion, 168

fire resistance, 45, 125, 129-30
grading, 45, 47, 53
screen tile, 49, 52
sizes and shapes, 46-48, 49, 50-52
Cleaning masonry, 397, 407-10, 447-48
Cleanouts, 367, 447
Codes and standards, 6, 218, 225
American Concrete Institute (ACI), 218, 225, 235, 251, 259-60
American National Standards Institute (ANSI), 171, 175, 210, 218-25
ANSI A41.1, 210, 218-25, 328-29
ANSI A41.2, 245, 248-49, 251-53, 255-56, 277-81
Basic Building Code, 125, 132, 171, 176, 190, 218, 245, 277
Brick Institute of America (BIA), 6, 171, 176, 211, 218, 225, 226, 229-33, 245, 254, 262-64, 277 282
National Building Code, 125, 132, 171, 176, 191, 218, 245, 277, 317
National Concrete Masonry Association (NCMA), 171, 210, 218, 225, 235-38, 242, 245, 250, 256-59, 277, 281-82
Standard Building Code 125, 129, 132, 171, 176, 191, 218, 245, 277
Uniform Building Code (UBC) 125, 129, 132, 171, 176, 190, 210, 211, 218, 225, 226, 245, 252-53, 260-62, 268, 272, 274-76, 277, 282-84, 317
Cold weather construction, 26, 400-401
Columns and pilasters, 171, 174, 244, 248, 251-53, 255, 261
brick, 232, 262-63, 347
concrete masonry, 236, 238, 240, 257-58, 259, 345
reinforcement, 244, 253, 255, 257, 259, 261, 263
Composite walls, 123, 222, 226
Compressive strength:
brick, 6, 31, 44, 52, 53, 228, 234, 261
CMU, 20, 23, 59, 64-65, 73-74, 228
grout, 103-4
masonry construction, 226, 227-29
mortar, 6, 94, 97, 98-99
Concentrated loads, 210-11, 224, 255, 259, 286, 298, 308, 310
Concrete masonry units:
admixtures, 18, 20
aggregates, 18-20, 58, 60-61, 73
absorption characteristics, 74, 92
acoustical characteristics, 19, 20, 61, 162-63
fire resistance, 19, 61, 63, 125, 128, 130
thermal characteristics, 19, 61, 168
types, 19, 61
weight, 18-19, 60-61
ASTM Standards, 20, 58-59, 60, 61, 71, 474
block, 60-76
equivalent solid thickness, 63, 75, 125, 128
hollow, 63, 74
loadbearing, 61, 63
non-loadbearing, 61
sizes and shapes, 66-67, 290, 365
solid, 61
brick, 58-59, 66
cast stone, 60
cellular concrete block, 60
cements, 20
color, 19, 20, 23, 25, 59, 76
coursing, 36, 342-43, 345
cracking, 23, 24, 65
custom designed, 21-22, 68-71
glazing, 24-25, 71
gypsum block, 60, 118, 129-30
lightweight, 24, 61, 66, 73, 75
materials, 18-21
methods of manufacture, 18, 20-26
modular coordination, 36, 66, 341-46
normal weight, 24, 61, 73, 75
pavers, 72-73
properties:
absorption, 59, 64-65, 73, 74
acoustical characteristics, 19, 20, 61, 74, 75, 162
compressive strength, 20, 23, 59, 64-65, 73-74
coring, 61-63, 74
durability, 64, 68, 74
expansion, 73, 74-75, 167-69
fire resistance, 19, 61, 75, 125, 128, 130
grading, 63
modulus of elasticity, 74
moisture content, 24, 58-59, 64-66, 340
shrinkage, 23-24, 64-66, 74-75, 95, 167-69
thermal characteristics, 19, 61, 74, 75, 149
sand-lime brick, 60
screen block, 66-67
textures, 18, 19-20, 21, 24-25, 66, 68-71
weight, 61, 66
Condensation:
analysis, 388-93
control, 123, 148, 333-34, 388, 390, 393-400
Conductivity, thermal, 138, 155, 158
Connections:
diaphragm, 208, 213, 259, 356
floor-to-wall, 208, 213-15, 224, 259, 260, 353-57
cast-in-place concrete, 214, 216, 354-57
precast concrete, 214, 216, 260, 356
steel, 214, 217, 354-56
wood frame, 215, 260
foundation, 179, 211, 356-57
wall intersections, 216, 223-24, 260, 343-46

Control joints, 24, 59, 75, 115, 167, 170, 186, 198, 346-47, 349, 350-51
Copings, 200-201, 337
Copper accessories, 105, 106, 109, 116
Corbelling, 2, 178-80
Corrosion, 105, 116
Cost factors, 134-36, 151, 283, 294, 413-27
Coursing:
 brick, 36-38, 342-43, 347
 clay tile, 36, 348
 concrete block, 36, 342-43, 345
Cracking, 23-24, 75, 95, 106, 166, 167, 170, 180, 226, 245, 272, 273, 284, 346-57
Curtainwalls, masonry, 175, 191-94

D

Damping, 243, 268
Deep wall beams, 198, 226, 283-84
Deflection:
 diaphragm, 212, 271-72, 273
 floor, 381
 lintel, 167, 175, 284, 288, 290
 shelf angle, 352
 wall, 176, 192, 226
Design methods:
 arches, 297-315
 beams, 277-83
 columns, 171, 174, 244, 248, 251-53, 255, 257-58, 259, 261-63
 empirical design, 8, 120, 217-25
 garden walls, 196-203
 lintels, 175, 180, 284-97
 pilasters (*see* columns)
 rational engineering, 8, 217-18, 225-76
 retaining walls, 316-24
 screen walls, 194-96
 serpentine walls, 198-200
 veneer:
  brick, 176-80, 181
  concrete masonry, 180-82
  stone, 182-86
 walls, 207-76
  partially reinforced, 8, 209-42
  reinforced, 8, 243-76
  unreinforced, 8, 208-42
Design theory, 207-16, 225, 243-48
Diagonal tension, 272, 273, 278-79, 282
Diaphragms, 207, 208, 212-13, 243, 268, 269-72, 273
Differential movement, 56, 106, 110, 115, 123, 166-70, 176, 186, 188-89, 194, 215-16, 290, 350-57
Dimension stone, 82, 84, 88
Domes, 2-3
Dry stack stone walls, 201-2
Ductility, 107, 243, 268
Dynamic thermal analysis, 138-60

E

Earthquake (*see* Seismic design)
Eccentricity, 207, 211, 225, 226, 230-31, 237-39, 255, 259, 262-63
Effective depth, 295, 297
Effective height, 232, 236
Effective thickness, 232, 236-37
Efflorescence:
 causes, 29, 96, 338, 399, 404-7
 cleaning, 404, 407-10
 prevention, 28, 96, 405, 406
Elastic deformation, 167, 169
Elasticity, 26, 54, 74, 96, 97
Empirical design, 8, 120, 171, 175-76, 217-25
 ANSI A41.1, 218-25
 limits, 8, 218-22, 225
Energy (*see* Thermal characteristics)
Engineered masonry, 8, 217-18, 225-76
Equivalent fluid pressure, 317
Equivalent solid thickness, 63, 75, 125
Expansion joints, 56, 115, 116, 169-70, 186-87, 189, 190, 318-19, 349-53, 377-78

F

Fasteners, 107, 113-15
FBA brick, 32-33
FBS brick, 32-33
FBX brick, 32-33
Fences (*see* Garden walls)
Fire clay, 10, 39, 100-101
Fireplaces, 202-6
Fire resistance, 124-34
 brick, 56, 100-101, 125-27
 building codes, 125, 129, 132
 clay tile, 45, 125, 129-30
 concrete masonry, 61, 75, 125, 128, 130
 equivalent solid thickness, CMU, 63, 75, 125, 128
 fire walls, 131-34
 ratings, 125-28
Flashing:
 materials, 115-16
 methods, 179, 181, 188, 384-85
Flexible anchorage, 56, 75, 110, 123, 188-89, 194, 215-16, 351-57
Flexural design, 277-97
Floor systems:
 brick, 381-83
 connections to, 208, 213-17, 224, 259, 260, 353-57
Foundations, 179, 196-99, 211, 226, 319-20, 332

G

Galvanized metals, 105-6, 109
Garden walls, 196-203
 brick, 198-201

concrete masonry, 196-98, 199-200
serpentine, 198-200
stone, 201-3
Gaudi, Antonio, 3, 5
Girders, 277-83
Glass block, 42-43, 118, 188-91
Glazed masonry units:
brick, 16, 35, 40
clay tile, 49, 50-51
concrete masonry, 24-25, 71
Grading:
brick, 30-32, 40, 43, 53
clay tile, 45, 47, 53
concrete masonry, 63
Granite, 77, 78, 84
Grout, 8, 26-29, 103-4
admixtures, 104
ASTM Standards, 103, 474
compressive strength, 103-4
curing, 103-4
ingredients, 26-29
mix designs, 103
placement, 103, 209, 244, 366-68, 447
space requirements, 103
Guide specifications (*see* Specifications)
Gypsum block, 60, 118, 129-30, 174-75

H

Headers, 38, 106, 107, 120-21, 123, 222-23, 359, 360
Heat-gain calculations, 134, 138-42
Heat-loss calculations, 134, 138, 142-43
Heat transmission, 134, 138-47, 153-60
Height-to-thickness ratios, 171, 217, 218-19, 222, 232, 238, 255, 258, 259, 261, 283
High-lift grouting, 244, 367-68
Hollow masonry units:
brick, 43-44, 118-21, 125
clay tile, 45-57, 118-21, 125
concrete block, 60-76, 118-21, 125, 128
h/t ratios, 171, 217, 218-19, 222, 232, 238, 255, 258, 259, 261, 283

I

Initial rate of absorption, 54-55, 74, 92-93, 94, 103, 104, 340-41, 445
Inspection, 95, 225, 229, 235, 444-48
Insulation, 134, 137
capacity insulation, 138
added insulation, 122, 123, 147-49, 150, 366
thermal inertia, 138-39, 141-47, 395-96

J

Joint reinforcement, 106-7, 167, 176, 180, 189, 192, 196, 198, 242, 258
Joints:
control, 24, 59, 75, 115, 167, 170, 186-87, 346-47, 349, 350-51
cracking, 23-24, 75, 94, 96, 123, 386
expansion, 56, 169-70, 186-87, 188, 189, 190, 349-51
moisture leakage, 123, 360, 361, 363, 386
pointing, 361
tooled, 360-61
tuckpointing, 411
types, 360
workmanship, 362-63

K

Kilns, 16-18

L

Lateral support, 171, 172-73, 174, 196, 218-19, 222, 225
Lightweight CMU, 24, 58, 60-61, 66
Lime, 10, 26, 28
Limestone, 77, 78, 84-86
Lintels, 175, 224, 284-97
brick, 180, 294-97
clay tile, 294-97
concrete, 288-89
deflection, 167, 175, 284, 288, 290
precast concrete, 289-90
steel, 286-88
Loadbearing:
masonry units:
brick, 52-57
clay tile, 45-46
concrete masonry, 61-63
walls, 6-7, 117-23, 207-76
brick, 229-35, 248-49, 251-56, 260-68
concrete masonry, 235-42, 248-53, 255-62
partially reinforced, 209-42, 243
reinforced, 209-10, 243-76
unreinforced, 207-42
Load distribution, 210-11, 240, 274-75, 285-86, 291, 298
Low-lift grouting, 244, 366-67

M

M-factor, 144-47
Manufacturing processes:
brick, 10-18
cement, 26-27
concrete masonry, 18, 20-26
Marble, 86-87
Masonry:
classifications/types, 6-8, 118-23
hollow, 121
loadbearing, 207-76
non-loadbearing, 118, 171-206
partially reinforced, 207-42

Masonry (*cont.*)
reinforced, 243-76
solid, 120-21
unreinforced, 207-42
Masonry bearing wall design theory, 207-16, 225, 243-48
Masonry bonding, 106, 107, 120-21, 123, 222-23
Metal accessories, 105-16
Metal anchors, 105-6, 110-13, 176-79
Metal stud walls, 176, 178, 179
Metal wall ties, 107, 120-23, 176, 222-23
Metallic oxides, 10-11
Mexican brick, 16-17, 34
Minimum reinforcement, 8, 209-10, 243, 244-45, 256, 258, 262, 263
Minimum wall thickness, 189, 196, 217, 219-22, 225, 255, 258, 261
Mixing water, 26, 29, 340
Modular coordination, 35-38, 341-46
Modulus of elasticity:
brick, 54
concrete masonry, 74
Modulus of rupture, 53, 54
Moisture content:
brick, 54-55, 340-41, 445
concrete masonry, 24, 58-59, 64-66, 340
Moisture control:
below-grade walls, 332-34, 386-88, 395
caulking, 115, 347, 386-87
cavity walls, 122-23, 148-49, 384, 393-94
condensation:
analysis, 388-93
control, 123, 148, 333-34, 388, 390, 393-400
efflorescence, 28-29, 96, 338, 399
flashing, 115-16, 179, 181, 188, 384-85
retaining walls, 318-19
silicone treatment, 399-400, 406
vapor barriers, 123, 150, 388-400
waterproofing, 332-34, 386-88
weep holes, 148, 180, 188, 318-19, 363, 384
Moisture expansion, 56, 123, 166-67, 169, 187
Monadnock Building, 3, 4, 120
Mortar:
admixtures, 29, 101, 226
air entrainment, 26-27, 92
ASTM Standards, 95, 97, 101, 103, 474
bond, 26, 27, 28, 92, 93, 94-95, 100, 180, 183, 208, 211-12
chemical resistant, 42, 101-2
compressive strength, 6, 94, 97, 98-99
corrosive effects, 105-6
durability, 96, 97, 99-100
extensibility, 95-96
ingredients, 26-29, 91
joints, 360-63, 411
lime in, 26, 28, 92, 96-97
masonry cement, 28, 97, 100, 212, 226
mix proportions, 95, 97, 98-100
mixing water, 29, 92-93, 340
portland cement, 26-27, 91, 97, 212, 226
properties, 26-29, 91-96
proprietary cements, 28, 97, 212, 226
refractory, 100-101
retempering, 92-93, 340
sand, 26, 29
shrinkage, 95-96
testing, 28, 29, 98
types, 97-100
Movement joints, 56, 75, 115, 167-70, 175, 186-88, 189, 190, 215-16, 346-47, 349-53, 377-78

N

National Building Code, 125, 132, 171, 176, 191, 218, 245, 277, 317
Noise control, 56-57, 74, 75, 161-66
Nominal dimensions:
brick, 35-37
concrete masonry, 66
Non-loadbearing construction, 6-7, 117, 171-206
masonry curtainwalls, 175, 192-94
panel walls, 175, 188-91
partition walls, 118, 172-75
veneer, 6-7, 118, 123, 175-94
brick, 176-80
concrete masonry, 180-82
stone, 182-86
Normal weight CMU, 24, 58, 60-61
Notations and symbols, 227, 471-72

O

Open end block, 67, 365
Openings in shear walls, 272
Oxides, 10-11

P

Painting masonry, 397-400
Panel walls, 175, 188-91
Parapet walls:
flashing, 170, 357, 385
movement, 190, 350
reinforcement, 170, 256
Partially reinforced masonry, 8, 191, 209-42, 243, 328-30
code requirements, 210, 218
design, 210-42
brick, 229-35
concrete masonry, 235-42
Passive solar design, 151-60
Pattern bonds:
brick, 358, 360
concrete masonry, 359

paving, 379-80
stone, 85
Paving, 371-83
bases, 72, 371, 373-78
bond patterns, 72, 379-80
brick, 40-42, 371-83
concrete masonry, 72-73, 371-83
stone, 87-88, 90, 377
structural floors, 381-83
Perlite insulation, 148
Pilasters, 171, 172, 219, 221, 225, 236-37, 248, 251-53, 255, 345, 347
Plain masonry (*see* Unreinforced)
Plasticity:
clay, 10-11, 13, 15
mortar and grout, 26, 28, 29, 92-93, 103, 169
Pointing, 361, 411
Prefabricated masonry, 284, 297, 402-3
Prism tests, 226-28

Q

Quality control:
mortar, 338-40, 445-46
workmanship, 94, 211, 225, 227, 361, 444-48

R

R-values, 134, 138-47
Racking test, 272, 273
Rational design methods, 6, 8, 217-18, 225-76
Refractory brick, 11
Reinforced masonry, 3-4, 6, 8, 117-18, 209-10, 243-76, 332
brick, 3-4, 6, 248-49, 251-56, 260-68
concrete masonry, 6, 248-53, 255-62, 331
Reinforcing steel, 117-18, 172, 209-10, 211, 229, 231
anchorage, 244, 253, 280-81, 282
ASTM Standards, 247-48, 475
clearances, 297
design tables, 193, 197, 246-47, 291-92, 295-96, 321-22, 324, 335, 336
function, 8, 243, 264
minimum required, 8, 209-10, 243, 244-45, 256, 258, 262, 263
placement, 365-66
spacing, 245, 246-47, 256, 258, 262, 263
Restoration, 38, 45, 52, 403-12
Retaining walls, 316-24
design, 320-24
drainage, 318-19
expansion joints, 318-19
loads, 317-18
Rigid insulation board, 148-49
Rowlock, 38, 359, 360
Rubble stone, 82, 85
Rudolph, Paul, 22, 71

S

Sailor, 38, 359, 360
Salmon brick, 33-35
Sample panel, 361, 430, 446
Sampling and testing, 475
Sand:
ASTM Standards, 29, 474
in mortar and grout, 26, 29
Sand-lime brick, 60
Sand struck brick, 15
Sandstone, 77, 78, 87-88
Saturation coefficient, 31, 44, 53-55
SCR:
Acoustile, 57
brick, 37
process, 364
Screen blocks, 66-67
Screen tile, 49, 52
Screen walls, 49, 66, 151-52, 194-96
Sculptured brick, 39
Seismic design:
box frame system, 208, 213, 243, 268, 269, 274
code requirements, 248, 258, 261
diaphragms, 268, 269-72, 273
dynamic forces, 227, 243, 268, 275
design methods, 272, 274-76
shear walls, 268-72
zones, 275
Serpentine walls, 198-200
Shading devices, 151-52
Shale, 11
Shear walls, 208-9, 212-13, 225, 233, 235, 240-41, 243, 268-72
Sheathing, 176-77
Shelf angles, 176, 185, 188, 191, 352-53
Shrinkage:
brick, 56
concrete masonry, 23, 64-66, 74-75, 167-69
mortar, 28, 29
Shrinkage cracking, 23, 24, 75, 106, 115, 166-67, 180, 186, 346-51
Silicone, 399-400, 406
Slate, 77, 78, 87
Slenderness, 225, 230-32, 236
Slump block, 21, 68
Solar heating, 151, 153-60
Solar screens, 151-52
Soldier, 38, 359, 360
Sound control, 56-57, 74, 75, 161-66
Spalling, 10, 169, 170, 352, 403, 404, 406
Specifications, 428-44
mortar and grout, 429-38
restoration, 441-44
stone, 438-41
unit masonry, 429-38
Split face block, 25, 68, 69-71

Standard Building Code, 125, 129-30, 132, 171, 176, 191, 218, 245, 277
STC ratings, 75, 161, 164-66
Stone:
- anchorage, 111, 113
- ashlar, 84, 85
- ASTM standards, 474
- classifications, 77-78
- cut stone (*see* dimension stone)
- dimension stone, 82, 84
- dressing, 80-83
- dry stack walls, 201-2
- fieldstone, 82, 84, 86, 87, 89
- finishes, 80-81, 83
- flagstone, 82, 84, 88
- granite, 77, 78, 84
- limestone, 77, 78, 84-86
- marble, 86-87
- patterns, 84, 85
- production, 79-84
- properties, 78-79, 84-88
- rubble, 82, 87
- sandstone, 77, 78, 87-88
- selecting stone, 88-90
- slate, 77, 78, 87
- travertine, 78, 86
- types, 79, 84-88
- veneer, 84, 86, 87, 88, 89, 182-86

Stretcher, 38, 359
Suction, 54, 74, 92-93, 94, 103, 104, 340-41, 445
Sun-dried brick, 16, 35
Surface clay, 11
Surface bonded masonry, 368-71
Swimming pools, 334-37
Symbols and notations, 227, 471-72

T

Tensile stresses, 8, 210, 211, 212, 213, 243, 244, 268, 281, 301, 307
Terra cotta, 52
Testing:
- ASTM standards, 475
- fire resistance, 124
- masonry prisms, 226-28
- mortar and grout, 28, 29, 98

Theory of bearing wall systems, 207-16, 225, 243-48
Thermal characteristics, 122, 134-60
- capacity insulation, 138
- conductivity, 138
- heat-gain calculations, 134, 138-42
- heat-loss calculations, 134, 138, 142-60
- M-factor, 144-47
- resistance, 134, 137
- storage capacity, 138-39, 150, 155
- thermal inertia, 138-39, 141-47
- thermal storage walls, 153-60
- Trombe walls, 154, 155-60

Thermal expansion, 56, 75, 123, 166-68, 187
Ties:
- allowable loads, 108
- materials, 105-6, 109-10
- spacing, 108-9, 222-23, 224
- types, 107-10, 222-23

Tile (*see* Clay tile)
Travertine, 78, 86
Trombe walls, 154, 155-60
Tuckpointing, 411

U

U-values, 134, 138-47
Uniform Building Code, 125, 129, 132, 171, 176, 190, 210, 211, 218, 225, 226, 245, 252-53, 260-62, 268, 272, 274-76, 277, 282-84, 317
Unreinforced masonry, 8, 207-42, 328
- empirical design, 4, 8, 120, 217-25
  - allowable stresses, 218, 220
  - ANSI A41.1, 218-25
  - h/t ratios, 171, 217, 218-19, 222, 232, 238, 255, 258, 259, 261, 283
  - lateral support requirements, 171, 172-73, 174, 196, 218-19, 222
  - wall thickness, 217, 219-22, 225
- rational analysis, 225-42
  - brick, 229-35
  - concrete masonry, 235-42

Used brick, 33-34, 41

V

Vapor barriers, 123, 150, 388-400
Vapor flow, 388-96
Veneer, 175-87
- adhered, 52, 100, 175
- anchorage, 176-79
- anchored, 52, 175-76
- anchors:
  - allowable loads, 112, 176, 177
  - spacing, 176, 178, 183
  - types, 107, 110-13
- design:
  - brick, 176-80, 181
  - concrete masonry, 180-82
  - stone, 182-86

Vermiculite insulation, 148
Vitrification, 17, 30

W

Wall copings, 200-201
Wall intersections, 216, 223-24, 260, 343-46
Walls, 6-7, 117-23, 171-203, 207-76
- cavity, 106, 107, 118-20, 122-23, 148-49, 166, 169, 223, 363, 384
- fire resistance ratings, 125-28
- h/t limits, 171, 217, 218-19, 222, 232, 238, 255, 258, 259, 261, 283
- lateral support, 171, 172-73, 174, 196, 218-19, 222, 225

loadbearing, 6-7, 117, 123, 207-76
brick, 229-35, 248-49, 251-56, 260-68
concrete masonry, 235-42, 248-53, 255-62
minimum reinforcement, 8, 209-10, 243, 244-45, 256, 258, 262, 263
minimum thickness, 189, 196, 217, 219-22, 225, 255, 258, 261
multi-wythe, 6-7, 36, 107, 117, 119-23
non-loadbearing, 6-7, 117, 171-206
masonry curtain walls, 192-94
panel walls, 188-91
partially reinforced, 8, 191, 209-42, 243, 328-31
partition walls, 118, 172-75
reinforced, 209-10, 243-76
retaining (*see* Retaining walls)
screen (*see* Screen walls)
shear (*see* Shear walls)
single-wythe, 6-7, 37, 106, 117, 118-19
STC ratings, 161, 164-66
unreinforced, 120, 207-42, 328
veneer, 6-7, 118, 123, 167, 169, 175-87
adhered, 52, 100, 171, 175
anchored, 52, 175-76
brick, 176-80, 181
concrete masonry, 180-82
stone, 182-86
Water retentivity, 74, 92-93, 94
Water struck brick, 15
Waterproofing:
basement walls, 332-34, 386-88, 395
caulking, 115, 347, 386-87
flashing, 115-16, 179, 181, 188, 384-85
foundation walls, 332-34, 386-88, 395
vapor barriers, 123, 150, 388-400
Weathering index, 30-32, 58-59, 64-65
Weep holes:
retaining walls, 318-19
walls, 148, 180, 188, 332, 363, 384
Wind loads, 123, 176, 188, 191, 198, 210, 211-14, 227, 241-42, 268, 274
Workability:
mortar, 92-93, 94, 100
stone, 78-79, 89
Workmanship, 94, 211, 225, 227, 360-71

Z

Zinc coatings, 105-6, 109, 111, 116